Van Sickle's Modern Airmanship

FIFTH EDITION

Van Sickle's Modern Airmanship

FIFTH EDITION

edited by

John F. Welch

Colonel, United States Air Force (Ret.)

VNR **VAN NOSTRAND REINHOLD COMPANY**
NEW YORK   CINCINNATI   ATLANTA   DALLAS   SAN FRANCISCO
LONDON   TORONTO   MELBOURNE

Van Nostrand Reinhold Company Regional Offices:
New York   Cincinnati   Chicago   Millbrae   Dallas

Van Nostrand Reinhold Company International Offices:
London   Toronto   Melbourne

Manufactured in the United States of America

Published by Van Nostrand Reinhold Company
135 West 50th Street, New York, N.Y. 10020

Published simultaneously in Canada by Van Nostrand Reinhold Ltd.

Library of Congress Cataloging in Publication Data

Welch, John F.
  Modern airmanship.

  Previous ed. by N. D. Van Sickle.
  Includes index.
    1. Aeronautics. 2. Airplanes—Piloting. I. Van
Sickle, Neil D. Modern airmanship. II. Title.
TL545.V3   1980   629.132′521   80-14137
ISBN 0-442-25793-7

Foreword

Since the first edition of *Modern Airmanship* appeared in 1957, the changes in each edition have kept up with the changing airman's world.

Now, jet power has virtually replaced propellers in military forces and in scheduled airlines. Near transonic speeds are routine in transport flying, and supersonic aircraft are regularly scheduled. Very High Frequency and Ultra High Frequency radio dominate communications and air route navigation aids. Air traffic control systems are based on computer assisted ground radar, aircraft identification and location indicators, and multiple discrete communications channels.

In aircraft, there are high precision gyro stabilized rate computing instruments and excellent quick switching radio. These, combined with high standards of training and pilot proficiency, permit traffic flow to safely reach the densities that have made commercial and private air transport the way to travel. Inertial navigation systems and weather radar have developed new expectations in flexibility, precision, and safety.

This Fifth Edition is largely the work of Colonel John F. Welch, a widely experienced military pilot, certified flight instructor, and charter pilot who on retirement from the Air Force put his flying and management skills into his own aircraft sales, flight instruction, and charter company. Colonel Welch, also an aeronautical engineer, has carefully examined the entire text and has revised it extensively to insure that it truly portrays modern airmanship in every way.

Neil D. Van Sickle
Major General, US Air Force (Retired)

Rapid City, South Dakota

Preface

More than ever, flying is a profession, whether the airman serves as an airliner crew member, a corporate or military pilot, an aircraft mechanic, a crop duster, a flight instructor, or in one of many other specialties. All these vocations demand long periods of training to a high degree of skill, and dedication to the good of the public served.

The person pursuing flying activities purely for pleasure or personal convenience must take an equally professional approach, for the air space is shared by all, and the safety of all depends on continual awareness and practice of safe flight procedures by every airman. The essence of this professional approach to flying is true airmanship. This new edition once again emphasizes the knowledge and skills which go to make up modern airmanship.

The task of editing this Fifth Edition was undertaken reluctantly, well realizing the immensity of the task. Not only is it a tremendous challenge to measure up to the standards set by previous editions, but changes in aviation during the past 8 years have dictated a new look at many aspects of the world of airmanship. At the same time, the foundations laid by the contributors to the first four editions have proved a sound basis on which to build, so this Editor is very grateful to them. Comments from readers of the Fourth Edition have been welcome and most helpful.

Special acknowledgment must be given to Myron W. Collier, who not only completely rewrote the chapter on navigation, but also provided valuable inputs to Chapters 5, 9, and 10. Neil D. Van Sickle's continued interest in the book has been appreciated as have his expertise and readily given help.

Finally, two ladies deserve special credit, Alberta C. Welch for her constant encouragement, and Juanita M. Firestone for her typing and retyping so much of the manuscript.

John F. Welch

Rapid City, South Dakota

Acknowledgments

Valuable contributions to this Fifth Edition of *Modern Airmanship* were made by a number of people not recognized in the Preface.

Don J. Ahrens, of Cessna Aircraft Company, who revised the chapter on Airplane Structures for the Fourth Edition, provided a new revision of the same chapter for this edition.

Dr. Harold E. Wright and Dr. William C. Elrod, of the Aero-Mechanical Engineering Department of the Air Force Institute of Technology made revisions and wrote new paragraphs for Chapter 4, Propulsion.

Don K. Halligan, of the U.S. Weather Service, made comments and suggested corrections for Chapter 6, The Atmosphere and Its Weather.

Dr. William H. King, Colonel USAF (MC), now retired, wrote a revised Chapter 7, Medical Aspects of Flight.

Joseph S. Davis of United Airlines, provided rewritten and new material which was incorporated in Chapter 10, in the section on Flying Large Multi-Engine Aircraft.

Col. Robert F. Hedrick of the South Dakota Army National Guard, reviewed Chapter 13, Helicopters, and made helpful comments and suggestions.

John F. Welch

Contents

Van Sickle's
Modern Airmanship

FIFTH EDITION

1

Modern Aircraft

*"I've chased the shout-
ing wind along, and
flung
My eager craft through
footless halls of air. Up,
up the long delirious
burning blue
I've topped the wind-
swept heights with
easy grace,
Where never lark, or
even eagle flew."*

High Flight
—JOHN GILLESPIE MAGEE, JR.

Magee's short poem expresses the exhilaration and joy of flying which are experienced at least sometimes by nearly every pilot. It is this feeling of freedom and elation that keeps a pilot flying for 50 years, but there is a great deal more to flying than flitting joyously about the sky.

The pilot takes his equipment and goes about the work of man, matching his knowledge and skill and the capabilities of the machine against wind, weather, and distance. His end products are people and goods carried, time and lives saved, distance made inconsequential, natural barriers overcome, and the world drawn closer. At the individual level, flying can be as great an art as music and dancing; it can also be drudgery. Putting the knowledge, skill, and dedication of the airman together for the benefit of all of us is the highest degree of modern airmanship.

The exponential growth of technology of the last hundred years has both helped and been helped by the science of aeronautics. The physical sciences do not stand alone; they interact continuously and are directed and modified by the moral fiber of the professionals involved in them.

There is hardly anyone today whose life is untouched in some manner or other by aviation. It has become the accepted standard for long distance travel, and its unique capabilities have caused a continuing expansion of its usefulness, so that it affects all phases of human existence.

Since the 1940's, air travel has almost totally displaced travel by train in the United States, and the present generation of travelers considers air travel the normal, practical way to go. They use not only the airlines, but air taxis, commuters, and self-piloted light aircraft as well.

Aircraft play an important part in community services, providing air ambulance, traffic surveillance and control, and assistance in law enforcement. Their part in these activities becomes more indispensable as time goes on.

Flying often saves lives. Critically needed blood may be sped through the night to a patient in surgery. A badly burned child may find his chance to live by being flown to a burn treatment center 1000 miles from his home. Thousands of war wounded owe their lives to rescue helicopters.

The reasons for the use of aviation are as varied as the users. At the top of the list is saving time. A busy executive can spend productive time at several widely separated locations in one day, and still be home for dinner. He can go from coast to coast in a few hours. The doctor can whisk his patient to a center providing urgently required treatment. The vacationer can maximize his time at his selected vacation spot. The farmer can cover hundreds of acres, instead of tens. Our ancestors took weeks of hardship to cross the ocean. We do it in hours, in luxurious comfort.

Because the only physical restrictions on the route of an airplane are terrain it may not be able to climb over, or weather it may not safely penetrate, it can truly travel the shortest distance between two points (a great circle on the earth's surface). Over land, this routinely allows a saving of 25 or 30% of the road distance. Many small aircraft can carry four people at miles per gallon rates comparable to those of small or medium size automobiles. They thus conserve fuel as well as time.

Many remote areas, particularly in the far north, can be reached much of the year only by air. It becomes the only way in for mail, supplies, medicines, and people. Surface travel that might require weeks over water or nearly impossible terrain is replaced by a few hours in the air.

For many people, aviation is a unique form of recreation, as spectators and as participants. Whether it's an aircraft show, a Sunday afternoon sight-seeing jaunt, a contest of flying skills, or coyote hunting, there is a host of enthusiasts.

A look at a few statistics reflecting the importance of aviation is enlightening. In the United States there are now nearly 190,000 civil aircraft. Some 2200 of these are airliners. The rest belong to the general aviation fleet. There are more than 780,000 active pilots, and new ones are learning to fly every day. In 1978 the airlines carried 262.4 million passengers in intercity travel, nearly three and a half times as many as in 1963. The general aviation fleet carries more than 100 million, traveling not just between the smaller towns, but to and from major population centers as well.

Modern Aircraft

The basic requirement of an aircraft is the capability to provide transportation through that all-pervasive medium contained within a few miles of the earth's surface, the atmosphere.

Different kinds of aircraft, as we shall see, have different capabilities, and different purposes require different capabilities. Certain capabilities exclude others, so that every design is a collection of compromises.

CATEGORIES OF AIRCRAFT

By official designation there are four categories of aircraft—lighter than air (balloons, airships), gliders (unpowered airplanes), rotorcraft (helicopters, autogyros), and airplanes. Not yet included, but being looked at very carefully, are hang gliders.

Balloons are at the mercy of the direction of the air flowing over the earth's surface, so they have little practical use for providing transportation between selected points. Airships once showed promise for long distance transportation and still have their enthusiastic supporters. The few nonrigid airships (blimps) now flying play very little part in modern air travel. Airships could be designed to carry heavy loads long distances, but certainly offer no speed challenge to modern jet transports. Lighter-than-air craft are accordingly not further considered in this book, but a typical sporting balloon is shown in Figure 1.1.

Gliders and *sailplanes* are marvelous sport vehicles, but they, too, have limited application in the normal transportation system. Aerodynamics and structural considerations, as well as many other aspects of flying, apply equally to them and to airplanes. Chapter 14 is devoted entirely to fixed wing, unpowered flight.

Rotorcraft are represented in largest numbers by helicopters. Basic aerodynamic principles apply to them, too. In the helicopter the main rotor is powered directly by the engine, and forward motion is obtained by tilting the rotor. Gyrocopters differ in that once in flight the main rotor is turned by the slip stream. Forward thrust is from a conventional engine and propeller, which usually are mounted in the rear. Gyrocopters have not achieved any large degree of use, so they are not further discussed in this book. On the other hand, helicopters have developed into highly successful, if specialized, transportation. They are the subject of Chapter 13.

Figure 1.1 Sport Balloon, *Courtesy Raven Industries*

By far the greatest number of aircraft are classed as *airplanes*. They depend upon fixed airfoils for lift and internal combustion engines to provide thrust, and are the main interest of this book.

Another way of categorizing aircraft is by their uses. These can be grouped under three major classifications—*airline transports, military,* and *general aviation*—reflecting their purposes.

AIRLINES

Airline transports are large in size but not in number. They specialize in carrying large numbers of people and high volumes of freight between large airports. Consequently they are large aircraft requiring long runways and specialized airport services.

The early growth of long distance public air transportation was very turbulent and much involved in politics because of mail contracts. Finally, in 1938, Congress passed the Civil Aeronautics Act. It established the Civil Aeronautics

Board which since then has had authority to *certify,* and grant routes and subsidies to, *air carriers,* while requiring particular stops and other service.

Certificated airlines must use large aircraft and offer sufficient scheduling to provide on-demand air transportation to the traveling public. Their aircraft must be certificated under very stringent Federal Aviation Administration (FAA) rules to afford the safest possible transportation to their passengers.

The airlines carry hundreds of millions of passengers annually among the 400 or so U.S. airports they serve. About 30 of these airports handle 70% of the airline passengers. The other 30% are distributed among the remaining 370 or so airports with scheduled airline stops.

A way of generally classifying airlines is according to the area served: The *regional* airlines serve a specific area. They connect with the *domestic trunks,* which in turn connect with the *international* airlines. The domestic trunks have the transcontinental routes, and the internationals offer service between points in the United States and those overseas.

Typical modern airliners are pictured in Figure 1.2.

MILITARY AIRCRAFT

In earlier years, the advantages and usefulness of flying to military forces were gradually established. Under the pressures of war, and more recently, threats of war, common effort has greatly accelerated the development of aeronautics. As a direct consequence, civil aviation has been able to make tremendous progress, using equipment, training, and knowledge first developed for military use.

Military aircraft have been in the forefront of military planning, tactics, and operations since the late 1930's. Their value was much debated after World War I, but their unique contributions became quite apparent early in World War II. During the latter conflict they achieved by their numbers and by the determination and skill of their crews what they lacked in capability and sophistication. With the advent of the jet engine and atomic and nuclear weapons, they have acquired awesome destructive power. So far the new weapons have been used only twice, but the jet engine has been well tested both for air lift and in combat, in more recent conflicts. National defense forces continue to rely on military aircraft as a vital means of maintaining peace and assuring national survival in the event of war.

Air Lift.   Military commanders from time immemorial have dreamed of being able to place men and material where they are needed when they are needed. The use of air lift has finally made their dreams come true. We can now transport personnel and supplies halfway around the world and put them on the battle line, in less than 24 hours. They can be placed where needed in a theater, and shifted from one point to another, entirely by air.

The major U.S. intercontinental and intertheater high-speed air lift is by the

Figure 1.2. Boeing jet passenger and freight transports were the first in use, and are widely used as to be fairly typical of all jet airline transports. There are, of course, many other fine aircraft in comparable classes.
TOP: Model 747B. Carries 374 passengers 5750 nm at 543 knots
TOP CENTER: Model 707-320B. Carries up to 189 passengers 5220 nm at 522 knots.
LOWER CENTER: Model 727-200. Carries 134 passengers 1480 nm at 522 knots.
BOTTOM: Model 737-100. Carries 119 passengers 1390 nm at 500 knots *(Courtesy of the Boeing Company)*

big jets, C-135, C-141, and C-5, all highly capable air transports (Figure 1.3). In addition, there is the Civil Reserve Air Fleet, which is made up of modern jet airliners modified for military purposes. These are in commercial service, but are earmarked for immediate military use when needed.

For intermediate distance transport within a theater, where airfields may be unimproved or almost nonexistent, the star performer is the multi-engine turboprop, typified by the C-130 Hercules (Figure 1.4). Special unloading systems make it possible to deliver a load of supplies to a given site without even touching down. Other propeller airplanes serve well in a similar capacity, and new

Figure 1.3. Large military transports.
LEFT: Lockheed C-5 Galaxy, for carrying large, heavy cargo items as well as passengers; over 75 tons useful load long distance at 450 knots.
RIGHT: Lockheed C-141 Starlifter; can carry 154 troops or 68,500 lbs. cargo 3675 nm at 490 knots. *(U.S. Air Force photograph)*

short-haul, short-field jet transports for the same purposes are under development.

The most versatile cargo and troop carrier in a battle zone is the helicopter. It can make nearly vertical ascents and descents, landing anywhere there is a level spot with sufficient clearance for its rotors. This concept has been well proved, in the rapid transportation of both men and materiel as needed within a combat area (Figure 1.5).

Bombers.   *Bombers* are an extension of artillery power. They can drop explosives and other destructive weapons on defensive positions and supply sources, storage depots, and transportation lines deep within antagonist territory, or they may be used in front line support. If they strike against the sources of enemy power, they are strategic bombers. During World War II they were sometimes used against the populations of large cities, with the hope that war production, transportation, and the will to wage war could be inhibited. Present day bombers

Figure 1.4. Lockheed C-130E Hercules Turboprop Logistic and Troop Transport. Cruises 2950 nm at 350 knots. Carries 92 troops or 64 paratroops or 74 litters, payload more than 20 tons. Shown here making a low altitude parachute extraction cargo delivery. *(U.S. Air Force photograph)*

Figure 1.5. Boeing Vertol CH-47C Chinook cargo and passenger helicopter. *(U.S. Army Photograph)*

with current weapons are potentially thousands of times as destructive. The bomber fleet of the U.S. is comprised primarily of the B-52, the newest models of which were delivered in 1962 (Figure 1.6).

Fighters.   The design mission of fighter aircraft is to establish and maintain control of the air space in a desired area. This might be a battlefield, a deployed naval fleet, a war theater, or the nation. These airplanes must be very fast, highly maneuverable, and heavily armed and fitted with the most sophisticated and deadly fire control systems that have been devised. They must also have defensive systems that can cope with the offensive systems of projected adversaries. Typical modern fighters are pictured in Figure 1.7.

Battlefield Support.   The *raison d'être* of an army is to take and hold ground. Today's ground forces are very dependent upon close air support in carrying out this mission. The small airplane provides a vantage point from which to view the ground action. Its pilot can spot and call out the positions of enemy forces, he can direct artillery fire, and he can mark enemy positions and direct ground sup-

Figure 1.6. Boeing B-52H Stratofortress, long range heavy, last built in 1962, still the U.S. Air Force's primary Strategic Air Command Bomber. *(U.S. Air Force photographs)*

Figure 1.7a. Northrup F-5E. *(Courtesy Northrup Co.)*

Figure 1.7b. McDonnell Douglas F-15. *(Courtesy McDonnell Douglas)*

Figure 1.7c. General Dynamics F-16. *(Courtesy General Dynamics)*

Figure 1.7d. McDonnell Douglas F-18. *(Courtesy McDonnell Douglas)*

Figure 1.7e. Grumman F-14. *(Courtesy Grumman Corp.)*

port fighter strikes against them. His work is becoming more and more dangerous as defensive weapons, such as shoulder launched heat seeking missiles, are developed.

The third dimension that air adds to ground combat is augmented by helicopters. They bring in fresh troops to where they are needed, provide battlefield surveillance, and, when armed, can attack enemy positions. In battlefield operations they are primary assault vehicles, for opening new salients and repositioning troops as needed. They move field commanders about quickly and evacuate the wounded to rear areas.

Close bombardment support for ground operations is provided by a variety of airplanes using various kinds of weapons. They must be good load carriers, with accurate sighting and delivery systems. For their own protection they are armored and carry defensive systems. Their fuel capacity should give them adequate orbiting time in the combat area. When working in close support of friendly forces, their strikes must be directed by a ground or airborne air liaison officer.

GENERAL AVIATION

Nearly 99% of civil aircraft belong to general aviation. (That's a catchall designation for everything except airliners and military aircraft.)

Aviation has developed into one of the most valuable tools of all time for business. It speeds executives and skilled people to the places they are needed in unimaginably short periods of time. Its greatest benefit to these people is freedom—freedom to concentrate on the things they need to do rather than on travel. And they can fly nearly anywhere, with more than 13,000 airports, seaports, and heliports from which to choose in the United States alone—not counting military and other airports barred to general aviation aircraft.

The purpose for which an aircraft is flown is one way of classifying both the aircraft and its use. Some of these are discussed in the following paragraphs.

Training.   Piloting an aircraft is not an art learned in a few hours. It requires an instructor skilled and knowledgeable both in flying and in teaching, and, for most efficient learning, an aircraft which is capable of performing all the required maneuvers, responsive to the pilot's inputs, and demanding of knowledge and skill to perform the maneuvers well. Examples of popular and successful airplanes used for beginning training are shown in Figure 1.8.

For more advanced training, larger, more powerful, and more specialized aircraft are required, the model determined by the particular need.

Personal.   Air transportation is the goal of the large majority of pilots. The most typical vehicle for them is a four-seat airplane which can travel from 125 to 200 mph (Figure 1.9). These craft can cover long distances economically, three or four times as fast as by automobile, and can land at almost any airport. There is no speed limit, except for the fastest of them when in the vicinity of airports. They are often equipped to fly in instrument conditions as well as at night. Halfway across the United States in one day is entirely practical.

The greatest use of general aviation aircraft is for *business travel.* The money value of people's time, the ever greater demand for efficiency, and competition for profitable business have made air travel a vital part of modern commerce. The aircraft used vary from two-place trainers to intercontinental jets, according to the needs of the firms and the airports to which they must fly. Particularly popular are the light twins, small turboprops, and small executive jets, examples of which are shown in Figures 1.10 and 1.11.

By no means all aircraft used for pleasure or for business are owned by the pilots or the companies who fly them. They can be rented by the hour to be flown by competent pilots, nearly everywhere.

Air Taxi and Charter.   These two terms are used interchangeably. Strictly speaking, an air taxi delivers the person(s) using it to a specific destination, with no further commitment to them, while a chartered flight stays with the user(s) for a longer trip, normally until return to the original point of departure. Charges for these services are based on distance traveled and waiting time of the aircraft and pilot. Aircraft employed range from the smallest to the largest.

Chartered-airlines-type flights for travel groups usually do not remain with the group, but take them to a point and return them from the same or another point to the original point of departure at an agreed-upon later date. Interim transportation is by other means, and the charter aircraft may meanwhile fly several other trips.

Figure 1.8a. Cessna 152.

Figure 1.8b. Piper Tomahawk.

Figure 1.8c. Bellanca Citabria.

Figure 1.9a. Gulfstream American Tiger.

Figure 1.9b. Piper Dakota.

Figure 1.9c. Beechcraft Bonanza.

Figure 1.9d. Cessna Centurion.

Figure 1.10a. Beechcraft Duchess.

Figure 1.10b. Piper Seminole.

Figure 1.10c. Gulfstream American Cougar.

Figure 1.10d. Cessna Skymaster.

Figure 1.11a. Beechcraft King Air.

Figure 1.11b. Gates Learjet.

Figure 1.11c. Rockwell Turbo Commander.

Figure 1.11d. Cessna Citatation II.

Figure 1.11e. Piper Cheyenne.

Commuters.   These are aircraft and crews flying over scheduled routes, operating much as the certificated airlines do. The routes they serve do not provide enough business to justify the use of large airplanes. The aircraft are chosen to meet the demands of the flying public. Most popular are twins with from six to 19 passenger seats. A typical commuter aircraft is shown in Figure 1.12.

Aviation serves extensively in the economical production of food, fiber, and building materials, to the benefit of all of us. Aircraft are used increasingly to fight insects, weeds, and plant disease; they plant and fertilize and increase the quantity and quality of food production. They spot and fight forest fires, and they help harvest the lumber needed for building new homes.

Agricultural Aircraft.   Many different aircraft have been used by "crop dusters" to aid farmers in producing crops to meet the ever-growing demands for food and fiber. Airplanes built 40 years ago, including the Stearman and the venerable DC-3, are still flown for this purpose.

Large fields are most economical for aerial agricultural operations, so the flat lands of the river deltas and of the great plains can profit the most from application of seed, fertilizer, insecticides, and weed killers by air.

As the old Stearmans have worn out, and the unique requirements of aerial application have increased, a family of specially designed agricultural airplanes has emerged, and some helicopters have also been modified for such work. All these must carry a heavy load of either liquid or dry materials, be able to apply the material evenly, fly relatively slowly, and be safe, responsive, and maneuverable. High altitude capability is seldom a requirement; most of them hardly ever go above a couple of hundred feet when working. They pose little collision hazard for other aircraft! Figure 1.13 shows typical "ag" aircraft.

Figure 1.12. De Havilland Twin Otter commuter and utility transport. Carries 19 passengers, plus crew, 750 nm @ 178 knots. *(Courtesy of De Havilland Aircraft of Canada Limited)*

Figure 1.13a. Bell Jet Ranger.

Figure 1.13b. Gulfstream American Ag Cat.

Figure 1.13c. Rockwell Thrush Commander.

Figure 1.13d. Piper Pawnee Brave.

Figure 1.13e. Cessna AG Wagon.

Forestry. Until recently the highest hills of our public forest lands were all capped by tall steel towers, atop which sat enclosed platforms providing radios, telephones, directional sighting equipment, and visibility in all directions. These were manned throughout the fire danger season, and when smoke was spotted, ground crews fought their way to the scene to battle the fire.

The growing use of airplanes and helicopters has brought a big change to forest fire spotting and fighting. Many of the towers now stand empty. The airborne observer can cover a much larger area, and spot fires more accurately. Often "smoke jumpers" can parachute in to stop a small fire before it becomes a big one. Highly skilled pilots, flying specially modified aircraft (slurry bombers), deliver heavy loads of borate solution to slow and stop the burning devastation of a forest. The most popular aircraft for this purpose are refitted large bombers and transports of World War II.

Fighting insects and diseases affecting trees being grown to supply our lumber needs is another use of spray planes. There simply is no other way to do the necessary job.

Some of the areas from which trees are now being harvested are very difficult to reach on the ground. It is even more difficult to bring the harvested logs out to the sawmill or to a point from which transportation is easy. Heavy lifting helicopters have proved to be the answer.

Fish and Wildlife.   Aircraft are used for spotting schools of fish for commercial ocean fishermen and help keep our lakes stocked with fish by dropping fingerlings grown in fish hatcheries. They are also used for getting deer and antelope counts in the prairie states. Small airplanes are excellent vehicles for hunting coyotes, and their aerial photographs document the range damage being done by prairie dogs.

Some of the larger cities contract each year for an aerial spraying service to control mosquitos.

Construction.   Builders have long talked jokingly of a "sky hook," to which they could attach their tackle to raise and position components of tall structures. Now they have such devices, in the form of helicopters. They have proved particularly useful in remote, rough areas, for positioning and assembling steel towers, specifically those used in building power lines. They can also minimize the time and man-hours required to bring in and assemble all kinds of structures in inaccessible areas.

Aerial Photography.   Aviation received a small "shot in the arm" in the 1930's when the Department of Agriculture, in connection with crop controls and soil conservation, took high altitude aerial photographs of most of the country. Updated photography still serves the same purposes. Special color films enable the experts to spot plant diseases and insect infestations. Geologists use aerial photography in identifying and plotting surface geological features, and in exploring for valuable minerals. Map makers stay up to date and accurate by working from continuous aerial photographic mapping operations. Road builders and other construction teams depend upon aerial photography for selecting routes and planning construction.

Observation and photography from the air for military purposes began over a hundred years ago and are now more important than ever.

Special cameras are used in specially modified aircraft for many aerial photography purposes. On the other hand, a simple hand-held camera may get the shot a motel or resort owner wants to advertise his enterprise. Ordinary news cameras provide excellent aerial coverage of natural disasters and other newsworthy events.

Off-shore Drilling.   Were it not for helicopters providing ready, rapid transportation between oil drilling platforms and the shore, the extensive off-shore petroleum production efforts of today would be a great deal more expensive and less efficient. This kind of operation has become quite specialized, carrying peo-

Figure 1.14. Energy exploration support.

ple, parts, and supplies back and forth much more rapidly than could surface vessels (Figure 1.14).

Aviation Safety

The early days of aviation were notoriously unsafe, and its reputation as a risky way to travel persisted long after it had become one of the safest means.

While the passenger miles traveled by commercial airlines have multiplied over and over, the passenger fatality rates have gone steadily downward. The passenger fatality rate per 100 million passenger miles in the 1944–1948 period was 2.01; by the 1969–1973 period, it had dropped to .09. All phases of aviation have experienced similar declines, as pilot knowledge and skill, and equipment, have improved.

CAUSES OF AVIATION ACCIDENTS

In most aviation accidents there is at least some element of human failure, which may range from failure to recognize a potential of danger, to deliberate risk taking.

Inexperience. The beginning pilot must be especially safety conscious because his knowledge and his skill level are both low. He may not recognize a developing hazardous situation or condition, he may not recognize the hazards in some maneuver he is trying to do, or he may not have the skill required to do what he wants to do safely or to recover safely from an unexpected event such as sudden strong wind gusts. These shortcomings are minimized by professional flight instruction, so that training to the private pilot level is usually completed without undue risk. As the new pilot continues to fly without serious incident, his self-confidence and complacency increase at an unjustified rate. Too often he becomes more and more hazardous as he approaches 500 hours. It is during this period that he should seek additional professional advice and instruction.

Overextending.   The pilot who convinces himself that he can fly 2 more hours without falling asleep, that he can reach that next airport on the fuel remaining, that the airplane can carry another 50 lb., or that the ceiling will not go as low as forecast is inviting an accident. Overextending may take the form of deliberate low flying, acrobatics without proper training or in a nonacrobatic airplane, or other show-off maneuvers, in violation of regulations. The regulations are designed to aid a pilot in setting safe limits to what he will do in an airplane. And they are just as important to the Boeing 747 pilots as to the new student pilot.

Weather.   Clouds, fog, rain, snow, sleet, hail, ice, heat, frost, cold, and wind all have their effects on flying. Under certain circumstances, any one of these or a combination of them can be a hazard to safe flight. Weather forecasters try to predict the conditions pilots will encounter, but their overall effectiveness is limited. First of all, they are powerless to change the weather. Secondly, weather observation points are limited in number, and information about conditions between observation points can only be estimated. Thirdly, the best efforts of forecasters are to no avail if the pilots do not use their services. Finally, forecasting remains more an art than a science, because so many factors of a given set of conditions are unknown.

It is no wonder, then, that many aircraft accidents are associated with unfavorable weather conditions. Too many of them, however, could be prevented by use of the information available, and by application of a thorough knowledge of how the atmosphere behaves. And there is a difference between inadvertently encountering unpredicted icing conditions and deliberate "scud running." The old safety principle of the 180° turn is too often violated.

Human Failure.   The tendency of the human being to make mistakes is recognized by putting erasers on pencils. Unfortunately, lack of knowledge or skill is often accompanied by lack of knowledgeable concern—the pilot doesn't recognize the danger until it is perhaps too late to make a safe recovery. The truly professional safe pilot minimizes human failures by being conservative, cautious, and alert, just as does the professional safe automobile driver.

Mechanical Failure.   Once the bugaboo of aviation, mechanical failure now occurs quite infrequently, and is often kept from being a serious problem by built-in redundancy. An Eagle Rock (late 1920's airplane) once made several forced landings going from Rapid City, SD to Atlanta, GA, but this author, in 12,000 hours of flying, has made only two precautionary landings away from an airport, both for only partial power failure.

Mechanical failures have been reduced in frequency by building in more reliability, by providing redundancy (back-up systems), such as both mechanical and electrical fuel pumps, and by high maintenance standards. Every small airplane must have thorough inspection for mechanical condition at least once a year; every airplane that flies commercially must be inspected at least every 100 hours of flying time. Any defect which shows up in a particular model of airplane or commonly used component may become the subject of an Airworthiness Directive (AD) issued by the Federal Aviation Administration (FAA). These directives require appropriate inspections and mechanical changes within a specified period of time or number of flying hours, reducing the chances that someone else will encounter the same problem.

Manufacturers of new designs and models, of both aircraft and components, not only try to incorporate as much safety as possible into their products, but also must submit them to the FAA for evaluation and testing under firmly established rules before they may be offered for sale. This process is called certification and is successfully concluded by the issuance of an Airworthiness Certificate for each aircraft built.

Growth of Aviation

In sheer numbers alone, aviation is making its greatest peacetime growth. In 1977, US factories turned out over 200 commercial transports, 900 helicopters, and nearly 17,000 general aviation airplanes.

Speed, load capacity, and equipment to fly in weather and at night have become more and more important to the users of airplanes. To meet these requirements, designs have become cleaner, engines have become more powerful, sizes have increased, radios and instruments have become more numerous and more reliable, and ground equipment for control of air traffic has become more complex. The numbers of pilots have continued to grow, too, but more dramatic than the total number is the increase in the number of instrument and multiengine rated pilots. Increased capabilities of airplanes have led to better qualified pilots. Better airplanes and better qualified pilots have steadily increased the safety of travel by air.

What of the Future?

The economic value of air transportation assures that it will continue to grow. This growth will be augmented by the increased need for fast, economical transportation as more and more industrial enterprises move away from the major population centers. This growth will not be accomplished without the solving of several major problems.

FUEL SHORTAGES AND FUEL EFFICIENCY

All modern transportation and industry are heavily dependent upon petroleum, for fuel, for lubrication, and as raw material for an enormous variety of products. Considering the rate at which petroleum is being used versus the rate at which it is being discovered, and the certainty that the total supply is finite, it is apparent that alternate fuels and new energy systems must be developed. Whatever the solutions, there obviously must be big changes in power plants and manufacturing materials. These changes will be expensive, but aviation will more than hold its own.

Present airliners produce 29 to 45 seat miles per gallon of fuel. Single engine general aviation aircraft deliver from 42 to 88 seat miles per gallon.

Automobiles may provide 100 or more seat miles per gallon, but seldom do, because occupied by only one or two people. They must go farther than airplanes between the same points because roads are not straight. Also, auto speeds are limited. By comparison, flying is more fuel and time efficient than driving.

POLLUTION

The swarms of cars in thickly populated areas, plus all the combustion products of modern industry, have caused serious health problems with the large volume of pollutants which they spew forth into the air. Aircraft are a very small part of the problem, because their relatively small numbers consume a minute percentage of total fuel used, their exhaust gases are much more widely dispersed, and their engines burn fuel more efficiently. Nevertheless, possible ways of reducing aircraft engine emissions are being investigated.

NOISE

It is most difficult to develop the high thrust needed to propel a large aircraft through the air without producing high noise levels, from both engine exhausts and propulsive equipment like propellers and jet nozzles. In this aspect, large aircraft are no better neighbors than a major railroad freight line.

Many airports were originally built some distance from population centers. Unfortunately, they often were perceived as being in excellent areas for industrial and residential development. When the runways were lengthened, when airplanes grew larger (and noisier), when flights became more frequent and jet engines entered the picture, residents began to think of the airports as much too noisy, and as undesirable neighbors. Some communities have placed restrictions on certain kinds of aircraft or hours of operations. Traffic patterns have been modified, as have operational procedures, to minimize the noise nuisance.

Chief noisemakers are the older jets. Newer designs have significantly reduced jet noise levels, and most jet engine installations will eventually be required to meet new lower noise level standards. The more common light aircraft with gasoline engines and propellers are much less of a problem.

Supersonic transports now flying the Atlantic produce their sonic booms harmleesly over the ocean. Their engines are only moderately more noisy than those of some subsonic airliners.

AIR TRAFFIC CONTROL

More aircraft flying means greater chances of midair collision. The basic method of avoiding such an event is "see and be seen," but tragic accidents have illustrated the shortcomings of this approach. It has been seriously proposed that all air traffic be precisely controlled by radar; near misses and accidents have shown that this does not work either. Furthermore, it becomes a physical impossibility as traffic grows.

The solution must be a system that will warn each pilot of the near presence of another aircraft, and either tell him where to look, which way to maneuver to avoid it, or, better yet, both.

Meanwhile, more stringent traffic control is being imposed, limiting the number and types of aircraft that may fly in a given area at one time. Carried to its ultimate conclusion, this approach would cripple air transportation, especially general aviation, without solving the problems. Other solutions must be found.

REGULATION

With increased control of traffic, regulation also increases. The FAA must be careful that any new regulations adopted are *necessary* and *the best available solution* to a safety or other problem.

Economic regulation of the airlines has sought to assure safe and adequate air service to the traveling public ever since the Civil Aeronautics Act of 1938. Unfortunately, in the view of many, monopolies, with unjustified higher prices, have resulted. Legislation in 1978 provided for gradual deregulation of the airlines and allowed greater competition for the more profitable airlines markets.

As they are allowed to do so, the airlines may very well decide the stop serving the unprofitable stops they previously had to accept in order to be awarded the routes they wanted. An accompanying growth of commuter airlines is to be expected.

The primary objective of regulation of air travel is the safety of the traveling public. Care must be taken to assure that regulation does not throttle the service, make it less available, or less safe and more expensive.

Aircraft Obsolescence

Properly maintained aircraft do not wear out. But they do outlive their original advantages as newer aircraft come along. The fighter planes of 20 years ago cannot survive if pitted against the newest ones. The older jet airliners cannot compete with the newer wide-body jets, which are not only faster, but safer and more economical to operate in terms of cost per seat mile.

The J-3 is hard to beat for slow sight-seeing, checking cattle, and hunting coyotes. But it lacks instruments and other capabilitities needed for modern pilot training. The Stinson 108 cannot compete with the new four-place aircraft as transportation for the modern businessman. So, while they may perform as well, and even look as good as new, progress passes them by. If they survive and avoid the scrap heap, they end up as someone's hobby. After all, that's what happened to horses.

2

Basic Aerodynamics and the Theory of Flight

A sound acquaintance with the principles of flight is an essential part of any modern airman's equipment. A special branch of the science of physics, aerodynamics, deals with these principles. It is, of course, possible to fly without speaking the language of an aerodynamicist. However, the airman, and particularly the professional airman, will derive much from a careful study of the concepts of aerodynamics and its mathematically expressed relationships. The subject cannot be covered in one chapter or even in one book; therefore the principal concepts, presented here, are highly condensed and require thoughtful reading.

Basic Principles

Before reading this chapter, an airman should have in mind the questions the chapter hopes to answer. Most basic of all is the question of what holds the airplane up in the sky. The explanation will require a discussion of some simple concepts of airflow which show how a wing produces lift. The breakdown of flow that produces lift is the explanation of an airplane's stalling.

Why is power needed? What are the controls, and how do they operate and affect the airplane? How do the characteristics and limitations of the aerodynamics influence performance? How does an airplane behave by itself, and how does this natural behavior affect the airman's job as a pilot?

How does the theory change for high-performance aircraft at speeds which approach or exceed the speed of sound?

The airman intuitively recognizes the relevance of these questions and should keep them in mind as he reads this chapter.

FORCES ACTING ON AN AIRPLANE: LIFT, WEIGHT, THRUST, AND DRAG

In Figure 2.1 the airplane is in straight and level flight. There are four net forces—*lift, drag, thrust,* and *weight*—acting on the airplane. By definition, the *lift* forces act perpendicularly to the relative wind or the line of flight; the *drag* forces act in parallel to the relative wind; the *thrust* forces usually act in parallel to the line of flight; while the *weight* always acts in the direction of gravity. *Relative wind* is the remote velocity (speed and direction) of air which strikes an airfoil (see Figure 2.11). In level unaccelerated flight the propulsion system must furnish a net thrust force adequate to balance the aerodynamic drag, which is the resistance of the air to the passage of the airplane. The lift force generated by the wing must be large enough to support the weight of the airplane.

VECTORS

For many quantities, both magnitude and direction are significant. They are called vector quantities and are represented schematically as arrows (Figure 2.2(a)). Velocity and force are the principal vectors of interest here. A vector may be the sum of several component vectors. Vectors are added by joining them tail to head. The resultant vector is represented by the line (arrow) from the tail of the first component to the head of the last as shown in Figures 2.2(b) and 2.2(c).

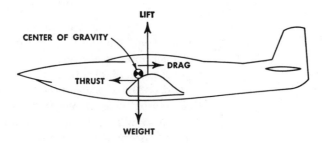

Figure 2.1. Forces acting in straight and level flight.

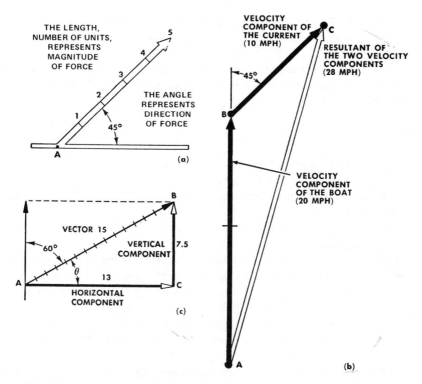

Figure 2.2. Vectors. (a) A force vector where Point A is the point of application. (b) Vectors may be added geometrically. For example, a boat whose velocity (speed in a certain direction) is represented by *AB* moves through a current having a velocity of *BC*. The boat's resultant velocity is the geometric sum, *AC*. (c) The vector having magnitude and direction *AB* may be conveniently divided into vertical and horizontal components equal to *CB* and *AC*. In trigonometric terms, $CB = AB (\sin \theta)$, $AC = AB (\cos \theta)$, and $CB = AC (\tan \theta)$.

FORCES AND MOMENTS

The effect of a force on an aircraft depends on its point of application. If the line of action of the force does not pass through the aircraft center of gravity (c.g.), a moment or torque is produced about the c.g., which tends to rotate the aircraft. A moment is given as a product of the force and the shortest distance between its line of action and the reference point about which it acts. The units are therefore those of a force times a distance, such as foot-pounds. When two equal and opposite forces act through separate points, they produce a pure moment or couple which does not change as the reference point is changed. In general, a force system can be represented at a reference point as a force and a moment.

These principles are well illustrated by a mechanic "pulling through" a propeller of a light plane. Consider a propeller 6 ft. in diameter with the mechanic pulling at one end. If the mechanic pulls straight down with a 50-lb. force, then the moment produced is $50 \times 3 = 150$ ft.-lb. about the prop shaft when the propeller is horizontal. If the propeller is inclined and the man pulls straight down, the moment arm is the horizontal distance from his handgrip to the vertical line through the propeller hub.

THE LAWS OF MOTION

Figure 2.1, again, shows an airplane in straight and level flight, with the four basic forces, lift, weight, thrust, and drag, in balance.

The actions and effects of forces are explained by Sir Isaac Newton's three laws of motion.

1. *A body remains at rest or moves uniformly in a straight line unless acted upon by a net force.*

Velocity is the combination of speed and direction. Acceleration is a change in velocity (speed or direction) and can be brought about only by a force which unbalances the existing forces. For unaccelerated flight, forces are equal in opposite directions. For example, thrust equals drag.

2. *The acceleration of a body is directly proportional to, and in the direction of, any net force acting on the body, and is inversely proportional to the mass of the body.*

This statement is usually expressed in equation form as:

$$F = \frac{W}{g} \times a, \text{ or } F = M \times a \qquad (2.1)$$

where F = the net force, lb.

W = weight of the body, lb.

g = acceleration due to gravity = 32.2 ft./sec.^{2}

a = acceleration of the body, ft./sec.^{2}

M = mass of the body defined as W/g, lb. sec.^{2}/ft.

This equation, when applied to an airplane, explains accelerations and decelerations in flight when thrust and drag or lift and weight are not equal. As noted above, any change in the velocity vector is an acceleration. A horizontal vector perpendicular to the velocity vector is required to perform a constant-altitude, constant-speed turn. In an airplane the force is produced by banking, so that a component of the lift vector acts horizontally to supply the centripetal acceleration.

3. *For every action there is an equal and opposite reaction.*

The third law explains how a propeller can produce a force sufficient to propel the airplane by accelerating the air, somewhat the same as the rower pushes against the water to accelerate it aft, thus propelling his boat forward. Similarly, the wing accelerates air downward to produce lift.

LOAD FACTORS

The term *load factor* is defined as the ratio of the lift developed by an airplane to its weight,

$$n = \frac{L}{W} \tag{2.2}$$

The simplest flight example is one in which lift is exactly equal to weight and is known as "1 *g*" flight; that is, the lift is exactly equal to the force needed to overcome the pull of gravity. A load factor of 1 exists in level flight. It is slightly greater than 1 in gentle turns or during the landing flare when the velocity direction is changing slowly. A more severe maneuver can impose a load factor substantially greater than 1 *g* on the airplane and crew. For example, it is 2 *g* in a level 60° banked turn.

INTRODUCTION TO AERODYNAMICS

Until the 1940's, the solution of subsonic flight problems dominated the field of aeronautics. Most design work was done assuming an incompressible medium (air), that is, the free air could not be compressed by a body moving through it. During the latter part of World War II, several cases of difficult control and buffeting were encountered at high speeds, especially in dives. It was soon discovered that these and other effects were related to the fact that air *is* compressible.

The range of speeds that must be considered for present-day flight is divided into four regimes. These are denoted successively as *subsonic, transonic, supersonic,* and *hypersonic.* Each regime denotes a speed range within which the aerodynamic design problems are generally similar.

AIRFOILS

An *airfoil* is any surface which is designed to obtain lift from the air through which it moves. In aerodynamic discussion, *airfoil* means a section of "wing" of infinite span (length); airplane fuselage and wing tip effects may be ignored in the study of a simple airfoil. NASA and other research agencies have developed and classified families of airfoils. Studies of the characteristics of these families have done much to increase our knowledge of aerodynamics, even though only a relatively small number of these airfoils are efficient lifting surfaces. Figures 2.3 and 2.4 illustrate typical airfoil shapes and airfoil terminology.

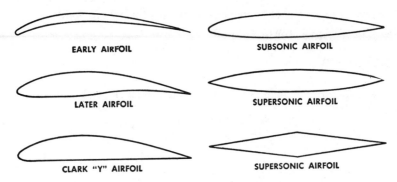

EARLY AIRFOIL   SUBSONIC AIRFOIL

LATER AIRFOIL   SUPERSONIC AIRFOIL

CLARK "Y" AIRFOIL   SUPERSONIC AIRFOIL

Figure 2.3. Typical airfoil cross sections.

Subsonic Aerodynamics

SUBSONIC FLOW

The theory of lift is based upon the forces generated between a body and a moving gas in which it is immersed. In subsonic flow, the velocity of particles of the gas never, at any point, exceeds the speed of sound. Air is the gas in which the aircraft is supported by the reaction of airflow around the airfoil and other parts' of the aircraft. Air is considered to be composed of a great number of small particles. In steady flow the path of each particle is called a streamline. Streamlines

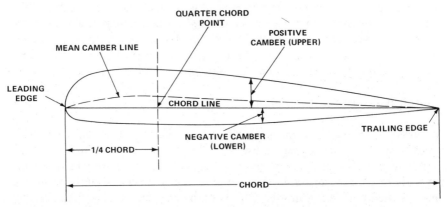

Figure 2.4. Asymmetrical subsonic airfoil cross section. In the symmetrical airfoil, the mean camber line and chord line coincide. Airfoil measurements are made with reference to fractions of chord length. Upper and lower camber are measured as perpendicular distances from the chord line to the upper or lower surface of the airfoil. The sum of the lengths of upper and lower cambers is the profile thickness.

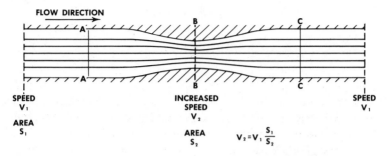

Figure 2.5. Steady flow of an ideal fluid through a restriction.

are everywhere tangent to the gas velocity, and either form closed paths or extend to infinity; no gas can flow across streamlines.

At speeds below about 260 knots, air can be considered incompressible. Under this assumption, air behaves no differently from water and is classified as a fluid. If it is further assumed that the effects of viscosity can be neglected, air is classified as an ideal fluid. Even with these seemingly severe restrictions, aerodynamicists have been able to obtain a great deal of practical information. In what follows, we will first concern ourselves with ideal-fluid aerodynamics and consider some basic principles that apply to this field.

CONTINUITY

The first concept to be considered is one involving a continuity principle. This principle expresses, in general, the requirement that fluid is neither created nor destroyed. For example, if in Figure 2.5 the illustration is considered to be a cross section of a tube with a restriction at *B-B,* the mass of fluid passing the restriction at *B-B* must be the same as the mass of fluid passing through the tube at *A-A* and *C-C.* To maintain continuity with an ideal fluid, the speed must therefore vary inversely as the cross-sectional area.

BERNOULLI'S LAW

In 1738 Daniel Bernoulli (1700–1783), a Swiss physicist, found that for an ideal fluid, one could write a very simple relation between the potential energy and the kinetic energy. The potential energy is represented by the pressure, and the kinetic energy is represented by the product of the fluid density and the square of the speed. Bernoulli found that along a streamline, the sum of these two energies is constant.

Mathematically this is written

$$P + \tfrac{1}{2}\rho V^2 = \text{constant} \qquad (2.3)$$

where P = the pressure
  ρ = the fluid density
  V = the fluid speed

This is a very important result, because it allows the aerodynamicist to calculate the pressure on a body once he knows the speed along the streamlines around the body. From the relationship above, we see that *where the velocity is lowered, the pressure must rise, and vice versa.* If we consider a streamline in Figure 2.5, we see that the pressure in the restriction must be lower than the pressure in the other parts of the tube. Furthermore, if flow conditions are such that the speed V becomes zero, the pressure P reaches a maximum, known as the stagnation point. From equation (2.3) it can be seen that the stagnation pressure is the constant on the right-hand side.

FORCES DUE TO FLUID FLOW

Jean le Rond d'Alembert (1717–1783), a French mathematician and encyclopedist, arrived at a famous quandary of fluid mechanics known as the "paradox of d'Alembert." He studied the forces that an ideal fluid exerts on solid bodies. His calculations showed that no net force would be exerted on the body by the fluid as the fluid moved past the body. For example, mathematically the flow past a circular cylinder of infinite length, illustrated in Figure 2.6, produces no net force on the cylinder. D'Alembert recognized that such results were in error, because experiments clearly showed that there was a net force, but he was never able to correct his calculations.

In the next century, Lord Rayleigh (1842–1919), in attempting to explain the swerving flight of a "cut" tennis ball, studied the flow over a rotating cylinder, as illustrated in Figure 2.7. He recognized that a real fluid, which is viscous, will adhere to the cylinder. The rotation thus imparts a circulatory component to the speed at the surface of the cylinder. In comparison with the flow in Figure 2.6, then, there is an increase in speed on the upper surface of the cylinder and a decrease on the lower surface. Bernoulli's law shows that there is a reduction of pressure on the upper surface and an increase of pressure on the lower surface,

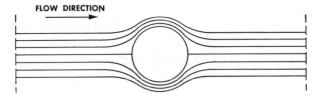

FLOW DIRECTION

Figure 2.6. Ideal flow past a circular cylinder.

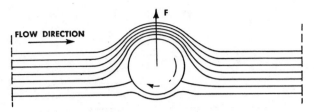

Figure 2.7. Ideal flow past a rotating cylinder.

producing a force F per unit cylinder length normal to the flow direction. A similar, though more complicated, flow pattern is caused by the spin of a tennis ball or a baseball; it is the unbalanced forces resulting from this spin that makes them curve.

CIRCULATION

Scientists soon recognized that the circulatory motion caused by the spin of Rayleigh's cylinder was an ingredient that had been left out of the theoretical models. This was pointed out independently by the English engineer, Frederick W. Lanchester (1876–1946), in the late nineteenth century, and early in this century by the German mathematician, Wilhelm Kutta (1867–1944), and the Russian mathematician and professor of mechanics, Nikolai E. Joukowske (1847–1921). Without this circulatory motion, or circulation, the flow about an airfoil would be as shown in Figure 2.8; there is no net force on this airfoil. However, there is also infinite speed at the trailing edge, where the flow is required to round a sharp corner.

Kutta and Joukowske propc ied that enough circulation be added to this theoretical model to correct the conditions at the trailing edge. Thus, while the average speed on the upper su face of the airfoil is increased, that on the lower surface is decreased, and the fl w over the trailing edge is smooth, as shown in Figure 2.9. Just as in the case c f Rayleigh's cylinder, the relative increase in the speed on the upper surface is acompanied by a reduction in pressure, the relative decrease in speed on the lower s irface causes a rise in pressure, and a net force F per unit length is calculated. Jo kowske was later able to demonstrate a remark-

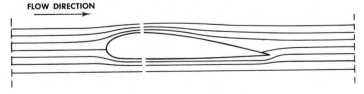

Figure 2.8. Ideal flow about an airfoil without circulation.

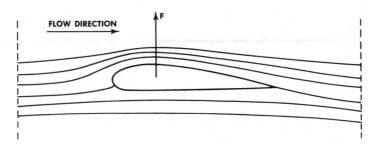

Figure 2.9. Ideal flow about an airfoil with circulation.

ably simple relation between the circulation around a two-dimension body and the aerodynamic force per unit length produced on it:

$$F = \rho V \Gamma \tag{2.4}$$

where Γ = circulation
ρ = fluid density
V = the speed with which the body moves through the fluid

This relation is useful because it permits calculating the aerodynamic forces directly from the circulation, rather than having to sum up pressure differences over the surface of the body.

Observed Circulation.   It is interesting to note that a very real justification for adding circulation to the theoretical model is a practical one—the flow picture of Figure 2.9 agrees quite well with experimental observations, as does the calculated force. In a real fluid, the circulation is induced by viscous forces acting near the airfoil surface. We do not have to spin the airfoil, as we had to spin the cylinder; the airfoil is shaped so that its linear motion alone suffices. Nonetheless, there is direct experimental evidence that circulation, or rotary motion, is produced around an airfoil as it is moved through a fluid. Figure 2.10 shows two views of an airfoil being started from rest in water. In both views a swirling of the water, called a vortex, is observed being shed from the rear of the airfoil. This vortex causes rotation, or circulation, in the water behind the airfoil. Theodore Von Karman[1] has related this flow to the lift generated by an airfoil with unparalleled clarity: "Now we must remember that, according to a fundamental principle of mechanics, a rotation, or more exactly a moment of momentum, cannot be created in a system without reaction. For example, if we try to put into rotation a body, such as a wheel, we experience a reaction tending to rotate us in the opposite direction. Or in the case of a helicopter with one rotor turning in one direction, we need a device to prevent the body of the craft being put into rotation in the opposite sense. Similarly, if the process of putting a wing section in mo-

[1] Theodore Von Karman, *Aerodynamics,* Cornell University Press, Ithaca, N.Y. 1954.

tion creates a vortex, i.e., a rotation of a part of the fluid, a rotation in the opposite sense is created in the rest of the fluid. This rotary motion of the fluid appears as the circulation around the wing section. In a way analogous to what we have seen in the case of the baseball, the circulation creates higher velocity (lower pressure) at the upper, and lower velocity (high pressure) at the lower,

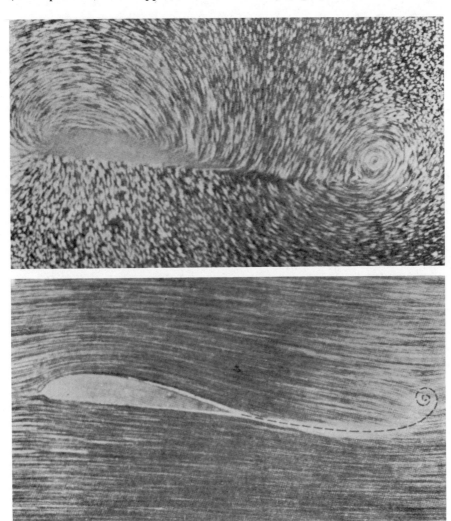

Figure 2.10. Starting vortex. Pictures of water flow around an airfoil. Lower: Camera at rest relative to undisturbed fluid. Upper: Camera moving with the airfoil. *(From T. Von Karman, Aerodynamics, Cornell University Press, Ithaca, N.Y., 1954. By permission from Applied Hydro- and Aeromechanics, by Prandtl and Tietjens, copyright by McGraw-Hill Book Company, Inc.)*

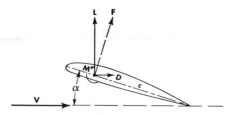

Figure 2.11. Airfoil notation.

surface of the wing. In this manner a positive lift is produced. This describes the basic principle of the way in which circulation, produced by vorticity, contributes to lift. It is further illustrated on page 44 in reference to airplane wings of finite span.

AIRFOIL FORCES

Conventional Airfoil Notation.   To simplify discussion of forces acting about an airfoil, standard notation is used as shown in Figure 2.11. The *chord, c,* is the usual reference axis. When written $\bar{c}$, it refers to the *mean aerodynamic chord (m.a.c.)* for the entire wing. *Angle of attack, α,* is the angle between the chord and the remote relative air velocity V, or *relative wind*. The aerodynamic forces acting on the airfoil are resolved into a force F acting at some point and a moment M acting about that point. The force F is further resolved into two components—a lift L acting perpendicularly to the relative wind, and a drag D acting in parallel to the relative wind. If we now refer to the force F in Figures 2.7 and 2.9, we see that it acts perpendicularly to the relative wind (or flow direction). Thus, the force F calculated by Rayleigh, Kutta, and Joukowske is all lift. In reality, there is also a drag, and to explain it we must turn to still another effect of viscosity that has heretofore been neglected.

BOUNDARY LAYER

An ideal fluid has no viscosity and therefore no ability to resist deformation due to a shearing force. Since viscosity is the property of a fluid that tends to prevent motion of one part of the fluid with respect to another, it follows that any real fluid which is viscous will resist the passage of a body immersed in it. Viscosity may best be visualized by thinking of the difference between a heavy oil and water; the oil is considerably more viscous than water.

The effects of viscosity may be seen easily if one considers a thin, flat plate immersed in a moving fluid. An ideal fluid would stream freely over the surface of the plate. However, any real fluid has a certain amount of viscosity, which will cause it to cling to the surface of the plate. Consequently the layer of particles nearest the plate will come to rest. The next layer of particles will be slowed down but not stopped. Figure 2.12 shows this effect. At the surface of the plate, fluid speed is zero. At some small measurable distance from the surface, the fluid is moving at freestream speed. This layer of fluid, within which viscosity induces a varying velocity, is called the *boundary layer*. Since it is usually relatively thin with respect to the thickness of an airfoil, it does not render lift calculations invalid. Typical boundary-layer thicknesses on an aircraft range from small fractions of an inch near the leading edge of a wing to the order of a foot at the aft end of a large aircraft such as the Boeing 747. This boundary-layer concept was introduced by the German engineer and professor, Ludwig Prandtl (1875–1953).

There are two different types of boundary-layer flow—laminar and turbulent (Figure 2.13). The laminar boundary layer is a very smooth flow, while the turbulent boundary layer contains swirls or "eddies." The English engineer and physicist, Osborne Reynolds (1842–1929), developed the basic relationships that enable us to determine which type of boundary layer exists in a given flow. His theories and experiments led to the development of a dimensionless number, called the Reynolds number, which can be used to determine the nature of flow along surfaces and about bodies. The Reynolds number may be expressed as:

$$RN = \frac{\rho V l}{\mu} \tag{2.5}$$

where ρ = fluid density
V = freestream speed
l = a characteristic length
μ = the coefficient of viscosity of the fluid

Examination of flow characteristics indicates that a transition from laminar to turbulent flow along a surface is dependent upon the Reynolds number. As shown in Figure 2.13, the laminar flow breaks down at some critical Reynolds number and becomes turbulent. The transition point depends upon the *surface roughness* and the *degree of turbulence* in the freestream, as well as on the terms

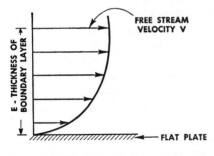

Figure 2.12. Boundary layer on a flat plate.

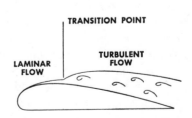

Figure 2.13. Boundary layer characteristics.

making up the Reynolds number. Typical values of the Reynolds number for aircraft range from 3,000,000 for a light plane to as high as 100,000,000 for the C-5A under certain flight conditions.

Another phenomenon associated with viscous flow is *separation*. Separation is said to occur when the flow breaks away from the body. In particular, this effect is predominant on airfoils at high angles of attack; leading-edge or trailing-edge separation results in extremely high drag and reduced lift. When separation has occurred over the upper surface, the wing is said to have "stalled." When stall has occurred, a further increase in the angle of attack produces a reduction rather than an increase in lift, and a sharp increase in drag.

It is important to remember that stall is primarily dependent upon angle of attack. An examination of equation (2.8) (page 59) shows that high angles of attack come about because of flight at low speeds or because of high load factors, such as occur in certain maneuvers. Thus an airplane will stall at many different airspeeds, depending upon the flight condition, but always at the same angle of attack in a given wing configuration.

The progression from laminar boundary layer to turbulent boundary layer and thence to flow separation gradually increases drag and ultimately destroys lift. Consequently much effort has been expended to control the boundary layer. Depending on circumstances, it may be desirable either to remove the boundary layer or to reinforce it.

Boundary layer *removal* has been accomplished experimentally by suction through a porous wing surface; this can serve to reduce skin friction drag and increase lift. Boundary layer *reinforcement* is accomplished by injecting air into the boundary layer through a porous wing surface; this serves to prevent separation, thus increasing lift. The precise location, power, and regulation requirements of boundary layer control devices make their application an extremely specialized operation, and quite expensive.

Profile Drag.   The presence of the boundary layer on an airfoil produces two types of drag. *Skin-friction drag* results from the tendency of the fluid to adhere to the airfoil surface, thereby producing shearing forces tangential to the surface in addition to the pressure forces discussed earlier, which are normal to the surface. *Form drag* produces the wake, a region of fluid of relatively low energy that trails to the rear of the airfoil. The surface pressure distribution is altered by the boundary layer and production of the wake, so that the resultant force is not all lift, as it would be in a fluid of zero viscosity, but also has a component in the drag direction. Airflow separation, which occurs over a larger area as angle of attack increases, produces high drag because it drastically increases the size of the wake. In practice, form drag and skin-friction drag on airfoils are lumped together under the name *profile drag*.

Pressure Distribution.   Figure 2.14 shows typical measured pressure distributions of an airfoil at two angles of attack. In the graphs, "relative pressure" means the difference between the pressure on the airfoil surface and the pressure

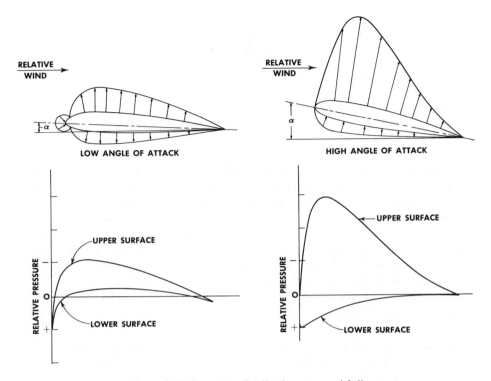

Figure 2.14. Pressure distribution on an airfoil.

in the undisturbed flow far ahead of the airfoil. Above each graph, the pressure distribution is indicated by arrows perpendicular to the airfoil surface. Arrows pointing away from the surface denote a negative relative pressure, or a pressure less than that in the undisturbed flow, while arrows pointing toward the surface denote the opposite effect. The lift developed by the airfoil is proportional to the area enclosed by the pressure-distribution curves, and from them it can be seen that most of the lift comes about from the reduction in pressure on the upper surface of the airfoil.

LIFT AND DRAG, FINITE SPAN WINGS

The discussion of lift and drag has, so far, been confined to the two-dimensional case, from which appears the picture of airflow in the mid-span region of an infinite span airfoil. Now a finite span wing must be considered to see what tip effects exist and how they affect the flow fields so far developed.

Figure 2.15 indicates the pressure distribution over a lifting wing. Because the pressure on the upper surface is less than the pressure on the lower surface, there is a tendency for the air to flow around the tips. In normal flight this flow is from

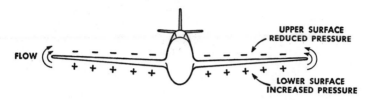

Figure 2.15. Spanwise pressure distribution.

lower surface to upper surface, from a high-pressure region to a low-pressure region. Prandtl introduced the proper theoretical model for this situation. The flow of air around the tips produces a system of trailing vortices, strongest at the tips, called a *vortex sheet*. A starting vortex, similar to the starting vortex discussed earlier in connection with the flow over an airfoil, terminates the vortex sheet downstream. A *bound vortex,* so called because it stays with the wing and produces the wing circulation, completes the model, which is sketched in Figure 2.16. The vortex sheet, with rotation in the sense shown by the arrows, induces a downward component of velocity in the flow called the *downwash field*.

Prandtl assumed that at any chordwise cross section of the wing, the force per unit span could be calculated from the airfoil theory. The relative wind for the airfoil would be given by the relative wind vector for the wing added to the downwash vector. This is illustrated in Figure 2.17. The resultant wind is seen to be rotated downward by a downwash angle, ϵ, and the force on the airfoil, perpendicular to the resultant wind, now has a drag component because lift and drag are defined as perpendicular and parallel to the relative wind. This drag component is called the *induced drag, D_i,* because it is induced by the downwash resulting from the vortex sheet. This drag, combined with the profile drag, gives the total drag for the wing.

WAKE TURBULENCE

The airman flying a light aircraft into an airport served by large commercial aircraft must heed carefully the controller's "Caution, wake turbulence." This extremely strong and dangerous turbulence is primarily caused by the system of

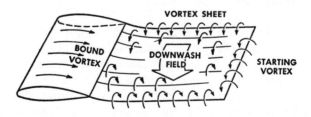

Figure 2.16. Ideal flow about a finite span wing.

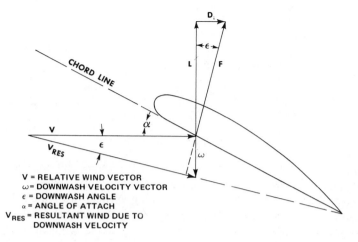

Figure 2.17 The origin of induced drag.

trailing vortices behind the aircraft. In actual fact, the vortex sheet soon "rolls up" into two concentrated trailing vortices that originate near the wing tips. These vortices are a particular problem near airports where large and heavy aircraft are flying relatively slowly.

The lift must always be near the weight of the aircraft. One way of looking at the creation of lift is to view it as the result of a rate of change of momentum imparted to the air by the wing. At slow speeds, the amount of air affected by the wing in 1 sec., for example, is reduced from what it would be at cruising speeds. To provide enough lift, then, the downwash must be correspondingly greater. This in turn means that the trailing vortices are strongest during this portion of the aircraft's flight. The trend toward larger and heavier aircraft, which land at approximately the same speeds as earlier aircraft, accentuates this problem. Singles, light twins, and DC-9's must avoid the 747's!

These trailing vortices are avoided by planning flight to stay above or well to one side of the flight path of the heavier airplane. One lands beyond the touchdown point of the large aircraft, and takes off before the rotation point of the large aircraft, turning aside to avoid flying through its wake.

Airport surface wind is important. A crosswind can blow the turbulence clear of the runway in a short time, but it can also blow it onto a parallel or intersecting runway. The strength and persistence of wake turbulence cannot be reliably predicted and is quite variable, due in large part to the variability of the weather.

ASPECT RATIO

Naturally, it is desirable to have the greatest possible lift with the least possible drag, or in other words, to make the ratio L/D_i as large as possible. To see how

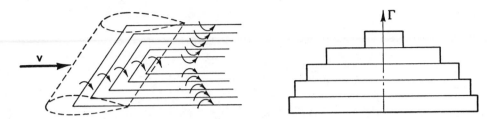

Figure 2.18. Vortex distribution.

this may be accomplished, one may look at the vortex pattern in a little greater detail. It is possible to approximate the vortex distribution on a wing as indicated in Figure 2.18. The bound vortex and trailing vortices of Figure 2.18 are shown as a series of "horseshoe" vortices, and the strength of the circulation is shown to vary from a maximum at midspan to zero at the wing tips. The resultant spanwise lift distribution approximates half an ellipse; the minimum induced drag is obtained when the spanwise distribution is elliptical. This distribution results in a constant downwash along the span.

One major factor in the determination of wing lift and drag characteristics is *aspect ratio*. Aspect ratio is defined as span squared divided by wing area:

$$AR = \frac{b^2}{S} \qquad (2.6)$$

or when the wing is rectangular, aspect ratio is span divided by chord:

$$AR = \frac{b}{c} \qquad (2.7)$$

where AR = aspect ratio
b = wing span
S = wing area
c = wing chord

Generally, the higher the aspect ratio, the smaller the downwash velocity, ω. Hence, L/D_i is improved. The use of high aspect ratio is limited by structural considerations; high strength is more easily obtained with low aspect ratio wings. Proper combinations of wing planform and wing twist can produce nearly elliptic pressure distribution on wings of low to medium aspect ratio.

Current design practice produces aspect ratios of 20 or higher for sailplanes. Some subsonic airplanes have aspect ratios approaching 12, although aspect ratios of 5 to 8 are more common.

GROUND EFFECT

Before continuing in the development of subsonic flight characteristics, we should recognize that the induced-drag characteristics of a wing are not the same near the ground as they are in free air. For operations such as take-off and landing, the effect of the ground on the flow pattern reduces the downwash velocity, and consequently the induced drag. The possible drag reduction varies from about 8% with a height above ground equal to semispan to as much as 50% at a height equal to $1/12$ span.

COMPLETE AIRPLANE DRAG

So far only wing lift and drag have been considered. For a complete airplane, it is necessary to consider the drag produced by the other component parts. The lift produced by components other than the wing is negligible. It is customary to refer to the drag of all parts not contributing to lift as *parasite* drag.

The correct determination of the total drag is an important phase in the design of an airplane since it is the first step in determining the power required to fly the craft.

The total drag of an airplane is obtained by summing up the drag of each part of the aircraft plus the drag resulting from the combination of these components. The airflow at interesections such as that of the wing and fuselage is often disturbed in such a way that the drag in the region of the intersection of the bodies is different from that which can be predicted by simply adding the drag of the parts. This is known as *interference drag,* a component of parasite drag. Interference effects can be favorable as well as unfavorable.

Exclusive of induced drag and interference drag, the drag of each aircraft component consists of two parts—*form drag* and *skin-friction drag.* Although it is usually quite difficult to separate the two, it must be remembered that form drag results from pressures induced on a component by the motion of air about it and depends on the wake characteristics, while skin-friction drag is a result of the viscous properties of the thin layer of air next to the surface of the component. The pressures and hence the form drag are largely a function of component shape, as shown in Figure 2.19.

Additional drag may also be caused during flight when control surfaces are deflected. And flaps, spoilers, or airbrakes may be used to add drag under certain circumstances such as rapid descent or landing approach.

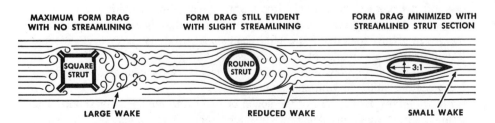

Figure 2.19. Effect of streamlining.

BUFFET AND FLUTTER

The pressure fluctuations associated with separated flow and turbulent wakes can create a feeling of "pounding" against some part of the aircraft. This effect is known as *buffet*. The separated flow may be due to ordinary stalling over local areas on the airplane (low speed buffet) or may be induced by a shock wave caused by the flow locally reaching sonic velocity (high speed buffet).

A loss in lifting force and an increase in the drag force are generally associated with buffeting. These effects may seriously limit the ability of the airplane to perform the missiion for which it was designed. Buffeting can further affect the airplane's gun or bomb platform stability, the peace of mind or fatigue of the pilot, crew members, and passengers, and, in extreme cases, the structural integrity of the airplane.

Flutter is an aeroelastic instability involving mutual interaction between the airstream and aircraft components such as the wing, the tail, control surfaces, or individual skin panels. When flutter occurs, the airstream interacts with the flexible motion of, say, the wing, such that any oscillation of the wing will rapidly increase in magnitude. Flutter is generally to be avoided at all costs, because it can lead very quickly to catastrophic structural failure. It is a fundamental design requirement that there be a given margin of safety between any flutter condition and the operating limits of the aircraft.

Supersonic Aerodynamics

In the preceding section we were concerned only with subsonic flow. Now we wish to consider flow at speeds greater than the speed of sound. World War II fighters attained speeds which produced, on critical points of the airplane, local flow velocities of Mach 1.0 and higher, although the airplane speed was less than the speed of sound. Many present-day airplanes are capable of exceeding the speed of sound in level flight.

For flight at low speeds, below about 260 knots, air acts as an incompressible fluid. However, as velocity increases, air density changes about the airplane and this effect becomes increasingly important. When flow velocities reach sonic speeds at some point on an airplane, the airplane's drag begins to increase at a rate much greater than that indicated by subsonic aerodynamic theory; subsonic flow principles are invalid at all speeds above this point.

Certain new definitions and concepts are necessary for dealing with air as a compressible fluid and with supersonic speeds.

Mach number is the ratio of the speed of motion to the speed of sound. The term comes from an Austrian physicist and philosopher, Ernst Mach

(1838–1916). An airplane traveling at the speed of sound is traveling at "Mach one" (Mach 1.0).

Increasing pressure is accomplished in supersonic flow by *shock waves* (compression waves).

Decreasing pressure is accomplished in supersonic flow by *expansion waves*.

SUPERSONIC FLOW CHARACTERISTICS

When an airplane flies at subsonic speeds, the air ahead is "warned" of the airplane's coming by a pressure change transmitted ahead of the airplane at the speed of sound. Because of this warning, the air begins to move aside before the airplane arrives and is prepared to let it pass easily. If the airplane travels at supersonic speeds, the air ahead receives no advance warning of the airplane's approach because the airplane is outspeeding its own pressure waves. Sound pressure changes are felt only within a cone-shaped region behind the nose of the airplane. Since the air is unprepared for the airplane's arrival, it must move aside abruptly to let the airplane pass. This sudden displacement of the air is accomplished through a shock wave.

The water-wave analogy furnishes a good physical picture of the subsonic "warning" system and supersonic shock formation. If one drops pebbles into a smooth pond of water, one each second from the same point, each pebble will produce a water wave moving outward with constantly increasing radius, as shown in Figure 2.20. This is similar to the pattern of sound waves produced by an airplane sitting on the runway before take-off. Even though one cannot see the airplane, its presence is signaled by these outward-rolling waves of engine noise.

Now suppose we move slowly over the pond, dropping pebbles at regular intervals. The picture of the waves is changed to that shown in Figure 2.21. Each pebble still produces a circular wave, but the circles are crowded together on the side toward which we are moving; the center of each succeeding circle is displaced from the prceding one by a distance proportional to the speed at which we are traveling over the water. This wave pattern is similar to the pattern of sound

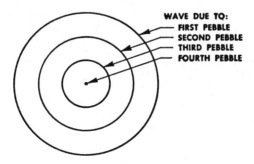

WAVE DUE TO:
— FIRST PEBBLE
— SECOND PEBBLE
— THIRD PEBBLE
— FOURTH PEBBLE

Figure 2.20. Stationary waves.

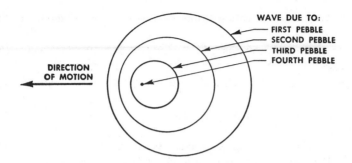

Figure 2.21. Waves for motion at subsonic speed.

waves around an airplane flying subsonically. The air ahead of the airplane is warned of the imminent arrival of the airplane, and the warning time decreases with increasing airplane speed. The warning time is zero when the airplane is flying at exactly sonic speed. The corresponding water-wave pattern is shown in Figure 2.22.

If we move across the water more rapidly than the water-wave speed, the wave pattern is markedly different from the patterns formed up until now. The smaller circles are no longer completely inside the next larger ones. Now all the circles are included within a wedge-shaped region as shown in Figure 2.23. This is similar to the sound-wave pattern for an airplane flying at supersonic speed. The airplane is, in fact, a continuous disturbance in the air rather than an intermittent one, as the pebbles falling regularly into the pond. Therefore, instead of several circles, there is an envelope surface of countless circles. The wedge on the surface of the pond looks like a section through the cone formed by an airplane in the air. Figure 2.23 indicates that there is considerable overrunning and interference between the wave circles; we might suspect from this that such interference will change the envelope shape for an actual airplane. This is, in fact, the case.

If the airplane is very streamlined and has a long, sharply pointed nose, then the air is not required to move a great distance suddenly in allowing the airplane

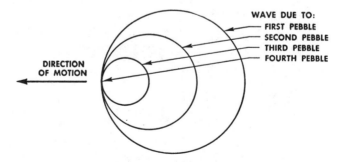

Figure 2.22. Waves simulating Mach 1.0 motion.

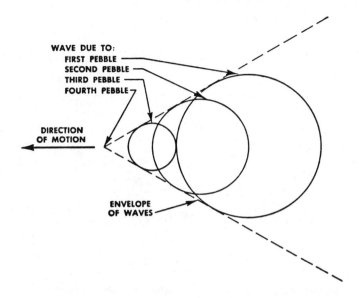

Figure 2.23. Waves simulating supersonic motion.

to pass. In this case the interference between sound waves is slight; the envelope is defined by the *Mach angle* μ. Figure 2.24 shows that the Mach angle is the angle whose sine is the speed of sound divided by body speed, or C/V. Thus, the Mach angle is 90° at a Mach number of 1.0, 30° at a Mach number of 2.0, and 10° at a Mach number of 5.75. This envelope is called a *Mach line* in two dimensions or a *Mach cone* in three dimensions.

TYPES OF SUPERSONIC WAVES

It is now apparent that waves are formed about any disturbance in a supersonic stream of air. The type of wave formed depends on the nature of the disturbing influence, which, in our case, is an airplane; its shape determines the location and characteristics of the waves formed. A wave of some type will exist whenever the air is required to change direction. The wave caused by a slight disturbance is defined as a Mach wave. Air passing through a Mach wave undergoes a small increase in temperature and pressure and a small decrease in velocity. The

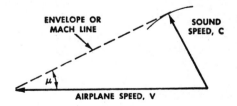

Figure 2.24. Mach angle.

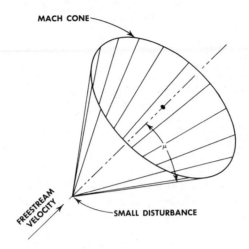

Figure 2.25. Mach cone.

Mach line envelope due to a small disturbance in three dimensions is conical in shape and is called a Mach cone (Figure 2.25). The envelope for a very thin wing is a wedge over most of the span bounded by a Mach cone at each tip as shown in Figure 2.26. The apex angle of these wedges and cones is the Mach angle, μ.

It cannot be overemphasized that Mach waves are associated only with very small or gradual changes in the flow direction of the passing air. The bodies that are small enough to produce Mach waves are in many cases too slender to be incorporated on an actual airplane. Many parts of an airplane must be too blunt and thick for Mach waves to form; instead, shock waves are formed. These shocks are formed by the interference of sound waves mentioned during the water-wave discussion. Shocks are like Mach waves in that the pressure and temperature of the air passing through are abruptly increased and the air velocity is decreased. However, the magnitude of these changes through a shock is many times greater than the magnitude of these changes through a Mach wave.

The difference in the magnitude of these changes is the essential difference between a Mach wave and a shock wave. Since the drag of an object is dependent

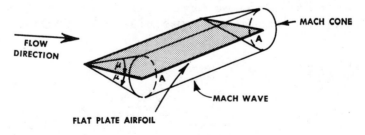

Figure 2.26. Mach wave and Mach cones.

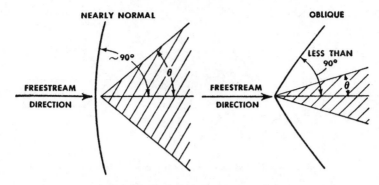

Figure 2.27. Types of shock waves.

upon the pressure on its surface, the drag caused by a shock is very high compared to that caused by a Mach wave on the same body. Fundamentally a Mach wave may be thought of as a shock of negligible strength, a shock through which air undergoes the smallest pressure, temperature, and velocity changes.

The magnitude of the changes in these properties is used to measure the strength of a shock. This strength is dependent upon the angle of the shock with the freestream and the freestream Mach number. Strong shocks are associated with high drag. The strongest shocks are normal shocks, so called because they stand at right angles to the freestream. All shocks standing at an angle of less than 90° to the freestream are called oblique shocks. Figure 2.27 shows examples of these two general cases.

The Mach wave and shock wave are compression waves. There is also the *expansion wave,* or fan, which has characteristics opposite to the compression wave. In passing through an expansion wave, air velocity increases, while temperature and pressure are reduced. Expansion waves occur where bodies begin to narrow, making more space available for the passing air to occupy. Figure 2.28 illustrates a typical expansion fan, or system of expansion waves. Since compression and expansion waves are opposite in nature, they tend to cancel each other when they intersect, and the shock's strength is reduced accordingly. Figure 2.29 shows a complete wave pattern on a double-wedge airfoil. The airfoil is

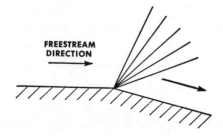

Figure 2.28. Expansion fan.

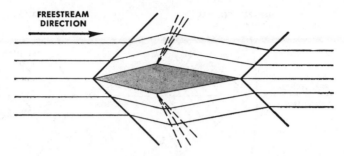

Figure 2.29. Wave pattern on a double-wedge airfoil.

Angle of attack = 0°.

at zero angle of attack. Shocks are formed at the leading and trailing edges while expansion fans occur at the surface angles or discontinuities. To obtain lift, the airfoil would have to have some finite angle of attack; the resultant wave pattern is shown by Figure 2.30 when the angle of attack is comparatively large.

The oblique shocks at the upper surface of the nose and lower surface of the tail are replaced by expansion fans in Figure 2.30. This is a graphic illustration of the effect of slope change on wave formation. In Figure 2.29, the turning of the flow required at the upper nose is concave, meaning that a rise in pressure and therefore a shock wave are required. A turning in the opposite, or convex, direction is required at this same point in Figure 2.30; this requires a drop in pressure and therefore an expansion fan. The situation at the lower trailing edge is similar.

AIRFOIL CHARACTERISTICS

The double-wedge airfoil used to illustrate wave patterns is convenient to study because the flow changes direction only at six definite regions. Another typical

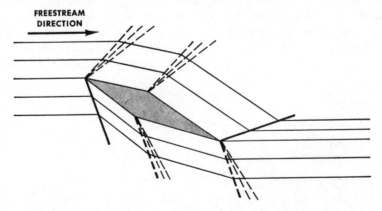

Figure 2.30. Wave pattern on a double-wedge airfoil at positive angle of attack.

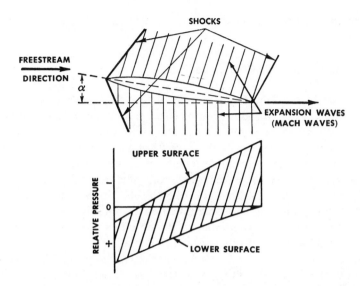

Figure 2.31. Pressure distribution on an airfoil in supersonic flow.

airfoil section might be as sketched in Figure 2.31. Here there are leading- and trailing-edge shocks, but the expansion is continuous over the entire surface between. The expansion waves intersect the leading-edge shock and progressively weaken it, thus making it a curved shock. The relative pressure, or difference between the airfoil surface pressure and the freestream pressure, is plotted as part of Figure 2.31. The pressure difference is proportional to the local inclination of the airfoil surface with respect to the freestream direction.

LIFT AND DRAG

As was pointed out earlier, the circulation theory of lift is not applicable in supersonic flow. Here, the upper and lower surfaces of the airfoil are, for all practical purposes, isolated from each other because of the inability of signals in the flow to propagate upstream. The pressure on the airfoil is determined by the freestream Mach number and pressure and local airfoil inclination to the freestream. The upper and lower surfaces contribute about equally to the lift. This is in contrast to the subsonic case, where most of the lift comes from the upper surface.

The drag of airfoils in supersonic flight is composed of three parts. These are *wave drag, skin-friction drag,* and *induced drag* or, as it is frequently called, *drag due to lift.* Wave drag is the drag at zero lift associated with the creation of shock waves or expansion fans in the flow. It comes from the pressure forces, or forces normal to the airfoil surface, as does form drag in subsonic flow. Drag due to lift comes from the alteration of the wave pattern as the angle of attack is changed from its zero-lift value. It is analogous to induced drag in subsonic flow, except that there is no induced drag for two-dimensional airfoils in subsonic flow.

WING (Finite span) CHARACTERISTICS

Remember that in subsonic flow, a wing of finite span experiences a three-dimensional flow which includes a vortex sheet, a downwash field, and induced velocities locally along the wing surface. This is not true in supersonic flow. In Figure 2.32, note that the pressure along the wing between the tip Mach cones is the same as for an airfoil of infinite length. Vortices produced within the tip Mach cones reduce the pressure from the airfoil value to zero at the tip, with the average lifting pressure in the tip region one-half the airfoil value. Thus, the influence of the tips is much less in supersonic flow than it is in subsonic flow. The drag due to lift is increased somewhat by the tip effect over its airfoil value.

If a wing with a planform other than rectangular is used, tip losses can be eliminated. The delta, or triangular, wing planform accomplishes this and can be illustrated by the two possible pressure patterns over a delta wing, depending on the relationship between freestream Mach number and wing leading-edge sweep.

In this case, Figure 2.33, the components of velocity perpendicular to the leading edge are subsonic, even though the freestream flow is supersonic. The lifting pressure is maximum along the leading edge and decreases rapidly toward the center of the wing. The average lift coefficient is less than would be obtained by a similar airfoil in subsonic flow.

In the other pattern, the wing leading edge lies ahead of the tip Mach cone. Figure 2.34 illustrates that the highest lifting pressure still exists along the wing

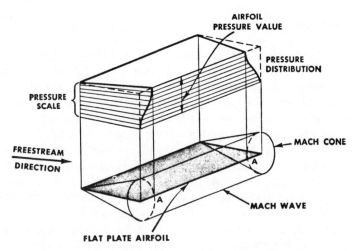

Figure 2.32. Supersonic pressure distribution on a rectangular wing.

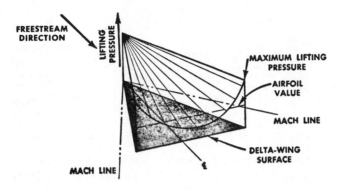

Figure 2.33. Pressure distribution on a delta wing (leading edge inside the Mach line).

leading edge. In this case, however, the lifting pressure remains constant at this peak value in the region between the leading edge and the Mach cone. Inside the Mach cone, the lifting pressure again decreases, but the wing's average lift coefficient is as high as can be obtained with an airfoil of a similar cross section. When the leading edge is outside, the wave drag is lower than airfoil wave drag. Wing wave drag reaches a maximum when the Mach cone lies along the leading edge.

Consider the effects of modifying the delta planform. If area is added at the trailing edge to make a diamond planform, it is being added where the local lift coefficient is low. In this case, the average wing lift coefficient is less than that obtainable with a delta planform. Conversely, cutting out area to give an arrow planform will increase the average lift coefficient.

BODY CHARACTERISTICS

This discussion pertains only to *bodies of revolution*. A body of revolution is one whose cross section perpendicular to its longitudinal axis is always circular (Fig-

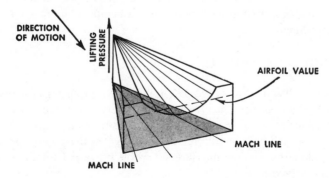

Figure 2.34. Pressure distribution on a delta wing (leading edge outside the Mach line).

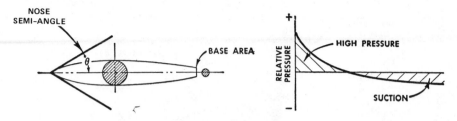

Figure 2.35. Pressure distribution on a body of revolution.

ure 2.35). Airplane fuselages are generally as nearly circular in cross section as volume requirements will permit. All studies, analytical and test, indicate that a parabolic longitudinal shape is desirable. The pressure distribution on a body of revolution at zero angle of attack shows a positive value at the nose lower than for an airfoil of the same "nose semi-angle" (Figure 2.35). Following this, the air finds more room in which to expand than in the case of the airfoil; it can fill the "ring" all around the body. The pressure drops so rapidly that the relative pressure returns to zero before the body slope has returned to zero, that is, before the body contour becomes parallel to the longitudinal axis of the body.

The expansion continues over the aft part of the body, and the relative pressure becomes more negative. However, the largest negative value of the relative pressure is limited by the occurrence of a complete expansion to a vacuum. Usually the positive pressure coefficient at the nose is larger than this maximum negative value.

The wave drag of bodies of revolution depends on their shape, angle of attack, and flight Mach number. Angles of attack other than zero usually do not cause a large drag increase, nor will they cause the body to produce much lift. Drag force increases with increasing Mach number just as does wing lift force. Body shape has a strong influence on body drag. The longitudinal lines should be parabolic; the exact equation of the lines is a function of Mach number. A good fineness ratio for low drag is a length to maximum diameter ratio of 8 to 12. Many bodies are designed so that the wave drag coefficient reaches its peak at very low supersonic Mach numbers, then drops rapidly until it begins to approach a minimum value at about Mach 2.

WING-BODY COMBINATIONS

In the study of subsonic drag (page 47), it was apparent that a factor was *interference drag,* resulting from the effect of angles between surfaces, such as wing and fuselage. This drag was reduced by the use of smooth fairings, thus avoiding sharp corners—in effect it was corrected "locally" at the point it occurred.

In supersonic airplanes, the interference problem is a much more critical one and cannot be solved locally.

We have seen that the ideal streamline shape is a body of revolution having a longitudinal parabolic curve. Another way of stating this is that if the cross-sectional areas of the ideal body, taken at even increments along its axis, were plotted, the result would be a parabolic curve.

Studies by Richard T. Whitcomb, NASA, have demonstrated that in supersonic aircraft the parabolic cross-sectional area distribution from nose to tail must be based on the complete airplane cross section, not just the fuselage cross section. This led to, and is illustrated in, the "Coke-bottle" shape of the fuselage of some early supersonic designs.

Lift-Drag Relationships

The preceding sections have discussed the means by which lift may be generated and the types of drag which must be considered in airplane design.

Lift and drag are normally expressed in coefficient form. These coefficients are dimensionless; that is, they are absolute numbers which are not associated with length, mass, or time.

The equations for lift and drag are:

$$L = C_L \rho \frac{V^2}{2} S \qquad (2.8)$$

where L = lift, lb.
   C_L = lift coefficient (which varies with angle of attack α)
   ρ = air density, slugs/cu. ft.
   V = speed, ft./sec.
   S = wing area, sq. ft.

$$\frac{\rho V^2}{2} = q = \text{dynamic pressure, lb. sq. ft.}$$

$$D = C_D q S \qquad (2.9)$$

where D = drag, lb.
   C_D = drag coefficient

Lift and drag are commonly plotted on curves similar to those of Figure 2.36. Notice that L/D, or C_L/C_D, reaches maximum value at some relatively low lift coefficient. Generally, designers attempt to have $(L/D)_{max}$ occur at the lift coefficient associated with maximum-range flight conditions. For a fixed airfoil, C_L, C_D, and L/D vary with α. In real life airplanes they are also varied by use of flaps and spoilers, which change airfoil shape and area.

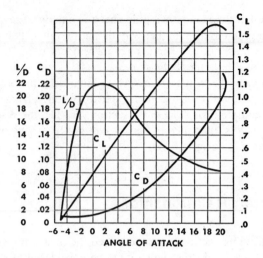

Figure 2.36. Aerodynamic characteristics of a typical airfoil.

PITCHING MOMENT

The pitching moment is the moment, M (Figure 2.11), which tends to rotate an airfoil about the pitch axis (Figure 2.38). As with lift and drag, it can be expressed in coefficient form:

$$M = C_M c q S \qquad (2.10)$$

where M = pitching moment, lb.–ft.
   C_M = pitching moment coefficient
   c = wing chord length, ft.

A moment must be defined with respect to a specific axis. For example, the moment may be summed about a line through the airplane c.g. parallel to the y-axis. In the longitudinal plane a reference point implies an axis so one simply says "about the c.g." A force acting through the reference point produces no moment about it. Given the moment about a point, the net forces can be placed at that point to represent the net force and moment.

Three reference points are of particular interest. The *center of pressure* (c.p.) of an airfoil is the point through which the net pressure force acts; that is, it is the point about which the pressure distribution (Figure 2.14) produces no moments. The *aerodynamic center* (a.c.) is the point at which the moment produced by the pressure distribution does not change with angle of attack. In considering the whole aircraft, the point about which there is no change in pitching moment as a function of angle of attack is called the *neutral point*.

In a cambered airfoil, the center of pressure moves along the chord as a result of various changes in angle of attack and relative wind. The moment of the c.p. about the a.c. is the pitching moment.

Performance

THE NEED FOR THRUST

Thrust is needed for airplane flight for two reasons: to overcome drag so that an airplane can fly at constant height and speed and to change the energy of the airplane as required by the *maneuver* at hand. The energy of the airplane is

$$E = (WV^2/2g) + Wh \tag{2.11}$$

where E = energy, ft.–lb.
W = weight of the airplane, lb.
V = speed, ft./sec.
g = acceleration due to gravity = 32.2 ft./sec.^{2}
h = altitude, ft.

The first term is called kinetic energy, and the second, potential energy. As an example of a mission requirement to change the energy, consider the landing of an airplane. The airman must reduce both the kinetic and potential energy. He does this by reducing thrust T until drag D is greater than thrust. The rate of change of the energy $\dot{E}$ is power and is given by

$$\dot{E} = (T \cos \alpha_t - D)V \tag{2.12}$$

where α_t is the angle between the thrust and the velocity vector.

The airman may exchange one type of energy for the other without changing the total energy. For example, a shallow dive from straight and level flight increases speed at the expense of altitude. The original flight conditions can be regained with very little change in energy by climbing back to altitude. The excess speed is returned to potential energy as the aircraft climbs.

In the discussion that follows, the relationships between lift and drag will make it possible to discuss the requirements for power under different flight conditions.

POWER REQUIRED

For straight and level flight at constant speed (V and h constant), the energy is constant, so thrust equals drag. "Power required" is a convenient way of specifying this equilibrium. It is given by the equation

$$P = \frac{DV}{550} \tag{2.13}$$

where P = power required, hp
  D = drag, lb.
  V = speed, ft./sec.

For jet- or rocket-powered aircraft, it is most convenient to work with thrust or drag force. In either case the lower curve of Figure 2.37 illustrates a typical speed-power required relationship.

Understanding fully the relationship between thrust, or power available, and total drag is vital to safe flying, particularly in large aircraft. Since parasite drag varies directly with the square of airspeed, it is relatively small at low airspeeds. Both lift and induced drag, however, vary directly with angle of attack. If sufficient lift is to be generated at low airspeeds, the wing must be at a higher angle of attack, resulting in a large induced drag, as is shown in Figure 2.36.

At low airspeeds, then, induced drag is a more important factor in determining thrust requirement than is parasite drag. As airspeed increases, a smaller angle of attack is required to produce the same lift, and induced drag decreases faster than parasite drag increases; their sum is less, and thrust required is less. Thus the faster the airplane goes, the less power is required, up to the point where parasite drag increases faster than induced drag decreases.

The power required at this point is the minimum required for level flight. For an airplane in this condition, it is the lowest point on a curve of power required

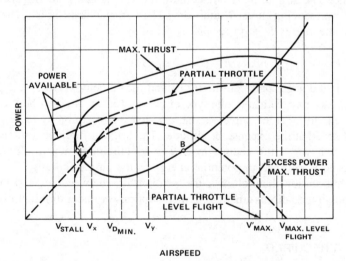

Figure 2.37. Typical aircraft power curve. V_x is the best angle of climb speed, V_y is the best rate of climb speed. The power available curve for a reduced throttle setting is lower and intersects the power required curve at a lower value for the V_{max} in level flight, V'_{max}.

versus airspeed, the lower curve in Figure 2.37. This minimum-power speed is also the maximum-endurance speed.

It is apparent that at a given level of power, such as line *A-B,* steady flight at either of two airspeeds is possible—one faster and one slower than the minimum point. Operation at the greater airspeed is normal and is known as "flying the front side of the power curve." After stabilizing at this greater speed, raising the nose will cause the airplane to decelerate to a somewhat lower airspeed and to climb.

Operation at the lesser airspeed is known as slow flight or "flying the back side of the power curve." Raising the nose causes the airplane to decelerate to a lower speed where more power is required to stabilize. If no more engine thrust power is available, the airplane can maintain the new speed only by losing altitude. Raising the nose still farther will result in a stall. Flying high on the back side must therefore be avoided during take-off, low-level flight, landing approach, or any other situation in which loss of altitude or a stall could be hazardous.

POWER AVAILABLE AND EXCESS POWER

Plotting the power available on the same graph gives further useful information. The difference between power available and power required is equal to the available rate of change of energy. For constant speed, this change corresponds directly to altitude change. The best rate of climb speed, v_y, is therefore where the difference is greatest. The maximum level flight speed is where the power-required and power-available curves are equal. The best angle of climb speed is v_x, where the tangent to the excess-power curve passes through the origin.

Subsonic Stability and Control

FUNDAMENTAL PRINCIPLES

An aircraft is designed to fly within a range of speeds and altitudes determined primarily by its weight, power, and structural strength. However, the aircraft must be stable and easily controlled to fulfill its design objectives. For stability the aircraft must be capable of holding a given flight condition once equilibrium has been established, and must tend to return to that condition if displaced by an outside force. For controllability, there must be controls that will allow the pilot to maneuver the aircraft safely and precisely. The determination of stability is precise, whereas the handling qualities which determine the controllability are judged subjectively and assigned a score by test-pilot rating.

An aircraft in flight is capable of six different types of motion. It may *translate*

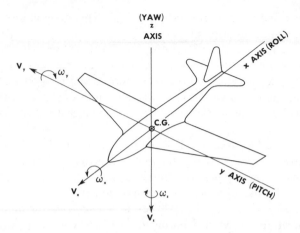

Figure 2.38. Stability axis system. *XZ* is the vertical plane (plane of symmetry), *XY*, the horizontal plane, and *YZ*, the lateral plane. V_x, V_y, and V_z refer to translation, and ω_x, ω_y, and ω_z refer to rotation with reference to the respective axes.

(move in a straight line) along any of three axes, or it may *rotate* about any one of these axes. Figure 2.38 shows the conventional stability axis system with its origin at the aircraft c.g., and indicates the direction of "positive" motion used for mathematical expression. Because of the symmetries of the aircraft, there are two types of motion: *longitudinal* motion and *lateral-directional* motion.

Longitudinal motion is in the plane of symmetry (*XZ*, Figure 2.38), and involves translation along the x- and z-axes, or rotation about the y-axis. During these motions, the symmetry of the airplane prevents any coupling to the lateral-directional motion.

Lateral-directional motion involves roll and yaw attitude and translation along the y-axis. These motions are coupled because rotation of the aircraft about either axis induces a moment of sufficient magnitude to cause motion about the other axis, and sideslip (translation along the y-axis) induces moments about both roll and yaw. (For example, an airplane will not bank without tending to yaw, nor yaw without tending to bank.)

The aircraft controls are designed to be safe and easy to operate so that a pilot can maneuver the airplane as required by the primary task for which it was designed. The purpose of the controls in an aircraft is to make it possible to navigate from one place to another. There is one main force, the lift, with which to perform maneuvers in conventional aircraft. (In VTOL and high-performance aircraft, the thrust may be so great that it is a primary force, too.) The lift magnitude depends on speed and angle of attack, while its direction depends principally on the orientation of the aircraft.

The controls which directly affect the lift are referred to as primary controls.

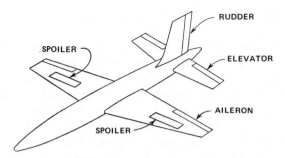

Figure 2.39. Control surfaces. Spoilers on top of the wings may assist or replace ailerons.

These include the elevators and the throttle, which are used to control the speed and angle of attack, and the ailerons (and sometimes spoilers), which control the angle of bank. A secondary control is the rudder, which controls the angle of sideslip and is used in keeping turns coordinated. Conventional controls are shown in Figure 2.39. The ailerons, elevators, and rudder principally produce moments about the roll, pitch, and yaw axes, respectively (although there is some coupling in roll-yaw), and the throttle controls the thrust.

AIRCRAFT CHARACTERISTIC BEHAVIOR

An airplane has dynamic behavior which is independent of the control an airman exercises over it. This inherent characteristic should be understood to appreciate the pilot's task in controlling the airplane.

Equilibrium is achieved when all of the forces and moments are in balance. In all but very small aircraft, small control surfaces are supplied to trim the aircraft so that the pilot does not constantly have to supply control forces or torques to maintain the equilibrium.

An aircraft is said to be *statically stable* if forces and moments are produced to return it to its equilibrium state when it is disturbed.

The airplane is *dynamically stable* if the motion that results from a disturbance dies out and the aircraft eventually returns to its equilibrium state.

A ball perched on top of a smooth hill is in a state of equilibrium that is both statically and dynamically unstable. In a valley, the ball oscillates about its equilibrium position. It is statically stable, but unless the oscillations die down due to friction, one cannot call it dynamically stable. Under some conditions of flight, an aircraft which is statically stable may be dynamically unstable, if oscillations that result from a perturbation increase in amplitude. Although static or dynamic instabilities are undesirable, they are not necessarily unacceptable, depending upon how fast they occur and what types of motion are involved.

LONGITUDINAL STATIC STABILITY AND TRIM

Gusts or other changes in the direction or velocity of the relative wind alter the angle of attack. A sudden gust changes the angle of attack, thus varying lift and drag (Figure 2.36). For static stability the increased lift occasioned by the increased angle of attack must also result in a nose-down pitching moment, or the nose will tend to rise and then continue to diverge farther and farther away from the original equilibrium. Therefore, the static stability of the airplane longitudinally is dependent upon the relationship of lift to pitching moment. The airplane longitudinal static stability is affected by the net contributions from the wing, fuselage, and tail.

First, let us see how the wing affects the static stability, that is, how its pitching moment about the aircraft c.g. (or the corresponding moment coefficient, $C_{M_{c.g.}}$) varies as a result of a sudden change in angle of attack. This change in angle of attack will cause a change in wing lift (or lift coefficient, C_L). In general, the moment coefficient will also change except if the lift is placed at the wing a.c. When placed at the wing a.c., the change in the moment about the c.g. is just due to the change in lift times the moment arm between the a.c. and c.g. A curve that relates $C_{M_{c.g.}}$ to C_L, as in Figure 2.40(a), shows how the pitching moment changes with change in lift. The solid C_L vs. C_M curve indicates a stable wing contribution to the overall airplane longitudinal stability, because a positive

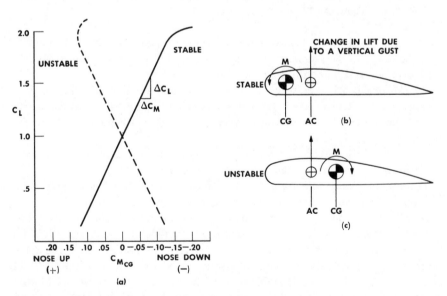

Figure 2.40. Relationship of wing lift and pitching moment to longitudinal stability. In (b) the nose-down pitching moment M is evident with the a.c. aft of the c.g. For changes in lift, the center of pressure acts generally at the quarter chord point, at the a.c.

increment of lift C_L will result in a negative, or nose-down, increment of pitching moment C_M as shown in Figure 2.40(b). Hence, moving the wing aft or the c.g. forward increases the wing's contribution to the aircraft static stability. As a result, in most airplanes the wing is placed so that there is a nose-down moment in equilibrium flight that must be trimmed by negative lift on the horizontal tail.

The horizontal tail is well aft of the c.g., so the principal changes in moment about the aircraft c.g. with angle of attack are due to changes in lift. An increased angle of attack produces an increase in lift, and hence a nose-down moment. The horizontal tail therefore produces a stabilizing contribution to the aircraft static stability.

In flying wings and in tailless airplanes, special techniques must be used. The a.c. must be behind the c.g., although this leads to trim problems. To obtain the noseup moment needed for equilibrium, reverse camber on straight wings and wing twist in swept wings (producing negative lift at the wing tips, which are behind the c.g.) are two techniques that can be used to obtain trim after the wing has been located so as to assure adequate longitudinal static stability.

The equilibrium or trim condition changes with airspeed. Supersonic aircraft must be designed to handle large changes in both the trim conditions and the static stability, as indicated by the slope of the lift-moment curves (see Figure 2.41).

The reference point for calculating aerodynamic moments is the airplane c.g. The location of the c.g. at which there is no change in moments as the angle of attack changes is called the *neutral point*. The static stability of an aircraft changes, therefore, with the changes in location of the c.g. due to payload and fuel variations. When the c.g. is in front of the neutral point, the aircraft is statically stable, whereas a c.g. location behind the neutral point would result in the aircraft being statically unstable. The neutral point-c.g. separation can be calculated from the slope of the aircraft $C_{M_{c.g.}}$ curve (Figure 2.40[a]), from the equation

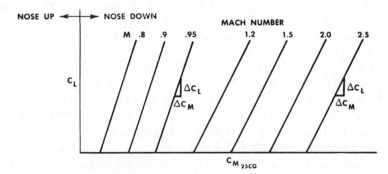

Figure 2.41. Pitching moment. In addition to the large trim changes required, these curves change slope as speed increases beyond Mach 1.0. Note: In illustration, change abscissa notation to Cm_{cg}.

$$neutral\ point - \text{c.g.} = \Delta C_{M c.g.} / \Delta C_L \qquad (2.14)$$

From this expression and the data of Figure 2.41, it can be seen that the neutral point of an airplane moves aft at supersonic speeds.

The range of allowable c.g. locations is an important design and operational consideration. This range is limited by stability and trim consideration. If the c.g. is too far forward, more negative left than can be produced may be required of the horizontal stabilizer to balance the nose-down moment produced by the wing. The absolute limit is when the horizontal stabilizer stalls. Also, excessive trim increases drag. More importantly, the pilot must have the ability to vary the tail lift if he is to control the aircraft. The amount of this variation is referred to as the *control authority* available. A minimum safe level of control authority, therefore limits the allowable forward c.g. travel, and stability limits its rearward travel. The airman meets these requirements by insuring that the airplane is loaded so that it remains within the airplane's moment and gross-weight envelope at all times.

A complicating factor at supersonic speeds is the change in the C_L vs. C_M curve as speed changes (Figure 2.41). The change in slope indicates that the neutral point shifts sharply aft as the airplane speed changes from subsonic to supersonic. This means that high elevator deflections must be employed to trim out the increased moment about the c.g., or the pilot must have a way to shift the c.g. aft to minimize trim drag. Presently, a c.g. shift obtained by fuel transfer is the most desirable solution. As shown in Figure 2.42, not only does the wing a.c. shift aft at supersonic speeds, but control-surface effectiveness, $C_{L\delta}$, decreases. Alternatively, variable-geometry aircraft, such as variable wing sweep on the F-111, could furnish an automatic method of trim control.

DIRECTIONAL-LATERAL STATIC STABILITY

The vertical tail provides the keel surface needed for adequate directional static stability, based on moments measured about the c.g. The tail size is not the only factor, however, and too large a tail would add unnecessary weight and drag, detrimental to airplane performance.

The neutral point for directional-lateral stability is analogous to the static longitudinal neutral point just discussed. The tail must be large enough to overcome the typically statically unstable effects of the fuselage and wing (although in lateral motion the wing does not play as important a role as it does in pitch). In directional-lateral behavior, the torques about yaw due to roll and sideslip make the discussion more complicated. The static effects of the coupling mechanisms should be understood to appreciate the types of natural motion that can occur.

The moment about yaw produced by sideslip may be accompanied by a rolling moment. If the moment tends to make the wing headed into the sideslip rise, the aircraft is said to have favorable or positive *dihedral* (see Figure 2.43). For ex-

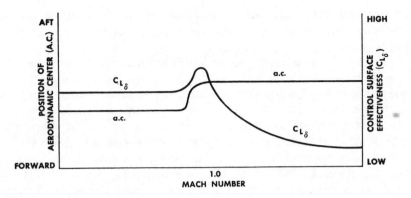

Figure 2.42. Control surface effectiveness.

ample, if the aircraft lowers its left wing but maintains heading control with the rudder, it will slip to the left. The rolling moment produced, as shown in Figure 2.43, tends to correct the low wing attitude and return the aircraft to its initial orientation. Positive dihedral is produced by several effects. It corresponds to the lateral neutral point being above the c.g. The aerodynamic center of the tail is usually above the c.g. and contributes positive effective dihedral. A swept wing on the left side of an aircraft slipping to the left meets the air more nearly perpendicularly and hence produces more lift. On the lee side, however, the right wing meets the air more obliquely and has less lift, thus producing positive effective dihedral. The most obvious effect is the geometric dihedral of the wing. Not as obvious is the effect of crossflow across the fuselage. In order to pass over the fuselage, the air flows generally upward over the upper part of the windward side of the fuselage and downward over the upper part of the lee side. A straight high-wing aircraft, therefore, experiences flow over its wings that resembles the flow over a wing with positive dihedral, as in Figure 2.43. A straight wing mounted at the bottom of an aircraft similarly has effective negative dihedral.

To obtain the stabilizing effect of dihedral, the whole airplane must be designed with positive dihedral, though the wing design is usually the principal factor.

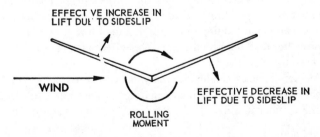

Figure 2.43. Rear view of dihedral effect (in a sideslip to the left).

In addition to these static effects, the angular velocity of an airplane produces moments and forces which are also important to the coupling in the directional-lateral behavior.

A yaw rate causes one wing to go faster than the other. Hence, the lift is increased on the faster wing and reduced on the slower one, causing a rolling moment. The yaw rate also produces a sideslip of the tail due to its tangential velocity perpendicular to the direction of motion. In addition to the side force that results, there is generally a rolling moment of the same sense as the effects of differential wing speed, because the aerodynamic center of the tail is above the c.g.

Rolling rates produce similar effects about the yaw axis. A rolling rate increases the angle of attack on the wing which is going down. The lift is increased, and it is accompanied by an increase in the induced drag. The difference in drag produces a yawing moment.

POWER EFFECTS ON CONTROL

The location of airplane engines can drastically influence the general control characteristics of an airplane. For a flight condition in which a large amount of excess thrust is available to the pilot, a sudden application of power can strongly influence the pilot's impression of controllability. If the airplane is arranged in such a manner that the thrust axis is displaced a considerable distance from the c.g., very rapid and positive control manipulations may be required in order to maintain trim for a steady flight path following a sudden increase in thrust. It is apparent, then, that careful attention must be given to the external geometry of the aircraft in order to insure its manageability.

Propeller-driven aircraft must allow for the effects of the yaw and roll forces produced by the propeller. Aircraft manufactured in the United States generally have propellers turning clockwise as viewed from the pilot's seat. Increasing the power output increases the torque reaction between the propeller and the engine-airplane combination, producing a left rolling tendency called the torque effect.

The slipstream is given a clockwise rotation, which impinges on the left side of the vertical tail surfaces, producing a left yawing tendency.

In an operation with substantial angle of attack, such as on take-off with a tail wheel airplane, or in a steep climb when the axis of propeller rotation is at an angle above the relative wind, the descending blade of the propeller is at a higher angle of attack than the ascending blade, producing more lift (thrust) on the right side of the propeller disk than on the left. This effect also causes left yaw and is known as "P-factor."

These three left turning tendencies may be compensated for in the design cruise regime by mounting the engine at an angle to the right, offsetting the leading edge of the vertical stabilizer to the left, or offsetting the rudder neutral to the right.

In a glide, when the engine is at low power or idling, the left turning tendency compensations for cruise flight will cause a right turning tendency.

Small aircraft must rely on the pilot's feet on the rudder pedals to compensate for turning tendencies in other than level cruise flight. The larger, more complex aircraft often have rudder and aileron trim adjustable from the cockpit, to assist the pilot in maintaining coordinated flight.

The propeller also behaves like a gyroscope. Precession causes a pitch-up effect when the aircraft turns or yaws left, and a pitch-down effect when it turns or yaws right. Pitch-up causes right yaw, and pitch-down, left yaw. These reactions are countered by use of the flight controls.

All these effects and correlations are reversed if engine rotation is counterclockwise.

LONGITUDINAL DYNAMIC CHARACTERISTICS

When there is more than one type of motion that can occur in a dynamic system, the different types are referred to as *modes*. There are two modes of motion that characterize longitudinal dynamics. One is a *short-period mode*. The pilot may typically notice this as a fast but well-damped oscillation about the pitch axis. He may also be aware of a variation in the *g* forces on the seat of his pants at the same frequency. There is almost no effect on airspeed during these oscillations. The short-period mode is excited by a sudden movement of the elevators or by a vertical wind gust. Its characteristics are the most important single factor determining the handling characteristics of an airplane.

In Figure 2.44, three aircraft have been disturbed in pitch. The amplitude of fluctuations through which they return to equilibrium has been plotted against time. Curve *A* shows how a statically unstable aircraft continually diverges. In curve *B,* the airplane is both statically and dynamically stable, but the return to equilibrium takes too long. It is not adequately "damped," so the pilot is faced with controlling an oscillation with a period of several seconds. Though the aircraft is stable, he may well feel he has a tiger by the tail. If there is too much damping, the pilot is given the impression of sluggishness. Too great a static sta-

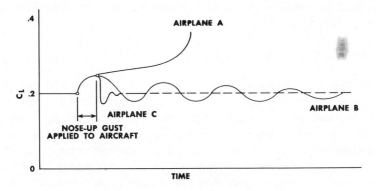

Figure 2.44. Stability in pitch.

bility (which makes the oscillations faster) requires excessive control forces to maneuver the airplane. Inadequate static stability makes the aircraft wander in pitch and oversensitive to trim. The airplane of curve C is acceptable because it returns to equilibrium in a relatively short time.

The second mode of motion is called the *phugoid mode*. The period of this typically lightly damped oscillation is about 25 sec. for light aircraft and increases directly with speed. Even when this mode is unstable, the pilot may be unaware of its existence because it is so slow. The phugoid mode is principally an exchange of potential and kinetic energy. The angle of attack is relatively constant, but the airspeed builds up in a shallow dive and increases the lift. The airplane climbs until the loss of lift starts the cycle again. The pitch attitude changes during the dive and climb.

DIRECTIONAL-LATERAL DYNAMIC CHARACTERISTICS

There are three modes that involve roll, yaw, and sideslip. The *dutch-roll mode* is similar to the short-period mode in pitch in that the directional-lateral static stability about the yaw axis is the restoring moment that produces the oscillation. The period is slightly longer, but there the analogy ends. The dutch-roll oscillations do not need to be as well damped because motions about yaw are not a primary mode of control. The motion is not limited to yaw but is coupled to the other axes so that roll and sideslip respond, too. Dutch roll is excited by a sudden application of rudder or by a lateral gust.

The *rolling mode* describes the rate at which a roll rate dies out. It is very fast, principally involves the rolling rate, but couples into yaw.

A long-period mode that is frequently unstable is called the *spiral mode*. Sideslip causes an aircraft to roll and yaw. If the yawing effect dominates, the aircraft turns into the direction of slip, and the nose lowers because the aircraft is banked. This aggravates the situation, and the airplane turns and dives at an increasing rate. When there is a lot of dihedral, the corrective rolling moment may right the aircraft and return it to straight and level flight (though on a new heading). The spiral mode is very slow and, as a result, is hardly noticed by the pilot just as is the phugoid mode.

CONTROL-SURFACE FUNCTIONS

The function of the elevators to pitch the airplane up or down is quite clear and straightforward, but their use in combination with the throttle deserves some emphasis. We have used the energy of the aircraft as a basis for discussing performance. It is very helpful in understanding the use of controls too. From equilibrium horizontal flight at constant thrust, elevator controls can exchange kinetic for potential energy but cannot change the total energy. Hence one can use the elevator to increase speed at the expense of altitude or to increase altitude at the expense of speed. The net force acting in the direction of the airplane velocity determines the rate of change of energy or net power (equation [2.11]). An increase in thrust from the trim condition increases the energy. The energy change can be effected in speed or altitude or both.

Energy management in landing is a good illustration. Maintaining the desired glide path requires careful management of the potential energy of the aircraft, which is accomplished by use of the elevator. Maintaining proper airspeed for desired performance (kinetic energy) is accomplished by use of the throttle. But when power is fixed, as during a full-power climb, or a power-off landing, airspeed (kinetic energy) is controlled by use of the elevator, and the resulting change of potential energy is accepted for whatever it is.

Some aircraft, particularly high-performance fighter types, use an angle of attack indicator to apprise the pilot of the aircraft performance on approach to landing and during other critical maneuvers; the indicator gives a direct reading of the item of greatest concern.

The basic purpose of the ailerons and rudder is to provide the forces and moments necessary for a heading change. The rudder affects yaw attitude. Its primary uses are to maintain zero sideslip during a turning or rolling maneuver, and to provide yaw control for crosswind landing or similar maneuvers. Lift provides the major turning force. The ailerons bank the aircraft to give a component of lift in the desired direction of turn (Figure 2.45). When deflected, they force the aircraft to roll because the up aileron reduces wing camber, reducing lift, while the down aileron increases lift by increasing camber. Spoilers may also be used to provide roll control. These are flaps on the top of the wing which are raised to interrupt the smooth flow of air over the airfoil, thereby reducing the lift. When an aileron reduces lift, it also reduces the induced drag. With a spoiler, however, the drag is increased. Ailerons produce adverse yaw because in initiating a left turn, for instance, a yawing moment to the right is produced, whereas spoilers may produce proverse yaw. In some aircraft the effects of ailerons and spoilers are combined to balance their yawing effect and to increase the aileron roll control authority at low speeds. Some aircraft, such as the B-52H, have only spoilers, not ailerons.

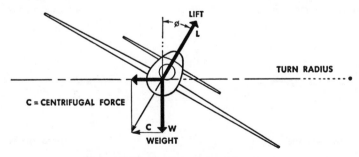

Figure 2.45. Airplane in level turn.

In various aircraft, conventional control surfaces are combined or augmented. Delta-wing aircraft combine elevator and ailerons, using flap-type surfaces at the wing trailing edge. They move together as elevators, differentially as ailerons. Ailerons and landing-flap controls may also be joined so that ailerons droop as flaps are lowered, improving low-speed control. The aileron and rudder may be interconnected to automatically coordinate the airplane in turns and thereby simplify flying, as in some light planes. Aircraft such as the Boeing jet transports or Boeing B-52D use spoilers together to reduce lift for rapid descent, and differentially in conjunction with ailerons to enhance ease of lateral control. Sailplanes also use spoilers to control descent.

CONVENTIONAL CONTROLS

As the speed of the air flowing over a control surface is increased, a given deflection of the surface against the airstream will produce greater and greater forces and moments, making the surface more and more difficult to move. Very large and supersonic aircraft have such high forces acting on the control surfaces that designers have been forced to use elaborate control systems in order to provide the pilot with sufficient mechanical advantage to move the surfaces.

Figure 2.46(a) illustrates a conventional type of system in which forces acting on the control surface are transmitted directly to the pilot's control stick. Figure 2.46(b) is the other extreme in design in which a hydraulic piston is connected to the aerodynamic control. When the pilot moves the control column, he actually positions a valve which allows high-pressure fluid to flow against one side of the hydraulic piston, moving the control surface against the airstream.

Figures 2.46(c) and (d) illustrate two other systems which allow the pilot's

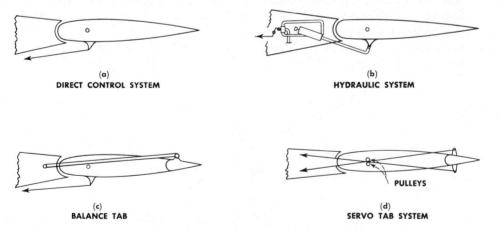

(a)
DIRECT CONTROL SYSTEM

(b)
HYDRAULIC SYSTEM

(c)
BALANCE TAB

(d)
SERVO TAB SYSTEM

PULLEYS

Figure 2.46. Typical control surface mechanisms.

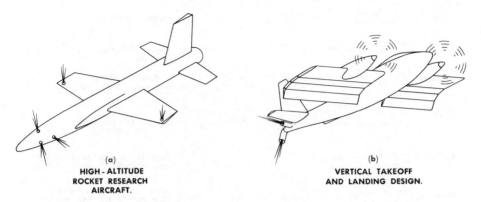

Figure 2.47. Reaction controls.

control column to be connected directly to an aerodynamic surface. The balance tab of Figure 2.46(c) is linked to the primary surface in such a manner that a displacement of the primary surface moves the small surface in the opposite direction, producing forces which assist the pilot in moving the primary surface. The servo tab of Figure 2.46(d) furnishes all the force applied to the primary surface. The control column is linked directly to the servo tab, and there is no linkage from the control stick to the main surface.

When aircraft operate at very low airspeeds or very high altitudes, the aerodynamic forces developed by conventional surfaces are too small to provide adequate control. Therefore reaction controls are used. These are either air jets which expel high-velocity air from the engine in variable amounts (Figure 2.47[b]), or miniature rocket engines as used to control the high-altitude research aircraft (Figure 2.47[a].

In addition to the primary controls, there are other controls which alter the aircraft characteristics. These include *speed brakes* and *high-lift devices*.

Speed brakes are hydraulically operated flaps which the pilot can project into the airstream to slow aircraft having the clean characteristics of jets without the retarding effects of windmilling props.

HIGH-LIFT DEVICES

For high speed and riding comfort in rough air, an airplane should have a high wing-loading; that is, for a given gross weight, the wing area should tend to be small. However, for take-off and landing the airplane should fly slowly and therefore have as large a wing area as is consistent with other factors of design for a given gross weight. Generally the compromise made between these two extremes results in a wing area too small to give satisfactory low-speed characteristics. Consequently, the wing must be modified by adding trailing-edge flaps, leading-edge slots or slats, or other high-lift devices.

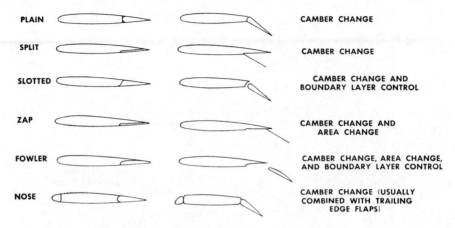

Figure 2.48. Various wing flap designs.

Trailing-edge flaps are probably the most commonly used high-lift devices. They increase the maximum lift coefficient attainable, which means that a given airplane can become or remain airborne at lower speeds with flaps extended than it can with flaps retracted. Some types of flap increase C_{Lmax} by increasing effective wing camber, while other types increase both area and camber. Flaps are controlled by the pilot and must be used judiciously because they increase drag markedly when fully extended. Figure 2.48 shows several types of flap.

Leading-edge slots and slats are provided in the leading edge of wings of some aircraft to smooth the airflow over the leading edge in such a way as to delay the stall at high angles of attack. Slots are fixed, while slats are mounted in such a way that at high angles of attack the decreased leading-edge pressure permits the slats to extend into the airstream, effectively increasing wing upper camber. They retract automatically when the aircraft reaches a lower angle of attack (see Figure 2.49).

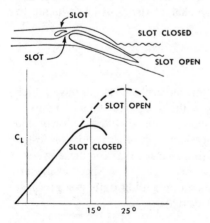

Figure 2.49. Typical aerodynamic characteristics with and without slot. (From *Airplane Performance, Stability and Control,* by Perkins and Hage.)

AUTOMATIC CONTROLS

Control systems may be designed to augment the natural stability of the airplane or to perform a mission automatically. In an automatic control system, for a quantity such as attitude, angular velocity, speed, or altitude, the signal is compared with the desired state and the error is used to deflect the control surfaces and make the aircraft behave as desired. In stability augmentation, one or more of the natural modes of the aircraft may need improvement. As discussed earlier, each mode involves certain variables but not all. For example, the short-period mode involves pitching motion and angle of attack, but not airspeed. To improve a particular mode, one senses one of the variables involved and applies it to a control that also affects the mode. For example, the short-period mode may have its damping improved by measuring pitch rate and applying it to the elevators. Dutch-roll damping can be improved in a similar way with a stability-augmentation system (SAS) that senses yaw rate or sideslip and uses it to control the rudder. A roll signal to the ailerons can overcome a tendency to spiral if the spiral mode is unstable.

In addition to SAS, mission-oriented autopilots can be designed to fly the airplane. They frequently work in conjunction with SAS and are subject to pilot override. Automatic attitude, heading, and altitude controls are common, and automatic landing systems have been developed. For a fighter aircraft, a maximum-performance turn at the limits of the airplane's structural capability can be performed by using an accelerometer to sense and maintain the maximum allowable *g* factor through elevator control. The possibilities are limited only by the ability to sense and by the intrinsic aerodynamic and structural limitations of the aircraft.

SUPERSONIC STABILITY AND CONTROL

The analysis of the stability and control characteristics of an airplane capable of flight at supersonic speeds is essentially the same as that used for subsonic aircraft. A supersonic airplane must be statically and dynamically stable; it must have a control system that gives the pilot accurate, safe control of the airplane throughout its flight regime. However, the magnitude of control forces and moments and the changes in these resulting from a displacement of the airplane about one or more of the stability axes are much greater at supersonic speeds than at subsonic speeds. Consequently, supersonic airplanes have power control systems to enable the pilot to move the control surfaces. The power control systems used on these aircraft can be designed so that the pilot has the same stick feel for a maneuver of given severity at all speeds despite the variation of control force with speed changes. In general, these airplanes also need three-axis stability-augmentation systems to provide satisfactory dynamic stability.

STRUCTURAL LIMITS ON PERFORMANCE AND STABILITY

The speed and maneuverability of an airplane are often restricted by structural criteria rather than by available power or aerodynamic characteristics. These structural limits are set by the strength and stiffness of the airframe and its components. Some aircraft can achieve level flight speeds that would destroy them.

Many normal flight maneuvers cause an increase in the loads on an airplane due to accelerations. The limit load factor is imposed to prevent loading any part of the airplane above its design strength. This limit varies from as low as 2 $g's$ for large transports and bombers to as high as 8.67 $g's$ for fighter-type aircraft. Another condition that determines aircraft strength requirements is gusts. A severe gust can impose high g loads so suddenly that the speed of airplanes is often limited to prevent structural damage due to gust loading.

Airframe flexibility, airplane weight distribution, and airplane aerodynamics can, singly or in combination, impose other restrictions. Buffet and flutter limits are a typical example of these restrictions. Flexibility can cause aileron reversal and wing divergence. Aileron reversal occurs when wing deformation due to air loads on the ailerons is great enough to nullify the effect of aileron deflection. Wing divergence occurs when the wing deformation due to air loads causes local wing loading to increase rather than decrease. This effect can be powerful enough to cause local failure in the wing. Both these effects can make it necessary to impose a high-speed limit.

Airframe flexibility problems are compounded by aerodynamic heating at supersonic speeds. At these speeds, skin temperatures are elevated well above ambient temperature. This means that stress analyses must be run on materials at extreme temperatures. To cope with this problem, the designer must resort to high-strength materials and limit high speed because of temperature.

AERODYNAMIC HEATING

Aerodynamic heating results from the conversion of kinetic energy to thermal energy. At the surface of an aircraft, the air is slowed to zero velocity. This means that freestream kinetic energy has been converted to thermal energy. The resultant stagnation temperature is a function of freestream temperature and Mach number. Most of the aircraft surface does not reach this stagnation temperature because the conversion of kinetic energy to thermal energy is not 100% efficient. The maximum resultant skin temperature, known as adiabatic wall temperature, is 85% to 95% as high as the stagnation temperature.

Radiation reduces skin temperature to less than adiabatic wall temperature; this effect increases as altitude and Mach number increase. Adiabatic wall temperature minus heat loss to radiation gives equilibrium temperature. This is the temperature to which an aircraft is designed.

Typical stagnation temperatures are about 260°F at Mach 2.0 and 1550°F at

Mach 5. These temperatures serve to explain the structural problems mentioned earlier and the need for air conditioning to provide a suitable environment for equipment and crew members. Space-vehicle designers have resorted to surface coatings which tend to ablate at extreme temperatures. The energy used to ablate the coating plus the insulating properties of the coating keep the vehicle primary structure at temperatures well below stagnation temperature, for the relatively short period of time required, *i.e.,* reentry into the earth's atmosphere.

There is another heating problem that is not due to aerodynamic effects—the effect of high-temperature engine exhaust gases on structure in the vicinity of the engine exhaust. Piston-engine exhaust location could be altered, but jet-engine exhaust cannot be diverted without a loss of thrust. Consequently, airplane structure must be of a material and design which can withstand this heating.

NOISE AND SONIC BOOM

The high-thrust jet engines used to power modern aircraft produce a sound level which tends to exceed human tolerance. Users of these aircraft equip their ground crews and mechanics with ear plugs or muffs to prevent physical and mental damage due to this noise. In addition to their effect on people, these sounds, or pressure waves, must be considered in the design of any part of the aircraft on which they impinge, and advances in engine design have reduced engine noises. Environmental restrictions imposed by political entities play a highly important part in new designs.

Supersonic flight is now commonplace to the public, but an associated phenomenon known as the sonic or supersonic boom has not always met with public acceptance. A boom will occur whenever a shock wave (pressure wave) emanating from an aircraft flying supersonically reaches an observer. According to the water-wave analogy, a supersonic boom is similar to a water wave moving past a floating leaf and causing the leaf to bob up and down.

There are many shocks emanating from an aircraft flying supersonically, but these usually interact and coalesce to form two main shocks—one from the nose and one from the aft end of the aircraft. For this reason, the *shape* of an aircraft has little or no influence on the number of shocks reaching an observer, although the *size* of the aircraft does. The bow and tail shock waves gradually diverge as they extend outward from the aircraft. This divergence is due to a slight difference in the propagation velocities of the two waves. The observer will hear two booms if the time interval between the passing of the two waves is of the order of .10 sec. or greater. If this time interval is much less than .10 sec., as would be experienced during a low-level pass of a fighter, the ear could not discriminate between the two waves and would hear only one boom.

The loudness of the boom is a function of the distance between airplane flight path and observer, Mach number, aircraft size and shape, atmospheric pressure, temperature, and winds. The factor having the strongest influence on loudness is the distance between observer and airplane flight path. The loudness, expressed as an increase in pressure above atmospheric, is inversely proportional to the three-quarter power of this distance.

Atmospheric temperature and wind gradients cause a bending or refraction of shock waves. Such conditions can produce a supersonic boom on the order of ten times the normal intensity. Other atmospheric conditions, such as clouds, cause a diffusion of the waves. Hence in some cases the shock-wave pattern from the airplane may become so distorted and attenuated that the boom is not even heard on the ground. Prediction of these effects under actual flight conditions is extremely difficult, though by precise control of airplane speed and flight path, shock waves may be focused at a point in space or on the ground.

Present and Future

There was once a move afoot to close the U.S. Patent Office, on the theory that everything that could be invented had been. Many thousands of patents have been granted since then.

In aerodynamics, too, valuable new ideas and mechanisms continue to come along, many from the research of the National Aeronautics and Space Administration (NASA), others from private research.

One of the more successful recent developments to come from NASA is the *supercritical airfoil* (Figure 2.50). By its shape it reduces the velocity of the airflow over the top of the wing, compared with that beneath the wing, and delays the transonic drag rise as speed approaches Mach 1.0. A downward cusp at the trailing edge also deflects the airflow downward more efficiently than do more conventional airfoils. The lower relative speed between the air and the top surface of the wing reduces drag due to skin friction, with consequent reduction in overall drag at a given speed. More speed is then possible at the same power. These characteristics make this type of airfoil attractive to designers of new high subsonic cruise speed aircraft. The supercritical airfoil is the result of the work of Richard Whitcomb of *area rule* fame and his associates at NASA.

Closely related to the supercritical airfoil, and developed by the same team, is the GA(W)-1 (General Aviation-Whitcomb, First Design) airfoil (Figure 2.51). It lifts better with less drag than conventional airfoils and is used in a number of new general aviation aircraft designs.

Figure 2.50. Supercritical airfoil.

Figure 2.51. GA(W)-1 airfoil.

Induced drag and wake turbulence are the result of circulation arising from the existence of a higher pressure on the bottom of the wing and lower pressure on the top; the air flows around the wing tip. Fences, and flat plates on the wing tips, have been tried with doubtful success. A more promising development, at least for larger aircraft, is winglets on the wing tips to counteract the pressure-difference-induced circulation. Such winglets might also use a supercritical airfoil, and provide additional lift.

Progress has indeed not come to a standstill in the art and science of aerodynamics.

Aerodynamics and Flight

As airplanes have become faster and more complex, the precision required of the pilot has increased correspondingly. Fortunately, the expanding knowledge of aerodynamics which has contributed to the design of high-performance aircraft has also come to the aid of the airman. His aircraft can do more than ever for him if he will handle it properly. To do this he needs only two basic tools. One is a conscientious desire to fly his aircraft as efficiently as possible; the other is a basic comprehension of airplane aerodynamics. An airman so equipped will better understand the data in his flight manual; he will use these data properly and thus get the most efficient and safe use of his aircraft.

3

Airplane Structures

An airplane, by its nature, must be an extremely efficient machine. In flight, its shape and contours must provide aerodynamic lift for overcoming gravity and for maneuvering and control, and yet they must present a minimum of area in order that the drag, or wind resistance, is a minimum. However, within this shape must be adequate space to house all the necessary equipment, propulsion units, and systems, as well as a comfortable environment for the crew, passengers, and baggage or cargo. An integral part of this shape, in fact its very contour, comprises the structure of the airplane. Since all items of the airplane must contribute to the total efficiency, it follows that the structure must have a high strength-to-weight ratio and be as simple as possible. It must safely resist all the loads imposed on the airplane from a variety of sources, supporting the aircraft in the atmosphere, maneuvering in three dimensions, and operating on the ground.

The airplane structure must be capable of withstanding much more load than that imposed by its weight alone. When a particular type of design is established, the designers provide structure according to strict standards established by experience and experimentation to insure safety. In general, airplanes are designed to withstand one and a half times the maximum expected loads.

The conventional structural elements are the fuselage, the wings, the stabilizers, the controls, and the landing gear. Regardless of the purpose of the airplane, these primary elements must all have high strength, low weight, safety, and an efficient aerodynamic shape.

Loads

External loads which act on an airplane come from the air—in the form of turbulence (gusts) or maneuvering loads—and from the ground during taxi, take-off, landing, or ground handling.

For convenience in fixing loads, moments, and directions, structural designers and aerodynamicists both use the same coordinate system (Figure 2.38) to depict the various axes of the aircraft. Thus, a rolling moment caused by aileron deflection would be about the x-axis; a pitching moment caused by elevator deflection would be about the y-axis; and a yawing moment caused by rudder deflection would be about the z-axis.

AIR LOADS

An aircraft in level, undisturbed flight is balanced, and the net lifting loads equal the weight of the airplane. This condition is only temporary, as the airflow in the atmosphere is usually changing in velocity and direction. Sudden changes in this airflow are felt as "bumps"—the result of turbulent air or, more precisely, gusts. The magnitude of the loads induced under these conditions varies inversely with the *wing-loading* (weight of the airplane divided by the square feet of wing area) of the airplane, and directly with the aircraft velocity and the velocity of the gust. That is, the greater the wing-loading, the less load developed; however, the greater the velocity of the aircraft, the greater the load becomes.

Maneuvering loads are created at any time the controls are operated in flight, resulting in a change in flight path. For example, in a sudden pull-up maneuver, the elevator is deflected upward, causing a down load on the horizontal tail. The nose rotates upward, the wing angle of attack increases, and the up load on the wing will increase. An opposite maneuver, of course, would impose loads in the opposite direction. Rudder deflection will form a sideward load on the vertical tail, resulting in yaw, and ailerons, when deflected, unbalance the loads on the wings, causing roll. In a well-coordinated maneuver, to change attitude and direction simultaneously, the forces created by all of the control surfaces are acting at the same time, each in proportion to its degree of deflection.

DYNAMIC LOADS

In addition to the air loads discussed previously, there exist loads resulting from the dynamic response of the structure. Any structure has a natural or resonant frequency of vibration. The structure of an airplane is elastic, *i.e.*, it deflects or twists under load and is subjected to aerodynamic forces; therefore aeroelastic oscillation of the component can develop. Excitation is caused by the airstream and is, thus, related to airspeed. At a certain velocity, the forcing frequency of the airstream will equal the resonant frequency of the component, such as a control surface, and cause it to vibrate. Since this occurs at the resonant frequency of the component, the amplitude of vibration will increase rapidly without an increase in forcing energy. The velocity must be reduced or destruction will result. This condition is known as *flutter* and may also be aggravated by another component sympathizing with the first—a phenomenon known as *coupling*.

Proper structural design provides for this type of load by "tuning" the airframe for certain frequencies of response. Further, control surfaces are statically balanced, and in some high-speed, complex aircraft, auxiliary dampeners are used. Any component can flutter if the speed is high enough and atmospheric conditions are right. During the structural design of the aircraft, the flutter speed is determined and must be well above the maximum design speed of the vehicle. It is therefore important that the manufacturer's specifications and operational limitations are followed. Specifically, control surfaces which are modified must meet the requirements stated in the manufacturer's manuals.

Divergence is another type of dynamic load. It explains the reasons for some of the structural design shapes.

The resultant wing load is generally far forward on the wing—usually about one-fourth of the chord length aft of the leading edge. This is normally forward of the elastic axis of the wing (the axis where load could be applied without twisting the wing). The load causes the wing to twist in a nose-up direction by an amount proportional to the torsional stiffness of the wing. If the wing lacked torsional stiffness, the twisting would be considerable. Twisting in the nose-up direction would increase the angle of attack of the wing and, therefore, the load on the wing. The additional load would further increase the angle of attack. If the wing twisted easily, the condition above would continue until the limits of the wing were exceeded—a phenomenon called divergence. The designer, therefore, provides adequate torsional stiffness for the airplane's specified operational limits.

Another load resulting from lack of torsional stiffness is *aileron reversal*. We have seen that the resultant wing load is usually forward of the elastic axis. When an aileron is deflected downward, it develops a load on its own surface from the airstream. This load is far aft of the elastic axis of the wing and causes a nose-down twisting action on the wing. As the wing noses down, it loses some of the lift gained by deflecting the aileron. If the wing lacks torsional rigidity, it is possible that the nosing down of the wing will cause the loss of more lift than is gained by deflecting the aileron. Then the wing will go downward with a down-aileron deflection—opposite the intended maneuver; hence, aileron reversal.

All of these various dynamic loads become a part of the considerations of the designer in developing an efficient and safe structure.

LOAD FACTOR

As mentioned previously, when an aircraft is in flight and a gust is encountered, a load results which is in addition to the normal weight of the craft. In level, unaccelerated flight, the wings support the weight of the craft plus any load needed to maintain equilibrium. A gust, when encountered, gives rise to an additional force. The gust attempts to change the path of the airplane, but because of its inertia, the airplane tends to remain on its original path. The force of the inertia, which acts so long as the acceleration is taking place, is algebraically added to the force of gravity, and a new total force or apparent weight will result. The ratio of this new force divided by the original weight of the aircraft is a nondimensional quantity called "load factor."

Gust loads in general are of shorter duration than maneuver loads, but their direction change can be much faster and sometimes will appear to be almost instantaneous. It is during these times of instantaneous change that the load factors produced will be highest.

The speed and wing loading of the airplane will have a direct bearing on the size of the load factor, whether the source of the acceleration be a gust or a maneuver. In general, the gust load factor increases as the speed of the airplane or the gust velocity increases, and decreases as the wing-loading increases. In maneuvers, the loads developed increase as the square of the velocity, that is, when the speed is doubled, the loads will increase four times.

In determining load factor by dividing the apparent weight or force by the actual weight, the result is a relative quantity for the particular airplane involved. For example, if a plane had a weight of 10,000 lb. and developed a load factor of 1.5, the resulting load would be $1.5 \times 10,000$ or 15,000 lb. The speed, the weight-lifting capabilities of the airplane, the specific usage of the vehicle, and the atmosphere all play a part in determining the limits to which an airplane is designed. Specific values of load factors are established for a variety of types of aircraft, and since these are the maximum limits an airplane is ever expected to encounter, they are usually referred to as *limit load factors*.

Recognizing that abnormal circumstances may result in sometimes exceeding these limits and that it is impossible to predict the different elements in nature with complete accuracy, it is necessary that a safety factor be incorporated in all aircraft design. The most common safety factor is 1.5. This safety factor provides a margin for unpredictable elements and insures good service life. It is important to note that this factor is applied to loads only, and not to the weights, speeds, or other limitations of the airplane. Examples of some of the limit load factors currently used are listed in Table 3.1. Notice that the load factors listed

TABLE 3.1. Limit Load Factors for Various Types of Aircraft.

Type	Performance or Mission	Positive Flight Load Factor
Fighters	Interception, aerial combat, aerobatics	5.3 to 8.7
Bombers	Long-range missions with relatively high payloads	2.0 to 3.7
Trainers	Instruction and practice in flight fundamentals or advanced maneuvers	5.7 to 7.3
Aerobatic	Sudden violent maneuvers, demonstrations	6.0
Utility	Agricultural dressing, training, cargo, general purpose	4.4
Liaison	Observation, transport, rough field operation, photography	3.8 to 4.4
Private	Business flying, training, pleasure	3.8
Passenger Transport	Airlines, large executive or military carriers	2.5
Cargo Transport	Commercial or military freight carriers	2.5 to 3.0

are sufficient for the craft to perform the mission or fulfill the purpose for which it is designed.

GROUND LOADS

These are the next important loading conditions to be considered. Each flight involves at least one take-off, one landing, and some taxiing. Once again the purpose of the aircraft will determine, to a large extent, the amount of time to be spent in the air and on the ground. Usually the landing loads, rather than take-off loads, govern the design of the gear attachment structure of an airplane, even though the allowable take-off weight may be higher than the landing weight. Primarily, designers must consider load conditions in level landing, tail-down landing, braked roll, side forces, and, on some special-purpose aircraft, barrier or hook loads for stopping the aircraft roll in a short distance.

Ground handling loads from towing, winching, jacking, and tiedown may present special problems in a particular design and are considered in all designs. Taxiing, turning (with brakes or steering system), and obstructions, such as chocks, curbing, or ground irregularities, also impose many special kinds and magnitudes of loads on the landing gear structure.

Descent velocities of the particular type of airplane, as well as the wing-loading in association with the shock-absorption characteristics of the landing gear struts and tires, will determine, in large part, the reaction at ground contact. Total reaction force divided by the weight of the aircraft is called the *landing load factor*.

Two types of inertia loads are involved in landing conditions. The first is *translational inertia*—the force resulting from the aircraft's downward momentum as ground contact is made. The second, *rotational inertia,* is developed when an aircraft touches down on its main landing gear first and begins to rotate toward its nose or tail wheel. The aircraft mass strives to resist this rotation with a counter inertia. The force from rotational inertia can produce airplane structural loads larger than those developed in flight.

MISCELLANEOUS LOADS

These are considered in the structural design of an airplane and come from a host of sources. Most aircraft operating at higher altitudes are pressurized internally, which usually results in a rather large load in the aircraft structure very similar to that of a pressure tank. Structure around doors, windows, and other cutouts becomes particularly critical in these designs. High-frequency vibratory loads have become rather commonplace in the modern airplanes equipped with turbine engines. Loads which are peculiar to transonic and supersonic craft as a result of shock waves influence the structural design of aircraft operating at those velocities. Loads due to buffeting at low speeds are fairly well known and are in the inertia or dynamic-load class. Vibratory loads in the frequency range of those produced by piston engines are still present on many aircraft and must be accounted for in a successful design.

In recent years a great deal of research has been performed in the area of crashworthiness, although there is still much to be learned about the human body and its resistance to large load factors under short time intervals. This ever-increasing knowledge is constantly being put to use in the design of structure, particularly in the cabin section, seats, and restraint systems. Structure is designed to slowly and progressively crush under high loads, slowing the absorption of kinetic energy and thereby reducing the magnitude of load factors. Restraint systems, consisting of a seat belt and shoulder harness, keep the occupant from moving forward under the deceleration forces. Careful design balance is required in order to minimize load factors developed and adequately contain the occupant within the seat and cabin space.

STRESS

Any load applied to a unit area of material is called a *stress* and will produce a deflection or deformation in the material, termed *strain.* There are *tension, compression, shear,* and *bending* stresses. While there are many other types of stresses, they are special arrangements of these basic ones.

Strain is directly proportional to stress as load is applied until the *proportional limit* is reached. Beyond that point strain may increase at a changing rate until *yield stress* is reached, but the part will return to original size and shape (strain will return to zero) when the load is removed. After yield stress is exceeded, the part is permanently deformed and will not return to original size and shape, although it may accept an appreciably greater load before failing completely at the *ultimate stress*.

A simple example of this may be shown in the action of a rubber band. As load is applied, the rubber stretches. The load applies a tensile stress to the rubber material while the actual amount of lengthening is the strain, usually measured as the number of inches of stretch per each inch of the material. Obviously, in the case of the rubber, the stretch is considerable, whereas in metals the amount of stretch could only be measured by precision equipment. All materials have elastic properties and below the yield stress (elastic limit), or the point of permanent set, will return to their original condition when the load is removed. The ratio of stress to strain, within the elastic limit of the material, is constant and is referred to as the *modulus of elasticity*.

Fatigue loading has become an important design consideration in all classes of aircraft. The increasing performance of modern aircraft and higher utilization

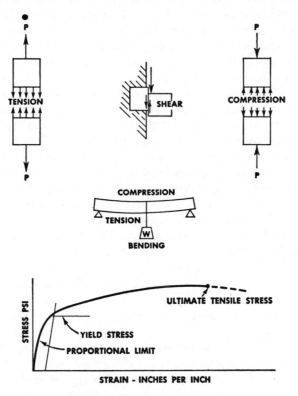

Figure 3.1. Stresses and strain. The lower sketch shows a typical stress-strain curve for aluminum.

rates have dictated the requirements for primary structure to approach infinite service life. Many structural areas, especially those subjected to highly concentrated cyclic loads, are designed by fatigue requirements. This is accomplished by designing for low stress levels to approach infinite life or by fail-safe techniques which deliberately provide multiple load paths within the structure.

Considerable research in recent years has provided data that are useful in simulating typical load spectra for the type of aircraft being designed. This permits accelerated testing of various airframe components or a complete airframe by subjecting it to repeated loadings simulating the mission profile of the aircraft, from takeoff through flight maneuvers, gusts, and landing. These tests substantiate the design and provide additional data for future designs.

Wings

The primary lifting airfoil of an airplane is the wing. While its planform may be widely varied, its function remains the same. Wings are attached to airplanes in a variety of locations, vertically and longitudinally. The terms *high-wing, low-wing,* and *mid-wing* all describe both airplane types and methods of wing attachment. Longitudinally, the location of the wing on a particular aircraft will be determined by the size and location of the mean aerodynamic chord of the wing and the center of gravity of the complete airplane. Other descriptive terms applied to wings describe their shape. Examples are *delta wings, swept wings, taper wings, elliptical wings,* and *rectangular* wings.

TYPES OF CONSTRUCTION

Wings may be either strut-braced or full cantilever. Many small aircraft use an external brace, or strut, to help transmit loads from the wing to the fuselage. Cantilever wings must resist all loads by means of structure within the wing contours.

Wings are designed with one or more main load-carrying members called *spars*. The most common arrangement is two spars, where the front spar is generally much more massive than the rear.

The *skin* may be made of metal and used as a primary load-carrying member, or the wing may be covered with plywood or fabric, although in only a few instances does one find wood and fabric used in modern aircraft.

If fabric is used as a covering, it is saturated with *dope*, a strengthening and filling fluid. Fabric covered wings have drag wires installed between the spars, running diagonally across the bays of the wing, to prevent fore and aft distortion. Such distortion may also be prevented by orienting the wing ribs diagonally so that they will carry loads normally carried by drag wires.

Dope causes the fabric to shrink and tighten as it dries, providing a smooth, firm surface.

DESIGN FEATURES

The shape and type of wing used on an airplane are determined by considerations other than structural. The airplane's primary usage will govern in choosing a structural design. Small, low-speed aircraft have straight, nearly rectangular wings. For these wings the main load is in bending of the wing as it transmits load to the fuselage, and the bending load is carried primarily by the spars. In high-speed aircraft, wings are often swept back. Sweepback imposes a very high torsion load because the resultant wing load is located far aft of the wing attachment to the fuselage. Depending upon the degree of wing sweep and its thickness, this torsional load might very well command the most importance in determining the particular structural design. Other items influential in the structural design of the wing are equipment to be housed within the wing, duct work and wiring, fuel cells, nacelles, and external loads. Cost of building and maintenance will also influence structural design. In fact, any existing airplane wing design represents a careful balance of performance, cost, fabrication techniques, weight, and strength.

STRESSED SKIN

The surface of the wing usually is covered with a sheet metal skin which has high strength and is employed as a primary load-carrying member. The skin is quite strong in tension and shear, and if stiffened by other members, may be made to carry some compressive load. The thickness of wing skins varies widely, depending upon the local stresses encountered in service. Thicknesses vary from as low as .016 in. in small aircraft to as much as .75 in. in the wings of heavy bombers. If weight reduction is of prime importance, the skins may be tapered in thickness so that the proper amount of strength is provided for local loads. Metal skins have the additional advantage that they are fairly rigid and hold their aerodynamic shape well.

CELLS

Because of the characteristic way in which loads are developed on the wing, a cell-type construction has important advantages. Ordinarily, wings have two spars which are joined by ribs and covered by metal skin which completes a *box*. Box construction, of course, utilizes the skin as a primary structural member. A cell is quite rigid and resists torsional deformation. The fact that aerodynamic demands limit the shape and space in which structure may be installed adds importance to cell structures. If the front spar is placed as close to the quarter chord

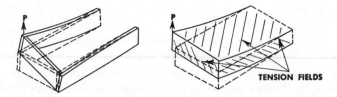

Figure 3.2. Deflection characteristics of a box beam versus those of a two-spar beam. Note how the tension fields in the skin of the box beam cause it to deflect as a unit.

line as possible, the air load will cause it to bend and will have little effect on the rear spar. The wing must twist if the front spar is to bend alone, but since the boxlike cell resists twist, the rear spar is loaded by the deflection of the front spar. Figure 3.2 shows the deflection pattern of the cell beam under the off-center load.

Because the cell allows a distribution of the load to other areas in the cell, an auxiliary spar added at a convenient location will also help support the bending load. It is therefore unnecessary to design a massive front spar in the limited space available, expecting it to carry the entire bending load. In a conventional two-spar cell-type wing structure, cells are also formed at the leading edge and trailing edges. Cells *A* and *C* in Figure 3.3 are important because they, too, resist torsion and bending, and contribute considerable stability to the spars. Since it is possible that creases or dents in the leading- or trailing-edge cells could cause premature collapse and failure of larger members, leading and trailing edges should always have proper maintenance.

It often becomes necessary to make cutouts in the cell skins to allow access for service or other reasons. These cutouts greatly weaken the cell, especially in its ability to resist twist, and relatively heavy frames or doublers are required around the cutouts. If a cutout is exposed to the airstream, or is in a highly stressed area, a cover is installed to complete the cell. The strength desired determines how the cover is attached.

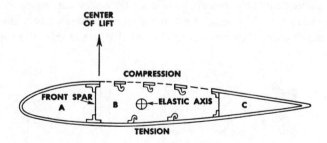

Figure 3.3. Typical wing cross section. Note that the skin on the compression side of the cell is considered ineffective in compression except for that portion attached directly to the stringer.

SPARS

Wing spars may be one of several types: simple formed or extruded channels or "I" sections; laminated sections utilizing heavy cap strips attached to simple formed members; truss type; and semitension field sections. One line of light aircraft uses a fairly thick-walled aluminum tube, which in some models also serves as the fuel tank.

Simple formed or extruded shapes have the advantage of being easy to build. Load calculations on these spars are well defined. This type of spar has its greatest use in small, low-priced utility airplanes. Though somewhat heavier, they cost less to produce.

Laminated spars are used where the outer flange loads are fairly high in relation to web (center section) loads, and it is necessary to add material to the outer fibers of the section.

Laminated spars permit the attachment of different sizes of cap strips to the basic spar along the span, thereby forming a lighter tapered effect. Another advantage of this build-up of parts is that a wing may be simply redesigned to support a heavier airplane model with perhaps no more than fabrication of heavier cap strips, which are attached to the basic spar.

Truss-type spars are occasionally found in larger wings using a fairly deep section. These are made up of hollow, rectangular tubing with welded joints, and arranged to form a truss. The truss consists of two large cap strips separated by vertical members, braced rigidly by diagonal members. These spars may be closely designed so that excess weight is not a problem, though they have very little torsional strength unless supported against twisting, and it is difficult to attach other structural members to them.

Semitension field spars are commonly used in aircraft structural design. The spar types previously mentioned, with the exception of the truss type, are called *web shear-resistant beams*. This means that the web material must be strong and stable in shear to withstand the shear load along the beam. The method of handling the beam shear load is what makes the semitension field spar differ from others. A semitension field beam is made up of heavy cap strips (usually extruded or milled), with a thin sheet metal web between them and vertical stiffeners spaced along the spar. As load is applied to the spar, the caps tend to

Figure 3.4. Typical spar cross sections. *A*, rectangular; *B*, channel; *C*, "I" section; *D*, built-up "I" section; *E*, truss; *F*, truss spar, side view.

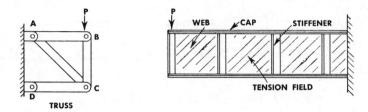

Figure 3.5. Tension field action in a spar.

hinge about their end attachments and allow deflection by sliding by each other. To prevent this deflection, a web is provided. The web must then absorb the shearing load caused by the tendency of the spar caps to pass over and under each other.

In a semitension field beam, this shear load is resisted by a tension field set up in the sheet metal web. The tension field runs diagonally across the web material, and if the load is large enough, diagonal buckles will appear in the web material. These buckles occur long before the ultimate spar load is reached and are not an indication of impending failure, since they disappear when the load is reduced. As its stability is not of prime importance, the web may be made much thinner than for web shear-resistant spars. The tension field developed in the web exerts a force that attempts to pull the spar caps together, hence the necessity for vertical stiffeners.

Figure 3.5 shows the similarity between a tension field beam and a truss. The frame is hinged at points A, B, C, and D. With a load applied as shown, members AB and CD would rotate about points A and D, and the frame would not support the load. Member AC, however, develops a tension load and prevents this rotation, enabling the truss to support the load. The semitension field beam acts in a similar manner, with the thin sheet-metal web replacing member AC. The buckle pattern in the thin web would be oriented the same as member AC for the same loading. If the load is reversed, the buckle pattern will also reverse. Since the web sheet is flat, it may buckle fairly severely, but provided the elastic limits have not been exceeded, it will still return to its original flat shape when the load is removed. Semitension field spars are generally the lightest in weight, but may be somewhat more expensive than some of the other spar types. To reduce weight further, the heavy cap strips are sometimes milled down to taper toward the wing tips where the spar loads are lower.

RIBS

Ribs are found in some form or another in all airplane wings and serve several purposes. They are placed at appropriate intervals along the wing span and transmit air load from the wing surface to the spars, act as formers to hold the airfoil

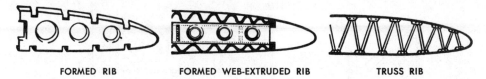

FORMED RIB   FORMED WEB-EXTRUDED RIB   TRUSS RIB

Figure 3.6. Wing ribs. (*Courtesy of US Air Force.*)

shape, stabilize the spars against twisting, distribute bending loads between the spars, close cells to complete the torque box, form barriers in the internal fuel installation, provide a point of attachment for other components such as landing gears, and serve many other needs. Ribs may be made of formed sheet metal, truss work, or a thin sheet metal web with cap strips attached. Generally, lighter aircraft have thin, formed sheet metal ribs with large holes ("lightening holes") cut into the web to conserve weight. Heavier aircraft usually have ribs either of the truss type or built-up type.

The particular type of rib will be determined by the purpose it must serve. For example, if a rib is to serve as one side of a fuel cell, it will have a solid web to withstand uniform internal pressure.

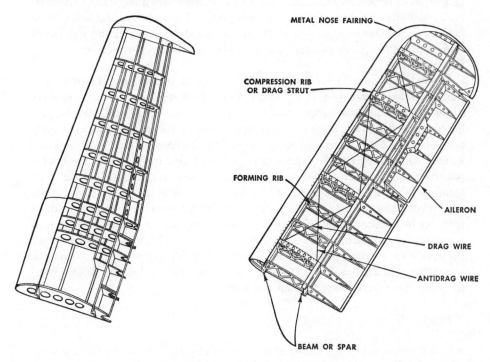

Figure 3.7. Typical light aircraft wing frames.

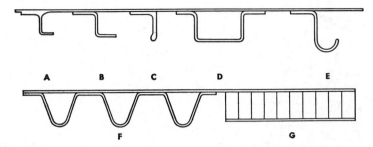

Figure 3.8. Typical methods of stiffening panels. *A*, integral; *B*, *Z* section; *C*, bulb angle; *D*, hat section; *E*, *J* section; *F*, corrugation; *G*, honeycomb core sandwich section.

STIFFENERS

Where high compressive loads are encountered, or where a shape must be held, stiffeners are attached to the skin. These are usually found fairly closely spaced on the upper wing surface, which normally is in compression, stiffening the compression skin to resist the induced bending loads. Where stiffening demands become extreme, the skin is usually reinforced by a corrugated panel or honeycomb sandwich instead of by individual stringers. Typical stiffened panels are shown in Figure 3.8.

ATTACHMENT OF AUXILIARY DEVICES

Wings are usually far more complex than we have seen so far, because to use the space available efficiently, much equipment must be housed within the wing or attached to it. Fuel cells break the continuity of the wing structure and may add high inertia or pressure loads. Engines and nacelles also disrupt the structural continuity and impose high local attachment loads, vibration, and torsion on the wing. Spoilers, flaps, ailerons, and dive brakes create high attachment loads and in addition often require openings or cutouts in the wing. If the landing gear is attached to the wing, provision must be made to support the high local loads. If gear is retractable, cutouts are necessary but will not allow permanent cover plates as would service access cutouts. Wires, tubes, ducts, controls, and other items must pass through various parts of the wing structure and further complicate the design.

For these reasons, a wing as designed is necessarily heavier than the ideal structure, but adds much to the utility of the completed airplane. Equipment not housed within the wing would have to be placed elsewhere in the airplane, perhaps resulting in even greater weight and performance penalties.

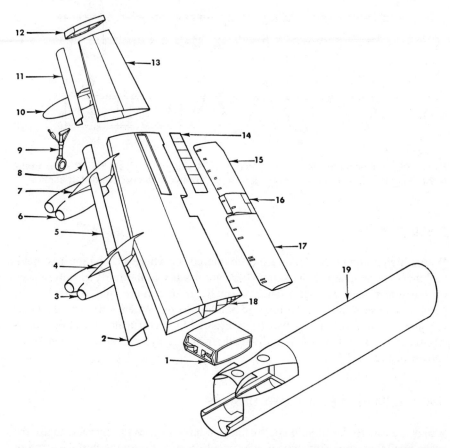

Figure 3.9. Jet bomber wing components. 1, center wing section; 2, wing leading edge; 3, inboard power plants; 4, nacelle strut; 5, wing leading edge; 6, outboard power plants; 7, nacelle strut; 8, wing leading edge; 9, wing landing gear; 10, external fuel tank; 11, wing leading edge; 12, wing tip; 13, outboard wing section; 14, spoilers (serve also as airbrakes); 15, outboard flap; 16, aileron; 17, inboard flap; 18, inboard wing section; 19, fuselage section.

FUTURE TRENDS

The demands on wings both as lifting devices and as places to house equipment seem to be increasing at a rapid pace. Wings are becoming more heavily loaded and more multipurpose in design. Several present airplane designs utilize the entire wing as a fuel cell and have additional external fuel pods or payload pods suspended below them. These increasing demands require a new structural style. To maintain a high degree of structural efficiency, wing structure in the future will make more extensive use of such features as semitension field spars and ribs, strong light metals, tapered spar caps, tapered skins, lightweight but rigid honeycomb sandwich panels, metal bonding, and chemically milled skins and stiffeners. Structural design, strength of materials, fabrication practice, and processes will certainly change in the future, and wing design will change with them.

Faced with the various demands on the wing structure, the designer would prefer to have the wing as thick as possible. This permits the use of less bending and torsional resisting material, because of lower stresses, and provides more space for equipment and fuel. Recent aerodynamic research has developed new airfoil sections, known as "supercritical" and "advanced technology" which permit the use of thicker wing sections with no penalty in performance and, in some cases, an increase in performance. It is rather apparent that these sections will be used to a large extent in the near future for aircraft of all sizes.

Data processing and the computer have provided the designer with tremendous new tools. Load, stress, and dynamic analyses can now be performed faster and far more accurately than before, permitting the designer to develop a stronger, yet lighter, structure.

Fuselage

The fuselage is the body of the airplane. Some of the features that all fuselages have in common are provisions for a comfortable, quiet environment for the crew and passengers, restraint systems and crashworthiness capability, attachments or carry-through structure for wings and empennage surfaces, instruments, controls, necessary equipment, possible fuel, baggage or cargo, engines, and landing gear. Perhaps the most distinct feature of the fuselage is a result of its purpose: providing space for payloads. The space required creates the need for comparatively large openings within the airframe in relation to its size. Around this space and function the fuselage is designed and built.

Fuselage structures of aircraft today can usually be classified as *truss, monocoque,* or *semimonocoque*. Truss or framework types of construction have steel-tube, aluminum-tube, or other cross-sectional shapes which may be bolted, welded, bonded, pinned, or riveted into a rigid assembly. The exterior is then covered with such materials as impregnated glass fiber, aluminum, or thin steel. Impregnated glass fiber is a glass cloth or mat, reinforced with epoxy or other resins, and sometimes is part of the primary structure. One light-aircraft manufacturer successfully uses metal-to-metal bonding of the airframe structure. Some airframe components, that is, fuselage halves and wing and tail sections, are being molded using glass fiber and epoxy. This results in a high degree of rigidity of the structure and very smooth skin surface with some penalty in weight.

The *monocoque* or *semimononocque* fuselage structure uses its covering or

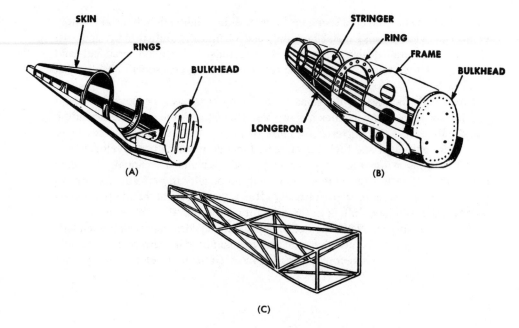

Figure 3.10. Fuselage structural styles. *A*, monocoque; *B*, semimonocoque; *C*, truss-frame. Note stringers used in semimonocoque structure. *(Courtesy of US Air Force.)*

skin as an integral, structural load-carrying member. Monocoque structure is a thin-walled tube or shell which may have rings, bulkheads, or formers installed within. It can carry loads effectively, particularly when the tubes are of small diameter. As its diameter increases to form the internal cavity necessary for a fuselage, the weight-to-strength ratio becomes more inefficient, and longitudinal stiffeners or stringers are added to it. The result is a type of structure known as semimonocoque. Use of this concept has enabled aircraft designers to use aluminum skins as light as .016 in. in thickness for the primary structure on airplanes as large as the modern light twins. Larger semimonocoque aircraft use progressively thicker skins and still maintain an equivalent stress level in the skin along with an equally good weight-to-strength ratio.

Fuselage cross sections may vary widely and are dictated by the type of aircraft designed. For aerodynamic reasons, the front area should be as small as possible. Fuselage cross sections may be a circle or an ellipse, a basic square or rectangular shape with large corner radii, or the shape that best fits the needs and constraints of the design. Pressurized cabins are usually circular sections, although other shapes have been successfully used. The outer contours of the fuselage along its length must be some form of streamlined shape.

Fuselage and wing structural design problems are similar. Because more curvature may, in general, be employed than on the wing—making the skin itself somewhat stronger for compressive or bending loads—and because of the greater cross-sectional area, the fuselage usually does not employ the spar-type structure found in the wing. Most fuselages can be considered single-cell structures, although when floorboards in a multilevel, larger aircraft are designed to bear structural loads, the fuselage is a multiple-cell structure.

The skins used in modern construction are particularly effective in tension but rather ineffective in compression. In addition, the fuselages of these craft are of large diameter or cross section when compared to skin thickness. Accordingly, a unique situation occurs when the structure is subjected to bending loads. One side becomes very much stronger than the other under bending loads, causing the neutral axis of the section to shift toward the strong or tension side (Figure 3.12). As the neutral axis shifts, the loads or stresses in the weaker side tend to become less and the loads or stresses become larger in the side that is more capable of withstanding them. Thus, a somewhat inherent structural equilibrium exists. In addition to this phenomenon, the longitudinal stringers form tension field webs, as explained in the discussion of the wing section, further improving bending-load-carrying capability.

Twisting loads on the fuselage are resisted primarily by the shear stresses in the skin, which are produced directly as well as by the development of tension fields in bays between stringers and formers or bulkheads (Figure 3.11).

Figure 3.11. Tension field action in a stressed skin structure. The tension field is set up by shear flow in the skin, caused by an extremely high torque on the fuselage tailcone. This semimonocoque specimen is under a load equal to twice the maximum expected load for this particular design. *(Courtesy of Cessna Aircraft Co.)*

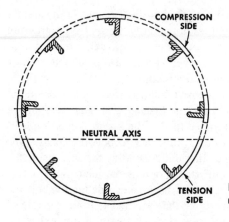

Figure 3.12. Cross section of fuselage under bending load.

Pressurized fuselages must bear an additional load, for the fuselage becomes a pressure tank. Even though the differential pressures are quite low in force per unit area, the area is large, and consequently so are the forces. Because of the type of construction (semimonocoque) used in most fuselages, each ring or former will develop stresses similar to the tension in a barrel hoop. Fatigue of joints due to fluctuating pressures, seals around control systems where they enter or leave the fuselage, door openings and seals, window frames and glass pressure, and—primarily—the need for a "fail-safe" structure to prevent explosive decompression if the skin is punctured from either the inside or outside while under pressure, are important in pressure fuselage design.

INTERNAL LOADS

High local loads are imposed on the fuselage internally. Seat and restraint system attachment loads are quite severe because of their limited area of application and their need to be adjustable. Cargo aircraft floors and tie-downs may be subjected to loads from extremely heavy objects concentrated in a very small area. Floorboard structure may be of the honeycomb variety, or sheet reinforced with corrugations, channels, or stiffeners of other types.

FUTURE TRENDS

Future trends of structural technique in fuselage construction will undoubtedly follow that of development of stronger, lighter metals and synthetic materials, processes of joining cheaply and efficiently, and new methods of forming intricate shapes. The aerodynamic heating resulting from hypersonic speeds has an important effect on the structural concepts of fuselage construction.

Landing Gear

TYPES

The basic types of landing gears are the *tail wheel* gear, the *tricycle* gear, and the *bicycle* gear (which utilizes a main gear under both the forward and the rear portions of the airplane). The *bicycle* landing gear usually has outrigger wheels near the wing tips. Skis, or combination wheel-ski types which will allow the craft to land on or take off from either a snow surface or a dry landing area, are widely used, especially in areas with long snow seasons.

Floats allow an aircraft to operate from water. Amphibious aircraft with wheels retracting into the hull, operate from either land or water. Float equipped aircraft may also have retractable wheels.

On aircraft operated at extremely high gross weights, tandem, dual-tandem, or multiple-tandem types of wheel arrangements may be found. Aircraft designed to operate from soft surface fields may also be equipped with tandem wheels.

Many light aircraft use a cantilever single-leaf or tapered tubular spring type main landing gear of steel or laminated fiberglass, which will store the energy of initial impact, thereby producing quite low load factors. Low maintenance, simplicity, and long service life characterize this gear type. Other types are hinged frames or hinged beams which use a shock cord, hydraulic cylinder, or a coiled or rubber-filled spring to store or absorb the shock. The torsion-bar method of shock absorption has been used successfully on some aircraft.

Crosswind gears are those in which the wheels are mounted with some degree of castering freedom to allow them to roll parallel to the aircraft direction on the ground when the aircraft is crabbed into the wind. The B-52 main landing gear can be preset before landing in a castered position to accommodate a crosswind.

Aircraft *tires* usually are made of conventional nylon or rayon cord, rubber-filled. Many modern tires on large and small aircraft alike are tubeless. They may be either the low or the high pressure type, depending upon the purpose of the aircraft and the space available for storage within the airframe if retracted.

Aircraft *wheel design* is determined by the type of tire desired, the method by which the tire and wheel assembly is attached to the remaining portions of the landing gear, and brake requirements.

Static and dynamic impact loads are both utilized in the design of tire and wheel combinations.

Brakes used on aircraft are determined by landing speed, thrust, and weight. They are used to assist in steering many aircraft. Because of the tremendous heat generated in slowing or stopping an airplane, the brake must be designed for rapid heat dissipation. The most popular brake on light aircraft is the single disc type. In this design the disc is attached to the wheel by bolts, gear teeth, clips, or

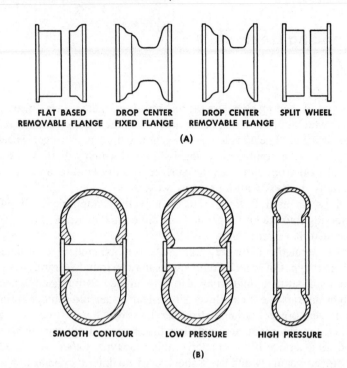

FLAT BASED   DROP CENTER   DROP CENTER   SPLIT WHEEL
REMOVABLE FLANGE FIXED FLANGE REMOVABLE FLANGE

(A)

SMOOTH CONTOUR   LOW PRESSURE   HIGH PRESSURE

(B)

Figure 3.13. Aircraft wheel and tire contours. (A) Wheels: (B) Tire contours.

other devices, and turns with the wheel. Small blocks, usually hydraulically actuated, press on this disc simultaneously from both sides to create the braking friction. Many other types are used, and the design is dependent upon the specific braking needs.

Because of the intense heat developed, brake materials are very important. Brake systems when heated by friction may begin to feel spongy. Hydraulic fluid pressure is the most popular actuating force.

Aircraft *axles* vary considerably in design. The conventional type for single wheel installation is similar to the front wheel spindles axles on automobiles. In the tandem wheel arrangement, the lower strut axle becomes a trunnion to which is attached the truck upon which the wheels are installed.

SHOCK (Oleo) STRUT

Shock-absorption mechanisms for all larger landing gears are of the air-oil hydraulic type. This type (Figure 3.14) simply dissipates energy by forcing hydraulic fluid through a small orifice. The rate of flow and the length of the stroke determine the efficiency of the shock strut and consequently determine to a large degree the load factor which will be developed when landing. When the aircraft

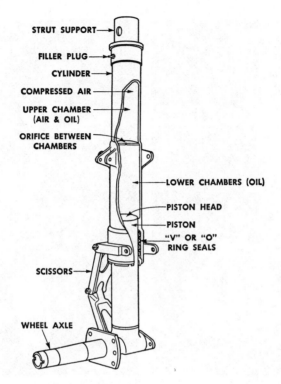

STRUT SUPPORT

FILLER PLUG

CYLINDER

COMPRESSED AIR

UPPER CHAMBER
(AIR & OIL)

ORIFICE BETWEEN
CHAMBERS

LOWER CHAMBERS (OIL)

PISTON HEAD

PISTON

"V" OR "O"
RING SEALS

SCISSORS

WHEEL AXLE

Figure 3.14. Basic landing gear shock (oleo) strut assembly.

is stationary or taxing, the weight is supported on a cushion of compressed air inside the upper chamber, above a piston. Oleo strut seals must be kept in good condition to prevent leakage. The inner tube must be reasonably free to move inside the outer tube and therefore would be free to turn independently. Consequently, some device must be added to prevent rotation on main or nonsteerable landing gears. *Scissors* (Figure 3.14) are used to permit vertical motion without rotation.

The upper trunnion or attachment of the landing gear strut to the airplane spreads the load out as much as possible to prevent high load concentrations which would require heavy structure or bring about a rather short fatigue life. On wing installations, as shown in Figure 3.15, the trunnion transmits load directly into heavy wing structure installed because of flight loads or for the power plant. This multiple use of aircraft structure increases its weight efficiency. Popular trunnion fittings used today are aluminum, magnesium, or steel forgings. Some aircraft, however, retain the frame-type weldments.

Most of the design features and the structure of the main landing gear will be applicable to the nose gear of an airplane. Brakes are not usually installed on nose wheels, and on most aircraft the nose gear carries a much lighter load than

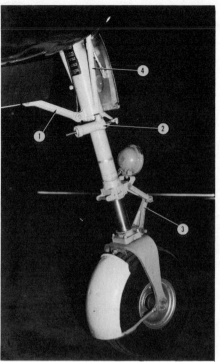

Figure 3.15. Typical tricycle-type landing gear installation on a low-wing airplane. (1) *Left*, side brace and down lock. (2) *Left*, drag brace trunnion. (3) *Left*, retraction link which also acts as down lock. (1) *Right*, drag brace down lock. (2) *Right*, shimmy dampener. (3) *Right*, scissors or torque links. (4) *Right*, side brace trunnion. This landing gear is retracted by a single electric motor acting through a system of bell cranks and push-pull tubes. *(Courtesy of Cessna Aircraft Co.)*

the main gear. This is primarily because the main gear is usually located at a position closer to the airplane center of gravity, and on landing the nose gear should not make contact until the main gear has absorbed the initial impact.

Nose-gear steering is a system which causes the lower strut to rotate within the outer strut or causes the entire strut to rotate in bearings mounted in the nose-gear attachment. Steering systems may be manual, either foot or hand controlled, but often are hydraulically actuated as in Figure 3.16. Many aircraft utilize positive steering for a limited number of degrees rotation and then allow free swiveling to enable the aircraft to turn about one main gear. Others are free swiveling, with no nose steering control. Because of this positive steering or comparatively free swiveling feature, many nose gears would oscillate at some rotational speed of the tire unless equipped with a shimmy damper. The most popular is the hy-

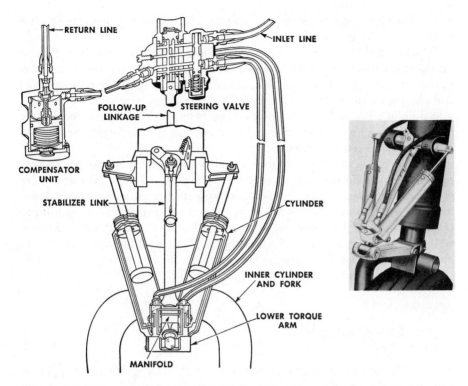

Figure 3.16. Nose-wheel steering mechanism. *(Courtesy of Bendix Aviation Corp.)*

draulic-type damper. On some craft the power-steering system serves also as the shimmy damper.

RETRACTION SYSTEMS

The performance demanded of many modern aircraft requires that the landing gear be retracted in flight to decrease drag. On low-wing aircraft, the main landing gear is often housed within the wing, and the nose gear retracts into the fuselage forward section. Other aircraft house the landing gear within the fuselage in order to keep the weight involved to a minimum. Large, high-performance aircraft—and, in general, most aircraft that cruise faster than approximately 160 knots—have retractable landing gear.

The power source for retraction systems may be hydraulic, electrical, electromechanical, air, or manual. The hydraulic system is the most popular, although many systems have electrical motors for each gear or an electromechanical system to operate more than one gear by utilizing push-pull tubes with a bellcrank or torque-tube system. The hydraulic pump may be electrical, engine driven, or air-turbine driven.

To realize fully the objective for which the gear is retracted, it is desirable that the opening for the gear in the structure be closed smoothly and tightly once the gear is retracted. To accomplish this, most retraction systems are equipped with doors that open to allow the gear to enter or exit, and that remain closed when the gear is up or down and locked. Some designers have eliminated the doors in the interest of saving weight and cost and have successfully used elastomer seals in contact with the retracted tires.

In a few cases a small performance loss has been accepted and the wheels simply retracted into the opening. Having the openings closed while the gear is down is advantageous because the tires will not throw foreign matter such as mud, oil, water, or ice into the wheel well during ground movement, and wing lift/drag characteristics will be better maintained for takeoff. The interior of the wheel well should be fairly smooth and reasonably well sealed off to prevent these same objects from becoming lodged on the internal structure should the wheel be spinning when it enters.

Empennages

The conventional arrangement of the empennage group (tail group) is an extreme aft location of a single *vertical fin* and *rudder, stabilizer,* and *elevator.* There are many variations in this arrangement: two surfaces mounted in a *V,* two or more vertical fins mounted on the horizontal tail, and the stabilizer mounted at various heights on the fin are several examples. Another concept, "flying tails," will be discussed with controls. There may not be a well-defined empennage group, as in delta-wing aircraft where the wing trailing edge must have features to simulate a horizontal tail.

Since the vertical tail must lift to either side and the horizontal tail must lift either up or down with equal efficiency, the airfoils are made symmetrical (upper and lower contours are the same). The control surfaces attached to the empennage are merely large-chord, plain flaps, usually amounting to about half the chord of the airfoil. Some of the more important considerations in determining empennage design are whether the airplane is single or multi-engine, the amount of fore-and-aft c.g. and trim range desired, and various stability and control requirements.

Structurally, empennages resemble wings but are normally much less complex and less integrated. They are usually full cantilever, metal-covered airfoils. Spars are quite often little more than formed sheet metal channels. In light business aircraft, the front spars may be formed by the skin. The rear spars are much heavier than the front spars—opposite the general arrangement used in wings—

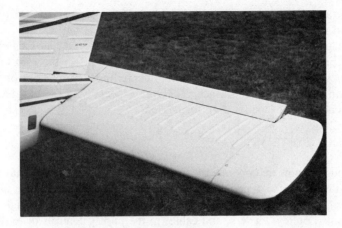

Figure 3.17. A flying tail, on a Piper Cherokee.

because the larger elevator and rudder flaps cause the center of pressure to be far aft on the airfoil, requiring strength farther aft.

Control Surfaces and Systems

Control surfaces are constructed much the same as wings, yet much simpler. Some aircraft make use of "slab" controls or "flying tails" in which the entire surface moves instead of just a flap. These may have either of two forms. With the *flying tail* the entire surface is rotated by the stick or wheel. With the *rotating stabilizer* the pilot rotates the stabilizer with his trim control. There

Figure 3.18. A rotating stabilizer, on a Piper Tri Pacer.

are important advantages to these movable surfaces—good control over stick forces and minimum drag profiles.

Dive flaps, speed brakes, or airbrakes must be designed and located so that they will provide high drag but will not appreciably affect the trim of the airplane. They may be located on the bottom centerline, on the sides of the fuselage, or on the wing upper surface. They are especially needed by jet aircraft, which do not have propeller drag for descent.

CONTROL SYSTEMS

The various systems that operate the control surfaces of an airplane have considerable influence on the dynamic characteristics of the control surfaces and on the structure of the airplane. Control surfaces are usually connected through a system of bellcranks, pulleys, levers, and cables to the pilot's controls or to the servomechanism used to operate the control. The method of actuating controls is widely varied among airplane designs. Light business and utility airplanes use direct pilot effort for operation of the controls, with conventional cable arrangements or push-pull rods making up the connecting system. Automatic pilots may also be connected into the system. Very large airplanes usually have servomechanisms to amplify the pilot's control forces, or in some cases, the pilot may operate a tab which actuates the control surface. When auxiliary controls such as trim tabs are installed, it is generally required that they be driven by a system separate from that used to drive the primary control; in the case of failure in the primary-control system, the airplane must be maneuverable and controllable by the tab. The tab will be operated in its normal direction when the primary control is free, but in a direction opposite of normal if the primary control is frozen in position.

For supersonic aircraft, a very complex electromechanical-hydraulic flight control system is used, because of extreme dynamic forces and variations of control displacements required.

The various components of a control system are selected on the basis of the load that they will be expected to withstand in operation. Also, control elements must be properly routed and cleared of the structure to prevent their becoming frayed or otherwise damaged. Cables are always installed with some degree of preload to prevent excessive looseness, to eliminate a portion of the cable stretch, and to aid in keeping them on the pulleys. All of these factors must be considered while designing a system that contains a very low amount of friction loss. Control system loads may become quite high and impose severe local loads on the airframe at attachments of bellcranks and pulleys; as a consequence, they affect the airframe design.

Good control system design avoids excessive deflection to reduce control-surface travel but still provide effective control force. All controls, since they are connected by means of elastic systems, have some particular *spring rate* or degree of sponginess. This characteristic could in part determine the flutter speed of the airplane. Preload, or rigging tension, has an influence on this system spring rate and should be checked periodically to see that it is up to the manufacturer's specification. Free play at the surface is another such influence and should be checked for excessive looseness periodically.

Some older small airplanes have interconnecting systems to simplify control of the airplane. Most of these designs include spring-loaded devices so that the pilot may override the system to execute such maneuvers as slips. In some instances, airplanes are equipped with bungee springs or control interconnects to meet certain stability requirements.

Provisions for Power Plants

As noted in other portions of this chapter, the power plant may be located in many different places on or in the airframe. Each location will dictate some particular design feature. Structurally, the power plant is a source of vibration, heat, concentrated mass, and sometimes undesirable thrust or drag forces. It necessitates many controls which in turn require cutouts in structure, as do fuel lines, oil lines, and electrical lines. All engines require some type of mounting through which mass, thrust, thrust reversing, torque, and vibration are transmitted to the rest of the airplane or are absorbed.

Various types of mounts have been designed. Among them is the welded-tube truss assembly (usually steel) which may be either the *bed* or *ring* type. The sheet-metal, built-up type is quite popular and will usually be equipped with some kind of steel leg for the final engine-to-mount attachment. With the advent of jet power plants, where propeller clearance is no longer a problem, the engines of multi-engine craft often are suspended below the wings (Figure 3.9) or are attached to the fuselage in pods (Figure 1.11[6]). These pods simplify the structural problems somewhat by removing the engine from other primary structures, and offer many safety advantages as well. The structural problem of pod suspension is largely one of pylon design with well-defined load paths and more straightforward attachment.

Large forgings have been used successfully as mounts for some engine installations. Internal-wing installations have been used on some jet aircraft. Usually these use the inboard portion of the wing-to-fuselage attachment area for their location. When this installation is used, or if the engine is mounted inside the fuselage, the immediate area around the engine will be covered with stainless steel, aluminized iron, or other material possessing greater flame and heat resistant capabilities than aluminum alloy alone. All openings in this firewall will be sealed to prevent flames from entering adjacent structures or compartments. When engines are located in the proximity of a structure that would be affected by temperatures of the engine itself or of its exhaust, the structure must be insulated.

Piston engines particularly are the source of undesirable, fatiguing airframe vibration which must be reduced or isolated. Vibration dampers are located at points of engine-to-mount or mount-to-adjacent-structure attachment. When the dampers are located in the engine-to-mount attachment, much of the vibration is isolated from the mount, which is spared this damaging fatigue. Although piston-engine vibration has been considered high-frequency, it is low compared to the frequencies of turbine-powered aircraft—which, fortunately, are usually accompanied by smaller amplitudes. Satisfactory design must accommodate these high frequencies, sufficiently isolate them, or control the stresses caused by them.

Joints

The large variety of joints found in a structure as complex as an airplane requires may different fastening jobs, each needing some particular type of fastener. The usual joining devices are: bolts, rivets, screws, rivnuts, cams, pins, spot welds, solder, braze, and adhesive bonds. Figure 3.19 shows most of the types of joints, classified generally according to the type of loading to be applied.

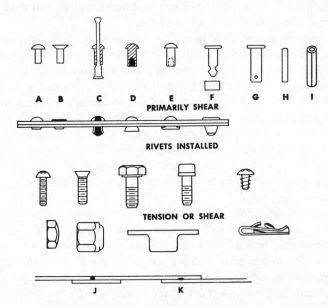

Figure 3.19. Typical aircraft fasteners. A and B are solid shank rivets; C, D, and D E typical blind rivets. F is a steel rivet with an aluminum collar. G is a steel pin-free fit. H and I are drive-fit steel pins. J is a typical metal-bond joint. K is a typical spot-weld joint.

Bolts are good in either tension or shear and are generally necessary where close tolerance must be held or where moving parts are involved.

Rivets are very common and are used where shear is the primary load. They are generally much cheaper to buy and install than any other fastener. The ordinary rivet has a button head and solid shank and is made of aluminum alloy. The shank is inserted in a hole, and the rivet is driven from the manufactured-head side, swelled into the hole by a bucking bar held at the end of the shank, and upset on the butt end to form another head called the shop head. If the back side of the work is inaccessible to form the shop head on the rivet, one of the several types of blind rivets shown in Figure 3.19 is used.

Screws are used in much the same applications as bolts, but where loads are lower. Two types of screws commonly used are the machine screw and the self-tapping screw. A machine screw has standard SAE threads and must be installed in a threaded hole or with a nut. Tapered, self-tapping screws may be installed directly in an undersize hole. Generally they are used in full-sized holes with sheet-metal nuts. They are mostly used where loads are relatively low or when frequent removal is necessary. In all joints secured by bolts or screws, precautions are taken to insure that the fastener will not vibrate or work loose in service by the use of locking devices such as washers, safety wire, cotter pins, and self-locking nuts.

Among the *special rivets* are tubular-steel rivets, used in firewalls because of their superior heat resistance. High-shear rivets have steel shanks with aluminum-alloy collars on the shop-head side. The high-shear type is shown in Figure 3.19. These rivets are installed where shear loads are very high and must be installed in fairly close-tolerance holes.

Pins are used only in shear applications and are popular because of their low cost, tight fit, and ease of installation. One popular type of pin fastener is the roll pin—a hollow steel pin with a longitudinal gap, driven into a slightly undersize hole.

Quick release fasteners are used extensively on cowling or other members which may require frequent removal. A sketch of these types of fasteners is shown in Figure 3.20. The latching screw has either a wing nut or a slot head, and the other end of the shank is formed into a cam which engages the wire and latches with a quarter turn of the screw.

Welding is, perhaps, the best known of the many nonmechanical fastening methods. Steel parts are particularly well suited for this type of joining. Welding of most steels may be accomplished easily, but other metals usually require special techniques. Aluminum welding is an example. Firstly, aluminum oxidizes easily in the presence of high heat so that a shield or stream of inert gas, such as helium, may be required to drive off the oxygen. Secondly, aluminum conducts heat so efficiently that it is dissipated away from the joint out into the part where the high heat could warp the part or remove its hardness.

Brazing and *soldering* are not fusion joints but rather obtain their strength from

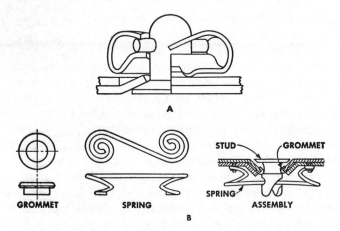

Figure 3.20. Special quick detachable fasteners. (A) Air lock fastener. (B) Dzus grommet and spring.

cohesion between the weld material and the parent metal. Brazing may be done with silver, copper, or bronze. Soldering may be done with silver or with the usual solder material made of an alloy of lead and zinc. Soldering and brazing are done with either hot irons or open flame, whereas fusion welding may be performed by flame or by an electric arc.

Spotwelding is a form of sheet-metal welding quite extensively used in aircraft structures. In resistance spotwelding, two or more sheets are joined at a spot by the simultaneous application of electrode pressure and heat from an electric current, which together fuse the sheets at the local spot. Advantages of this type of weld are its low cost and smooth surface appearance. Spotwelds withstand shear load and some degree of tension. Since they have a low resistance to peeling loads, rivets are generally installed at each end of a row of spotwelds. If the members must be separated for replacement of parts, the spots may be drilled out as if they were rivets and the parts replaced, using rivets instead of spotwelds. Other methods are ultrasonic and fusion spotwelding, for joining of thin and thick materials, respectively.

Adhesive bonding is a common method of joining materials, rapidly gaining popularity in aircraft structural design. New bonding materials offer a high degree of strength and reliability. There are three general classes of adhesive resins presently in use—*thermoplastic, thermosetting,* and *elastomeric.* Thermoplastic resins soften with the application of heat and must be maintained under pressure to effect a bond. They are low in cost but also fairly low in strength. Thermosetting resins, when heated, develop full strength and harden. This type usually requires both heat and pressure to effect a bond, which is stronger than that made with the other resin types. Some of the resins in the group are *catalyst-setting,* that is, a catalyst is added to the resin to accelerate curing. Elastomeric resins are polymeric rubber compounds used where a high degree of flexibility is required.

For metal bonding, as in spotwelding, the surfaces must be very clean and free of oxide, grease, and oil. Special inspection techniques may be required to insure safety, and precision equipment is often necessary to obtain structural bonds. Bonded joints, which have a very good shear strength, are used in applications where this feature is important. Many of the resins are also good in pure tension.

Bonded metal joints have many important advantages: (1) The stress is distributed over a wide area so that little concentration develops. (2) Members are kept free of holes which raise stresses or create concentrations. (3) Large panels may be secured by a single bond joint, thus eliminating hundreds of small fasteners. (4) The resin between the sheets acts as an insulator to prevent corrosive electrolytic action when electrically dissimilar metals are joined. (5) The exterior surface is quite smooth, giving a high degree of aerodynamic efficiency.

Sandwich construction is a special application where metal bonding is very desirable. When resin is used to bond the outer sheets to the core material, it will flow down on the core to form a fillet at each joint. The formation of the fillet yields more contact area for a greater bond strength (Figure 3.21).

Bonding is rapidly growing in popularity with airframe designers. This type of fastening will appear in greater quantities in the future. Weld-bonding is a system combining the methods of spotwelding and bonding and is a recent development in the field of joining metals. With the advent of metal bonding, weld-bonding, and other new innovations in joint design yet to come, airframes will become increasingly safer, stronger, lighter, and more durable. High-frequency vibrations

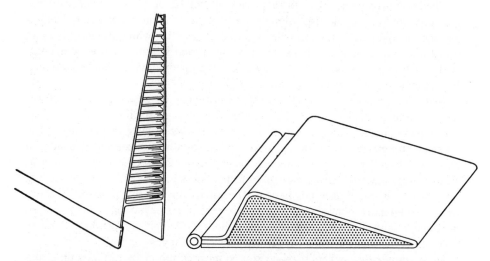

Figure 3.21 *Core construction*. The sketch at left shows a honeycomb core structure. Such panels are coming into common use in aircraft. The sketch at right shows how plastic foam core can be used, in control surfaces, for instance.

and fatigue loading require joint designs that distribute load in a favorable manner.

Resins will, undoubtedly, be increased in strength, and fabrication practices will be simplified in the future so that bonding will be available at low cost for all aircraft.

Materials

In order to obtain a high strength-to-weight ratio and a high degree of safety, materials used in aircraft structures usually are of a higher quality than those found in other structures. Some of the materials and alloys commonly found in aircraft are steel, aluminum, magnesium, stainless steel, bronze, brass, plastics, and glass fiber. Titanium is being used in increasing quantities, as is boron and graphite filament composite structure.

Aircraft structural materials are selected for their particular properties, which vary considerably from one to another. A few of the properties that influence selection are: strength, weight, rigidity, hardening ability, corrosion resistance, ductility, toughness, resistance to heat (or cold), notch sensitivity, and resistance to abrasion and wear. Certain of these characteristics are very important in aircraft structural design. For example, a ductile material is much more desirable than a brittle material, because failure in a brittle material is sudden and complete, whereas in a ductile material the part yields long before the ultimate load is reached and gives warning of the approaching failure. A deformed part can thus be discovered during an inspection of the airplane and can be replaced. The job assigned to the part in the airplane structure determines which property is most important. Tough material is used where shock and vibration are encountered; heat-resisting material is used for firewalls; and a material which can be hardened is used where high strength is necessary.

Materials are formed into useful shapes primarily by cutting, bending, forging, extruding, casting, welding, stretching, grinding, machining, and rolling. For making aircraft structural members, materials are usually chosen to give the greatest strength-to-weight ratio. For this purpose, alloys of metals are used. An alloy which will respond to some of the known methods of hardening by heat treatment is selected. As the hardness of the material is increased, its strength is also increased. Steel responds best of all known alloys to this process. Because of this excellent response, it is possible to strengthen steel so that for a given job it can be "lighter" than aluminum—even though it weighs almost three times as much in equal-volume units—by using very thin gauges.

Steel, aluminum, and *magnesium* are the most common aircraft structural materials because they each possess a large number of the desirable properties previously listed. *Glass fiber* and *glass cloth,* when impregnated with a thermosetting resin, are used in wing tips, fillets, ducts, radomes, spinners, and in many other ways—particularly where loads are light or where contours make metal forming difficult. Acrylic *plastics* are used in making canopies, windows, and windshields. They have a fairly high strength, are easily formed when heated, and exhibit excellent optical qualities. The urethanes or foaming plastics have recently come into use in the structure of light airplanes. The plastic is foamed in to place between two sheet metal panels, forming a sandwich structure (Figures 3.8(g) and 3.21). Panels made in this way are light in weight, very rigid, strong, and reasonably inexpensive.

Stainless steels are used because of their corrosion resistance and superior resistance to heat. Although stainless steel is heavy (almost three times an equal volume of aluminum), it is often used as an aircraft structural material. The weight is not considered excessive because it can easily be made two to three times as strong as aluminum.

Paper is a rather unusual aircraft structural material, but is sometimes used as a core material in a sandwich of two sheet-metal panels, to form floorboards or shelves. The floorboards absorb a large amount of high-frequency noise, as well as providing a strong, rigid support for structural furnishings or cargo.

Aluminum alloys can be hardened and are quite ductile and fairly corrosion-resistant. Generally, a thin coating of pure aluminum is rolled onto the outside of the alloy sheet to provide corrosion protection. This "cladding" normally amounts to about 5% of the total thickness of the sheet (Figure 3.22).

Magnesium is one of the lightest of structural metals, but is not as strong as aluminum in equal volume. Magnesium has roughly the same strength-to-weight ratio as aluminum, but is only two-thirds as rigid. It also must be protected from corrosive atmosphere and flames.

Bronze and *brass* have rather limited use in the actual structure of an airplane. These materials appear mainly in bearings, fasteners, cable turnbuckles, and springs.

Certain materials are used to perform special duties. An example is *ceramics,* also called "cements," "ceramels," "intermetallics," and "refractory" metals. In general, these materials are a combination of ceramics and metals which are formed in sintering processes using powdered materials. The ceramics impart a high heat resistance, and the metals furnish the strength. They were developed for special uses where high strength at high temperatures is the goal.

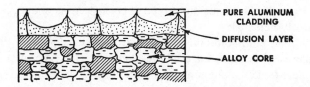

Figure 3.22. Cross-sectional diagram of Alclad.

Titanium is being used in supersonic aircraft, due to its high strength, light weight, and retentiveness of strength at high temperatures.

Lithium is proving valuable when alloyed with aluminum, by increasing the modulus of elasticity, yet it reduces weight, when compared to some alloying elements.

Boron and *graphite* filaments in *composite* structures present high strength with light weight. At the present time, however, the cost is prohibitive for general application, and their use is confined to large aircraft.

Almost every time an aircraft structural material is mentioned, the question of its resistance to *corrosion* arises. For an airplane to possess utility and have a long, safe life, it must be resistant to ordinary corrosive atmospheres. An airplane, once delivered to the customer, may be flown to any spot on earth; since the parts of the airplane are exposed to the local conditions, they must be protected. Each material has its own needs in this respect. Metals may be attacked by marine atmosphere and especially by urine, oxygen, chemicals, industrial atmospheres, exhaust gases, and gun smoke. Steels generally tend to rust and must be protected by paint or a plating of a corrosion-resistant metal such as cadmium. Pure aluminum is quite resistant to most types of corrosion, but aircraft structural members made of aluminum alloys will tend to corrode. The plating of pure aluminum on these alloys to impart corrosion resistance has already been mentioned. This "Alclad" material, if heat-treated several times, will lose its resistance to corrosion because the pure aluminum layer will eventually alloy with the core material when subjected to the high heat-treating temperatures. Zinc chromate primer is often applied for corrosion resistance on aluminum and magnesium. Aluminum alloys may also be protected by a process known as anodizing, in which a thin layer of oxide is caused to form on the surface.

Aluminum-alloy corrosion will usually appear in one of the following ways. First (and least serious) is surface pitting, which while presenting a disturbing appearance, does not seriously weaken the material and is easily discovered. Figure 3.23 shows this type of corrosive action.

The second type of corrosion is galvanic action, which occurs at a joint between two or more electrically dissimilar metals, such as two different alloys of aluminum or aluminum and steel. Every metal has an electrical potential; if two metals having differing electrical potentials are joined, a partial electric cell is formed. When moisture and acids from the atmosphere enter the joint, the cell is complete and current will flow. This cell eats away the metal and may seriously weaken the joint. This is an insidious type of corrosion since it attacks vital parts and is not easily detected.

A third type of aluminum corrosion is intergranular corrosion, in which the grain boundaries of the core material are attacked. This type of corrosion may occur when heat-treated aluminum alloy is improperly quenched. As the boundaries are destroyed, the grains lose their bond to each other, and the part loses its

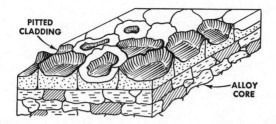

Figure 3.23. Typical surface appearance of superficially corroded unpainted aluminum sheet.

strength. Intergranular corrosion is serious because of the damage and the difficulty of locating it.

The airplane manufacturer and material supplier are constantly guarding against the possibility of corrosion and usually help the customer in solving a particular corrosion problem when an unusual use is planned for the airplane. Corrosion protection, though vital to the airplane, is usually difficult and expensive, so the corrosion properties of a particular material frequently play a major role in its selection.

Plastic components are subject to damage from temperature—either high or low—and from the action of certain solvents. The damage will usually be a progressive embrittlement with subsequent checking at the surface called *crazing*. The transparent plastics may become clouded and lose much of their optical efficiency unless protected from heat and dust.

The future holds much hope in the field of aircraft structural materials. New metals are being applied to structural problems, either alone or alloyed with other metals; new alloys are being discovered; new processes are being developed; and new methods of arranging structural parts are being designed.

Strength levels obtained from conventional materials today seem quite ordinary, but 20 years ago they would have been the dream of metallurgists. Similarly, the metallurgist's dream of today will be a reality to designers 20 years hence. Structural steel used in buildings has an ultimate tensile strength of about 55,000 psi. Even now, aircraft structural steels are being produced with an ultimate tensile strength of about 350,000 psi. This is indeed a very high strength—but metallurgical technology is already predicting ultimate tensile strengths of 1,000,000 psi in the near future. Theoretically, only 1/1000 to 1/100 of the perfect strength of metals is being used at present; undoubtedly the million psi level will be achieved.

Composite structure, that is, multimaterials, show much promise in permitting the designer to optimize the structure and obtain a better balance of strength to weight. During fabrication, the structure can be assembled by laminations of glass cloth, boron or graphite filaments, epoxy, and similar materials, with both

BUILT UP METHOD

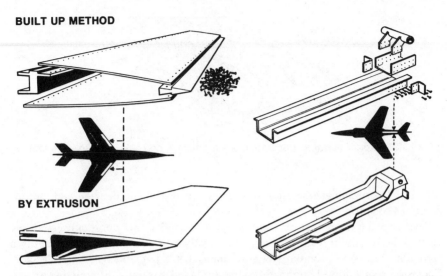

BY EXTRUSION

Figure 3.24. Extrusions in aircraft fabrication. Cost of construction can be reduced by use of extrusions to replace built-up parts. *Left:* Old and new leading edges. *Right:* Old and new longerons.

uni- and multidirectional fibers. Thus, the designer can approach ideal structural requirements based on the stress paths in the part or assembly. The stress paths may also be controlled. The use of these materials will increase in the future and result in reducing their cost. Technology in aircraft structural materials has a long, difficult development process ahead, but gains are being made at a gratifying rate.

Similarly, the use of newer methods of manufacture of aircraft parts shows great promise in holding down the costs of construction. Conventionally, many small parts are assembled into an aircraft, using hundreds of man-hours and many thousands of rivets. Resin bonding and the use of large honeycomb panels, which require no rivets, are noteworthy advances in this area. Another is the use of aluminum extrusions for forming complex parts, illustrated in Figure 3.24.

The constantly increasing costs of building aircraft and the healthy competitive atmosphere in the industry assure that the benefits of new materials and improved methods of construction will quickly reach the market place and the user of aircraft.

4

Propulsion

The early desires of man to fly were thwarted
for lack of a suitable means of propulsion.
The oldest well-authenticated sketches of
heavier-than-air flying are those of machines
which were designed by Leonardo da Vinci,
who conceived of three different flying machines in the fifteenth century. His devices,
like most earlier flight ideas, called for
human muscles to provide the motive power.
After da Vinci, four centuries passed before
the Wright brothers developed an engine sufficiently light to power an airplane built from
their advanced aerodynamic theories concerning propellers and airfoils.

During these four centuries many efforts
were made to produce engines light enough
for dynamic flight. Stringfellow developed
engines of sufficient lightness to support a
properly designed dynamic flying machine
and exhibited models in London in 1842.
Unfortunately, the comprehension of aerodynamic principles had not progressed far
enough to permit the trial of his engines in an
airplane. Even by 1875 the lightest engines
made weighed over 80 lb. per hp. By comparison, the engine designed and built by the
Wright brothers for their first flight weighed
only 13 lb. for each of its 12 hp.

Propulsive means were used with limited
success in man-carrying balloons. The first

dirigible (1852) was powered by a 3-hp steam engine weighing 330 lb. This engine turned a three-bladed propeller, 11 ft. in diameter, and made possible the first demonstration of some control over direction of flight.

A 9-hp electric motor weighing 25 lb. per hp enabled the airship "La France" in 1884 to demonstrate considerable capacity in powered flight.

Following the success of the Wright brothers in developing the first true airplane power plant, engine development progressed rapidly. The Gnome "monosoupape" air-cooled engine of World War I fame was developed in France in 1908. This engine was unique in that the crankcase and cylinders rotated with the propeller about the stationary crankshaft. Good cylinder cooling was obtained by this arrangement, but engine lubrication was a difficult problem. By 1913 it was producing 160 hp. Major developments by British and American designers of this period were in liquid-cooled engines. Two famous engines of this type were the eight-cylinder OX-5 which produced 90 hp and the 12-cylinder Liberty engine which developed 400 hp. Both engines were used in a number of different applications through the 1920's.

The eight-cylinder liquid-cooled Hispano-Suiza also powered many World War I airplanes and was the first to incorporate cast aluminum-alloy cylinder blocks with steel liners.

The 1920's saw the development of radial engines, followed in the 1930's by small air-cooled, horizontally opposed engines. In the 1920's, also, the development of the controllable-pitch propeller was a tremendous advance and opened many new possibilities for airplanes of greater weight and speed. The next milestone, of course, was the perfection of the gas-turbine engine into a useful airplane power plant.

By the end of World War II, reciprocating engines were producing as much as 3500 hp and eventually 3800 hp, and weighed less than 1 lb. per hp. This was the peak of development for large engines of this type. But they could not compete with the gas-turbine engine which offered far more power output for each pound of installed weight.

Theory of Propulsion

A dynamic flying machine must be set in motion by a force called thrust, which is a result of momentum change of a gas stream. The momentum change can be provided by a propeller, a helicopter rotor, flow through a gas-turbine, or a rocket engine. Gliders and sailplanes are originally set in motion from the thrust provided by a launching or towing device. Continued flight is maintained by descending in rising air currents.

With an airplane, the thrust force must be equal to total airplane drag to maintain level flight at any given speed or altitude. Since airplane drag increases with speed, greater thrust is required for an increase in speed. Thrust force must also be greater than total drag when the airplane is accelerating, as at take-off, or when the airplane is climbing. Generally, an airplane has reached its maximum speed or altitude when all available thrust is being utilized. However, in many current high-altitude jet airplanes, the amount of thrust available to meet take-off acceleration and high-altitude requirements is somewhat greater than that required to obtain maximum permissible speeds at lower altitudes, because of airframe structural limitations.

In the case of rocket-powered vehicles, thrust is utilized in a somewhat different manner. A large thrust force is used for a short initial period to gain a very high speed. After the rocket motor has depleted its fuel, the vehicle continues on its trajectory by means of the kinetic energy imparted to it by the high initial thrust. Once outside the earth's atmosphere, space vehicles and satellites will continue in motion indefinitely since there are no friction forces to absorb their energy, unless they are pulled back again by the earth's gravitational field or are "captured" by another heavenly body.

Propellers

The function of a propeller is the conversion of engine shaft torque (turning force) into thrust. In the early days of propeller design, it was thought that thrust was obtained by the rear face of the propeller blade pushing against the air. Later it was learned that propeller blades should be considered as small, rotating airplane wings with the shape of the airfoil being of primary importance. Therefore the upper surface of the airfoil, or the front rather than the back of the propeller blade, should be given primary design consideration.

Propellers may be installed in two different ways. In the "tractor" installation, the propeller "pulls" the airplane through the air, and the thrust bearings in the engine are designed for thrust loads in this direction. "Pusher" type propellers are installed aft of the wing and push the airplane forward.

PROPELLER THEORY

Two theories are recognized on the operation of a propeller in converting engine shaft torque to thrust. The momentum theory, developed by R.E. Froude, deals with the energy change to the air mass acted upon by the propeller. The other, and the most commonly used in determining propeller performance, is the blade-element theory developed in practical form by Stefan Drzewiecki.

In the Drzewiecki theory, each propeller blade is considered as being composed of an infinite number of airfoils (called blade elements) joined end to end, forming a shape similar to a twisted airplane wing. Because the propeller is rotating about the engine shaft centerline, each blade element will be rotating in a different arc. The greater the distance of each blade element from the shaft centerline, the larger is the arc it circumscribes and thus the greater the distance it must travel for each revolution. Each blade element, therefore, is moving at a different velocity, which is greatest at the blade tip. If each blade element is to operate at maximum L/D, the angle of attack α for each element must diminish as the distance from the hub increases. This gradual decrease in blade angle gives the propeller blade its twisted appearance.

Since the circumference of a circle is $2\pi r$, each blade element rotates through a circular path of $2\pi r$ during each revolution, with r as the distance from the center of rotation, expressed in feet. If the propeller is turning at n revolutions per sec., the linear velocity of each blade element in its individual plane of rotation must be $2\pi rn$ ft. per sec. In flight, the propeller is moving forward at the same speed as the airplane, as well as rotating about its own axis. This forward velocity must also be expressed in feet per second to keep all terms of the equation in common dimensions. (The conversion is: mph x 1.467 = feet per second, where 1.467 = [5280 ft. per mile/3600 sec. per hour]). Thus, the total velocity of each blade element is the vector sum of the two velocities: the forward velocity V and the rotational velocity $2\pi rn$. This may be shown graphically as in Figure 4.1, for each blade element.

As shown on the diagram, the angle of attack α is equal to the fixed blade angle β, minus the angle ϕ which is determined by airplane velocity V and propeller rotational velocity $2\pi rn$ (or $\pi\,dn$ since diameter $= 2r$).

The flow of air about the propeller blade creates lift and drag forces in the same manner as airflow about an airplane wing. The resultant of these forces can be divided into two components: the thrust component acting in the direction of the propeller axis, and the torque component. The torque component acting in the plane of rotation resists the rotation of the propeller and is the component that must be overcome by the engine shaft torque. Similar to the airplane wing, the magnitude of the thrust and torque components for the propeller depends upon angle of attack and velocity.

Coefficients for thrust and torque could be plotted against angle of attack, but

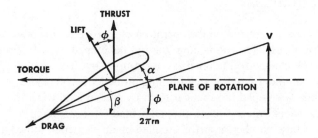

Figure 4.1. Vector diagram of propeller blade element.

since α is equal to fixed blade angle β minus variable angle ϕ, it is more convenient to plot in terms of angle ϕ. The usual procedure is to plot these coefficients against the tangent of ϕ or against V/nd. Such plotted diagrams, together with the blade angle β of any of the blade elements, define the operating conditions of the propeller as a whole. The blade angle generally given is the one for the blade element three-fourths of the distance from the propeller axis to the tip.

Graphs of thrust coefficient C_T, and torque or power coefficient C_p, plotted against V/nd, are prepared by the propeller manufacturers for various families of propellers. With reference to such graphs, values for C_T or C_p may be obtained for any chosen value of V/nd. Power required to drive the propeller and forward thrust delivered can then be determined by multiplying C_T or C_p by air density ρ and certain variables consistent with the method used in the chart preparation. Usually, thrust $= \rho C_T n^2 d^4$, and power required $= \rho C_p n^3 d^5$.

PROPELLER EFFICIENCY

The efficiency of a propeller in converting engine power to thrust is obviously of tremendous importance to airplane performance. The efficiency of a propeller during cruising flight is as important in determining the range of an airplane as is the efficiency of the engine in converting fuel energy to shaft power.

Propeller efficiency is expressed as the ratio of *thrust power* delivered to *engine power* required to turn the propeller. Thrust power is thrust force T(lb.) multiplied by airplane velocity V(ft. per sec.), or TV (ft.-lb./sec.). The power required to turn the propeller may also be expressed in ft.-lb./sec. by multiplying engine shaft horsepower by 550, since one horsepower is 550 ft.-lb./sec.

$$\text{Efficiency,} \quad \eta = TV/P = VC_T n^2 d^4 / C_p n^3 d^5 = (C_T/C_P)(V/nd). \qquad (4.1)$$

As noted previously, propeller thrust is expressed in terms of V/nd. Therefore, efficiency may be expressed in terms of V/nd and plotted as shown in Figure 4.2.

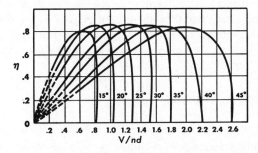

Figure 4.2. Propeller efficiency vs. V/nd for various angles of β. (N.A.C.A. Tech. Rept. 642.)

In the simple fixed-pitch propeller, blade angle β is fixed. Each curve of the chart shown by Figure 4.2 represents efficiency for one angle of β.

A study of Figure 4.2 shows that for a fixed propeller blade angle β, high efficiency is obtained in only a very narrow range of airspeeds, and that a propeller designed for good cruise performance will be deficient for take-off, climb, low-speed flight, and high-speed cruise. The airspeed at which maximum efficiency occurs is called the "design airspeed" for the propeller.

The advent of the controllable-pitch propeller solved these difficulties extremely well. Provided with a means for varying pitch angle, the pilot could select the optimum angles for take-off, for climb, for efficient cruising, or for high-speed flight. In fact, modern airplane performance really began with the development of the controllable-pitch propeller.

Fixed-pitch propellers are currently used only on small, light airplanes where cost and lightness of installation are major considerations. The speed range of such airplanes is limited by the power output of the engine as well as by the propeller.

Propeller Tip Speed.   The velocity of the blade elements has little effect on efficiency until velocities near the speed of sound are approached. When the velocity of blade elements close to the tip approaches the speed of sound, the effect of compressibility increases power input requirements at no increase in thrust, thereby reducing efficiency. Propeller tip speeds are normally limited to less than .8 times the speed of sound due to compressibility effects in airflow over the blade. To prevent tip speeds from exceeding these limitations, many engines are equipped with reduction gears. This arrangement permits lower propeller rpm than the engine rpm required for efficient engine power output.

Thrust at Take-Off.   In addition to providing efficient thrust during cruising conditions, the propeller must be able to convert total engine power to thrust for take-off. This is necessary to provide rapid acceleration to flying speed and to keep take-off distances within reasonable limits. In Figure 4.2 it will be noted that when V/nd is zero ($V = 0$), efficiency is also zero. Obviously this is not correct or an airplane would never start in motion. Because of this apparent fallacy in the Drzewiecki theory, the efficiency curves are shown as dotted lines at low values of V/nd, indicating that they should not be used in the low-speed range.

Charts of thrust output versus shaft horsepower are provided by propeller manufacturers for use in computing thrust output during the take-off phase of flight.

PROPELLER TERMINOLOGY

The theoretical advance of the propeller blade per revolution is determined by blade angle β. This advance per revolution is called geometric pitch. However, air is not a solid medium, and a certain amount of slippage occurs. The effective pitch is the actual distance the propeller advances each revolution. The difference between geometric pitch and effective pitch is called "slip."

Fixed-Pitch Propeller.   This is a propeller with blade angle fixed.

Adjustable-Pitch Propeller.   This is a propeller on which blade angles can be adjusted on the ground with propellers not turning. This is especially convenient for take-offs where maximum thrust is required at the sacrifice of cruise efficiency (take-off on a hot day from a high-altitude airport, or for preparing an airplane for float operations).

Constant-Speed Propeller.   This is a controllable-pitch propeller with a speed governor which maintains a selected rpm by automatically changing the blade angle regardless of airspeed or engine power. Propellers of this type used on multi-engine aircraft also provide for full "feathering" of the blades in flight to prevent "windmilling" of the propeller; this reduces drag when the engine has failed.

Feathering.   The means mechanically increasing pitch angle until the blade is turned approximately to the direction of flight.

Reverse Thrust.   This indicates thrust in the direction opposite to flight. Nearly all larger multi-engine airplanes utilize constant-speed, full-feathering, reversible-pitch propellers. The pitch-change mechanism is designed to permit the pilot to select negative blade angles immediately after landing, thus creating reverse thrust for rapid airplane deceleration. This feature saves brakes on heavy airplanes and permits safe landings on icy runways where wheel braking is extremely ineffective.

Windmilling Drag.   Perhaps the most undesirable action in the use of propellers as a thrust-producing device is the drag that results from propeller windmilling. If engine power is lost in flight, the airflow over the propeller blades will cause the propeller to rotate and, in turn, rotate the engine. When the blade angle is great enough to be commensurate with the flying speed, propeller efficiency will remain high and the drag will not be appreciably greater than engine load. However, if a failure occurs in the pitch-change mechanism and the blade angle is reduced, excessively high rotational speeds may result. The low blade angle and resultant negative angle of attack of the blade airfoil element cause a tremendous reduction in propeller efficiency. This means the load required to turn the dead engine is supplied by an inefficient propeller, and the resulting windmilling drag can become dangerously high. In such instances it is necessary to reduce airspeed or altitude to reduce propeller rotational speed. But it is important to remember that minimum airspeed for adequate directional control of a multi-engine airplane will be increased above that required if the propeller of the failed engine can be feathered. However, unless actual failure occurs to the pitch-change mechanism, the propeller speed governor will maintain rpm as it was before engine failure, and the feathering system, if installed, will feather the propeller to stop windmilling rotation altogether. Even on a nonfeathering propeller, if propeller control remains after engine failure, drag can be substantially reduced by selecting low rotational speed (high blade angle).

PROPELLER SELECTION

Analysis and investigation required for proper propeller selection depend upon the type of operation of the airplane involved. Careful investigation is required to insure that all conditions of take-off and flight are met with the lightest and most efficient installation. The propeller manufacturer provides charts for propeller performance under all conditions of flight. Ideas as to the relative importance of the various factors influencing propeller selection may vary among airplane designers. One may choose a three-bladed propeller for a specific type of airplane, while another would choose a four-bladed propeller for superior performance.

The selection of propellers for the smaller airplanes designed to operate at low and medium altitudes and airspeeds under 200 mph is fairly simple. The propeller manufacturer lists propellers that are designed for use with engines of a given horsepower rating and often lists the make and type of airplane on which the propeller may be used. Fixed-pitch propellers for a specific engine are usually obtainable in a choice of two or more pitch angles. A blade of higher pitch may be chosen for better cruise performance. Controllable-pitch propellers for small airplanes are designed by the manufacturer for application on specific engines. Sufficient pitch change is usually designed into the control mechanism to take care of all speed and altitude conditions of a number of airplanes using the specific engine.

PROPELLER CONSTRUCTION AND INSTALLATION

Propellers made of forged aluminum alloy blades are in general use (Figures 4.3 and 4.4). This type of construction offers durability and strength as well as good manufacturing control of airfoil shape, all of which are especially important in larger sizes. In service, small nicks and scratches may be polished out without impairing the strength or performance. If the propeller is bent by some misfortune, often it can be straightened by a properly equipped repair facility.

The simple adjustable-pitch propeller has rounded blade shanks which are secured by collars to the hub. By loosening the securing bolts in the collar, the blades may be rotated to the desired blade angle (pitch angle) and the bolts retightened.

Figure 4.3. Fixed-pitch metal propeller. *(Courtesy of McCauley Industrial Corp.)*

Figure 4.4. Hartzell constant-speed, full-feathering propeller, installed on a Piper Apache airplane. *(Photo by editor.)*

Constant-speed, controllable-pitch propellers are used on many small and medium-sized personal and executive airplanes of over 180 hp. Such propellers are shown in Figures 4.4 and 4.5. Engine oil under pressure provides pitch control.

In the type shown by Figure 4.5, the twisting moment of the blade, resulting from centrifugal force, tends to decrease blade angle; it is offset by oil pressure from the governor, acting against the forward side of a piston in the hub, the piston being connected through the linkage to the blades. The flyball governor is mounted on the front case of the engine and is rotated by gears driven by the engine crankshaft. The gears also drive the governor oil pump. If the propeller tends to increase in speed above the rpm selected by the control in the cockpit, the governor will open the oil passage and admit more pressure to the piston. This action will increase blade angle slightly and restore rpm to the selected value. Conversely, if the propeller tends to decrease in speed, the governor will drain pressure from the piston, permitting the centrifugal twisting movement of the blades to reduce blade angle and increase speed. Changes in engine power (within limits), airspeed, and air density are compensated for by automatic selection of the proper blade angle for the condition, the propeller speed remains constant. The rpm at which constant-speed regulation is desired is selected by the control in the cockpit. This control is directly connected to a spring acting against the centrifugal force of the flyballs in the governor. Increasing or decreasing the

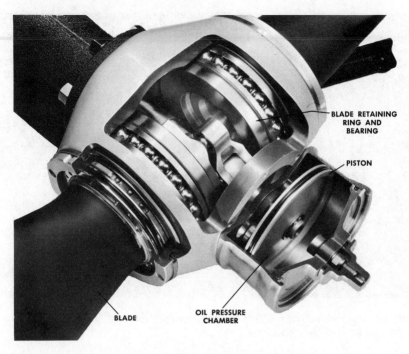

Figure 4.5. Cross-sectional view of one type of contant-speed propeller mechanism. *(Courtesy of McCauley Industrial Corp.)*

spring tension requires a new position for the governor flyballs, which in turn repositions the valve to admit oil to the piston in the propeller hub.

Another type of controllable-pitch propeller is somewhat similar in principle to the propeller described above, although different in operation (Figure 4.4). Counterweights connected to the blade shanks in the hub tend to increase blade angle. The force of the counterweights is offset by oil pressure acting against a piston. Regulation of speed is obtained by controlled oil pressure in the reverse manner to the previously described propeller. Increased blade angle will result when oil pressure is reduced, and vice versa. Feathering may be accomplished automatically in some models of this propeller by the action of a spring assisting the counterweights in rotating the blades to the feather position when all oil pressure is bled from the piston.

Constant-speed feathering and reversing propellers are used on large turboprop and piston-engine transport airplanes. In addition to constant-speed control, a synchronizing system may be used to maintain equal rpm on all engines.

Two different principles of operation are employed in the propeller-pitch control systems used predominantly on large multi-engine airplanes. Electrical motor power for pitch-change actuation is used in the Curtiss electric propellers, while oil pressure is utilized in Hamilton Standard, Hartzell, and McCauley propellers.

In the Curtiss propeller, a motor rotates the blades, through a gear train, by means of segment gears attached to each blade root. When the motor is not running, the blades are held in position by a brake that releases only when power is applied to the motor. Electrical power is transmitted to the rotating propeller mechanism through slip rings attached to the propeller shaft. Electrical limit switches mounted in the rear housing accurately control the low, high, feather, and reverse blade-angle limits. In addition to the electrical low-pitch limit switch, a mechanical low-pitch stop is provided on many models to prevent dangerously low blade angles in the event of electrical brake failure. This mechanical stop is retracted when the pilot selects reverse actuation during landing. All propellers are maintained at the same rpm by the synchronizer located in the airplane fuselage.

The Hamilton Standard hydromatic propeller is used on many commercial and military multi-engine airplanes. Oil pressure acting against either side of a piston, plus the centrifugal twisting movement of the blades, provides the force for blade pitch change.

The force of the piston is transmitted through cams and rollers to the drive gear which is meshed with the gears affixed to each blade root. Constant speed is maintained by the engine-driven, double-acting flyball governor, which meters governor oil pump pressure to either side of the piston as required.

Figure 4.6. Hamilton standard hydromatic propeller. Constant speed, full feathering and reversing, installed on a Lockheed C-130B turboprop transport. *(Courtesy of US Air Force.)*

Jet Propulsion

The principle of jet propulsion is very familiar and very simple. The jet or rocket engine moves the airplane or missile forward by reacting to the momentum of a mass of air or gaseous matter accelerating out the rear (nozzle) of the engine. The recoil of a rifle and the reactive thrust of a fire hose are typical examples of this principle. Sir Isaac Newton explained jet propulsion in 1690 when he propounded the third law of motion: "Every action produces a reaction, equal in force and opposite in direction." Thus the reaction to the change of momentum of the gases escaping from the jet engine nozzle creates the thrust force which propels the vehicle.

From Newton's law it can be seen that the thrust force is not the result of the jet gases pushing against the air behind but instead is the reaction to the change of momentum of the escaping gases. This fact is demonstrated by rocket motors, which produce full thrust in outer space beyond the normal atmosphere. (Unlike air-consuming turbojet and ramjet engines, rockets carry their own oxidizer supply for combustion and do not depend upon the oxygen in the atmosphere.)

ENGINE TYPES

All engines employing the principle of jet propulsion are jet engines. However, common terminology denotes jet engines which carry their own oxidizing agent as rocket motors. Engines which draw in air, heat it by burning fuel, and expel it from the nozzle are commonly termed jet engines. The various types of jet engines and rocket motors, with their principles of operation, are discussed later in the chapter. The following symbols, followed by the model number of the engine, are used by the Armed Services to identify the various types of jet propulsion engines.

J = Turbojet   R = Rocket
PJ = Pulsejet   T = Turboprop
RJ = Ramjet   TF = Turbofan

POWER MEASUREMENT

Jet thrust force is measured in pounds as is the thrust force of a propeller.

The magnitude of jet gross thrust force is determined by the mass flow of air or gaseous matter multiplied by the velocity with which it escapes from the nozzle. This may be expressed by:

$$F = \frac{W_a}{g}V_j = MV_j \qquad (4.2)$$

where F = thrust force, lb.

M = mass flow rate = W/g

V_j = jet velocity, ft. per sec.

W_a = weight flow, lb./sec.

g = constant of proportionality = $32.2 \dfrac{\text{lb.} M - \text{ft.}}{\text{lb.} F - \text{sec.}^2}$

Because aerodynamic equations require thrust to be expressed as a force, in order to equate thrust requirements against drag and acceleration forces, there is no purpose in attempting to express jet-engine output in terms of horsepower. Though it is relatively meaningless, jet-engine "power" can be determined by multiplying thrust force times true airspeed in feet per second and dividing by the horsepower constant.

$$hp = (\text{jet thrust} \times \text{TAS} \times 1.467)/550 \qquad (4.3)$$

Therefore, when true airspeed is 550 ft. per sec. or 375 mph, the jet thrust and jet "power" are the same. At speeds below this value, jet "power" is less than jet thrust in pounds. In the case of the turboprop engine, where both propeller thrust and jet thrust are utilized, it is sometimes desirable to convert jet thrust into horsepower. By adding this value to the horsepower rating of the engine, the total propulsive output of the engine can be expressed as "equivalent shaft horsepower" (eshp). Assuming a propeller efficiency\of 80%, we have

$$\text{eshp} = \left(\frac{\text{jet thrust} \times \text{TAS} \times 1.467}{550 \times .80}\right) + \text{shaft horsepower} \qquad (4.4)$$

Propulsive Efficiency

PRINCIPLES AND LIMITS

The efficiency with which a propulsion system is utilized in propelling the airplane or missile is called "propulsive efficiency." Thorough understanding and investigation of propulsive efficiency are necessary for successful design, when comparing one propulsion system with another. This is especially true when comparing propeller propulsion with jet propulsion.

Propulsive efficiency may be expressed very simply by

$$Efficiency = 2V_a/(V_J + V_a) \qquad (4.5)$$

where V_a = airplane velocity, ft. per sec.

V_J = jet velocity (or propeller slipstream velocity), ft. per sec.

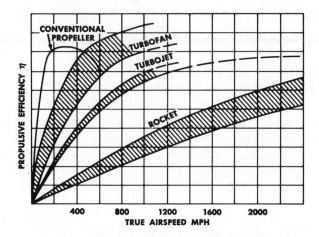

Figure 4.7. Comparison of propulsive efficiency. *(Courtesy of Hamilton Standard.)*

In the case of propeller-driven airplanes, propulsive efficiency very closely approximates propeller efficiency (Figure 4.2) because the propeller moves a large mass of air at relatively slow speeds. During cruising conditions, the air behind the propeller is moving at approximately the same speed as the airplane; hence, V_a and V_J are approximately equal. In contrast, the thrust-producing gases escaping from the pure jet engine nozzle are extremely high in velocity, and relatively smaller in mass flow.

High-altitude operation must be considered in propulsive efficiency analysis. Because of compressibility effects on the propeller airfoil, propeller efficiency decreases with both airspeed and altitude. This factor limits subsonic propellers to approximately 500 mph true airspeed and 40,000 to 45,000 ft. altitude.

Since jet thrust is accomplished at high jet velocities, the higher the airplane speed, the greater is the propulsive efficiency. Current turbojet engines are limited to approximately 2000 mph and 90,000 ft. altitude. Turbo-fan engines provide an extremely good compromise between propeller and turbojet engines. Higher propulsive efficiency is available for take-off and is maintained to speeds and altitudes far greater than are possible with propellers. For efficient operation the ramjet is bounded between 1500 mph at sea level and 3000 mph at 100,000 ft. altitude. Because the rocket engine carries its own oxidizer supply, its high-speed performance is not limited by atmospheric considerations; however, its high exhaust velocity results in low propulsive efficiencies for low speeds. A comparison of propulsive efficiency for several propulsion systems is shown in Figure 4.7.

TAKE-OFF PERFORMANCE

During take-off, the propeller offers some advantage compared with the pure jet. Not only does the propeller move large masses of air at a more advantageous velocity for greater propulsive efficiency, but the slipstream of the propeller flowing over the wing increases the effective lift of the wing. A turboprop engine, comparable in physical size to a pure jet engine, will develop shaft horsepower approximately equal in rating to the thrust rating of the pure jet engine. Using an approximate conversion factor of 2.0 to convert shaft horsepower to thrust for take-off, it can be seen that the turboprop engine will develop about 2.0 times as much take-off thrust. The newer high-bypass-ratio turbofan engines approach the performance of propellers for take-off and, in the larger sizes, provide thrust levels far greater than are possible with propellers of any reasonable diameter.

Increased take-off thrust for pure jet airplanes is obtained by several methods. A method used extensively in fighter airplanes, called afterburning, is the burning of additional fuel in the engine tailpipe. While this method requires a high fuel consumption during operation, it can result in as much as a 50% increase in take-off thrust. Water, or water-alcohol mixtures, injected into the engine compressor or burners on turbojet and turbo-fan engines also can provide increased thrust for take-off; as much as 15% to 25% increase can be developed by this method. The use of afterburning or water injection for increased take-off thrust is termed augmentation. Further details of augmentation are discussed later in the chapter.

Those flight vehicles with a thrust-to-weight ratio less than unity require a runway. These vehicles are accelerated along the runway until the lift surfaces provide sufficient lift for flight. A helicopter is set in motion by the lifting force of its rotating wing (rotor). Forward velocity is obtained by tilting the aircraft to provide a forward component of the lifting force. In an effort to obtain the speed and lift force advantage of the fixed wing, coupled with the vertical take-off and landing advantage of the helicopter, the VTOL (vertical take-off and landing) aircraft has been developed. Several design concepts have undergone study and test. With the turboprop-powered tilt wing and tilt engine, either the entire wing (or sections thereof), with the engines, is tilted, or the engines only are rotated to the vertical position for take-off and landing. The thrust force provided must be greater than the entire weight of the airplane. Separate turbo-fan lift engines and vectored main engine jet exhaust have been used on some experimental aircraft. A successful example of vectored main engine jet exhaust is the British Harrier jet fighter.

Reciprocating Engines

Reciprocating aircraft engines operate on the same principle as the crude engine which powered the Wright brothers' airplane. The advancement to present-day, high-powered engines has been the result of improved efficiency, construction, and reliability. The vast improvement in these factors is clearly evident. Compare the Wright brothers' 13 lb. per hp with larger modern engines weighing less than 1 lb. per hp. Each cylinder on a large modern engine will produce more than ten times the horsepower of the entire Wright brothers' engine, and do it for many hundreds of hours before overhaul is necessary.

PRINCIPLES OF OPERATION

The term *reciprocating engine* is derived from the action of the pistons moving back and forth within the cylinders in a reciprocal motion as the crankshaft rotates. The term *internal combustion* denotes the action of burning fuel within the engine to provide the forces that develop the engine power output.

All reciprocating engines in general use for airplane propulsion are internal-combustion engines operating on the *Otto cycle* (Figure 4.8).

The Otto cycle principle is named for Nicholas A. Otto, who built the first successful engine operating on this principle in 1876. Another description commonly applied to engines operating on this principle is "four-stroke cycle, spark ignition." The basic power-producing elements of a reciprocating engine are the *cylinder, piston, crankshaft,* and *connecting rod* (see Figure 4.8). The piston receives the force of the expanding gases within the cylinder and transmits the force to the crankshaft by means of the connecting rod. The crankshaft is caused to rotate by the force of the connecting rod and transmits the turning effort (torque) to the propeller. One complete cycle consists of four strokes of the piston—two inward strokes (toward the crankshaft) and two outward strokes. The completion of one cycle, therefore, requires two complete revolutions of the crankshaft. On multicylinder engines, the pistons are connected to the crankshaft in a manner such that each cylinder completes its cycle at a different time relative to propeller position. This results in smoother operation, since the power stroke of each piston occurs at a different time.

The four strokes of the Otto cycle are called *intake, compression, power,* and *exhaust.* The relationship is shown by Figure 4.8. When the piston is at the very end of the stroke, away from the crankshaft, it is said to be at "top dead center." The term "bottom dead center" means the opposite end of the stroke or the one closest to the crankshaft.

The intake stroke begins with the piston at (or near) top dead center. At this point, the intake valve is opened by the valve mechanism, and the intake stroke of the piston draws the fuel-air mixture (produced by the carburetor) into the cylinder from the intake manifold. (In supercharged engines the fuel-air mixture is forced into the cylinder under pressures greater than atmospheric.) When the piston reaches the end of the intake stroke, the intake valve closes and the compression stroke begins. (Fuel-injection engines operate similarly except that the fuel is injected in the proper amount under very high pressure directly into the cylinder during the compression stroke.) As the piston moves toward the top of the cylinder, the enclosed combustible mixture is compressed to a small volume. An instant before the piston reaches top dead center, the spark plug is energized, igniting the fuel-air mixture. While burning takes place, the piston passes top dead

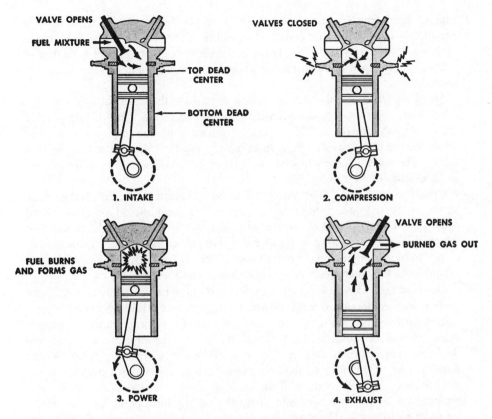

Figure 4.8. The events in the Otto cycle or four-stroke cycle spark ignition engine.

center and power stroke begins. The burning of the fuel-air mixture greatly increases the temperature and pressure of the gases. These high pressures act against the piston, which transmits its force by means of the connecting rod into the crankshaft, creating the turning torque of the crankshaft. When the piston reaches the bottom of the cylinder, at the completion of the power stroke, the exhaust valve opens and the upward movement of the piston in the exhaust stroke expels the burned gases from the cylinder. At the completion of the exhaust stroke, the intake valve opens and a new cycle begins.

Valve Timing. Because the intake and exhaust valves open and close only once for two revolutions of the crankshaft, the mechanism which actuates the valves is driven at one-half crankshaft rpm. The time at which each valve opens and closes, relative to piston and crankshaft position, varies with different engines. The amount of supercharging and the rpm at which the engine is to derive its greatest power or efficiency determine the inertia and velocities of inrushing combustible mixture and outgoing burned gases which, in turn, determine the valve timing for which optimum engine power is realized. On some supercharged engines, the intake valve is opened before the exhaust valve is completely closed to insure complete expelling of the burned gases. This timing is called "valve overlap." During the development of an engine, a great deal of testing is performed to determine optimum valve timing, among other considerations.

Spark Advance. This term is applied to indicate the angular position of the crankshaft, relative to top dead center, at the moment spark ignition is applied to the combustible mixture. A 15° spark advance means that the spark is applied when the crankshaft is 15° from reaching top dead center on the compression stroke. The spark is nearly always applied before top dead center, as this results in better engine performance.

Compression Ratio. The ratio of the cylinder volume at the end of the intake stroke to the cylinder volume at the end of the compression stroke is called *compression ratio*. The greater the ratio, the smaller is the relative volume to which the air-fuel mixture is compressed. Greater compression (within limits) results in higher engine power output for a given engine size, but opposed to this are other factors such as spark-ignition timing and fuel-burning rate.

Detonation. Detonation will occur if the burning rate of the fuel is not compatible with compression ratio and spark plug advance. Holding a fuel/air mixture at sufficiently high temperature and pressure for a sufficient time produces instantaneous self-ignition. The leaner a mixture, the slower its burning rate, and the more subject it is to self-ignition (detonation). It is necessary to control the burning rate of the fuel so that complete burning takes place during several degrees of crankshaft rotation. If the burning rate is too slow, the pressure and temperature in the cylinder build up to the self-ignition point, and the entire remaining mixture will explode instantaneously, causing detonation. When applied to automobile engines, detonation is often referred to as "pinging" or "knocking."

Preignition. This will occur with an effect similar to detonation if a "hot spot" (improperly cooled valve, carbon particle, etc.) in the cylinder causes ignition of the fuel prior to the normal firing of the spark plugs.

Octane Rating. Fuels for Otto cycle engines are rated according to their ability to remain at sustained high temperatures and pressures without causing detonation, by comparing them with a pure hydrocarbon, called octane. This hydrocarbon has an assigned rating of 100. Thus, a fuel with an octane rating of 80 has inferior detonation characteristics compared to pure octane, while a fuel rated at 130 has better characteristics. The comparison tests are conducted in a special single-cylinder engine in which the compression ratio may be varied.

FUEL-AIR MIXTURE

After fuel and air are mixed in the desired proportions in the carburetor, the mixture is forced by throttled pressure into each cylinder through pipes or passages in

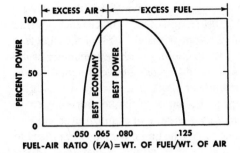

Figure 4.9. Effect of fuel-air mixture on power output.

the intake manifold. With fuel injection, the airflow through the carburetor signals the injection pumps to meter the proper amount of fuel directly into the cylinder. The ratio of fuel mixed with the air is called "fuel-air ratio" (F/A) and has a very pronounced effect on the operation of the engine. If the mixture is extremely "lean," low in fuel content, the mixture may not burn at all or may cause "backfiring." Backfiring is caused by incomplete combustion in the cylinder so that, when the intake valve next opens, the still-burning fuel will ignite the fresh mixture in the intake manifold. If the mixture is moderately lean, but still not the proper ratio, detonation may result at high power settings. Excessively rich mixtures cause loss of power as well as uneconomical fuel consumption. However, excessively rich mixtures, which provide a cooler fuel/air charge, are sometimes employed on a supercharged engine for short-period operations at high power, such as is used for take-off. Power loss due to the rich mixture is more than offset by forcing a greater weight of fuel-air mixture into the cylinders. The objective of the overly rich mixture in this instance is faster combustion at the higher pressures, while preventing the detrimental higher temperatures within the cylinder and the attendant prevention of detonation. The effect of fuel-air mixture on power is shown by the curve in Figure 4.9. The mixture entitled "best economy" is the mixture that yields the most hp per lb. of fuel burned. On some supercharged engines, the fuel-air mixture is not enriched during high-power operation. Instead, the engine is operated with "best power" fuel-air ratio, and detonation suppression is obtained by substituting a mixture of water and alcohol to cool the fuel/air charge, in place of the normally-excess fuel. This practice is done only on very high-powered engines and is called "antidetonant injection," or ADI.

PERFORMANCE

Only a fraction of the total heat energy of the fuel burned is converted to useful horsepower. The remainder of the heat energy is lost through the exhaust gases, the engine cooling system, and the friction of the moving engine parts. The ratio of useful energy to total energy of the fuel consumed is called *thermal efficiency*.

The best reciprocating Otto cycle aircraft engines are about 34% thermally efficient. In some instances, discussed later, utilization of some of the energy in the exhaust gases has resulted in higher values. Approximately 10% of the total fuel energy is lost to engine friction, 10% to 15% to cooling. and the remainder of the loss to the exhaust gases. Friction loss is the difference between actual horsepower developed, called *indicated* horsepower (ihp), and useful shaft horsepower, called *brake* horsepower (bhp). The ratio of brake horsepower to indicated horsepower is called *mechanical efficiency*. (Mechanical efficiency = bhp/ihp.) As shown by later discussion, not all of the brake horsepower is used to drive the propeller. A portion may be used to drive pumps, generators, superchargers, etc., depending upon installation.

The indicated horsepower developed by a reciprocating engine may be expressed simply as the foot-pounds of work done on all pistons in 1 min. divided by 33,000. (The value 33,000 ft.-lb./min., or 550 ft.-lb./sec. is an established standard for 1 horsepower.)

$$\text{ihp} = PLAN/33,000 \tag{4.6}$$

where P = mean effective pressure, psi [1]
    L = length of stroke, ft.
    A = area of piston, sq. in.
    N = number of power strokes per min.

The horsepower expressed by the above equation is indicated horsepower and does not account for friction losses of the moving engine parts.

Because friction losses and mean effective pressure would be very difficult to calculate, actual engine horsepower is determined by operating the engine while coupled to a measuring device called a *dynamometer*. A dynamometer does not measure horsepower directly; instead it measures the turning force of the crankshaft. This turning force is called *torque* and is measured in pounds-feet. Thus, if an arm were connected to the crankshaft, 200 lb.-ft. of torque will exert a force of 200 lb. at a right-angle distance of 1 ft. from the centerline of the crankshaft. At a distance of 2 ft. the force would be 100 lb., etc. (The expression *pounds-feet* for torque should not be confused with *foot-pounds* for work, which is the product of force times distance in the direction the force is acting.)

Once torque is determined, the useful horsepower delivered by the engine shaft is easily calculated. Since $2\pi r$ is the circumference of a circle, the work per revolution is $2\pi r$ times torque, and

$$\text{hp} = \frac{2\pi \times \text{radius (ft.)} \times \text{lb.} \times \text{revolutions/min.}}{33,000 \times \text{ft.} \times \text{lb./min.}}$$
$$= \frac{2\pi \times \text{torque} \times \text{rpm}}{33,000} \tag{4.7}$$

---

[1] Average cylinder pressure during the power stroke.

The horsepower output thus determined is called brake horsepower.

For the purpose of comparing the relative performance of various engines, the original equation, hp = $PLAN/33,000$, is used. Since all parts of the equation are known except mean effective pressure P, this value is calculated. The resulting pressure is now called brake mean effective pressure (bmep), since it is calculated from measured or brake horsepower.

$$hp = (\text{bmep}LAN/33,000$$
$$\text{bmep} = hp \times 33,000/LAN \qquad (4.8)$$

Since N is the number of power strokes per minute, and one power stroke occurs in each cylinder every two revolutions, N = no. of cylinders $\times$ rpm/2. Converting the powerstroke L to inches and multiplying by piston area and the number of cylinders, the total piston displacement in cubic inches is determined. The equation now becomes

$$\text{bmep} = \frac{hp \times 33,000}{L/12 \times A \times \text{no. of cylinders} \times \text{rpm}/2}$$
$$= \frac{hp \times 792,000}{\text{rpm} \times \text{piston displacement}} \qquad (4.9)$$

Thus, engines (especially the larger sizes) are compared according to total piston displacement and maximum allowable bmep.

POWER REGULATION

From the last equation it is apparent that only two variables determine the power output of a particular engine. These factors are bmep and rpm. The magnitude of bmep for an engine, at any constant rpm, will be determined by the amount of fuel-air mixture pumped into the cylinders during the intake stroke. The greater the weight of mixture charge, the greater is the internal cylinder pressure resulting from the burning fuel.

The weight of the mixture charge will be determined by the density of air in the charge. Therefore, all the factors that influence charge air density will also influence the engine horsepower output. The density of the charge air is determined by atmospheric pressure and temperature, position of the throttle valve in the carburetor, and if supercharging is used, the amount of air compression in the supercharger. The throttle valve provides the basic control for charge air density and is regulated by the throttle lever.

When the throttle is wide open, the charge air density is determined only by atmospheric pressure and temperature and the amount of supercharging. Since air density decreases with altitude, the power of the unsupercharged engine will also decrease as the airplane climbs to higher altitudes. The effect of altitude on

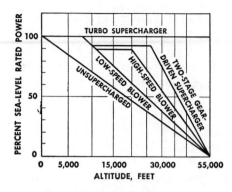

Figure 4.10. Engine performance at various altitudes for several methods of supercharging.

engine power output and the effect of supercharging are shown in Figure 4.10. All curves below the top line reflect wide-open throttle. Partial closing of the throttle has the same effect on the density of the air charge as an increase in altitude.

Charge air density not only varies with pressure altitude, but varies considerably with temperature and humidity. Even on small airplanes, the effect of these variables on engine power cannot be ignored. The effect of temperature and altitude both on engine output and on aircraft performance must always be taken into account. Engine power output varies as the square root of the ratio of standard day temperature to observed temperature, where both are converted to absolute temperature.

$$\text{corrected hp} = \text{chart hp } \sqrt{(460 + T_s)/(460 + T_a)} \qquad (4.10)$$

where T_s = standard day temperature = 60°F
  T_a = observed temperature, °F

A fairly accurate approximation that may be used is 1% correction for every 10°F temperature variation from standard day temperature.

Humidity corrections are made by determining *absolute* humidity (not relative humidity) of the atmosphere at the particular moment and using the manufacturers' power correction charts.

The engine manufacturer designates various power settings at which each type of engine may be operated for short periods and at which the engine may be operated continuously. For large-engine operation, *take-off* or *emergency* power may be held for a period of only 5 min. The designation *rated power* on military engines, or METO power (maximum except take-off) on commercial engines, is the maximum power permitted for continuous operation. Both of these powers require a rich fuel-air mixture. The maximum cruise power permitted under lean fuel-air mixtures is called *maximum cruise*. The altitude at which rated or METO power requires wide-open throttle is called *critical altitude*. Above this altitude engine power will reduce because of reduced atmospheric density. Obviously, the amount of supercharging will determine the critical altitude, as shown by Figure 4.10.

POWER INDICATION

Revolutions per minute and bmep are the primary variables affecting horsepower output of any particular engine. Revolutions per minute may be readily indicated to pilot or flight crew by means of the engine instrument called the *tachometer*, which measures the rotating speed of the crankshaft. On small airplanes this is often the only power-indicating instrument.

The other variable, bmep, cannot be measured. However, pressure of the fuel-air mixture in the inlet manifold can be measured on a pressure gage. Since this pressure is the primary influence on bmep, its measurement serves the purpose well. The pressure in the intake manifold is called *manifold pressure* (mp or map), and the measuring instrument is called the *manifold pressure gage*. The manifold pressure gage always registers total pressure above absolute zero (vacuum) since the function of the gage is the indication of air density in the manifold (Figure 5.1).

With the two main variables indicated to the pilot, the engine and airplane manufacturers can prepare airplane performance charts in terms of rpm and map for maximum power limitations, a variety of cruising powers, and other operating conditions.

Another power-indicating instrument is the *torquemeter*. This is a torque-measuring device built into the nose of the engine, and it measures the amount of turning effort of the crankshaft being delivered directly to the propeller. The advantage of this direct indication is that all factors which affect power output will be reflected in the torquemeter reading.

SUPERCHARGERS

Supercharging increases power output by forcing more fuel-air mixture into the cylinders than could be normally aspirated at ambient atmospheric pressure. Because superchargers increase the weight and cost of the engine, they are not often used on engines below 200 hp. Also, airplanes utilizing engines of smaller horsepower generally operate at altitudes that do not require supercharging. The compressor in nearly all the makes of superchargers operates on the centrifugal principle. This type of compressor permits high-speed operation and attendant lightness of construction relative to airflow delivery.

Superchargers may be designated by two categories: single-stage and two-stage (or auxiliary-stage). Single-stage denotes a single compressor; two-stage signifies two compressors in series. The single-stage compressor is usually contained within the engine case and is driven by a system of gears directly from the crankshaft at several times the rotating speed of the crankshaft. It is usually the engine *blower*. The total pressure at the outlet divided by the total pressure at the inlet is called *blower ratio*.

Auxiliary-stage supercharging is supplied by a separate gear-driven supercharger or by an exhaust-gas-driven turbosupercharger. Its chief advantage is its use of exhaust gas energy to provide the large power requirements for air compression at high altitudes rather than taking this power from the engine crankshaft.

Auxiliary-stage supercharging often requires the employment of an *intercooler* to remove some of the heat of compression from the supercharged air. The intercooler exchanges the heat of compression with atmospheric air directed through passages in the intercooler. Cooling of the air not only increases its density, assisting the compressor, but prevents fuel-air mixture at excessively high temperature from entering the cylinders, thereby helping to prevent detonation.

A number of engines in the 160-hp to 500-hp class are now using single-stage turbosupercharging to provide high engine power up to above 20,000 ft. altitude. In some installations intercooling is provided, and a portion of the turbosupercharger output is used for cabin pressurization.

PERFORMANCE CHARTS

Charts furnished by engine manufacturers show power output for various operating conditions. Performances of smaller engines used with fixed-pitch propellers are usually shown by charts similar to Figure 4.11. A *propeller load curve* represents the power that will be absorbed by a specific fixed-pitch propeller at any given rpm. In Figure 4.11, curve *A* is a propeller chosen to provide full power at take-off rpm (the usual case). If full engine power is not needed for acceptable take-off performance, a propeller of greater blade angle may be chosen, resulting in full throttle before maximum allowable rpm is reached. This propeller, which will permit higher cruising speeds, is shown by curve *B*. The curves marked *full throttle* indicate maximum output of the engine at any given rpm and would require a controllable-pitch propeller of extensive blade angle variation to absorb power. Normally, the blades of a controllable-pitch propeller will contact the low-pitch stops at an intermediate rpm and follow a prop load curve to idle rpm. The power absorbed by a fixed-pitch propeller increases as the cube of the rpm change.

$$hp_2 = hp_1(rpm_2/rpm_1)^3$$

The performance charts for engines incorporating superchargers, and to be used with constant-speed propellers are plotted to show altitude performance as well as sea-level performance. A typical chart of this type is shown by Figure 4.12 for one rpm line.

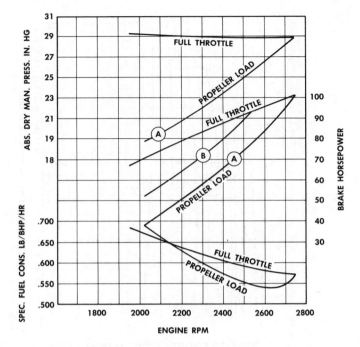

Figure 4.11. Typical sea-level engine performance chart.

ENGINE COOLING

A substantial portion of the total energy of the burning fuel is lost as heat within the engine. In addition, considerable heat is created by the friction of the moving engine parts. This heat must be removed to maintain metal temperatures within structural limits. Engine heat dissipation, especially from the cylinder head and valves, is also necessary for proper control of the fuel-air mixture burning rate. Heat is removed by the cylinder cooling system and by the circulating oil system, which removes internal engine heat; the oil is then cooled.

Air-cooled engines, which are used almost exclusively today, are cooled by passing air directly over the cylinders and bringing it into contact with the cooling fins. The cooling fins are necessary to present a large surface area for transfer of heat to the cooling air since air is not efficient in absorbing heat. The size and location of the cooling fins and the amount of air circulation required about the various parts of the cylinder play a very important part in the success of the air-cooled engine as an airplane power plant. To direct the cooling air around each cylinder for uniform cooling, a system of *baffles, dams,* and *hoods* fabricated from sheet metal is closely attached to portions of the cylinders. These baffles can be seen on some of the engine illustrations on the following pages. Note the

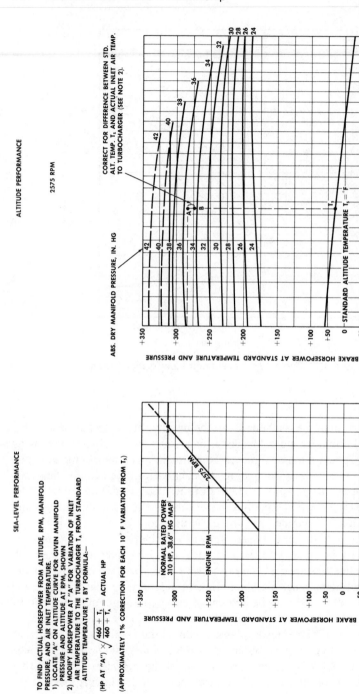

Figure 4.12. Typical turbosupercharged engine performance chart. *(Courtesy of Avco Lycoming.)*

greater depth of the cooling fins on the cylinder heads, inside of which the combustion takes place.

A portion of engine heat is dissipated through the circulating oil system. Besides providing lubrication, the oil removes heat from the bottom of the pistons and absorbs the heat caused by friction. The oil is then pumped through oil coolers where the heat is transferred to atmospheric air. On some small engines, the lower part of the crankcase, or oil *sump,* provides sufficient heat transfer without requiring a separate oil cooler.

Cooling Drag. Because heat loss represents a substantial portion of the total thermal energy output of the engine, the drag imposed on the airplane to provide adequate cooling airflow must be carefully considered. The larger the engine, the greater is the cooling requirement. Thus, cooling drag receives considerable attention from both the engine manufacturer and airplane manufacturer, and flight tests are accomplished to determine the most satisfactory installation.

Cooling drag is equal to the loss in velocity energy of the air mass required for cooling. This may be expressed as:

$$\text{cooling drag (lb.)} = M(V_a - V_E) \qquad (4.11)$$

where M = mass airflow = W/g
$\quad V_a$ = airplane velocity
$\quad V_E$ = exit velocity of cooling air leaving airplane
$\quad W$ = weight of airflow, lb./sec.
$\quad g$ = gravitational acceleration = 32.2 ft./sec.^{2}

This equation represents total drag loss to the airplane and includes the aerodynamic losses of the ducting system that conducts the air to and from the engine.

Regulation of Cooling Airflow. In small airplane installations, the cooling air inlets and exits are usually fixed. The inlet area can be seen directly behind the propeller in Figure 4.4. The cooling air exits from the engine compartment through the openings on the bottom of the nacelle. Optimum areas are determined during flight testing by measuring cylinder temperatures and oil temperatures. In larger airplanes where speed ranges and power variations are greater, the cooling-air exit areas are adjustable from the cockpit for control of cooling airflow. The adjustable portions of the exit on air-cooled engines are called *cowl flaps* since they form part of the engine cowling.

CARBURETION

The measurement of airflow for combustion and the metering of the proper amount of fuel to obtain the desired fuel-air ratio are called *carburetion.* As discussed previously, if the cylinders are not supplied with the proper fuel-air mixture for each operating condition, engine performance will be penalized. For this reason, carburetors have become rather complex, especially on high-power engines.

Principle of Operation.   Every carburetor operates on the same basic principle, that is, the measurement of airflow and the metering of fuel. Every carburetor contains an *air venturi* as the airflow measuring device (or part thereof) and a *throttle* for regulating the amount of airflow. The principal differences in carburetors are the methods employed for metering fuel.

Types of Carburetors.   The *float-type carburetor* (Figure 4.13) of most small-horsepower engines is similar in operation to automobile carburetors. The fuel is led from the float chamber, in which a constant fuel level is maintained by the float mechanism, through the fuel metering jets to the *throat* of the air venturi. When air is flowing, the static pressure in the throat is less than the atmospheric pressure in the float chamber, with the amount of static pressure reduction proportional to the airflow. As the airflow increases, the static pressure decreases, which in turn causes increased fuel flow. By proper design of the air venturi and the metering jets, the proper fuel-air mixture will be obtained for most engine operating conditions. During *idle operation,* the airflow through the venturi is not sufficient to provide the rich mixture needed for idle. To overcome this, additional fuel is added through an idle enrichment jet below the throttle. This jet is ineffective at higher airflow. To provide the *power enrichment* required for high-power operation, an auxiliary jet is opened by a linkage connected to the throttle, admitting additional fuel at higher throttle angles. An *accelerating pump* is also connected to the throttle linkage to provide an additional charge of fuel during rapid throttle opening. *Altitude compensation* is provided on some float-type carburetors, to reduce metered fuel flow commensurate with

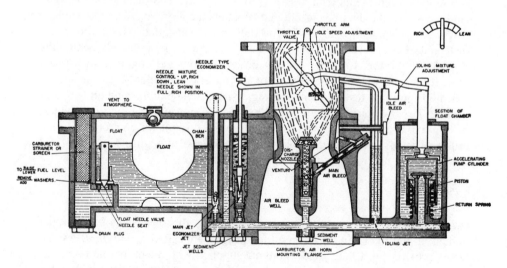

Figure 4.13. Float-type aircraft carburetor.

reduced atmospheric air density and to maintain desired fuel-air ratios. This additional control is required because the air venturi does not measure the true weight flow of air. Manual mixture control for cruise economy and compensation for density altitude changes are provided.

Pressure carburetors, sometimes called automatic carburetors, are used on many of the large radial engines. The larger fuel flow requirements of high-power engines do not permit the simple metering system employed in float-type carburetors. Fuel is pumped under constant pressure into the pressure carburetor, and the utilization of this pressure to force fuel through the metering jets in proportion to airflow provides the principal means of fuel-air mixture control. The fuel pressure on the metering jets, called *unmetered fuel pressure,* is regulated in proportion to airflow by means of a valve connected to diaphragms positioned in accordance with airflow through the air venturi. After passing through the metering jets, the fuel is led through discharge nozzles below the throttle. On some engines the metered fuel is directed through hollow passages in the supercharger impeller before escaping to be mixed with the air.

Pressure carburetors generally provide two mixture settings, auto-rich and auto-lean. The auto-rich setting, which provides richer fuel-air mixtures, is used under all operating conditions except stabilized low power cruise. The richer fuel-air mixture insures lower temperatures in the cylinder to avoid excessive cylinder head temperature, detonation, and burned exhaust valves. The auto-lean setting may be manually selected for increased economy at constant low-power cruise operation, but only after cylinder head temperature has reduced to a safe level so that detonation does not occur with the leaner fuel-air mixture. Altitude compensation, an acceleration pump, and power enrichment are always provided on the pressure carburetor.

Direct fuel injection is used on many of the horizontally opposed engines over 200 hp. A very simple low-pressure system is shown diagrammatically in Figure 4.14. Fuel is fed from the supply tank, to a positive-displacement-engine-driven fuel pump which supplies fuel flow in proportion to engine speed. An electric pump provides positive pressure for starting.

By arranging a calibrated orifice and a relief valve in the pump discharge, pump delivery pressure is also maintained proportional to engine speed. Fuel is then fed through the fuel-air control unit which is linked directly to the throttle where fuel flow is properly proportioned to airflow for correct fuel-air ratio. From the control unit, fuel is delivered to the fuel discharge nozzles, one for each cylinder. These are installed in the cylinder heads outside each intake valve. An air "bleed" or vent arrangement is incorporated in each nozzle. This helps to vaporize fuel and by breaking the high vacuum which exists at idle speed, maintains the fuel lines solidly filled, for instant acceleration of the engine.

For high-altitude operation with turbocharged engines, an aneroid bellows is added to the fuel pump to control orifice calibration, and the fuel nozzles are fed with ram air.

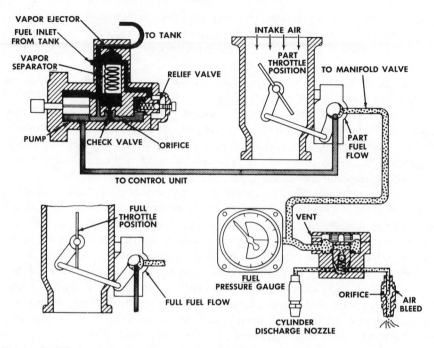

Figure 4.14. Schematic diagram of fuel injection system. *(Courtesy of Teledyne Continental Motors.)*

Ice. Carburetor icing may stop the engine completely or seriously restrict the power output. The float-type carburetor has the worst icing characteristics, since the fuel is evaporated, cooling the air charge, immediately above the throttle valve, in the narrowest air passage, where reduced pressure also reduces temperature. In moist air, the combined cooling effects will cause ice to form at atmospheric temperatures considerably above 32°F. The pressure carburetor and fuel injection reduce this type of icing. However, under normal atmospheric icing conditions, impact ice may form on the throttle valve and on the air pressure tubes leading to the fuel-control diaphragm. Carburetor and induction system icing are prevented by heating the induction air, using heat from a muff around the exhaust system. External (turbo) supercharging greatly reduces the likelihood of intake icing.

TYPES OF ENGINES

The air-cooled, horizontally-opposed engine is used almost exclusively in new normal and utility category aircraft today. Larger aircraft in production rely almost entirely on turbine engines for their propulsion.

Horizontally-Opposed Engines. The smaller air-cooled engines, up to 500

Figure 4.15. Utility aircraft engine. Continental Model I0-520-A, six-cylinder, direct-drive, weight 475 lb., fuel injection, uses 100/130 fuel. Rated 285 hp at 2700 rpm for take-off at sea level. Displacement, 520 cu. in. *(Courtesy of Continental Motors Corp.)*

hp, are now constructed with opposed cylinders similar to the engine shown in Figure 4.15.

This cylinder arrangement lends itself to compact, rigid construction with a small frontal area. Four-, six-, and eight-cylinder engines are used with various horsepower ratings from 100 to 500. Many of the medium and larger sizes employ single-stage turbosupercharging, and some have propeller-speed-reduction gearing. Magnetos are used for ignition, and electrical starting is customary. The oil systems are usually self-contained. Cylinders may be replaced individually for easy and low-cost maintenance.

Engines of advanced technology and design have been developed to combat the encroachment of turboprop engines in this power range. These engines utilize a much higher operating rpm (4000 to 5000), resulting in reduced engine weight per horsepower, greater operating smoothness, and reduced noise, both in the cabin and outside. Engines of newer design have achieved weight ratios of slightly above 1 lb. per hp in the four- and six-cylinder size, to slightly less than 1 lb. per hp in the eight-cylinder class.

Many design features are used to reduce engine complexity, weight, and cost. One manufacturer has developed a system of torsional vibration control, eliminating the need for pendulum crankshaft dampers, and therby sharply reducing vibratory torque in the crankshaft, propeller gearing, propeller, and accessory drives. This also permits the use of lighter, slower, and quieter propellers. Perhaps most significantly, it reduces stresses in all engine components, permitting lighter weight in all structural members.

Reduction gearing for propeller drive is accomplished by spur gears, which

Figure 4.16. Light training aircraft engine. Lycoming O-235, L-2C, 235 cu. in. displacement, developing 115 hp at 2700 rpm for take-off at sea level, burning 100 octane fuel, installed in a Grumman American Aviation AA-1C. (*Photo by editor.*)

also drive the camshafts. While this limits the range of reduction to one-half engine speed, the resulting 2000 to 2500 rpm is about right for the power range and propeller diameter.

An additional significant feature of these engines besides reduced weight is a reduction of 25% to 30% in cooling air pressure. Since cooling drag in light twin-engine airplanes is a large percentage of total drag, the improvement is significant.

These horizontally-opposed engines rely more and more on turbosupercharging ("turbocharging") for high-altitude performance, providing greater selection of cruise altitudes and high true airspeeds (Figure 4.17).

Radial Engines. The radial aircraft engines which have served aviation so long and faithfully became famous with the Wright "Whirlwind" J-5 which powered the *Spirit of St. Louis* from New York to Paris with Charles A. Lindbergh in 1927. These engines have the cylinders arranged radially about the engine case to accommodate the greater number of cylinders required (Figure 4.18).

The horsepower ratings range from 220 hp to about 3800 for take-off. They have one to four rows of 7 (or 9) cylinders each. Although the radial engine presents a rather large frontal area (an aerodynamic disadvantage), the short crankshaft coupling results in great strength and lightness, and arrangement of the cylinders in this manner permits greater utilization of cooling air.

Figure 4.17. Six-cylinder opposed aircraft engine. Teledyne Continental Model Tiara T6-285, geared drive dual turbocharged. Rated 285 hp at 4000 rpm for take-off at sea level. Displacement 406 cu. in., weight 402 lb. (*Courtesy of Teledyne Continental Motors.*)

Figure 4.18. Twin-row radial engine—R-2800. (*Courtesy of Pratt & Whitney Aircraft Co.*)

The largest radial engine in use today is the 28-cylinder R4360, which produces up to 3800 hp for take-off with water injection. The cylinders are arranged in four rows of seven cylinders each. The R4360 was developed toward the close of World War II and was chosen for a number of late-war and postwar multi-engine airplanes. It is the last and largest of the big aircraft reciprocating engines. Very few new aircraft are now being built with radial engines, and almost all of these are for agricultural purposes. A detailed discussion of their features may be found in earlier editions of *Modern Airmanship*.

ENGINE CONSTRUCTION

Horizontally-opposed engines are constructed with a single case split at the centerline for access to internal parts—crankshaft, bearings, gearing, etc. The cylinders are bolted to the case and may be removed and replaced individually.

One of the new generation high-speed fuel-injection engines with rear-mounted turbosuperchargers is shown in Figure 4.17. Accessories are side mounted for ease of removal for maintenance. An additional 20 bhp for driving other accessories, such as air conditioning or spray pumps, is available from the two four-bolt pads shown at the rear of the engine.

Turbojet and Turbofan Engines

The principle of jet propulsion was recognized very early in aviation development as possibly the best method for obtaining greater and greater speeds. The practical application of the principle proved very difficult, however, because of the extreme inefficiency of early engine design.

In 1930, Frank A. Whittle was granted an English patent for a turbojet engine design, but it was not until 1937 that intensive work was begun on his design in England. The Germans initiated turbojet development in 1936 along with other jet engine and rocket motor programs. This work was carried on in an intensified manner, and the first turbojet-powered airplane flight took place in Germany in 1939. The airplane was a Heinkel HE178 powered by a Heinkel HE 53B engine. Meanwhile, the Whittle engine program progressed rapidly in England. In 1941 the fourth Whittle engine, Model W1, delivering 855 lb. of thrust, propelled a Gloster E-28 airplane to 339 mph at 20,000 ft. In this same year, the General Electric Company agreed to develop turbojet engines for the U.S. Army Air Forces from plans of the Whittle W-2-B engine. The first American flight took place in 1942 when a Bell Aircraft P-59, powered by two General Electric 1A engines of 1300-lb. thrust each, flew successfully. By 1944, production was started on the 4000-lb. thrust J-33 engine for the P-80 airplane. During 1944, the first axial-flow compressor turbojet engine was successfully tested by General

Figure 4.19. The first turbojet engine built in the United States. (*Courtesy of General Electric Co.*)

Electric. The first turbojet engine built in the United States is shown in Figure 4.19.

PRINCIPLES OF OPERATION

The operating principle of the turbojet engine is quite simple. There are five basic parts: inlet, compressor, combustor, turbine, and nozzle. The inlet (normally considered part of the airframe) converts the ram-air pressure to static pressure. The compressor handles large quantities of air, compresses it to high pressure, and delivers it to the combustor. Here, the air is divided into two streams, one passing through the combustion zone where a near stoichiometric mixture burns to products of combustion at very high temperature, around 4000°F. The other airstream is used for combustor cooling and then is mixed with the combustion products to produce a gas temperature of 1600 to 2400°F. The hot gases subsequently flow through the turbine, where part of the energy is extracted to drive the compressor and accessories. The gases are then passed through a duct (where additional fuel may be added, called afterburning) to the nozzle, which functions to greatly increase the momentum of the gas by expanding it to high exit velocity. The change of momentum from inlet to nozzle exit produces most of the thrust force which propels the airplane. The fuel adds some mass flow, and water or a water-alcohol mixture may be introduced to add mass flow for additional thrust for take-off.

The efficiencies at which the basic parts of the turbojet engine operate, particularly the turbine and compressor, are very important to thrust output. Since approximately two-thirds of the energy available in the hot gases is absorbed by the turbine to drive the compressor, one can easily see the necessity for high efficiency of both the turbine and the compressor. Early models failed because of inefficient turbine and compressor designs, coupled with inferior materials of construction. Since thermodynamic laws reveal that extractable energy increases with an increase of both pressure and temperature, designers strive to develop compressors that will provide even greater pressure ratios and turbines that will withstand higher temperatures. It is the improvement in these two components that has led to the tremendous development in turbojet engines since the first flight in 1939.

MAJOR PARTS

Turbojet engine construction is relatively simple compared to a reciprocating engine. The major parts of typical turbojet engines are shown by the cutaway drawing in Figure 4.20.

Compressor. The centrifugal compressor was used in early engines but is seldom used in new designs except in combination with an axial flow compressor. In the centrifugal compressor, air enters the rotating impeller at the center through inlet guide vanes. Rotating at high speed, the impeller compresses the air by centrifugal action. The air passes from the rim of the impeller through a diffuser, which reduces the velocity and increases static pressure, and then enters the combustor.

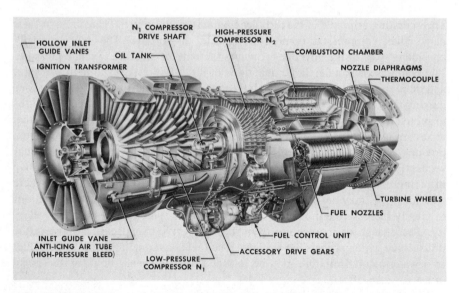

Figure 4.20. Principal parts of a dual-rotor axial-flow turbojet engine. (*Courtesy of Pratt and Whitney Aircraft Co.*)

The axial-flow compressor is used on nearly all turbojet, turbofan, and turboprop engines today. The distinct advantage of this type of compressor is its small diameter. As can be seen in Figure 4.20, many "stages" can be added together without increasing the diameter of the engine. As shown by the illustration, the compressor blades are very similar in appearance to the blades on the turbine wheel. The combination of a row of rotating blades (rotor) and a row of stationary blades (stator) comprises a stage. The stator blades are provided for the purpose of decelerating the air, while increasing static pressure and directing the air at the proper angle into the succeeding row of rotor blades. As the air passes through each stage, the pressure and temperature are increased. Air passes from the last stage of compression through a diffuser section which reduces the angular and axial velocity and increases static pressure. From the diffuser, the compressed air passes into the combustion chambers, where heat of combustion is added, prior to its entering the turbine.

The engine shown in Figure 4.20 employs two compressors, each driven independently by its own set of turbine wheels and at different speeds. This has been accomplished by making the drive shaft for the rear compressor hollow and by rotating the shaft for the forward compressor within it. Advantages of this arrangement are greater ease in starting and controlling high-pressure-ratio engines, and higher speed with attendant smaller size for the rear compressor. As a result of this design, the engine performance envelope is enlarged. Some modern designers have gone to three compressors to permit more flexibility in operation. Engines of these types are often called two-spool or three-spool engines, respectively. High-pressure compressors with up to 17 stages in a single rotor are used in some modern turbojet or turbofan engines. Variable geometry (stator vane angle in several stages) is employed in these compressors to permit the necessary flexibility of operation.

Combustion Chamber.   The performance of the combustion chamber, often called "burner," is very important to satisfactory engine operation. The fact that the burner must operate properly over a wide variation of airflow adds to the complication of design.

Combustor (burner) designs are of three types: (1) can type, (2) can-annular type, and (3) annular type. Early turbojet engines employed can-type chambers where the compressor discharge is divided evenly among a number of individual can burners with perforated liners (Figure 4.19). The can-annular design has an annular flow channel inside which individual perforated cans are used to control the primary combustion of fuel and air (Figure 4.20). Most modern turbojet and turbofan engines use the annular design, which has an annular flow channel with an annular liner in which fuel is introduced at several discrete locations (Figure 4.31). As shown by the figure referenced, these combustors surround the engine in order to provide a uniformly distributed gas flow to the turbine. Combustion takes place within the "liner," which is perforated to permit the entry of the portion of air used for combustion. The remainder of the airflow, called the diluent, passes between the liner and the jacket to cool the liner, and then enters the liner toward the rear. At the aft end of the combustion chamber, the two airstreams have been thoroughly mixed to control turbine inlet temperature. Fuel is sprayed into the forward end of the liner. The fuel is scheduled for start-up and run conditions. Spark plugs for starting combustion are usually installed in two opposite locations, and all chambers are interconnected with tubes for flame propagation. When combustion is thoroughly underway, the sparking of the plugs is discontinued. For this reason, any interruption in fuel flow thereafter will cause the engine to "flame out," requiring restarting in flight, an "air start."

Turbine Wheels. Perhaps one of the most outstanding achievements in turbojet engine progress is the development of alloys to withstand the temperature and stresses encountered in turbine wheels. Not only is the turbine wheel required to withstand high stresses due to centrifugal and axial loads, but it must do this when exposed to gases at temperatures of 1600 to 2400°F.

As shown in Figure 4.20, the turbine wheel is composed of a number of individual blades. The blades, which provide the turning force, are individual parts keyed or slot-mounted to the heavy rim of the wheel. The turning force of the turbine wheel is imparted to the compressor by a shaft suitably supported by ball and/or roller bearings. The ability of the turbine wheel to withstand the high gas temperatures is enhanced by cooling air that is bled from the compressor. The turbine-entrance guide vanes (turbine-nozzle diaphragm in the illustration) are usually hollow so that a portion of the cooling air may be directed through them to facilitate cooling. New high-performance engines also have cooled hollow turbine blades. Variable geometry turbines, which use an adjustable nozzle diaphragm to improve the engine performance over a wide range of operating conditions, are under development. This will also have an impact on the aircraft inlet design since it influences the quantity of air flowing through the engine.

Large compressors may require more than one turbine wheel to provide sufficient power to drive the compressor. Split compressor engines likewise require more than one turbine wheel. If more than one turbine wheel is required, a row of stationary guide vanes, often called the nozzle diaphragm, affixed to the engine case, is positioned between each wheel and the next.

Exhaust Cone. The exhaust cone and the inner cone aft of the turbine wheel play an important part in engine performance. By controlling the relative diameters of the inner and outer cones at each station moving aft, a predetermined area for exhaust gas passage is established.

Tailpipe and Nozzle. The tailpipe is attached to the end of the exhaust cone. Frequently the entire tailpipe is called the nozzle, but the nozzle actually is the opening in the end of the tailpipe. Tailpipes will vary in length, dependent upon the installation. In pod installations where the engines are mounted on struts below the wing, a very short tailpipe suffices. Engines installed in the fuselage or wing root may require tailpipes of some length. Special tailpipes are required for engines using afterburning. This is discussed under augmentation.

Sound Suppression.   Both the engine inlet and exhaust nozzle are sources of noise pollution. The high frequency component of the noise radiating from the engine inlet, which is generated by the compressor (and the fan on turbofan engines), can be attenuated. Noise reduction can be achieved by the use of acoustical liners in the inlet duct, by lowering compressor and/or fan tip speed, and by modern design techniques applied to blade profile design. Noise from the exhaust streams of turbojet and turbofan engines includes components of blade aerodynamic noise, combustion noise, and jet noise produced by turbulent mixing of the high velocity stream with the atmosphere. The first two of these may be attenuated by use of acoustical liners in the flow ducts similar to the inlet treatment. The jet noise can be reduced by design technique applied to the nozzle (such as using a multi-lobe nozzle) and by reducing the velocity gradient between the jet and the surrounding air. The low-velocity airstream between the jet and the surroundings in the case of the turbofan provides a large reduction in shearing velocity and a substantial reduction in jet noise.

Thrust Reverser.   Thrust reversal to assist wheel braking during the landing roll is accomplished by stopping the aft flow of the jet and directing it partially forward in the direction of airplane movement. This feature is provided on all large commercial transports. It is almost a necessity when landing on extremely wet runways on which wheels can hydroplane, or on icy runways. It greatly reduces brake wear when used during normal runway conditions, and has opened many shorter runways to jet operations.

To prevent the hot gases from entering the compressor inlet of the same or adjacent engines, the reverser directs the jet blast outward at a suitable angle from the nacelle. Because of this angle, and losses due to turning of the gases, thrust available in the reversed direction is about 40% to 50% of forward thrust for the same engine rpm.

Two major types of thrust reversers are in general use. In one type the primary exhaust gases and fan air are directed forward by means of turning vanes. Aft direction of the gases is blocked by means of clamshell doors or segmented blocker doors swung into position when the cowling is translated to expose the turning vanes. A diagrammatic sketch is shown in Figure 4.21. This type is used on the Boeing 747, Mc Donnel Douglas DC-10, and Lockheed L-1011.

The other main configuration is the target type, which swings a portion of the cowling aft and outward to direct the gases forward. This system is used on engine installations where the fan air and the primary air are mixed inside the engine or tailpipe and then exit either through a single nozzle or through separate coplanar nozzles. "Coplanar nozzles" are those in which fan air and primary air exit at the rear but through separate nozzles. The fan-air nozzle usually surrounds the primary nozzle. A photograph of this type, used on the DC-9, the Boeing 737, and the C-141 is shown in Figure 4.22. The Boeing 727 also uses a target type, but the aft direction of the gases is blocked by an internal set of clamshell doors (Figure 4.23).

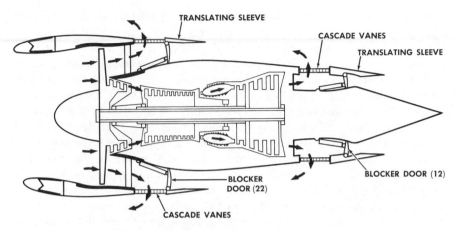

Figure 4.21. Short-duct turbofan thrust-reverser principle, utilizing turning vanes and blocker doors. (*Courtesy of The Boeing Company.*)

Accessories. The accessories, such as fuel pumps, ignition coils, or transformers used for engine operation, are mounted on the outside of the engine or in the mounting strut, and are driven by the engine shaft through gear trains. Other accessories necessary for airplane operation, such as electric generators and alternators, hydraulic pumps, and vacuum pumps, may also be mounted on the engine and driven by gear trains.

Figure 4.22. Target-type thrust reverser used on single-nozzle and coplanar-nozzle installations, such as the Boeing 737. (*Courtesy of The Boeing Company.*)

Figure 4.23. Target-type thrust reverser. Boeing 727 with Pratt and Whitney JT8D engine. Side panels are open, exposing engine for maintenance. (*Courtesy of The Boeing Company.*)

Another method for driving airplane accessories mounts the accessories in the airplane wing or fuselage where small turbines drive them by means of compressed air bled from the engine compressor. On some engines they are mounted below and on the sides of the engine-compressor case.

Oil System. A very simple oil system is required since there are few moving parts in the turbojet engine. The circulating oil provides lubrication for the shaft bearings and accessory drive gears, as well as being used to remove heat from the bearings. Fairly low pressure is required to force the oil into the bearings and gears. After passing through the bearings and gears, the oil is collected and scavenge pumps direct it back to the oil tank. Before it reaches the tank, it is pumped through heat exchangers to remove the heat picked up in the bearings. On some engines, the heat in the oil is exchanged with incoming fuel (before the fuel reaches the nozzles) in the heat exchanger provided for this purpose. Larger engines may require an air-oil heat exchange, in addition, since a fuel-oil heat exchanger may not remove all the heat as required.

Fuel System. With the exception of the fuel control system, the fuel system (fuel supply) is quite simple. Fuel is pumped from the airplane supply tanks into an engine-driven fuel pump where the pressure is raised to several hundred pounds per square inch to insure adequate atomization in the nozzles. After pressurization, the fuel enters the distribution manifolds to which the nozzles are connected. Some American engines employ a dual manifold system and a double nozzle to control atomization at starting rpm. During starting, a spring-loaded valve called a flow divider directs the fuel into the smaller manifold until rpm and fuel pressure rise above predetermined values. In this interval, fuel is fed from the small manifold to the smaller orifice in the nozzle to provide adequate atomization at low fuel flow. As fuel pump pressure builds up and fuel demands increase, the flow divider changes position, permitting fuel to enter the large manifold and be atomized through the large orifice in the nozzle. On other engines, low flow atomization is controlled by means of valves in the nozzle itself. British engines often obtain atomization in quite a different manner. Fuel enters the nozzle at relatively low pressure and is atomized within the nozzle by a jet of air.

At engine shutdown, the fuel supply from the pump is quickly cut off by shutoff valves actuated when the throttle is closed. However, the residual pressure in the fuel manifold will still cause a small flow of fuel until the pressure bleeds off. To prevent this fuel from entering the combustion chambers and creating an uncontrolled fuel-air mixture, or being dumped overboard and polluting the environment, the residual fuel is drained into an auxiliary reservoir.

PERFORMANCE

The magnitude of the thrust output of turbojet and turbofan engines is determined by the mass flow through the engine and by the velocity of the mass leaving the jet nozzle and fan. This has been expressed earlier in the chapter as

$$F = \frac{W_a}{g} V_j = MV_j \tag{4.2}$$

This equation represents the thrust output when the engine and airplane are at rest. When the airplane and engine are in motion, air is being rammed into the engine compressor inlet, tending to retard the forward motion. This retarding force is called *ram drag,* and its magnitude is determined by the mass airflow through the engine and the airplane velocity:

$$\text{ram drag} = \frac{W_a}{g} V_a = MV_a$$

The net thrust (F_n) for the engine in motion is, therefore, the gross thrust output minus the ram drag:

$$F_n = \frac{W_a}{g} V_j - \frac{W_a}{g} V_a$$

$$F_n = \frac{W_a}{g}(V_j - V_a) = M(V_j - V_a)$$

(4.12)

and

$$V_j = \frac{gF_n}{W_a} + V_a$$

Equation (4.5) may be reexpressed as

$$\text{Propulsive Efficiency} = \frac{2V_a}{\frac{gF_n}{W_a} + 2V_a}$$

Net thrust is equal to the product of mass airflow and the change in velocity through the engine. Actually, the mass flow of gases from the nozzle is increased by the amount of fuel added for combustion. However, since the airflow is about 70 times greater than fuel flow, the mass of the added fuel is neglected in basic equations.

Airflow.   Factors that determine compressor airflow will also affect thrust output. For turbofan engines, the discussion applies to *total* airflow (fan + compressor). Compressor airflow, measured in lb. per sec., is determined by the size of the compressor (and fan, if turbofan), compressor rpm, airspeed, and air density. Compressor airflow varies extensively with rpm, as exemplified by the fact that thrust output at idling speeds is only 6% to 10% of thrust output at maximum rpm. Figure 4.24 shows the relationship of rpm with compressor airflow and thrust. Compressor airflow also increases with airspeed because more air is rammed into the compressor inlet. This increases the *gross* thrust output of the engine but also increases the ram drag. The effect of airspeed on gross thrust, ram drag, and net thrust is shown by Figure 4.25.

The design of airplane ducting systems leading the air into the compressor plays an important part in compressor airflow. Multi-engine airplanes with the engines mounted in the wing or in pods below the wing will have relatively short inlet ducts. Fighter-type aircraft with engines mounted in the aft portion of the fuselage may require relatively long inlet ducts. With both types of installation, great care, usually requiring extensive testing, is required to insure that the maximum amount of air will enter the engine with a minimum of duct losses. This is particularly true at take-off speeds and during high-speed flights. The efficiency with which the compressor inlet ducts scoop in the air and conduct it to the compressor is designated by several terms—*ram recovery, ram efficiency,* or *recovery factor*. The effects of ram efficiency on engine performance are shown by Figure 4.26(A).

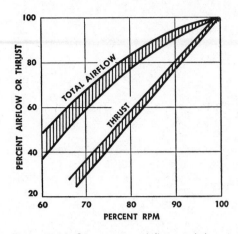

Figure 4.24. Compressor airflow and thrust vs. rpm at zero airspeed.

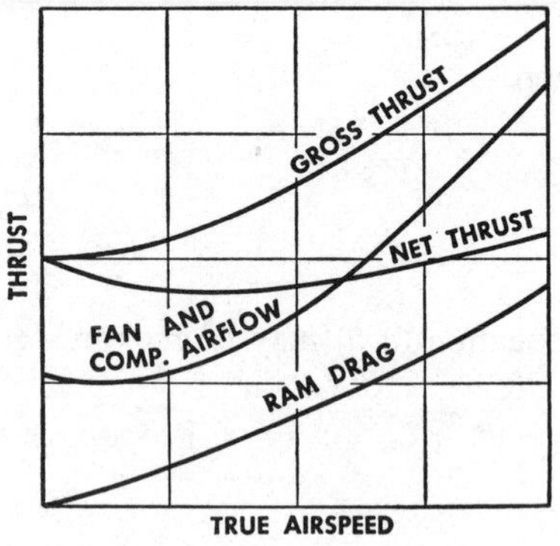

Figure 4.25. Thrust and ram drag vs. airspeed.

For supersonic flight, the engine air inlet ducting must provide for reducing the flow of ram air to subsonic velocity. This requires careful design and usually dictates variable inlet geometry.

Because air becomes less dense as ambient air temperature rises, as well as with increase in altitude, compressor airflow and engine thrust also decrease. The effect of reduced air density and the resulting reduction in engine thrust must be considered carefully by the designer when selecting the proper size engine for a given airplane, for performance at take-off and at high altitude. Figure 4.26(B) shows the effect of atmospheric temperature on engine performance, and that of altitude is shown by Figure 4.27. An increase in atmospheric temperature at altitude from the standard day conditions used for the plot in Figure 4.27 will further decrease altitude performance.

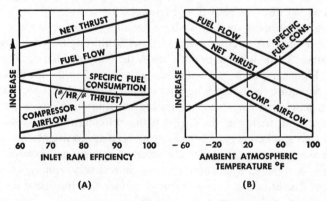

Figure 4.26. Effect of (A) inlet ram efficiency, and (B) atmospheric temperature on engine performance at constant airspeed.

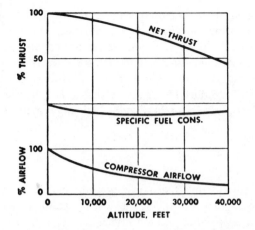

Figure 4.27. Effect of altitude on engine performance at constant airspeed.

Primary jet velocity (V_j) is determined during design of the engine. At any given engine rpm, jet velocity depends on the temperature of the hot gases in the tail-pipe (and the pressure if the gases leave the nozzle at subsonic velocity). The pressure is controlled by the area of the nozzle through which the gases escape. The energy available in the hot gases is greater at higher temperature; therefore it is desirable to maintain tailpipe temperatures as high as possible. The factor that determines the highest allowable temperature is the turbine, since this is the part subjected to the greatest stresses. Because the turbine temperature limitation is approximately the same in all engines, turbojet engine jet velocity will be ap-proximately the same, and jet nozzle area will correspond to the mass flow for any given engine at full rpm. Tailpipe temperature is often referred to as *exhaust gas temperature* (EGT).

Fan air exhaust velocity is determined during the design of the engine by the number of stages desired and the pressure rise across each stage. The nozzle area is generally fixed to result in some flow at a given altitude, airspeed, and com-pressor rpm.

Primary jet nozzle area is adjusted prior to take-off to obtain maximum tailpipe temperature and jet velocity on engines that operate at fixed maximum compres-sor speed. Since air density decreases with increase in ambient air temperature, compressor airflow will decrease. Therefore, a smaller nozzle area is required to maintain maximum tailpipe temperature and pressure. Some engines, particu-larly two-spool engines, correct for thrust loss due to high ambient temperatures by adjusting the compressor speed control, permitting an increase in rpm to maintain airflow. Although jet velocity decreases with reduced compressor air-flow, resulting from lower rpm or higher ambient air temperature, it does not diminish appreciably as altitude is increased, even though compressor airflow by weight does diminish.

When airspeed and rpm are held constant, compressor and fan airflow diminish with altitude, as shown in Figure 4.27. However, the reduced atmospheric pressure that causes the reduction in compressor airflow also acts on the nozzle. The reduced atmospheric pressure behind the nozzle tends to increase the velocity of the gases escaping from the nozzle by permitting greater gas expansion within the tailpipe. The reduced compressor airflow and the greater expansion of the hot gases tend to offset each other, and jet velocity remains substantially constant.

FUEL CONSUMPTION

Turbojet engines have higher fuel consumption, relative to their output, then have reciprocating engines. However, the cost per gallon of jet fuel is less than reciprocating engine gasoline, and the higher fuel consumption of the turbojet engine as an airplane range factor is greatly offset by its light weight compared to a reciprocating engine and propeller installation. Also, jet airplanes are able to cruise at much higher altitudes and take advantage of reduced airplane parasite drag.

Current intermediate turbofan engines consume about 20% to 30% less fuel than the best axial-flow jet engines. The new generation turbofan engines of very high bypass ratio (5 to 6) yield fuel economies 50% to 60% better than the best turbojet engines. These engines approach the ultimate reached in the large reciprocating engine specific fuel consumption. These increased bypass engines have been made possible by improved compressor design and by the use of turbine wheel cooling and materials capable of withstanding higher temperatures.

AUGMENTATION

This term applies to methods for increasing thrust during takeoff in addition to that delivered by the basic engine. Two such methods are in common use. Afterburning is used predominantly on fighter airplanes and supersonic transport installations for thrust augmentation. As the term implies, fuel is added to the hot gases in the tailpipe after they have passed through the turbine wheel. The burning of this additional fuel greatly increases the gas temperature and correspondingly the jet velocity, resulting in as much as 50% increase in thrust. Because the hot gases pass through the tailpipe at high velocity, it is necessary to introduce the fuel through flameholders in the tailpipe similar, in some respects, to the flameholders employed in the combustion burners. The increase in gas temperature from afterburning causes the gases to expand, which tends to increase tailpipe pressure and turbine wheel temperatures. When the afterburner is turned on, the nozzle is automatically opened to a new area schedule in order to accommodate the required mass flow. To permit complete combustion of the afterburning fuel, a greater length of tailpipe is required than for a normal engine

Figure 4.28. Supersonic low bypass ratio axial-flow engine with afterburner. Pratt and Whitney TF-30 used on F-111A fighter aircraft. Develops 20,000-lb. thrust with afterburner. (*Courtesy of Pratt & Whitney Aircraft Company.*)

without afterburning. A typical afterburning engine is shown in Figure 4.28. Note the length of the tailpipe and the variable area nozzle.

Another method for obtaining augmentation is water injection, which has produced up to 25% additional thrust. One desirable feature of this method is the small increase in engine weight over the basic engine, compared with considerable weight increase for afterburning. Also, water injection equipment causes little or no change in normal engine performance, whereas afterburning equipment causes a slight increase in fuel flow during normal (afterburner off) operation because of the blocking effect of the flameholders in the tailpipe. To obtain the augmentation created by water injection, the water is introduced into the compressor or is forced through nozzles into the combustion chambers, or both. When introduced into the compressor, the water cools the air by evaporation, permitting the compressor to deliver more mass flow of air at maximum rpm. To offset the additional airflow, a signal is sent to the fuel control, increasing fuel flow slightly to heat the additional air. Thrust is increased by the increased mass airflow, plus the mass of the added water and fuel.

When water is introduced into the diffuser section of the combustor, the resulting increase in compressor airflow is less than by direct compressor injection. However, large amounts of water injected into the combustion chambers, plus the slight increase in compressor airflow, result in a substantial gain in thrust. The injection water must be distilled (mineral free) to prevent mineral deposits from forming on the compressor or turbine blades, which would seriously reduce their efficiency.

TURBOJET CONTROLS

Since the fundamental operating principle of the turbojet engine is to increase the velocity of air passing through it by increasing the temperature of the air, the fundamental requirement of the engine control is that the correct amount of fuel be provided for heating the air. This must be accomplished in a very precise manner for the wide range of airflow, rpm, altitude, airspeed, and ambient temperature.

In addition, the control must meter in the proper rate to provide controlled acceleration from low to high rpm. Turbojet controls are designed with a single thrust lever to select the desired thrust output.

The turbojet control meters fuel at any selected rpm by measuring compressor airflow. This is done by slightly different methods by various engine manufacturers. One system measures compressor discharge total pressure (static pressure plus velocity pressure) only; another measures compressor discharge pressure but also measures compressor inlet temperature and pressure and combustion chamber pressure (which is substantially the same as compressor discharge). By any of these means, changes which influence the compressor airflow, such as ambient temperature, airspeed, or altitude, automatically call for a corresponding change in fuel flow. Rpm is maintained constant at each selected speed by a mechanical governor directly driven by the engine shaft.

For any tendency of the engine to exceed or drop below the selected rpm, the governor will reduce or increase fuel flow to keep rpm constant. Selected rpm does not necessarily remain constant as altitude is increased, however. At higher altitudes, the reduced air density has an adverse effect on the burning characteristics of the fuel-air mixture as well as affecting compressor and combustion chamber operation. To prevent the burning process from suddenly blowing out, either during rpm acceleration or sudden changes in airspeed, an altitude bias is integrated into the governor control. This bias causes the governor to call for higher and higher idle rpm as altitude is increased, regardless of sea-level idle throttle position. Idle rpm, therefore, will be considerably higher at altitude than it is at sea level.

Engine acceleration rate from a lower to a higher rpm must be carefully regulated automatically by the control at all altitudes. Compressors, particularly axial-flow compressors, have a characteristic known as "surge" or "stall." This condition results from air separation on the compressor blades in a manner similar to stall and separation on an airplane wing. During rpm increase, usually in the medium speeds, separation followed by stall will occur if the compressor discharge pressure (combustor pressure) is permitted to build up too rapidly as a result of increased fuel flow.

The control, therefore, must be designed to meter the exact amount of fuel which will permit the engine to accelerate as rapidly as possible but not so fast as to cause a stall. This rate of change of rpm also varies with altitude, being much slower at high altitude. On some engines an additional control is added to reduce compressor stall tendencies and increase acceleration rate. This control opens bleed ports in the compressor which "waste" some of the compressor air by dumping it overboard during acceleration through the critical rpm region. As soon as the critical rpm region is passed, the control automatically closes the bleed ports.

Compressor stall on some very high pressure ratio compressors is controlled by varying the angle of the stator vanes in several stages of the compressor. This control may be used in combination with compressor bleed. The engine shown in Figure 4.30 has variable stator vanes.

TURBOJET AND TURBOFAN OPERATION

Turbojet and turbofan operation is relatively simple. An individual throttle lever and starting switch for each engine are the only controls necessary in the cockpit. Movement of the throttle positions the mechanism within the automatic control on the engine, which then regulates fuel flow to establish an rpm in relation to the throttle position. In installations where augmentation is used, a separate switch may be employed to "arm" the afterburning or water injection electrical control circuit. Operation of the circuit is then controlled by switches actuated when the throttle reaches near to the wide-open position. A fixed "detent" position of the throttle is usually provided to facilitate setting of the throttle at the proper ground idle position. In some installations, another throttle position below the idle detent is marked for engine starting. Moving the throttle to the fully closed position closes a fuel cutoff valve on the engine. A separate start lever provides this function on some engines.

On many engines, thrust output is indirectly indicated to the pilot by the tachometer which measures engine rpm. Because different engines will operate within different rpm ranges, the tachometer indicator is usually graduated in percent of maximum rpm for the particular engine. The tachometer permits the pilot to select cruising thrust on multi-engined airplanes. On dual-rotor engines, where the two compressors rotate at different rpm, each tachometer indicates the speed of only one rotor. This does not provide a satisfactory indication of relative thrust output, since the ratio of the two compressor speeds (hence the thrust output) may vary from engine to engine.

Thrust output is determined on some dual-rotor installations by measuring tailpipe total pressure at the nozzle inlet. This pressure may be displayed on a gauge in the cockpit. In most installations, the tailpipe total pressure is divided by compressor inlet total pressure and the ratio presented on a gauge in the cockpit. The latter system, called Engine Pressure Ratio (EPR), has the advantage of eliminating altitude density effect on the indication.

The only other engine instruments used in turbojet installations are fuel pressure, oil pressure, oil temperature, exhaust gas temperature, and in some cases, fuel rate of flow. Exhaust gas temperature indication is especially necessary during maximum engine rpm operation and certain other operating conditions to insure that excessive temperatures are not imposed on the turbine wheel. If limiting exhaust gas temperatures are reached before maximum rpm is attained, the pilot must not advance the throttle farther since serious damage to the turbine wheel can result.

Starting of turbojet and turbofan engines is relatively simple. The starter motor on turbojet engines is geared to the engine rotor shaft. On some engines the starter is electrically powered; on others it may be operated by compressed air. In the latter, the starting turbine wheel shaft is connected to the main engine starting gears. To start the engine, the starter is engaged, which commences rotation of the compressor (high-pressure compressor on dual-rotor engines). When the compressor has reached an rpm at which it will deliver sufficient airflow for combustion, the throttle is opened to admit fuel, and combustion takes place. The starter continues to operate until self-acceleration speed is obtained. At this point, it disengages automatically. A spark ignition system is used during starting and may be actuated during flight for restart or for operation in icing conditions which might cause flameout.

Types of Engines.   Because jet propulsion requires high speed, high altitude flight for greatest propulsive efficiency, the major jet engine design effort has been concentrated on the larger thrust output machines. However, several smaller models, with 2000 to 4000 lb. thrust, one of which is shown in Figure 4.29, have been developed for use in high speed target airplanes, business aircraft, and jet training airplanes. A majority of the larger turbojet engines developed in recent years fall within the 10,000 to 25,000 lb. thrust class (without augmentation).

The axial flow JT-3C engine (J-57) is used on several commercial and many military airplanes. Thrust ratings vary from 12,000 lb. augmented, 13,500 lb. with water, to 18,000 lb. with afterburning.

The J-79 (CJ-805) engine is unique because it uses a single 17-stage compres-

Figure 4.29. Small turbojet engine. The General Electric CJ 610 (J-85) produces 2950 lbs. take-off thrust. The J-85 produces 4300 lbs. thrust with afterburner and eight-stage compressor, to 5000 lbs. thrust with nine-stage compressor. This basic engine powers the T-38, Aero Commander, Lear Jet, F-5A, and several foreign aircraft. *(Courtesy of General Electric Company.)*

Figure 4.30 Single-rotor J-79 turbojet engine. This engine uses 17 stages of compression with stator vane angle control for the first seven stationary blade rows. The J-79 is used on a number of subsonic and supersonic fighter aircraft and develops 18,000 lb. thrust with afterburner. (Afterburner not shown in illustration.) *(Courtesy of General Electric Co.)*

sor with variable vanes in the first seven stationary blade rows, developing 16,000 to 18,000 lb. thrust with afterburning. This engine is used on several supersonic fighter airplanes and is shown, without afterburner, in Figure 4.30.

The foregoing are typical examples of turbojet engines in current use. Generally, new development is centered around the turbofan or bypass engines.

Turbofan engines are being used exclusively on all new commercial and military subsonic transports. As noted previously, these engines were developed to provide greater propulsive efficiency than turbojets by increasing total airflow and reducing velocity of the combined jets. The additional airflow does not pass through the combustion chambers and turbine wheels but is ducted aft around the engine case. The fan air may be exited at the forward part of the nacelle through a separate nozzle as shown by Figure 4.33 or may be conducted around the entire engine case and exited either through a common nozzle as in Figure 4.23 or through separate coplanar nozzles at the rear as on the C-141.

Medium bypass ratio engines are in widespread use on a number of large transports. The JT-3D (TF-33) was developed from the J-57 engine. This turbofan powers the Boeing 707, the later McDonnell Douglas DC-8 series, and the C-141 Starlifter. Take-off thrust is rated at 18,000 to 19,000 lb. for the 15-stage commercial engines, and to 21,000 lb. for the 16-stage military TF-33 version.

The 13-stage JT-8D-5, developing 12,500 lb. thrust, and the JT-8D-7, with 14,500 lb. thrust, are used on the Boeing 727 and 737 airplanes and the McDonnell Douglas DC-9. In this engine the fan air is conducted around the main engine case by an outer case and exits with the primary gases through a common nozzle.

The new generation of high bypass ratio turbofan engines is exemplified by the Pratt and Whitney JT-9D. A cutaway view of this engine is shown in Figure 4.31, and a front view of the 8-ft.-diameter fan is shown in Figure 4.32.

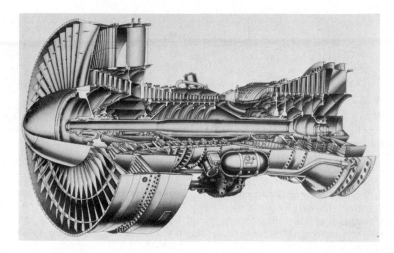

Figure 4.31. Cutaway view of the JT-9D high bypass ratio dual-rotor turbofan engine. It produces 47,000 lb. thrust at take-off and has a weight-to-thrust ratio of .187 lb. per lb. of thrust. *(Courtesy of Pratt & Whitney Aircraft Co.)*

Figure 4.32. Front view of JT-9D turbofan engine showing 8-ft.-diameter fan. Installed on Boeing 747. *(Courtesy of The Boeing Company.)*

Take-off thrust is 47,000 lb. with water injection. The 15 stage compressor operates at a compression ratio of 24, and the single-stage fan has a pressure rise of 1.56. The specific fuel consumption is .36 lb. per lb. of thrust per hr., and the dry uninstalled weight ratio is .187 lb. per lb. of thrust. All values are for a standard day at sea level. This engine powers the Boeing 747 and some models of the DC-10. The General Electric CF5-50 (TF-39) engine, which is used on the Mc-Donnell Douglas DC-10 and the C-5A, is somewhat similar in design and performance, as is the Rolls Royce RB-211 used on the Lockheed 1011 airbus. Growth versions of these engines in the 55,000-lb. range are under development.

Smaller high bypass ratio engines (5000 lb. thrust and under) are in developmental stages but have strong competition from new turboprop engines in this power range. Even smaller engines are in the design and testing stages.

TURBOJET AND TURBOFAN POWER PLANT INSTALLATION

Common terminology has classified engine installations into two types, the "buried" and the "pod." As suggested by the title, a buried installation is the enclosure of the engine (or engines) within the body or wing structure. When the engine is mounted in a nacelle which is separated from the wing or body by a supporting strut, the term *pod installation* is applied. One of the advantages of the pod installation is the exposure of the entire engine for maintenance by simply removing the side panels. The pod installation also provides safety in case of engine fire since the engine is remote from the basic airplane structure.

Figure 4.33. Turbofan engine nacelle. JT-9D engine installation on the Boeing 747 airplane. *(Courtesy of The Boeing Company.)*

A buried engine installation is used for the center engine in the Boeing 727 commercial aircraft in Figure 1.2. A pod-mounted installation with the engine mounted below and forward of the wing is shown in Figure 4.33. This is a short-duct turbofan installation with the fan exhaust at the forward end of the nacelle. Side mounting at the tail section of the fuselage is used on the Lear Jet, North American Sabreliner, DC-9, BAC-111, Boeing 727, Grumman Gulfstream II, Lockheed Jetstar, and others.

Turboprop Engines

Turboprop engine development actually began in 1926 when the English scientist Griffith proposed an axial-flow compressor and turbine engine for driving a propeller. After wind tunnel tests, development was discontinued because of the economic depression and scientific reluctance to accept the engine as having any practical significance. Following the successful development of the turbojet engine, interest was again directed toward the development of the turboprop engine, and successful flights in England and the United States were made with turboprop installations in 1945. For several years following these flights, the majority of effort was devoted to improving turbojet design in the United States, and turboprop development lagged. More emphasis was placed on turboprop development in England and France during this period, and the turboprop-powered English Viscount 630 was placed in commercial passenger service in 1950. Turboprop engine development during recent years in the United states has been directed toward smaller engines of a few hundred horsepower. These engines are now appearing on numerous medium-weight twins such as the Beech King Air, Piper Cheyenne, Mitsubishi MU-2 and others.

PRINCIPLES OF OPERATION

Basically, the turboprop engine is a gas turbine with most of the heat energy converted to shaft power by the turbine wheels, leaving very little for jet reaction. The basic parts of the engine compressor, combustion chambers, and turbine wheels are the same in both engines, except that larger or more turbine wheels are used in the turboprop engine.

In most turboprop or gas turbine engines of recent design, separate turbines are used to drive the compressor and power output shafts. This type of design is called "free turbine." In the cutaway picture shown by Figure 4.34, the two turbines, each with two stages, are indicated. In all of the earlier larger turboprop engines, the compressor and propeller gearing were driven by a single shaft, losing a great amount of flexibility and efficiency offered by the free turbine design.

Turboprop Controls.   A significant difference exists between turboprop and

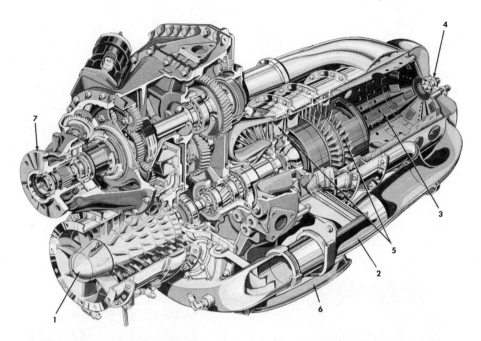

Figure 4.34. The Allison Model 250 turboprop engine. This engine utilizes six stages of axial-flow compression coupled with a single-stage centrifugal compressor. It develops 418 eshp. 1, Compressor: six axial stages and one centrifugal stage of the compressor. 2, Air transfer tubes: air from compressor to combustion section through two air transfer tubes. 3, Combustor: single combustor. 4, Fuel nozzle. 5, Turbines: two-stage axial turbine, which drives compressor; second two-stage axial turbine, which drives power-output shaft. 6, Exhaust: twin exhaust ducts. 7, Propshaft flange: accessories gear case and propeller reduction gear box. *(Courtesy of Allison Division of General Motors.)*

turbojet engine controls. In the turbojet engine, fuel is metered as required in proportion to airflow to maintain constant rpm, or is metered in the proper amount to effect changes from one rpm to another. As mentioned previously, the control must be designed to meter fuel within precise limitations to prevent overheating of the turbine wheels, compressor surge, or burner blowout. The control must also correct for varying conditions of altitude, air temperature, and airspeed for a given power lever position.

All of the control requirements of the turbojet are present in the turboprop engine, with additional requirements imposed by the propeller. By means of the blade pitch-angle control, the speed and power absorption of the propeller can be varied independently of the normal engine control. Propeller controls are generally different for the free-wheel engine and the direct-drive engine. Usually the propeller rpm control and engine speed control are separate for the free turbine installations. The windmilling drag with flame-out for this type of engine is only

Figure 4.35. A 3750-shp turboprop engine. Gas turbine versions of this engine are used in several helicopter installations. *(Courtesy of Lycoming.)*

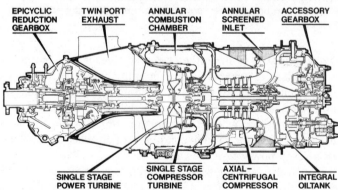

BASIC CONFIGURATION RETAINED
THROUGH TO PT6A-50

Figure 4.36. PT-6A engine. This engine is available in several power ranges including twin pack with a single reduction gear assembly. It powers a large number and variety of aircraft throughout the world. *(Courtesy of United Artist of Canada Limited.)*

Figure 4.37. Allison 501-D13 turboprop engine with separately mounted reduction gearing. This engine is used in the Lockheed Electra transport; it develops 3750 eshp at take-off. The T-56 version at 4900 eshp powers the C-130. *(Courtesy of Allison Division of General Motors.)*

about 25% that for the direct drive. In addition, power section acceleration is extremely rapid since the propeller mass is not directly connected to it.

For direct-drive turboprop controls, the engine is operated within a narrow range in the high-rpm region for all flight conditions. This operation minimizes the effect of propeller inertia on power changes. Fuel is metered into the engine in relation to blade angle and desired power output. The propeller rpm governor is integrated with the fuel control and monitors fuel flow to cause rpm to remain within the established limits. With this type of control, flight idle rpm is generally quite close to maximum rpm. Thus, during landing approach conditions, engine rpm will be high and propeller blade angle low. In some instances the blade angle is so low in proportion to flight speed that negative thrust will result and the propeller will windmill, feeding a small amount of power into the engine shaft. This action makes the turboprop installation substitute as an air drag brake.

Since idle rpm is only slightly less than maximum rpm, a rapid increase in power requires extremely rapid change in blade pitch. This is accomplished on one type of propeller by a system of clutches and gears which obtain their large power requirements directly from the engine shaft.

Windmilling drag in direct-drive turboprop installations is much more severe than with reciprocating and free turbine engines. If flame-out should occur in the engine, the large power requirements of the compressor will be absorbed by the propeller, creating a serious airplane control condition, especially with low blade angle settings during approach. To prevent this condition, a device is incorporated in the propeller drive which immediately clutches in the propeller feathering drive (pitch increase) when negative thrust rises above a preset value. The blades are then driven at a very high pitch change rate toward the feather position.

TYPES OF ENGINES

Because of light weight and simplicity of construction, turboprop engines are being developed in a wide variety of power ranges. Turboprop and turbine engines ranging from 200 to several hundred hp in small airplanes and helicopters are in widespread use. Continued improvement in design is resulting in engines with fuel consumption rates approaching those of the reciprocating engine.

TURBOPROP POWER PLANT INSTALLATIONS

Similar to the turbojet, the simplicity and small diameter of turboprop engines permit light and aerodynamically clean installations. A representative installation is shown in Figure 1.11(A).

TURBINE ENGINE CONSIDERATIONS

Turbine engines for aircraft have far outclassed reciprocating engines in producing thrust for flight. They have, from the first, shown superiority in lightness and size, especially in frontal area (and cooling drag).

Their high rate of fuel consumption has been a problem from the first, but that has slowly given way to advances in materials and special cooling techniques allowing higher operating temperatures. Finally, large high bypass ratio engines have been able to challenge the best of the large reciprocating engines in lb. thrust per lb. fuel per hour.

An original disadvantage of short engine life has been overcome gradually as designs, materials, manufacturing processes, and just plain knowledge have improved.

After initial problems were solved, the engine life of turbines has gone beyond anything dreamed of for reciprocating engines, and they run for thousands of hours with only periodic "hot section" inspections.

Rocket Motors

Rocket motors are the oldest man-made propulsion devices, having been used by the Chinese to propel incendiary weapons as long ago as 1232 A.D. The *propellant* fuel used in these early rockets is believed to have been some form of slow-burning black powder.

Rocket development for the next several hundred years was devoted to purely military application, except for some use to propel lifelines to ships in distress. Slow-burning black powder was used in all rocket motors during this period. Toward the close of the nineteenth century, the possibility of rocket motor power for aircraft propulsion led a number of investigators throughout the world to propose various types of rocket motor power plants. In the United States, Dr. Robert H. Goddard conducted intensive studies in rocket motor development from 1909 until his death in 1946. Early experiments made Dr. Goddard realize that desired speed and endurance could not be achieved with solid propellants of the types used in earlier motors. He therefore devoted most of his efforts toward the development of liquid-propellant motors, and a successful flight of a liquid-propellant machine was made in 1926. The fuel he used in this motor was liquid oxygen and gasoline. The American Rocket Society, formed in 1932 by a group of amateurs interested in Dr. Goddard's work, perfected a number of new types of motors, and flew various models.

Similar experimentation with liquid-propellant rocket motors was carried on in other countries during this period, much of it inspired by Dr. Goddard's writings. Using Dr. Goddard's work as a basis, Germany developed the V-2 rocket propelled bomb employed against England late in World War II. This machine had a total weight of 28,500 lb., of which 10,800 lb. was liquid oxygen and 8400 lb. was fuel (alcohol). The fuel was consumed in approximately 1 min., producing 56,000 lb. of thrust, which accelerated the machine to supersonic speed in a high-altitude trajectory. The range of the V-2 was 150 to 180 miles. Design details of the V-2 rocket motor were important reference materials in later liquid-propellant rocket motor development in the United States.

PRINCIPLES OF OPERATION

Rocket motors produce thrust by ejecting a mass of hot gases at high velocity from a nozzle, in a manner similar to that of the jet engines described earlier in the chapter. The principal difference is that jet engines carry only the fuel, drawing oxygen from ingested air for combustion. The jet engine thus achieves an air-fuel ratio of about 70 to 1.

This means that only $1/70$ of the total mass flow need be carried with the aircraft. In rocket motors, the mass of gaseous material ejected from the nozzle is composed entirely of the products of combustion of the oxidizing agent and fuel, both of which are carried with the aircraft. The weight penalty involved in transporting the entire jet mass limits the duration of combustion in rocket motors to a relatively short period. Since thrust is the product of jet mass and jet velocity, it is essential to obtain very high jet velocities. Therefore, the rocket fuels used must burn at extremely high temperatures, to produce the required thrust.

Rocket motors have been classified generally according to the two types of propellants used—*solid propellants* and *liquid propellants*.

Solid-Propellant Rocket Motors.   Solid propellants contain a mixture of oxidizer and fuel in one substance. Usually the constituents are pressed into a dense solid form, with the shape of the form being a factor in the burning rate of the charge. Various types of nitrocellulose smokeless powders are often used to form solid-propellant charges. Combinations of other chemicals are also used.

Solid-propellant charges are themselves classed into one of two types—*unrestricted burning* or *restricted burning*. The unrestricted burning charge will burn from all exposed surfaces, with the geometrical shape of the charge determining the burning area. Usually the burning time is quite limited. The restricted burning charge has some surfaces coated with a burning inhibitor to limit the burning area and provide a longer burning period.

In solid-propellant rocket motors, burning of the charge is started by a pyrotechnic igniter which is set off with electric current.

The storage area for the charge forms the combustion chamber, and the gases resulting from combustion are discharged through a nozzle at the aft end. Jet velocities range up to 7000 ft. per sec. This type of rocket motor is used in military missiles.

Liquid-Propellant Rocket Motors.   Liquid propellants are carried in tanks separate from the combustion chamber and are fed into the chamber at a controlled rate during the combustion process. A *monopropellant* contains an oxidizing agent and a fuel as a single liquid. A *bipropellant* system carries the fuel and oxidizing agents in separate tanks and provides for controlled mixing within the combustion chamber. The latter type is the principal system in use today. In the bipropellant system, the fuel and oxidizing agents are forced through feed lines, by means of pumps or tank pressure (or a combination), into suitable nozzles in the combustion chamber where burning takes place. Because of the extremely high combustion temperatures, the fuel is normally circulated through a jacket or liner surrounding the combustion chamber, to provide cooling of chamber surfaces before the fuel enters the nozzles. Several combinations of liquid propellants are in general use. Common among these is liquid oxygen with either kerosene or hydrogen. Combustion-chamber pressures vary from 300 to 3000 psi, jet velocities from 7000 to 10,000 ft. per sec., and combustion chamber temperatures range from 4000 to 7500°F. Some propellants such as nitric acid analine (fuel) are extremely toxic and require special handling precautions, reducing the practicability of their use. With hypergolic propellants, ignition is spontaneous (as with nitric acid and gasoline, analine, or jet fuels); with other combinations, spark or other types of ignition may be required to start combustion. This type of rocket motor is used primarily for space launches.

Impulse.   To compare rocket motor size and relative performance, a rating of *impulse* is used. *Specific impulse* (I_{sp}) is defined as the thrust produced per pound of propellant burned per second; *i.e.:*

$$I_{sp} = \frac{\text{lb. Thrust}}{\dfrac{\text{lb. Mass Propellant Consumed}}{\text{sec.}}}$$

$$= \frac{\text{lb. Thrust} \times \text{sec.}}{\text{lb. Mass Propellant}} \tag{4.13}$$

Total impulse (I_t) is defined as the product of the thrust and the firing time.

Solid fuel rocket motors have been used successfully to produce additional thrust for take-off by military aircraft. The rocket motors are attached to the fuselage, actuated for take-off, and then jettisoned when their fuel has been exhausted. This use is called *jet assisted take-off* (JATO), or *rocket assisted take-off* (RATO).

5

Aircraft Instruments and Avionics

The efficiency and utility of modern aricraft are largely dependent on the ability of instruments to depict accurately what the aircraft is doing in flight, and how well its power plants and components are functioning. In additioh to instruments, a variety of radio equipment is required to permit efficient navigation. The combination of these two, and the use of electronics in instrument design, is called *avionics*. To understand the roles of flight instruments, regard them in these categories: *control instruments, performance instruments, and navigation instruments*.

Control Instruments

These instruments permit the pilot to control how the aircraft flies by showing him directly and immediately both the *attitude* of the aircraft with respect to the natural horizon, and the *power* being delivered by the propulsion system. In other words, they permit him to control its performance. They are the *attitude indicator* (or horizon indicator) and a power indicating instrument—in propeller aircraft usually the *manifold pressure gauge and tachometer;* in turboprop air-

Figure 5.1 Instrument panel, Beech Bonanza V35B. Top row, left to right: clock, airspeed indicator, attitude indicator, altimeter, manifold pressure gauge and tachometer. Second row: ADF indicator, turn coordinator, horizontal situation indicator (HSI) and vertical speed indicator. Fuel gauges and engine instruments are grouped in lower center panel. Radio equipment, top to bottom: audio selector panel, dual communication transceivers, dual VOR receivers, transponder and automatic direction finder (ADF). Second VOR course deviation indicator (CDI) is located out of view behind control wheel. Three-light marker and DME distance read-out are located above attitude indicator. Second altimeter is located extreme right. *(Courtesy of Beech Aircraft Corporation)*

Figure 5.2. Cessna 421 panel.

craft, the *torquemeter;* and in turbojets, the *tachometer,* the *exhaust pressure ratio indicator,* or the *exhaust total pressure indicator.*

ATTITUDE INDICATOR

The attitude indicator is mounted directly in front of the pilot. There is good reason for this location. It is the only instrument which serves as a direct substitute for the natural horizon, and shows immediately whether the aircraft is straight and level, climbing, diving, or banking. It exactly duplicates the natural horizon seen through the windshield.

SUCTION-DRIVEN ATTITUDE INDICATOR

This instrument is very similar to the ones which permitted the development of attitude instrument flying as it is known today. It serves well all over the world in thousands of aircraft and is widely used in general aviation aircraft.

The attitude indicator may be considered a small horizontal metal bar, visible through the glass face of the instrument. It is kept parallel to the earth's horizon by the *rigidity in space of a universally mounted vertical gyroscope.* The base of the universally mounted gyroscope may be repositioned at will, without changing the relationship between the plane of the rotor and the earth's surface, once the gyro is spinning at the proper speed.

Suction System. Power to spin the rotor of air-driven instrument gyros is provided by the aircraft suction system (Figure 5.4).

The vane-type pump, consisting of sliding vanes on an eccentrically mounted rotating shaft in the pump case, is driven from the engine accessory section. At 1000 rpm or more, it gives a maximum suction pressure of 10 in. of mercury (Hg). The suction relief valve is adjusted to the required suction pressure, usually

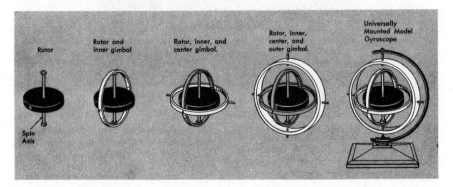

Figure 5.3. Mountings of the gyroscope. Great density, high spin rate, and the least possible friction in bearings and mountings are determinants of gyroscopic efficiency. *(Courtesy of US Air Force.)*

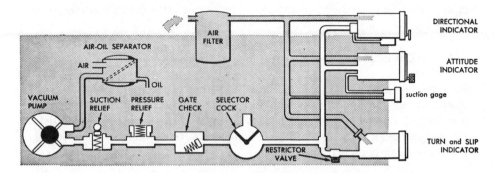

Figure 5.4. Typical aircraft suction system. *(Courtesy of US Air Force.)*

3.8 to 4.2 in. Hg. The gate check valve protects the system from engine back-fires, and the pressure relief valve releases the pressure which would be built up in such cases. Most multi-engine aircraft have suction pumps on more than one engine; either can be selected in the event of failure of one.

Attitude Indicator.   In the attitude indicator in Figure 5.5, the air sucked out of the case is replaced by air channeled through the instrument in such a way that it is eventually forced against buckets carved in the rotor, spinning it at about 15,000 rpm.

Figure 5.5 Suction-driven Attitude Indicator. This instrument provides the pilot with an artificial horizon when in IFR conditions or in reduced visibility. Pitch angle markings are shown above (10°) and below (10° and 20°) horizon line. Angle of bank markings are at top. Color coding is used, top half blue and bottom half dark with grid lines, to present a more natural presentation to the pilot. Most single and light twin-engined general aviation aircraft use this type attitude indicator. *(Courtesy Edo-Aire)*

The air spilling off the rotor is channeled down through the cylindrically shaped rotor housing and out through four vertical slits cut 90° apart in the wall of the housing. After the air departs from the housing, it is drawn out of the case, thus completing the cycle. If air under pressure were forced in and against the rotor, the effect would be the same. Thus the man-made horizon is gyroscopically kept parallel to the earth's horizon. One has but to attach the rear view of a miniature aircraft in front of this horizon, with its wings fixed horizontally in the case (parallel to the actual aircraft's wing), and any rolling motion of the aircraft will be detected by the miniature aircraft appearing to dip its wing into the horizon. In order to determine bank precisely, a scale is provided on the upper half of the circular instrument face and is marked at 30°, 60°, and 90° increments from top center.

The horizon will displace above or below the miniature aircraft to reflect the respective nose-high or nose-low pitch attitude of the aircraft. A knob below the face of the instrument permits the pilot to adjust the height of the miniature aircraft to align it with the horizon on the face of the instrument for continuous flight reference at different pitch attitudes and different pilot sitting heights.

Precession, the term given to any tilting or turning movement of the spin axis of a gyroscope, is the result of friction in the gryo mounting bearings, or the deflection of the spin axis by an externally applied force. In flight instruments, this force is the result of friction, of turning the aircraft, or of changing airspeeds. The rate at which the gyro precesses varies inversely with its speed of rotation, and directly with the applied force. As shown in Figure 5.6, a force (A) which attempts to *tilt* the spin axis will cause it to *turn;* one which attempts to *turn* the

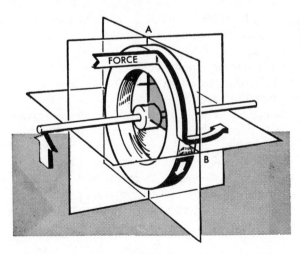

Figure 5.6. Effect of precession. An external force applied to the rim of a spinning gyro will cause it to precess in a direction 90° ahead in the plane of rotation and in the direction of the applied force. *(Courtesy of US Air Force.)*

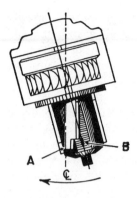

Figure 5.7. Action of pendulous vanes. The vane A is un-covering the slit B. On the opposite side, the vane is cov-ering the slit. As the gryo erects from the force of the air jet, the vanes cover the slits equally, equalizing the jet force. *(Courtesy of US Air Force.)*

spin axis (B) will cause it to *tilt*. Gyroscopic *drift* is precession caused by internal friction in the gryoscope assembly.

Friction, however small, will cause the gyro in an attitude indicator to precess away from vertical. An automatic erection device is installed to counteract pre-cession error. Hanging from above the four vertical slits through which air leaves the bottom of the rotor housing are four pendulous vanes, which half cover the slits when the gyro is erect (Figure 5.7). Should the gyro's spinning axis tilt away from vertical, that is, no longer be perpendicular to the earth's surface, the gyro housing would tilt with it. But since the vanes hang vertically under the influence of gravity, their relative movement causes the greater opening of one slit and a corresponding narrowing of the slit located on the opposite side of the housing. The unequal escape of air which results is, in effect, a jet blast from the largest opening, which drives the bottom of the gyro housing away from the force of the blast. The resultant movement or precession of the bottom of the gyro forces it back to an erect position. When erect, the air escapes equally and no unbalanced force exists.

Centrifugal force due to turns, and acceleration or deceleration, will displace the vanes and introduce a false precession. Thus, during turns (coordinated or other-wise) and during periods of acceleration and deceleration, the horizon bar will be displaced slightly in pitch or bank. These errors are seldom greater than 3° or 4° and are of minor concern to the pilot who is adept at using the instrument.

Operating Limits.   The safe operating limits of the typical air driven in-strument are 60° of pitch and 100° of bank. Exceeding the limits of the in-strument will cause it to strike against its mechanical limits and "tumble." As explained above, this is a violent and uncontrolled precession. It can result in cracked, flattened, or loosened bearings, any of which causes excessive friction and increased precession.

ELECTRIC ATTITUDE INDICATORS

The inherent limitations of suction-driven attitude indicators made essential the development of indicators which could cope with wider performance ranges in acceleration, speed, temperature, altitude, and maneuverability, as the capabilities of modern aircraft expanded. The first widely used electrically driven attitude indicator was a self-contained unit, the J-8.

The vertically mounted gyro is an electric motor, driven at 21,000 rpm by 115-volt, 400-cycle ac current. The gimbal mounting, a yoke and pivot assembly, supports the horizon bar, which can move up and down through an arc of 27°. A kidney-shaped partial sphere forms a black background for the instrument face. On the upper lobe is painted a black bullseye and the word *DIVE;* on the lower lobe is painted a white bullseye and the word *CLIMB*. These words, when under the nose dot of the miniature aircraft reference bar, represent about 60° pitch. The bank index pointer is attached to the yoke and pivot assembly and is free to rotate through 360°.

The dial face of the attitude indicator is marked with 0°, 10°, 20°, 30°, 60°, and 90° of bank and is used with the bank index pointer to indicate degrees of bank left and right.

The J-8 incorporates an erecting mechanism which maintains the gyro's spin axis vertical. Errors due to false erection in turns, acceleration, and deceleration can be as much as 5° of pitch or bank; however, they are usually less than 5°. As soon as the sensing device senses true gravity again after a maneuver, the errors are corrected rapidly.

Designed originally for fighter-type aircraft, the gyro is nontumbling, and the instrument has a "pull to cage" knob with which the pilot can erect the gyro in 30 sec. after power is available, with enough rigidity for an instrument take-off. An "off" flag appears whenever current to the instrument is interrupted; it disappears whenever the current flow is adequate. It is spring-loaded down, and its appearance in flight indicates current failure only.

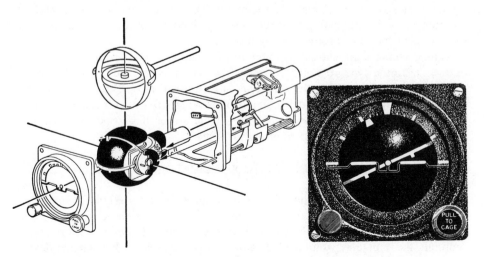

Figure 5.8. J-8 self-contained attitude indicator. *(Courtesy of US Air Force.)*

Operation.   The J-8 permits 360° rolls and loops without tumbling the gyro, and the expanded motion of the horizon bar provides sensitive pitch indications in near-level flight attitudes. When the aircraft attitude exceeds 27° of pitch, up or down, the horizon bar is held in extreme position, and the reference changes to the painted bullseyes. At 90° of pitch, as in going inverted at the top of a loop, the sphere rotates quickly through 180°. As soon as the aircraft departs the vertical, the instrument again indicates the true attitude of the aircraft. This rapid rotation of the ball is controlled precession, not tumbling.

Both the angle between the horizon bar and the miniature aircraft, and the bank index pointer, show the bank attitude of the aircraft.

The J-8, with a small face, serves well as the only attitude indicator in a limited size instrument panel. In aircraft with a larger panel, it or one similar to it is used as a stand-by instrument, providing reliable indications even in turbulence-caused extreme attitudes, which upset or tumble other types of attitude indicators.

SYNCHROS

When convenience or necessity dictates location of a sensing unit far from the instrument panel, its information can be suitably displayed on the panel by using synchros.

A *synchro* is an electrical motor system which controls electrically the angular position of one shaft by the angular position of another remotely located shaft (Figure 5.22). Simple synchros have two components: the transmitter and the receiver. The transmitter, often called a synchro generator, consists of a rotor which has a single winding, and a three-pole stator made up of windings displaced 120°. The rotor is supplied from an ac source and is coupled directly to the controlling shaft, as in a small electric motor. The voltages induced in the stator windings, by the alternating field set up by the windings of the rotor, are representative of the rotor position at any instant.

These voltages are carried by cable to the field windings of the synchro receiver, a similar synchro motor. Its rotor, free to turn, is coupled to a shaft, similar to that which positions the needle in an instrument face. The receiver rotor assumes an angular position depending on the voltages from the transmitter. The entire system is called a *servomechanism* if the torque has to be increased or amplified to turn the receiver rotor and its shaft.

REMOTE VERTICAL GYRO ATTITUDE INDICATORS

By separating the gyro assembly from the cockpit indicator, designers were able to eliminate the restrictions imposed by mechanical connection between the two and the space limitations behind the instrument panel. For high-performance aircraft, the remote gyro attitude indicator is currently used, with its greater flexi-

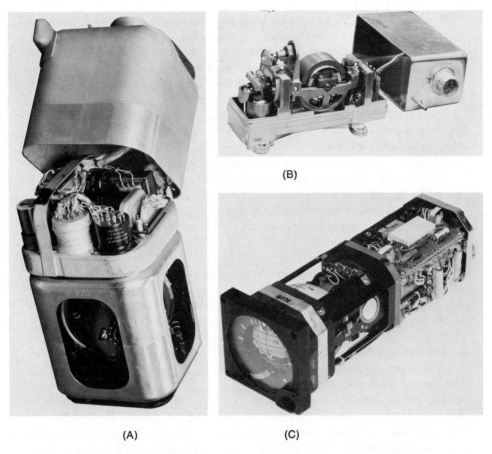

(B)

(A) (C)

Figure 5.9. Remote attitude indicator components. The control assembly, opened, is at (A); the rate switching gyro is at (B); and the indicator is at (C). The amplifier is not shown. In a sealed case, it is about the size of the control assembly. *([A] and [B] courtesy of Lear Instrument Division. [C] courtesy of Bendix Corp.)*

bility of mounting and greater accuracy. These systems consist of a displacement gyroscope, a rate gyroscope, and one or more amplifier-indicator combinations.

The displacement gyro, usually called the *control assembly,* is a hermetically sealed, gas-filled unit, displayed in Figure 5.9(A), which contains a 2°-of-freedom vertical gyro element, an automatic starting cycle circuit, an internal gravity erection system, roll and pitch erection cutout switches, and a provision for external application of fast erection when desired. The vertical gyro establishes the vertical reference line from which roll or pitch deviations of the aircraft are measured. Driven by 400-cycle, three-phase 115-volt ac, it turns about 22,000 rpm and is mounted in gimbals which give 360° of freedom in roll and 82° of freedom in dive or climb.

The erection mechanism, which holds the spin axis of the gyro vertical, consists of torque motors and rings on the gyro gimbals. These motors are engaged by the switching system whenever the gyro is not vertical, and the appropriate motor acts on the gimbal ring to return the gyro to vertical.

The separate rate switching gyro, or *rate gyro* (Figure 5.9[B]), which is mounted with its spin axis parallel to the aircraft lateral axis, senses turns; when the turn is more than 15° per min., the erection mechanism is interrupted so that no false erection of the vertical gyro takes place during a normal turn. A time delay eliminates nuisance switching in and out of the erection system due to rough air and during small turns faster than 15° per min.

When ac and dc power are applied simultaneously, as when aircraft power is turned on or a ''fast erection'' switch is activated, a thermal relay or fast erection switch allows faster operation of the torque motors, and the gyro is erected in 2 min. There is no manual erection device. Without the fast erection switch, erection requires up to 15 min.

The *amplifier* is also a sealed, gas-filled unit, which amplifies the error and rate signals from the indicator, and produces an output to run the servo motor in the indicator which responds to the servos in the control assembly. The amplifier has two independent amplifier channels, one for the roll axis and one for the pitch axis. Amplifiers may be separate units, or may be in the same case with the indicator.

The *indicator* (Figure 5.9[C]), hermetically sealed and gas-filled, contains two transistorized amplifier channels and two servo assemblies, one for the roll and one for the pitch axis.

The attitude indicator consists of the instrument face, calibrated in degrees of bank, a fixed miniature aircraft reference, an ''OFF'' flag which shows whenever the control unit is getting insufficient current, the reference bar and sphere, and a pitch trim knob which adjusts the horizon reference electrically by vertical placement of the reference bar.

Attitude indicators are now being widely used such as that in Figure 5.10, in which the attitude of the aircraft is indicated by orientation of the movable sphere with respect to the fixed miniature aircraft. The horizontal centerline of the sphere moves below the aircraft symbol for climb, and above it for dive. Pitch angle is indicated by reference marks on the sphere, and bank angle by the bank-angle scale around the upper face of the instrument and a movable bank index or ''sky pointer'' which rotates with the sphere.

The pitch trim knob electically positions the sphere to provide for pitch trim corrections relative to the fixed miniature aircraft.

Turn and acceleration errors have been reduced from those experienced in the J-8 so that turn error is virtually eliminated, and acceleration error is generated at the rate of .8° to 1.8° per min. versus the 3° to 6° error of the J-8.

Future Trends.   Remote gyro systems have the greatest opportunity for development for application to all aircraft. At present, scales, lighting, and sym-

Figure 5.10. MM-3 attitude indicator. Pitch trim knob is in the lower right corner of the indicator. The "OFF" flag indicates no power. *(US Air Force photograph.)*

bols are simplified and improved. Color is being used to make them more natural and easier to use without error and with reduced fatigue. They are incorporated into integrated flight director systems. Gyro inputs from inertial guidance systems are being used, and video displays of complete instrument systems are being evaluated.

POWER INDICATING INSTRUMENTS

The *tachometer* is the basic power indicating instrument in light planes having fixed-pitch propellers; it is also basic in many jets, with the only difference being that the jet engine tachometer measures percent of maximum rpm rather than actual rpm.

There are two basic types of tachometer system: mechanical and electrical.

The *mechanical* instrument consists of flyweights working against a spring, in either tension or compression. The flyweights are mechanically connected to the engine so as to revolve at a speed proportional to the engine speed. The collar to which the weights are attached is free to move up and down the shaft against the force of the spring. As the weights revolve around the shaft, centrifugal force causes them to move away from the shaft. This movement away from the shaft is proportional to the speed of rotation and is resisted by the compression of the spring. For each speed there will be a state of equilibrium when the centrifugal force is exactly equal to the spring compressive force. For a given speed there will be a certain displacement of the collar up the shaft. As the collar is connected to the indicator, this displacement, acting through a lever and gear train, shows the speed of the engine in revolutions per minute. In tachometers for light planes, a counter is added which records engine time in hours and hundredths. It is calibrated by assuming an average rpm, and is accurate at some specified rpm.

The *electrical tachometer* consists of an ac generator and a dc indicator. The

Figure 5.11. Color in engine instruments. Tape instruments shown here for a twin jet are colored in the usual green ⁄⁄⁄ for operating range, yellow ∷∷∷ for caution, and red ▒▒▒ for danger. Exhaust Pressure Ratio, Exhaust Gas Temperature, Tachometer, Fuel Flow, Oil Pressure, and Oil Quantity are all shown. Red flag on Fuel Flow indicates the instrument is inoperative. Instant readability is provided by color-contrasting the tapes. *(Courtesy of US Air Force.)*

generator is a simple inductor-type generating mechanism. The speed of the engine determines the frequency output of the generator.

The indicator on the instrument panel is a small millivoltmeter. A small transformer, built into the case, is designed so the core saturates at the voltage output of the generator when the engine speed is less than 100 rpm. Thus the voltage in the secondary coil is dependent on the frequency of the primary coil. The ac voltage is rectified to dc by a small unit placed in the indicator. The displacement on the millivoltmeter then indicates the speed of the engine in revolutions per minute. (See Figure 5.1.)

Jet engine tachometers are primary power-indicating instruments. They indicate the speed of rotation of the turbine rotor, and because of the high turbine speeds (as high as 40,000 rpm), they indicate *percent of* maximum speed. These tachometers are electrical instruments consisting of two units, indicator and generator, connected by electrical cables. Large turbine engines having two rotors use two tachometers, N1 for the front rotor and N2 for the rear rotor (Figure 4.20). The scale on the indicator ranges from zero to 110%. Indication is by two needles, one a vernier of the other, to facilitate accurate readings at slow speeds for engine starting and to enable the pilot to adjust power with greater precision.

The generator that actuates the tachometer is located on the engine accessory section and is driven by the main engine rotor shaft through a gear train. When the shaft of the generator rotates, it supplies a three-phase electric current, the frequency of which is directly proportional to shaft speed. This current is conducted to synchronous motors in the tachometer indicator which indicates the related speed in percent of rpm.

In the event of aircraft electrical failure, the tachometer will continue to function since it generates its own electricity. The scales are marked in green and red, with the upper operating limit normally set at 100% rpm.

MANIFOLD PRESSURE GAGES

The most accurate measure of power in piston-powered aircraft equipped with variable-pitch propellers is the manifold pressure gage. It measures the pressure under which the fuel-air mixture is supplied to the intake manifold. In conjunction with the engine tachometer, which indicates the engine speed (constant for a given takeoff, climb, cruise, or approach power range set by the propeller control), the manifold pressure gage is the most easily read and accurate measure of engine power change. It is calibrated in inches of mercury (Figure 5.1).

There are many types of manifold pressure gages manufactured. One type uses a curved, flexible, hollow tube (Bourdon tube) which is flexed by changes in pressure. It measures pressure in the engine blower section or the carburetor air intake and transmits it to the indicator by a capillary tube. A second uses a bellows or evacuated chamber, one end of which is sealed with a diaphragm which electrically or mechanically indicates pressure on an indicator.

Manifold pressure gages indicate the station ambient air pressure when engine and airplane are stopped. In flight, with the engine stopped, they indicate air pressure at the flight level plus the pressure induced in the engine intake manifold by the ram-air effect of airspeed. For this reason, when an engine fails, the manifold pressure gage will not go to zero, but will generally indicate about 30 in. Hg (at sea level). Note the reading in the left-hand dial on the center panel in Figure 5.2. The engines are stopped.

TURBINE ENGINE POWER MEASUREMENT

The two most sensitive indicators of power change in turbojet engines are the tachometer, previously described, and the engine *pressure ratio (EPR) indicator*. The EPR is described on page 167. The ratio is mechanically computed and

Figure 5.12. Exhaust pressure ratio indicators. *Left:* Round-dial indicator. The set knob permits setting in the digital readout a desired referenced value, as for take-off or climb, with which the index marker coincides. *Right:* Tape-type instrument, showing the exhaust pressure ratio for all four engines. This is used on the Air Force C-141. *(Courtesy of US Air Force.)*

transmitted electrically to the indicator, which is calibrated in nondimensional units. A set knob on the instrument face permits a limit index to be set to maximum allowable or desired EPR reading, as for take-off.

Turboprop engine power output is measured by the *torquemeter*. The torquemeter measures power available at the propeller shaft in pounds per square inch of torque oil pressure. This pressure is developed by pistons of the torque measuring system which reflect the changing pressures as engine reduction gears advance along the propeller shaft. A pressure transmitter, through synchronous motors, transmits the pressure indication to torquemeters, one for each engine, to the pilot's instrument panel.

This Falcon 10 panel displays complete, dual King KFC 300 Integrated Flight Director/Autopilot systems teamed with dual King KNR 665 10-waypoint RNAV which provide digital keyboard frequency management for all RNAV/NAV/COMM/ADF/Transponder functions.

Figure 5.13(A). Boeing 747 control cabin. Safety dictates simplicity, cleanness, and physiological accessibility. All but pilots' control, performance, and navigation instruments, pilot-essential controls and COM/NAV, lighting, and engine switches are located on the flight engineer's panel in the TWA cockpit. *(Courtesy of The Boeing Company.)*

Figure 5.13(B). Boeing 747 flight engineer's panel. Fuel, heat, electrical power, air conditioning, cabin pressure, and engine condition instruments and controls are here, aft of the copilot. Circuit breaker panel is overhead. *(Courtesy of The Boeing Company.)*

193

Performance Instruments

The performance of an aircraft in a given attitude and with a certain power is indicated by the *airspeed indicator, Mach indicator, heading indicator, altimeter, vertical velocity indicator,* and *turn and slip indicator.* In high-performance aircraft, the *angle of attack indicator* has come into general use.

PITOT-STATIC SYSTEM

Altimeters, airspeed indicators, and vertical velocity indicators all operate by sensing differential air pressures around the aircraft. Ram air, that is, air flowing by at aircraft speed, compared with static air, actuates airspeed indicators. Static, or still air at the pressure existing at the aircraft altitude, actuates altimeters. Changes in static air pressure actuate vertical velocity indicators.

Ram air is sensed by the *pitot head,* or *"pitot tube."* Impact pressure is taken from the pitot head through pressure lines either to the instrument or to the air data computer. The pitot head is located on the leading edge of a wing, the vertical stabilizer, or the nose section, on a shaft which projects it into undisturbed air. On supersonic aircraft, it is mounted on a shaft or "boom" of sufficient length to project ahead of the nose shock wave.

Static pressure is vented through small holes in the side of the fuselage and led through lines to instruments or to an air data computer. On most aircraft using a flush-mounted static source, there are two vents, one on each side of the fuselage. They are connected by a Y-type fitting to compensate for any possible impact pressure that might occur on one of the vents from rapid changes in attitude, such as slips, skidding turns, and rolls. A new pitot tube for transonic aircraft is in use with an aerodynamic shape such that the airflow at any speed insures static air pressure over ports in the tube head, about 2 in. aft of the tip. This system eliminates the installation error described below.

The pitot chamber is affected by the impact pressure of the air on the open pitot-tube tip as the aircraft moves in flight. The static chamber, if it is contained in the pitot head, is vented through small holes on the top and bottom of the tube to free, undisturbed air. The accumulation of ice and water are prevented by a heating element in the pitot head which the pilot turns on before entering areas of visible moisture; clogging of the static ports may be offset by an alternate source vented into the cockpit in unpressurized aircraft or into a fuselage area outside the pressure vessel in pressurized aircraft.

When switching to the "alternate source," one should expect to observe the altimeter and airspeed move slightly higher than normal, and the vertical speed indicator momentarily indicate a climb.

These effects result from the normally slightly reduced pressure in the fuselage caused by the relative wind past openings (mostly small) in the fuselage.

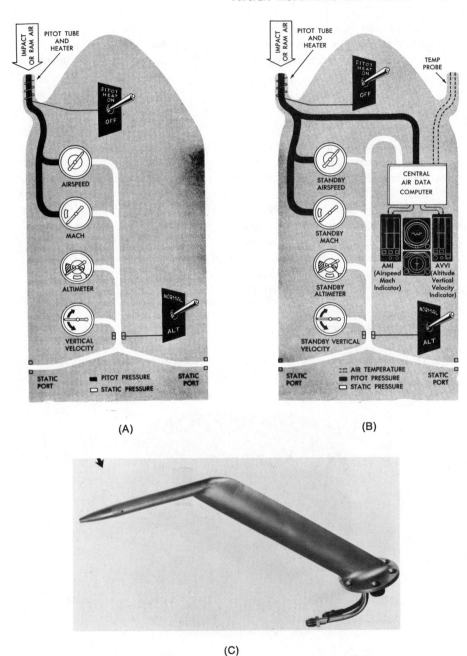

Figure 5.14. Pitot-static system. (A) Conventional round-dial instruments. (B) Integrated flight instruments. (C) Compensated pitot-static tube as used in the C-141. *(Courtesy of US Air Force.)*

Static-pressure sensing systems in high-speed subsonic or transonic vehicles are greatly affected by shock waves. Consequently they are typically placed in booms which put them forward of the fuselage outside the area of the disturbed air.

AIR DATA COMPUTERS

Higher-performance aircraft are now customarily equipped with flight and navigation instruments which do not take information directly from outside air sources, but receive electrical impulses from a *central air data computer* (CADC). The computer receives ram air, static air, and temperature information and converts it to electrical signals. Airspeed, vertical velocity, altitude, and outside air temperature all use this source (Figure 5.14). While all the errors introduced in instrument aneroids and bellows are present in the CADC, they are reduced to a minimum by the elimination of mechanical linkages in electrically operated instruments. The CADC is designed or selected for a particular aircraft installation, further minimizing error.

AIRSPEED INDICATOR

The airspeed indicator measures the difference between static air pressure and ram air pressure. Ram air is led to the inside of a hollow flexible bellows. Static

Figure 5.15. Airspeed indicator. The present wide range of airspeed performance requires a drum or other vernier device to permit precise airspeed readings during slow flight for approach and landing. The small figures are Mach numbers. The striped needle is the variable Mach limit needle. *(US Air Force photograph.)*

air is vented into the instrument case and surrounds the bellows. The pressures are equal on the ground at rest, but as the aircraft accelerates as, for example, on take off, the ram-air pressure increases. The relatively greater pressure in the bellows expands it, and through a system of gears and levers, moves the airspeed indicator needle or drum in the face of the instrument, registering airspeed in knots or in statute miles per hour.

Jet aircraft airspeed indicators also have a maximum allowable airspeed needle. It is used in a manner similar to the "red line" marking on airspeed indicators in piston aircraft, to show the maximum safe speed of the aircraft in which it is installed. The difference is that the maximum allowable needle is a movable "red line" and points out the maximum safe indicated airspeed for that aircraft at all altitudes, whereas the stationary red line marking is accurate only for sea level on a standard day. A second sealed bellows expands under the lighter atmospheric pressures of higher altitudes. This expansion is linked to the maximum allowable needle to cause it to decrease its indications with an increase in altitude. The higher the altitude, the farther it descends, constantly showing the maximum allowable indicated airspeed at increasing altitude.

There are three general errors characteristic of airspeed indicators:

Installation error is the error caused by the difference between the actual and the theoretical pressure differential developed by the pitot-static system.

Compressibility error results from "packing" of air in the pitot tube at high airspeeds, and results in indications of higher than normal airspeeds. This error is usually not more than one or two knots, but increases sharply to as much as 10 knots just below the speed of sound, about Mach .96. It decreases rapidly after the aircraft is slightly above the speed of sound.

Air density error is a variation between true and actual speed through the air, resulting from the effect of variations in altitude and temperature. It is determined and corrected by the use of a flight computer, or a true airspeed indicator, which in addition to the normal bellows contains aneroid and temperature diaphragms to compensate for changes in air density.

As a result of these errors, various types of airspeed are identified as follows:

Indicated airspeed (IAS) is that read directly from the face of the indicator.

Calibrated airspeed (CAS) is IAS corrected for installation error. It may be obtained from a table in the aircraft's performance data, and is sometimes given by a card in the cockpit.

Equivalent airspeed (EAS) is CAS corrected for compressibility error; it may be taken from aircraft performance charts, or from an aircraft performance computer for the particular type.

True airspeed (TAS) is CAS (or EAS in transonic aircraft) corrected for air density at flight level, and determined from outside air temperature and indicated pressure altitude, either by using a computer (Figure 11.7) or by a true airspeed indicator which must then be corrected for installation error and compressibility.

Ground speed is TAS corrected for wind effects (page 616).

Figure 5.16. Combined airspeed-Mach indicator. *(Courtesy Sperry Flight Systems.)*

MACH INDICATOR

This instrument indicates the ratio of aircraft speed to the speed of sound in the air at flight altitude. It is essentially an airspeed indicator which has an aneroid diaphragm to sense static air pressure. No temperature diaphragm is necessary because both true airspeed and Mach are dependent upon air temperature and in the ratio, air temperature values are cancelled. By mechanical means the differential air pressure of the airspeed indicator and the static pressure of the aneroid are combined to provide an indication of Mach number on the face of the instrument.

Combined airspeed-Mach indicators are provided for aircraft in which panel space is at a premium. Airspeed, Mach, and limiting Mach are displayed. A knurled knob is provided to permit setting a desired reference speed.

ALTIMETERS

The altimeter measures the height of the aircraft above a given reference, both for maintaining terrain clearance and for aircraft separation. The pressure altimeter is a simple barometer which measures atmospheric pressure and displays indicated altitude in feet above a preselected reference. As an aneroid barometer, it actually measures the weight of air lying above it at any given altitude.

Because atmospheric density at various altitudes is constantly changing from the lapse rates of Figure 6.4, due to changes in temperature and pressure, the altimeter must be designed to some arbitrary standard. The standard is the lapse rates in Figure 6.4.

Figure 5.17. Altimeters.

Altimeter Setting.   In order to account for deviations from the standard day at the surface, the altimeter has a knob by means of which the pilot can set barometric pressures from 28.10 to 31.00 in. Hg into the *Kollsman window* (Figure 5.17). If 29.92 were set in the Kollsman window, and the aircraft were on an airport at the seashore on a standard day, the altimeter would read 0 ft., the elevation of the airport above sea level. But if it were not a standard day, and the barometric pressure were, for example, 30.20, the altimeter—reading altitude above a standard datum plane of 29.92—would show the aircraft to be 280 ft. lower than sea level. If the pilot then turned the setting knob until the altimeter indicated 0 ft. 30.20 would appear in the Kollsman window.

Indicated altitude is that shown by the instrument. It is altitude above sea level pressure when the current barometric pressure, or *altimeter setting,* is in the Kollsman window, assuming a standard lapse rate.

Pressure altitude is indicated altitude above mean sea level when the barometric pressure setting in the Kollsman window is 29.92 in. (1013 mb).

True altitude is actual altitude above mean sea level.

Density altitude is pressure altitude corrected for temperature. It is used for predicting aircraft and engine performance, for take-off, climb, and landing. The computer or charts are used to calculate it. On most computers, it can be read directly when calculations for true airspeed are made.

Absolute altitude is the height above the surface of the terrain over which the pilot is flying. It can be determined only by a radio altimeter (Figure 5.46). One can estimate it fairly closely by subtracting the height of the ground, read from the map, from true altitude; however, the reading will err by the amount of error in the map and by the degree to which the air (through which the pilot is flying) deviates from the standard lapse rates. It can vary so greatly due to pressure systems aloft and wind effects that the FAA requires 2000 rather than 1000 ft. clearance above the highest terrain in mountainous areas.

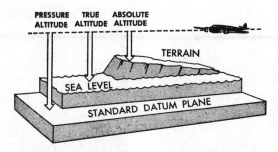

Figure 5.18. Types of altitude. Sea level and the standard datum plane are the same only when barometric pressure equals 29.92 Hg. *(Courtesy of US Air Force.)*

In international flight, there are three distinct altimeter setting symbols. They may be measured in inches of mercury or in millibars:

QNH   Standard altimeter setting as known in the United States, measuring existing surface pressure reduced to sea level at that point. (Pressure altitude above mean [average] sea level, "MSL".)

QNE   Always 29.92 in. or 1013.2 mb.

QFE   Actual surface pressure, not reduced to MSL. This causes the altimeter to read altitude above the reporting point.

Because the altimeter is affected by temperature and pressure both in the air and on the ground, a study of deviations from the standard shown in Figure 6.4 will show that:

When flying from an area of higher pressure to one of lower pressure, the altimeter, if not reset, will place the aircraft at an altitude LOWER than indicated.

When flying from an area of higher temperature to one of lower temperature, the altimeter will place the aircraft at an altitude LOWER than indicated.

Altimeter Error.   Air traffic density, high rates of climb and descent of modern aircraft, and instrument approach procedures which require precise altitude all make altimeter error of vital importance. These errors consist of *mechanical errors* in the instrument itself, which are compensated for by those altimeters which derive their information from a central air data computer; *installation error,* which is error introduced by faulty airflow around the static ports; and *reading error,* in which the pilot misreads the altimeter.

Installation error is corrected by the design of compensating pitot tubes (Figure 5.14), and by compensation provided by the CADC. Reading error is reduced by the design of the altimeter presentation. The three-pointer altimeter (Figure 5.17[A]) is the most difficult to read. To reduce error, it may have a rotating disc on the face which uncovers a brightly hatch-marked sector, as shown in Figure 5.17(A), when below 16,000 ft.

The vertical scale type (Figure 5.44) is the easiest to read. Among round-dial altimeters, the "counter-drum-pointer" type is the easiest to read (Figure 5.17[B]). It derives its information from a CADC and presents it on counters which indicate 10,000 and 1000 ft., a drum which indicates 100 ft., and a pointer which indicates tens of feet. It is based on a servo-pneumatic mechanism in which the pneumatic portion is based on evacuated bellows which expand and contract under different pressure altitudes, as in the customary altimeter. This system acts as a stand-by system. Normally the drum, counter, and pointer are driven by servos from the CADC. These altimeters are also capable of transmitting encoded information to the aircraft transponder and thus to Air Traffic Control.

VERTICAL SPEED INDICATOR (VSI)

This instrument indicates the rate of climb or descent in feet per minute by measuring the rate of change of atmospheric pressure.

The VSI employs a thin metal bellows similar to the one in the airspeed indicator, one side of which is connected to the static pressure line. The airtight case of the indicator is also vented to the static pressure line, but through a restricted passage called the "calibrated leak." This opening is extremely small and of a uniform specific size. With the opposite or free side of the bellows connected by levers and gears to the indicator needle, any expansion or contraction of the bellows will cause a corresponding downward or upward movement of the needle. In level flight, the same static pressure fills the case and bellows. There is no movement of the bellows and the indicator needle zeroes. In a climb or descent, however, a differential pressure is created, and there is a corresponding expansion or contraction of the bellows, with the corresponding indications by the needle.

In a descent, for example, the aircraft descends into greater atmospheric pressure; the denser air fills the bellows instantly. This denser air would also fill the instrument case instantly, but the calibrated leak restricts free flow and in effect traps lighter pressure in the case. The relatively greater pressure in the bellows causes it to expand, and the indicator needle registers a descent. But the instrument is also measuring the rate of descent because the faster the descent is made, the harder it is for pressure to exchange through the calibrated leak. This naturally creates a greater pressure differential and greater travel of the indicator needle.

As long as the aircraft is descending at a constant rate, the pressure differential will be constant, and it is merely a matter of scaling the instrument dial in feet per minute correspondingly. When the aircraft levels off from the climb or descent, the indicator needle lags behind. This is the result of the time interval that it takes for the pressure to equalize again through the calibrated leak. Experienced instrument pilots expect the lag and disregard the indicator during this settling-down period.

The primary use of the vertical speed indicator is determining rate of climb or descent. However, it is also an excellent trend instrument for maintaining level flight, since entry into climbs and descents from level flight is indicated instantly.

Inertial-Lead Vertical Speed Indicators (*IVSI*).   These instruments are designed to reduce VSI lag. In them, two small accelerometers (Figure 5.19) sense

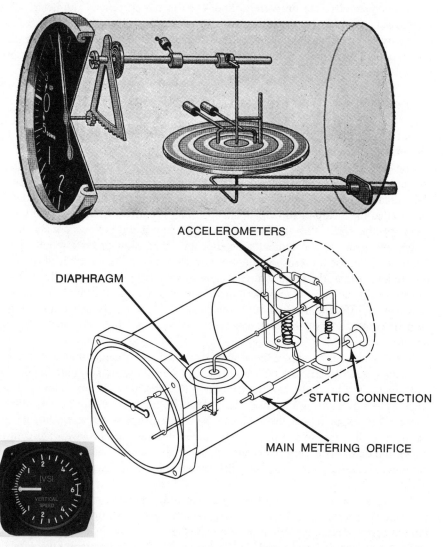

ACCELEROMETERS

DIAPHRAGM

STATIC CONNECTION

MAIN METERING ORIFICE

Figure 5.19. Cutaway view of vertical speed indicators. *Upper:* The standard model. *Middle:* The "instantaneous" or "inertial" model, employing accelerometers. *Lower:* A full-face photo. *(Courtesy of Teledyne.)*

the initial acceleration in climb or descent. They provide a response before the indicator can respond to static pressure change, giving an "instantaneous" reaction, and enhance vertical speed control when used properly with the attitude indicator, during wings-level climb and descent. Because the accelerometers turn with the aircraft, they induce an error during turn entries. Outside a .7 *g* to 1.4 *g* load factor, the accelerometers are inhibited, and the IVSI acts as a normal VSI.

HEADING INDICATORS

This term is applied to any gyro-stabilized heading reference instrument. A *directional gyro* is a heading indicator which is gyro-stabilized, its gyro spun by pressure, suction, or electric power. It is not magnetic but may be aligned to the heading of a magnetic compass or any other directional reference. The term is also used to mean a gyro in a gyro-stabilized magnetic compass system which provides "short-term" directional stability, preventing compass oscillations, and connected by synchros to a cockpit indicator. Such a system is said to be in the "directional gyro mode" when it is detached or unslaved from the magnetic reference and set to an independent reference such as a stand-by magnetic compass or a runway heading.

Directional gyros of any type depend upon the gyroscopic property of rigidity in space. They are subject to errors caused by gyro drift, by real precession caused by aircraft maneuvering forces, and by apparent precession. *Apparent precession* affects universally mounted gyros which are translated rapidly over long distances at high speed. As the earth rotates, or as the gyro is translated long distances, the plane of rotation changes from the viewpoint of an observer on earth. For example, a gyro rotating in a horizontal plane at the pole would appear to be rotating in the vertical plane at the equator, since its plane of rotation in space does not change as it is transported through 90° of earth curvature.

SUCTION-DRIVEN DIRECTIONAL GYRO

The rotor spins in the vertical plane. Fastened to the vertical gimbal, at right angles to the plane of rotation, is a circular compass card. The case of the instrument, and the airplane, simply turn about the compass card. The aircraft suction system draws air from the case of the instrument; air coming in to replace it is directed at buckets on the periphery of the gyro rotor, turning it at about 18,000 rpm. Precession is corrected by two parallel jets which strike at the center of the rotor when it tilts from the vertical; the precessive force returns it to vertical. A caging and heading set knob is provided. It permits rotation of the card to a desired heading indication by means of gears and a friction clutch.

The limits of this instrument are 55° of pitch and roll. The gyro, striking stops, will tumble and the compass card will spin when these limits are exceeded. The aircraft must then be returned to level flight, and the gyro caged and reset. The

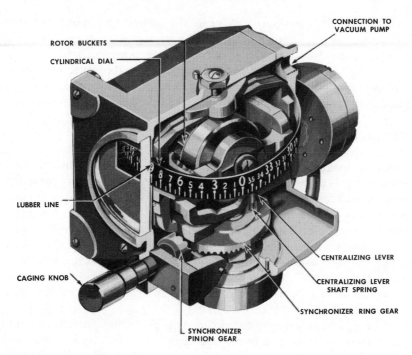

ROTOR BUCKETS

CYLINDRICAL DIAL

CONNECTION TO
VACUUM PUMP

LUBBER LINE

CAGING KNOB

CENTRALIZING LEVER

CENTRALIZING LEVER
SHAFT SPRING

SYNCHRONIZER RING GEAR

SYNCHRONIZER
PINION GEAR

Figure 5.20. Cutaway view of suction-driven heading indicator. *(Courtesy of US Air Force.)*

gyro should be caged during maneuvers which exceed its limits, in order to avoid undue wear on rotor and gimbal bearings.

REMOTE INDICATING GYRO-STABILIZED COMPASSES

These compasses, known also as *gyrosyn compasses,* have been developed for a wide variety of applications, but all have four basic components: the *remote compass transmitter,* the *directional gyro,* and *amplifier,* and the *heading indicator.*

The remote compass transmitter is the direction-sensing device of the system. It is usually located in a wing tip or other magnetic interference-free location.

It contains a *flux valve,* which is an iron core shaped like a three-spoke wheel with a heavy rim, broken between spokes. Each spoke of this wheel is wound with primary and secondary coils. Alternating current from the compass inverter creates a reversing magnetic field through the primary coil. This wheel is suspended horizontally from its hub in pendulous fashion so that as the aircraft pitches, it remains horizontal; an oil bath restricts its fluctuation. It cannot rotate in its housing and turns with the aircraft. As long as the aircraft is on a fixed course, the earth's magnetic lines of flux have a constant effect, and the field of the primary coil is constant. When the aircraft turns, the field changes, and a current is induced in the secondary coil. These signals from the secondary coil are amplified in the amplifier, and synchronous motors align the *heading indicator* with the magnetic heading of the aircraft.

The heading indicator is stabilized by the directional gyro, which may be either a remote unit or integral with the heading indicator.

The gyro compass systems have operating limits of 85° of pitch and roll, and are free of northerly turning error, oscillation, or swinging. Since inherent gyro drift is continuously being corrected by being slaved to the magnetic compass transmitter, no resetting is required unless the system is one which can be unslaved for operation in the directional gyro mode.

When a slaving cutout switch is provided for "DG mode" operation, a setting switch is provided; there is also an *annunciator,* a left-right needle which, when centered, indicates that the compass transmitter and gyro are properly synchronized when reslaved.

In today's era of specialized flying, it is impossible to expect a single type of compass system to satisfy the many varied commercial, general aviation, and military operational requirements. Specialized models are available to meet the

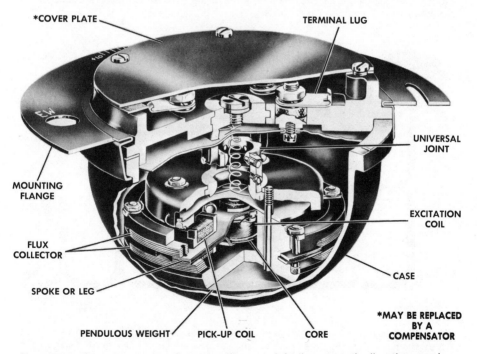

Figure 5.21. Gyrosyn compass flux valve (flux gate.) As the magnetic direction-sensing element, the flux valve is the heart of the remote indicating-type compass. *(Courtesy of Sperry Gyroscope Co.)*

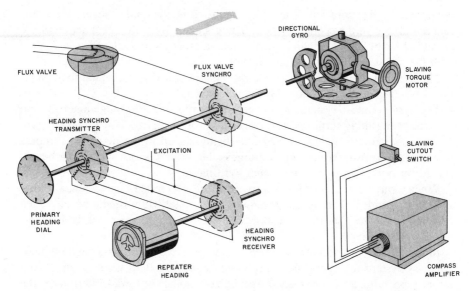

Figure 5.22. Schematic of basic components of the gyrosyn compass system. *(Courtesy of Sperry Gyroscope Co.)*

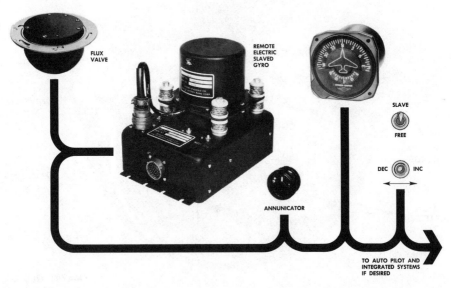

Figure 5.23. Gyrosyn compass system designed for business and executive aircraft. This system uses a solid-state amplifier, may be operated in the DG mode, and can provide for more than one indicator, or for directional reference for an auto-pilot system. It provides a drift-free heading reference, self-maintained to ±1°. Self-erecting in the slaved mode. *(Courtesy of Sperry Phoenix Co.)*

needs of auto-pilots, high-performance aircraft, and transpolar operation. Very high accuracy and automatic corrections for latitude are the features of the more advanced systems.

MAGNETIC COMPASS

This old and familiar instrument, in use for the last 600 years, is the simplest and least precise aircraft heading reference. It is the basic heading reference in most single and light twin aircraft, and serves as an emergency back-up and cross-check instrument for more complex systems in the more sophisticated aircraft.

The panel or independently installed magnetic compass contains a circular framework with two small parallel mounted magnets whose north-seeking poles point in the same direction, encircled by a numbered card. This framework is suspended at its top by a post set in the bottom of the case. The circular framework is restrained from vertical movement, but is free to rotate and may tilt to a certain degree in any direction. The card has letters for cardinal headings, and

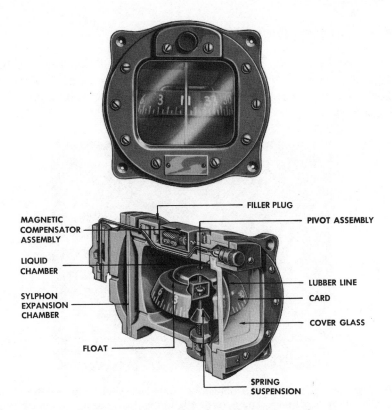

Figure 5.24. Cutaway and front view of panel-type magnetic compass.

each 30° is represented by a number, from which the last zero is omitted. Between the numbers, the card is marked for each 5°.

The card is visible through a window, with a vertical lubber line between the card and the window. The numbers on the near, visible side of the card correspond to the direction with which the case is aligned. The case, or bowl of the compass, is filled with acid free white kerosene, which dampens excessive oscillations, lightens, by buoyancy, the weight of the card frame on the pivot, provides lubrication, and prevents rust within the case.

Magnetic needles tend to point downward in middle and high latitudes, aligning themselves more completely with the earth's magnetic lines of force (flux) as they flow downward to the surface at the magnetic pole. This tendency, called *magnetic dip,* induces fluctuating errors in the compass during acceleration or deceleration, and during turning.

Acceleration-deceleration errors are most evident on east-west headings, and turning errors most evident on north-south headings. Errors are also caused by aircraft electrical equipment, radios, or even by a camera lightmeter placed near the compass.

TURN AND SLIP INDICATOR

The turn and slip indicator, often called the turn and bank indicator, was the first gyroscopic instrument. It now serves primarily to indicate a need for yaw trim, though it is a primary instrument in sailplanes and of value as back-up bank control in airplanes.

Figure 5.25. Electric turn and slip indicator mounted in a single degree of freedom gimbal. When the aircraft turns, the gyrorotor is turned, causing it to tilt left or right an amount proportional to the rate of turn. Mechanical and damping linkage is located between the rotor and instrument face. *(US Air Force photograph.)*

SLIP SINGLE-NEEDLE-WIDTH TURN

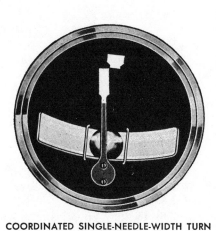

COORDINATED SINGLE-NEEDLE-WIDTH TURN   SKID SINGLE-NEEDLE-WIDTH TURN

Figure 5.26. Three indications of a turn and slip indicator in a left turn. *(Courtesy of US Air Force.)*

It consists of two parts, a turn needle and an inclinometer, or "ball." The turn needle indicates a turn when the gyro, horizontally mounted with its spin axis parallel to the aircraft lateral axis, tilts in precessive response to a turn or yaw of the instrument case. The gyro is usually electrically driven. Its rotational axis is parallel to the lateral axis of the airplane (a line through the center of gravity parallel to the wings). Turning the airplane causes precession of the gyro, restrained by small springs. The motion of precession is transferred to the turn needle by suitable linkage. The needle is held centered by the springs when the aircraft is not turning. The force of precession and the consequent amount of needle deflection vary directly with turning rate.

The *turn needle* indicates the rate (number of degrees per second) that the aircraft is turning, regardless of bank. It is calibrated for a single-needle-width deflection to indicate a *"standard rate"* of either 2 or 4 min. for an aircraft turn through 360°. With a full-needle-width deflection, the 2-min. needle indicates a rate of 3° per sec., and the 4-min. needle indicates a rate of 1½° per sec.

The inclinometer, or *ball,* restrained only by a stabilizing fluid in its tubular race, shows whether the aircraft is slipping or skidding, that is, the relationship between the angle of bank and the rate of turn.

Only gravity and centrifugal force act on the ball. During a coordinated turn these forces are in balance and the ball remains centered. In a skid, the rate of turn is too fast for the bank angle, and excessive centrifugal force moves the ball toward the outside of the turn. In slips to the inside of the turn, the cause and effects are reversed.

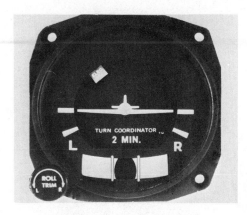

Figure 5.27. Turn coordinator.

When the aircraft is flying level but out of trim longitudinally, the ball will be off center; a rudder trim adjustment is needed and will center it.

A better pictorial presentation is provided by the turn coordinator (Figure 5.27). It is basically the same instrument as the turn and slip indicator, but the needle has two ends, representing wing tips, and is positioned horizontally, pivoting about its center.

ANGLE OF ATTACK INDICATOR

This instrument has become of primary importance in maneuvering high-performance aircraft, particularly those which experience violent maneuvers or great changes in gross weight.

Angle of attack is defined as the angle formed between the chord line of the wing and the relative wind (Figure 5.28). In its simplest form, the small stall warning vane which actuates a buzzer or light on the instrument panel of light planes is essentially an angle of attack indicator. Called a "stall warning" device, it operates when the airspeed is so low, related to load and relative wind, that airflow gets under the vane, lifts it, and closes the warning circuit.

At some angle of attack, every wing will stall. This stall angle will always remain the same for that particular wing, regardless of the gross weight, attitude, airspeed, or "G" load. High-lift devices, such as flaps, in effect change the wing, changing the wing area and chord line, so that use of flaps results in a different wing, a different stall speed, and different stall angle of attack.

When you consider that each wing has the same stall angle of attack for each maneuver, that is, final approach, climb, maximum range, maximum endurance, steep turns, etc., regardless of gross weight, it is apparent why the angle of attack is valuable information, eliminating the guessing associated with differing airspeeds for these maneuvers, under different weight conditions and "G" loadings. Since a lift requirement results from weight supported by the wing, the

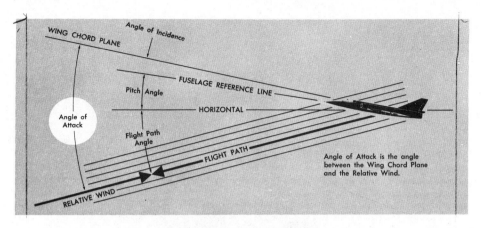

Figure 5.28. Angle of attack.

airspeed must be increased to maintain the same angle of attack at higher gross weights and at increased "G" loading. The only exception is that angle of attack indications may not be accurate during take-off roll and at lift off while in ground effect.

ANGLE OF ATTACK SENSORS

Both vanes and slotted probes which sense differential pressures are used along the fuselage where airflow disturbance is a minimum. Both may transmit a signal either directly to the instrument or to the central air data computer.

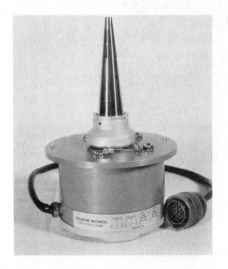

Figure 5.29. AOA—Sensor and indicator.

Figure 5.30. Standard instruments, Lockheed 1011 Airline Transport. *Left group:* Dual-pointer radio magnetic indicator (RMI); altitude director indicator (ADI); and horizontal situation indicator (HSI). *Right group:* Counter-drum-pointer encoding pneumatic-servo altimeter; combined airspeed-Mach indicator; true airspeed indicator; vertical-speed indicator; and outside air temperature indicator. All air data derived from a CADC. *(Courtesy of Sperry Flight Systems Division.)*

Cockpit indicators of angle of attack vary. They may be graduated in actual angles, units, symbols, or percent of lift being used, or as a "fast-slow" indication. Both round dials and vertical tapes (Figure 5.44 and 10.3) are used.

Navigation Instruments—Aircraft Radio

The third group of instruments includes the visual omni range (VOR) course deviation indicator (CDI), the radio magnetic indicator (RMI), the radio compass or automatic direction finder (ADF), radar transponder, marker beacon, and flight director. Glide slope indicator is usually included in a CDI. Indications may be combined in a single instrument, *e.g.*, marker beacon lights in the CDI, heading, VOR, and glide slope indications in the Horizontal Situation Indicator (HSI), and attitude, performance, and navigation in the flight director.

Navigation instruments complete the picture by showing the position of the aircraft relative to fixes it passes en route, and relative to destination, including both radio fixes and inertial navigation data, if the aircraft is so equipped. In essence, they provide information the pilot needs for guidance of the aircraft to the desired destination. All but inertial inputs are dependent on ground and aircraft radios, operating according to the characteristics of the particular stations.

Radio is so essential to modern aircraft that it is literally as much a part of the aircraft as the power plant. Modern air traffic control has developed to meet the needs of increased traffic congestion of high- and low-performance aircraft under all kinds of weather. It has, of course, been entirely dependent upon the flexibility and efficiency of air-to-ground radio communications equipment, radio aids to navigation, and the skill of the airmen using them.

RADIO ENERGY CHARACTERISTICS

While most people have some notion of radio theory, it is important for airmen to understand the following basic ideas.

Radio waves, although lower in frequency, are similar to light and heat waves and can be reflected, refracted, and diffracted. The velocity of travel is approximately 300,000 km (186,000 miles) per sec. in free space. When the wave amplitude is plotted against angle or time, the curve is normally a sine function (Figure 5.31).

A *cycle* is the complete process of starting at zero, passing through two maximums of opposite direction, and returning to zero again.

Frequency is the number of cycles completed in 1 sec. One cycle per second is the basic unit of measurement, the *Hertz*. The terms *kilohertz* (1000 cycles, abbreviated kHz) and *megahertz* (1000 kilohertz, abbreviated mHz) are applied in discussing radio frequencies.

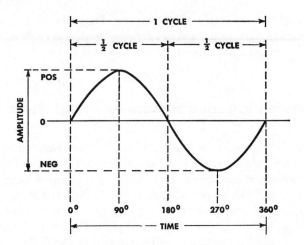

Figure 5.31. Radio wave characteristics.

Wavelength is the distance traveled by a wave during one cycle. This distance is normally expressed in meters.

Modulation is the process of creating a variation in the amplitude, frequency, or phase of the radio wave with another waveform, such as speech, to transmit intelligence.

Bandwidth is the amount of frequency spectrum required to transmit the desired intelligence. For example, a television signal, which transmits both visual and audio information, has an average bandwidth requirement of 6 mHz. Other types of information, such as voice or music, require a bandwidth of 3 to 30 kHz, depending upon the quality or fidelity desired.

Audio waves and the associated frequencies are those detectable as sound by the average human ear. The velocity of travel in air is approximately 1180 ft. per sec. or 663 knots. The frequency range extends from 20 to 15,000 Hertz (Hz), subject to variation in hearing of different individuals.

Radio communications of many types are conducted in a frequency range of 10 kHz to 30,000 mHz. This overall frequency range is divided into bands, which assists in identifying general frequency characteristics and the equipment which applies. These bands are classified as follows.

VLF	Very low frequency	10 to 30 kHz
LF	Low frequency	30 to 300 kHz
MF	Medium frequency	300 to 3000 kHz
HF	High frequency	3 to 30 mHz
VHF	Very high frequency	30 to 200 mHz
UHF	Ultrahigh frequency	200 to 3000 mHz
SHF	Superhigh frequency	3000 to 30,000 mHz

TABLE 5.1. Characteristics of Radio Waves.

	Frequency						
	3 kHz	30 kHz	300 kHz	3 mHz	30 mHz	300 mHz	30,000 mHz
Designation	VLF	LF	MF	HF	VHF	UHF	SHF
Maximum range (miles)	World-wide	3000	5000	12,000	Line-of-sight (sporadic long ranges)	Line-of-sight	Line-of-sight
Propagation	Ground wave and D-layer reflection	Ground wave and D-layer reflection	Ground wave and E-layer reflection	F-layer reflection	Sporadic ionosphere reflection	Sporadic atmospheric reflection	Sporadic ducting, some atmospheric absorption
Applications	Communication, long-range navigation	Communication, navigation	Communication, navigation	Communication, navigation, control, medical	Communication, navigation, television, control, relay, radar, industrial, medical	Communication, navigation, television, control, relay, radar, medical	Communication, navigation, control, relay, radar, industrial, nuclear resonance

Basically, a radio frequency signal or wave is generated by equipment known as a transmitter. This signal is then coupled to a suitable antenna which in turn radiates the energy into space. Once radiated, the signal is detected or intercepted by equipment known as a receiver.

A radio signal that is not modulated is known as a "continuous wave" (cw). In order to transmit intelligence, however, the signal must be altered in some manner and these alterations decoded at the receiver. One of the most simple methods of transmitting intelligence is to interrupt the signal at different rates and sequences to conform to a code for letters, numbers, and punctuation. The Morse code is, of course, an excellent example of this method of transmission.

Once some form of modulation is applied to the "continuous wave," it then becomes known as a "carrier." If the modulating signal varies the amplitude of the carrier, this is called "amplitude modulation" (AM). "Frequency modulation" (FM) indicates that the carrier is varied in frequency. "Phase modulation" (PM) indicates that the phase of the carrier is varied. The most common form of modulation used in aircraft radio transmitters is AM.

To intercept a signal and reproduce the transmitted intelligence, the receiver must be tuned to the correct radio frequency, amplify an overall bandwidth equivalent to that of the transmitted signal, and provide for the appropriate type of demodulation. "Demodulation" is the reverse process of modulation and performs the task of extracting the intelligence from the carrier.

General Characteristics of Propagation.   A nondirectional radiator (antenna) in free space will radiate radio frequency energy in all directions. Part of the energy travels along the earth's surface and is called the "ground wave." The balance of the energy is radiated into space and is referred to as the "sky wave." Particles in the atmosphere, the earth itself, terrain features, and structures all act to absorb or attenuate radio frequency energy. The amount of attenuation increases with distance from the radiating source.

Ionospheric Reflection.   The ionosphere is composed of a number of ionized gas layers that exist above the surface of the earth. The height and density of these layers vary with the amount of radiation received from the sun. Thus a pattern of variations arises between daylight and darkness, with the changing seasons of the year, and during periods of sunspot maxima. Certain radio frequencies, particularly those in the HF band, are subject to reflection by one or more of the ionosphere layers with only slight attenuation. This creates a condition in which the sky wave of an HF radiator is reflected back to the earth at some distant point with appreciable strength. The distance from the transmitting location to the first point of return to the earth is called the "skip distance." Since the earth is also a reflective surface, multiple skips are not uncommon.

Interference.   Anything that acts to degrade the quality or intelligibility of a signal at the receiver location may be classed as interference. *Man-made interference* is that caused by poor power connections or an arcing motor brush. *Natural interference* is caused by thunderstorms, charged rain or snow particles, and general atmospheric noise. *Mutual interference* is caused by other transmissions on or closely adjacent to the same frequency. When deliberate, this is called "jamming."

General Characteristics of Various Frequency Bands. There are many variations within a designated band, and the range or direction of transmission may be limited or enhanced by factors of power, antenna configuration, and type of modulation. However, general characteristics may be summarized as follows:

VLF: Almost entirely ground wave. High power and large efficient antennas can provide world-wide coverage. This is used by Omega, a recently developed long-range (world-wide) navigation system.

LF: Primarily ground wave. Used for homing beacons. Subject to sky-wave reflections during dark hours and unstable sky-wave reception at dawn and dusk. Vulnerable to natural interference.

MF: Primarily ground wave. Standard broadcast stations are included in this band and can be used as homing beacons. Subject to medium- and long-range sky-wave reflections during hours of darkness, especially during winter.

HF: Ground wave attenuates rapidly. Reflected sky wave is normally employed and useful for communications up to 12,000 miles or more, depending on ionospheric conditions. It is used mostly by long-range aircraft for position reporting, receipt of weather information, change in flight plans, etc., to and from distant ground stations. It may also be employed by light aircraft operating beyond VHF and UHF facilities. Bush pilot operations are an excellent example.

VHF: Practically no ground wave. Line-of-sight range via sky wave. Normally penetrates ionosphere as opposed to strong reflection. This is the principal air-to-air and air-to-ground frequency band for civil aviation. Voice communications and navigational aids are included. Since line-of-sight is a basic restriction, the useful range increases with an increase in aircraft altitude, up to about 200 nautical miles or 300 km.

UHF: Frequency characteristics closely parallel those of VHF. Utilization of this band for air-to-air and air-to-ground communications was brought about by the need for more channels and the severe congestion existing in the allotted portion of the VHF band. Most airport facilities and virtually all FAA stations now have both UHF and VHF capabilities. UHF is used primarily by the military.

SHF: Very little application to aircraft radio at the present time. Line-of-sight characteristics subject to some atmospheric absorption.

ASSIGNMENT OF AERONAUTICAL FREQUENCIES

In the United States, VHF and UHF communications and navigation frequencies are assigned generally as follows:

Visual Omni Range (VOR) and ILS Localizer	108.00 to 117.9 mHz
VHF Communications	118.00 to 135.95 mHz
Instrument Landing System (ILS) Glide Slope	328.60 to 335.40 mHz

All odd-tenth mHz frequencies beginning with 108.1, that is, 108.3, 108.5, etc., through 111.9, are localizer (LOC) frequencies. The even-numbered-tenth frequencies from 108.00 through 112.00 and all frequencies higher through 117.9 are VOR only. Seventy-five megahertz is used solely for marker beacons and Z markers.

UHF frequencies between 225.0 and 400.2 are used for air traffic control, and are assigned differently from time to time depending upon the development of air traffic control facilities and traffic loads. They are assigned and changed so as to provide the least likelihood of mutual interference between nearby stations.

Nondirectional radio beacons use low frequencies between 200 kHz and 550 kHz.

Radio frequency assignments change. Current ones are listed in the FAA-published "Airman's Information Manual," and on current navigation charts, as well as in nongovernmental publications, such as Jeppseson J-Aid.

VHF OMNIRANGE (VOR)

VOR is the primary instrument navigation system because its frequencies avoid the effects of static, and because it can give a detailed, precise display of the aircraft's position by easily read instruments.

Principle of Operation. Two VHF radio signals are transmitted from the same facility. One is constant in phase throughout 360° of azimuth, while the other, a variable-phase signal, is transmitted in a rotating signal pattern. Rotating at 1800 rpm, this variable phase combines with the constant phase to provide a signal which varies uniformly throughout 360°. The phase differential is oriented with magnetic north, so that aircraft receivers can measure the phase differential electronically and present it visually to the pilot in the cockpit.

The infinite number of courses which radiate from the VOR station are called radials. They are identified by their magnetic bearing outbound from the station. Thus, regardless of your heading, if you were momentarily due east of the sta-

TABLE 5.2. Line-of-Sight Distances for VHF Reception.

Aircraft Altitude (ft.)	Approximate Transmission Range (nautical miles)	Aircraft Altitude (ft.)	Approximate Transmission Range (nautical miles)
100	12	4,000	70
200	15	5,000	80
400	25	8,500	100
600	30	10,000	115
800	35	12,500	125
1,000	40	15,000	135
2,000	50	17,500	145
3,000	65	20,000	160

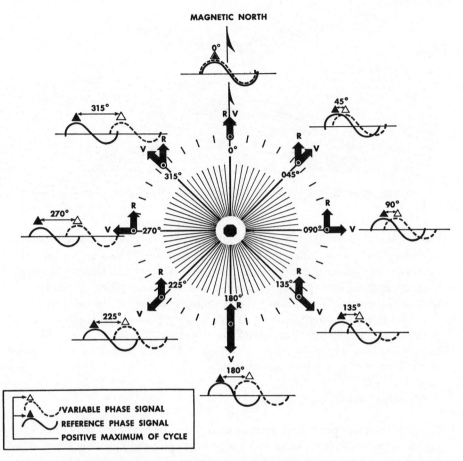

Figure 5.32. VOR phase angle relationship. Signals are in phase at magnetic north and vary elsewhere around the station. *(Courtesy of US Air Force.)*

tion, you would be on the 90° radial. If inbound on this radial, your course would be 270°. Because VOR uses frequencies in the VHF band, signals are limited to line-of-sight reception, and the usable range varies with altitude.

Station Classification. VOR stations are classified according to the altitude and interference-free distance they are able to serve, as shown in the following table:

Class	Altitude	Distance
H	Up to 45,000	130 nm.
H	Above 45,000	100 nm.
L	Up to 18,000	40 nm.
T	Up to 12,000	25 nm.

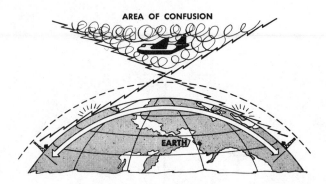

Figure 5.33. Area of confusion resulting from co-channel interference.

Identification.   VOR stations may be identified by a coded three-letter station identifier, a voice identifier, or a combination of both. Stations on the same frequency are spaced to avoid interference. However, with the increased density of installations, it is possible, at certain locations and altitudes, to receive two stations with approximately equal signal strength. The resulting "area of confusion" is shown in Figure 5.33. You can recognize it by oscillation of the instrument panel indicator and by an aural whistle. Eliminate it by selecting different stations along the route. Eventually .05 mHz frequency intervals are planned, to reduce interference problems.

TACAN, VORTAC, AND VOR-DME/T

The need for more precise all-weather navigation for both civil and military flying has required the development of equipment to give the pilot a direct picture of his bearing and distance from a station. The result is TACAN (tactical air navigation) equipment and its near relatives, VORTAC (co-located VOR and TACAN) and VOR-DME/T, a co-located VOR and DME portion of TACAN.

TACAN-equipped aircraft can receive both bearing and distance information from TACAN and VORTAC stations. An aircraft equipped with VOR and DME equipment can receive bearing and distance information from VORTAC stations, and distance information from TACAN stations.

Flight procedures for using VORTAC are generally the same as those used for VOR and TACAN. TACAN ground equipment consists of receiver-transmitter combinations (transponders) and rotating-type antennas for transmission of bearing and distance information. TACAN stations, called beacons, have a practical receiver-limited range of 195 nm., and are identified by a coded signal which is repeated every 30 sec. TACAN and VORTAC information appear in the cockpit on the navigation instruments and flight director systems.

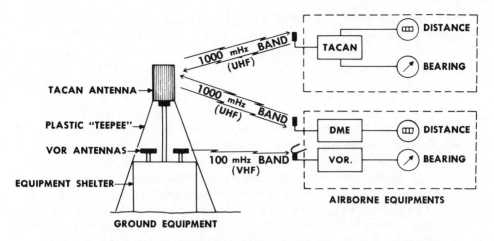

Figure 5.34. Utilization of VORTAC equipment and frequencies. *(Courtesy of US Air Force.)*

TACAN BEARING INFORMATION

As far as the pilot is concerned, the information might be coming from a VOR, except that TACAN's signal pattern is somewhat more complex and gives greater bearing accuracy. In addition, the use of UHF frequencies makes it less vulnerable to terrain effects. Using UHF frequencies, TACAN requires a much smaller ground-station antenna. TACAN antennas consist of a vertical wire omnidirectional signal transmitter around which rotate two concentric cylinders. These cylinders have embedded in their walls "parasitic elements" (vertical wires), which distort the signal pattern into fine and course phases as the cylinders rotate. One stage of the phases is oriented with magnetic east. The airborne receiver measures the time differential between the phases of the signal, giving bearing information which is accurate to within ±1°.

Aircraft Radio Equipment

The versatility of modern aircraft and the necessity for a high order of navigation precision and communication clarity and reliability have resulted in the development of the extensive airways and traffic control systems which now exist world-wide. To use this system, aircraft communications and navigation equipment have been developed in a pattern which follows aircraft cost and use. It may be grouped generally into systems for utility and pleasure aircraft, executive or business aircraft, and airline transports. Military equipment embraces all classes, depending upon the complexity of the mission. There is a wide variety available for each class. The distinction between classes is a matter of versatility, reliability, precision, and cost.

COMMUNICATIONS AND NAVIGATION EQUIPMENT

Because the selection of this equipment depends so much on the locale of aircraft operations, the aircraft use, and the probable pilot capability, a precise list is inappropriate. However, the network of airways in the United States is such that even while not following airways, reliance unpon VHF omnirange-type stations as the principal VFR radio aid to navigation is adequate. An ADF is helpful in areas (Canada, Alaska, Mexico, mountainous terrain) where VHF stations are unavailable, or where HF nondirectional beacons and radio broadcast stations are generally used. A popular and typical VFR installation is shown in Figure 5.35.

For IFR capability the installation shown in Figure 5.36 is considered minimal. It is not only essential to be able to communicate easily and navigate precisely. One must be able to do so efficiently in the air traffic control environment if the aircraft is to provide acceptable utility in most parts of the United States.

Combined COM/NAV equipment consists of the COM transceiver (transmitter and receiver), and the NAV receiver. Both receive voice broadcasts, using either headsets or cockpit speakers.

From this basic idea, in which VHF communications and navigation are com-

Figure 5.35. VFR radio package.

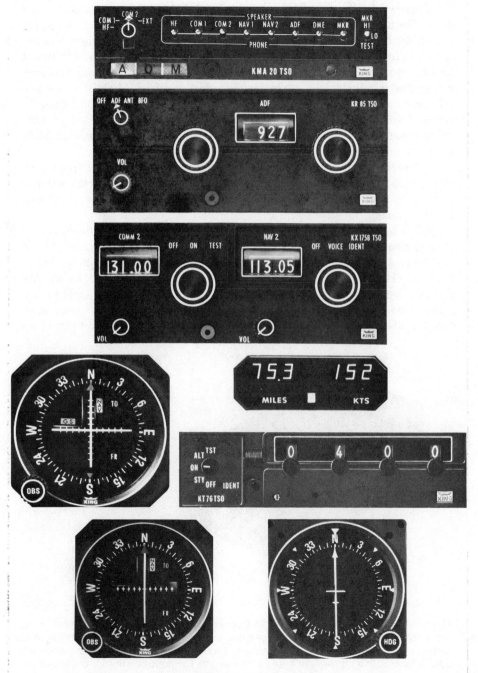

Figure 5.36. IFR radio package.

bined, grows the concept of separate and dual NAV/COM receivers and control units. The more versatile units have refinements which increase their reliability and utility: Manual and automatic squelch circuits improve readability by reducing background noise. Transistors instead of vacuum tubes are used to reduce heating and size, and to lengthen service life. Remote controls and indicators permit dual installation and use by multiple crew members. Modular design permits quick removal for test and repair of circuits. Frequency controls are designed into the circuits to prevent audio frequency distortion and to limit interference caused by unstable radio frequencies. Selectivity is increased: Many more crystal channels are available and they can be turned faster.

UHF radios are used for communications to greatly expand the frequencies and to reduce static. Separate audio control panels permit the pilot to receive or transmit on more than one radio at a time, or to select various receivers for monitoring or tuning.

UHF is used for TACAN and DME, and for military communications.

Long-Range Communications.   Aircraft flying over the ocean or remote areas of the world use high frequency (HF) for communications because of the limited range of VHF and UHF frequencies. HF radio signals (3 to 30 mHz) travel great distances when radiated. The signal leaving an HF transmitter splits into two segments. One part remains near the earth's surface and is called a *ground wave,* traveling about 100 miles. A second travels into outer space, but is reflected back

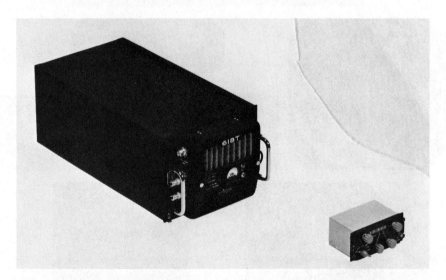

Figure 5.37 High Frequency (HF) Radio Operating in the 3 to 30 mHz frequency band, these radio waves travel long distances because of ionospheric skip, enabling voice communications to be carried over several thousand miles. The receiver-transmitter (left) is located in a convenient area of the aircraft. The control unit (right) is located in the cockpit. *(Courtesy of Collins Avionics Division, Rockwell International)*

by the ionosphere. These reflected signals are called *sky waves* and are responsible for long-range communications.

The ionosphere's reflective properties vary with time and season. Night time results, for example, in an increase in reflectability, which is why a distant AM commercial broadcast station can often be heard at night but not by day. Also, long-distance properties of HF can cause disappointing or nonexistent voice communications due to skip and interference.

Improved performance can usually be gained by using higher frequencies (10 to 30 mHz) during the day and lower frequencies (3 to 10 mHz) at night. At these times, the respective frequencies are more stable.

The current development and eventual deployment of satellite communications in the VHF and UHF bands will likely decrease reliance on HF.

Navigation receivers and indicators are more complex, have more positive test and warning flag systems, and are combined with directional indicators. The indicators also give more stable indications of bearing and ILS position. They are combined with glide slope receivers to permit full ILS use. Indicator and control head dials are designed and illuminated with emphasis on human factors affecting error.

Figure 5.30 shows the *RMI,* which consists of a rotating compass card, a double-barred bearing pointer, and a single-barred bearing pointer. The rotating compass card operates from the aircraft's master compass system and is independent of the VOR receiver. The compass card rotates as the aircraft turns so that the magnetic heading of the aircraft is always under the index at the top of the instrument.

The double-barred bearing pointer gives magnetic bearing from the aircraft to the VOR station to which the receiver is tuned. If the aircraft is headed directly toward the station, the head of the double-barred pointer and the magnetic heading of the aircraft will both be directly under the index at the top of the instrument. Should the aircraft be turned 90° to the right, the compass card will rotate 90° to the left, and the double-barred pointer will also rotate 90° because it always points toward the station. Thus, the magnetic bearing from the aircraft to the station is always shown under the head of the double-barred pointer.

The single-barred pointer operates in the same way. It is connected to an ADF or may be used with a dual VOR receiver installation, or a TACAN receiver.

The course deviation indicator of Figures 5.35 and 5.36 has a course set knob, the course selector window, a TO-FROM indicator, and a course deviation indicator (the vertical needle). It may also have OFF warning flags, a marker beacon light, and a glide slope indicator (horizontal needle). Some instruments also have a heading pointer, which shows the heading of the aircraft with respect to the selected course.

The course knob can be used to set any bearing to or from the station. Once a bearing is set, the TO-FROM indicator tells whether the selected course, if flown, would take the aircraft closer to or farther from the station. The course needle is unaffected by the heading of the aircraft. It shows position relative to the course set by the selector knob. If the aircraft is turned to the heading of the course selected, the needle shows the relative position, right or left, of the bearing. With the needle centered, the aircraft is on the bearing or its reciprocal, no matter what the aircraft heading.

The dots in a horizontal plane on the face of the instrument represent deflection on either side of the on-course position. A full-scale deflection indicates 10° or more from the selected course.

The glide slope indicator is operative only when an ILS frequency is selected; it will be centered with the alarm flag displayed when the receiver is tuned to VOR frequencies.

The two red alarm flags on the CDI and glide slope indicators are spring-loaded to the OFF position and operate independently of each other. When the flag disappears from an individual indicator, it shows you that the appropriate indicator is receiving a strong enough signal for reliable indications.

When a VOR is transmitting an abnormal or erroneous signal, the station identification is not transmitted. In this case the receiver may be receiving an erroneous signal strong enough to keep the alarm flag from showing. Therefore, the indicator is reliable only when the alarm flag is not showing *and* the station identification is being received.

DISTANCE MEASURING EQUIPMENT (DME)

To determine distance, an airborne set transmits an interrogation pulse which the ground station receives and answers with a pulsed reply. The airborne equipment measures the time interval between the interrogation and reply; this interval is in proportion to the airplane's distance from the station. In order to prevent an airborne receiver from mistaking ground replies intended for other aircraft, the repetition rate of each airborne transmitter is wobbled or varied slightly to give it an instantaneous repetition rate which differs from that of other aircraft in the area.

When the receiver is tuned to a station, it first must go through a "search mode," in which it looks for ground-station replies which are effectively synchronized with its own repetition rate. This searching process may take up to 20 sec., and during this time the range indicator distance drums rotate rapidly through the whole range. When a synchronized signal is found, range-gate circuits lock onto it, the set goes into its normal operating "track mode," and the distance drum stops at the distance to the station. If the TACAN DME signal is momentarily cut off, as when the aircraft banks and screens the receiving antenna, the range indicator will show the same distance for 10 sec.; beyond that time, the receiver reverts to the search mode and starts a new search cycle.

DME is accurate within ±600 ft., plus .2% of the distance being measured. Because DME measures slant range, the error is greater close to the station. At 12,000 ft. and 10 nm., the error would be only 720 ft. If the aircraft crossed the

Figure 5.38. DME receiver.

station 12,000 ft. above it, however, the minimum distance indication would be 2 miles (6080 ft. per nm.).

Recent electronic developments in DME receivers have reduced search time to near zero and provide greater accuracy. Read out is electronic, with no moving parts.

THE AUTOMATIC DIRECTION FINDER (ADF)

The ADF is a homing and direction-finding radio using the transmissions of commercial broadcast stations, and nondirectional radio beacons (NDB). Geographical locations, identification, and frequencies of these stations are listed in aeronautical publications and on aeronautical charts.

Operating in the low frequency band (190 to 1750 kHz), the ADF is susceptible to atmospheric disturbances and other natural phenomena which can affect the accuracy and reliability of the bearings indicated. Consequently, the ADF is not used as a primary aid to navigation except where no other aids exist. In the United States it is used primarily as a supplemental radio aid, providing bearings to outer compass locators and nondirectional radio beacons. NDB's are useful in locating, and in making instrument approaches to, smaller low-density airports.

Limitations.   Problems inherent in ADF operation (radio frequency phenomena) are:

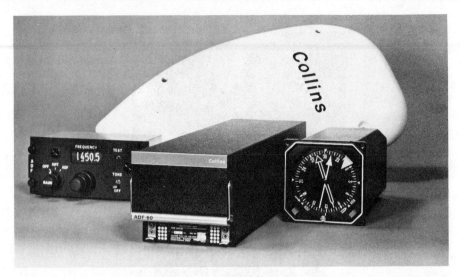

Figure 5.39. Automatic direction finder (ADF) components. *Left:* control unit for selecting desired station frequency and operational controls. *Middle:* receiver unit which can be mounted in any convenient location. *Right:* bearing indicator. Unit shown is an RMI indicator. A bearing indicator with a single needle and a fixed or manually rotatable compass card may be used. *Background:* the solid state loop antenna also contains the sense antenna, simplifying installation. (*Courtesy Collins Avionics Division, Rockwell International.*)

(1) *Station Overlap:* Occasionally more than one station will be assigned the same frequency. Certain conditions (night effect, for example) may cause signals from such stations to overlap. Sunspots and atmospheric phenomena may distort reception and also cause overlap.

(2) *Electrical Storms:* In the vicinity of an electrical storm, an ADF pointer may point to the center of the storm.

(3) *Night Effect:* This is particularly strong just after sunset and just before sunrise. An ADF pointer may swing erratically at these times.

(4) *Mountain Effect:* Radio waves reflecting from the surfaces of mountains may cause the ADF pointer to fluctuate or show an erroneous bearing.

(5) *Coastal Refraction:* Radio waves may be refracted (bent) when passing from land to sea or when moving parallel to the coast.

Equipment.   Advancing technology has resulted in easier use by the pilot and increased reliability, by minimizing many of the operational problems associated with earlier ADF's, including those listed above.

An ADF consists of a *receiver, control unit, loop antenna, sense antenna,* and *bearing indicator.* The receiver and control unit may be combined in one panel unit, and the sense antenna may be combined within the loop antenna, simplifying installation.

Receiver. New ADF's use solid state electronic design, decreasing weight and size, and increasing reliability. Digital tuning with crystal control eliminates manual fine tuning.

Control Unit. This unit contains all the controls for operating the ADF. The *function selector* is used to select: OFF, ADF, ANT, and BFO. The ADF position provides normal ADF operation. Signals from both the loop and sense antennae are used to determine and display the bearing to the station relative to the aircraft heading. ANT is selected for best audio quality when making positive identification of a station or for program enjoyment. The ANT position of the selector switch provides for maximum signal reception, bypassing the directional future. BFO (beat frequency modulation) is selected when using an unmodulated signal. In the United States station transmissions are modulated, and this position is not used. To identify an unmodulated station, BFO is selected and the code underlying the tone will be heard. This is extensively used by European stations.

The *tuning knobs* are used to select the desired frequency. With frequency selected, the function switch in the ANT position, and volume turned up, the station is identified by checking its code call letters. With station identified, the function switch is set on the ADF position, and volume is turned down if desired. The indicated bearing is then usable for navigation purposes.

The bearing indicator may be built in as part of the receiver/control unit with a fixed or rotatable card or may be separately located, as with an RMI. The needle shows the angular position of the loop antenna in relation to the longitudinal axis of the aircraft. The top index represents the nose of the aircraft. The bearing indicator always indicates relative bearing from the nose of the aircraft.

Loop Antenna. This is the directional element of the ADF system. A loop

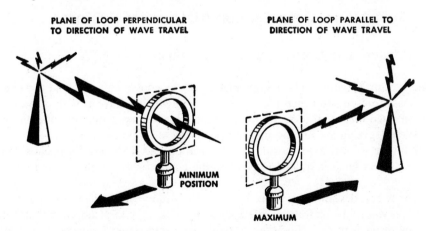

PLANE OF LOOP PERPENDICULAR TO DIRECTION OF WAVE TRAVEL

PLANE OF LOOP PARALLEL TO DIRECTION OF WAVE TRAVEL

MINIMUM POSITION

MAXIMUM

Figure 5.40. Minimum and maximum positions of loop antenna. The minimum (null) position is the loop's position during ADF operation. The modern ADF loop antenna is totally solid state with no moving parts. Electrically, it functions in the same manner as the earlier rotatable type loop shown above.

antenna gives maximum reception when the plane of the loop is parallel to the direction of electromagnetic wave travel. Minimum reception or a *null* is produced when the plane of the loop is perpendicular to the direction of wave travel. The needle of the bearing indicator is synchronized with the null of the loop, that is, the needle points perpendicularly to the plane of the loop.

Sense Antenna.   When the loop is in the null position, the radio station received is on a line perpendicular to the plane of the loop. However, the station may be on either side. The inability of the loop to determine which of the two possible directions is correct is called the 180° ambiguity of the loop. The 180° ambiguity is resolved by the use of a nondirectional *sensing antenna*. When the signal strength pattern of the sense antenna is superimposed on that of the loop antenna, there is only one null position of the loop. These combined signals energize a phasing system which operates the ADF indicator needle in synchronization with the loop, indicating bearing to the station.

Development of the ADF system has progressed in concert with contemporary technology. The receiver unit is totally solid state, employing crystal tuning. The rotatable loop antenna has become a solid state flat plate (¾ in. thick) antenna with no physical moving parts. The external wire sense antenna (clothesline) has often been concealed in the aircraft structure, and recent technological developments have combined the sense antenna within the structure of the loop antenna. These and other refinements have improved performance and reliability, assuring the ADF of a viable place in aircraft navigation for many years to come. A discussion of ADF procedures appears in Chapter 9.

Advanced Navigation Instrument Systems

PICTORIAL-SYMBOLIC PLAN VIEW SYSTEMS

This system (Figure 5.41), a combination of the course indicator and the gyro compass, presents in visual planform the position of the aircraft with respect to the selected course. It greatly simplifies the pilot's computation and orientation problems, giving a clear and instant orientation at all times.

With the VOR receiver tuned to a VOR station, the NAV flag disappears. The course select knob is rotated to position the course arrow on the desired radial heading.

The lateral deviation bar represents the VOR radial and indicates its position with respect to the aircraft by reference to the miniature airplane symbol etched on the instrument face. Each dot of deflection from center indicates 2° aircraft displacement from the radial, and the expanded scale bar an additional 6°.

To intercept a radial, the pilot turns the aircraft until the miniature aircraft is headed toward the bar at an acceptable interception angle; in Figure 5.41 the

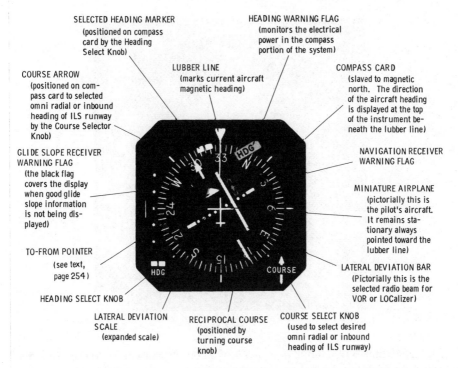

SELECTED HEADING MARKER
(positioned on compass
card by the Heading
Select Knob)

HEADING WARNING FLAG
(monitors the electrical
power in the compass
portion of the system)

COURSE ARROW
(positioned on com-
pass card to selected
omni radial or inbound
heading of ILS runway
by the Course Selector
Knob)

LUBBER LINE
(marks current aircraft
magnetic heading)

COMPASS CARD
(slaved to magnetic
north. The direction
of the aircraft heading
is displayed at the top
of the instrument be-
neath the lubber line)

GLIDE SLOPE RECEIVER
WARNING FLAG
(the black flag
covers the display
when good glide
slope information
is not being dis-
played)

NAVIGATION RECEIVER
WARNING FLAG

MINIATURE AIRPLANE
(pictorially this is
the pilot's aircraft.
It remains sta-
tionary always
pointed toward the
lubber line)

TO-FROM POINTER
(see text,
page 254)

HEADING SELECT KNOB

LATERAL DEVIATION BAR
(Pictorially this is the
selected radio beam for
VOR or LOCalizer)

LATERAL DEVIATION
SCALE
(expanded scale)

RECIPROCAL COURSE
(positioned by
turning course
knob)

COURSE SELECT KNOB
(used to select desired
omni radial or inbound
heading of ILS runway)

Figure 5.41. Pictorial-symbolic course indicator. (*Courtesy of Collins Radio Co.*)

aircraft is heading 330° to intercept a course of 300°. With no drift, the aircraft would by flying along the radial when the bar was centered and the miniature airplane aligned over it on a heading of 300°. The TO-FROM pointer indicates the course which, if selected and flown, will lead to the selected station.

While the bar is centered, the difference between the selected heading under the course arrow and the actual heading under the lubber line is the correction for wind drift.

With a glide slope receiver, glide slope information is shown on the left of the instrument by a glide slope arrow. This instrument, with integrated data, is called a horizontal situation indicator (HSI).

INTEGRATED FLIGHT DIRECTOR SYSTEMS

The attitude indicator gives the pilot a forward-looking view of his aircraft's position. The RMI and CDI combined in the pictorial-symbolic plan view system gives him a downward-looking, or planform, view of his position relative to a VOR radial. By putting these instruments one above the other, the pilot's cross-check problem is greatly simplified, but he still must practice to retain the fine

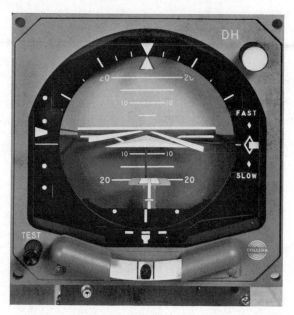

Figure 5.42a.

judgment which will tell him *how much* of a correction to make to remove a displacement from his desired position and altitude or rate of climb or descent.

The "director" portion of the system determines this for him. In the flight director mode, a symbolic airplane as in Figure 5.43 (or in other models, cross bars) derives its displacement from center from a computer which computes the

Figure 5.42b.

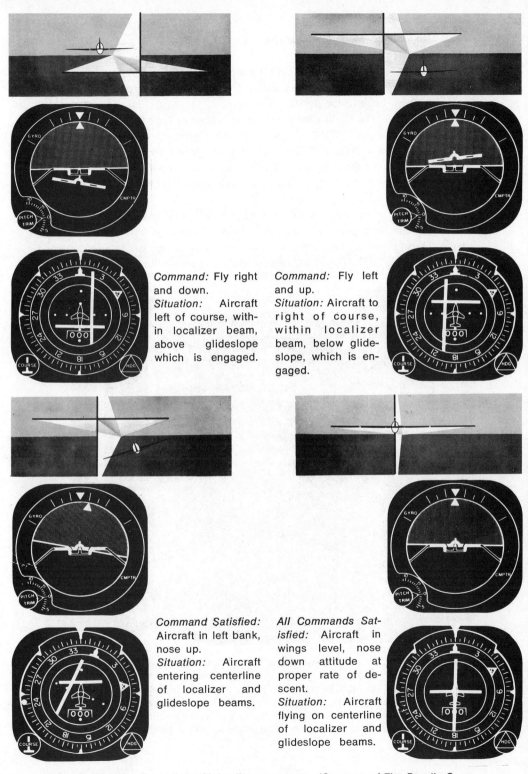

Command: Fly right and down.
Situation: Aircraft left of course, within localizer beam, above glideslope which is engaged.

Command: Fly left and up.
Situation: Aircraft to right of course, within localizer beam, below glideslope, which is engaged.

Command Satisfied: Aircraft in left bank, nose up.
Situation: Aircraft entering centerline of localizer and glideslope beams.

All Commands Satisfied: Aircraft in wings level, nose down attitude at proper rate of descent.
Situation: Aircraft flying on centerline of localizer and glideslope beams.

Figure 5.43. Indications of the flight director system. *(Courtesy of The Bendix Corp., Eclipse-Pioneer Div.)*

rate of return to a desired or command flight condition. The pilot merely changes pitch or bank attitude to center the symbol. By keeping it centered, the aircraft is gradually flown back to the desired condition.

Components.   The flight director system consists of four basic parts: an attitude director indicator (ADI), a course deviation indicator (CDI) or horizontal situation indicator (HSI), a flight instrument amplifier, and the flight steering computer. These components receive inputs from the aircraft's radio navigation (VOR-LOC) system, from its vertical and rate gyros, and from its compass system.

Commands may be put into this system in level flight by setting an "altitude hold" function on a given altitude and setting the course arrow on a desired course. During an ILS, the commands are put in by tuning the ILS frequency, and by switching to the ILS mode of the system. Because the command steering

Figure 5.44. High-performance integrated flight instrument panel. This panel appears in the F-105, F-106, and the F-111. (A) Angle of attack-airspeed indicator. (B) Stand-by airspeed indicator. (C) Attitude director indicator. (D) Horizontal Situation indicator. (E) Vertical speed-altimeter indicator. (F) Stand-by altimeter. Note that in this ADI, the turn and slip indicator is mounted in the lower rim, and that the steering bars are the cross-pointer type. The attitude indicator horizon is tipped because there is no power on the instrument.

indication of the command bar in the ADI is more sensitive than the displacement information presented by the attitude indicator and the CDI or HSI, a relatively small attitude change will center the command bar, correcting the displacement at a smooth rate.

The flight instrument amplifier, a transistorized unit, amplifies the attitude, heading, and command signals to operate the servomechanisms in the ADI and CDI or HSI.

The flight steering computer is the "brain" of the system. It contains the transistorized computer circuits needed to use input signals from the gyros and radio receivers for computation of a combined command for display on the ADI.

HIGH-PERFORMANCE AIRCRAFT INSTRUMENT PANELS

The flight director system is coupled with vertical reading performance instrument tapes in Figure 5.44. The result is a system in which the pilot is concerned with matching only horizontal and vertical lines. "Command" means the value used as a guide for some flight condition, as take-off, climb, or cruise. The pilot sets this value. The vertical scale indexes show the degree to which the aircraft is meeting the desired values. Developed originally for supersonic fighters, this system is now used also in high-performance subsonic aircraft such as the C-141.

Vertical-Tape Engine Instruments. With the multitude of engine power and condition data that is vital to modern multi-engine aircraft, and the relatively limited time in which a pilot must make critical judgments based on readings of those instruments, the vertical-tape presentation is considered superior. This is clearly illustrated in Figure 5.45 in which the eye immediately detects the difference in values in the vertical instruments, while the round dials, even though all appearing alike, are reading quite differently.

RADIO ALTIMETERS

Possible errors in barometric altimeters ordinarily limit minimum instrument approach altitudes to 200 feet. The radio altimeter (Figure 5.46) makes a safe instrument approach to 100 feet above the ground. Operating below 2500 feet, the radio altimeter has an accuracy of $\pm2\%$ from 500 feet down to 80 feet, and $\pm1\%$ below 80 feet.

An FM transmitter sends continuous wave energy through a wide-beam antenna and receives reflected energy from the ground through another wide-beam antenna. The signal undergoes an apparent frequency change dependent upon aircraft altitude, when it is reflected. The apparent change is measured and converted into signals to drive the indicator on the pilot's panel. The signal can also be used to drive other indicators, such as a flight director, and a digital read out may be provided.

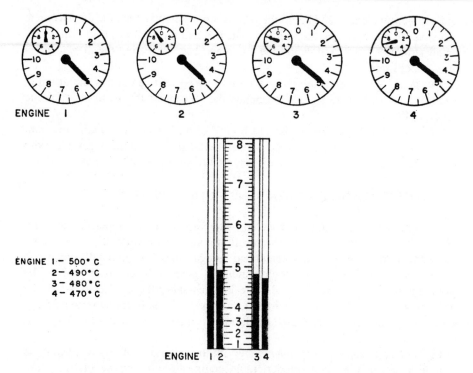

Figure 5.45. Comparison: round dial vs. vertical scale display. This illustration shows only one element of the complex information indicated in the jet engine panel of Figure 5.13. The simplicity and clarity of multiple critical indications are evident. *(Courtesy of The Bendix Corp., Eclipse-Pioneer Div.)*

Terrain Alert.   The pilot may set in an altitude above the terrain, and when the aircraft reaches it during descent, a visual and/or aural signal is activated.

From 200 feet on down, the signal may be used to activate a rising runway symbol in the flight director, more closely simulating the outside environment.

THE AUTOMATIC PILOT

In large aircraft and small, automatic pilots relieve the pilot of the task of maintaining course and often altitude, and provide more precision and accuracy for instrument approaches. The auto-pilot normally can do a better job of flying than can the human pilot, because it senses variations in heading, pitch, and trim more quickly than he can and applies more precise corrections. It reduces the work load of the pilot(s) and allows more attention to navigation and radio, and more time to look out for other aircraft. It can also be tied into navigation systems, making course following automatic.

Figure 5.46. Radio altimeter.

Operation. The basic idea of auto-pilot operation is this: A vertical gyro, such as the remote gyro for the attitude indicator, provides a stable reference platform for pitch and roll axes; a directional gyro, as for the heading indicator, provides a stable directional reference. To each of the three flight surfaces, elevator, aileron, and rudder, there are attached servo motors capable of moving the control surface. When the aircraft deviates from the attitude and heading in which the pilot placed it when he engaged the auto-pilot, synchros in the gyros signal the servo motors through a central coordinating computer to correct the deviation by a movement of the control surface. This is the basic auto-pilot. It may be designed around any or all of the three axes; some light plane auto-pilots are designed to maintain only the roll, or lateral, axis stable.

There are complications. A trim servo is required to insure that the elevator trim reflects shifting loads in the aircraft under various flight conditions. Without it, the elevator could gradually become held out of alignment under great stress, and a sudden disengagement of the servo could mean violent dive or climb.

To provide smooth operation, rate gyros (sensors) are required in the pitch, roll, and yaw axes to detect the rate at which the aircraft is deviating from its set attitude. This permits the auto-pilot to govern the rapidity and force of the correction, preventing over and under controlling.

Controls have to be provided to the pilot so that while the auto-pilot is engaged, he can make smooth turns to selected headings and can remain level, climb, or descend.

The basic auto-pilot system senses and controls attitude. To enable the pilot to select and fly at an altitude, a barometric sensor is provided in the more sophisticated systems. It signals the elevator servo and elevator trim servo whenever the aircraft deviates from the null established by setting an altitude.

When the pilot wants to follow a certain course, as an OMNI radial, a computer can be introduced which signals the auto-pilot to turn, to intercept, and to follow the radial. This function is expanded in many auto-pilots to permit automatic low approaches with ILS and glide slope receivers, controlling both course and altitude.

Controller.   The auto-pilot controller contains the various controls for operating the auto-pilot system. Included are: engage button, pitch control (climb-descend), turn knob, and other functions such as heading hold, capture, and track (the latter two modes provide automatic interception and tracking along a selected radio course). A single controller is often used for operating both the auto-pilot and flight director.

Using the Auto-pilot.   A pilot who studies and practices with the auto-pilot can get a great deal more from it and provide a smoother and more pleasant ride for passengers. Proper trimming of the aircraft before engaging the auto-pilot and

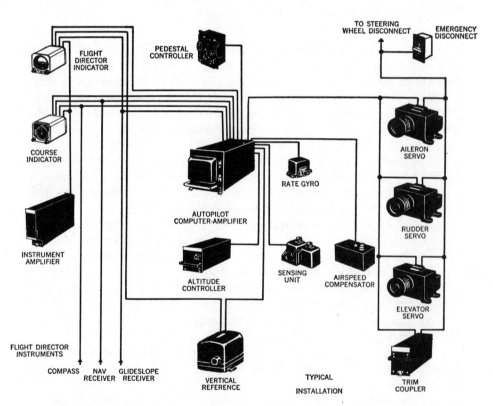

Figure 5.47. Auto-pilot system schematic. *(Courtesy of Collins Radio Corp.)*

monitoring aircraft trim after engaging are important. In strong updrafts and downdrafts, an excessively high or low airspeed can result with the altitude hold engaged. In turbulence, autopilots try to maintain too fixed an attitude, resulting in rough movements of the flight control surfaces and placing a high stress load on the aircraft and passengers. Modern auto-pilots often have a *soft ride mode* (lower auto-pilot sensitivity) for use in heavy turbulence. It is generally wise to hand fly the aircraft in such turbulence with emphasis on a level attitude and lesser concern for altitude.

TRANSPONDERS

The rapid increase in the number of aircraft utilizing the airspace and the increasing number of these aircraft operating in instrument flying conditions have rendered the basic radar system inadequate to provide separation while ensuring maximum safety and minimum traffic delays. To correct this problem a radar beacon system has been developed. It greatly improves the effectiveness of the radar system.

Air traffic control leans more and more heavily on the airborne radar beacon or transponder. The transponder responds to the interrogation of a ground interrogator used in conjunction with ATC radar, so that the blips on the controller's radar scope, representing certain aircraft, are reinforced and identified by the coded signal. This particular signal, transmitted as a reply, is selected by the air traffic controller, who assigns a given code to the aircraft by voice, for example "8883 Lima, squawk 1203." Individual aircraft responding to the same code may be identified by transmitting an identifier signal on the controller's request, "8883 Lima, ident."

By this method also, a small aircraft whose surface will not provide a suitable radar reflection on the controller's scope can be identified by its reply signal which appears on the scope at the aircraft's location. Because the controller then knows the aircraft's precise location, he can advise the pilot, thus providing an even more precise position than VOR with DME.

The transponder may also respond with the aircraft's altitude if equipped with an altitude encoder. An altitude encoder is required in Group I Terminal Control Areas and for flight above 12,500 ft., 2500 ft. or more above ground level (AGL) in the contiguous 48 states. Altitude encoding is Mode C, set by selecting "ALT" or "Mode C" on the control panel. This feature is discussed in more detail later.

The Air Traffic Control Radar Beacon System (ATCRBS) consists of three main components.

a. *Interrogator.* Primary radar relies on a signal transmitted from the radar antenna site, reflected or "bounced back" from an object (aircraft). This reflected signal is then displayed as a "target" on the controller's radar scope. No "on board" equipment is required by the aircraft for this procedure.

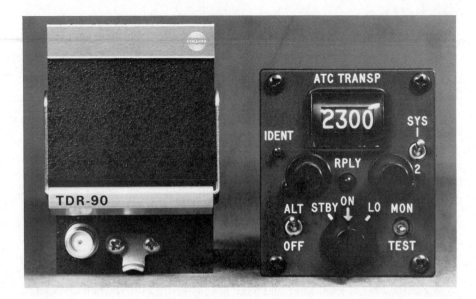

Figure 5.48. Transponder. This airborne radar beacon is a vital component in the air traffic control system. Each aircraft is assigned a specific four digit code when operating under instrument flight rules (IFR), enabling the ATC controller to maintain positive identification of the aircraft. The receiver transmitter above (left) is located in the avionic section of the aircraft, and the controller (right) is located on the instrument panel. *(Courtesy Collins Avionics Division, Rockwell International.)*

In ATCRBS the interrogator, a ground-based radar beacon transmitter-receiver, rotates with the primary radar antenna and transmits discrete radio signals which repetitiously request all transponders (on board aircraft), on the mode being used, to reply. The replies received are then mixed with the primary radar returns, and both are displayed on the same radar scope (Figure 5.49).

b. *Transponder*. The airborne radar beacon transmitter receiver automatically receives the signals from the interrogator and selectively replies with a specific pulse group (code) only to those interrogations being received on the mode to which it is set. These replies are much stronger and independent of a primary radar return.

c. *Radar scope*. The radar scope used by the controller displays returns from both the primary radar system and the ATCRBS. These returns (targets) are used by the controller in the control and separation of air traffic.

A part of the ATCRBS ground equipment is the decoder. This equipment enables the controller to assign discrete transponder codes to each aircraft under his control. Normally, only one code will be assigned for the aircraft's entire flight. This equipment is also designed to receive altitude information (Mode C) from aircraft that are equipped with an encoding altimeter.

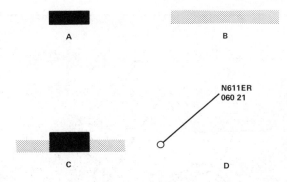

Figure 5.49. Radar targets.

ENCODING ALTIMETERS

The transponder provides great flexibility for the air traffic controller but is lacking in one important dimension—altitude. The encoding or servo altimeter in association with the transponder provides this vital link. It may be an additional feature of the primary altimeter, or a separate unit with no read out in the cockpit. One design is shown in Figure 5.50.

With an encoding altimeter and transponder, the altitude at which the aircraft is flying is converted in 100-ft. increments to electronic signals and, through the aircraft's transponder (Mode C), is sent to the ground interrogator/decoder along with the normal beacon response.

The ground receiving equipment analyzes these signals and converts them into alphanumeric data on the controller's radar screen.

With altitude information, the controller may sequence a greater number of aircraft through the airspace, increasing safety and significantly decreasing delays due to air traffic saturation. Mode C altitude information greatly reduces altitude verification requests as well.

As air traffic increases, the requirement for all aircraft to have altitude reporting capability may become mandatory.

TRANSPONDER OPERATION

Transponders must have 4096 code capability. For normal VFR operation, code 1200 is set in the window, and altitude encoding (Mode C) may also be used. The transponder selector switch is turned to STBY (Standby) for warm up, then turned to ON or ALT at take-off. It is turned back to STBY or OFF after landing. For instrument flight and at any time the pilot is using radar guidance or assistance, the code is set as instructed by ATC. The accuracy of the altitude encoder may be checked with the radar facility.

When requested to "Ident" by the radar controller, the pilot pushes his

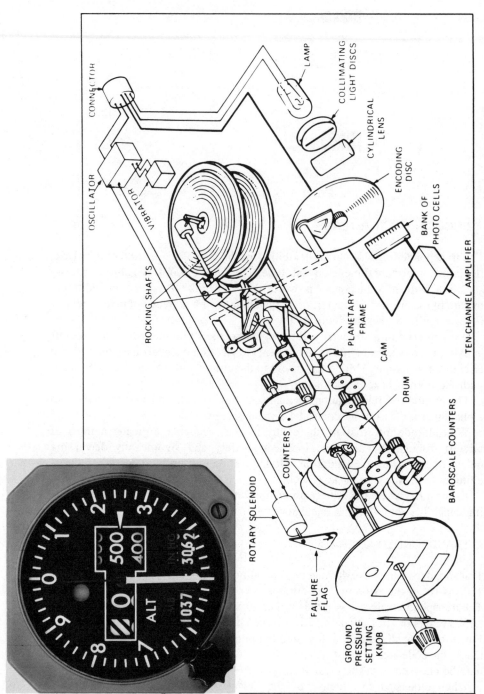

Figure 5.50. Schematic view of an encoding altimeter. Altitude is converted to electronic signals in 100 ft. increments. (*Courtesy Smiths Industries, Inc.*)

IDENT button. A distinctive return or blip then appears at the location of the aircraft on the scope, for about 1 min.

USES OF ATCRBS

VFR.   The transponder is set on 1200 unless otherwise instructed, and altitude encoding is selected if it is available. This enables ground controllers to assist air traffic by providing traffic separation and vectoring arriving and departing aircraft so as to promote a smooth flow of traffic. The controller will warn of other radar targets and give vectors.

IFR.   The transponder is set as directed by ATC, and altitude encoding is selected if the aircraft is so equipped. The same code may be used from take-off until arrival in the destination terminal area. Position reporting is virtually eliminated. Some assistance in avoiding severe weather is provided.

Emergency.   When the pilot has an airborne emergency, he sets the transponder emergency code, 7700, in the window. (In case of radio failure only, he squawks 7700 for 1 min. then goes to 7600.) Most radar facilities are now equipped to recognize 7700 as an emergency signal, and so signal the operator.

Transponder Phraseology.   The following phraseology is used in connection with operation of the transponder:

"Squawk 7352"—Set in code 7352 in transponder.
"Squawk Ident"—Activate the identification feature of the transponder. Instructions could be "Squawk 7352 and Ident."
"Squawk Stand-by"—Turn selector switch to "Stand-by."
"Squawk Altitude"—Turn selector switch to ALT or turn on Mode C, if so equipped.
"Squawk Mayday"—Set code 7700 in transponder.
"Squawk VFR"—Set code 1200 in transponder.

GROUND PROXIMITY WARNING SYSTEM

All air carrier aircraft (airlines) and many large corporate jets are equipped with ground proximity warning systems (GPWS). They are used to help prevent inadvertent flight into terrain.

The GPWS, using inputs from equipment already installed in most modern aircraft, automatically and continuously monitors the aircraft's flight path with respect to the terrain at all radio altitudes between 50 and 2450 ft. If the projected flight path would imminently result in impact, the system issues a warning to the flight crew.

Warnings are both visual and audible. The visual warning is a red light in the cockpit, labeled PULL UP. The audio warning, heard through the flight deck speaker and interphone, is a loud whooping signal plus the spoken words "Pull Up" repeated continuously.

GPWS Pictorial Diagram

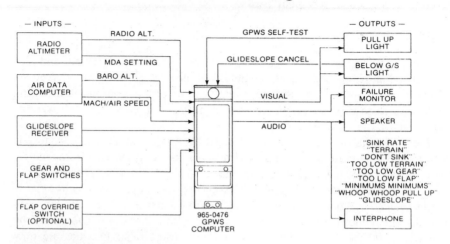

Figure 5.51. The ground proximity warning system is a monitoring system to prevent inadvertent flight into the ground. Inputs from several sources, as shown above, provide a warning only when an actual terrain emergency exists, remaining silent during all normal flight situations. The GPWS computer is shown at left. *(Courtesy Sundstrand Data Control, Inc.)*

If the aircraft deviates below the glide slope during an ILS approach, an amber light in the cockpit labeled BELOW G/S is illuminated, and the spoken word "Glide slope" is repeated on speaker and interphone. If the glide slope deviation continues to increase and altitude continues to decrease, the glide slope voice repeats faster and at a higher audio level.

Figure 5.52. Weather radar.

AIRBORNE WEATHER RADAR

Large aircraft have been equipped with airborne radar for a number of years. Recently, solid state technology has reduced the size, weight, power requirements, and cost, enabling virtually all twin engine aircraft and larger singles to be radar equipped.

Radar does not see clouds, thunderstorms, or turbulence directly. Rather, it detects water droplets within clouds and thunderstorms. It is these water droplets that provide the clue to potential turbulence, heavy rain, and hail.

Components.   Airborne weather radar consists of an *antenna* enclosed within a fiberglass *radome* (located on the wing or nose of the aircraft), a *receiver-transmitter,* and the *radar scope,* located in the instrument panel (Figure 5.52).

ANTENNAS

The typical weather radar antenna sweeps back and forth horizontally, covering 60° on either side of the nose of the aircraft.

Antenna Tilt.   The radar antenna can be tilted up and down by a tilt control located on the radar scope. The tilt angle will vary with flight altitude and range being used. At low altitudes, the antenna should be tilted up slightly to remove most of the ground return (ground clutter). If tilted too high, the radar beam will pass over the top of distant weather and will not show it on the scope. At higher altitudes, the antenna should be tilted slightly downward. If tilted too far down, again distant thunderstorms will not be displayed and the scope will become cluttered with undesirable ground returns. The important thing is for the radar beam to be directed in the area (altitude) through which the aircraft will pass. (The projected radar beam is narrow, usually about 5° vertically and horizontally.)

Antenna Stabilization.   Many airborne radar installations have a gyro stabilized antenna. This permits the antenna to remain fixed in the attitude of radar beam projection while the aircraft is turning or pitching up and down, as in turbulence or in maneuvering around weather cells. Information from the aircraft's regular gryo system usually is utilized for this purpose.

RADAR OPERATION

The radar transmitter fires a burst of electromagnetic radiation through its antenna system. If sufficiently large water drops are within this coverage, a portion of this energy is reflected back and displayed as bright returns on the radar scope. The shape and relative position of these displayed returns enable the pilot to make required course corrections to avoid the potentially dangerous weather.

Gain Settings.   The gain control should be set at a predetermined level and not adjusted up or down to improve the echo return. If the setting is too high, returns will appear relatively stronger, and if too low, relatively weaker. A low but adequate gain setting enables the operator to develop judgment in evaluating weather returns based on comparative data. Some radar units are now equipped with automatic gain control.

It is good procedure when approaching a weather area to tilt the antenna up and down slightly to determine if any significant weather is indicated at higher or lower levels than the selected antenna setting. For ground mapping purposes, the antenna is tilted down for maximum ground return. Some radar units have special circuitry for ground mapping.

Interpretation of Imagery.   Success with airborne weather radar lies primarily with the skill of its operator. Recent electronic developments have made his task easier.

The radar image (echo) seen on the scope is in direct correlation with raindrop size. The larger the drops, the better reflectors they are and the stronger the radar return. The pilot is not necessarily concerned with flying through rain, but rather is more concerned with turbulence. It is the correlation of water droplets with turbulence that makes the radar an invaluable tool for turbulence avoidance.

Strong vertical currents are structural components of .thunderstorms and will readily support large rain drops. Hence, strong weather radar echoes are indicative of turbulence, the degree indicated by the strength of the echo.

Contour and Determining Turbulent Areas.   To aid in the identification of intense moisture concentration and likely areas of turbulence, airborne weather radars are equipped with contour circuitry. Selecting contour causes the radar to *blank out* returns above a fixed degree of power or brightness. These ''punched out'' areas represent the heaviest water concentration. When in the contour mode, an evaluation is made of the area around the blanked out portion.

Figure 5.53. Weather—Normal—Contour.

If a very thin surrounding return is displayed, a *steep gradient* is indicated with a high probability of intense turbulence. These areas are avoided by a wide margin. Inasmuch as areas of heavy turbulence tend to lie where the contour gradient is steepest, rather than in the region of most intense echo return, accurate evaluation of severe weather requires a constant comparison of normal and contour weather depiction (Figure 5.53).

Recognizing and Avoiding Severe Weather.   Adroit weather radar evaluation requires skill and experience on the part of the operator. The first step is to attend a radar school to become acquainted with operational procedures and to learn what radar returns constitute potentially hazardous weather.

Hooks, fingers, scalloped edges, and *nodules* are usually associated with heavy turbulence, including hail. Extreme turbulence can inflict severe damage or total destruction on even the largest aircraft. Flying in areas of heavy hail can do considerable damage to any aircraft. Therefore, radar returns displaying any of the above configurations should be avoided by considerable margin.

The margin of avoidance should always be as great as possible; however, the following minimums apply:

(1) Avoid any cell by 5 miles if the temperature is above freezing and by 10 miles if the temperature is below freezing. When flying above the freezing level, where ice crystals and snow are likely to prevail, accurate cell identification is more difficult. When penetrating an area of storms and passing between two thunderstorms, a minimum corridor of 10 to 20 miles is required, depending upon the temperature and altitude. If the radar return is very intense with hooks and scallops, this distance should be increased accordingly.

(2) Avoid by 10 miles or more any storm which is changing shape rapidly.

(3) Never fly under an overhang from a mature cumulonimbus (thunderstorm) cloud. In this area, hail is likely to be falling.

(4) Never pass directly over a thunderstorm if it can be avoided. If unavoidable, a minimum of 5000 ft. should be maintained over the top. Turbulence can often extend above the visible top of a thunderstorm.

Airborne weather radar, although invaluable for detecting adverse weather areas, is not a substitute for a complete understanding of weather phenomena and their effect on safe flight. A preflight weather briefing is essential, often enabling the pilot to avoid those areas of greater weather potential. Even with weather radar, severity of a weather system may preclude safe passage by an aircraft.

Remember, radar cannot see turbulence, only water droplets. It is of no value in detecting clear air turbulence (CAT). Only through the skill of the pilot operating quality radar equipment can a weather map be constructed as the aircraft flies through stormy skies.

Miscellaneous Avionics

ALTITUDE ALERT SYSTEM

An altitude alert system (AAS) provides the pilot with a visual and aural warning as he approaches a preselected altitude during climb or descent. If he deviates from the set altitude, the warning signals operate again.

High rates of climb and descent in jet aircraft, at the very time when cockpit demands are heavy on the pilot, led the Federal Aviation Administration (FAA) to require this system on all civil jet aircraft.

This system is dependent upon an input from an aneroid device which determines altitude from atmospheric pressure. The radio altimeter, in contrast, is referenced to the height above the terrain (absolute altitude). Thus, each has its place in the cockpit, complementing the other.

When descending from FL 410 (41,000 ft.) to 10,000 ft. (example), the pilot sets 10,000 in the altitude alert selector (Figure 5.54). As the aircraft passes

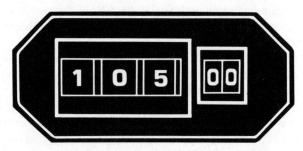

Figure 5.54. Altitude alert. This system is required of all United States certified civil jet aircraft. A warning is provided when approaching or inadvertently deviating from a preselected altitude. *(Courtesy Smiths Industries, Inc.)*

through 11,000 ft. (1000 ft. above the selected altitude), an aural signal sounds through the speaker or pilot's headphones and a panel light illuminates, warning the pilot of the approach of his desired altitude. At 300 ft. above the selected altitude, the light goes out. A deviation of more than 300 ft. from the selected altitude causes reactivation of the warning signals.

ANTENNAS

Discussion of aircraft radios is incomplete without a discussion of antennas. The finest engineered and manufactured aircraft radio will deliver less than satisfactory performance if a poorly designed, improperly located, or improperly installed antenna is used.

In large airline or military type aircraft, antenna design is a part of the initial design concept of the airframe. In small general aviation aircraft, antenna selection and location are often given little critical attention, and radio performance suffers.

Factors considered in the design of an antenna include: *reciprocity,* an antenna's ability to transmit as well as receive; *polarization,* the orientation of the electrical field component of the electromagnetic field; and *radiation pattern,* a graphic representation of the variation in direction of the radiated pattern in receiving and transmitting.

The major types of VHF and UHF antennas are as follows.

VHF Communications. The frequency range is 118–136 mHz, vertically polarized, and the antenna appears in the form of a blade or rod (Figure 5.55A). It should be located on the top or bottom of the fuselage, away from other an-

Figure 5.55a. VHF COM antenna.

Figure 5.55b. VOR/LOC antenna.

Figure 5.55c. DME/Transponder antenna.

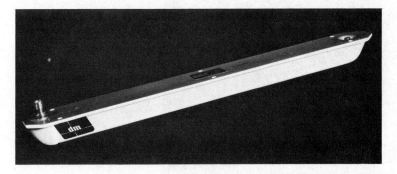

Figure 5.55d. Marker Beacon antenna.

tennas for best performance, because of potential interference of one antenna with another in close proximity.

VOR-LOC. This antenna is used for VOR and ILS localizer receivers in the 108–118 mHz frequency range, horizontally polarized. It appears as a ''V'' or blade-type balanced loop antenna (Figure 5.55B), generally located high on the vertical stabilizer. It is often flush-mounted in large aircraft.

DME and *Transponder.* This is used with either a DME or a transponder in the 960–1220 mHz frequency range and appears as a small blade (2¼ in.) or spike, mounted on the bottom surface of the aircraft (Figure 5.55C). A UHF antenna is a somewhat longer blade.

Other specialized antennas such as marker beacon (75 mHz) and glide slope (329–335.2 mHz) should also be carefully selected and installed for maximum performance (Figure 5.55D).

AIRBORNE RADIOTELEPHONES

An extensive network of ground stations (over 45) enables the businessman to maintain his nation-wide communications capabilities while he is airborne, through airborne radiotelephone (Figure 5.56).

The system consists of a receiver-transmitter, antenna, and control units. An example is the Wulfsberg Electronics, Inc. ''Flitefone III,'' pictured below.

Receiver-Transmitter. The R/T unit is completely solid state, with no life limited components, ensuring maximum reliability and performance, and weighs about 7 lb. It has 13 receiver channels, operating in a frequency band of 454.675–454.975 mHz, and 12 transmitter channels in a band of 459.700–459.975 mHz. In this UHF frequency spectrum, transmission and re-

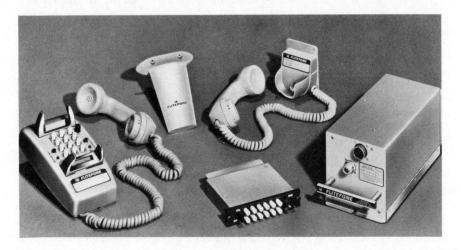

Figure 5.56. Radiotelephone.

ception are line-of-sight. Consequently range at 5000 ft. is about 100 miles, increasing to 300 miles at 40,000 ft.

Antenna.   A small rod or blade-type antenna is mounted to the underside of the aircraft.

Control Unit(s).   A "touch tone" type telephone is completely self-contained and includes illumination of selected channel, volume control, off switch, and interphone system for communicating with the flight deck.

Air-to-Ground Calls.   After turning on the set and allowing 1 min. for warm up, remove the phone from the cradle and select the nearest ground station by the appropriate channel button. Listen for dial tone, then flash the operator by depressing the operator button. When the operator answers, give the aircraft telephone number, city of registry, and number you are calling. For billing convenience, it is recommended that calls be charged to a telephone credit card.

Ground-to-Air.   Determine the ground station most likely to be within range of the aircraft. Dial the operator and ask for the mobile operator in the city having the selected ground station. Reaching the desired mobile operator, give your number and the telephone number of the aircraft you wish to call. It is necessary that the aircraft unit be "on" if contact from the ground is to be successful.

Operation.   When contact with the desired party has been made, full-duplex operation allows a normal conversation as with any desk phone. There are no push-to-talk requirements.

NAVIGATION SYSTEMS

Development of microprocessors and lightweight computers has led to the advancement of electronics which makes possible very sophisticated, highly accurate navigation systems. Some of these are discussed in detail in Chapter 11.

Future Trends

One hundred years ago communications were by voice and writing; 80 years ago, transportation depended on steam, wind, and animal power; and only 50 years ago most farming was by use of animal power. The vast changes that have occurred in recent history can only be termed an explosion of technology.

The same vast changes in aviation, particularly in aircraft instrumentation and avionics, have been condensed into a much shorter time. In just 25 years, radio tubes have been replaced by transistors, then integrated circuits, then large scale integrated circuits, with the promise that ultimately small chips will accommodate over 20,000 active elements. The space program has no doubt made a major contribution to this progress. The modern DME, which simultaneously computes distance and time to the station and ground speed, and the digital clock, which provides time accuracy undreamed of a few years ago, are only two examples of recent advancements.

It appears that progress in aircraft control and communications is not leveling off, but continuing to snowball. We can now simultaneously observe events occurring thousands of miles apart; we are getting closer and closer to being there.

6

The Atmosphere and Its Weather

The atmosphere is defined by Webster's Dictionary as "the whole mass of air surrounding the earth; . . . a surrounding influence or environment." This is the airman's world. In this world of surrounding air, thinness and buoyancy permit measuring distance in minutes rather than miles, and beauty, enchantment, and freedom are the pervading influence. Only in the airman's world can one view horizons encompassing land and sea, mountain and prairie, storm and calm, even peace and war, at a single glance.

Chemical Structure of the Atmosphere

The ancient Greeks considered air to be one of the four basic "elements"—the others being fire, earth, and water. As we know, air is not an element in the chemical sense but a complex mixture of gases, like apples, oranges, and bananas thrown together in a fruit bowl. The basic unit of each gas is the molecule, far too small to be seen even through the most powerful electron microscopes yet developed. The molecules of oxygen, nitrogen, and other atmospheric gases maintain their separate identities as they exist side by side.

In the lowest 40 to 50 miles of the atmosphere, the relative proportion of each gas in the mixture stays remarkably constant. Nitrogen and oxygen, by volume 78% and 21%, are the most abundant gases. Argon, an inert gas, makes up about 1%. Only .03% is carbon dioxide, yet this part is vital to life, since plants require it for photosynthesis. In addition, carbon dioxide, like water vapor, is a powerful absorber of the sun's radiation, thus moderating temperature. Traces of helium, krypton, and neon are also present. Water vapor is the most variable part, ranging from values as high as 5% by mass in moist tropical air to almost 0% in regions of intense cold and great heights. In addition to its gaseous parts, the atmosphere contains various types of small, solid particles, such as sea salt and dust. Since the condensation of moisture in the atmosphere occurs on these solid particles (*condensation nuclei*), they are nearly as important as the gases.

Although nitrogen and oxygen compose 99% of the atmosphere up to altitudes of a thousand miles or more, the forms of the gases do not remain constant. At some distance above the earth, ultraviolet radiation from the sun causes some of the molecules to dissociate from the molecular state to the atomic state. At lower levels where the atmosphere is relatively dense, the atomic state is rare because recombination takes place about as rapidly as dissociation.

When an atom and a molecule of oxygen combine, ozone is formed. Most of the ozone occurs in a layer about 15 miles thick, between 10 and 25 miles from the surface of the earth, although it extends from as low as 5 to as high as 35 miles. This layer is frequently called the ozonosphere.

While there is only a minute quantity of ozone present, it so efficiently absorbs certain ultraviolet wavelengths that it greatly warms the upper part of the ozone layer. This makes the temperature at 30 miles altitude about the same as that at the earth's surface. Ozone also protects the lower layers from excessive short-wave radiation which would kill nearly all forms of life on the earth.

Above the ozone layer, air density decreases until recombination eventually does not occur, leaving both oxygen and nitrogen in atomic form.

Physical Properties of the Atmosphere

While the airman should understand the atmosphere's chemical structure, he deals more directly with its physical properties and structure. The primary physical properties which determine structure are pressure, temperature, humidity, and radiation.

PRESSURE

Atmospheric pressure at any point is equal to the weight of the air column of unit cross-sectional area above that point. It can be expressed either in terms of actual

TABLE 6.1. Standard Atmospheric Pressure at Sea Level Expressed in Various Units.

1013.25	millibars (mb)
14.70	pounds per square inch (psi)
760	millimeters of mercury (mm Hg)
33.90	feet of water
29.92	inches of mercury (in. Hg)

weight, such as pounds per square inch (psi), or in terms of the height of a column of some other fluid, such as inches of mercury, which exerts the same pressure. At sea level the average pressure of the air is 14.7 psi. The same pressure is exerted by a column of mercury 760 mm. (29.92 in.) in height. In meteorology the unit of pressure most commonly used is the millibar (mb), which is the pressure exerted by a force of 1000 dynes/cm.^{2}. These equivalents are shown in Table 6.1.

TEMPERATURE

While the temperature of a substance may be considered as the degree of its hotness or coldness, measured on a relative scale, the kinetic theory better explains this physical property of gases.

The molecules of a gas move constantly. Their average speed is high at higher temperatures, low at lower temperatures. The temperature at which all motion of molecules stops is defined as absolute zero.

The two thermometer scales in common use today are the Fahrenheit and Celsius. Gabriel Daniel Fahrenheit, a German physicist, developed a scale which selected as 0° the temperature on a particular winter day in Danzig. This gave a boiling point of 212°, a freezing point of 32°, and an absolute zero of −459.69°.

A French scientist, Rene Antoine de Reaumur, devised a more scientific scale in 1731, using alcohol as his liquid and the freezing point of water as 0°. From this idea was developed the centigrade (now called Celsius) scale, with the freezing and boiling points of water equal to 0° and 100°, respectively, and an absolute zero of −273.16°.

HUMIDITY

The quantity of water vapor contained in the atmosphere is relatively small. Without this moisture, however, most life as we know it could not exist. Moisture is the basis of rain, snow, thunderstorms, clouds, fog, frost, and dew; by its absorption of solar and terrestrial radiation, it affects the temperature of the air.

The capacity of air for holding moisture increases with increasing temperature. Air is termed *saturated* when it is in equilibrium with a flat water surface at the same temperature and pressure as the air. In other words, such a water surface will neither evaporate into nor gain water from saturated air. The actual water vapor content of an air parcel can be described in a number of ways. Most commonly used by meteorologists are *mixing ratio, dew point,* and *relative humidity.*

Mixing ratio is the ratio of the mass of water vapor to the mass of dry air in the parcel, usually expressed in parts per thousand. Dew point is the temperature to which air must be cooled to reach saturation at its original pressure and mixing ratio. It is also the temperature at which a smooth surface would first show traces of condensation or dew. Further cooling would cause condensation of a part of the vapor as dew—hence the name *dew point.* Relative humidity is the ratio of the actual mass of water vapor in the air to that which would exist if air at the same temperature and pressure were saturated. It is expressed in percentage. The difference between the actual temperature and the dew point temperature is called "spread," and is an indication of relative humidity. A 0° spread corresponds to 100% relative humidity.

The ability of air to hold moisture falls off rapidly with decreasing temperatures. Assuming a surface pressure of 1000 mb, the mixing ratio of saturated air at a temperature of 20°C (68°F) is 15 parts per thousand. This value falls 50% with a cooling of only 10°C. At a temperature of -37°C, the saturation mixing ratio is down to 1/100 of its original value.

Water vapor is added to the air mainly through evaporation from wet surfaces and transpiration from plant life. Since these are low-level sources, the amount of moisture, like temperature, tends to decrease with altitude.

RADIATION

All motions in the earth's atmosphere, all forms of weather, and even life itself owe their origin to energy received from the sun. This energy is transmitted to the earth in the form of electromagnetic waves. Electromagnetic radiation occurs over a wide spectrum of wavelengths. At one end of a complete spectrum are the extremely short waves, including the cosmic rays, gamma rays, and X rays. Next come the ultraviolet, visible light, and infrared rays. Finally at the other end of the spectrum are the longer waves used in radar, television, and radio broadcasting. Solar radiation is concentrated in only part of this spectrum: visible light and parts of the ultraviolet and infrared bands.

The quantity of solar radiation received at the outer layers of the earth's atmosphere on a unit of surface in a unit of time is called the *solar constant.* Its value has been calculated to be 1.94 calories/cm.²/min. After this radiation is absorbed, heat is liberated. A small fraction (about 5%) of the incoming solar radiation (*insolation*) is absorbed by the ozone layer. The next barriers are the cloud layers, which cover about half the sky. The *albedo* (reflecting ability) of the clouds is about .80, so 40% of the total insolation is reflected back to space without releasing its heat. Another 15% is absorbed by the various atmospheric constituents, especially by water vapor in the lower levels; most of the remaining 40% is absorbed by the earth.

These figures include the small amount of insolation reflected either back to space or down to earth by the process known as scattering. Light waves are scattered, or reflected in all directions, when they strike very small particles. If the particles are smaller than the wavelength of light, as is true for molecules of air and many condensation nuclei, the shorter wavelengths (blue light) are scattered most. This selective scattering gives the sky its characteristic blue color. At very high altitudes, where the molecules of air are widely separated, scattering becomes negligible and the sky appears black. Particles larger than the light waves, such as fog, cloud, and precipitation droplets, scatter nonselectively and give the sky a whitish appearance.

All this absorbed radiation must necessarily be radiated back to space, since otherwise the earth would be growing progressively hotter. The average wavelength of radiant energy varies inversely with the temperature of the emitting body. The earth, which is relatively cool, radiates principally in the long-wave infrared band. The atmosphere is not nearly so transparent to these long waves as it is to the short-wave solar radiation. Water vapor, clouds, and (to a smaller degree) carbon dioxide are all good absorbers of rays in the infrared spectrum. Thus the energy is trapped, and the lower layers of the atmosphere are warmed. Glass has the same property of being transparent to short waves and absorbing long waves; the resulting temperature rise is known as the *greenhouse effect*.

Although for the earth as a whole the terrestrial radiation must equal the absorbed insolation, there is a marked discrepancy between income and outgo at certain latitudes. In equatorial regions, where the sun is most nearly overhead, there is an excess of insolation, while in polar regions there is a deficiency.

Physical Structure of the Atmosphere

Atmospheric structure may be seen as comprising a series of layers, each with its own characteristics. Several nomenclature systems have been devised for identifying the various layers. Perhaps the most common, and certainly the simplest, divides the atmosphere into four layers of alternately decreasing and increasing temperature—the troposphere, the stratosphere, the mesosphere, and the thermosphere—plus a boundary layer between the atmosphere and space called the exosphere (Figure 6.1). The layers in which the temperature decreases with height tend to be unstable and turbulent, while those in which the temperature increases with height tend to be stable and smooth. Within the thermally bounded layers are two regions that have unique physical properties—the ozonosphere, which coincides with the stratosphere, and the ionosphere, which overlaps the mesosphere and thermosphere (Figure 6.1).

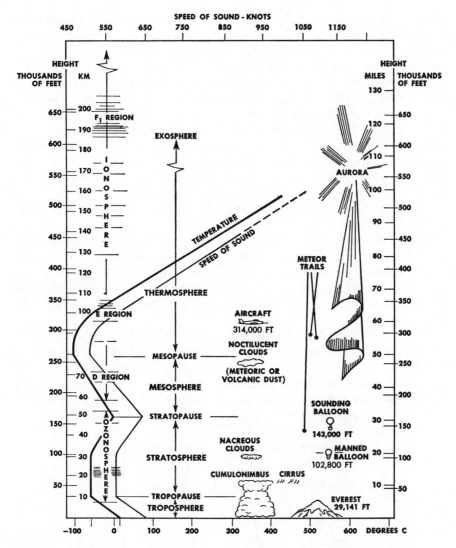

Figure 6.1. Phenomena and structure of the atmosphere. Radio waves are, in general, absorbed in the *D* region, and refracted at the *E* region.

THE TROPOSPHERE

The lowest layer and the one most familiar to man is the troposphere. The name is derived from the Greek *tropos,* meaning "turn" and refers to the great amount of overturning of the air in this layer. Convective currents rapidly and continuously carry air parcels upward and downward. Nearly all of the varied activity termed *weather* takes place within this layer. The troposphere is characterized by a decrease in temperature with height at the rate of about 3½°F per 1000 ft. until a leveling-off point is reached at the tropopause. The change in temperature with height is called the *temperature lapse rate*. If the temperature within a layer is constant, the lapse rate is said to be *isothermal*. An *inversion* is a layer in which the temperature increases with height.

The troposphere extends to greater elevations in warm regions than it does in cold, reaching 50,000 ft. or more at the equator, and only 25,000 ft. at the poles. This difference results in the phenomenon of temperature reversal at high altitudes. At the surface of the earth, polar air is much colder than tropical air. However, the temperature of polar air decreases with height only up to its low tropopause, while that of tropical air continues falling until its much higher tropopause is reached. At some altitude, therefore, tropical air becomes colder than polar air. Typical temperature-altitude soundings for polar and tropical air are shown in Figure 6.2. Here the north-south temperature reversal occurs near 35,000 ft.

Another important characteristic of the atmosphere, particularly noticeable in the troposphere, is its decrease in density with altitude (Figure 6.3). Though the troposphere contains less than 1% of the total volume of the atmosphere, it contains over 75% of the total mass. In fact, half of the total mass is contained in the lowest 18,000 ft. Similarly, the pressure exerted by the air decreases with altitude.

The distinction between air pressure and air density is important. As described in other chapters, the two have independent effects on aircraft engine operation and on the human body. Density is a measure of mass contained in a unit volume, and is often expressed in the number of slugs in a cubic foot. Pressure is a measure of force exerted on a surface of unit area, and may be expressed in pounds per square foot (lb./ft.^{2}) or pounds per square inch (psi).

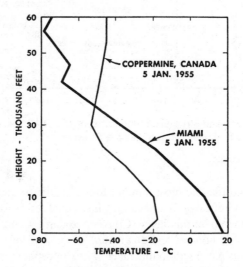

Figure 6.2. Reversal of the north-south temperature relationship at high altitudes.

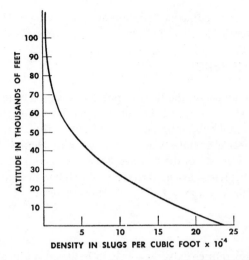

Figure 6.3. Decrease in density of the atmosphere with altitude. Density in slugs per cubic foot × 10⁻⁴.

THE STRATOSPHERE

Above the troposphere, extending to great altitudes, lies the stratosphere. An isothermal layer usually extends from the tropopause to perhaps 10 or 20 miles, at an average temperature of $-56.5°C$. Above this point the temperature increases, reaching a maximum value at about 30 miles.

Scientists estimate that the average peak temperature is $-2.5°C$, not far below the earth's average surface temperature. This temperature peak near 30 miles altitude is called the *stratopause*. It is close to the top of the ozone layer. Even though the total quantity of ozone in the atmosphere is small and the great bulk lies at lower levels, its great ability to absorb short-wave radiation is the primary cause of the high temperature at this level.

Since the source of moisture is at the surface of the earth, one would expect to find few clouds in the stratosphere. In general, this is true. However, it is not unusual for the tops of large "thunderheads" to be found there also. More surprising is the fact that clouds have been found at altitudes ranging from 70,000 to 90,000 ft. above ground. These rare mother-of-pearl clouds, so called for their iridescence, are believed to be composed of tiny globules of ice.

THE MESOSPHERE

Above the stable stratosphere is a turbulent layer called the mesosphere, in which the temperature falls rapidly with height. It reaches an average minimum value of $-92.5°C$ near 50 miles. The upper boundary of the mesosphere is called the mesopause. The rare noctilucent "shining-at-night" clouds, which are probably composed of meteoric dust and ice crystals, occur near the mesopause.

THE THERMOSPHERE

Like the upper stratosphere, the thermosphere is a region of increasing temperature with height. Here are found the aurorae and most of the meteor trails. The high temperatures in the thermosphere are at least partly due to the absorption of some of the shortest ultraviolet waves by oxygen molecules. There is still sufficient air in the lower part of the thermosphere to cause significant drag and frictional heating on vehicles passing through it. As shown in Figure 6.1, there is no temperature boundary to the top of the thermosphere.

THE EXOSPHERE

The very high temperatures existing above 350 miles result in a very rapid movement of the atoms and molecules, some of which achieve sufficient velocity to overcome the force of gravity and escape into space. This fringe region between the atmosphere and outer space is termed the exosphere. Both the lower and upper boundaries of the exosphere are diffused and poorly defined. Generally the limits are considered as lying roughly between 350 and 600 miles, although some authorities put the upper limit above 1000 miles.

THE IONOSPHERE

An important effect of radiant energy from the sun is to ionize air. That is, electrons are dissociated from neutral air molecules and atoms, producing positively charged ions and free electrons. The most effective agent for this process is solar ultraviolet radiation. Most of the ultraviolet is blocked from reaching the lower atmosphere by the stratosphere's ozone. Also, at lower altitudes the density of the air is so great that ionized particles rapidly meet and recombine. In the mesosphere and above, however, solar ultraviolet radiation acts with full force, and density is so low that charged particles have a long lifetime. This deep layer in which substantial numbers of freely moving charged particles are present is termed the ionosphere. It extends between about 35 and 300 miles altitude.

The main effect of the ionosphere upon the aviator is its ability to refract, reflect, or absorb radio waves. This property makes long-distance radio communication possible through reflection of the longer radio waves between the earth and the ionosphere. Since the ionosphere is created and maintained by solar radiation, its characteristics vary markedly with the passage of day and night and with varying solar activity. Predictions of its radio propagation activities are routinely made by the National Oceanic and Atmospheric Administration (NOAA) of the Department of Commerce, which contains the National Weather Service.

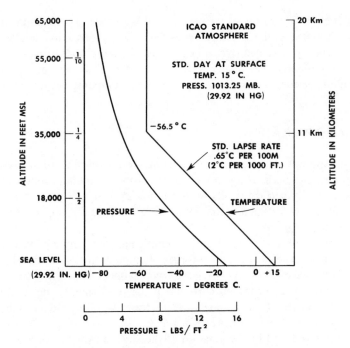

Figure 6.4. The standard atmosphere, lower levels, with standard day at the surface.

THE STANDARD ATMOSPHERE

Atmospheric properties at any level vary greatly with location and time of year, and because of moving weather systems. But for many practical purposes it is desirable to have a universally accepted model atmosphere. For example, altimeters which depend upon pressure should all behave the same. Such uniform standards are also essential for aircraft design.

The most generally accepted standard atmosphere is the ICAO Standard Atmosphere shown in Figure 6.4. This is the U. S. Standard Atmosphere adopted in 1962. Supplementary standards for higher altitudes and varying seasons and latitudes were published in 1966. These give the aircraft designer and operator valuable guidance on the nature of his environment.

Circulation of the Atmosphere

The most striking feature of the atmosphere to airmen is its unceasing motion and change. Clearly this constant activity requires some outside source of energy to maintain it. This is provided by the sun, whose radiant energy acts most effectively at low latitudes where the earth's surface is perpendicular to the sun's rays. Conversely, the sun illuminates the polar regions at low angles so that each unit of energy is spread over a larger area. Because both tropics and poles can lose heat to outer space equally well, the net effect is a constant input of heat to the earth and atmosphere near the equator and a compensating deficit near the poles.

GENERAL CIRCULATION

If the earth were not rotating, unequal heating would cause a simple system of convection currents with rising motion in the tropics, sinking motion over the poles, and northerly surface winds in mid-latitudes, capped by southerly winds aloft. (See Figure 6.5.) Such a circulation would efficiently transfer heat from equator to pole.

The rotation of the earth significantly complicates this simple picture. Air set in motion by heat-generated pressure forces attempts to follow a straight course as viewed from space. However, the earth rotates under this moving air, so that to an earthbound observer the air currents appear to deviate to the right (in the Northern Hemisphere). To us it appears as if a force is pulling the air to the right. This apparent force is termed *Coriolis force*.

Coriolis force induced by the rotation of the earth makes the simple one-cell circulation of Figure 6.5 impossible. Instead, a far more complex circulation is used by the atmosphere to transfer heat from south to north. Near the equator and pole, prevailing easterly winds are found. The subtropical easterlies are termed trade winds and are among the steadiest winds on earth. In mid-latitudes the

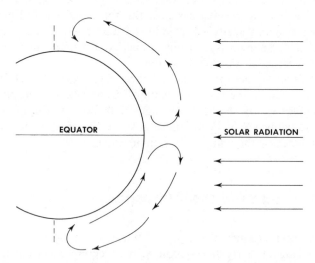

Figure 6.5. Atmospheric circulation on a nonrotating earth.

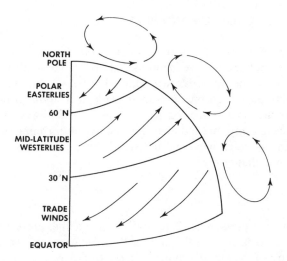

Figure 6.6. Observed general circulation of the atmosphere.

prevailing wind is westerly, but the dominant circulation feature is the continual series of vortexes or eddies which move from west to east in the westerly winds. For this reason, any depiction of the average circulation, such as Figure 6.6, bears little resemblance to the weather of any individual day.

GEOSTROPHIC WIND

Another remarkable and useful consequence of the earth's rotation is a relationship between the pressure distribution and the wind known as the *geostrophic wind law*. Any parcel of air always experiences a force directed from high pressure toward low pressure. When we increase the pressure in a bicycle pump, the air in the hose is forced out. But on the earth a moving parcel of air experiences an apparent Coriolis force to the right of its direction of motion. However, we also observe that moving streams of air experience relatively little acceleration or deceleration in the course of a few hours. This is possible only if the wind and pressure cooperate so that the pressure force and Coriolis force almost exactly balance each other.

This state of balance generally prevails in the atmosphere. In the Northern Hemisphere, large-scale winds almost universally blow so that high pressure lies to the right of the direction of motion, balancing the deviating Coriolis force. In the Southern Hemisphere, the direction of Coriolis force is reversed, and high pressure is found to the left of the wind. In the tropics, the earth's surface is parallel to its axis of rotation, Coriolis force is virtually absent, and the winds are poorly related to the pressure distribution. Near the earth's surface, the force of friction is also significant, and winds turn somewhat toward low pressure.

LOW-LEVEL CIRCULATION AND PRESSURE

The distribution of average pressure on the earth is consistent with these rules and the average air circulation we mentioned above. A "trough" of low pressure lies over the equator between the two trade wind belts, while to the north and south large subtropical high-pressure areas mark their northern boundary. The polar high-pressure area suggested by the polar easterlies is much less pronounced, but the stormy mid-latitudes are marked by traveling low- and high-pressure systems in which wind and pressure behave much as suggested by the geostrophic wind relationship.

WINDS ALOFT

These migrating low-pressure areas (or cyclones) and high-pressure areas (or anticyclones) are associated with the bulk of the weather experienced in mid-latitudes where most of the world's flying is done. But to aviators, the winds above the earth's surface are of almost equal interest. The typical wind flow in the troposphere can be represented by a ring of subtropical highs bounded to the north by a belt of westerlies surrounding a weak polar low (Figure 6.7). The westerlies are by no means constant. They meander around low-pressure troughs and

Figure 6.7. Average pressure and wind distribution from 20,000 to 40,000 ft.

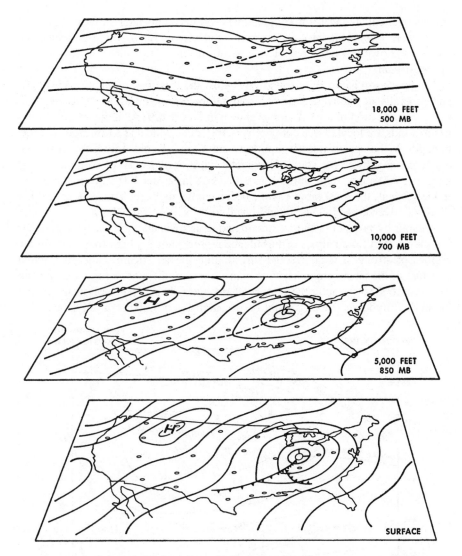

Figure 6.8. Relationship between winds aloft and surface pressure patterns.

high-pressure ridges which are generally associated with the surface cyclones and anticyclones mentioned above. A typical relationship is shown in Figure 6.8.

THE JET STREAM

A remarkable feature of the upper air was discovered during World War II when high-flying bombers encountered westerly winds of nearly 200 knots on bombing attacks over Japan and Europe. Systematic observations later showed that these winds were organized into a semicontinuous narrow band which encircled the globe, as shown in Figure 6.7. This narrow current of fast-moving air is termed the *jet stream*.

The jet stream is typically found just below the tropopause at about 35,000 ft. and generally lies above a zone of strong temperature contrast (a front). A schematic cross section of a jet stream is shown in Figure 6.9. As indicated, the maximum winds are located in a core perhaps 100 miles wide and a few thousand feet in depth. Wind speeds fall off rapidly to the north, or cold side of the flow, and much less rapidly to the south. Horizontal wind changes can be as great as 100 knots per 100 miles to the north or as little as 25 knots per 100 miles to the south. In the vertical, the most rapid change occurs above the core. Thus a pilot seeking to avoid adverse headwinds should head north or toward colder air.

A major by-product of the jet stream and its associated wind shears is clear air turbulence (CAT). Analysis of turbulence reports shows a high frequency in the regions of strongest wind change surrounding the jet stream. Since the actual structure of the jet stream consists of many narrow fingers of strong wind, it is no easy task to locate exactly the regions of strongest shear. Hence pilots traversing the high-altitude jet stream should be prepared for turbulence at all times.

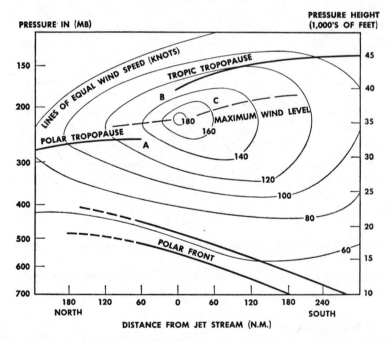

Figure 6.9. Polar front and cross section of a jet stream looking downwind. Most probable CAT is at *A, B,* and *C.* Cirrostratus clouds frequently found in area *C. (Courtesy of US Air Force.)*

STRATOSPHERIC WINDS

With the advent of supersonic transports, the weather of the stratosphere becomes of increasing interest. Above the tropopause, the effect of the small weather-producing disturbances of the troposphere rapidly dies out. At altitudes of 70,000 ft. or more, atmospheric pressure and density are less than 3% of their surface values.

The circulation alternates between radically different summer and winter situations. The summer stratosphere is dominated by a warm high-pressure area over the pole, surrounded by a broad, weak belt of easterly winds. In fall, this pattern gradually reverses itself, and by winter the polar high has been replaced by an intense low with temperatures of −80°C or lower. This low is bounded by intense polar night westerlies which frequently reach velocities of several hundred knots. These westerlies undulate in a slowly moving fashion, seemingly independent of the tropospheric turmoil below. In spring the return of the sun to polar latitudes produces an abrupt transition to the summertime stratospheric circulation.

The Growth of Clouds

MOISTURE IN THE ATMOSPHERE

Water is the only compound which naturally occurs on our planet in all three phases—solid, liquid, and gas. In the atmosphere, it is present chiefly as the gas, water vapor, but its occurrence in the form of water droplets or snowflakes is of greater interest to the aviator. Most of the atmosphere's moisture is concentrated in the lower troposphere because its source is at the ground and also because the cold air at higher levels holds but little moisture.

The moisture content of air can be described in many ways. The concept of *saturation* is essential to understanding these measures. Air is saturated when it is in equilibrium with a water surface. That is, the air contains sufficient water vapor that no more can be evaporated into it at the existing temperature.The saturation water content varies strongly with the temperature. For approximately every 20°F (11°C) increase in temperature, the water-holding capacity of the air is doubled. Relative humidity and dew point, described on page 257, are measures of saturation.

CONDENSATION PROCESSES

In order for the water vapor in the air to condense into liquid or solid form, the air must be brought to its saturation point either by cooling or by adding water. There must also be present a suitable particle or surface to serve as a base for condensation. In absolutely clean air (which is never found in nature), water molecules in the air would coagulate together only by accident, and water vapor concentrations many times "saturation" would be required for condensation.

In the real atmosphere, however, small particles of dust or salt are always present, and condensation starts to build cloud droplets on these when the air's water content only slightly exceeds saturation. The number and kind of these "condensation nuclei" have a great effect on the character of clouds. Over the continents dust particles are plentiful, and clouds consist of quite small droplets. On the high seas, the air is clean, and moisture condenses on a small number of large salt particles, forming clouds with large wet drops.

The majority of clouds and fogs are produced by cooling the air to its saturation point, that is, reducing the temperature to the dew point. This can occur by passage over a cooler surface or by loss of heat through radiation. The most important cooling process, however, is lifting. As will be discussed later, rising air cools at a steady rate but keeps its moisture content. Eventually saturation is reached.

On occasion clouds or fog form without cooling by addition of moisture. For example, rain falling from higher clouds can saturate lower layers, producing low clouds or fog. Sometimes dry air moves over warmer moist surfaces and "steam fog" is formed.

The development of precipitation is a complex and imperfectly understood process. Droplets, falling at different speeds, probably collide and coalesce into larger drops. These grow further and break apart, starting a chain reaction. In continental air the presence of ice crystals appears to be crucial in forming con-

Figure 6.10. Formation of clouds due to expansion and cooling of air lifted over terrain features. *(Courtesy of US Air Force.)*

densation nuclei for the requisite large drops. In any event, continual influx of moist air and cooling by rising motion are essential to produce any significant amount of precipitation. An individual cloud at any insant holds a surprisingly small quantity of water.

Stability

To a pilot, the stability of his aircraft is a major concern. A stable aircraft, if disturbed from its intended altitude or flight path, will return to normalcy of its own accord. An unstable aircraft, however, will continue to deviate away from level flight (Figure 2.44.)

The same phenomenon is observed in the atmosphere. Air normally flows in horizontal paths because any deviation up or down is resisted by strong buoyant forces. When these forces diminish or vanish, the air is said to become unstable, and vertical air currents are possible with little input of energy. The pilot experiences these vertical currents through turbulence or the clouds and weather they produce.

ADIABATIC LAPSE RATE

We have seen that temperature decreases with altitude at a normal lapse rate of 2°C (3½°F) per 1000 ft.

If a parcel of air is forced to rise (is *lifted*) in the atmosphere, it encounters

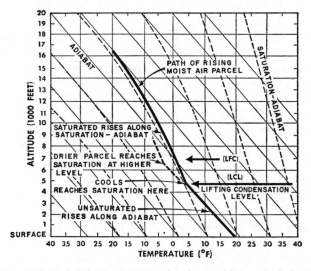

Figure 6.11. Adiabatic diagram showing stability calculations.

lower pressure and expands. If no heat is added or removed (adiabatic conditions), it will cool as it rises. This *adiabatic lapse rate* (adiabat) is about 3°C (5½°F) per 1000 ft. of ascent.

If the air contains water vapor, it will eventually cool to a temperature at which the air is saturated, and condensation will begin. The level at which this first occurs is called the *lifting condensation level* (LCL). Upon condensation, the heat which was required to evaporate the water originally is released. Hence rising air cools at a much slower rate when condensation is taking place. This rate is termed the "moist" or *saturation adiabatic lapse rate* and is about 1½°C per 1000 ft.

These processes can be best studied and depicted on an adiabatic diagram such as shown in Figure 6.11. Different versions of these charts are used by the various meteorological services, but all have coordinates of temperature and pressure (or altitude) and are overprinted with a pattern of lines showing the dry and saturation adiabiatic lapse rates.

STABILITY AND INSTABILITY

The density of air depends upon its temperature and pressure. At constant pressure, a parcel of cold air is denser than the warm air around it and will tend to sink. If lifted, it will tend to return to its original level as long as it cools faster than surrounding air. The air is then *stable*.

If the cooling rate of a lifted air parcel is less than the temperature lapse rate of the surrounding air, the lifted parcel will become warmer and lighter than its surrounding air. It will then continue to rise spontaneously. When this condition prevails, the air is said to be *unstable*.

Frequently such a comparison will show a sounding whose slope lies between the dry adiabat and the saturation adiabat. This represents *conditionally unstable air,* because a lifted parcel will be stable until it reaches the LCL, and then will become unstable. The point at which it becomes unstable is the *level of free convection* (LFC). If the parcel were lifted a small distance above the LFC it would become warmer than the surrounding air and be accelerated freely away from its original location. Lifting action which sets off this free acceleration might result from forced ascent over a hill (orographic lifting) or from a warm air mass riding over a cold, dense air mass (frontal lifting). If lifted above the level of free convection, cloudiness and precipitation usually result (Figure 6.11).

The concept of moisture and stability is very important to those who fly, as Table 6.2 will show.

EFFECTS OF STABILITY AND INSTABILITY

The atmosphere's stability helps to determine what type of clouds will form. For example, gradual ascent of stable moist air up a sloping surface will produce

TABLE 6.2. Characteristics of Air According to Moisture and Stability.

	Dry Stable Air	Moist Stable Air	Dry-Un-stable Air	Moist Unstable Air
Visibility	Poor in haze	Poor in haze and fog	Good	Good except in precipitation
Cloud type	None	Clouds in stratified layers— *stratus*	None	Clouds of vertical development— *cumulus*
Precipitation	None	Steady	None	Showers
Flying conditions	Smooth	Smooth	Turbulent	Turbulent
Type of icing	None	Mainly rime	None	Mainly clear

layered, or *stratus* clouds with little vertical development or turbulence. Conversely, clouds forming in unstable air show a puffy, turbulent character. In them, release of latent heat adds to the intensity of vertical motion.

Highly stable air greatly inhibits the exchange of low-level air with the free-flowing air at higher levels. Pollutants generated at low levels can accumulate, producing haze or smog. Thus poor visibility is common in areas where large-scale descent of air produces warm layers aloft. Some of the effects mentioned are conveniently summarized in Table 6.2.

Cloud Types

Clouds are significant to the flyer as obstructions to vision and as sources of potentially hazardous weather. However, they also aid him by acting as sensitive indicators of air motion, turbulence, and weather.

Although dozens of cloud types are recognized by meteorologists, a much smaller number of categories is sufficient for practical aviation. Four general classes are: low clouds (below 6500 ft.), middle clouds (6500 to 16,500 ft.), high clouds (above 16,500 ft. and generally below 45,000 ft.), and clouds with extensive vertical development. Within these families, we can distinguish two main subdivisions: lumpy or *cumiliform* clouds formed by turbulent local air motions in unstable air; and smoothly layered or *stratiform* clouds produced by gradual uplifting of whole layers of fairly stable air.

Dark clouds which produce precipitation are modified by the word *nimbus*. Middle cloud types are denoted by the prefix *alto*, while high clouds which are generally formed of ice crystals are prefixed by *cirro*. Thus a puffy, fair-weather, low cloud is termed "cumulus," and a thunderstorm is "cumulonimbus." Smooth-layered clouds at different levels are called "stratus," "altostratus," and "cirrostratus."

Cloud forms offer valuable clues to flying conditions inside them. As we mentioned earlier, lumpy cumiliform clouds suggest turbulence, while smooth-layered stratus clouds indicate smooth air. Dark clouds, particularly those with considerable vertical extent, have a high liquid water content. Above the freezing level, these water droplets will be supercooled, ready to freeze on contact with any foreign body such as your aircraft.

Air Masses

Although the atmosphere is in constant motion, there are large regions where the air is relatively stagnant for a number of days. In these conditions air can take on quite uniform properties over a large area. Such a body of homogeneous air is termed an *air mass*.

SOURCE REGIONS

The geographical area in which an air mass acquires its characteristic properties is termed its source region, and the air mass generally carries the name of this region. The two principal types of air mass are *polar* (P) and *tropical* (T). Further

TABLE 6.3. Summer Flying Weather Conditions in Various Air Mass Types.

Air Mass	Clouds	Ceilings	Visibilities	Turbulence	Surface Temp. (Degrees F.)
cP (near source region)	Scattered cumulus	Unlimited	Good	Moderate turbulence up to 10,000 ft.	55 to 60
mP (Pacific Coast)	Stratus, tops 2000 to 5000 ft.	100 to 1500 ft.	½ to 10 mi	Slightly rough in clouds, smooth above	50 to 60
mP (east of Pacific)	None except scattered cumulus near mountains	Unlimited	Excellent	Generally smooth except over desert regions in afternoon	60 to 70
mT (east of Rockies)	Stratocumulus early morning; cumulonimbus afternoon	500 to 1500 ft. A.M.; 3000 to 4000 ft. P.M.	Excellent	Smooth except in thunderstorms, then severe turbulence	75 to 85

TABLE 6.4. Winter Flying Conditions in Various Air Mass Types.

Air Mass	Clouds	Ceilings	Visibilities	Turbulence	Surface Temp. (Degrees F.)
cP (near source region)	None	Unlimited	Excellent (except near industrial areas, then 1 to 4 miles)	Smooth except with high wind velocities	−10 to −60
cP (southeast of Great Lakes)	Stratocumulus and cumulus, tops 7000 to 10,000 ft.	500 to 1000 ft., 0 over mountains	1 to 5 miles, 0 in snow flurries	Moderate turbulence up to 10,000 ft.	0 to 20
mP (on Pacific Coast)	Cumulus, tops above 20,000 ft.	1000 to 3000 ft., 0 over mountains	Good except 0 over mountains and in showers	Moderate to severe turbulence	45 to 55
mP (east of Rockies)	None	Unlimited	Excellent except near industrial areas, then 1 to 4 miles	Smooth except in lower levels with high winds	30 to 40
mP (East Coast)	Stratocumulus and stratus, tops 6000 to 8000 ft.	0 to 1000 ft.	Fair except 0 in precipitation area	Rough in lower levels	30 to 40
mT (Pacific Coast)	Stratus or stratocumulus	500 to 1500 ft.	Good	Smooth	55 to 60
mT (east of Rockies)	Stratus or stratocumulus	100 to 1500 ft.	Good	Smooth	60 to 70

significant classifications are *maritime* (m) and *continental* (c) because of the importance of the differing moisture content.

In the Northern Hemisphere, polar air masses are formed in northern regions where the air is cooled from below. Typically, large-scale sinking motion is also present. Hence these air masses are cold, stable, and dry. In contrast, tropical air masses are warmed from below, producing less stable lapse rates. If they form over the ocean, convection can carry surface moisture high above the surface. Tropical air masses are thus fertile sources of clouds and weather. In middle latitudes mixing of air by the action of moving weather systems prevents the formation of unique air mass types. These regions are the meeting place of air masses from north and south, as we shall see.

AIR MASS WEATHER

The weather changes associated with moving systems are complex and difficult to forecast. There still are many occasions in which an identifiable air mass overlies a large area of the United States producing characteristic and fairly uniform weather. These characteristic weather types are summarized in Tables 6.3 and 6.4. Air mass types are indicated in the tables as c for continental, T for tropical, P for polar, and m for maritime origin.

Fronts

The circulation of the atmosphere moves air masses from their source regions and brings them into contact. Thus the differences in temperature, moisture, and stability which distinguish one air mass from another tend to be concentrated in narrow transition zones. These surfaces along which air masses meet are termed *fronts*.

THE POLAR FRONT

The boundary between the two principal air masses—polar and tropical—is known as the *polar front*. We can roughly describe this as a continuous transition zone encircling the globe. It is by no means uniform in character. In some regions terrain features obscure the transition. In others, the circulation tends to disperse the temperature gradient, producing a gradual transition between polar and tropical air.

For much of the year a picture rather like Figure 6.12 could be drawn. This chart, typical of winter conditions, illustrates a number of frontal characteristics. Polar air, colder and heavier than tropical air, pushes underneath it and extends further south at the surface than aloft. Also evident are a number of areas where polar air bulges to the south or warm air pushes north. These waves are created by air circulation around moving high- and low-pressure systems in the westerlies.

Although the polar front is the dominant feature of weather charts in mid-latitudes, other fronts may be found between air masses of the same general source region. The necessary conditions for front formation are simply the existence of dissimilar air masses and an airflow pattern which concentrates their differences into a narrow zone.

COLD FRONTS

Fronts are classified by their motion. A front which moves so that cold air replaces warm air at the ground is called a *cold front*. The leading edge of the ad-

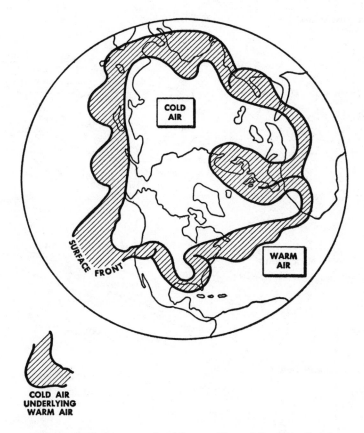

Figure 6.12. An example of the position of the polar front.

vancing cold air has a fairly steep slope of 1:50 to 1:150. A slope of 1:50 means that at a point 50 miles back from the surface position of the front, the front would be found 1 mile high.

Because of this steep slope, rapidly moving cold fronts are associated with rapidly rising warm air in a narrow band. This produces clouds of considerable vertical development in a narrow strip with frequent showers and thunderstorms. The weather pattern classically associated with cold fronts is shown in Figure 6.13(A).

In the Northern Hemisphere, cold fronts are generally oriented from northeast to southwest and move to the southeast. Ahead of the front, winds are southerly, and fair-weather clouds are prevalent. As the front approaches, clouds thicken, the pressure falls, and the wind increases. Showers and thunderstorms are encountered around the time of frontal passage, which is signaled by a shift of the wind to the northwest, a drop in temperature, and a marked rise in pressure. With a strong, fast-moving cold front, clearing is rapid after frontal passage.

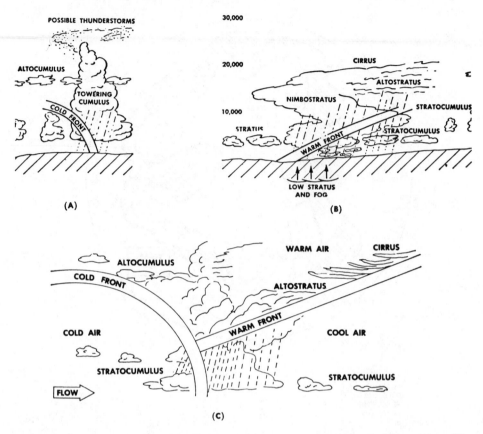

Figure 6.13. Classical concept of frontal weather. (A) Cold front. (B) Warm front. (C) Occluded cold front. In an occluded warm front, the cool air would be riding up on the slope of warm air.

WARM FRONTS

A *warm front* is the leading surface of an advancing mass of warm air. The slope of the frontal surface is generally much flatter than in a cold front. The warm air rises over the cold air mass much more gradually, producing a widespread area of clouds and precipitation. Rain falling out of the warm air mass into the lower cold air may saturate it, causing widespread stratus and fog. With ground temperatures below 32°F, freezing rain may result. A model picture of warm-front weather is shown in Figure 6.13(B).

STATIONARY FRONTS

On occasion, the circulation pattern will maintain a front's intensity, but will not give it a significant motion toward either the warm or cold air masses. Such a nonmoving front is termed a *stationary front*. The weather associated with it typically resembles warm-front weather. Stationary fronts are often produced on the trailing southwestern ends of cold fronts which advance into the Gulf Coast region of the United States and stagnate. Persistent fog, drizzle, and stratus are found on the cold side of the front.

In the northwestern United States during the winter months, there is frequently a stationary front oriented northwest-southeast. It is the boundary between the very cold, dry cP air masses from over central and western Canada, and the warmer mP air from over the northern Pacific. Temperature differences across the front are very pronounced, and it typically moves back and forth with each new air mass. Quite strong inversions may exist in the area of the front, and low clouds with fog, light snow, or drizzle may persist for days at a time in the colder air.

FRONTAL WAVES AND OCCLUSIONS

The air masses bounding fronts are set into motion by moving circulation systems in the westerly wind patterns of middle latitudes. These systems are seen on weather charts as centers of high and low pressure with winds circulating around them in agreement with the geostrophic wind law. In the Northern Hemisphere, the airflow is clockwise around a high and counterclockwise around a low. The reverse directions are observed in the Southern Hemisphere. In either hemisphere, circulation around a low is called *cyclonic,* around a high *anticyclonic*.

Low-level air tends to spiral inward in lows and outward in highs. Thus lows are regions of converging and rising air, while highs are characterized by diverging and sinking air. As a result, strong fronts and massive cloud systems are primarily found in low-pressure areas.

Lows and their associated frontal systems go through a typical life cycle as shown in Figure 6.14. In the figure, the dark line indicates a front. In standard meteorological practice, pointed pips on this line pointing toward warm air indicate a cold front, rounded pips pointing into cold air show a warm front, while alternating pointed and rounded symbols denote a stationary front. When circulation around a low-pressure area on the front begins, cold air is forced south to the west of the low, and warm air is moved north to the east, forming cold and warm fronts. Deepening of the low continues this process, producing the frequently observed pattern shown in Figure 6.14(C).

The cold front generally moves faster than the warm front, and eventually overtakes it near the low-pressure center (Figure 6.14[D]). There is generally sufficient air mass contrast along this line of convergence to define an *occluded* front (denoted by mixed pointed and round pips on weather charts). A cross section of an idealized occluded front viewed from the south is shown in Figure 6.13(C). In this example, the air behind the cold front was colder than the air ahead of the warm front. The cold air thus slides in under the cool air, lifting a tongue of warmer air aloft and forming a *cold-front occlusion*. The opposite situ-

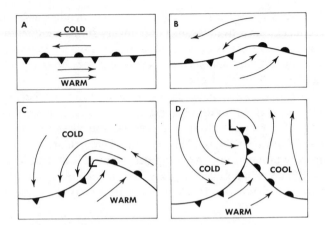

Figure 6.14. Life cycle of a frontal wave. (A) Undisturbed front. (B) Initial wave. (C) Fully developed wave. (D) Occlusion.

ation is also possible: Air to the east of the low may be colder than that to the west. The occlusion then appears as a continuation of the warm front and is termed a *warm-front occlusion*.

Occluded fronts are generally accompanied by an extensive band of clouds and precipitation curving back into the low-pressure center.

Tropical Weather

While the weather of middle latitudes is dominated by the effects of traveling cyclones and anticyclones, tropical weather generally follows a rather routine course. In place of the violent alternations between northerly and southerly winds with accompanying weather and temperature changes, we find in general a simple daily cycle of clouds and weather superimposed on a cycle of wet and dry seasons. For this reason, a close study of local and regional climatology is particularly rewarding to the aviator in tropical regions.

The weather systems which disturb the regular course of events in the tropics are not nearly as well defined and understood as those of middle latitudes. Here we can but briefly mention some of the outstanding features.

EASTERLY WAVES

In the easterly trade wind belts, wavelike disturbances traveling from east to west are often observed. Typically these are preceded by northerly winds and some clearing of the normal trade-wind cumulus clouds. As the crest of the wave passes, heavy showers with tops rising to 15,000 ft. or more are observed, and the wind shifts to a southerly direction.

THE EQUATORIAL TROUGH

The trade winds of both hemispheres converge toward a band of low pressure near the equator. This equatorial trough, sometimes known as the *intertropical convergence zone,* is a region of increased cloudiness and weather. It is generally displaced from the equator toward the summer hemisphere and is subject to north-south oscillations.

HURRICANES

The most spectacular features of tropical weather are the intense cyclonic storms called hurricanes in the Atlantic or typhoons in the Pacific. These storms originate over warm tropical waters in the late summer and early fall.

Initially, they appear as a cyclonic air circulation around a weak low-pressure center. These tropical depressions may increase in intensity as they drift westward, and may intensify until winds reach hurricane force (65 knots). On reaching the western Pacific or Atlantic, the storms generally curve northward, where they bring extensive rain, winds frequently over 100 knots, and great damage. Although these storms are far less frequent than the cyclones of middle latitudes, their immense strength and unpredictable shifts in direction make them the most feared weather disturbances of our earth.

Weather Services

Aviation weather needs the world over have made observation, reporting, and analysis of weather one of the truly international activities of man.

In the United States, the National Weather Service provides basic weather service. Supported by the weather information networks of the Air Force, Navy, NASA, Coast Guard, and FAA, it operates the World Meteorological Center (WMC) near Washington, D.C. This is the analysis center for all data, and issues the basic analyses and forecast charts which are used by civil and military forecasters as source material for their briefings. Forecasts for individual terminals are produced by regional centers or the terminal stations themselves, where WMC's output is also interpreted by local forecasters for individual flights.

Through Flight Service Stations (FSS) and Weather Service Offices (WSO), pilots can obtain the information and guidance necessary for safe and efficient flight. Knowing how to obtain and use this information is an absolutely essential element of modern airmanship.

WEATHER OBSERVATIONS

While there are many diverse sources, the most abundant and useful weather reports are those recorded regularly by professional weather observers through-out the world. At many active airfields, observations are taken every hour or whenever significant changes occur. These aviation-oriented reports are distrib-uted rapidly by teletype circuits to other aviation terminals. Observations for major international airports, for example, are distributed world-wide, in but a few minutes.

The pilot is most likely to encounter the Aviation Weather Reports transmitted on regular schedules over civil and military circuits to teletype units in airfield weather stations or Flight Service Stations. A little study of the typical example given below, and queries in the local weather station should enable the aviator to understand the bulk of these reports. Understanding the terminology used is im-portant because, for example, the difference between an *estimated* and a *mea-sured* ceiling could be critical in pilot decision making.

"MIA 8 SCT E15 BKN 50 OVC 2R-F 122/54/52/0910/989/BINOVC"

MIA: *Station identification symbol:* This identifies the reporting station on all weather teletype transmissions. When it is followed by *SP*, the weather reported is a special observation, rather than the normal hourly observa-tion, because of a significant change in the weather since the last hourly report.

AMOS immediately following the station identification symbol indicates that the observation was taken and transmitted by an unmanned Automated Meterological Observation Station.

The sky condition group: Cloud heights are reported to the nearest 100 ft. and transmitted in hundreds (8 = 800 ft. etc.). Cloud amounts are determined in tenths of sky cover and encoded as:

CLR: Clear—Less than 1/10 sky coverage.

SCT: Scattered—not more than 5/10 total sky cover.

BKN: Broken—more than 5/10 but not more than 9/10 sky cover.

OVC: Overcast—more than 9/10 sky cover.

X: Obscuration—interference to vision such that the sky cannot be seen.

A "–" sign may precede any of the above to denote thin coverage. A thin cov-erage (one that can be seen through) does not constitute a ceiling, and there will not be any indicated method of height determination preceding it.

The method by which cloud cover height is determined is indicated by one of the following letters:

M: Measured
B: Balloon

A: Aircraft report

E: Estimated

R: Radar or RAOB

W: means the ceiling is indefinite, the number following it indicates a best estimate of its effective height.

The group in the example

"8 SCT E15 BKN 5Ø OVC 2R-F"

indicates the following multiple cloud layers: scattered clouds at 800 ft.; a broken layer of clouds estimated at 1500 ft.; an overcast cloud condition at 5000 ft.

It is important to recognize that clouds are reported *as seen by the ground observer*. The coverage reported at a level reflects all the clouds the observer can see *between the surface and that level*. A report of 15 SCT 30 BKN, for example, could represent 4/10 of clouds at 1500 ft. and 6/10 coverage at 3000 ft. On the other hand, the same report could result from a judicious arrangement of 4/10 at 1500 ft. and only 2/10 at 3000 ft.

2R-F: *Visibility in statute miles and the weather restricting it.* The visibility reported is 2 miles, and the visibility is restricted by light rain and fog. Some of the common weather symbols are:

R: Rain (steady)

RW: Rain shower (W indicates showery type precipitation)

T: Thunderstorm

S: Snow

L: Drizzle

ZL: Freezing drizzle (Z indicates freezing on contact)

A: Hail

IP: Sleet

F: Fog

GF: Ground fog

H: Haze

K: Smoke

D: Dust

BD: Blowing dust

N: Sand

There are many of these symbols, and they may be found listed in any weather station. They may be grouped together to show multiple weather conditions.

Other symbols are:

122/: *The sea level pressure.* This is the barometer reading in the weather station reduced from the station elevation to standard sea level. It is a pressure reading in millibars. This example, 122, decodes as 1012.2 mb.

54/52: *The temperature and dew-point temperature in degrees Fahrenheit* (54°F temperature/52°F dew point in this example).

0910: *The surface wind*. The first two digits indicate the true (not magnetic) direction from which the wind is blowing. The last two digits indicate speed in knots. In the example, the wind is blowing from 90° at 10 knots. Variations in the wind report might be:

00 Calm
G   Gusts (followed by the peak speed)
Q   Squalls (followed by the peak speed)

989/: *The altimeter setting in inches of mercury*. This example is decoded as 29.89 in. of mercury.

The remainder of the report might carry special cloud data, maximum or minimum temperatures, radar reports, NOTAMS (notices to airmen) in coded form, abbreviated remarks descriptive of the weather (such as BINOVC—breaks in the overcast), coded forecast groups from Navy stations, etc.

Only a small part of the weather report has been covered here. The most essential items to the flyer are clouds, visibility, precipitation, surface temperature, wind, temperature-dew point spread, and altimeter setting.

Pilot Reports.   Surface-based observations give valuable information on conditions at and near airports. However, conditions between observing stations and weather phenomena aloft can best be observed by the pilot. Because of this, flight controllers and weathermen eagerly solicit weather reports from aircraft.

On teletype summaries, pilot reports are indicated by the code word PIREP and contain the following information:

1. Originating station designator
2. The word PIREP and the filing time (GMT)
3. Location or extent of the reported weather
4. Time of observation (local time)
5. Weather condition (turbulence, icing, cloud tops, weather, etc.)
6. Altitude
7. Type of aircraft

SPECIAL WEATHER OBSERVING TECHNIQUES

Increasingly, there are better capabilities for observing weather in remote areas. Three of the more valuable means are AMOS, Airborne Weather Reconnaissance, and Weather Satellite.

AMOS.   Not all established airports, even with published instrument approaches, have qualified aviation weather observers. Many of these, as well as locations where there is not an airport, are equipped with Automated Meterological Observation Stations (AMOS). These report temperature, dew point, wind direction and velocity, altimeter setting, peak wind during the previous hour, and the time at which it occurred.

OTHER INFORMATION

NOTAMS.   Notices to airmen, of a temporary nature, are also placed on the hourly sequence reports, using abbreviations not too difficult to read after gaining familiarity with them. A NOTAM begins with an arrow (→), followed by the station symbol, and appears immediately after the weather report.

For example:

"JMS 200 –BKN 10 124/37/32/2903/986"

"→JMS 03/06 JMS 3-21 RWY LGHTS OTS"

Translation:

Jamestown 20,000 ft. thin broken, visibility 10 miles, pressure 1012.4 millibars, temperature 37, dew point 32, wind 290 at 3 knots, altimeter setting 29.86. Jamestown Notice to Airmen, sixth one for March, Runway 3 and 21, runway lights out of service.

WEATHER RADAR OBSERVATION

Raindrops, snow, sleet and hailstones reflect radio waves and hence can be "seen" by radar. Large raindrops reflect much more energy than small ones. These large drops require strong updrafts to generate and support them. Hence the brightest radar echoes are associated with hazardous convective storms, making radar an invaluable aid to the aviator.

Airborne radar is discussed in Chapter 5. The weather services operate many weather radars throughout the United States. They permit local forecasters to detect and track thunderstorms in an area of as much as 250 miles radius. Verbal descriptions of radar echoes are regularly transmitted to a Weather Bureau unit at Kansas City, Missouri, where summaries and maps are prepared. These are transmitted to weather stations nationwide and give a quite complete picture of significant weather throughout the nation.

Weather Reconnaissance.   The Air Force, Navy, and Weather Bureau all operate specially instrumented aircraft in their respective areas of interest to obtain detailed current reports of the existence and intensity of adverse weather. This is particularly important for coastal storm and hurricane warnings, where frequent, precise, specific information is essential.

Weather Satellites.   Both the United States and the USSR have launched a number of weather satellites. These circle the earth at an altitude of several hundred miles in orbits designed to be sun-synchronous. That is, they pass over each part of their path at the same local time.

Two basic types of pictures can be received from weather satellites. Certain satellites will transmit their latest picture whenever interrogated by a ground station. Facilities throughout the world can receive pictures from these Automatic

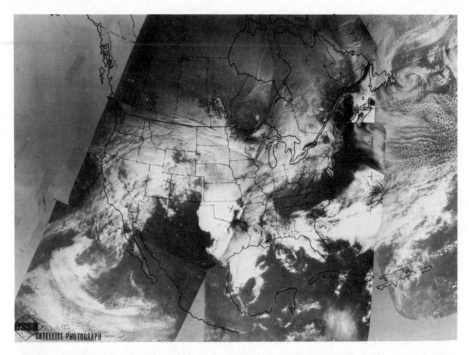

Figure 6.15. Mosaic of APT weather satellite pictures. *(Courtesy of National Environmental Satellite Center.)*

Picture Transmission (APT) satellites. A mosaic of a number of these pictures is shown in Figure 6.15. The light areas are cloud masses, while land and sea are about equally dark. These pictures are particularly valuable in forecasting weather for transoceanic routes or for isolated island stations.

Other weather satellites store a number of pictures in on-board tape recorders and transmit a large volume on demand to special NASA installations. These data are then processed by computer at the National Environmental Satellite Center at Suitland, Maryland, to form global pictures of cloud distribution. For example, Figure 6.15 shows weather systems over most of the United States, information obtainable from no other source.

Weather Analyses and Forecasts

Weather observations are of maximum use to the meteorologist and his customers only after a coherent picture of the weather has been drawn from them. A knowledge of the recent past, climatology, forecasts, and judgment all are used to produce these analyses. From them, the future course of events can be projected by a mixture of scientific knowledge, empirical rules, and experience.

METHODS OF PREPARATION

In addition to the basic NMC analyses and forecasts, specialized analyses and forecasts for world-wide military operations are made by the Air Force Global Weather Central (AFGWC) at Offut Air Force Base, Nebraska, and the US Navy's Fleet Numerical Weather Central at Monterey, California. All three centers are supplied with data by teletype and by the Air Force's Automated Weather Network which furnishes overseas weather data collected world-wide. AFGWC and NMC also receive satellite data from the national satellite system.

The weather centrals use electronic computers to decode, edit, and sort these data. Computers also analyze the upper-air data in preparation for "numerical"

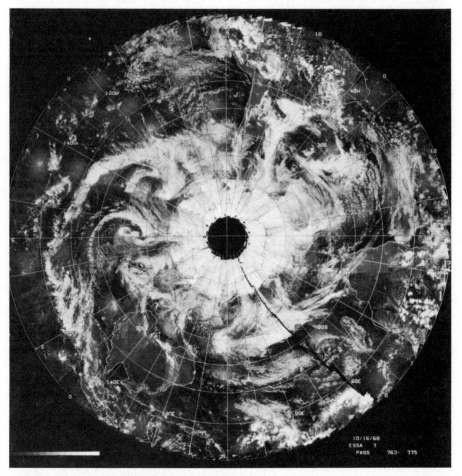

Figure 6.16. Satellite cloud patterns. Computer-processed pictures of cloud patterns over the Southern Hemisphere. Composite of data observed by Essa-7 in several orbits. *(Courtesy of National Environmental Satellite Center.)*

forecasting. This comparison of data characteristic of selected vertical atmospheric columns at a given time with typical models, is the simulation of the atmosphere's behavior through numerical calculation based on physical laws.

WEATHER SERVICES FOR AVIATION

The NMC is the source of information, analyses, and forecasts, principally used by other weathermen. These, in turn, can serve the pilot. Aviation weather specialists in the Weather Bureau's Flight Advisory Service Centers prepare and distribute terminal and area forecasts. They are responsible for areas roughly coinciding with Air Route Traffic Control Centers of the FAA.

In the WSO (Weather Service Office) or FSS (Flight Service Station), most pilots find their weather information. The briefer here is not necessarily only a briefer. He may have observing and forecasting duties, too, and it is likely that his experience in the area can be of considerable assistance in solving local problems. In addition, he can request and expedite the occasional greater-than-normal requirements which pilots may have. He is also responsible for distributing hazardous weather warnings, known as SIGMETS and ADVISORIES FOR LIGHT AIRCRAFT.

High-Altitude Forecast Centers, which still serve mainly airline jet operations, provide forecasts, mostly in graphic form. Seven of these centers cover the area from the Philippines and Japan east into Europe, and from the North Pole to the equator. Much of the product of the United States center, located with NMC, goes directly to airline operations offices.

The National Severe Storm Forecast Center (NSSFC) is located in Kansas City, Missouri. The Air Force's Air Weather Service has a hazardous weather office at Offut Air Force Base. It furnishes similar products for military operations; the two provide mutual support and are fully cooperative.

The National Hurricane Center in Miami, Florida, uses information from Air Force and Navy aerial reconnaissance, satellites, and surface observations to locate, track, and forecast the movement of hurricanes. They distribute forecasts and warning bulletins by aviation weather circuits, and by any other available means, including news media.

With the aid of weather satellites, it is almost impossible for a tropical storm to reach damaging size without detection. When one is located, it is christened with a suitable name, and regular bulletins are disseminated on its progress. If it approaches land, reconnaissance aircraft of the Air Weather Service and the US Navy are regularly dispatched to determine its exact location and intensity.

This system obviously requires an extensive distribution network. The FAA distributes weather and aviation information on a series of circuits to more than 500 terminals. The Air Force Continental Meterological Data System (COMEDS) collects information and distributes it to military installations. The Navy operates radio circuits from shore bases to ships at sea, both sending and receiving information. All these systems have facsimile networks used to send a wide variety of charts, discussed later, throughout the system.

AIRCREW BRIEFING

Military and airline operations offices insure that their aircrews use all the information available, presented in the form most useful for the flight program they conduct. The individual pilot or other interested airman has a briefing service available to him through the WSO or FSS nearest him, by telephone, and by radio via FSS.

He will find information available on almost any aspect of flight in the form of charts, reports, and forecasts in chart form and in writing; nevertheless, it is important to know what to ask for, and what to expect.

The hourly teletype sequence reports and special teletype reports provide the most up-to-date overall picture of the surface weather, and that portion of weather aloft that can be seen from the ground. Checking several succeeding hourly reports provides a good picture of how the weather is changing, or how a weather system is moving. Special reports pinpoint times of frontal passage, warn of rapidly deteriorating weather, and so on.

Pilot reports help fill in the gaps between ground observer points, which often are more than 100 miles apart.

Other data available include facsimile Weather Depiction Charts, Surface Weather Charts, Constant Pressure Charts, Winds Aloft Charts, and Radar Summary Charts. Similar charts provide surface weather forecasts. Teletype forecasts include Terminal Forecasts, Area Forecasts, and Winds Aloft Forecasts.

In contrast to the pilot who asks for and gladly accepts the assistance of the trained and experienced briefers of Weather Service or Flight Service is the pilot who enters the facility, barely says "Hello," heads for the reports and charts on display, and proceeds to brief himself.

He would be much better prepared for the flight he is about to make if he would seek out the expertise of the persons whose primary business it is to help him. They can combine all the information available to provide the best possible picture of the conditions he can expect to encounter.

WEATHER DEPICTION CHART

One of the most popular briefing aids, this chart is prepared at NMC every 3 hours beginning with data observed at 0100Z. Because the data are at least 1 hr. 20 min. old by the time the chart is posted in the WSO or FSS, and because these charts are based on *observed* data only, pilots should check local conditions.

Weather Depiction Charts contain precipitation and other important phenomena, sky coverage, visibility if less than 6 miles, and cloud bases in hundreds of feet above terrain up to 20,000 ft. Using this chart in conjunction with the hourly

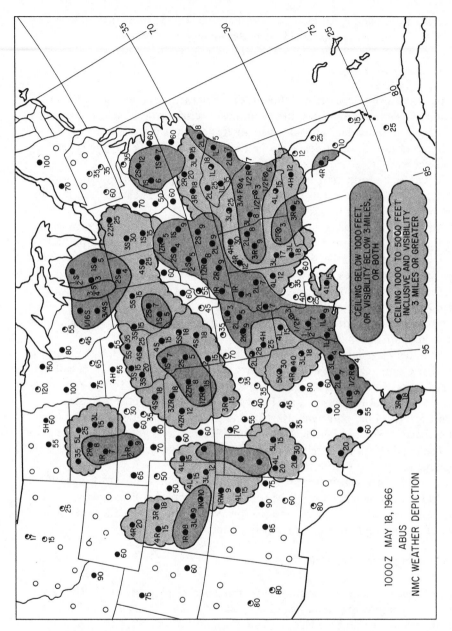

Figure 6.17A. Weather Depiction Chart.

STATION MODEL	TOTAL SKY COVERAGE	SIGNIFICANT WEATHER
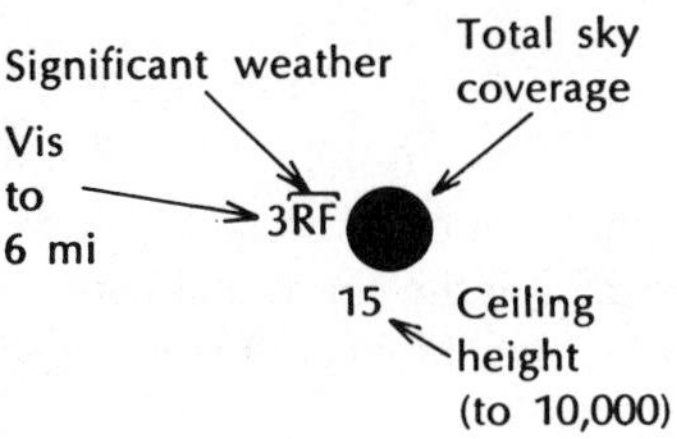 Significant weather — Total sky coverage / Vis to 6 mi — 3RF / 15 Ceiling height (to 10,000)	 ○ Clear ◐ Scattered ◕ Broken ● Overcast ⊗ Obscured	T Thunderstorm ZR Freezing rain ZL Freezing drizzle R Rain S Snow E Sleet L Drizzle W Shower H Haze K Smoke F Fog GF Ground fog ≢ Clouds topping ridges
ANALYSIS:		
Ceiling and visibility classes:	Outline:	Color code:
(1) Ceiling below 1000 ft. or visibility below 3 miles, or both.		Red
(2) Ceiling 1000 to 5000 ft. inclusive and visibility 3 miles or greater.		Blue

NOTES:

(1) Precipitation will always be entered when reported as occurring at the station at the time of observation, except that no more than two symbols for significant weather will be entered.

(2) Obstructions to vision other than precipitation will be entered when visibility is reduced to 6 miles or less, except that if two forms of precipitation are reported, the other obstructions to vision will not be entered. Haze and smoke will be omitted if precipitation is reported.

(3) Visibility values of 6 miles or less will be entered in miles and fractions.

(4) All ceiling heights will be entered in hundreds of feet.

(5) The analysis of the ceiling-visibility classes will be based on the reports themselves and is not meant to include possible between-station orographic effects, or even systematic interpolations between stations.

Figure 6.17B. Symbolic notation used on Weather Depiction Charts. *(A and B reprinted from M. W. Cagle, "A Pilot's Meteorology," Third Edition, Van Nostrand Reinhold, New York, 1970.)*

reports and the Surface Weather Chart is an excellent briefing technique for shorter flights.

SURFACE WEATHER CHARTS

This is the most informative single chart available to the pilot or forecaster. The data are observed almost simultaneously for the entire area of the chart, and with NMC's computer methods, the chart is no more than 1½ hr. old when it reaches

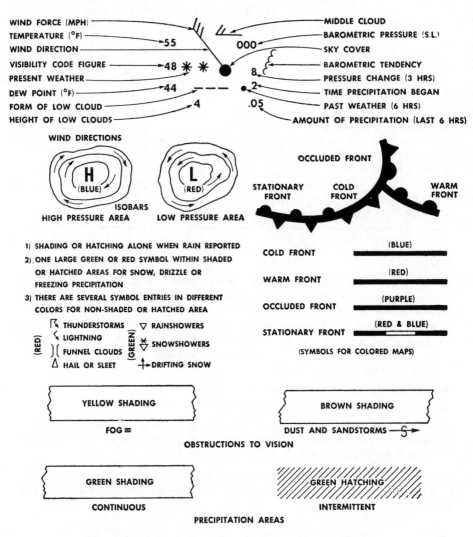

Figure 6.18A. Legends and symbols found on weather map.

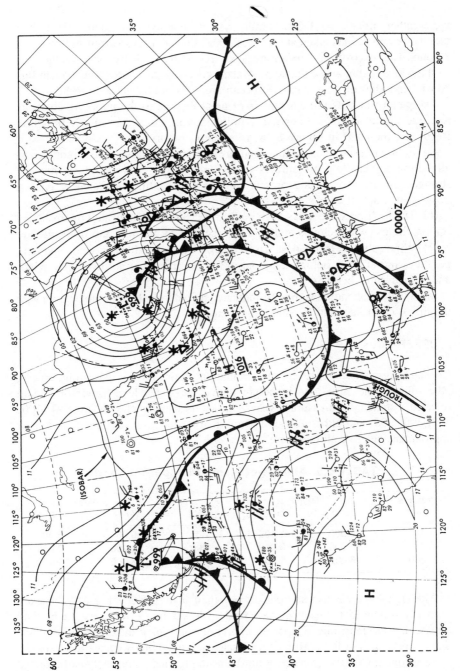

Figure 6.18B. Surface weather map.

the weather office over the facsimile system. For the contiguous states, it is based on observations taken at 0000Z, 0300Z, 0600Z, etc. and transmitted every 3 hours. Charts covering smaller areas are frequently made by local weathermen to show local detail and influences. A chart covering the United States including Alaska and Hawaii, and extensive ocean areas, is published every 6 hr.

The pattern of pressure systems is of great interest to the pilot. All lines are *isobars*, which are lines connecting points having equal barometric pressure. Regarding the chart in Figure 6.18, it is interesting to note:

(a) Following the geostrophic wind law mentioned earlier, the winds generally blow so that low pressure is to the left looking downwind. Winds are strongest in areas of strong pressure gradients, identified by closely packed *isobars*, or lines of constant pressure.

(b) The high-pressure areas have generally clear skies, indicated by unshaded station circles, while clouds and weather accompany the low and its fronts.

(c) Winds shift as fronts pass. In Kentucky we find southerly winds and temperatures in the 50's. Behind the front in Illinois, there are northwesterly winds and temperatures below 40.

(d) Weather phenomena may be indicated by colored shading (not shown with the map), color supplied by station personnel.

UPPER-AIR CHARTS

Constant-pressure charts reveal the circulation of air aloft by indicating, among other things, the variations in height above mean sea level (msl) of certain selected "standard" pressure levels. The levels usually chosen are 850, 700, 500, 300, 200, and either 150 or 100 millibars (mb). Standard charts and their approximate relative altitudes are:

TABLE 6.5. Altitude of Standard Upper-Air Charts.

Pressure (mb)	Height (feet)	(meters)	Pressure (mb)	Height (feet)	(meters)
850	5,000	1,500	300	30,000	9,000
700	10,000	3,000	200	40,000	12,000
500	18,000	5,500	100	53,000	16,000

Although some weather stations prepare their own high-level charts, these are normally produced for each of the standard levels by NMC every 12 hr. and transmitted via facsimile. Upper air charts not only give information on winds and temperatures at flight altitudes, but also serve as vital forecast tools for the meteorologist. Circulation patterns aloft are simpler and more easily extrapolated in time than those of the surface layers. Small-scale weather systems are steered by the broad currents of the upper air, and the life cycles of individual storms are closely tied to upper-air patterns.

On these charts, data are plotted at the locations of rawinsonde (radar wind sounding) stations. Winds are indicated by the shaft of an arrow pointing in the wind direction toward the station circle. A dark triangular flag indicates 50 knots, while barbs and half barbs indicate 10 and 5 knots, respectively.

The temperature in degrees Celsius is plotted to the left of the station circle, and the dew-point spread is plotted below it. To the right of the station appears the height of the pressure surface in meters, with the first or last digit omitted to save space. The lines are used in analysis, called *isolines* (lines of equal value), represent pressure, temperature, moisture, and wind speed.

Contour Lines.   Contour or height lines connect points which represent equal altitudes above sea level based on the pressure level of the constant-pressure chart. These are drawn for intervals of 60 meters on charts below the 300-mb surface, and for intervals of 120 meters on the 300-, 200-, and 100-mb charts. Winds blow parallel to the contour lines with lower heights to the left and higher heights to the right.

Isotherms.   These short, dashed lines connect points of equal temperature. They are normally drawn for even 5°C intervals.

Moisture Lines.   Isolines of dew point, usually called moisture lines, connect points of equal dew-point temperature. On the higher-level charts, generally 500 mb and higher, moisture lines are not drawn since the low temperatures at these higher levels not only prevent much moisture from existing, but tend to make measuring difficult.

Isotachs.   On the higher level charts, generally 500 mb and higher, dotted lines of equal wind speed, called isotachs, are often drawn. They relate only to speed and do not indicate direction. Heavy broken lines with arrowheads are indicative of jet-stream locations.

Fronts.   Fronts are indicated in the same symbolic manner as they are on surface charts, provided that the frontal surfaces extend to the pressure levels involved. It is important to remember that fronts are practically always displaced geographically at different levels because of frontal slope.

PROGNOSTIC CHARTS AND SEVERE WEATHER SERVICES

The atmosphere is subject to the same natural laws which govern all other material bodies. If we know its state at one moment and know all the forces acting upon it, its future behavior should be as well determined as that of the sun and moon. This capability has been partly realized today with the aid of modern data-gathering and analysis systems, and global analyses of the atmosphere can be produced in reasonable time. Computers produce useful forecasts out to 5 days once each day by numerically simulating the behavior of the atmosphere. Numerically produced prognostic charts for most of the standard analysis levels are transmitted by facsimile to weather stations. All stations receive NMC products, while Air Force and Navy stations also receive products from the military weather centrals.

Prognostic charts are similar in content to actual analyses. Contours of pressure or height, isotherms, isotach, etc., are presented, but station data are naturally absent. The surface forecast charts are produced manually, because of low-level weather's local influences, although the forecasters rely heavily upon numerical forecast guidance.

Severe Weather Forecasting. Special analysis and forecast techniques are required to predict the probability of occurrence of small severe weather phenomena such as thunderstorms and tornadoes. For this reason, both the National Weather Service and the Air Weather Service of the U.S. Air Force have established specialized centers for severe storm prediction. The National Weather Service's National Severe Storms Forecast Center at Kansas City, Missouri, prepares advisories on severe weather for public and aviation use. These are issued by local field offices which also disseminate local warnings based on radar, public observations, pilot reports, and similar sources.

The AWS Military Weather Warning Center is located with the Air Force Global Weather Central at Offutt Air Force Base, Nebraska. It issues warnings on hail, gusty winds, severe thunderstorms, and tornadoes both for areas and for certain individual military installations. Its principal product is a daily teletype map showing areas of expected severe weather.

Severe weather forecasts have become increasingly reliable over the years, and now merit the most careful attention of pilots and others considering the traverse of an area in which severe weather is forecast.

WINDS ALOFT CHARTS.

Planning for efficient, safe, and economical flight requires the use of winds aloft in predicting ground speed, determining fuel required, and in flight, being able to determine practicable alternate routes and destinations. Because the winds aloft are determined more by the pressure systems than by terrain effects, they are generally more reliable, either as observed, or in forecasts.

NMC plots the data it receives at 0000, 0600, 1200, and 1800Z using the barb and pennant arrows described for the constant pressure charts. In facsimile form, these reports are transmitted in three sets of four charts for Lower Levels (second level above the surface to 10,000 ft.), Intermediate Levels (14,000 to 30,000 ft.), and Upper Levels (35,000 to 60,000 ft.). Here again, remember that these are *observed* data, and may be 2 to 9 hr. old.

RADAR SUMMARY CHARTS

These summaries of all radar reports in the contiguous states are transmitted 16 times a day by facsimile. They include severe weather warnings as well and are an essential briefing aid during the thunderstorm season.

FORECASTS

Forecasts may be mental, by the pilot looking at the weather office reports or the sky, or may involve the most sophisticated skills of the Weather Service and the weather centers of the armed forces.

They are much more detailed and specific than 12- to 36-hr. prognostic charts, and thus, while more valuable in flight planning, are more vulnerable to error in details. They may relate to areas, terminals, or to specific routes on request.

Area Forecasts.   Area forecasts are issued at 0040Z and 1240Z, with valid times from 0100Z to 1900Z and from 1300Z to 0700Z the following day. They provide a further outlook for the periods 1900Z to 0700Z the following day and from 0700Z to 1900Z the following day, respectively. Area forecasts are issued by Weather Service Centers located in Chicago, Kansas City, Salt Lake City, etc., each covering several states. A synopsis of current weather is given, with a forecast of the expected changes during the forecast period. This is followed by state-by-state forecasts of significant clouds and weather. Kansas City's forecast, for instance, is for Wyoming, Kansas, Colorado, Nebraska, South Dakota, and North Dakota. For example (from the 1240Z forecast 0300Z Monday to 0700Z Tuesday, outlook 0700Z Tuesday to 1900Z Tuesday):

WY CO. . . .
NO SIGCLD. AFT19Z OCNL 120SCT VRBL BKN WITH ISOLD TSTMS MOSTLY INVOF MTS. TSTMS DSPTG BY 05Z. CB TOPS 350. OTLK. . .
VFR

Translation:

For Wyoming and Colorado. . . . .
No significant clouds. After 1900Z occasional 12,000 scattered, variable broken with isolated thunderstorms, mostly in the vicinity of mountains. Thunderstorms dissipating by 0500Z. Cumulus cloud tops 35,000 ft. Outlook VFR.

Terminal Forecasts. Terminal forecasts are issued three times a day, at approximately 1000Z, 1500Z, and 2200Z. Each is for the period until the next forecast. The cloud cover, weather, wind direction, and velocity (if above 8 knots) are forecast throughout the next period (up to 12 hours), followed by the general outlook beyond that period. The forecasts are grouped by states.

For example:

ID 032244. . . . .
BOI 032323 80 SCT 150 SCT V BKN CHC C60 BKN RW–/TRW–G 35. 07Z 100 SCT 200 SCT OCNL WIND 1310. 17Z VFR

Translation:

Idaho, forecast issued the third of the month at 2244Z.

Boise, for the period beginning on the third at 2323Z. 8000 scattered, 15,000 scattered variable broken, chance of ceiling 6000 broken in light rain shower or light thundershower, gusts to 35 knots. Beginning at 0700Z, 10,000 scattered, 20,000 scattered, occasional wind 130° at 10 knots. VFR after 1700Z. (Winds are not forecast unless expected to exceed 8 knots.)

Area and terminal weather forecasts are amended if the weather outlook changes before the end of the period. For example:

BOI FT
BOI FT AMD1 040923 0855Z C7∅ OVC OCNL C 45 BKN RW–/TRW–
G35. 17Z VFR

Translation:

Boise terminal forecast, amendment 1, at 0923Z on the fourth. Beginning at 0855Z, ceiling 7000 overcast, occasional ceiling 4500 broken in light rain showers or light thundershowers, gusts to 35 knots. VFR after 1700Z.

Winds Aloft Forecasts. The airflow aloft (winds aloft) is predictable to the extent of knowing reasonably well, from current flow patterns, pressure systems, and past history, what the likely trends are. Radiosonde data from the many stations around the country are fed into a computer in the NMC in Washington. The output is a prediction of winds and temperatures aloft above each of about 26 selected stations around the country. These are transmitted by teletype to Weather Service Offices and Flight Service Stations and are available to pilots for flight planning. They are for altitudes above mean sea level, beginning at 3000 ft. (where ground level is less than 3000). Data used are for 0000Z and 1200Z soundings, forecast valid times are for times 12 and 18 hr. later. The data are for planning use for a period before and after each valid time, and forecasts are for 3000, 6000, 9000, 12,000, 18,000, 24,000, 30,000, 34,000, and 39,000.

An excerpt, for example:

FDUS 1 KWBC 040540
DATA BASED ON 040000Z
VALID 041200Z for USE 0600–1500Z.   TEMPS NEG ABV 24000

FT	3000	6000	12000	18000	24000	39000
ABR	2022	2216+23	2513+09	2814−09	3022−21	314357

Translation:

Wind aloft forecast for the US, forecast number 1 for the day, from Weather Central, issued on the fourth of the month at 0540Z. Data are based on soundings taken at 0000Z on the fourth. Valid time is 1200Z on the fourth. For use from 0600Z to 1500Z. Temperatures are negative above 24,000. Forecasts at 3000, 6000, 12,000, 18,000, 24,000, and 39,000 *(excerpted by editor)*.

For Aberdeen (South Dakota), forecast winds for the period are: at 3000 ft., 200° at 22 knots; 6000 ft., 220° at 16 knots, temperature 23°Celsius; 12,000 ft., 250°, 13 knots, 9°Celsius; 18,000 ft., 280°, 14 knots, −9° Celsius; 24,000 ft., 300°, 22 knots, −21°Celsius; 39,000 ft., 310°, 43 knots, −57°Celsius.

AIRMETS.   These are forecasts of weather of significance to light aircraft pilots and are issued when wind or weather conditions may affect operations of light aircraft. They include such things as low level turbulence, strong winds below 3000 ft. above the surface, and IFR weather conditions.

Example:

SLC WA 040425
040425–041025
AIRMET ALFA 1 NV WRN ID OCNL MDT TURBC BLO 14∅ CONDS CONTG BYD 1025Z.

Translation:

Salt Lake City Weather Advisory issued at 0425Z on the fourth.
For the period 0425 to 1025Z. Airmet Number A-1 by Salt Lake City. Nevada and western Idaho, occasional moderate turbulence below 14,000 ft., conditions continuing beyond 1025Z.

SIGMETS and CONVECTIVE SIGMETS.   These forecast conditions of significance to all aircraft, such as icing, severe turbulence, and thunderstorms. They also are issued for specific time periods and use codes and abbreviations similar to those used for AIRMETS.

Fog

Unlike other weather hazards a pilot is likely to encounter, fog presents little or no hazard during the en-route portion of the flight. Fog, however, is a hazard during take-offs and landings, since it restricts the visibility near the surface of the earth.

FOG FORMATION

The three factors that favor the formation of fog are high relative humidity, very light wind, and condensation nuclei.

Since fog is composed of liquid water, condensation is necessary, so high relative humidity is obviously of prime importance. In fact the relative humidity must be very near 100%. The natural conditions which bring about high relative humidity (or saturation) are frequently cited as being fog-producing processes: evaporation of additional moisture into the air, or cooling of the air to its dew point. A high relative humidity may be noted on the hourly teletype reports by observing the spread (difference in degrees) between temperature and dew point. Fog rarely occurs when the spread is more than 4°F. A very light wind or no wind is generally favorable for fog formation. Light winds provide a mixing action which extends the depth of cooling near the surface and thereby increases the depth of the fog. Finally, there must be condensation nuclei suspended in the air. These nuclei provide a base around which moisture condenses. Smoke, dust, and salt particles are the most common forms of nuclei found in the atmosphere. Although almost all regions of the earth contain sufficient nuclei to permit fog formation, certain regions (such as industrial areas) have a marked abundance of them (smoke particles and sulfur compounds). These regions frequently have fog with greater than normal spreads between temperature and dew point, and the resulting fogs tend to be more persistent.

FOG DISSIPATION

Turbulence, resulting from strong winds, causes fog to dissipate by mixing warmer or dryer air from higher layers of the atmosphere with that of the surface, thus widening the temperature-dew point spread. Dissipation may also be accomplished by heating of the fog layer, either from the radiation of solar heat or from adiabatic downslope motion, both of which result in a widening of the temperature-dew point spread, so that the fog evaporates, or "burns off."

FOG TYPES AND CHARACTERISTICS

One common type of fog encountered by the pilot is *radiation fog*, which is formed by radiational cooling. After sunset, the earth radiates heat, gained during daylight hours, to the atmosphere, and by early morning the temperature at the surface may drop as much as 20°. Since the dew point normally changes only a few degrees during the night, the result is a decrease in the temperature-dew point spread. If the radiational cooling is great enough and other conditions are favorable, radiational fog forms. This type of fog is most likely to form under the following conditions:

1. Clear skies (maximum radiational cooling).
2. Sufficiently high dew point (so that temperature may cool to it).
3. Low, steady wind less than 10 mph.

These conditions are most frequently realized when a land region is under the influence of a high-pressure cell.

A second type of fog is called *advection fog* (Figure 6.19). Advection fog is very common along coastal regions and is formed by the movement of moist air over a colder surface. One example of this fog is *sea fog*, which is formed (as the name implies) over water. Cold ocean currents, such as are present off the coast

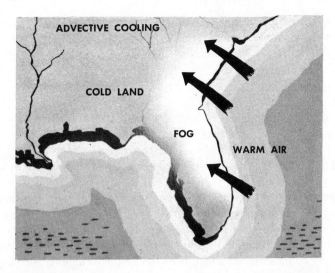

Figure 6.19. Advection fog is formed as warm moist air moves over cold ground. *(Courtesy of US Air Force.)*

of San Francisco, serve to cool and condense warm moist air which comes from seaward. This fog may be carried inland by the wind and is frequently intense. Advection fog frequently forms during periods of relatively strong winds despite the apparent dissipating effects of the turbulence generated by strong winds.

A third type of fog, which is called *upslope fog,* is formed by the transport of air up a rising land surface. As the air rises, it cools by expansion (adiabatic cooling) as a result of the decrease in pressure. Fog results when the cooling is sufficient to bring the temperature down to the dew point. The most common fog of this type is formed on the eastern slope of the Rocky Mountains by the westward flow of air from the Missouri Valley. Sufficient wind to support continued upslope motion is a necessity for this type of fog formation. However, if the wind is too strong, the fog may be raised from the surface and occur as low stratus.

All other types are classified under the general term *evaporation fogs.* Included within this group are the frontal fogs which are fairly common in the winter months and are associated most commonly with slow-moving systems. They are always found in a cold air mass under warm, moist air. Precipitation from the warm air falls through the colder air, and evaporation sufficient to produce fog may take place. In the case of a warm front, such fog will be prefrontal; postfrontal fog is associated with a cold front.

Steam fogs occur where cold, stable air flows over a water surface several degrees warmer; the intense evaporation of moisture into the cold air results in fog when and if saturation is produced. Conditions favorable for this type of fog are light winds, clear nights, and stable air. These conditions are common around lakes and rivers in the fall of the year.

ICE FOG

Ice fog is in a separate category and is formed of tiny ice crystals that have sublimated directly from the vapor state—frozen water vapor. It is very fine, misty, and dangerous. Its danger is in the speed of formation (extremely rapid) and in the fact that it is encountered primarily just where one doesn't want it, near centers of habitation.

This type of fog is a hazard in the Arctic regions. It may be expected to form from clear, cold air at temperatures between $-20°F$ and $-50°F$. The trigger action for this type of fog is a great deal like others, a nucleus of some impurity upon which the ice will gather. In the Arctic regions, these condensation nuclei are not too prevalent and are found downwind of towns, highways, and (most unfortunately) around airports.

It is not unusual for an entire airport to fog in within 10 min. after an aircraft engine has disturbed the fine balance between fog and no fog. Vertical visibility in ice fog is usually good; horizontal visibility may be zero.

Thunderstorms

Since approximately 44,000 thunderstorms occur daily over the surface of the earth, a pilot can expect to encounter thunderstorm areas frequently. A knowledge of how to avoid them in planning and flying, or how to penetrate them as a last resort, is basic airmanship.

A certain combination of atmospheric conditions is necessary for the formation of a thunderstorm. These factors are: unstable air of relatively high moisutre content and some type of lifting action.

Thunderstorms generally have the same physical features. They do differ, however, in intensity, degree of development, and associated weather phenomena such as hail, turbulence, and electrical discharges. They are generally classified according to the manner in which the initial lifting action is accomplished. Two general classifications are currently used—*frontal* and *air-mass*.

1. Frontal types	2. Air-mass types
Warm front	Convective
Cold front	Orographic
Prefrontal (squall line)	Nocturnal

Figure 6.20. Warm-front thunderstorms. *(Courtesy of US Air Force.)*

FRONTAL THUNDERSTORMS

The *warm-front thunderstorm* is caused when warm, moist, unstable air is forced aloft over a colder, denser shelf of retreating air (Figure 6.20). Owing to the extreme shallowness of the warm frontal slope, the air is lifted gradually. The level of free convection will normally be reached at isolated points along the lateral surface of the front aloft, and therefore warm-front thunderstorms are generally scattered. They are extremely difficult to identify visually because they are obscured by other clouds. The warm-front thunderstorm is dangerous because the pilot may fly from an area of relatively smooth instrument flight into an area of great turbulence in a matter of seconds. A thorough preflight study of the stability chart and weather radar charts with the weather forecaster forewarns the pilot of the existence of possible thunderstorm area. Presence of showery precipitation in the warm-front weather pattern also is an indication of cumulus or thunderstorm activity.

The cold-front thunderstorm is caused by the forward motion of a wedge of cold air into a body of warm, moist, unstable air (Figure 6.21). Cold-front storms are normally positioned along the frontal surface aloft in what appears to be a continuous line. The problem of recognition is negligible because most cold-front storms are partly visible to the pilot approaching the front from any direction. The bases of these storms are usually closer to the surface than the bases of warm-front thunderstorms.

Prefrontal or squall-line thunderstorms are the result of the lifting action that takes place at a squall line, normally great enough to force the warm air up to the level of free convection, thereby producing a line of thunderstorms along the

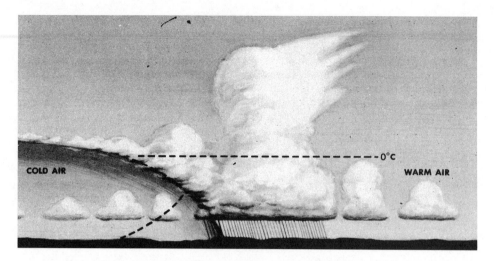

Figure 6.21. Cold-front thunderstorms. *(Courtesy of US Air Force.)*

squall line (Figure 6.22). These thunderstorms are normally similar to but more violent than the cold-front thunderstorms. Tornadoes are often associated with the most violent of squall lines.

AIR-MASS THUNDERSTORMS

Air-mass thunderstorms have two things in common: They form within an air mass and they are generally isolated or scattered over a large region.

Convective Thunderstorms.   Convective thunderstorms may occur almost

Figure 6.22. Squall-line thunderstorms. *(Courtesy of US Air Force.)*

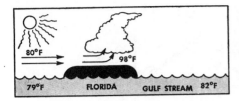 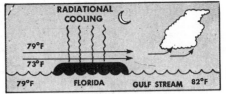

Figure 6.23. Convective-type thunderstorms.

anywhere in the world, over land or water (Figure 6.23). Formation of these storms is caused by solar heating of areas of the land or sea, which in turn provides heat to the lower layers of the air and results in rising currents of the warmed air. Land-type convective thunderstorms normally form during the afternoon hours at about the time the earth is receiving maximum heating from the sun. If the circulation is such that relatively cool, moist, unstable air is passing over the land area, heating from below will cause convective currents and result in towering cumulus which may continue to develop into thunderstorm activity. Dissipation normally occurs during the night or early morning hours after the land has lost its heat to the atmosphere.

Sea-type convective thunderstorms form in the same manner but at different hours. They generally form during the night after the sun has set and dissipate during the later morning hours. Examples of both types of convective thunderstorms may be found around the Gulf States. Clockwise circulation around the Bermuda high transports moist air over the land surface of these states during the entire day. The flow is off the water and across the land during the hours of sunlight. Since the land surface is considerably warmer than the air, the air is heated from below. Convective currents result, and the common afternoon thunderstorm follows after continued convective activity. Toward sunset, the land surface gives off heat rapidly to the air, and a balance between the land and the air is gradually reached.

At about the same time that the thunderstorms which have formed over land begin to dissipate, new ones begin to form over the sea surface. Thus it may appear that the thunderstorms move from land to sea. Although this may occur to a limited extent, basically the daytime and nighttime thunderstorms form as a result of the different heat-retaining properties of water and soil. Water is not subject to such rapid temperature changes as is land and therefore does not give off sufficient heat to the air to produce extensive convective activity until after sunset. During the night, however, a large-scale loss of heat from the water surface to the atmosphere occurs, and thunderstorms are formed by essentially the same process that transpires during the formation of late afternoon thunderstorms over land.

Orographic Thunderstorms.   As the name implies, orographic thunderstorms form in mountainous regions where the air traverses a long or steep slope (Figure

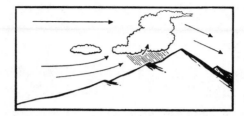

Figure 6.24. Orographic thunderstorms. *(Courtesy of US Air Force.)*

6.24). A good example of this type of storm is frequently found in the northern Rocky Mountain region. When the circulation of the air is from the west, moist air from the Pacific Ocean is transported to the mountains and forced upward by the slope of terrain. If the air is unstable, the upslope motion may cause thunderstorm activity on the windward side of the mountains. Such thunderstorms frequently form in a long, unbroken line similar to the cold-front type and persist as long as the upslope circulation continues. From the windward side of the mountains, identification of orographic storms may sometimes be difficult because they are obscured by other clouds, usually somewhat stratiform. From the lee side, identification is positive; the outlines of each storm are plainly visible. Don't fly through an orographic thunderstorm because this type of storm, almost without exception, enshrouds a mountain peak or a hill. For the same reason, don't fly under this storm unless the opposite side of the area is clearly visible, and then have plenty of terrain clearance.

Nocturnal Thunderstorms.   This is a type of thunderstorm which occurs later at night and in the early morning hours in the Central Plains area of the United States from the Mississippi Valley region westward. Its occurrence is explained as follows: A weak warm frontal surface overlies the area affected in such a manner that a relatively moist layer of air exists aloft. Nighttime radiation from the moist layer causes extreme cooling of the air at this level. The layer of air which has cooled settles in the early morning hours and forces unstable surface air aloft to initiate thunderstorm activity.

STRUCTURE OF THUNDERSTORMS

The fundamental structural element of the thunderstorm is the unit of convective circulation known as a convective cell. A mature thunderstorm contains several of these cells, each varying in diameter from 1 to 5 miles. By radar analysis and measurement of drafts, it has been determined that the circulation in each cell is generally independent of that in surrounding cells in the same storm. Each cell progresses through a cycle which usually lasts several hours. In the initial stages of cumulus development, the cloud consists of a single cell, but as the development progresses, new cells may form and older cells dissipate in a continuous and unbroken process. The life cycle of a thunderstorm consists of three stages: (1) the *cumulus stage,* (2) the *mature stage,* and (3) the *dissipating* or *anvil stage.* It should be noted that the transition from one stage to the next is not an abrupt one, but that the borders (with respect to time) of each stage blend gradually into the next.

STAGES OF A THUNDERSTORM

Cumulus Stage. The common cumulus cloud is the initial stage of all thunderstorm formation. It begins when some form of lifting raises moist unstable air to the level of free convection. Actually, only a very small percentage of cumulus clouds ever become thunderstorms, but the ordinary cumulus activity is always the basic or initial stage of the storm (Figure 6.25[A]). The chief distinguishing feature of the cumulus or building stage is the updraft that prevails

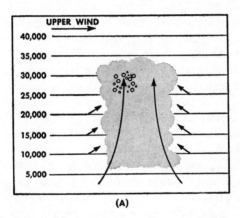

(A)

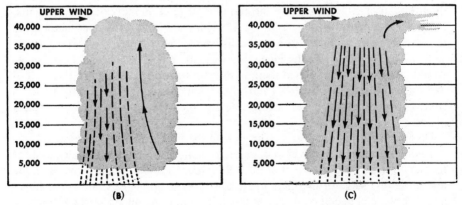

(B) (C)

Figure 6.25. (A) Cumulus stage of the thunderstorm. (B) Mature stage of the thunderstorm. (C) Anvil or dissipating stage of the thunderstorm.

throughout the entire cell. The speed of the updraft may vary from a few feet per second (fps) to as high as 100 fps in well-developed cumulus clouds. As the updraft builds a cumulus cloud to higher altitudes, small water droplets are formed. The greatest number of droplets form near the freezing level and grow in size by collision with other droplets. Usually rain does not fall during the cumulus stage because the raindrops are carried upward or remain more or less suspended by the upward air currents which exist throughout the cloud.

The Mature Stage. The occurrence of rain at the ground generally indicates the transition from the cumulus stage to the mature stage. By this time the cell has usually attained a height of 25,000 ft. or more. After the number of raindrops has increased to such an extent that they can no longer be supported by the updrafts, they begin to fall from the cloud and exert a drag on the air within the clouds. Such a drag is a major factor in the formation of a downdraft which characterizes each convective cell in the mature stage. The downdraft is evident at the surface as strong and gusty horizontal outflow which results when the downward flow of air is forced to move horizontally because of the solid lower barrier presented by the surface (Figure 6.25[B]). Downdrafts begin to form in the middle region of the cloud and gradually increase in both horizontal and vertical directions. Downdrafts in thunderstorms tend to be less gusty and smaller in speed than updrafts but frequently are as strong as 40 fps. In contrast, updraft speeds of 100 fps are not uncommon early in the mature stage. When one considers that the downdraft is immediately adjacent to the updraft in each cell, and that several cells are continuously forming and dissipating, it is apparent that severe turbulence may be expected when flying through a thunderstorm.

Mature cells generally extend to altitudes of 30,000 to 50,000 ft. It is in this stage that hail is frequently formed when a cell extends to great heights and/or has very strong updraft speeds. The "hard" rains which are often characteristic of thunderstorms also occur during the mature stage of each cell. In contrast to the life span of 10 to 15 min. of the cumulus stage of an individual cell, the mature stage usually lasts for 15 to 30 min. Again, one must remember that a given thunderstorm is composed of several cells which continuously form and dissipate so that the life of an entire thunderstorm is usually much longer than these figures would tend to indicate. When there is increasing wind with height, the updraft axis may tilt and rain may fall out of the updraft without impeding it. Such storms spawn hail and other severe weather; they are often found on the great plains.

Dissipating or Anvil Stage. Throughout the life span of the mature cell, the downdraft continues to develop both horizontally and vertically as the result of momentum and the drag effect of the falling precipitation. At the same time, the updraft is in a stage of dissipation. As this process progresses, the entire lower levels of the cell become an area of downdraft, and the upper levels become an area of little or no vertical motion. A downward current throughout the cell is the chief characteristic of the dissipating stage of a cell (Figure 6.25[C]).

Since the source of air necessary to maintain a downdraft is gradually reduced by the dissipation of the updraft, and since the descending motion effects a drying process, the entire structure now begins to dissipate. The dissipation stage may be identified at the surface by the gradual abatement of precipitation and finally complete cessation. At the same time the lower levels of the cloud frequently tend to become stratiform in appearance. A further identifying feature of this stage of the development process is the formation of the characteristic anvil top which results from the strong horizontal winds aloft blowing the upper portions of the cloud forward of the main portion. The life span of the dissipating stage is considerably more variable than that of the other stages and is dependent upon the strength of the original updrafts and the amount of moisture which the air contains.

VERTICAL DEVELOPMENT

The height of thunderstorms is a primary concern of pilots whose responsibility it is to determine an optimum flight altitude. Prior to the advent of radar analysis, it was difficult to give accurate estimates of cumuliform cloud tops, because of the general presence of more or less stratiform clouds in the lower levels.

Measurements of the vertical extent of thunderstorm activity were made by personnel of the original Thunderstorm Project in 1946 by using radar equipment with a range-height indicator. It was found that the closest correspondence between the radar-measured top and the actual top occurs during the cumulus stage, because ice crystal tops in the mature and anvil stages do not give good radar echoes.

Storms with heights of 50,000 ft. or more were measured in less than 10% of the cases observed. The greatest frequency of storms measured had heights between 25,000 and 29,000 ft. The average of all heights measured was 37,000 ft.; the maximum height recently observed was 66,000 ft.

Another factor indicated by draft research was the relative strength of updrafts and downdrafts. Updrafts appear to be of consistently greater magnitude than downdrafts. Thus, a good general rule with regard to flight is that an aircraft making a penetration of a thunderstorm will emerge at a higher altitude than that at which the penetration was started. Although this may not hold true for the dissipation stage of the development, the ability to maintain a given altitude is relatively easy in this stage because of the lesser speed and lesser gustiness which characterized the downdrafts.

Turbulent motion (gusts) within the cellular circulation pattern of thunderstorms has a considerable effect upon an aircraft in flight. In fact, the actual severity of a storm, so far as flying is concerned, is dependent upon the intensity and frequency of these gusts.

Gusts are both vertical and lateral, and all contribute to the dangers of turbulence to be expected. Gusts have been measured as high as 260 fps up and 160 fps down. The absolute value of gusts is not as significant as the frequency, which has been recorded at 40 fps up and 40 fps down in $^1/_{10}$ sec. The evidence indicates that gust intensity increases with altitude clear to the top of storms.

The forecast or observed tops of thunderstorm cells give the most important single indication of the severity of the storm. The greater the vertical development, the stronger will be the up and down drafts and turbulence, and the chance for hail.

Flight techniques in severe turbulence are discussed on pages 314–316.

WEATHER WITHIN THE THUNDERSTORM

Rain.   The pilot, upon entering any thunderstorm, may expect to encounter considerable quantities of liquid moisture which may not necessarily be falling rain. Liquid water in a storm may be ascending if it is encountered in a strong updraft; it may be suspended seemingly without motion, in extremely heavy concentrations; or it may be falling to the ground. If it is falling, it is rain in the usual sense of the word. Rain, as normally measured by surface instruments, is associated with the downdraft. This does not preclude the possibility of the pilot's entering a cloud and being swamped, so to speak, even though rain has not been observed at the surface. Rain will be found in almost every case of penetration of a fully developed thunderstorm. Liquid water, in heavy concentration, has been encountered at 40,000 ft., at temperatures far below freezing. There have been instances in which no rain was reported, but in these the storm probably had not developed into the mature stage. Altitudes above the freezing level showed a sharp decline in the frequency of rain of any intensity.

Hail.   Hail can be found virtually anywhere near or within a thunderstorm or its overhanging anvil. Although the likelihood of hail is closely related to the vertical development and intensity of the storm, there is no infallible method of forecasting hail or determining whether an individual storm contains hail. Because wet hailstones reflect radio energy very strongly, bright radar echoes are well correlated with hail.

Lightning.   According to pilot reports substantiated by subsequent ground inspection during the project, lightning strikes occurred in less than 2% of all thunderstorm penetrations. In general, damage was minor, being limited to small punctures in the aircraft skin. Radio failure caused by lightning occurred only when the antenna was damaged. This was rare. A maximum frequency of strikes occurred at the 16,000-ft. level, and the next highest frequency was at 26,000 ft. However, strikes may occur at any level, if conditions are favorable.

Icing.   On more than 50% of all traverses in the Thunderstorm Project, icing was encountered at 20,000 ft. Mostly rime ice, it never accumulated to the degree that it impaired safety, but this is attributed to the short exposure time in these traverses. Icing is considered most likely at or above the freezing level; the strong vertical currents and water at high levels indicate it should be expected at any level.

Snow.   The maximum frequency of moderate and heavy snow occurred at the 20,000- and 21,000-ft. levels. Snow, in many cases mixed with supercooled rain, was encountered at all altitudes above the freezing level. This was apparently of considerable concern to the pilots, for it presented a unique icing problem; wet snow packed on the leading edge of the wing and resulted in the formation of rime ice. This was particularly true close to the freezing level, and on occasion this altitude had to be abandoned because of the rapidity at which ice accumulated on the aircraft.

First Gust.   A significant surface hazard to approach, landing, and take-off associated with thunderstorm activity is the rapid change in wind direction and speed immediately prior to thunderstorm passage. Strong winds at the surface accompanying thunderstorm passage are the result of the horizontal spreading-out (divergence) of the downdraft currents from within the storm as they approach the surface of the earth. The total wind speed is a result of the downdraft divergence plus the forward velocity of the storm cell. Thus, the speeds at the leading edge, as the storm approaches, are ordinarily far greater than those at the trailing edge, and the effect of the approach of a thunderstorm is to cause a sudden and marked increase in the surface winds at a station. This initial wind surge as observed at the surface is known as a "first gust." The speed of the first gust is normally the highest recorded during storm passage and may vary in direction by as much as 180° from the surface wind direction which previously existed. First-gust speeds increase to an average of about 15 knots over prevailing speeds, although gusts with a total speed of more than 90 knots have been recorded. The average change of wind direction associated with the first gust is about 40°; this value is not considered particularly significant, as the wind shifts occurring with thunderstorms over a station may be as much as 180° in direction, depending on the orientation of the thunderstorm with respect to the station.

Altimeter Errors.   Rapid and marked surface variations generally occur during the passage of a thunderstorm. The typical sequence of these variations is for an abrupt fall as the storm approaches, an abrupt rise accompanying the rain, and a gradual return to normal as the rain ceases. The importance to a pilot of such pressure fluctuations is that they may result in significant altitude errors even though the altimeter setting which he uses is current almost to the minute. In the investigations by the Thunderstorm Project, it was found that if a pilot had landed during the rain, in approximately 24% of the storms studied, he would have been approximately 60 ft. higher or lower than his altimeter indicated, using an altimeter setting given him a few minutes earlier. *Two cases of the altimeter reading more than 140 ft. too high were noted.* A pilot should realize that any altimeter setting given him during or immediately before a thunderstorm is highly unreliable.

AIRBORNE WEATHER RADAR

Airborne radar, designed primarily for assisting flight crews in avoiding severe weather, has come into common use in light and medium twins and in airliners. Its features and uses are discussed in Chapter 5.

Flying in Severe Turbulence

Pilots are concerned with turbulence in storms at lower levels, with Clear Air Turbulence (CAT) in mountainous areas, and with CAT in jet-stream areas. While CAT is considered generally less severe than thunderstorm turbulence, both are highly dangerous, can come as a surprise, and have resulted in the loss of bombers, transports, and utility aircraft flying at all altitudes.

CLASSIFICATION OF TURBULENCE

Though adjectives describing turbulence are highly subjective and therefore imprecise, Table 6.6 is the generally accepted classification.

Low-Level Turbulence. This turbulence may be considered as random eddies in the air, caused by convection from ground heating and from wind over rough terrain.

The least low-level turbulence occurs in the early morning, in stratus cloud conditions, and in light winds.

Wake Turbulence. Every airplane creates turbulence as it flies, because lift depends upon downward acceleration of the air through which it moves. The resulting air motion is a spiral from each wing tip. Viewed from behind the airplane the left vortex rotates right, the right one, left. Between them, there is a downward flow of air. The strength of these vortices varies directly with the lift that is required and inversely with the speed of the airplane. Most severe vortices are generated by large aircraft (such as the Boeing 747) at slow speeds. The vortices settle slowly, and if they reach the earth's surface, they spread apart. A light wind could therefore keep a vortex directly over a runway. They are avoided by staying above the airplane's flight path, or to one side. Strong vortices may persist for several minutes.

Clear Air Turbulence. CAT is the turbulence in clear air, usually associated with high altitude, that can produce a violent "chop" or washboardlike effect. It is frequently a surprise. Remote thunderstorms and dry convection can contribute to its presence. It is patchy and transitory; these patches average 2000 ft. deep, 20 miles wide and 50 miles long; they are elongated in the direction of the wind and are difficult to locate.

CAT is likely found near mountain waves (Figure 6.26) and jet streams. The

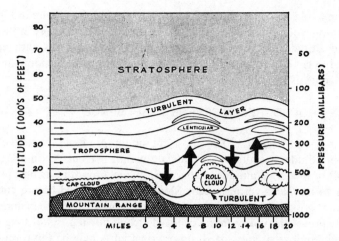

Figure 6.26. Mountain-wave turbulence. *(Courtesy of US Air Force, Aerospace Safety.)*

intensity of mountain-wave CAT will vary directly with wind speed across a mountainous area. The most intense CAT is found in a strong jet stream oriented perpendicularly to a mountain ridge; this produces a very strong mountain wave. Most frequently, it is found at 30,000 ft. in winter and 34,000 ft. in summer, whether in relation to mountain waves or jet-stream wind shear. CAT is a winter hazard, with three or four times more occurrences in winter than in summer, and more than twice as many occurrences with a cirrus overcast than with a broken condition.

Remember that mountain waves and their turbulence are found not only in high mountains and high-altitude flying. They occur downwind of *any* obstacle, including an isolated hill a few hundred feet high. The stronger the wind, the larger the effect, and while there are a variety of other meteorological conditions which affect the strength of the wave and its turbulence, this is a good rule of thumb: *If the wind at the altitude of the top of the obstacle is 20 knots or more, there will be a noticeable wave turbulence*. This turbulence will be greatest in the rotor zone downwind from the obstacle, as shown in Figure 6.26, and will be at a maximum at about the same altitude as the top of the obstacle.

Avoid flying downwind of peaks at about their altitude. This point of maximum turbulence will be at the same place no matter what the wind speed. The rotor zone gets larger in diameter as the wind speed increases, and more severe turbulence will be encountered, but it does not change position. In extreme cases it can reach the ground. This will be indicated by blowing dust clouds.

The rotor zone will be indicated by a roll cloud looking like wispy cumulus if there is moisture in the air, and this can be used to help avoid it. However, waves can be just as strong when the air is very dry, with no roll clouds or lenticular clouds to mark their position.

You can avoid wave turbulence by keeping in mind the picture in Figure 6.26 and adjusting your flight path to avoid the turbulent areas. Fly well above or below the altitude of the obstacle, or if this is not possible, turn upwind or downwind to get out of the turbulent zone. The rising or descending air between the rotor zones is usually very smooth.

FLIGHT PROCEDURES IN SEVERE TURBULENCE

Straight-Wing Aircraft. The most important principle is to fly *attitude.* Allow the altitude and airspeed to deviate within reasonable limits, correcting with small or moderate attitude changes. Slow the aircraft to its best penetration speed, or to its maneuvering speed. Slow to desired speed before entering the turbulence if possible, and in piston aircraft use a higher RPM than normal cruise to provide power flexibility in the event of strong down gusts of long duration. With relatively low aircraft speed, up or down gusts may be felt for a relatively long period of time, which could result in undesirable changes in altitude. *Don't "chase" the performance instruments.*

It is generally best to fly manually, particularly the elevators, and do not try to maintain altitude by trimming. Be sure the "altitude hold" function of the auto-pilot is OFF. The auto-pilot "altitude hold" function can establish excessive tail loads and severe out-of-trim conditions for level flight.

High-Performance Swept-Wing Aircraft. There are two general concerns: imposing excessive structural loads, and allowing the aircraft to get into undesirable extremes of attitude. It has long been generally accepted that a rough airspeed limit slower than cruising is less likely to result in structural damage. Experience indicates, however, that structural damage has not occurred unless there has been a severe change in attitude and a subsequent combination of stresses resulting from both the recovery maneuver and the severe turbulence.

Flexible wings and high wing loadings, considered with the effect of strong, frequent, longitudinal, lateral, and vertical gusts, all of which affect the angle of attack of the wing, make the importance of severe maneuvers much more significant.

In large swept-wing aircraft, it is now considered essential to penetrate turbulence at higher than cruise speeds, because (1) the aircraft at low speed is closer to the stall buffet and high drag, tempting the pilot to make undesirable thrust changes, (2) trim changes due to thrust changes at low speed are greater, compounding control problems, and (3) the airplane can be laterally and directionally upset at low speeds in severe turbulence. In these aircraft, higher speeds are now specified, and pilots should fly to the high rather than the low side of the target speed. Refer primarily to the specific speeds prescribed in the operating instructions for the specific aircraft.

Attitude is all important, but it is better to do nothing than to attempt control attitude too rigidly, exceeding moderate control inputs.

TABLE 6.6. Classification of Turbulence.

DEFINITION: ALTHOUGH AIRCRAFT REACTIONS TO A GIVEN TURBULENCE CONDITION VARY WITH AIRSPEED, WING-LOADING, ATTITUDE, PILOTS EXPERIENCE AND ABILITY, ETC., THE DESIGN CRITERIA OF ALL TRANSPORT, NORMAL, UTILITY, AND ACROBATIC-CATEGORY AIRCRAFT REQUIRE ESSENTIALLY IDENTICAL LIMIT GUST LOADS, EXPRESSED IN TERMS OF POSITIVE AND NEGATIVE ROUGH AIR GUSTS.

THE CLASSIFICATION OF TURBULENCE IS BASED UPON AIR MOVEMENT ALONE, NOT UPON AIRCRAFT SIZE OR TYPE.

CLASS OF TURBULENCE	TYPICALLY FOUND IN ASSOCIATION WITH THESE CONDITIONS	EFFECT
EXTREME (Gusts ± 50 fps)[1]	RARELY ENCOUNTERED. USUALLY CONFINED TO THE STRONGEST FORMS OF CONVECTION AND WIND SHEAR. IN OR NEAR THE ROTOR EFFECT OF A STRONG MOUNTAIN WAVE. +50 KT. IN SEVERE THUNDERSTORMS (LARGE HAIL 3/4", ALMOST CONTINUOUS LIGHTING, ETC.) USUALLY IN SQUALL LINES RATHER THAN ISOLATED THUNDERSTORMS.	AIRCRAFT IS VIOLENTLY TOSSED ABOUT AND IS PRACTICALLY IMPOSSIBLE TO CONTROL. MAY CAUSE STRUCTURAL DAMAGE.
SEVERE (Gusts ± 35–50 fps)[1]	IN ASSOCIATION WITH EXTREME TURBULENCE. + 50 KT. FOUND UP TO 150 MILES LEEWARD OF THE RIDGE. 25 TO 50 KT. FOUND UP TO 50 MILES LEEWARD OF THE RIDGE, FROM RIDGE LEVEL UP TO SEVERAL THOUSAND FEET ABOVE. IN AND NEAR MATURE THUNDERSTORMS AND OCCASIONALLY IN TOWERING CUMULIFORM CLOUDS. NEAR JET STREAMS.	AIRCRAFT MAY BE MOMENTARILY OUT OF CONTROL. OCCUPANTS ARE THROWN VIOLENTLY AGAINST THE BELT AND BACK INTO THE SEAT. UNSECURED OBJECTS ARE TOSSED ABOUT.
MODERATE (Gusts ± 20–35 fps)[1]	IN ASSOCIATION WITH EXTREME AND SEVERE TURBULENCE. + 50 KT. FOUND AS MUCH AS 300 MILES LEEWARD OF THE RIDGE. 25 TO 50 KT. FOUND AS FAR AS 150 MILES LEEWARD OF THE RIDGE. IN, NEAR AND ABOVE THUNDERSTORMS AND IN TOWERING CUMULUS. NEAR JET STREAMS, IN UPPER TROUGH, COLD LOW, AND FRONT ALOFT SITUATIONS. AT LOW ALTITUDE (USUALLY BELOW 5000 FT.) WHEN SURFACE WIND EXCEEDS 25 KNOTS, WITH STRONG THERMALS, AND COLD AIR ADVECTION.	OCCUPANTS REQUIRE SEAT BELTS AND OCCASIONALLY ARE THROWN AGAINST THE BELT. UNSECURED OBJECTS MOVE ABOUT.
LIGHT (Gusts ± 5–20 fps)[1]	IN MOUNTAIN AREAS EVEN WITH LIGHT WINDS. IN AND NEAR CUMULUS CLOUDS. AT LOW ALTITUDES WHEN WINDS ARE NEAR 15 KNOTS OR WHERE AIR IS COLDER THAN UNDERLYING SURFACE.	OCCUPANTS MAY BE REQUIRED TO USE SEAT BELTS, BUT OBJECTS IN THE AIRCRAFT REMAIN AT REST.

[1] Furnished by the editor.

U.S. Weather Bureau

Control pitch solely with the elevator, *never* with trim. As a general rule, do not change thrust. It is much better to accept large variations in altitude, so long as adequate terrain clearance is maintained.

Disengage the auto-pilot to avoid excessive out-of-trim conditions. If equipped with a yaw damper, keep it engaged. Do not "chase" performance instruments.

THUNDERSTORM PENETRATION

While the principal problem in thunderstorms is severe turbulence, the following items are also essential:

Anti-Icing: Be sure this equipment is full ON.

Seat Belts and Loose Items: Be sure all are are secured.

Cockpit Lights: Turn them high to avoid blindness by lightning contrast.

Power: Set at whatever is desired before entry. Then use this as a basic value. Higher rpm provides power flexibility in piston aircraft, and reduces the rate of prop ice accretion.

Gear and Flaps: Do not extend them. They serve only to make the airplane less efficient aerodynamically.

Turns: Avoid turns if possible. If necessary, make them shallow.

Expect the experience to be an upsetting one with many surprises. Have yourself and your airplane prepared as far in advance as possible.

Icing

This hazard has long been most dreaded, particularly in slower aircraft flying at lower and intermediate levels. Icing hazards dictate constant alertness and substantial respect at all times, and in all aircraft. *Structural* ice forms externally and reduces aerodynamic efficiency; ½ in. of ice on some airfoils can reduce lift by 50%. Ice can also completely mask outside vision, reduce effectiveness, limit operation of gear, flaps, and brakes, disrupt the pitot-static system, and destroy radio communication. Ice forming on the propellers, then breaking off, can damage the airframe.

Internally, ice forms in jet intakes and carburetors, seriously restricting airflow. It may accumulate on jet inlet structure, then break off and damage compressor blades.

RESULTS OF ICE ACCRETION ON AIRCRAFT STRUCTURES

The presence of ice on an airfoil disrupts the smooth flow of air over the airfoil and thus decreases lift and increases drag, with a resulting increase of the aircraft's stalling speed. Ice on the propeller results in a loss of thrust. The addition of ice to the various parts of the aircraft may cause vibration, placing added stress on these parts. This is especially true in the case of the propeller, which is very delicately balanced. Even a very small amount of ice, if not distributed evenly, can cause great stress on the propeller and engine mounts. The danger of added weight is not too great under ordinary circumstances, if too much of the lift and thrust are not simultaneously lost. Weight does become an important factor, however, in critically loaded aircraft, and in conditions of heavy clear icing.

FACTORS NECESSARY FOR ICE FORMATION

The first factor necessary for structural ice formation is the presence of liquid moisture visible to the naked eye.

Secondly, free-air temperatures of freezing or lower are necessary, although observations have shown that ice may form on a static object when the free-air temperature is as high as $+4°C$, as a result of evaporational and adiabatic cooling. When an aircraft is in flight, the heating from skin friction and the impact of water droplets cause the temperature of the skin to rise. Generally, the heating and cooling effects cancel each other; thus, structural ice can be considered possible at $0°C$ or below.

At temperatures less than about $-2°C$ or $-4°C$, the frequency of ice formation decreases gradually with lowering temperature, and the most severe icing is encountered generally between $0°C$ and $-15°C$; however, under unusual circumstances dangerous icing conditions have been encountered at temperatures less than $-15°C$.

FORMS OF VISIBLE LIQUID MOISTURE

Clouds are the most common form of visible liquid moisture; however, many clouds at temperatures below freezing do not give rise to ice formation. The pilot does not have any way of knowing which clouds will present an icing situation, but he must be aware that the possibility exists if the temperature is below $0°C$. Icing is more likely in turbulence than in smooth air.

Freezing rain may be encountered in otherwise clear air below a cloud deck. When warm moist air is forced to rise over a colder air mass, a frontal inversion will exist. Below this inversion, icing conditions are frequently encountered. In being forced aloft, the air may be sufficiently cooled to produce saturation and precipitation. The raindrops falling into the cold air may freeze upon contact with the aircraft if the temperature is at freezing or below. Freezing rain is probably the most dangerous form of ice because it can build to great proportions in a matter of minutes and is extremely hard to break loose.

TYPES OF STRUCTURAL ICE

The type of ice that will form depends on four factors:

1. The free-air temperature.
2. The surface temperature of the structure.
3. The surface characteristics of the structure (configuration, roughness, etc.).
4. The size of water droplets.

Clear Ice (Glaze).   Clear or glaze ice is considered to be the most serious of the various forms of structural ice. It is most often encountered in regions of large, supercooled (liquid at below-freezing temperatures) water droplets or in an area of freezing rain. This condition most often presents itself over mountainous terrain and in regions of unstable weather conditions, in and below cumuliform clouds (Figure 6.27). Clear ice is dangerous because of the great amount of freezing water available and the high rate of accretion—in addition to this, it is very tenacious.

Rime Ice.   Rime ice is generally encountered in stratiform clouds, that is, stable weather conditions. The small water droplets of a stratus cloud freeze immediately upon impact with the aircraft surface. Rime ice builds up slowly by comparison to clear ice. The greatest resultant danger is the added drag and serious deformation of the airfoil, destroying lift, caused by the very rough ice surface created. Rime ice is relatively easy to break loose by conventional methods. It should be realized that the various icing forms and attendant hazards may often occur in any and all combinations if the weather situation allows. For instance, ice resulting from freezing drizzle is usually a combination of rime and clear ice.

Frost.   Frost is a hazard very often underestimated. It is likely to form during the night when surface temperatures are below freezing. Frost forms in clear air

Figure 6.27. Icing over mountains. *(Courtesy of US Air Force.)*

by sublimation (change of state directly from gas or vapor to solid ice) when the moist air comes into contact with a very cold surface.

Frost may form in flight when descending into warmer (but still freezing) more moist air, or when flying from a very cold air mass to a warmer air mass. Frost causes added drag and offers a very real hazard at lower, critical air speeds. For this reason, the aircraft should be checked thoroughly, and frost should be removed prior to take-off. *Any* frost is too much forst. It *must* be removed before take-off. Another frost hazard is the restriction to visibility caused by frost on windshield surfaces.

ANTI/ICING AND DEICING AIDS

An anti-icer is an item of equipment which prevents the formation of ice, and a deicer is an item of equipment which eliminates ice after it has already formed. Anti-icing and deicing equipment can be divided into three general classes:

1. *Mechanical:* Deicing boots which are used on wings, tail assembly, and radio mast.

2. *Chemical:* Anti-icing fluid and paste. Anti-icing paste is used on propellers, and anti-icing fluid is used on the propeller, windshield, carburetor, and jet engine accessory section.

3. *Thermal:* Electrical heat and exhaust or compressor heat. Electrical heat is used on the pitot tube, electrical wingboot, wing-tip tanks, and propellers. Exhaust or compressor heat is used on the wings and tail assembly, windshield, intake air, canopy, and carburetor.

Anti-icers should be used before or immediately upon entering an icing zone and operated continuously until the aircraft is out of the icing zone.

Deicing boots should be used intermittently and only after an appreciable thickness of ice has accumulated upon the boot. Deicing boots are most useful in eliminating rime ice, but are not so reliable in removing clear ice. They should not be used during take-off or landing.

The "hot-wing" or exhaust-heated wing may be considered as either an anti-icer or deicer, depending upon the time at which the unit is put into operation. It is usually most effective as an anti-icer.

FACTORS INFLUENCING THE RATE OF DEPOSIT OF ICE ON AIRCRAFT

The concentration of liquid water is a definite factor determining the rate of deposit; the more liquid available, the more pronounced and rapid the deposit.

Diameter of the Droplets of Water. When a wing moves through the air, the air is deflected at the leading edge. If water droplets are present in the air, they tend to move with the airstream. The smaller the drops, the greater their tendency to follow the airstream; and the larger the drops, the more they resist this deflecting influence. Therefore, the large drops (small deflection) are collected more easily than the small drops (large deflection).

Airspeed. As an airplane goes faster, it strikes more water droplets, and ice accumulates more rapidly. As speed continues to increase, however, skin friction increases the temperature of the surface, and the water droplets cannot stick to it when it is above freezing. At a true airspeed of 380 knots, the temperature rise is 15°C. Supercooled water droplets are unlikely at 14°F (-10°C) and below, so an airplane flying at 400 knots or faster is not likely to accumulate structural ice.

Other Factors. Research has added another group of factors that affect the structural icing rate of an aircraft. The shape of the airfoil is importnat. The thinner the cross section of the airfoil, the less tendency there is for drops to bypass the surface under certain conditions. The smoothness of the airfoil surface also serves to affect the rate of icing. An airfoil that is physically or aerodynamically unclean presents a greater surface area to catch the freezing droplets. Another important factor is the temperature difference between the airfoil and the surrounding (ambient) air. Finally, turbulence in clouds supports larger droplets; ice will accumulate more rapidly.

PRECIPITATION ICE FORMS THAT MAY BE ENCOUNTERED

Hail. Hail develops in highly turbulent thunderstorms. Water drops, which are carried upward by vertical currents, freeze into ice pellets, start falling, accumulate a surface of water, and are carried upward again; the newly added water freezes. A repetition of this process increases the size of the hailstone. Hail is a warm-weather phenomenon and can be produced only by strong vertical currents. It does not lead to the formation of structural ice, but it can cause serious physical damage to the aircraft skin. Hail is frequently encountered in clear air, falling from the anvil cloud in front of a thunderstorm. Baseball-size hail is not unknown.

Sleet. The initial stage of sleet formation results from water passing directly from the vapor to the solid state (sublimation) and forming very small grains or pellets of ice. This early stage of the formation process usually occurs at a considerable distance (approximately 8000 to 10,000 ft.) above the freezing level. As these minute ice particles begin to fall from the cloud, they grow by impact with the supercooled water droplets contained in lower levels of the cloud. Variations in the height of the freezing level will result in variations of the altitude through which the sleet particles will fall before reaching this level. This results in varying sizes and textures of the sleet particles which may be encountered. Sleet is essentially a cold-weather phenomenon which results in the formation of structural ice only when it is mixed with supercooled liquid water. When sleet is encountered in flight, it is not wise to climb because frequently there are associated freezing rain areas above.

Snow. When condensation takes place at temperatures below freezing, water vapor changes directly into minute ice crystals. These crystals grow to form a single snowflake. Dry snow does not lead to the formation of structural ice, but wet snow frequently does.

INDUCTION SYSTEM ICING

Carburetor ice constitutes a very real problem that results from a process quite different from that by which structural ice is formed. It can, in fact, be formed under conditions in which it is impossible for structural ice to be formed. If the humidity of the free air is high, carburetor ice may occur with temperatures as high as 21°C. The ice results from the marked cooling due to two separate processes which occur during carburetion: (1) vaporization of the gasoline and (2) the decrease of pressure in the venturi-shaped throat of the carburetor. The greatest temperature drop is caused by the first of these.

Induction system ice may form in various places and on various parts of the induction system. It may form in the air scoop, in curves of the induction system, at the discharge nozzle, in the venturi, or on the throttle (butterfly) valve. One of the best indications of induction system icing is an otherwise unexplained loss of manifold pressure which disappears when carburetor or induction air heat is applied, if it is applied early enough after the indication is detected.

Ice can also form on the intake air filter. Carburetor or induction air heat will not remove it but will provide an alternate source of air, allowing continued flight.

JET AIRCRAFT ICING PROBLEMS

Ice formation on airframes of turbojet aircraft differs from that on piston-engined aircraft, mainly because of the cleaner design and much higher speeds of jets.

As the operational speeds of aircraft increase, there is an associated increase in the importance of aerodynamic heating in the prevention of airframe icing. Airspeed and altitude are the primary factors to be considered in determining the amount of aerodynamic heating.

For a constant indicated airspeed, the amount of aerodynamic heating decreases with increasing altitude. At a constant altitude, the amount of aerodynamic heating is greater for higher airspeed. Aerodynamic heating also varies with position on the wing; it is greatest on the wing's leading edge and decreases to a minimum just behind the wing shoulder or mid-chord. For this reason, it is quite possible for an aircraft to experience *runback-icing* on the wing at slightly higher temperatures and speeds than those which permit icing on the leading edge.

It is not always true that the high speeds and altitudes which characterize turbojet performance keep them out of icing conditions.

When operating above 30,000 ft. aircraft are not likely to encounter icing.

However, during operations at lower altitudes, jet aircraft may experience icing weather while climbing, cruising, descending, or making final approach. The amount of ice accretion will depend upon the attitude and true airspeed when in an icing situation.

JET-ENGINE ICING

Because any surface which is subjected to the direct flow of air may collect ice, jet engines experience icing both externally and internally. Inlet lips, accessory housing domes, islands and island fairings, compressor inlet screens, inlet guide vanes, and compressor blading are all vulnerable. Other surfaces—such as duct-splitter plates and openings for boundary-layer bleed-off and accessory cooling purposes—may experience icing, though not directly in the airstream.

Compressor-inlet screens are now retractable, removable, or heated both before take-off and during flight in most jet aircraft. The inlet guide vanes of an axial-flow jet engine are also critical icing components. When the inlet screen has been removed, the centrifugal flow-type engine is, by contrast, relatively unsusceptible to icing. Rear screens of centrifugal type engines are not susceptible to icing because of their proximity to warm portions of the engine.

Ice accumulation decreases the performance efficiency of turbojet engines. As ice builds up, there is a reduction in the total pressure at the compressor inlet, which results in reduced airflow, reduced thrust, and an increase in fuel consumption. Although the tail-pipe temperature may rise when airflow decreases due to ice accumulation, it is unwise to depend upon tail-pipe temperature rise as an indication of icing in turbojet engines.

While damage due to icing in centrifugal flow engines is unknown, it may occur in axial flow types. Structural damage does result from the shedding of large ice accumulations from components ahead of the compressor inlet when no screen is used. Small pieces will pass harmlessly through the engine, but a large piece of ice may lodge in the space between rotor and stator blade stages or may strike a rotor blade and be deflected into the guide vanes or stator blades, causing damage. Ice formed in the compressor during outside parking or storage of an aircraft may result in damage if the ice is not removed before attempting to start.

Prolonged operation of jet engines in supercooled fog (fog composed of water droplets at below-freezing temperatures) can result in intake icing with possibly hazardous consequences.

Although not directly associated with existing atmospheric conditions, other than low temperatures, fuel-screen icing may occur in turbojet engines. Jet fuel can contain a large amount of water vapor, even though there is no liquid water in it. When the fuel cools, the water condenses into minute water droplets. When the fuel cools below 0°, the droplets become supercooled but remain in the liquid state. As the fuel flows through the fuel filter, the droplets strike the solid parts of it and freeze instantly, adhering to it. As this ice builds up, it reduces, and then stops, the fuel flow, causing loss of engine power due to fuel starvation.

Fuel filter heating and use of fuel icing inhibitor minimize this fuel icing problem.

Induction icing problems in axial flow engines are being met by direct heating of the critical components of the engine inlet. General practice for jet engine anti-icing is:

1. Use of retractable screens which can be retracted before icing conditions are encountered.
2. Use of hollow, heated vanes utilizing heated compressor discharge air or combustion chamber hot gas as the heat transfer medium.
3. Use of hollow, heated construction similar to guide vanes or the use of electrical surface heating.
4. Use of double wall construction with hot air, hot gas, or hot oil between the walls, or use of electrical surface heating.

The most efficient and successful method of wing anti-icing is the thermal system of hot air heating the wing with a chordwise flow of heat. This system prevents ice by completely evaporating water droplets by the heating of the leading edges back to about 15% of the chord.

The heat is obtained by use of compressor bleed air taken off the last compressor stage, which has the necessary operating heat of 121°C to 260°C (250°F to 500°F), and pressure of from 20 to 100 psi. This air is channeled through ducts to the internal leading edge and flows back inside the wing to dissipate out the trailing edge.

Loss of power thrust during the thermal heating operations is approximately 8%, with a corresponding percentage of increase of specific fuel consumption. Also there is an increase in tail-pipe temperature, due to the increased fuel/air ratio in the combustion section.

POLAR ICING REPORTS

A summary of pilot reports of icing encountered north of 60° north latitude provides the following information:

About 10% of pilots surveyed had encountered icing. Severe icing most often occurred in December. February had the maximum number of icing incidents and September, the minimum. Rime icing was the kind most often experienced and was encountered down to −40°C. The maximum number of encounters occurred within a few degrees of the freezing temperature of water. Rime ice, clear ice, and mixtures of the two were experienced.

Icing occurred in summer as well as in winter. Icing was reported at temperatures as high as 7°C, and since the main study, has been reported in a severe form down to −55°C.

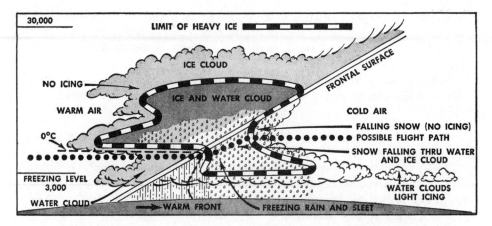

Figure 6.28. Icing under a warm front. *(Courtesy of US Air Force.)*

WEATHER STATION AIDS FOR DETERMINATION OF ICING REGIONS

Whenever a pilot plans a flight, he should utilize the weather charts and other facilities, including the knowledge of the qualified forecaster on duty, to aid in anticipating possible icing regions. Some of these facilities are listed and explained below.

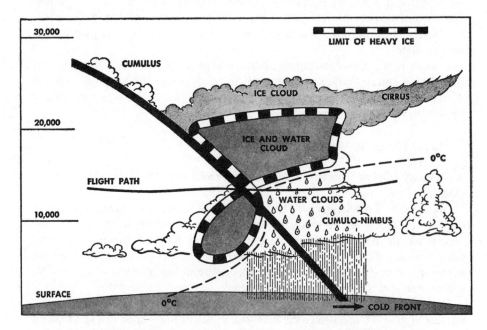

Figure 6.29. Icing zones along a cold front. *(Courtesy of US Air Force.)*

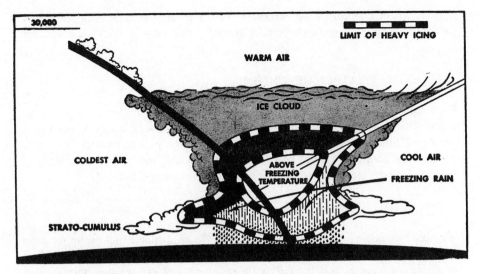

Figure 6.30. Icing zones along an occluded front. *(Courtesy of US Air Force.)*

Weather Maps. The various surface and constant-pressure weather maps indicate the positions of fronts, thus indirectly marking regions of possible aircraft icing. For example, the region of the cloud system of any warm front (Figure 6.28) presents a potentially dangerous icing situation in an associated region of below freezing temperatures. This region may extend over a great area, vertically and horizontally, and severe icing conditions may be present at any point in it. In the smooth or stratiform clouds, icing tends to be predominantly rime in character. Look for more rapid accumulation of rime ice if turbulence is present.

In cold fronts, prefrontal squall lines, and air-mass thunderstorms, clear ice usually is formed as a result of the turbulence and the presence of large water droplets (Figure 6.29).

Occluded fronts (Figure 6.30) present an icing hazard which may be considered to be about the same as in a warm front—particularly in northern regions such as Alaska and the North Atlantic. The extensive cloud cover may necessitate considerable flying time in clouds, with the type, severity, and zones of icing determined by considering each case separately.

AIR WEATHER SERVICE AND NATIONAL WEATHER SERVICE WRITTEN FORECASTS

Both types of weather forecasts give information on the height of the freezing level and regions in which the pilot can expect ice formation. Since it is often possible to fly in clouds at temperatures below freezing and not pick up any ice, the forecast may seem to be in error. These forecasts, then, simply indicate regions where conditions are favorable for ice formation. The pilot should be prepared to cope with ice and consider himself fortunate if he does not encounter it.

TELETYPE WEATHER REPORTS

Pilot reports (PIREPS) of icing will often be included in teletype weather reports—PIREPS collection or individual special reports—and may be used as a source of much useful preflight planning information by other pilots.

The pilot planning a flight should note areas where the free-air temperatures are favorable to ice formation and where clouds are expected.

PHILOSOPHY FOR AIRMEN

Weather forecasting, in spite of long study of the atmosphere and the application of computers and the latest technology, must be regarded by airmen as more of an art than a science. It is limited by the relatively few observation points, most of which are on the surface, and the almost infinite number of variables affecting weather.

The airman, as he flies, can help the timeliness and accuracy of the information available, and thus his fellow fliers, by reporting his observations of weather he encounters along his route.

7

Medical Aspects of Flight

Since earliest times man has been involved in a constant struggle to adapt to his environment.

The most significant environmental conditions in aviation, other than the aircraft itself, are the extreme changes in atmospheric pressures and temperatures as altitude increases. While these naturally occurring phenomena of the atmosphere have remained constant, their effects have become more pronounced as aircraft have reached greater heights.

The performance of powerful, present-day aircraft greatly exceeds the physiological capabilities of man unless he is supported by elaborate equipment such as oxygen masks, pressurized cabins, anti-g suits, ejectable seats, and at very high altitudes, pressure suits. To perform effectively within his environment, the flyer must be fully aware of his own physiological and psychological limitations and the functioning and use of his equipment. This chapter will discuss some of the more important medical aspects of flight, the health care of the flyer, and the role of the flight surgeon.

Hypoxia

Oxygen, essential to human life, is even more critical than food or water. Man literally lives from one breath to the next in that death rapidly occurs if he is not constantly supplied with oxygen by the process of respiration. Complete lack of oxygen, known as *anoxia,* is promptly fatal. It is seen in such conditions as suffocation, drowning, strangulation, and certain types of poisoning. Much more common, however, is partial lack of oxygen, which is termed *hypoxia.* The threat of hypoxia is a serious and constant hazard at high altitude. The early symptoms are insidious, and the victim may actually have a feeling of well-being even as more serious effects are developing. Because its onset is without pain or other warning symptoms, hypoxia is the most important physiological hazard of high-altitude flying.

Modern military and civil aircraft are equipped with pressurized cabins and oxygen breathing equipment to maintain sufficient oxygen pressure for the crew and passengers. If this equipment is maintained and used properly, the flyer is adequately protected against hypoxia. To insure that they appreciate the need for this equipment, aircrews must understand something of human physiology and how respiration takes place. Basic physiological training on this subject is required of all USAF aircrew members, with periodic refresher courses.

CHARACTERISTICS OF THE ATMOSPHERE

The air is a mixture of gases, nitrogen and oxygen being the principal components. Up to extremely high altitudes, these two gases, with minute amounts of several others, are present in constant proportions. As indicated in Table 7.1, nitrogen constitutes nearly four-fifths and oxygen about one-fifth of the total volume of dry air. At sea level and for about 10,000 ft. upward, the 20% of oxygen in the air is adequate for human breathing because the total pressure of the air (and hence the partial pressure of oxygen) is relatively high. With increasing altitude, however, there is a corresponding decrease in atmospheric pressure and less partial pressure of oxygen. Atmospheric pressure varies from 760 mm. of mercury at sea level to about one-half this amount at 18,000 ft. (379.4 mm. of Hg) and to less than one-eighth this amount at 50,000 ft. The second column in Table 7.2 illustrates the progressive decrease in pressure as altitude increases.

TABLE 7.1. Components of the Air and Percentages at Sea Level.

Nitrogen	77.14	Neon	.0012
Oxygen	20.69	Helium	.0004
Argon	.93	Water vapor	1.2
Carbon dioxide	.03	Hydrogen	.01

TABLE 7.2. Total Atmospheric Pressure and Oxygen Partial Pressure at Various Altitudes.

Altitude (ft)	Atmospheric Pressure (mm Hg)	O_2 Partial Pressure (mm Hg)
Sea level	760	160
5,000	632	126
10,000	523	105
15,000	429	85
20,000	349	70
25,000	282	56
30,000	225	45
35,000	179	36
40,000	141	29
45,000	111	23
50,000	87	18

The atmospheric pressure at any altitude is the sum of the pressures of the various component gases, including water vapor. Since the atmosphere is 20% oxygen, the partial pressure of oxygen at sea level is 20% of the total atmospheric pressure, or 160 mm. Hg. But at 50,000 ft., the oxygen partial pressure is 20% of the total pressure at that altitude and amounts to only 18 mm. Hg.

HOW MAN BREATHES

During the process of respiration, oxygen is extracted from the air which passes in and out of the lungs. Inspiration is the active phase of muscular expansion of the chest by which air is drawn into the lungs. Expiration is a passive phase, of relaxation of chest muscles and the diaphragm, by which air is exhaled. With each breath, oxygen must be extracted from the inspired air and absorbed by the red blood cells, and carbon dioxide must be given up by the blood to the lungs to be exhaled. Both of these exchanges are accomplished through the membranous walls of innumerable minute lung sacs adjacent to tiny, equally thin-walled blood vessels. Pressure differences across these thin walls cause the required oxygen to pass in, and waste carbon dioxide to pass out. The important requirement for the transfer of oxygen into the blood is sufficient oxygen partial pressure in the air, which progressively decreases with increase in altitude.

Oxygen is carried to the tissues by the hemoglobin inside each red blood cell. Each cell of the body must have an adequate and continuous supply of oxygen to survive. There is, for practical purposes, no capacity to "store" oxygen in the tissues. Certain tissues of the body are more sensitive than others to oxygen deprivation. These include the retina of the eyes and the brain. Therefore, a flyer's vision, level of consciousness, and ability to function are quickly impaired in any hypoxic environment.

MANIFESTATIONS OF HYPOXIA

As hypoxia develops with increasing altitude, the blood carries decreasing amounts of oxygen to supply the body tissues. Because of its considerable adaptability, the body is able to compensate fairly well for this oxygen lack for the first several thousand feet of altitude. Going higher, mental and motor coordination deteriorate steadily until collapse occurs.

An important feature to remember about hypoxia is that its onset is usually unnoticed by the affected person. There is no pain or discomfort, and the individual has no insight into his deteriorating condition.

The lowest altitude at which hypoxia is of practical importance to the flyer is 5000 ft., at which level night vision is measurably diminished. Except for this effect, the mild hypoxia experienced up to 10,000 ft. has little practical significance. Above 10,000 ft., however, the results of hypoxia rapidly become serious. These effects include loss of visual acuity, mental dullness, confusion, loss of judgment, failure of muscular coordination, and eventually unconsciousness. The nail beds and lips may be noted to have a dusky hue, due to their perfusion with poorly-oxygenated blood.

The effects of hypoxia are readily demonstrated in low-pressure training chambers, as students observe each other's reactions and note their own symptoms while performing mathematical calculations. As the hypoxia is relieved by the administration of oxygen, vision improves dramatically along with mental clarity.

DURATION OF CONSCIOUSNESS

A convenient and practical way to evaluate the risk of hypoxia at any altitude is in terms of the average time of useful consciousness for subjects without oxygen, that is, the interval during which the subject can perform some purposeful activity, as in the event of an in-flight emergency, after being deprived of oxygen. Man's performance at different altitudes without oxygen has been studied by means of controlled experiments. Under hypoxic conditions, the time of useful consciousness is also influenced by the amount of physical activity, method of exposure, and whether or not oxygen has been in use just prior to the episode. The time of useful consciousness under various conditions at different altitudes is indicated in Table 7.3., e.g., in the event of loss of cabin pressurization in a modern jet airliner, cruising at 35,000 ft., the time of useful consciousness would be approximately 30 sec. if sitting quietly, and less with any physical activity.

ALTITUDE TOLERANCE

Flyers differ somewhat in their susceptibility to lowered oxygen pressure, and even the same individual may show varying susceptibility from one day to the

TABLE 7.3. Time of Consciousness Under Various Conditions of Hypoxia and Physical Activity[a].

Altitude (ft)	O₂ Rapidly Disconnected, Sitting Quietly	O₂ Rapidly Disconnected, Moderate Activity	Rapid Decompression, Sitting Quietly
22,000	10 min	5 min	—
25,000	3 min	2 min	2 min
28,000	1½ min	1 min	1 min
30,000	1¼ min	¾ min	¾ min
35,000	¾ min	½ min	½ min
40,000	30 sec	18 sec	23 sec
65,000	12 sec	12 sec	12 sec

[a](Adapted from Air Force Manual 52-13).

next. Extreme examples of tolerance to hypoxia are seen among inhabitants of very high altitudes and among experienced mountain climbers. After considerable time spent at high altitudes, the body makes physiological adjustments which allow more effective use of the available oxygen. In Asia and South America, entire communities are found at altitudes above 10,000 ft. One of the highest permanent settlements in the world is a sulfur-mining community in Chile at an elevation of 17,500 ft. In attempts to scale Mount Everest, mountain climbers have spent several days as high as 25,000 ft. without oxygen. In 1953, Sir Edmund Hillary removed his oxygen mask on the summit of Everest at an alti-

Figure 7.1. Low-pressure chamber for simulating high-altitude flights. *(Photo taken at the USAF School of Aerospace Medicine, Brooks Air Force Base, Texas.)*

tude of 29,002 ft. for 10 min. These examples show that under special conditions some men are capable of considerable adaptation to altitude.

However, the intensive training required to produce this high degree of physiological conditioning has not been practical for aircrews. Although peak physical conditioning is highly desirable in all flyers, it still does not permit them to fly safely at high altitude without oxygen equipment. For practical and safe flying operations, 10,000 ft. is the upper limit, above which some type of oxygen breathing equipment (or cabin pressurization) must be used.

PREVENTION OF HYPOXIA

The several devices used to prevent hypoxia in flight are all designed to increase the amount of oxygen absorbed by the blood in one of the following ways:

a. Increase the percentage of oxygen in the inspired air (constant-flow or demand oxygen mask).

b. Increase the pressure of oxygen in the inspired air (pressure-demand oxygen mask).

c. Increase the pressure of air in the aircraft cabin (cabin pressurization).

d. Combine the pressure-demand oxygen mask with mechanical pressurization of the body (partial pressure suit).

e. Pressurize the entire body with oxygen in a sealed suit (full pressure suit).

Oxygen systems are designed to maintain an alveolar oxygen pressure as close to the sea level equivalent as possible. Aircraft oxygen systems consist of the containers in which the oxygen is stored, the regulators which regulate the oxygen flow, and the masks through which the oxygen is delivered to the individual user.

Oxygen Storage Systems.   Oxygen is carried in cylinders or containers mounted in the aircraft. Three types of oxygen storage systems are currently utilized in the US Air Force: low pressure, high pressure, and liquid oxygen. In the low pressure system, the oxygen is stored in yellow, light-weight cylinders at pressures below 450 psi. The pressure should never be allowed to drop below 50 psi, to prevent moisture from accumulating in the cylinders and contaminating the system. The low pressure system, in the form of portable oxygen assemblies, provides freedom of movement within many nonpressurized aircraft at high altitudes. In the USAF high pressure system, storage cylinders are color-coded green and are normally filled to a pressure of 1800 to 2200 psi. Liquid oxygen (LOX) systems are the most widely used storage systems in USAF aircraft, because of savings in both space and weight. A LOX converter is used to store the oxygen in a liquid state at a temperature of $-297.4°F$ and to convert it to gaseous oxygen as needed. The standard LOX converter is a double-walled container with a vacuum between the walls. It will expand each liter of liquid oxygen into 860 liters of gaseous oxygen. The expansion is controlled, and the gaseous oxygen is maintained at a constant operational pressure. In large aircraft the LOX

TABLE 7.4. Oxygen Equipment Required for Various Altitudes (U.S. Armed Forces).

Altitude	Oxygen Required	Oxygen System Required
Sea Level to 10,000 ft	Physiological Zone No supplemental oxygen required	None
Above 10,000 ft	Physiological Deficient Zone Oxygen recommended for all occupants Required for pilots	Continuous Flow or Demand Oxygen System
13,000 ft	Oxygen required for all occupants	Continuous Flow or Demand Oxygen System
25,000 ft	62% supplemental oxygen required	Continuous Flow Oxygen System operational ceiling
35,000 ft	100% oxygen	Diluter Demand Oxygen System operational ceiling
Above 40,000 ft	100% oxygen with positive pressure	Pressure Demand Oxygen System
Above 50,000 ft	Breathing pressure and body counter-pressure necessary	Partial or Full Pressure Suit

converter is designed to produce an operational pressure of 300 psi in order to permit portable oxygen assemblies to be refilled in flight. In small aircraft not utilizing portable assemblies, the operational pressure is 70 psi.

Oxygen Delivery Systems. The oxygen delivery system consists of a regulator which meters the percentage and flow of oxygen, and a mask which fits closely over the flyer's nose and mouth. There are three types of delivery systems currently used in the US Air Force: continuous flow, diluter demand, and pressure demand. Each system can be installed as a fixed (aircraft mounted) or portable system. The type of equipment required for various altitudes is indicated in Table 7.4.

In the continuous flow system, the regulator delivers oxygen to the user continuously whether he is inhaling or exhaling. This delivery system is utilized in transport aircraft to provide oxygen for passengers in the event supplemental oxygen is required (*i.e.,* in the event of failure of cabin pressurization). It is designed primarily as an emergency or "get me down" system. Generally, oxygen is delivered through the distribution lines to a number of continuous flow masks when the cabin altitude exceeds 10,000 ft. A continuous flow system is not normally used for extended periods of time above 25,000 ft.

The diluter demand system delivers oxygen only during inhalation, that is, on demand. It is designed to mix ambient, or cabin air, with oxygen in a proper proportion for the particular altitude. A lever attached to the regulator also offers a selection of 100% oxygen at any time desired, however. A diluter demand system is not used for prolonged periods above 35,000 ft.

The pressure demand system is most commonly used by crew members of

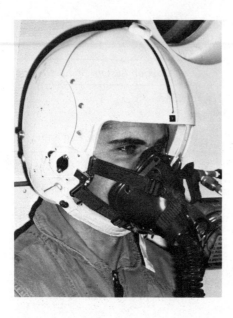

Figure 7.2. Demand oxygen mask (USAF type MBU-5/P) and flying helmet (USAF type HGU-2/AP). *(Taken at USAF School of Aerospace Medicine, Brooks Air Force Base, Texas.)*

high performance aircraft and has both the demand and diluter capability discussed above. At higher altitudes it will deliver oxygen to the lungs at pressures greater than that of the cabin environment. If cabin pressure is lost above 30,000 ft., oxygen will be delivered to the lungs under a positive pressure to insure an adequate alveolar oxygen pressure. The pressure demand system is designed for short duration use at the higher altitudes and has an operational ceiling of 43,000 ft.

Pressure Suits. The capability of modern aircraft to operate safely at altitudes in excess of 45,000 ft. has made it necessary to design personal flying equipment which will provide the required protection against the reduced barometric pressures at these higher altitudes. Pressure suits are designed to create an artificial atmosphere around the body by introducing oxygen pressure into the helmet and suit. This pressure is controlled in the helmet by a regulator which meters the correct breathing pressure and in the suit by a controller to counterbalance the breathing pressure. This delicate balance is necessary to afford the aircrew member maximum breathing comfort and body protection at altitudes approximating the near vacuum of space.

Mechanical Effects of Barometric Pressure Changes

Although generally considered to be a mass of solids and liquids, the human body also includes a significant amount of gas, largely air and its compo-

TABLE 7.5. Increase of Volume of Dry Gas with Altitude.

Altitude (ft)	Number of Atmospheres of Pressure	Relative Gas Volume
Sea level	1	1
18,000	½	2
28,000	⅓	3
33,000	¼	4
42,000	⅙	6
48,000	⅛	8

nent gases. Some of these are dissolved in the body fluids, as, for example, nitrogen, oxygen, and carbon dioxide in the blood. In addition, air is found as a free gas within the lungs, gastrointestinal tract, the middle ear, and the sinuses. These gases, dispersed throughout the body, react to variations in atmospheric pressure (resulting from altitude changes) according to well-known physical laws.

During changes in pressure, the behavior of free gas, such as the air in the body cavities, is determined by Boyle's law. This law states, that *when the temperature remains constant, the volume of a gas varies inversely as the pressure.* This characteristic of a gas is readily demonstrated in a low-pressure chamber by observing a partially inflated balloon which gets larger in proportion to the decrease in pressure. Thus at 18,000 ft., where atmospheric pressure is half its sea level value, gas volume is doubled. Table 7.5 indicates the approximate relative increase in dry gas volume at different altitudes. The symptoms produced by these trapped gases during flight can be very painful and distracting. These phenomena have been implicated in the cause of a number of aircraft accidents.

GASTROINTESTINAL GAS

The intestinal tract normally contains gas derived from swallowed air and from the action of the digestive process. The amount of this gas is variable among individuals, and from day to day in the same person. Like all other gases, it conforms to Boyle's law and so expands with ascent of the aircraft. Flyers are familiar with the sensation of abdominal bloating which is sometimes experienced during fast ascents to high altitude. Normally, this is not noticeable until an altitude of 15,000 to 20,000 ft. is reached, but the onset of distress will be determined, to some extent, by the amount of gas and the rate of ascent. Usually the healthy flyer is able to expel enough of the expanded gas to relieve the condition.

Flyers should avoid gas-forming foods in meals taken before and during flight.

Although certain foods such as beans, cabbage, and fresh bread are notorious in this respect, there is considerable individual variation from person to person. Any type of irritation within the gastrointestinal tract may produce discomfort which is not necessarily directly proportional to the amount of gas present. For this reason foods difficult to digest or those which are highly seasoned should be avoided.

BAROTITIS

Consistent with Boyle's law, air in the cavity of the middle ear expands and contracts with changes in atmospheric pressure. During altitude changes, if pressure in the ear is not readily equalized with the outside air pressure, pain and temporary deafness is produced in the affected ear. This condition has been given the name of *barotitis* because it is caused by changes in atmospheric pressure. It occurs frequently in persons who fly and is one of the most common medical complications seen by flight surgeons, affecting experienced aircrews as well as passengers.

In order to understand how barotitis develops, it is necessary to have some knowledge of the anatomy of the ear and Eustachian tube (Figure 7.3). Structurally, the organ of hearing and equilibrium is composed of three parts, an *outer ear*, a *middle ear*, and an *inner ear*. The outer ear includes the external carti-

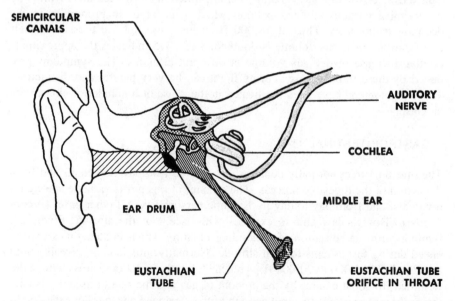

Figure 7.3. Structure of the ear and Eustachian tube. *(Drawn by Charles J. Shaw, Chief of the Gunter Branch Graphics Section, Air University Library, Gunter Air Force Base, Alabama.)*

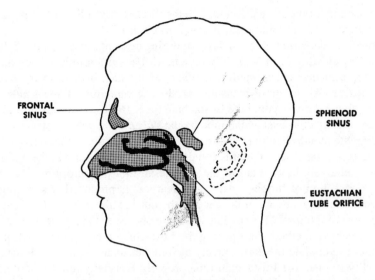

FRONTAL
SINUS

SPHENOID
SINUS

EUSTACHIAN
TUBE ORIFICE

Figure 7.4. The sinuses. The maxillary and ethmoid sinuses are not shown in this drawing. *(Drawn by Charles J. Shaw, Chief of the Gunter Branch Graphics Section, Air University Library, Gunter Air Force Base, Alabama.)*

laginous portion and the *auditory canal.* The latter extends inward and terminates at the thin, membranous *eardrum.* The auditory canal, the middle ear, and the inner ear are surrounded by bones of the skull.

The middle ear is a small air-filled cavity separated from the auditory canal by the eardrum. It contains three small bones which serve as sound conductors. Barotitis involves this small enclosed cavity of the middle ear. The inner ear is composed of delicate, membranous structures occupying a small chamber within the temporal bone. It has the dual function of hearing and assisting in the maintenance of balance. The narrow duct leading from the middle ear to open into the back of the throat is called the *Eustachian tube* (Figures 7.3 and 7.4). By this means the middle ear communicates with the outside air so that normally any pressure difference between the middle ear and the atmosphere is equalized. This tube remains collapsed except at intervals when it is briefly opened by the act of swallowing or yawning. At such times, small throat muscles come into play, causing it to open momentarily.

As the air in the middle ear expands during ascent, it is forced out through the Eustachian tube at frequent intervals of altitude. A pressure differential of approximately 15 mm. Hg in the middle ear cavity is sufficient to produce this effect. This is manifested by the familiar faint popping sensation in the ears during ascent. Ordinarily no difficulty is encountered on ascent because pressure equalization occurs automatically. On descent, however, the situation is reversed, and air in the middle ear decreases in volume. When this happens there is a tendency for the opening of the Eustachian tube in the throat to react like a flutter valve so that ventilation of the middle ear is much more difficult.

Normally, ventilation of the middle ear during descent can be accomplished by frequent swallowing or yawning. Experienced flyers often relieve such an "ear block" by a maneuver resembling a stifled yawn which becomes very effective after practice. All of these measures produce contraction of the small throat muscles, thus briefly opening the Eustachian tube. If these methods fail, air may be forced through the Eustachian tube by the *Valsalva maneuver,* which consists of closing the mouth, holding the nose and blowing steadily with moderate force.

As descent progresses, if ventilation does not occur, pressure in the middle ear becomes relatively lower than in the outside air and the eardrum is forced inward. A sensation of fullness, pain, and deafness is produced. As the pressure differential increases between the ear and the outside air, there is a critical point (approximately 90 mm. Hg) beyond which this pressure cannot be overcome by any of the above methods for opening the Eustachian tube. For this reason, it is important to ventilate the ears actively at frequent intervals during descent.

Inability to open the Eustachian tube properly is usually due to the common cold or other upper respiratory infection. The resultant swelling of the tissues in the throat and nose closes the small orifice of the tube. Other frequent causes of such interference are nasal allergy and overgrowth of tissue around the orifice. The eardrum is retracted to varying degrees, and fluid collects within the middle ear, producing temporary deafness. Under unusual conditions when a very high pressure differential develops, there may be rupture of the eardrum.

In order to prevent the development of barotitis, airmen should not fly while suffering from upper respiratory infections or other conditions which interfere with ventilation of the middle ear. If there is difficulty in equalizing pressure, the aircraft should be brought down at as slow a rate of descent as possible. In severe cases, reascent followed by very gradual descent may be necessary.

BAROSINUSITIS

Barosinusitis is pain in the sinuses caused by the same mechanism as barotitis, that is, by changes in pressure. Like the middle ear, the sinuses are air-filled cavities, each with a single opening, and are similarly affected by pressure differences. Though less common than barotitis, barosinusitis is equally painful, and relief is more difficult because of structural differences. The sinus cavities, located within the bones of the face and skull, are grouped about the nasal passage, one of each pair on either side of the midline (Figure 7.4). Each drains into the nasal cavity by a narrow opening through which equalization of pressure takes place during altitude changes. The mucous membrane lining of the sinuses is continuous with that of the nose and throat. If the openings are closed by inflammation and swelling of the membrane, a pressure differential develops within one or more of the sinuses. Pain is most common in the forehead, but may be felt in the face or even in the back of the head, depending upon which of the sinuses is involved.

As with the Eustachian tube, obstruction is more likely to occur on descent than on ascent. Unlike the Eustachian tube openings, however, the small sinus apertures cannot be controlled by the individual, since they are merely rigid openings in bone. Failure to ventilate the sinuses is more likely during rapid descents. When pain occurs, descent should be stopped if possible, and level flight maintained until the pressure is equalized. If severe pain persists, reascent is necessary to relieve it. Relative negative pressure causes fluid to be drawn out of the mucous membrane linings and to collect in the cavities. The same precautions should be followed as for the prevention of barotitis.

DECOMPRESSION SICKNESS

Decompression sickness is a condition developed at reduced atmospheric pressure and manifested by pain in the joints ("bends"), nervous system impairment, or pain in the chest with difficult breathing ("chokes"). It is caused by the formation of gas bubbles (mostly nitrogen) in the tissues and body fluids. The same conditions prevail and similar symptoms result when deep sea divers are suddenly decompressed by being brought to the surface too rapidly. Before the use of pressurized cabins in aircraft, decompression sickness was a frequent occurrence among aircrews flying above 25,000 ft. It is still a potential hazard in the event of loss of cabin pressure at high altitude or any unpressurized flight above 18,000 ft. (one-half sea level pressure). Also, it is seen occasionally in individuals undergoing high-altitude indoctrination in low-pressure chambers.

EVOLVED BODY GASES

To understand the mechanism of decompression sickness, further consideration of the behavior of gases is necessary. Because of constant absorption of oxygen and release of carbon dioxide in the lungs, these dissolved gases are maintained at relative equilibrium within the body. These two gases occur in both physical and chemical combination and actively enter into the body metabolism. This is not true, however, of nitrogen, which is inert and is therefore found only in physical combination. Physically combined nitrogen makes up about 80% of the gas that is dissolved in the tissues. Since it is held in solution solely by physical pressure, its amount varies according to the atmospheric pressure.

This is in accordance with Henry's gas law which states that *with a constant temperature the quantity of gas which goes into solution is proportional to the partial pressure of the gas concerned.* Therefore, at any altitude, dissolved gases are present in the body in the same percentage as they occur in the air at that altitude. As altitude increases, gases tend to come out of solution in proportion to the alteration in pressure and to maintain equilibrium with the atmosphere.

If the altitude attained is less than 18,000 ft., the nitrogen dissolved in tissues and blood is held in solution and eliminated by the lungs. If, however, the pressure is reduced to less than one-half that at sea level, nitrogen pressure exceeds that which can be held in solution and is evolved as bubbles. This phenomenon is similar to the bubbling which occurs when the lid of a carbonated beverage is released rapidly. Because it is inert, nitrogen is the principal component of such bubbles. As these bubbles enlarge, they may obstruct the very small blood vessels and produce various symptoms because of interference with the circulation in local areas of the body. There is considerable evidence that symptoms are also due to the direct pressure of bubbles around small nerves. Bubble formation is more likely to occur in tissues of the body where circulation is poor and where nitrogen is most abundant. Fat has a greater affinity for nitrogen than does any other tissue, and fat people are most likely to have decompression sickness.

The most frequent symptom of decompression sickness is pain in the joints, known as bends. Any joint may be affected, but the knee and shoulder are most often involved. Onset is usually fairly sudden, and pain tends to become worse as long as the individual is at high altitude. Pain is often severe enough to force the flyer to descend. Symptoms occurring in the chest are known as chokes, referring to a sharp, burning pain accompanied by shortness of breath and coughing. Skin symptoms include itching, burning, warm or cold sensations, and a mottled appearance of the skin. The nerves may be temporarily affected so that the individual has partial loss of vision, inability to speak, or partial paralysis.

Very rarely, decompression sickness may be fatal or result in permanent impairment. Most cases, however, are painful and temporarily incapacitating, but leave no permanent effects. The pain of bends usually subsides shortly after descent. Chokes and manifestations of nerve involvement, although rare, are of much more serious import. Decompression sickness is unusual below 25,000 ft., but becomes common above 30,000 ft. It has actually occurred as low as 18,500 ft., however.

PREVENTION OF DECOMPRESSION SICKNESS

Even apparently mild cases of decompression sickness must be considered potentially serious. The only satisfactory immediate treatment is recompression to ground level. In low-pressure chambers this is easily done, but in actual flight, rapid descent may not be feasible for operational reasons. The incidence of decompression sickness increases directly with the following factors:

    a. The pressure altitude.
    b. The total time spent at altitude.
    c. The amount of exercise performed.
    d. The age of the subject.
    e. The degree of overweight of the subject.

Experience indicates that the higher the altitude and the longer the flight, the greater will be the incidence and severity of symptoms.

Subjects who perform active physical exercise at altitude are more likely to develop symptoms. Analysis of data on aircrew trainees in low-pressure chambers shows that susceptibility to decompression sickness increases in overweight individuals. This then is an added reason for air crewmen to maintain themselves in good physical condition.

Other than pressurization of the aircraft cabin, the best prophylactic measure against decompression sickness is denitrogenation. By breathing 100% oxygen for a period of time, the dissolved nitrogen is removed from the tissues prior to the subject's reaching high altitude. Since the partial pressure of nitrogen in the lungs is greatly diminished by this procedure, nitrogen diffuses outward from the blood and tissues and is exhaled.

If as much as one-half of the body nitrogen is lost by denitrogenation, significant protection is afforded. This amount can be removed by prebreathing 100% oxygen for 1 hr. prior to flight. After the first hour, nitrogen is lost at a much slower rate. The body's nitrogen elimination rate is, however, not uniform for all tissues. Fatty tissue, for example, contains much more nitrogen, and its blood supply is much less than muscle. Thus, despite the total body loss of nitrogen within a given period of oxygen breathing, parts of the body may still have excess nitrogen sufficient to cause bubble formation.

Decompression sickness symptoms which do not clear up by recompression to ground level must be treated by chamber compression to greater-than-sea-level pressures. The techniques and equipment used are similar to those used for the treatment of divers with the same disorders.

In view of the value of compression therapy in the treatment of decompression sickness, the US Air Force and Navy have a number of strategically placed compression or hyperbaric chambers in operation. The Air Force has a medical treatment team at the School of Aerospace Medicine, Brooks Air Force Base, Texas. It is available on 24-hour call and will provide advice to assist in the treatment of affected airmen.

SCUBA DIVING

If one has participated in SCUBA[1] or surface supplied diving, with the associated increased pressure, before flying, decompression sickness may occur at relatively low altitude. The additive effect of the further decrease in pressure when flying soon after surfacing from a dive may be sufficient to cause symptoms. Under such circumstances decompression sickness may occur at altitudes well within the capabilities of light aircraft. After surfacing from a dive, sufficient time should be allowed to reestablish ground-level nitrogen equilibrium

---

[1] Self-contained underwater breathing apparatus.

before flying. For maximum safety after a dive to any depth, individuals should not fly to any altitude for at least 24 hours.

Pressurized Cabins and Rapid Decompression

The use of pressurized cabins in aircraft has partially solved the problems of hypoxia, barotitis, barosinusitis, and decompression sickness by reducing the peak pressure altitude of flight and by avoiding rapid pressure changes. The flyer is surrounded by an artificial atmosphere under greater pressure than the outside air. The oxygen partial pressure is thus also raised so that more oxygen is made available for absorption by the lungs. For example, when the true altitude of the aircraft is 30,000 ft., the cabin pressure altitude may be only 8000 ft. so that breathing oxygen is not necessary. Since the entire body is pressurized, it is protected against the effects of free and evolved gases as well as hypoxia. Pressurization up to a cabin altitude of 10,000 ft. may be used in lieu of oxygen breathing equipment. For cabin altitudes from 10,000 to 42,000 ft., oxygen equipment must be used in conjunction with cabin pressurization. Oxygen equipment is also required to offset the hazard of sudden loss of cabin pressure, that is, rapid decompression.

PRESSURIZED CABINS

Isobaric control of pressure is a system in which the cabin is maintained constantly at a fixed pressure throughout the flight, regardless of flight altitude. In *differential control* of cabin pressure, the cabin pressure varies with the flight altitude according to a given ratio. Isobaric control may be used until the aircraft reaches the altitude where its maximum allowable structural pressure differential exists. Above this critical altitude, differential control is required.

In most present cargo aircraft, the cabins are relatively large as compared to the surface area which is likely to undergo mechanical rupture or perforation. This feature minimizes the hazard from rapid decompression. The pressure differential for this type of aircraft is 6.55 psi. The most recent interceptors are provided with a pressure differential of 5 psi. In the latter, up to a flight altitude of 31,000 ft. the cabin "altitude" is maintained at 12,500 ft. by isobaric control. Above this flight altitude the constant differential of 5 psi is used. Under combat conditions this is reduced to 2.75 psi as a precaution against sudden decompression (over a wide pressure differential), which is more likely to occur in combat. Because of the relatively small size of the cabins in some of the jet bombers, the same precautions are necessary and a similar pressurization schedule is used.

RAPID DECOMPRESSION

If pressure is suddenly lost in flight, there is a forceful equalization of cabin pressure with that of the outside air as cabin air is rapidly expelled. The magnitude of this decompression, and, hence, the physiological effect on the flyer, is determined by:

  a. The size of the cabin defect.
  b. The altitude of the aircraft.
  c. The volume of the cabin.
  d. The amount of pressure differential.

The smaller the cabin, the larger the defect, and the greater the pressure differential, the more rapid is the rate of decompression. Extremely rapid loss of cabin pressure is rightfully termed *explosive decompression.*

Physiological Effects. With the sudden equalization of pressure, there is a forceful blast of air outward through openings and passages in the aircraft. At such times, crewmen near these openings have been swept out of the aircraft. For this reason personnel at their stations in pressurized cabins should always have their seat belts fastened.

In addition, the free gases within the body cavities suddenly expand and are partially expelled. The decompression which occasionally occurs in flight is usually relatively slow so that body gases escape without dangerous internal pressures being developed. The middle ear and the sinuses are ventilated without difficulty because the greater pressure inside the cavities forces open the Eustachian tube and the sinus orifices.

If excessive, expanding gases within the intestinal tract may cause pain, but since this organ is normally capable of considerable stretching, serious injury is not likely to occur. The same is true of the lungs; the lung gas is normally expelled through the trachea with little resistance, but breath holding during a rapid decompression could cause catastrophic lung rupture. Immediately following decompression, the flyer is exposed to the risk of decompression sickness and hypoxia. As previously discussed, these two hazards are in direct proportion to the flight altitude. As shown in the fourth column of Table 7.3, the time of consciousness is unusually brief following rapid decompression, that is, hypoxia develops more quickly than after simple deprivation of oxygen.

Aircrews should maintain oxygen masks in a state of readiness at all times when flying in pressurized aircraft. In case of rapid decompression, oxygen equipment will be needed immediately (if it is not already being used), and descent of the aircraft will be necessary under many circumstances. A practice rapid decompression in a low-pressure chamber is carried out as part of the altitude indoctrination of all US Air Force aircrews.

Very-High-Altitude Emergency Equipment

For aircrews flying above 43,000 ft., some type of emergency protection must be available in case of cabin decompression. At up to 43,000 ft. a well-fitted face mask with 100% oxygen at 18 mm. Hg. pressure gives adequate protection for a short time, even if cabin pressure is lost. Above this altitude, though, the breathing pressure required to prevent hypoxia is too high for any but short duration emergency use. Breathing oxygen under pressures greater than 18 mm. Hg rapidly becomes prohibitive because of leakage around the mask, respiratory fatigue, and interference with the circulation of blood to the heart and lungs. By the time 63,000 ft. is reached, atmospheric pressure has diminished to 47 mm. Hg. At this altitude water boils at man's normal body temperature (98.6°F).

Since the body is approximately 70% water, its gases come out of solution at 63,000 ft., and the blood is said to "boil." In order to survive at such altitudes if cabin pressure is lost, the flyer must have a garment which will give him some degree of pressurization. Breathing pressures required for this necessary pressurization exceed the limits of human tolerance unless counterpressure is supplied to the surface of the body by a tightly fitted suit. Such a suit also prevents interference with circulation and supports the chest against the required high breathing-pressures. There are two types of pressure suits in use: partial-pressure suits (mechanical pressure applied to the body) and full-pressure suits (the entire body enclosed in a controlled environment). The latter is the most common today and is exemplified by NASA's lunar suits.

THE PARTIAL-PRESSURE SUIT

This suit is a tightly fitted garment which completely covers the body and limbs. It is worn uninflated and inflates automatically when cabin pressure is lost. Each suit is carefully fitted to the individual flyer by means of laces and zippers. *Capstan tubes* are attached to the back, arms, and legs by a series of small straps. When these tubes are inflated, the closely-placed straps are tightened, thus tightening the suit to give pressure over the body and limbs (Figure 7.5).

THE FULL-PRESSURE SUIT

Although adequate as high-altitude emergency protective equipment, the partial-pressure suit has certain limitations for prolonged wear. The wearer has some restriction of movement, limitation of vision, and discomfort when the suit is worn for long periods. A better system for overcoming these difficulties and for furnishing the flyer with a livable atmosphere is a full-pressure suit in which the wearer is surrounded by a layer of pressurized oxygen or air. In developing such

Figure 7.5. High-altitude partial-pressure suit, helmet, and gloves. Note snug fit of suit even though capstan tubes are not inflated. *(Taken at School of Aviation Medicine, USAF Aerospace Medical Center, Brooks Air Force Base, San Antonio, Texas.)*

Figure 7.6. Full-pressure suit, helmet, boots, and gloves. A ventilation-insulation garment is worn next to the pilot's body beneath the suit. Curved facepiece of helmet extends laterally to afford maximum *visibility. (U.S. Air Force photograph.)*

a suit, further problems of overheating, weight, and lack of mobility are encountered. The full-pressure suit used by US Air Force crews is shown in Figure 7.6. This suit, with helmet, gloves, and boots is a complete full-pressure outfit which insures safety, mobility, and comfort of crewmen at very high altitude. It maintains a complete envelope with ventilation, pressurization control, breathing oxygen, and communications.

Extremes of Temperature

In the relative comfort of cabin accommodations in modern aircraft, a flyer is generally not concerned with the extremely cold temperatures which exist at high altitudes. The environmental temperature at 20,000 ft., for example, is approximately $-22°F$ and at 30,000 ft., $-58°F$. The effects of these temperatures on the human body are greatly increased by the effects of rapid movement of air which would also be encountered in flight. Thus, frostbite or cold injury was a serious problem in many World War II aircraft. It remains a hazard in the event of high altitude bailout or ejection, during which cold injury can quickly occur if adequate protective clothing is not worn. Parachute opening at high altitudes prolongs the exposure to cold and to hypoxia and is therefore more hazardous than a free-fall descent, with parachute opening at a lower altitude.

Heat stress may also be encountered in flight operations. When sitting on the taxiway enclosed by a glass canopy, the flyer is subject to the "greenhouse" effect. The military flyer's required protective clothing and equipment tend to accentuate the buildup of body heat. Malfunctioning of a cabin heating system in flight has been known to lead to heat stress, contributing to a fatal aircraft accident.

Vision in Flight

From the earliest days of aviation, the flyer's vision has been regarded as a most vital part of his physical equipment for flights. In flying, constant demands are made on the vision in order to avoid obstacles, to judge distance, to read colored signals and maps, to study the terrain, and to read flight instruments within the cockpit. Elaborate and careful tests of vision are a prominent part of every physical examination for military and commercial flying, and failure to meet high visual standards is a common cause for rejection of candidates for flight training. It is absolutely necessary that a candidate starting his flying training have perfect vision. In later years some defects of vision will inevitably de-

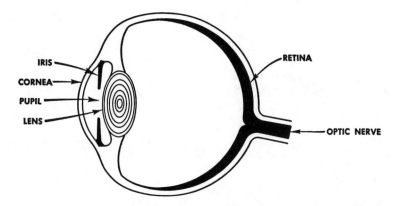

Figure 7.7. Diagrammatic section of the eyeball.

velop with increasing age. This loss of visual ability is expected and is acceptable after the flyer has accumulated considerable flying experience, which compensates to some extent for visual defects.

STRUCTURE AND FUNCTION OF THE EYE

When an object is seen, an image has been formed at the back of the eye and then transmitted to the brain. To be seen, objects must, of course, be illuminated in some way. This light is reflected from the object and enters the eye through the cornea, the outer transparent layer of the eyeball. Behind the cornea is the iris, an opaque layer with a central opening, the pupil, through which light enters. The lens is a semisolid, transparent disc behind the iris. A delicate muscle regulates the size of the pupil, enlarging it in dim light and contracting it in bright light. As light enters the eye and passes through the lens, it is focused by the latter to produce an image on the retina. The retina is the receiving portion of the optic nerve which is arranged to form a thin lining over the back of the eyeball. This is the nerve of vision which transmits the image to the brain. The multiple, minute nerve endings making up the retina are of two different types, differentiated by their structure and function (Figure 7.8).

The small central area of the retina is composed largely of tapered nerve endings called cones. These are most effective when illumination is good, and they are required for maximum visual acuity. They are also necessary for the discrimination of color, but function very poorly under low illumination. They account for what is called central vision which is effective only in good light. In the outer or peripheral portion of the retina, the nerve endings are mostly long, cylindrical rods which give less effective visual acuity. These rods give us our peripheral vision upon which we depend for seeing at night or in very dim light.

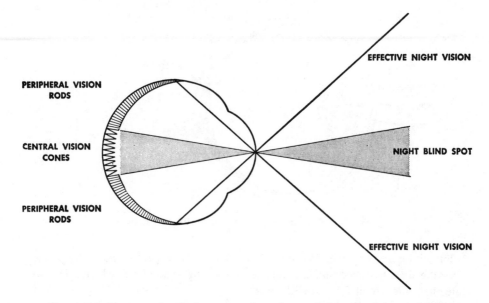

Figure 7.8. Diagram of retinal nerve endings for central and peripheral vision.

VISUAL ACUITY

Visual acuity is the ability to perceive the shape and detail of objects. This is the function which is measured when a flyer is asked to read a series of progressively smaller test letters at a standard distance of 20 ft. The smallest letters that can be read correctly indicate the visual acuity. If the subject is able to read letters of the prescribed size at a distance of 20 ft., he is said to have 20/20 visual acuity, which is considered to be normal. This visual test is the single most common cause of failure of the physical examination for military flying. Studies on military pilots of World War II indicate that of all visual factors, visual acuity is most closely related to success in air-to-air combat.

Regardless of whether an individual's visual acuity is good or poor, the clarity of his vision will be influenced by several characteristics of the object viewed, such as distance from the viewer, size and contour, movement, the amount of light reflected, and contrast with the background. Also of obvious importance are the amount of light and the condition of the intervening atmosphere in the field of vision.

Since visual acuity is best with central vision and diminishes steadily toward the periphery of the retina, not everything in the field of vision is seen with equal clarity. At a few degrees off center, visual acuity is significantly diminished, and therefore we see objects most clearly by looking directly at them. Also, if the eye is in motion, nothing but an indistinct blur is seen. As a pilot constantly scans the sky for other aircraft, he must carefully examine small, individual patches of sky rather than sweeping his gaze over wide areas. In this way, for search of any particular area of the sky, the gaze is immobile and directed at the area being scanned.

To combat eye fatigue caused by glare, sunglasses should be worn while scanning from aircraft. The most desirable lenses are those which are tinted to eliminate most of the ultraviolet and infrared light, but not light within the visible spectrum. This is best achieved by tinting sunglasses and helmet visors a neutral shade such as currently used by the US Air Force.

DEPTH PERCEPTION

Depth perception is the ability to judge distance. It is important in making landings, in formation flying, and in avoiding obstacles in flight. It is essential to the success of military aerial refueling operations, a specialized type of formation flying.

Several factors enter into an estimation of distance, some involving only one eye, and others requiring both eyes. The value of the latter is derived from the fact that the two eyes are at slightly different angles from the object. In flying situations, however, most of the distances are so great that this angle is too small to be of importance. In the critical phases of landing, factors such as relative size of objects seen, relative motion, familiarity with the airfield, and atmospheric conditions can also influence the pilot's visual performance.

At the speeds of jet aircraft, no human visual capability is sufficient to avoid a midair collision if two aircraft are indeed on a collision course. Hence arises the necessity for control of flight paths by ground radar as well as developmental work on equipment, to be mounted within the aircraft, which would automatically take corrective action in order to avoid a collision.

NIGHT VISION

Man does not see at all in the absence of light, but such conditions of total darkness are rare. If there is any light at all from the moon or stars, it can be picked up by peripheral vision, since the rods in this part of the retina are sensitive to light intensities as low as 1×10^{-6} candlepower. Under most conditions moonlight provides enough illumination for central vision, but starlight alone is rarely bright enough. The ability to see at night can be greatly improved by understanding and applying certain techniques. The individual flyer's night visual ability can actually be increased by practice. If the rods of peripheral vision are exposed to strong light even briefly, their sensitivity is temporarily destroyed. For this reason, avoidance of strong light must begin well in advance of a night flight; several minutes are required to regain rod sensitivity after it is lost.

Dark Adaptation.   After rod function is interrupted, the period during which sensitivity is slowly regained is known as the period of dark adaptation. This process takes place in each eye independently. The eyes become partially dark-adapted very quickly, as when one enters a darkened theater; only a brief period is required before the aisle and seats can be clearly seen. Dark adaptation is nearly complete after 30 to 40 mins. Rod function is temporarily destroyed most quickly and completely by white light, and dark red light exerts the least amount of influence on dark adaptation.

By wearing dark red goggles before a night flight, flyers whose mission requires effective night vision can continue to read or work in a lighted room without interfering with their dark adaptation process. Otherwise they must stay in darkened surroundings for 30 min. prior to flight. During the preflight check of the aircraft, as little light as possible should be used. Red flashlights, red cockpit lights, and red luminous instrument dials are used in aircraft for this reason. However, one disadvantage of red light is its interference with color perception. Some white light may still be required for certain uses such as the reading of colored maps. An instant of exposure to white light makes it necessary to start over with dark adaptation. If such exposure is unavoidable, one eye may be closed to preserve the function of its rods.

Eccentric Vision.   Normally we look directly at an object, utilizing central vision in order to see it more clearly. For night vision this habit must be altered. Since central vision is ineffective under low illumination, the flyer should not look directly at objects at night. They are seen much more clearly if vision is directed slightly away, rather than directly toward them. Eccentric vision, that is, gazing about 10° off center, should be practiced as a valuable means of improving night vision. In this way, the peripheral rods are used instead of the central cones. This effect can be demonstrated by counting a cluster of very faint lights in the distance at night. By looking slightly away from center, more lights can be clearly seen than by looking directly at them.

At night, objects are seen by contrast with their background, that is, they must be darker or lighter than their surroundings. Aircraft are more easily seen if they are above and silhouetted against the lighter sky. Since rod vision is not sensitive enough to see small objects, aircraft at night are seen best from above or below, where they present a larger outline.

Other handicaps to night vision are the lack of color discrimination and depth perception, for which the rods are of little use. Without colors, normally familiar objects may be difficult to identify. Also, because of poor depth perception, a small aircraft near at hand may be mistaken for a larger one far off. Thus at night it is important to properly interpret such visual clues as are available. The ability to do this improves with practice and night-flying experience.

Effects of Hypoxia.   Night vision is reduced by lowered oxygen tension, exposure to carbon monoxide, and deficient diet. As previously mentioned, night vision is unusually sensitive to hypoxia. When flying at night at an altitude as low as 5000 ft., instrument dials may blur slightly if oxygen is not used. At 12,000 ft. there is a marked reduction, and at 16,000 ft. nearly complete loss of night vision can occur because of hypoxia. Exposure to carbon monoxide reduces the oxygen-carrying capacity of the blood and so has an effect on night vision similar to that of hypoxia of altitude. A flyer who smokes steadily during or even prior to flight is certain to impair his night vision.

The ability of the eyes to adapt to darkness depends upon an adequate amount of vitamin A in the diet. This vitamin is found in many commonly consumed foods such as green and yellow vegetables. Therefore, vitamin A deficiency is not likely to occur under normal conditions. Taking excessive amounts of vitamin A (above the normal requirement) is of no benefit to night vision.

VISUAL ILLUSIONS AT NIGHT

The night flyer has very few available visual references with which he can orient objects outside the aircraft. For this reason he is susceptible to certain visual illusions.

If one stares at a stationary light in a darkened room it will soon appear to be moving about from side to side in wide, irregular arcs. This phenomenon, in which a fixed light appears to be moving, is called autokinetic movement, or "stare vision." It may also occur during night formation flying. This has been offered as an explanation for certain night accidents in which an aircraft is seen to break away from a formation, go into a steep dive, and crash without any apparent attempt to recover. It seems likely that a pilot who can see nothing but the taillight of the lead aircraft and stares fixedly at it may leave the formation and even lose control of his aircraft because he tries to follow the apparent movement of the light. The exact cause of autokinetic movement is not known. It can be prevented by constantly shifting the gaze from the light to some reference point on the aircraft.

COLOR VISION

The discrimination of color is entirely dependent upon the cones of central vision. Individuals who have below-average ability to differentiate color are said to be "color blind." A better term is "color defective" inasmuch as total color blindness (caused by an absence of retinal cones) is rare. Defective color vision is usually hereditary, does not change throughout life, and involves both eyes. The condition is not related to other visual defects. About 10% of the male population has some degree of defective color vision. The most common type of defect involves red-green perception, wherein these colors cannot be discriminated and are seen as a neutral shade. There is also another variety of red-green deficiency wherein the colors can be identified, but appear as a darker shade. Good color vision is essential in aviation in order to interpret colored navigation lights on aircraft, airdromes, and obstructions, for the reading of maps and the recognition of terrain features, and for colored emergency signals such as flares, smoke grenades, or standard light signals from a control tower. Any degree of defective color vision is disqualifying for military flying. In civil aviation, there are provisions to issue medical certificates with appropriate limitations to some individuals with mild color vision deficiencies.

Effects of Acceleration

Another important stress which the human body encounters in flight is that of accelerative forces. Acceleration is produced by any change in either the speed or direction of the aircraft, as in take-offs, landings, dives, or turns. It was once thought that speed alone might cause serious bodily damage, but it now appears that this fear was unfounded. As long as the rate of motion remains constant, man is able to function normally at even the astronomical speeds of space flight. However, if speed is increased or decreased or if the direction of flight is altered, there may be significant physiological effects.

In modern aircraft, because of their great speeds and maneuverability, acceleration is frequent and may be extreme. When this occurs, accelerative forces are brought to bear upon both the aircraft and its occupants. Aircraft designers and flyers know that every aircraft has a limit to the accelerative forces which it can safely sustain. If this limit is exceeded, there may be structural failure of the aircraft, and it is said to have been "overstressed." As might be expected, flyers themselves can be physiologically overstressed, if accelerative forces are excessive. Moreover, in the design of advanced aircraft, man may be the limiting factor in the performance of the man-machine system. Aeromedical research has therefore devoted considerable time and effort to the study of means to increase man's tolerance to these forces.

The standard unit of measurement of accelerative forces is called the "g" (for gravity). A man at rest is normally subject to the pull of gravity (1 g) equal to his own weight. But, in a pull-out from a dive, a flyer may sustain 3 to 6 g, and momentarily weighs three to six times his normal weight.

An accelerative force may also be described in terms of the means by which it is produced, its direction relative to the axis of the human body, and the length of time during which it acts.

Linear acceleration is produced when there is an increase or decrease in speed along a straight line, as during take-off, or in using an upward ejection seat. (A decrease in speed is called deceleration.) Radial acceleration is produced by a change in the direction of movement, as in recovery from a dive. Angular acceleration occurs in an aircraft in a spin, with changes in both speed and direction of movement. With relation to the body position of a flyer normally seated in an aircraft, g forces may be described as positive if acting in a head-to-foot direction, or negative if acting in the opposite direction. Positive g's are commonly experienced in turns and pull-ups from dives. They can be accurately reproduced in the laboratory by the use of large human centrifuge devices. Negative *g*'s occur when an aircraft arcs over, following a climb. *G* forces may obviously act in different directions if one is seated sideways or, as in the case of our astronauts, reclining at right angles to the accelerative force of blast-off. Accelerative forces may also be described as sudden (of less than 1 second's duration) or prolonged. In a single in-flight maneuver, several types of *g* forces may occur.

The effects of a *g* force on the human body depend not only on the magnitude of the force (the number of *g*'s), but also on its duration and its direction. In experimental situations, with adequate body support and proper positioning, man has withstood very high *g* loads for brief durations (measured in milliseconds) without serious injury.

The short-duration high *g* acceleration is not routinely experienced by the flyer but represents a stress to which he may be exposed in emergency situations. As such, sudden acceleration or deceleration is of primary concern in designing aircraft cockpits, seats, safety belts, helmets, and other protective equipment. Proper use of the equipment, body positioning, and adequate body support can also be essential in avoiding injury. Improper body positioning, for example, can increase the risk of a vertebral fracture in ejection. Rearward-facing seats obviously offer better body support for passengers in crash landings. Unfortunately, medical recommendations on this point have for years been ignored.

Prolonged acceleration, on the other hand, is a daily experience of every flyer. Figure 7.9 illustrates the duration and magnitude of accelerative forces produced by various flying maneuvers. Most of the accelerations experienced in flying are caused by a change in the direction of the aircraft, with the force acting in a head-to-foot direction (positive *g*).

The principal physiological effects of positive *g* are due to disturbances of blood supply, the temporary molding of soft tissues, and increase in the weight of all parts of the body. During a typical pull-up from a dive, for example, a pilot is forced down into the seat, his face sags, his limbs become heavy, and all his movements are markedly restricted. Manipulation of controls is difficult because of the immobility of arms and trunk. Blood is forced away from the head and into the lower parts of the body, the heart being unable to pump enough blood upward against this unusual force. As the effective blood pressure in the head (and hence the oxygen supply) is reduced, the first symptom is a progressive loss of vision. Dimming of vision or "grayout" usually occurs at 3.5 to 4.0 *g*, with loss of vision beginning peripherally and closing in toward the center of the vision fields. "Tunnel vision" refers to the preservation of the central area of vision, directly in front of the eyes, prior to a complete loss of vision or "blackout" which may occur at 4.0 to 4.5 *g*. Without protective maneuvers or equipment, the individual will suffer loss of consciousness, because of the deprivation of oxygen supply to the brain, at 4.5 to 6.0 *g*.

It should be remembered that "blacking out" indicates only loss of vision and not loss of consciousness. Between the time when the maximum *g* force is sustained and the actual occurrence of blackout, there is a time lag of 2 to 3 sec., *e.g.*, if vision is lost just at the start of the climb after pulling out of a dive, consciousness is retained and the flyer can carry out most required actions. Vision will return within a few seconds after acceleration ceases.

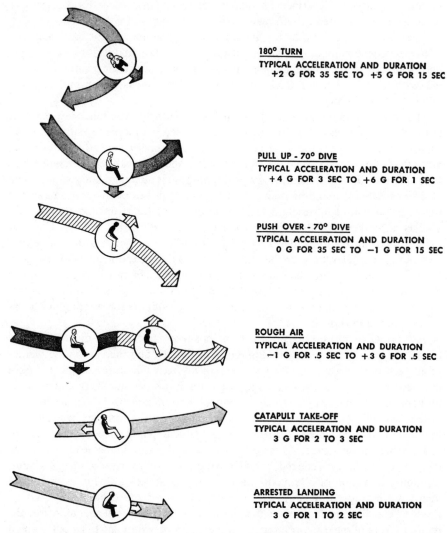

180° TURN
TYPICAL ACCELERATION AND DURATION
+2 G FOR 35 SEC TO +5 G FOR 15 SEC

PULL UP - 70° DIVE
TYPICAL ACCELERATION AND DURATION
+ 4 G FOR 3 SEC TO +6 G FOR 1 SEC

PUSH OVER - 70° DIVE
TYPICAL ACCELERATION AND DURATION
0 G FOR 35 SEC TO −1 G FOR 15 SEC

ROUGH AIR
TYPICAL ACCELERATION AND DURATION
−1 G FOR .5 SEC TO +3 G FOR .5 SEC

CATAPULT TAKE-OFF
TYPICAL ACCELERATION AND DURATION
3 G FOR 2 TO 3 SEC

ARRESTED LANDING
TYPICAL ACCELERATION AND DURATION
3 G FOR 1 TO 2 SEC

Figure 7.9. Accelerations produced by flying maneuvers. The dark-stippled portions of the arrows indicate that part of the maneuver which produces positive acceleration, the cross-hatched portions indicate negative acceleration, and the light-stippled portions indicate sagittal acceleration. *(Courtesy of Engineering Department, Douglas Aircraft Company, Inc., El Segundo, California.)*

Going into a dive from level flight, the accelerative force is in a direction opposite to that of gravity (foot-to-head) and thus briefly neutralizes the normal force of gravity. For this short time the flyer is in a state of "weightlessness" or O_g. Performance is not adversely altered if the pilot's seat belt is tight. Following this, the effects of negative g prevail, so that blood tends to be forced toward the head. The small blood vessels become distended, the face is flushed, and there is a sensation of fullness in the head and eyes. If negative g becomes excessive, there may be minute hemorrhages in the face and eyes. Becuse of this, the limit of tolerance to accelerative force acting in this direction is much lower than for positive g forces. Some pilots have reported a "red out" of vision during negative acceleration. Unlike the effects of positive g, which progress through several stages before becoming serious, negative g effects are sudden in onset and quickly become serious. Fortunately, sustained negative acceleration is infrequent in controlled flight. It does occur in a push-over into a steep dive, during buffeting in rough air, and in downward seat ejection.

Hence, the effect of g forces on man, as well as his tolerance to them, is a function of time, magnitude, and the direction in which these forces act on the body. The average flyer can safely withstand an acceleration of $10\,g$ for 1 sec. if the acceleration is in a positive or transverse direction, but would be in danger if this force were applied in a negative direction.

INCREASING HUMAN TOLERANCE TO ACCELERATIVE FORCES

The most frequent and toublesome effects of acceleration in aircraft are those resulting from sustained positive g by its reduction of the effective blood pressure in the head. Anything that opposes the tendency of blood to collect in the lower part of the body will raise the flyer's tolerance for these positive g forces. Short, stocky individuals generally have a higher tolerance than tall individuals, probably due to the diminished distance from the heart to the brain. In the same individual, g tolerance may vary from day to day because of fatigue, emotional state, overindulgence in alcohol, hypoxia, temperature, and possibly other factors.

In general, blood flow and blood pressure are beyond any sort of conscious control. To a limited extent, however, movement of blood toward the lower portions of the body and the fall in blood pressure can be retarded by performing certain body maneuvers, assuming special postures, or by wearing *anti-g* suits. Contracting, or tensing, the muscles of the lower extremities, with a simultaneous prolonged exhalation of air from the lungs will increase g tolerance for several seconds. These maneuvers require demonstration and preflight practice with an experienced instructor. Improper performance of the maneuver, *e.g.*, taking a deep breath and holding it, has occasionally led to lowered g tolerance and loss of consciousness in inexperienced student pilots. The average pilot can improve his g tolerance by 1 to 2 g through these maneuvers. Special subjects with intensified training in the human centrifuge have increased their tolerance to

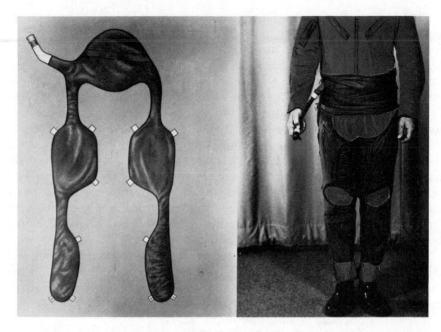

Figure 7.10. U.S. Air Force anti-g suit. The tightly fitted suit has an abdominal bladder and paired bladders over thighs and calves. As the wearer experiences g forces, these bladders inflate automatically to supply counter-pressure to legs and abdomen, thus raising g tolerance. *(U.S. Air Force photograph.)*

as much as 9 *g*. Since an important factor in human tolerance is the direction of the force relative to the body's long axis, changing body position will readily increase tolerance. If the pilot leans forward from the waist and at the same time raises his knees, his tolerance is raised by 2 to 3 *g*. The effectiveness of crouching in this way was well demonstrated by pilots during World War II. Assumption of the reclining, or prone, position places the head at approximately the same level as the heart, and hence the gain in tolerance is considerable. There are obvious practical limitations on the use of this position while retaining control of the aircraft, but the use of an automatically reclining seat remains a possible feature of future aircraft. The full prone position has been used successfully in space flight launches, to enable NASA astronauts to endure 10 to 12 *g* during lift-off.

A closely fitted *anti-g* suit which operates automatically during high accelerations will materially increase man's tolerance. The calves and thighs are constricted and the abdomen compressed by means of air bladders which are inflated whenever acceleration exceeds 2 *g*. The effect is to prevent pooling of blood in the lower extremities and abdomen and to raise blood pressure. The degree of inflation provided varies in direct proportion to the magnitude of the *g* force. Tolerance to positive acceleration can be raised by as much as 3 *g* by using some *anti-g* suits, but bladder pressure required to give this much protection is so high as to be uncomfortable. The pressures commonly used raise the blackout level by 1 to 2 *g*. A flyer may further increase the protection offered by his suit by straining or crouching. None of these measures offers any protection against negative *g* forces.

VIBRATION

An additional type of accelerative force encountered in aviation is that of vibration. Its effects range from mild fatigue to complete incapacitation and severe pain. Helicopters are known to produce continuous, low frequency, large amplitude vibrations that will lead to significant fatigue among crew members and passengers. Military pilots on high-speed, low-level flights in moderate turbulence have experienced fatigue with accompanying deterioration of performance. As far as possible, pilots should avoid continuous turbulence, even if minimal. Helicopter and crop-dusting pilots should plan on shorter flights with longer rest periods between them, if other fatigue-provoking factors are known to coexist.

Spatial Disorientation

Man's physiological mechanisms for maintaining his equilibrium, or sense of balance, are designed to function well on the ground. Maintaining equilibrium in the air is much more difficult. There is ample opportunity for confusion or misinterpretation of information, which can lead to sensory illusions, or spatial disorientation, in flight. Despite the design of modern instruments to aid the aviator, and emphasis on this subject in pilot training programs, spatial disorientation continues to be one of the leading causes of aircraft accidents— both civilian and military.

Whether in flight or on land, the sense of balance by which man is able to stand, walk, sit, swim, or fly is derived from three sources: visual, muscular, and vestibular. These three senses work together to keep the body properly positioned relative to the earth and the immediate environment. In the healthy person, these senses are more than adequate to maintain balance on the earth's surface. In the air, certain imperfections of these balancing mechanisms are manifest. Even when all are functioning, they may not insure that the flyer will maintain his equilibrium under all conditions of flight. If the effect of one of the senses is lost (for example, vision may be of little use on a dark night), equilibrium is seriously compromised. Man's balancing senses, then, are primarily designed to be used on the ground, to supplement each other, and to accommodate for only relatively gradual changes in position. In the air, not only may one or more of the triad lose its effectiveness, but man is subject to rapid and violent changes of direction in three dimensions.

On the ground or in flight, the eyes are the most accurate of the three means of maintaining equilibrium. In flight, however, even with good visibility, the eyes are not as reliable as on the ground because they lack valid visual references. While flying on a clear day, the pilot may see the earth below, other aircraft in the formation, and most importantly, the distant horizon. Aerial equilibrium is easily maintained by watching the ground and the horizon as long as these can be seen. Under these conditions the eyes act as a check on the other two balance senses and may even make corrections for them.

Because of weather conditions or darkness, the horizon and the ground below may be lost to view for variable periods so that the visual sense is ineffective. Even if the horizon remains visible, it may change suddenly and radically due to the maneuvering of the aircraft. Visual reference either outside the aircraft (ground, horizon) or inside the cockpit (flight instruments) is absolutely necessary for aerial equilibrium. Visual reference to the earth is most useful when there is a clearly defined horizon.

The muscle sense of equilibrium comes from minute nerve endings in the muscles, skin, tendons, ligaments, and joints. These are stimulated by pressure and tension in the feet, legs, or buttocks when man stands, walks, or sits. The muscle sense is effective enough that man is able to stand or walk with his eyes closed and still maintain balance. Muscle sense measures only the up and down components of movements and so is helpful chiefly when movement is in the vertical plane.

By means of his muscle sense, the experienced flyer can identify many movements of the aircraft by the pressure of the seat and the cockpit floor. For example, this pressure increases when going into a climb and decreases as descent is begun. Thus, although the pilot can tell that he is in a turn, spiral, or zoom, he cannot differentiate between these maneuvers without visual reference to flight instruments or to the horizon. To some extent, position is "felt" in this way and has given rise to the expression "flying by the seat of the pants." Besides its failure to record circular motion, the muscle sense has another weakness. During flight without benefit of visual references such as instruments, the tops of cloud layers, or the horizon, vision alone does not always differentiate between the normal effects of gravity and similar sensations which are produced by accelerative forces. In turbulent air or during aerobatics, many erroneous and conflicting impressions of equilibrium are thus sent to the brain.

The vestibular apparatus or organ of balance in each inner ear consists of a system of three small semicircular canals and two otolithic membranes. The canals are arranged at right angles to each other and terminate in saclike enlargements, which contain fluid and sensory hairs. The otolithic membranes consist of sensory hairs which are weighted with tiny calcium carbonate granules for perception of changes in acceleration in a straight line, including gravity.

When the head is moved, the fluid within the canals in each inner ear moves in the same direction. This stimulates the sensory hairs and gives them an appropriate impression of motion which is transmitted to the brain. Because of the special geometric arrangement of the canals in three different planes, there is a corresponding movement of fluid in at least one pair of canals each time the head is moved. Owing to the inertia of the fluid, there is a brief pause before its rate of flow equals the motion of the head. Likewise, when the head motion is stopped, the fluid continues to flow for a short period. This continued motion of the fluid causes the same sensation as actually moving the head in the opposite direction, so that the individual feels that he is turning in the opposite direction.

In prolonged turns the fluid movement catches up with the movement of the canal, and the hairs are no longer stimulated. This gives an erroneous impression to the brain that turning has stopped. Therefore, the semicircular canals are unable to perceive a turn after it has been continued for a few seconds, particularly if it is made gradually. If a rapid turn is abruptly stopped, the fluid continues to move and gives an impression of turning in the opposite direction.

In addition, the otoliths enable one to sense the magnitude and direction of *g* vectors, *i.e.*, which way is up. Under accelerative forces one may receive an incorrect sensation of what is up. Attempts to orient the aircraft based on this information may be disastrous.

The vestibular apparatus is not stimulated by movement at a constant velocity. That is, there must be some change in either speed or direction (acceleration). Since flying straight and level at a constant speed produces no acceleration, the vestibular sense is not stimulated. If accelerations are not of sufficient magnitude, they are below the threshold of sensitivity of the vestibular sense and so will be overlooked. This occurs during a gradual maneuver of the aircraft in which the rate of acceleration is minimal. If at the same time the flyer cannot see his surroundings, he may reach a dangerously high degree of rotation without knowing it.

Having left the earth, the flyer loses much of the value of his muscle sense because it is no longer reliable. During blind flying conditions he also loses the advantage of his visual sense except by reference to instruments. In this situation, while relying on the vestibular sense, he may be confused by it because it is excessively stimulated by *g* forces, or because small accelerations produced by aircraft maneuvers may not be perceived.

An awareness of the types of sensory illusions which can occur is essential to the aviator. The following are examples of illusions commonly encountered during flight.

An illusion of relative motion occurs when the motion of one aircraft is falsely interpreted as motion of the adjacent aircraft in the formation; *e.g.*, if the flight leader maintains a straight course and the rest of the flight veers to the left, the leader may appear to have turned to the right. A false horizon illusion occurs when a pilot flies over an inclined cloud bank. If this takes place gradually, the pilot is misled and changes the flight attitude of the aircraft by aligning it with the cloud bank. Thus, although flying in a wing-low attitude, he believes his aircraft to be horizontal. One other visual illusion, autokinetic movement, was discussed in connection with night vision.

Because the accelerations which they produce are below the threshold of sensitivity of the vestibular apparatus, certain motions of the aircraft are frequently overlooked.

"Leans" of the Same Direction.   If the aircraft rolls abruptly to the left and it then slowly resumes horizontal flight, the pilot may be aware of the tilt but not the recovery. He feels as though the plane is tipped to the left and corrects this sensation by leaning far to the right. He may persist in doing this for a short time even though his instruments are indicating horizontal flight.

"Leans" of the Opposite Direction.   If the aircraft is very slowly rolled to the left, the pilot is not aware of this and believes his aircraft to be horizontal. A rapid recovery of the aircraft to true horizontal is perceived as a tilt to the right. In this instance the flyer leans to the left in order to correct the sensation.

Illusions of Pitch.   The average person can be tilted upward 20 degrees and downward 10 degrees without being aware of it when this is done slowly and without visual reference. In turbulence, if the aircraft pitches upward less than 20 degrees and then recovers slowly, the sensation of pitch will persist. The pilot feels as though the aircraft is climbing steeply, even though it is flying straight and level. If pitching is downward, a false sensation of diving will persist after a slow recovery to horizontal. The pilot may react to this sensation by pulling back on the stick and so go into a steep climb. A sensation of climbing excessively may be experienced after take-off. This condition has resulted in multiple crashes, particularly during night take-offs over water, with no visual references available. The impulses from the otoliths, combined with those from the skin and muscle, produce a sensation of an excessively steep climb-out. Corrections for the false perception have resulted in fatal dives into the ground or water.

Sensation of Climbing While Turning.   In a properly banked turn, acceleration tends to force the body firmly into the seat in the same manner as when the aircraft is entering a climb or pulling out of a dive. Without visual references, an aircraft making a banked turn may be interpreted as being in a climbing attitude, and the pilot may react inappropriately by pushing forward on the control column.

Sensation of Diving While Recovering from a Turn.   The positive *g* forces sustained in a banked turn are reduced as the turn is completed. This reduction in pressure gives the flyer the same sensation as going into a dive and may be interpreted in this way. He may overcorrect by pulling back on the control column and cause the aircraft to stall.

Sensation of Diving Following Pull Out from a Dive.   The accelerative forces on the body during the pull-out from a dive are reduced after recovery is complete. This reduction in *g* force may be falsely identified as originating from another dive.

Sensation of Opposite Tilt While Skidding.   If skidding of the aircraft takes place during a turn, the body is pressed away from the direction of turning. This may be falsely perceived as a tilt in the opposite direction.

Illusions which are the result of stimulation of the visual and vestibular senses are called oculogyral illusions. These result when conflicting impulses from the eyes and the semicircular canals are transmitted to the brain.

The Coriolis Phenomenon.   This is a severe loss of equilibirum which occurs when the pilot moves his head in one plane of rotation while the plane is moving in another. Movements of the head in flight should be gradual and deliberate, particularly if the plane is not in straight and level flight. Improper design of cockpit instruments, which require the pilot to look downward and to the extreme left or right, has led to this occurrence in many instances. For example, if during a spin the pilot moves his head forward or backward, an additional pair of semicircular canals is stimulated and extreme dizziness and nausea are suddenly produced. The consequence of such an unusual reaction in flight is apparent.

Sensation of Reversed Rotation.   If a rotary motion persists for a short period and is then discontinued, there is a sensation of rotation in the opposite direction. After recovery from a spin to the left, there is a sensation of turning to the right. In attempting to correct for this, the pilot puts the aircraft back into the spin to the left. Flyers have given this illusion the sinister name of "graveyard spin."

Severe disorientation has been reported by light-aircraft pilots flying into the setting sun with an idling or windmilling propeller, and by helicopter pilots whose rotors came between their eyes and the sun. It is not necessary to look directly at the light source to become disoriented, as reflected light from the periphery of the visual fields will also cause this phenomenon. The frequency of interruption which will produce disorientation varies from individual to individual. This problem can be avoided by not allowing the propeller or rotor to come between the eyes and any strong light source.

PREVENTION OF SENSORY ILLUSIONS

Although sensory illusions in flight occur infrequently, they are extremely important because each is a serious threat to the flyer's safety. They are most likely to occur during flying under conditions of poor visibility but may occur at any time. All pilots are susceptible, but those who have developed and who maintain proficiency in basic instrument flying are least vulnerable.

All pilots must remember that there is only one way to maintain orientation in flight, and that is through use of a valid visual reference. This may be the presence of reliable horizon outside the aircraft, or in many cases the use of flight instruments (Chapter 5). The more a pilot understands how sensory illusions are produced, the more he will come to rely on the visual presentation of flight instruments to recognize and disregard misleading sensations and the less likely he will be to develop spatial disorientation.

Airsickness

Airsickness is nausea, vomiting, and a general feeling of discomfort caused by the motion of an aircraft in flight. It is essentially the same as other forms of motion sickness (car, swing, train, and seasickness) in its cause and manifestations. Its primary cause is recurrent disturbances of equilibrium through overstimulation of the vestibular sense. The up and down components of motion appear to be most important in producing this condition. The vertical "bumping" motion of the aircraft in turbulent air rather than yawing or rolling is the usual cause of symptoms. In addition, psychological factors probably are important in some people. Apprehension and anxiety of any kind increase the tendency to become airsick.

OCCURRENCE

Nearly everyone is susceptible if turbulence is extreme and prolonged. Experienced flyers rarely become airsick while they are actually operating the aircraft, but they may while riding as passengers. There is a very low incidence of airsickness among civilian airline passengers, because of the precautions taken by the airlines to make passengers comfortable, such as reclining seats and avoidance of turbulent air. In military flying, airsickness has been much more frequent. An analysis of 2080 flying cadets showed that 5.7% were airsick on their first flight, but only 1.1% were affected after they had made ten flights. While up to 40% of students may become sick at some time during training, less than 1% of Air Force students fail their training because of persistent airsickness. There has been some effort to rehabilitate individuals with recurrent or persistent airsickness, with limited success.

PREVENTION

Airsickness is difficult to avoid completely in military operations because combat aircraft are not built for comfort, and missions must be flown regardless of turbulence. The following measures may be taken to prevent or control symptoms of airsickness:

   a. A reasonably gradual introduction of the flying student to the motion of aircraft—particularly in aerobatics.

b. Selection of the most favorable flight route to avoid turbulence.
c. Avoidance of violent and unnecessary maneuvers.
d. If possible, choosing a position in the aircraft near its center of gravity where there is less motion.
e. In turbulence, lying down with eyes closed. (Shifting visual references cause increased susceptibility.)
f. Avoidance of strong odors or hot stuffy air in the aircraft.
g. Avoidance of overindulgence in food or alcohol prior to flight.
h. Fixing the gaze on a stable visual reference outside the aircraft, preferably the horizon.
i. The taking of antimotion-sickness drugs by susceptible passengers prior to flight.

It is possible, to some extent, for most persons to become accustomed to aircraft motion and so to develop some resistance to airsickness. This is commonly observed in flying students after they have made several flights. The use of preventive drugs by aircrews is not an acceptable practice because of certain secondary effects such as visual disturbances and drowsiness. In special instances, drugs may be prescribed temporarily by a flight surgeon for relief of airsickness in a beginning student.

Disturbance of the Circadian Rhythm

The jet age and long-distance airline travel have given importance to another aspect of human physiology, that of the circadian (from the Greek: *about one day*) rhythm, or the body's "biological clock." There is established in everyone a certain hormonal pattern related to the customary hours of sleep and wakefulness. At modern air speeds the aircrew member or passenger traveling west to east or vice versa may cross several time zones in a single flight, and find his "clock" out of sequence with local time upon arrival. Significant fatigue, difficulty in concentration, and other symptoms result. Digestive symptoms may result from eating at an unaccustomed time. For the aircrew member on an extended mission, but with return to his home base scheduled within a few days, it probably is best to maintain a pattern of sleeping and eating corresponding to the home schedule. This means disregarding the local time, sleeping in daylight hours, etc.

For an extended stay at a distant point, it is advisable to allow one day for adjustment to each hour of time change. Adequate rest before the flight and avoidance of any planned activity during the immediate postflight period, will facilitate the adjustment.

Noise in Flight Operations

Noise is unpleasant sound due to acoustic waves of scattered frequencies and intensities. In aviation, noise is a problem both during flight and on the ground. Higher noise levels for longer periods of time are experienced by ground maintenance crews than by aircrews. Nevertheless, flyers are subjected to noise of sufficient intensity and duration to produce significant effects, including partial hearing loss after repeated exposures.

THE MECHANISM OF HEARING (See Figure 7.3)

The outer ear collects sound waves from the air and conducts them inward to the middle ear, causing the eardrum to vibrate. This, in turn, sets up motion in the small bones of the middle ear by which vibrations are transmitted to the cochlea in the inner ear. From here, stimuli are sent through the auditory nerve to the brain where the vibration is perceived as sound. This is the perception of sound by air vibrations. Sound can also be perceived by vibrations being conducted through the bones of the skull to reach the auditory nerve.

MEASURING SOUND

Every sound has two measurable components, pitch (tone) and intensity (loudness). Two sounds may have the same pitch but different intensities, or vice versa. The human ear can hear sounds only within a certain range of intensity and a certain range of pitch. Since sound is a wavelike phenomenon, its pitch is measured in cycles per second or Hertz (Hz). Sound intensity is measured in units called decibels (dB). Audible tones for man vary from 20 Hz to 20,000 Hz. Human hearing is most acute for sounds between 500 Hz and 5000 Hz, which is the usual range of the human voice. Audible intensities range from 10 dB to 140 dB. The whispered voice averages 20 to 30 dB and the normal spoken voice, about 40 dB. There is no upper limit to audible sound intensity, but above 140 dB sound can actually be felt as a vibration in parts of the body other than the ears. Jet aircraft at full power setting can produce ground level noises as high as 150 to 170 dB.

CHARACTERISTICS OF AIRCRAFT NOISE

In-flight noise from propeller-driven aircraft is produced by the propeller, the engine and accessory power plants, engine exhaust, the ventilating system, and aerodynamic noises (slipstream). The propeller is the greatest contributor to the total noise intensity, producing intensities of 90 dB to 130 dB, primarily in the lower frequencies (below 300 Hz).

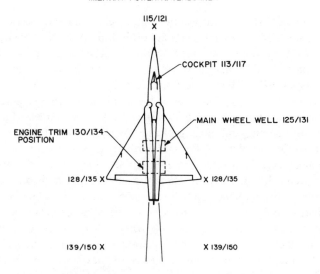

F–106 A AIRCRAFT   J 75–P–9 ENGINE

OVERALL SOUND PRESSURE LEVELS(DB RE .0002 DYNE/CM²)

MILITARY POWER/AFTERBURNER

115/121
X

COCKPIT 113/117

MAIN WHEEL WELL 125/131

ENGINE TRIM 130/134
POSITION

128/135 X   X 128/135

139/150 X   X 139/150

Figure 7.11. Sound intensities produced by jet aircraft. Noise levels are shown for various points around an F-106A aircraft. Highest intensities are found near the tail. *(Drawn by John Cole, Aerospace Medical Research Laboratories.)*

The two principal sources of noise from jet aircraft are the engines and the aerodynamic noise. Jet noise differs from that of propeller aircraft in that all the frequencies of the audible sound range are represented at equal levels of intensity. The proportion of the total jet noise from aerodynamic sources is much greater than that for propeller aircraft. This proportion becomes still higher with increasing speed but tends to diminish with altitude. Jet noise gives a sensation of smoothness and does not seem to be as loud, even though the total decibel values may be the same as for propeller aircraft. Since most noise originates outside the cockpit, its intensity within the cockpit is greatly reduced by the walls of the cabin. Pressurized aircraft are particularly well insulated in this respect because of the tightly enclosed cabin.

EFFECTS OF NOISE ON THE FLYER

Noise and In-Flight Communication.   Perception of voice and radio communication is made difficult in flight by interference of the aircraft noise. When subjected to two similar sounds, the ear detects the louder sound more clearly and may overlook the lesser sound. When an undesirable sound thus interferes with the perception of a desired sound, the effect is called "masking." Jet noise has a greater masking effect on the sound of the human voice than propeller aircraft noise of equal intensity. This is because the greatest proportion is within the speech range, that is, at the same frequencies as the most common sounds of the human voice.

In addition to the masking effect of noise, the decreased barometric pressure at altitude adversely affects communication. With diminished air density, the spoken voice does not produce the same sound pressures as at sea level. Aircraft radios are built with additional power to correct partially for this loss in speech intensity. Another difficulty in radio voice communication at high altitude is the effect of pressure breathing. Reversal of the respiratory cycle required in pressure breathing has an inhibitory effect on speech. Also, it is masked to some extent by frequent opening and closing of the oxygen mask valves.

Hearing Loss Due to Noise.   Partial hearing loss is a frequent finding in the periodic medical examination of aircrew members. Typically, the loss initially involves the higher frequencies, of which the subject may be unaware. At more advanced stages, the frequencies for human speech reception become involved. Once established, noise-induced hearing loss is irreversible. There is no medical treatment or auditory device which is of benefit. It is, however, entirely preventable. Personnel entering a career in aviation, or aircraft maintenance, should be instructed in the use of simple ear defenders whenever working in a hazardous noise area. The US Air Force has established a formal Hearing Conservation Program designed to reduce noise hazards, to educate personnel regarding appropriate preventive measures, and periodically to check the hearing ability of those occupationally exposed to noise.

The danger of hearing loss varies not only with the intensity and pitch of the noise, but with the duration of exposure and the time intervals between repeated exposures.

Fatigue Due to Noise.   Work done in the presence of noise is followed by a degree of fatigue which is out of proportion to the work itself. Since noise interferes with concentration, more energy is expended in performing in a noisy environment than in a quiet one. Operating an aircraft is a work situation which demands sustained concentration in a noisy environment. Aircraft noise is an important factor contributing to the overall effect of flying fatigue.

NOISE PROTECTION IN FLIGHT

Aircrews may be protected against noise by the wearing of insert-type ear defenders or flight helmets. Ear plugs of synthetic rubber, plastic, cotton, and cotton-wool, which insert into the external ear canal, are effective and practical. Impregnating cotton or wool with paraffin increases their ability to block sound. Plastic or rubber plugs are of most value when individually fitted to the ear canals. All types of ear defenders tend to protect against high-frequency noise more effectively than against noise of low frequency. Properly fitted soft plastic ear plugs may reduce noise intensity by 20 dB in the lower frequencies and by 40 dB in the high frequencies. In high noise levels, ear defenders do not mask out speech and radio signals even though they do reduce the total noise. The wearing of a well-fitted flying helmet, which is padded with sponge rubber, eliminates almost all possibility of bone conduction of noise as well as much of the air-conducted noise. Maximum protection is afforded by wearing ear defenders in combination with a helmet. However, this is impractical with plastic or rubber ear defenders, because they may so completely seal the external ear canal that a partical vacuum develops inside the canal during rapid descents. The wearing of a standard flight helmet will reduce the overall sound level inside the cockpit of a modern jet fighter from 100 to 80 dB at normal cruise conditions.

Aviation Toxicology

The flyer must guard against exposure to certain toxic materials which may be encountered during flight operations. Carbon monoxide poisoning may occur from fuel, heater, or exhaust system leaks and could result in incapacitation of the aircrew member. Carbon monoxide is a colorless, odorless gas which competes with oxygen for space available in the hemoglobin of red blood cells. Unfortunately, carbon monoxide has a 300 to 1 advantage in its affinity for hemoglobin and thereby has the effect of depriving body tissues of their oxygen supply. A small but significant source of carbon monoxide may be carried aboard by the flyer himself. Cigarette smoking, before or during flight, introduces carbon monoxide into the blood stream and can appreciably lower the individual's altitude tolerance.

Crop-dusting pilots are continually at risk of poisoning from their toxic cargo. An individual engaged in this hazardous occupation should be knowledgeable of the specific properties of the particular chemical being carried and should take measures to avoid exposure. Many crop-dusting compounds are readily absorbed through the intact human skin as well as by the respiratory route. They may affect the nervous system to cause nausea, headache, weakness, and eventually convulsions. A number of fatal accidents among crop-dusting pilots has been attributed to exposures which could have been prevented.

There are also other toxic exposures which, being more insidious, can be even more dangerous. *Medications* are a far too common and serious hazard for the uninformed or careless flyer. Medications used to treat respiratory or gastrointestinal illnesses may adversely affect the airman's visual acuity, depth perception, reaction time, and state of alertness. Many nonprescription drugs which are in everyday use for the common cold have side effects which can seriously compromise the flyer's effectiveness. The only safe policy for the flyer is to take drugs only when prescribed by his flight surgeon or a physician trained in aviation medicine, and to accept periods of temporary grounding during illnesses as being in his own best interests. In recent years a very limited number of drugs which are useful in treating certain chronic diseases among aging flyers have been approved for long-term use by the US Air Force, but only after appropriate medical evaluation and waiver action, and under the continuing supervision of the flight surgeon. A research program to evaluate specific drugs for safety in flyers is being conducted at the US Air Force School of Aerospace Medicine.

In recent years as piloting of private aircraft has become more accessible to the general public, aviation accidents are being linked to misuse of alcohol with increasing frequency. The effects of alcohol are greatly intensified by even the small degree of hypoxia experienced at 5000 ft. of altitude. Thus one or two cocktails can seriously impair judgment and coordination. *The combination of alcohol and then flying in the same 24-hour period may be deadly.*

Medical Care of the Flyer

The individual considering a career in aviation will encounter medical requirements and medical personnel at an early stage. Whether in a military service or civil aviation, an appropriate physical examination and certification of fitness is prerequisite to entering any flying training program. The strictness of medical standards may vary somewhat from the situation where an individual is applying to take private flying lessons at his own expense, to that where several hundreds of thousands of dollars of public funds are to be invested in a military pilot training program. It is obviously desirable to select the best qualified candidates, and those least likely to develop medical problems in the future, for such a program.

Medical standards are relaxed somewhat for applicants for navigator's training and for the periodic examinations required of trained, experienced aircrew members. In many instances, for trained flyers, it is possible to request and obtain a waiver for a medical defect, after an appropriate evaluation.

Other than the need to select the best qualified individuals for military programs, the rationale behind medical standards for flying is based upon consideration of flying safety. Any medical condition which would compromise a pilot's ability to control the aircraft would obviously constitute an unacceptable risk to the individual and to his fellow crew members or passengers. For example, an individual suffering from the disease epilepsy, with the possibility of recurrent convulsive episodes, would not be acceptable for flying training. Moreover, a history of significant head injury might constitute a risk of developing a convulsive disorder in the future and not be considered acceptable. A defect of a bone or joint which would interfere with bailout or the use of an ejection seat would not be acceptable for military flying. Conditions which would be aggravated by exposure to high altitudes, such as certain hemoglobin abnormalities of the red blood cells, are not compatible with flying duties. Any condition which would interfere with the free passage of air to the nose and middle ear would require correction. Any condition giving difficulty or requiring frequent use of medication (*e.g.*, asthma) would not be accepted. There are even limits for such simple measurements as height and weight due to the fact that aircraft and personal equipment are built for an average size adult and not for the extremely tall, short, large, or small. The special senses of vision and hearing are vital to the performance of flying duties, and only persons completely normal in these senses are suited to enter into flying training.

Once an individual has been successfully trained as a pilot, the required medical examinations focus on factors related to retention or longevity rather than selection. It is obviously in the flyer's best interest to undergo periodic checks in an effort to detect any disease early and, hopefully, to correct or to prevent any more serious manifestation.

In the aging pilot population, considerable attention is given to the detection of possible heart disease, which is the leading cause of death among men in the US and other countries. The occurrence of a heart attack in a pilot during flight would obviously have disastrous consequences. One such incident accounted for the loss of 83 lives in 1966. There have been few occurrences in the US Air Force, perhaps due to an excellent program for the detection of latent disease. This involves the recording of electrocardiograms (tracings of the heart's electrical activity) on each flyer annually, beginning at age 35, and review of all records at a centralized repository located at the School of Aerospace Medicine. Many flyers are referred to the School by their base flight surgeons for the evaluation of difficult or obscure medical problems at the Aeromedical Consultation Service. Similar services for the evaluation of flyers have been established by the US Army and Navy.

In the light of present knowledge concerning the risk factors for the development of heart disease, it is unfortunate that more active preventive measures are not taken. This is particularly true in a select group such as the flying population, where health maintenance is vital to each individual's career. It is believed that such simple measures as maintaining a high level of physical conditioning, avoiding a high intake of fatty foods in the diet, and abstaining from the use of tobacco would markedly decrease the risk of heart attack.

The Role of the Flight Surgeon or Air Surgeon

The flight surgeon is a physician who has taken postgraduate training in aviation or aerospace medicine. He may be employed by a military service, a civil governmental agency, or a commercial airline, and thus has responsibilities to the command or management of the organization as well as to his patients. His duties include serving as the primary physician for flyers and the conduct of required periodic examinations. He is the medical adviser to the organization's commander or president on matters affecting the health and morale of his personnel. He is responsible for preventive medicine programs, which seek to detect or to control environmental hazards in the air or on the flight line. He may be a fellow crew member who is required to participate regularly in aerial flights in order to maintain familiarity with flight operations and the problems of flyers. In the event of an aircraft accident, he serves as a member of the investigating team, seeking to identify any human factor which could have been involved in the cause of the event. His basic concern is not to "ground" his patients, but rather to keep them flying. In many situations, he is the flyer's best friend.

FAA MEDICAL EXAMINER

Most of the pilot population is not employed by an organization which has a professional air surgeon on its staff. The Federal Aviation Administration (FAA) requires that these individual pilots show medical fitness for crew member positions at the beginning of flight training, and thereafter at intervals ranging from 6 months to 2 years.

To ascertain and certify to such medical fitness for flight crew duties, the FAA selects and designates as Aviation Medical Examiners certain physicians in private practice. These Examiners must have qualified by special courses and other training for such designation. They are not employees of the FAA and so may charge reasonable fees for their services. They must submit their findings to the FAA Air Surgeon and may routinely issue waivers only for such conditions as the requirement to carry or wear corrective eye lenses while flying. After evaluation of doubtful or disqualifying physical conditions found by a Medical Examiner, the Air Surgeon may issue a waiver allowing the examinee airman's privileges with some or no restrictions.

When possible, the airman should have as his personal physician a designated Medical Examiner. When that is not possible, he should make sure his physician understands that he does fly as a crew member, and considers the effects and side effects of any medical condition the airman may encounter, and of any treatment or medication the physician may prescribe.

8

Basic Flight Techniques in Light Aircraft

This chapter presents the maneuvers and techniques applicable to the operation of light aircraft in visual flight conditions.

Each aircraft has its own set of flight characteristics, and the underlying purpose of flight training must be to develop skills and safe habits that are transferable to any aircraft. If the pilot learns to fly light planes with precision and develops safe habits in them, he will find the transition to heavier aircraft much easier than he probably expects. His time in light airplanes will be more enjoyable and safely spent, too, although the forgiving nature of light training airplanes may conceal dangerous habits for many hours of flight.

Preparing the Aircraft for Flight

PREFLIGHT INSPECTION

Even for the "simple" light plane, and even with good maintenance service, the wise pilot will form the habit of making a thorough preflight inspection with the aid of a printed checklist prepared by the manufacturer or the operator of the aircraft. The approved operator's manual or the owner's manual for the aircraft expands the list of specific items to be examined during the interior and exterior inspections.

The interior inspection assures that required documents are aboard, and that all switches and controls are properly positioned in preparation for the exterior inspection.

The exterior check is made during a complete circuit about the aircraft. Some items of the check are common to all aircraft—security of control surface fittings and tire inflation, for example. Any aircraft, however, has at least one item or two on its recommended preflight inspection that is peculiar to that aircraft alone because of design or operational experience. The reward for meticulous inspection habits is safe, pleasant, and inexpensive flying; the penalty for carelessness is always too great.

COCKPIT PROCEDURES BEFORE ENGINE STARTING

Fire is a remote but constant possibility during any start, and a fire extinguisher should be positioned near the aircraft or installed in the cockpit. Set the seat and rudder pedals, if adjustable, so that full rudder and elevator movement can be obtained, and so that all controls in the cockpit can be reached without strain. Move the flight controls through their full ranges, checking for freedom and correct movement. Adjust and fasten the shoulder harness and seat belt. The shoulder harness should be locked at this time if all controls can be reached after locking. If the harness is of the inertia reel type that locks automatically in case of impact, it can be left unlocked manually during all flight unless a crash landing appears likely.

The cockpit now can be readied for the start, still following the checklist. As nearly as possible the inspection should start at one point in the cockpit and proceed in an orderly manner around it. As an example, one might proceed in the following manner:

Radios—OFF
Lights and other Electrical Equipment—OFF
Carburetor or Induction Air Heat Control—OFF
Mixture Control—SET FOR START
Throttle—SET FOR START
Fuel Valves—ON, recommended tank(s)
Cowl Flaps—OPEN
Master Switch(es)—ON

Starting

STARTING SEQUENCE

Improvements in modern light-plane engines make starting no more difficult than starting an automobile; however, procedures vary considerably according to the type of engine and fuel system. The following information is general and should supplement the instructions of the manufacturer.

The propeller control should be in full high rpm (low-pitch) position for all starts. Changing pitch angle too quickly after a cold start can deprive the engine of needed oil pressure.

Safety Precautions During Starting. Before cranking the engine, clear the area 360° and call "clear." If a fire guard is standing by the aircraft, make sure he acknowledges. Have your feet on the brakes in the event the parking brake system is inoperative and keep one hand near the throttle and mixture controls ready for immediate adjustment.

Horizontally opposed engines equipped with a venturi-type carburetor are started with the mixture in the RICH position and the throttle partially open. If the engine is cold, the manual primer should be used prior to start. Amount of priming varies from one to two strokes in cool temperatures to eight strokes when the temperature is well below freezing. During extreme conditions it may be necessary to have the primer pulled out so that, as the engine begins to fire, additional fuel can be supplied by slowly pushing in the primer knob. The starter-ignition switch used in some aircraft operates exactly like that in an automobile. To crank the engine, the switch is turned to the START position. As the engine fires it will automatically disengage from the starter. At this time release the switch and adjust the throttle to the desired rpm.

Horizontally opposed engines equipped with a fuel-injection system are started with the aid of an electrical boost pump, since the engine-driven fuel pump will not supply sufficient fuel until the engine is operating at approximately 500 rpm or above. The mixture is set at RICH, and the throttle is partially opened. Use of the manual primer is necessary only in very cold weather. Actuate the boost pump just before cranking, and check the fuel flow to be within limits. Do not delay cranking the engine, or excess fuel will accumulate in the intake manifold, which will cause flooding and a possible fire hazard. After the engine is running smoothly, turn the boost pump off and adjust the throttle to the desired rpm.

Larger engines of the radial or inverted inline type are normally started by breaking down the start procedure into separate steps.

First. Make sure the start will not damage the engine. Rotate the engine through one complete cycle (two or three revolutions) without introducing fuel.

At the first evidence that the starter may stall, the engine should be checked for presence of oil or fuel in the inverted cylinders. A start with fluid in the cylinders may produce a condition known as hydraulic lock, which can bend or break a piston rod. (*Note:* Horizontally opposed engines are not subject to hydraulic locks from oil, although a partial lock can occur from overpriming.)

In addition to the check for hydraulic lock, turning the engine improves lubrication at the first of the start.

Second. Provide a combustible fuel-air mixture to the cylinders. Set mixture and prime as required or recommended.

A proper air mixture must be provided for primer fuel; to do this, adjust the throttle to the position which would produce about 1000 or 1200 rpm. This position varies from one aircraft to another. A cold engine requires less throttle at the start than a hot engine, because the cold fuel does not vaporize as well.

As described above, light-aircraft engines with fuel injection and no impellers are started with the mixture in RICH position, boost pump ON, and without the use of primer except in very cold weather. The larger engines also *can* be started in this manner, and there are certain advantages to the method. If the battery is weak, the current drain by the primer pump is eliminated. If the engine has a blower (single-stage supercharger), the fuel from the carburetor will be vaporized as it enters the cylinders more than will primer fuel put directly into cylinder or manifold. The danger of the latter method is that the possibility of induction systems fires is much greater.

Third. Provide ignition for the combustible mixture. One does this by turning the ignition switch to BOTH. The engine should fire within one or two revolutions.

Fourth. Adjust the engine to normal combustion. If the start is accomplished on an initial charge of manually introduced primer fuel only, move the mixture control to RICH position as soon as the engine definitely fires; otherwise, use more priming.

Most electric primers are capable of supporting combustion at 1000 to 1200 rpm. There is no reason for hurrying the transition to mixture control fuel, and the throttle may be adjusted slightly to provide smooth combustion before the transition. This procedure will generally prevent engine backfire, which can damage the induction system.

Check manual primers carefully after the start to ensure that they are locked closed, and check that electric primer switches are in the OFF position. Otherwise, leaking primer fuel will be drawn into the engine, causing an excessively rich mixture.

Starting Difficulties and Malfunctions. If the mixture is decidedly too lean or too rich, no start can be obtained. Raw fuel draining out of the exhaust or out of the impeller section drain valve under the engine indicates the rich condition. Discontinue priming. If the engine is of the horizontally opposed type, set the throttle FULL OPEN, the mixture at FULL LEAN, and continue cranking to clear the rich mixture from the cylinders. When the engine first fires, return the throttle and mixture to the starting position.

When the engine does not fire at all and no raw fuel appears, the probable difficulty is that the mixture is too lean. Additional prime and a check of fuel quantity and flow will usually solve the problem.

A mixture that is slightly too rich will permit the engine to fire weakly on one or two cylinders. Black exhaust smoke may be seen. With priming discontinued, the engine should "catch" after another revolution or two.

A mixture that is slightly too lean will permit the engine to fire intermittently, possibly with backfiring. Additional priming will help the engine to "catch." On large engines with injection-type carburetors and a continuous prime for the start, the difficulty probably is in the throttle being too far advanced. Since the priming is continuous already, the obvious correction is to retard the throttle slightly.

Sometimes when hot, fuel injected engines start best with the mixture control initially set in IDLE CUT-OFF.

If no start is obtained after a reasonable interval (10 to 30 sec. of operation, depending upon starter specification), discontinue the start by cutting off fuel and disengaging the starter. Turn the ignition switch off *after* the engine stops turning, so that no combustible fuel mixture can remain in the cylinder after the engine stops. Check for hydraulic lock before another starting attempt.

Engine fires during start usually result from overpriming and may occur in the intake or exhaust manifolds. The first corrective action is to set the mixture at FULL LEAN, boost pump OFF, discontinue priming, throttle FULL OPEN, and continue cranking. If the engine starts, an induction or exhaust fire will be blown out the exhaust. If the fire persists, discontinue cranking, turn the ignition and master switches OFF, and evacuate the aircraft. Fires not controlled by these methods usually are the result of leakage from fuel lines or accumulated fuel in the engine compartment.

HAND CRANKING

Some very light aircraft do not have electric starters. Larger aircraft sometimes must be started when the battery is very weak and no external power supply is available or when the electric starter is inoperative. In such cases the propeller can be swung by hand. Engines up to 500 hp have been started by this method.

To keep the job safe, strict adherence to procedures is required. On first approaching the aircraft, the prop man ascertains that the wheels are chocked and calls, "Switch off." The pilot checks the master and ignition switches and replies, "Switch off." The prop man calls, "Gas off," and the pilot checks and replies. The fuel may be shut off either with a valve in the fuel line or with the full lean position of the mixture control if the carburetion system is so equipped. With the fuel off, the propeller is pulled through two or three revolutions to check for liquid in the cylinders.

Figure 8.1. Hand cranking.

The prop man then calls, "Gas on and throttle cracked." The pilot assures that a supply of fuel will be available for the start. If a carburetor mixture control is used, it is set to RICH. Two or three strokes of the primer are normally used if the engine is cold. Priming may be unnecessary for a warm engine. The throttle is "cracked" from the idle position to the position normally producing about 1000 to 1200 rpm. The pilot then replies, "Gas on and throttle cracked."

The prop man has positioned one of the propeller blades at a convenient height where stiffness of rotation indicates that a cylinder is just short of a compression stroke. He assumes the stance shown in Figure 8.1. He should stand just clear of the propeller arc, since standing too far clear will cause him to lean forward dangerously unbalanced. His fingers should rest across the top of the blade, but should not grip the blade. The prop man calls, "Switch on," and the pilot replies *before* turning the ignition switch to BOTH position. He will usually turn the master switch on at this time also, although the master switch might be left off until after the start if the aircraft is normally started with the electric starter. This would prevent possible inadvertent operation of the starter while the prop man was handling the propeller.

The prop man calls, "Brakes and contact." The pilot applies brakes before replying. The prop man spins the propeller with a hard, quick, snapping motion, stepping back with the foot that is raised for balance. If no start is obtained, the switches are turned off, the propeller is rotated to clear the cylinders of fuel from the starting attempt, and the starting procedure is repeated. Since engine rotation is limited to one or two compression strokes with this starting method, success depends upon introduction of a correct fuel-air mixture before the start. Take care not to overprime.

Minor deviations of the procedure given above may be necessary for certain types of carburetion systems. The important thing is to duplicate as closely as possible the ideal starting procedure for the engine with the electric starter, adding safety conditions for protection of the prop man.

ENGINE WARM UP

After start, operate the engine at a setting between 1000 and 1200 rpm that produces smoothest operation. Watch oil pressure during warm up, and if it does not register an increase within 30 sec. after the start, shut down the engine for investigation. Up to a minute may be required to attain normal oil pressure after a cold start. The engine should be running smoothly before taxiing.

Use of idle rpm with a cold engine may produce fouled spark plugs. During the warm up, engine instruments and aircraft equipment can be checked. Save engine performance checks until just before take-off. They are more valid then, and the propeller blast is less unpopular when created off the parking ramp.

Taxi Technique. Taxiing is the movement of the aircraft on the ground under its own power. At a tower-controlled airport, first obtain taxi clearance from ground control. This is done by radio or, in the event of communications failure, by means of light signals from the tower (Figure 8.2).

In congested areas, ground crewmen should provide initial taxi guidance. If wing-tip clearance is doubtful, a "wing-walker" should accompany the aircraft.

Normal taxi speed is about that of a brisk walk—somewhat slower in congested areas and faster on open taxiways. The engine power required may vary from idle to 1500 rpm, depending on wind, the slope of the surface, and the aircraft.

Movement of the rudder pedals turns the aircraft, acting through the aerodynamic force of the rudder surface and, on most present-day aircraft, through a steerable nose or tail wheel. The effectiveness of the rudder increases along with the airflow over its surface, whether that airflow comes from increased propeller rpm, increased taxi ground speed, or increased headwind. As the effectiveness of the rudder surface increases, that of the steerable nose or tail wheel decreases. To keep as much pressure on the tail wheel as possible, hold the stick full back during taxi, especially when using engine power. The only exception to this is when taxiing with a tailwind that significantly exceeds taxi speed. Forward stick then will act to keep the tail down.

With the tricycle gear, elevator technique is less important during taxiing. Because the taxi attitude is approximately level with the ground, increased airflow tends to strike the top and bottom of the horizontal stabilizer almost evenly. Any tail-up tendency that does occur puts *more* pressure on the nose wheel, thus

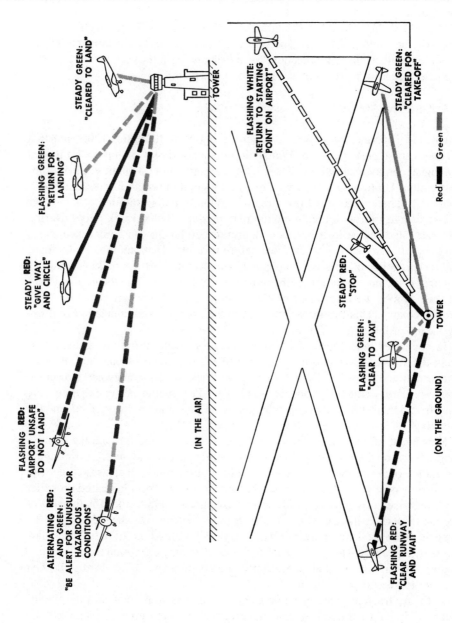

Figure 8.2. Light signals from control tower to aircraft.

increasing its steering effectiveness. Leave the control stick in neutral during taxi in tricycle-geared aircraft, except with a strong tailwind. As with a tail-wheeled aircraft, hold the stick forward in this situation to keep the tail down and prevent the aircraft from tipping.

A strong crosswind will try to lift the upwind wing during taxiing. Its effect will be greatest on very light aircraft with a high center of gravity, a large dihedral, and a narrow width of main landing gear. Holding aileron control into a quartering headwind and away from a quartering tailwind will help hold the wing down, for reasons substantially the same as those during flight. With a strong quartering tailwind, additional caution must be exercised.

Some high wing tricycle gear aircraft become unmanageable in a strong cross or tailwind and cannot be safely taxied in such conditions without wing and tail walkers to help hold them down.

Steering nose or tail wheels usually can be steered to about 15° to 20° of arc in either direction, after which point they are manually or automatically disengaged to permit tigher turns. With or without a steerable wheel, the brakes must be used for tight turns. The brake pedals usually are mounted on each rudder pedal, and they may be used selectively or in concert. Because braking for the turn slows the aircraft, anticipate the need for additional power and apply it before turn entry. Be careful to coordinate brakes, nose wheel direction, and power to avoid nose wheel side loads.

At any time brakes are used to slow the aircraft, retard the throttle first. Modern aircraft brakes are so efficient that carelessness leads easily to abuse. The pilot's feet should rest with heels on the floor and toes only on the rudder pedals (Figure 8.3), unless the need for strenuous braking justifies sliding the balls of the feet up on the brake pedals. Use intermittent rather than continuous application for prolonged braking, so that the brakes may have a chance to cool.

The engine cowling obstructs straight-ahead visibility for taxiing in some tail-

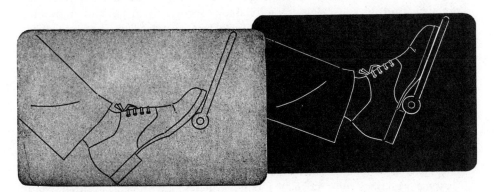

Figure 8.3. Normal position of pilot's foot. *Left:* on rudder. *Right:* on brake. *(Courtesy of US Air Force.)*

wheeled aircraft, and constant S-turning is necessary. Tricycle-geared aircraft, in which the pilot can easily see straight ahead over the nose, need not be S-turned.

PRE-TAKE-OFF CHECKS

Make final checks of the engine as close to the take-off runway as practicable. Brake the aircraft to a halt directly into the wind, with the nose wheel straight. By this time, the engine, if air-cooled, should be at normal operating cylinder head temperature. Complete the pre-take-off check list, including controls free, fuel selector on proper tank, flaps and trim set for take off, and proper carburetor heat setting.

The test of a power plant prior to takeoff may include some or all of the following checks as recommended by the manufacturer.

Idle Speed.   With the throttle closed, the engine must operate smoothly at the idle rpm recommended for the engine.

Magneto Grounding.   Switch the ignition off momentarily while at idle rpm. If the engine does not stop firing while the switch is off, the following check of the ignition system cannot be valid. The aircraft should be returned for correction of the difficulty, and everyone near warned to keep clear of the propeller.

Propeller Governing.   At a specified rpm, usually about 1800, retard the propeller control to the specified position. The result must be the response specified in the checklist.

Power Check.   The power check on supercharged engines measures the ability of the engine to produce a specified rpm with the introduction of fuel-air mixture into the intake manifold at outside barometric pressure. The unsupercharged light-plane engine always will produce slightly less than outside barometric pressure in its intake manifold at full throttle, because, while at rest, air intake passages open to the atmosphere are at atmospheric pressure. The suction needed to deliver air to the cylinders fast enough lowers the pressure in these passages slightly at full throttle. With a controllable propeller, the maximum allowable rpm should result from full throttle; additional rpm is evidence of faulty propeller governing. This check is not applicable to the light plane without a controllable-pitch propeller; however, satisfactory rpm at full throttle should be checked during the initial part of the take-off roll.

Ignition Check.   The ignition check, often called the "magneto" or "mag" check, tests the ability of each one of the two separate ignition systems to operate the engine at a selected power setting. The test usually is conducted at an engine rpm specified by the manufacturer, and this value may be as low as 1700 or as high as 2400. For supercharged engines with controllable-pitch propellers, the check should be made at the rpm achieved in the power check.

At the prescribed rpm, move the ignition switch to the R (right) position. Note the drop in rpm after the tachometer reading has stabilized. Return the switch to BOTH and again allow the rpm to stabilize. Repeat for the L (left) position. The drop in rpm on either independent system should not exceed the specified value.

An excessive drop in rpm during the check may indicate magneto or plug troubles; or it may be evidence of carburetor, valve, or other malfunctions that are not apparent when both ignition systems operate. When reporting an excessive rpm drop, mention whether it was slow or fast, and whether it was smooth or rough. This will assist maintenance diagnosis.

A marginal rpm drop may be caused by plugs fouled during extended idling. A fouled plug condition may be cleared by operating the engine at take-off power for a few seconds. Avoid lengthy operation at high power and lean mixture to "clean" spark plugs. A satisfactory reading might be obtained because a faulty spark plug became sufficiently overheated to ignite fuel whether it was functioning properly or not. Such a "glow plug" might result in preignition, detonation and loss of power during takeoff.

Cruise Fuel-Air Mixture. For aircraft equipped with a controllable-pitch propeller and a mixture control having RICH and NORMAL settings, adjust the throttle to a specified rpm (usually about 1700 rpm and not above that used for the ignition check in any case). As the mixture control is then moved from RICH to NORMAL, the engine speed should increase about 25 to 75 rpm, indicating a proper relationship between the two positions. This also provides a check against overrich idle mixture if the carburetor has an idle jet system that affects mixture in the cruising range.

Other Tests.   Scan the engine instruments continually during all performance checks. Other checks, such as of oxygen or hydraulic systems, may be required for particular aircraft or flights. Perform all engine checks as expeditiously as possible, because ground operation at high power settings produces uneven cylinder cooling and adds to engine wear.

Check flight controls for freedom of movement, and accomplish all other items on the pre-take-off checklist.

The Elementary Skills of Flight

AIRCRAFT CONTROL

The aircraft can be considered controllable when the pilot is able to produce desired airspeed, altitude, and direction of flight. The pilot reads these three primary conditions of flight in the cockpit from the airspeed indicator, altimeter, and some form of directional indicator. Together these three most basic flight instruments can present to the pilot the results of lift, thrust, drag, angle of attack, and similar aerodynamic factors discussed in Chapter 2. Despite their theoretical adequacy, however, the instruments are not used for primary reference during flight under visual conditions. In fact, the object of considerable practice during early flight training is to learn to spend as little time looking at cockpit instruments as possible, except for quick glances to check aircraft performance.

Pilots with their "heads in the cockpit" are a serious hazard in dense air traffic, and there are many maneuvers that require careful attention to the ground track of the aircraft or its attitude relative to the horizon.

FLIGHT BY ATTITUDE

The most important single reference that can be substituted for instrument watching is aircraft attitude, that is, visually establishing the aircraft's attitude with reference to the horizon. Attitude is the angular difference measured between an aircraft's axis and the line of the earth's horizon. Pitch attitude is the angle formed by the longitudinal axis, and bank attitude is the angle formed by the lateral axis (Figure 8.4). Since measurements from a line can be only two-dimensional, the vertical axis is pointedly ignored in relation to the horizon. Rotation about the vertical axis ("yaw" or "crab") is termed an attitude relative to the flight path of the aircraft, but not relative to the horizon.

INTEGRATED FLIGHT

Flying by aircraft attitude alone is obviously not the total answer when any degree of precision is required. The pilot must have available some means of determining aircraft performance. He does this by crosschecking the flight instruments to see if altitude, airspeed, and bank are as desired. If a deviation is

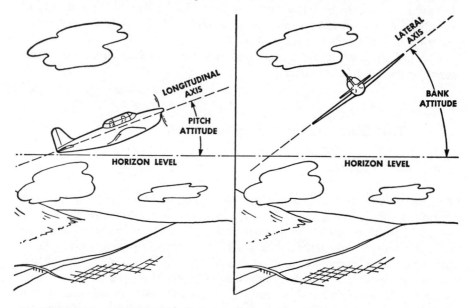

Figure 8.4. Aircraft attitude. Note that the angle of attitude is formed by the line of the axis and not in rotation about the axis.

noted, the pilot then makes a pitch or bank change to correct the situation. This method of aircraft control is known as integrated flight. More simply stated, it is the use of outside references supported by flight instruments to establish and maintain desired flight attitudes (Figure 9.1).

Although integrated flight becomes second nature with experience, the beginner must make a determined effort to master the technique. The following fundamentals should be constantly applied.

First. Establish and maintain attitude by positioning the aircraft in relation to the horizon. At least 90% of the pilot's attention should be devoted to this end, along with clearing the area to avoid other aircraft. When pitch or bank is rechecked and found to be other than desired, an immediate correction should be made to return the aircraft to the proper attitude. By continuously rechecking attitude and making immediate smooth corrections, the pilot will avoid any substantial deviations from the intended heading, altitude, or flight path. When the pilot's seat is to one side of the fore and aft centerline of the fuselage, he will not have a symmetrical view through the wind screen, but an off-center one, much as he does in an automobile.

Second. Monitor performance of the aircraft by frequent visual scanning of the flight instruments. With practice, it takes only a quick glance at the instrument panel to spot a trend away from the desired readings, and another glance to see when a correction has achieved the desired results.

Third. If the performance of the aircraft indicates a need for a correction, a specific amount of correction must first be determined, then applied with reference to the outside horizon, for example, by raising the nose ¼ in. farther above the horizon, or applying 5° left bank. The most common error made by the beginning student is to make a pitch correction while still looking inside the cockpit. Stick pressure is applied but the student does not know if the pitch changed 2 in. or 2 ft. As a result, level flight is not achieved, because pitch corrections made in this manner are either excessive or insufficient.

Pitch versus Angle of Attack. Pitch attitude is not the same as angle of attack. Angle of attack is related incidentally to pitch attitude, because raising the nose leads to lowered airspeeds and a high angle of attack. Lowering the nose leads to higher airspeeds and a low angle of attack.

The pilot must understand angle of attack before he can understand why this wing works as it does—why it cruises most economically at a certain airspeed, climbs most rapidly at another airspeed, and "stalls" or refuses to fly at other airspeeds, depending on flight conditions. An angle-of-attack meter in the cockpit would be a valuable asset, and the stall warning device found in most modern aircraft is a step in that direction. If angle of attack were completely instrumented, however, the pilot still would use the pitch attitude of his aircraft to approximate the condition of climb, descent, or level flight that he wanted. Attitude is an efficient measuring device—immediate in its indications and presented many times larger than any instrument.

The difficulty with attitude is that it must be considered in relation to engine power, airspeed, drag condition, and aircraft weight (both static weight and g loading) before the pilot can use its indications. Certainly the largest part of learning to fly is in the acquisition of ability to calculate from the conditions mentioned just what is going to happen to the airspeed (or airfoil angle of attack) of an airplane at any selected pitch and bank attitude. The calculating job is not as difficult as it may appear. A motorist approaching a hill in an automobile solves the same sort of problem. He knows that in his new pitch attitude he must increase power or settle for less forward speed. As he climbs the hill, he keeps in mind the trend of his speed as well as the indication at any moment. If he is carrying several passengers, he may notice that the "old bus lacks its usual zip." His eye is not fixed on the speedometer, because he has learned to estimate his speed within reasonable limits by the sound of his engine and the airstream around the car.

An aircraft provides many ways, other than the airspeed indicator, by which the pilot can learn what is happening to his condition of flight. The light plane is especially communicative in its reactions. The ability to sense flight condition is often called "feel of the aircraft," but senses in addition to "feel" are involved. The most important are the following.

Sounds.   The open-cockpit biplane is all but extinct, and with it have gone the wires and struts that would sing out the airspeeds as precisely as any visual indicator. The air that rushes past the modern canopy or cabin is often screened by careful soundproofing, but it still can be heard. When the level of that sound increases, the airplane is picking up airspeed, and for some reason.

The power plant may loaf in a glide, roar in a dive, hum contentedly in cruise, or labor in a climb; the amount of this noise that can be heard will depend upon how much the slipstream masks out. The relation between slipstream noise and power plant noise helps the pilot estimate both his airspeed and the trend of his airspeed.

Sight.   Attitude has been touched upon so far as though it were a static thing. It is not always so. When the pilot changes his direction of flight, the rate of change provides one indication of what is happening to airspeed and some of the other aerodynamic essentials. The rate of change is especially meaningful in conjunction with sensations the pilot can *feel*.

Feel.   The term *feel* of the aircraft basically is correct. There are three sources of *feel*, each highly important to visual flight.

One element of feel is in the pilot's own body as it responds to forces of acceleration. The g (gravity) loads imposed on the airframe are felt by the pilot with his famous vestibular organ, "the seat of his pants." Accelerations force him down into his seat or raise him against the seat belt. They affect his stomach in a manner familiar to any elevator passenger. Radial accelerations, as they produce slips or skids of the airframe, shift the pilot from side to side in the seat. These forces need not be strong to be useful. When the pilot can sense any change at all, he has an immediate index of control effectiveness.

A second element of feel, and one that provides direct information concerning airspeed, is the response of the stick and rudder controls to the pilot's touch. The control surfaces move in the airstream, and there they meet resistance proportional to the speed of the airstream. When the airstream is fast, the controls are stiff and hard to move; but only slight movement is needed to obtain results. When the airstream is slow, the controls move easily; but they must be deflected a greater distance. Note the use of the word air*stream*. Only the ailerons, outside the propeller arc, respond directly according to the speed of the aircraft through the air. The elevator and rudder are influenced also by the wash of air driven back by the propeller. Their response to pressures, then, must be considered cautiously—especially at slow speeds.

The third sort of feel, in the literal sense, comes to the pilot through the airframe. It consists principally of vibration, either from the power plant or from airflow. An example is the buffeting and shaking that should precede a stall in a well-designed aircraft.

Feel of the Aircraft—A Restatement.   Perhaps now, in summary, it can be said how the pilot gets the heading, altitude, and airspeed he wants from his airplane. He does it by obtaining an attitude and power setting that his past experience tells him are about correct. First, of course, he has to judge whether his desired situation is a practical one. Here, again, he must base his judgment on past experience—on what he has learned about his airplane's capabilities and limits. It is this need of "experience" which makes the most competent veteran pilots seek a check flight in a strange aircraft with a pilot familiar with its characteristics. This is particularly true where stepping from a small to a large, or a slow to a fast aircraft.

The pilot's main concern is airspeed, not only in terms of knots but also in terms of how much reserve airspeed he must have to support his climb or the loads of accelerations. His air sense must be able to balance the trend of his airspeed against the requirements his maneuver creates for airspeed.

Each maneuver or operation in the air offers its own problems, and one maneuver may demand closer attention to stick and rudder coordination than it does to airspeed. The knowledge of this, too, is part of "air sense."

STRAIGHT AND LEVEL FLIGHT

Flight at a constant altitude and heading, with all aerodynamic forces in balance, is "straight and level" flight. All flight maneuvers basically are deviations from the central straight and level reference, and the pilot must become acquainted with the attitude of his aircraft at straight and level flight before he can do other maneuvers precisely.

Level Flight.   Proper pitch attitude for level flight comes after a sort of trial-and-error process. Level flight can be approximated by placing the top of the engine cowling down slightly below the horizon about where experience on the other flights dictates. Airspeed selected governs pitch attitude required.

The altimeter will indicate whether altitude is being lost or gained. When the altimeter is stable, the pilot notes the point in his wind screen where the horizon appears to rest. If his sitting height is the same on the next flight, and he levels off at the same indicated airspeed, the horizon will be very near the same spot.

Adding power in level flight will cause nose-up, reducing power will cause nose-down. This exaggerated initial effect occurs because the first reaction is a changed force of downwash on the horizontal tail surfaces. If the new power setting is retained, a new attitude and a new airspeed will result, and retrimming of the elevator may be desirable.

All these complications emphasize the need for crosschecking the altimeter frequently during the first of the level-off. Except when near the ground, the pilot can determine gain or loss of altitude precisely only by checking his altimeter.

Straight Flight.   An aircraft is turned by banking and not by steering boatlike with the rudder. The best way for the pilot to keep his airplane headed straight, then, is to keep the wings level. Bank control in straight and level flight requires no trial-and-error process, but it does need some close watching. The surest way for the pilot to check that the wings are level is to look out at the wing tips and see that the horizon appears an equal distance above each tip (or below each tip for high-winged aircraft). The student pilot will have to check his wing tips frequently until he learns to detect small degrees of apparent tilt in the horizon ahead of him. Eventually one gets the idea that the aircraft must be in a bank if the horizon line appears out of its normal alignment over the nose.

Keeping the airplane aimed at a point on the horizon and holding a constant heading on the heading indicator are two ways of achieving straight flight.

When the wings are level, the pilot can expect his aircraft to fly at something close to a constant heading. If it is to be truly constant, rudder control must be trimmed or held to correct for any yaw tendencies. As stated above, the rudder is not used to turn the aircraft, but at the low airspeeds of the light plane it *can* be used for that purpose.

The rate of turn caused by wings-level yaw during normal flight usually is almost imperceptible, and that is the exasperating thing about it! It can be detected by a changing heading, or by slightly out-of-center position of the ball in the bank indicator.

TRIM CONTROL

One of the duties of the aeronautical designer is to arrange his aircraft weight and structure so that the flight controls (rudder, aileron, and elevator) are streamlined when the aircraft is cruising straight and level at normal weight and loading. If the airplane is flying out of that basic condition, one or more of the controls is going to have to be held out of its streamlined position. The holding duties are too tiring for a busy aviator, and so small movable *trim tabs* have been provided at the trailing edges of the control surfaces. These tabs, deflected in one direction, hold the primary control in the opposite direction through aerodynamic force. Because of their low power and relatively small range of speeds, not all light planes have a complete set of trim tabs adjustable from the cockpit; some have only elevator trim, and the rudder and aileron trim quite frequently are adjustable only from the ground. If all three controls are present, a definite sequence of trim makes the job easier when trimming for a protracted condition of flight.

Elevators should be trimmed first, since any attempt to trim the rudders at varying airspeed is impractical in propeller-driven light planes because of the built-in torque corrections. Once a constant airspeed has been established, hold the wings level with the control stick while rudder pressure is trimmed out. Finally, adjust aileron trim to relieve any lateral pressure. To avoid overcontrolling with trim adjustments, hold the airplane in the desired attitude with primary controls. Then apply trim to relieve hand or foot pressure. Because proper trimming decreases the drag of control surfaces, airspeed will increase. As it does, repeat the trim cycle until the airspeed stabilizes.

No special sequence need be followed for occasional trim applied during maneuvers or power adjustments to relieve temporary pressures. The more experienced a pilot becomes, the more frequently will one hand be found among the trim tabs. But trim must not be used as a primary control. Trim should be adjusted during steady flight conditions, rather than during changing ones.

AIRCRAFT STABILITY

One of the customary early lectures from pilot instructor to pilot student runs like this: "You don't have to balance this machine in flight as though it were on the point of a pin. It *wants* to do the right thing! We will go up, and I will trim for level flight. Then I will push into a dive or pull into a climb, and I will release the stick. The aircraft will do a few dives and climbs and gradually work its way back to level flight." He doesn't say much about directional stability. They go up, and they do the experiment, and it works out pretty much as the instructor said (Figure 8.5). He does have to help the wings back to level now and then, but "this aircraft is slightly out of trim, and the changing airspeed affects the torque rigging of the vertical stabilizer."

The instructor is taking the easy way out of a problem. Most present-day aircraft are stable in respect to pitch. Generally speaking, for a given trim condition the aircraft tends to find and hold a constant airspeed. It will keep that airspeed in a climb, a dive, or in level flight, depending upon how much power is used. Aircraft are not so stable directionally. An aircraft disturbed into a slight bank by a gust will tend to return to a wings-level position, *provided little or no turn develops first.* Once the aircraft has started turning, the turn will tend to increase. The outside wing in the turn will travel faster than the inside wing, will produce more lift, and will cause an overbanking tendency. At the same time,

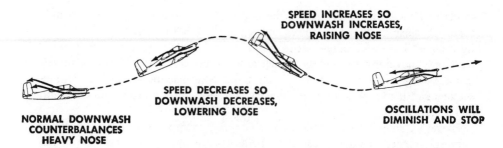

Figure 8.5. Stability in level flight.

loss of vertical lift from the banked airfoil will cause the nose to drop, further increasing speed and tightening the turn. The end result, if allowed to ensue with disinterest by the pilot, is a tight spiral down.

The most important thing for the student pilot to learn about aircraft stability is that the airplane is not going to fall out of flight in any direction so rapidly that normal attention will not provide plenty of time to correct.

Climbing and Descending

Climb and descent are extensions of level flight. The differences are in the amount of power being used and the attitude being maintained. The airplane's performance depends upon these two factors, and is translated into airspeed and rate of climb or descent. *Attitude + power = performance*.

When power is fixed, such as full throttle for takeoff and climb, idle for power-off glide, or some selected value in between, the airspeed is controlled by the pitch attitude of the airplane. When power is variable and pitch is fixed, as for level flight or a selected angle or rate of climb or descent, the throttle controls the airspeed.

With power at a reduced or idle setting, a given attitude results in a particular combination of airspeed, rate of descent, and glide distance. Increasing the pitch attitude will slow the aircraft, but will not increase glide distance, and can lead to stall with consequent abrupt loss of altitude. One simply cannot stretch a glide.

If pitch is too high, full power cannot maintain flying speed, and the aircraft again will stall. It can also descend with partial or full power, the rate and airspeed being determined by the pitch attitude.

Changing power changes pitch by varying the downwash on the horizontal tail surfaces. Adding power causes a pitch-up tendency, reducing power, pitch-down.

TECHNIQUE IN THE CLIMB

The mistakes of a new student pilot in learning to enter a climb usually do not stem from ignorance of his difficulties but rather from overconcern with them. If he would trust the pitch attitude and power setting demonstrated by his instructor, he probably would find them very close to correct.

The attitude selected usually will not be exactly correct because of variances caused by altitude, sitting height, individual aircraft performance, and the pilot's memory. Some minor corrections will therefore become necessary, but should be made only after the airspeed has stabilized or its trend carefully analyzed. At this time a small pitch correction, usually no more than ¼ in. to ½ in., may be made with reference to the outside horizon.

In a constant-airspeed climb, the student should not think of the necessary correction as an adjustment to his airspeed indicator, but rather as an adjustment to the pitch attitude. The difference is vital because the student *can* succeed in keeping attitude constant.

Level-off from a climb is a matter of adjusting pitch attitude and power to produce the desired level flight performance. Depending on the rate of climb, the pitch attitude can be reduced to stop the altimeter increase beginning anywhere from 10 to 50 ft. below the desired altitude. The throttle is left at the climb power setting until cruising speed is attained and then is set at the power which will maintain that speed. Once level flight and cruising speed are established, the trim is adjusted to relieve control pressures, and the mixture control is set for cruise conditions.

SOME PRACTICAL CONDITIONS OF CLIMB

It is important to know the approximate power settings and pitch attitudes for at least the following conditions of climb.

Best Rate of Climb (V_y). The best rate of climb is obtained at approximately the same airspeed for any power setting. It is the airspeed at which the most excess POWER is available over that required for level flight (Figure 2.37). This condition of climb, with full engine power, should be used when the pilot wants to gain the most altitude in the least amount of time. The best rate of climb speed is usually about 1.4 times stall speed clean.

Best Angle of Climb (V_x). The airspeed for steepest angle of climb produces the most distance up for the least distance forward and is less than that giving the best rate of climb. It is determined by flight test and should be found in the owner's manual or the approved flight manual for the airplane. The best angle of climb is of value in clearing obstacles after a minimum run take-off, and a properly qualified and prepared pilot knows the correct speed for his aircraft weight and configuration. In a multi-engine aircraft, if V_{mc} (minimum controllable airspeed with a failed engine) is greater than V_x, then V_{mc} must be used for climbing over obstacles. The best angle of climb airspeed is usually about 1.2 times stall speed clean.

During protracted climbs at low airspeed and high power settings, the pilot must watch his engine instruments for evidence of overheating. Cylinder head temperature usually is the limiting factor, although design of cowling facilities may be such that excessive oil temperature is encountered first.

With unsupercharged engines the pilot will have to advance his throttle during climbs and retard it during dives if he wants to keep a constant engine power setting. This becomes especially important during prolonged descents, when the increased power obtained at lower altitudes can cause excessive manifold pressures in engines with controllable-pitch propeller installations.

SOME PRACTICAL CONDITIONS OF DESCENT

The pilot should learn the approximate power settings and pitch attitudes for at least the following conditions of descent.

Descent at Minimum Safe Airspeed.   This nose-high, power-assisted descent condition is useful principally for skirting obstacles during a landing approach to a short runway. Considering a margin for pilot technique and the possibility of wind gusts, the minimum safe speed for approach to landing is a little closer to normal approach speed than to the power-on stall speed of the aircraft. Some margin must be left because of excessive power required to produce acceleration at low airspeeds. With a 10% reduction in airspeed at the critical point, a 100% increase in engine horsepower might be required to keep the rate of descent constant. If the engine already is being operated at 75% power, only the exchange of altitude for airspeed will restore the balance.

Partial-Power Descent.   The normal method of losing altitude is to descend with engine power slightly above idle, usually with 15 to 20 in. of manifold pressure for aircraft equipped with controllable-pitch propellers. Airspeed may vary from cruise to that used on downwind leg in the traffic pattern; however, do not interpret this wide range in airspeed to permit erratic pitch changes. Select the desired airspeed, pitch, and power combination, and attempt to keep them constant. The moderate airspeed and the use of some engine power help to keep the engine from cooling excessively in the descent, reducing possibilities of plug fouling or carburetor icing. Normal approach speed is about 1.3 times stall speed.

Glides.   Gliding flight is flown with the throttle closed and at the airspeed giving the most favorable airfoil lift-drag (L/D) ratio (Figure 2.36). This results in traveling the most distance forward for the least altitude lost. Practicing glides at this airspeed improves the pilot's ability to cope with an engine-failure situation and benefits the traffic pattern since it is the same airspeed and pitch attitude used for the power-off landing approach. If a protracted glide is made without power, the engine must be "cleared" every minute or so by advancing power momentarily to at least the cruising setting.

Dives. There is no sharp dividing line between a glide and a dive. As a general rule, if a pilot's primary visual reference is below the horizon, the maneuver may be considered a dive. Some engine power is essential during a protracted dive, for the same reasons mentioned in connection with glides. The pilot should know the limiting airspeed permissible in dives with his aircraft and avoid that speed by a wide margin. If the engine rpm exceeds allowable limits in a dive, the proper corrective action is to place the throttle in "idle" position and raise the nose of the aircraft to place a load upon the propeller.

Level Turns

The turn is so fundamental a maneuver in all phases of flying that a clear understanding of what happens to an airplane in a turn is essential to the modern airman.

WHAT DO THE CONTROLS DO?

There are four controls, and following is what the pilot needs to know about each of them:

The ailerons bank the wings and so determine the rate of turn at any given airspeed.

The aerodynamicist says: "For a given airspeed and rate of turn, there is one correct angle of bank." This same principle is familiar to any motorist who has tried to negotiate a corner with greater speed than the road engineer had in mind. The pilot does not have to compute what his bank angle should be at his airspeed (except for the special conditions of instrument flying) because he does not have to make good a precise rate of turn. He banks slightly (about 15°) for a shallow turn, more (about 30°) for a medium turn, and 45° or more for a step turn. If he must arrive at a certain heading at a certain spot, he will watch his progress and increase or decrease the rate of turn as needed to correct his first judgment.

The indicated airspeed at which an aircraft will stall in the turn increases sharply after the bank angle exceeds 45° (Table 8.1). The bank angle is secondary—the direct cause is the elevator control—but bank angle provides the pilot with a measuring stick of what he is demanding from his aircraft. He should not try to select steep angles of bank unless he has sufficient power, or altitude to lose, to keep the required airspeed in the resulting turn.

The elevator moves the nose of the aircraft up or down, that is, "up or down" as the pilot sits, and perpendicular to the wings. Doing that, it sets

TABLE 8.1. Effect of Banking Angle.

Angle of Bank (°)	Load Factor ("g" Load)	Percent Increase in Stalling Speed
0	1	0
20	1.065	3.0
40	1.31	14.4
60	2.00	41.4
80	5.76	140.0
90	Infinity	Infinity

the pitch attitude in the turn and "pulls" the nose of the aircraft around the turn.

When the wing of the aircraft is banked, lift at a right angle to the plane of the wings will have a "vertical" component opposing gravity and a "horizontal" component parallel to the horizon. The farther the wing is banked, the greater will be the percentage of total lift exerted horizontally, and the less will be the percentage of total lift exerted vertically. If the aircraft is to maintain a constant altitude in the turn, total lift will have to be increased to provide enough for both jobs. Arranging for the needed extra lift is the job primarily of the elevator, although throttle can assist by providing thrust to overcome the additional induced drag, and increased downwash on the horizontal tail surfaces.

By increasing the back stick pressure in a bank, you increase the angle of attack, thus increasing the total lifting ability of the wing. With exactly the right amount of back stick pressure, at any airspeed and bank attitude, the "vertical" component of lift will be just enough to keep the altitude constant. Any further back stick pressure will result in a climb; any less will result in a descent. To increase the rate of turn (usually expressed in degrees per second), the airplane must be banked steeper and more back pressure applied. To decrease the rate of turn, the bank must be shallowed and less back pressure applied.

The throttle provides thrust which may be used for airspeed to tighten the turn or to provide climb.

The throttle should be advanced as the turn is entered, if it is to be more than a shallow bank turn.

If you leave the throttle constant during a turn, it can be dismissed as a control. Should you advance the throttle, you will notice the response almost immediately from lift, from both the vertical and the horizontal components of lift. This is due to the same effect considered in climbs and dives—the immediate increase in flow of downwashed air from the wing on the horizontal stabilizer.

The rudder offsets any yaw effects developed by the other controls.

And that is all it does. Expect to apply rudder pressure when you add or subtract power, when losing or gaining airspeed, and when using the ailerons to bank the wings. In the last instance the rudder is counteracting yaw from *aileron drag*.

The ailerons, elevator, throttle, and rudder are operated together to achieve smooth turns, and for entry into and exit from turns. This is coordinated flight.

MAINTAINING THE TURN

Some major difficulties of the turn are overcome when you realize that creating a turn and keeping one constant are two individual problems.

Once the required angle of bank for the desired rate of turn is obtained, no more aileron is required in that direction. In fact, the aircraft may have a tendency to keep increasing its own bank. The outside wing, traveling faster than the inside one, creates more lift. To keep the bank constant, you may hold just a little top aileron—opposite to that used for the turn.

With the aileron almost streamlined, aileron drag difference disappears. The only requirement for rudder pressure, then, is that created by engine "torque effect" at the reduced airspeed. Only in steep or slow climbing turns does that become a problem. Generally speaking, a skid will result if uncoordinated rudder is held into a turn, and a slip will result if uncoordinated rudder is held away from a turn. Only by crosschecking the turn and slip indicator for a centered ball can coordinated flight be guaranteed.

Leaving the Turn. To roll out level on a particular heading, begin the rollout in advance of that heading. Start the aileron pressure back in the direction of level flight. Reduce back stick pressure, which in a rapid rollout may have to be reversed to avoid climbing out of the turn. Reduce power to cruise setting.

Some Tips on Learning Turns.. No substitute has been invented for practice. Before a student pilot can practice turns intelligently he *must* know the functions of his controls in turn so well that action follows need, without cogitation interfering! To develop such response, try entering turns and then exaggerating the application of one of the controls. This should be practiced at a safe altitude with the instructor present, although it is not necessary to be roughshod or to continue the exaggeration too long.

Try turns, leaving out one or more of the essential controls, and note what happens to the aircraft. Instructors should allow the students to attempt some fairly tight turns in which they are permitted to push on the rudders alternately. The sensing of a slip or skid is one of the most difficult skills to acquire, and training should include full-fledged slips and skids with some *g* loads pushing one into the seat. The student cannot learn slips and skids from light pressures on what doctors insist is one of the least sensitive parts of the anatomy!

COORDINATION MANEUVERS

There are many varieties of coordination maneuvers. All involve rolling into some sort of turn in one direction and then rolling into the same type of turn in the opposite direction. They all provide good practice in learning to handle any airplane and consist merely of applying coordinated rudder and aileron so as to roll from side to side, at the same time holding the nose on a point of the horizon. They are being properly performed when the nose stays on a point and when there is no sensation of roughness, slip or skid.

This maneuver can be modified to allow the airplane to turn as much as 30° right or left, rolling back into a snakelike succession of coordinated turns in opposite directions about a line on the ground.

CLIMBING OR DESCENDING TURNS

The most difficult part of climbing and descending turns probably will be in use of the rudder. On entering a climbing left turn, for example, the turn entry will require left rudder; but the diminished airspeed at the climb power setting reduces the requirement for left rudder and eventually may require right rudder to counteract torque effect at low airspeed.

Now that we have considered the techniques of flight, let's put them to use.

Take-off Technique

If an aircraft is taxied fast enough to reach flying airspeed and is allowed to assume a shallow climb attitude, it will become airborne. The take-off may be considered in three phases: *ground run, transition to flight,* and *climb-out.* These divisions are not exactly precise, because the effectiveness of certain controls changes with the progression of the take-off. Since the take-off involves both ground and flight operation of the controls, you must be able to make the transition from the ground functions of the controls to the flight functions with smoothness and coordination.

GROUND RUN

Tower clearance (where there is a control tower) is necessary before the aircraft may occupy the runway, but such clearance does not relieve the pilot of responsibility to check the pattern visually. Landing aircraft always have priority.

The wise old pilot lines up on the very end of the runway for take-off, and he does not use a taxiway intersection or a slightly downwind runway to save taxi distance. A trite aviation saying bears the point out: "Two things you can't use—the altitude above and the runway behind."

When the aircraft is aligned in take-off position with nose or tail wheel straight, release the brakes and open the throttle smoothly to the recommended take-off setting. Do not use reduced power settings for take-off to "save" the engine. Occasional operation at full power helps to keep spark plugs clean and firing properly. Rated take-off power will get the airplane to safe altitude and airspeed quickly; and if the engine is not functioning properly, the difficulty can be detected early enough to permit safe discontinuance of the take-off. Smooth throttle advance is essential, not only to avoid engine wear and plug fouling, but also to avoid directional control problems during the first of the take-off run.

Chapter 2 discusses the effects of the forces generated by the propeller and engine combination. When power is applied for take-off, torque and rotational wash (and P-factor on tail draggers) tend to turn the aircraft left. This tendency is compensated for by applying right rudder sufficiently to overcome it. Both the turning tendency and the correction by rudder are strongest at full throttle, and rudder effectiveness increases with forward speed.

The rudder must also compensate for crosswind, more so if the aircraft does not have nose or tail wheel steering. A left crosswind requires more right rudder, a right crosswind will reduce the need for right rudder and may even require left rudder.

Hold the elevator in neutral on a smooth runway, so that pressures developing with increasing airspeed may be sensed naturally. During rough field operation some back stick pressure is needed to avoid nosing down as the result of shocks during the roll. If a crosswind exists, the upwind wing is held down with the aileron by moving the control against the wind.

TRANSITION TO FLIGHT

As airspeed increases on take-off roll, the first control to become effective is the rudder, which becomes the primary directional control. As lift-off speed is approached, the elevator becomes effective and establishes pitch attitude for the departure from terra firma. Once the aircraft is airborne, the ailerons become effective.

There are several ways of determining that the aircraft is nearing flying speed. Most of these were mentioned in the discussion of "air sense." Like a motorist, you can check the "speedometer" or airspeed indicator. The runway to each side becomes more blurred. Flight controls at the start of the run yielded to light pressure and required large movement to produce effect. Now the controls offer resistance, and only slight movement is required. On tail-wheeled aircraft, the tail has a tendency to rise of its own accord. Engine noise appears to lessen as the slipstream is heard rushing past fuselage, wings, and landing gear. Forward acceleration, felt conspicuously at the start of the run, decreases. All this has happened within 7 to 10 sec., so that there is little wonder that the beginning student sometimes is overwhelmed by the barrage of changing sensations. After a few flights, the sensations become familiar; and their correlation informs him that take-off speed is near.

As the aircraft accelerates, the elevator becomes able to hold the aircraft in the shallow climb attitude from which it will soon leave the runway. With tricycle gear, lift the nose wheel slightly from the runway as soon as moderate back stick will do the job. The aircraft then will be in take-off attitude, and the nose gear will be relieved of high-speed runway shock.

Tail-wheeled aircraft rest on the ground in an attitude more nose-high than needed for climb. As speed increases, lift on the tail section causes it to rise. When this natural development places the nose of the aircraft in take-off attitude, maintain that attitude with stick pressure as required. Once the correct attitude is found and maintained, there is no difference of technique between tricycle and tail-wheeled aircraft.

The light plane can be made to leave the ground in a wide variety of pitch attitudes. The correct attitude is approximately that to be used during the normal climb; thus there are only slight pitch adjustments to be made between the later part of the ground run and climb to altitude. Make no effort to ''pull'' the aircraft into flight with increased back stick pressure, as it will fly off most safely without such assistance as soon as flying speed is reached.

If no crosswind exists, hold the wings level through the ground run and initial flight. Directional control is a function solely of the rudder until the wheels leave the runway.

CROSSWIND TRANSITION TO FLIGHT

Aileron pressure was applied into the wind at the start of the crosswind ground run. This initial correction was a mechanical one, because the ailerons lacked effectiveness at slow speed and the full weight of the aircraft on its tires gave sufficient traction to prevent drift effects from the crosswind. As the airspeed provides increased aileron effectiveness, so does the increasing wing lift act to relieve weight from the landing gear. By the time flying speed is near, the now-responsive ailerons hold the wings level. With too much aileron, the aircraft will bank into the wind and may leave the ground hazardously wing-low. With too little aileron, the force of the wind may lift the upwind wing and cause the aircraft to skip downwind across the runway as soon as traction of the tires becomes light enough to permit. Because the skipping is the more difficult of the two alternatives to control, most pilots prefer to keep extra aileron pressure into the wind during the later part of the ground run. They accept a bank into the wind as the aircraft leaves the ground and level the wings when safely airborne. In a stiff crosswind, the aircraft is held on the runway with slight forward control pressure until airspeed is slightly above normal take-off speed. The aircraft then may be pulled cleanly into the air to avoid side skipping.

CLIMB-OUT

The third phase of the take-off is that portion between the leaving of the runway and the exit from the traffic pattern. If the runway surface is rough, the aircraft may bounce a few times before definitely becoming airborne. The pilot concentrates on keeping directional control and the take-off pitch attitude.

Maintaining direction over the runway immediately after lift off in a crosswind is simplified by relaxing the rudder pressure that was used for directional control for the take-off run. The aircraft will yaw into the wind just about the right amount to maintain the desired ground path.

Raise the retractable landing gear when you are certain of remaining airborne. The raising of the gear will produce only a slight trim change for most aircraft, but this change must be resisted with control pressure to keep take-off pitch attitude. If flaps are used for takeoff, they generally are retracted when airspeed is slightly less than that used for climb. Raising the flaps on most aircraft will affect trim more than retracting the landing gear, and definite back stick pressure usually is required to prevent settling from the climb path as the flaps come up.

At full power, the take-off pitch attitude should tend to produce more than the desired climb airspeed. When that speed is reached, set the throttle to the normal climb power setting, and adjust pitch attitude as necessary to hold constant the desired airspeed.

Always follow a definite sequence when adjusting power in aircraft equipped with controllable-pitch propeller and fuel-air mixture controls (Figure 8.6). To increase power, advance the mixture first, followed by propeller and then throttle. To decrease power, retard the throttle first, followed by propeller and then mixture. This sequence is necessary to provide adequate cooling in high power ranges and to prevent the introduction of excessive pressures into the cylinders at low rpm.

Exit the traffic pattern when past the end of the runway and at a safe altitude and airspeed. This is usually done by making one 90° turn followed by an imme-

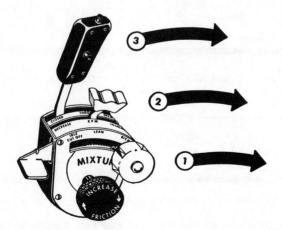

Figure 8.6. Sequence in power adjustment. 1. Mixture must always be prepared for rpm. 2. Rpm must always be prepared for throttle. 3. Do not exceed the recommended manifold pressure for rpm.

diate 45° turn in the opposite direction. The first of these turns is made in the same direction as the final turn for landing on the same runway, unless local procedures specify otherwise.

OTHER FACTORS AFFECTING TAKE-OFF DISTANCE

Few light planes require more than 1500 ft. of runway for takeoff under average conditions, so a short runway seldom is a deterrent to flight.

Table 8.2 illustrates the effect of five important variables upon take-off distance. The table is based on data extracted from the US Air Force Flight Handbook for the T-34A trainer. Similar data, in one form or another, are available to the pilot of any civil or military aircraft.

The principal factors affecting the length of take-off roll are altitude, temperature of the ambient air, aircraft weight, headwind, obstacles, and runway surface. Table 8.2 indicates that thin air experienced at high altitudes and high temperatures can multiply the required take-off distance by as much as three

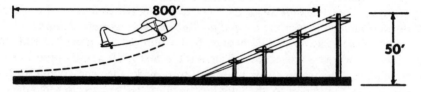

Hard surfaces: Concrete runway, highway landing strips, macadam and asphalt, packed snow or ice.

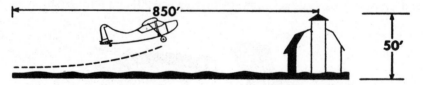

Medium surfaces: Sod turf, cinders, dry dirt runway, dry field roads, or cow pastures.

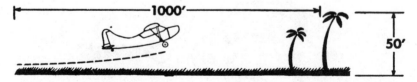

Soft surfaces: Muddy field, soggy turf, soft snow or sand, loose gravel.

Figure 8.7. Effect of runway conditions on the take-off of a typical light plane over a 50-ft. obstacle. Take-off flaps down 30°, no wind, G. W. 2050 lb. at sea level. *(Courtesy of US Air Force.)*

TABLE 8.2. Take-off Distances in Feet.

(10° Flaps, Hard Dry Runway Engine: Continental 0–470–13)

Gross Weight of Aircraft	Pressure Altitude	23°F (−5°C)				60°F (+15°C)				130°F (+55°C)			
		Zero Wind		30-Knot Wind		Zero Wind		30-Knot Wind		Zero Wind		30-Knot Wind	
		Ground Run	Clear 50 ft.	Ground Run	Clear 50 ft.	Ground Run	Clear 50 ft.	Ground Run	Clear 50 ft.	Ground Run	Clear 50 ft.	Ground Run	Clear 50 ft.
2900 lb	S L	910	1310	300	510	1130	1610	390	660	1690	2360	630	1020
	1000	1020	1460	340	580	1270	1800	450	750	1880	2620	720	1160
	3000	1270	1810	450	740	1580	2220	585	960	2360	3250	950	1500
	5000	1600	2230	590	960	1980	2730	770	1220	3040	4150	1280	1950

Source: USAF Technical Orders.

times. Although indicated take-off speed will remain the same, the true airspeed and hence the ground speed, will be increased. The time required to reach the necessary safe airspeed will also be increased because of decreased engine power and propeller efficiency at the higher altitude.

Not considered in Table 8.2, but significant, are the runway surface (Figure 8.7) and runway slope. Careless consideration of all of these factors has brought many a pilot to grief.

SHORT-FIELD TAKEOFF

If a short-field take off is necessary, revised technique will depend on how marginal the distance is. To go "all the way," drain fuel down to that actually needed, remove baggage and even radio equipment, consider pilot weight, and so on. If the wind is contrary to the otherwise best take-off direction, it may be a good idea to wait for better winds. Partially deflating tires will assist if the surface to be used is soft. Use all possible runway, including over-run surface.

The role of wing flaps in short-field operation has been the subject of much hangar talk. Part of the answer lies in the type of flap employed. *Simple flaps* or *split flaps* increase both drag and lift throughout their travel—mostly drag after about 50% of travel. The lift is needed in order for the aircraft to leave the ground at slower speed; but the drag is an unwelcome obstacle to the obtaining of that speed.

The best answer to "how much flaps for short field?" is in the recommendations of the manufacturer for the particular type of aircraft. This figure, established by flight tests, is seldom more than 30° of flap travel for simple or split flaps. *Fowler* or other *lift-type flaps* reach their optimum at greater extension, but this type of flap is not often found on the light aircraft.

Because the lift of the flaps is needed only at the moment of takeoff, it might seem advantageous to leave them retracted during the ground run until just before take-off. The possibility of the mishandling of controls or flap actuator malfunction makes this a hazardous proceeding of questionable value.

When the aircraft is aligned for a short-field takeoff, be sure you have full power—with satisfactory instrument readings—before releasing brakes. It is important that the stick be held full back at this time if the aircraft is a tail-wheeled type. There is little danger of nosing over if the brakes are firmly held, and no forward motion allowed. Should the aircraft start to creep forward, or the tail start to rise, release the brakes entirely and begin the take-off roll while applying the remainder of the power quickly but smoothly. The danger of nosing over on a full-power run-up occurs in tail-wheeled aircraft when the airplane is allowed to roll a few feet and *then* brakes are reapplied.

Leave the elevators approximately in neutral, and avoid imposing a take-off attitude by back stick pressure until just before reaching flying speed or until the tail naturally arrives at the take-off attitude. Attempts to raise the tail of tail-wheeled aircraft too early will only create extra drag through the deflected eleva-

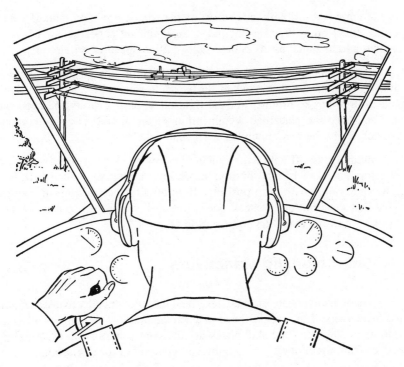

Figure 8.8. An old trick of the trade. If in doubt about getting over those wires, aim slightly under them at first. If you have enough speed when you get there, you can pull over. If you haven't the speed, YOU GO UNDER.

tors. "Feel" of the aircraft is important in leaving the runway. Just under flying speed, move the stick back rapidly but smoothly. When clear of the runway, retract the gear and lower the nose slightly to pick up climbing airspeed. If the objective is to clear obstacles as well as to get airborne quickly, you must know the speed at which the aircraft will deliver its highest angle of climb. Note that here the best *angle* of climb speed, which may be substantially less than the speed for the best *rate* of climb, is the required speed. If the speed for the steepest climb is not available, however, it is safer to err on the high airspeed side. Altitude lost in mushing along too slow cannot be regained, whereas that lost by too much climb speed may be regained partially in a zoom over the obstacle (Figure 8.8). The flaps set for take-off never retracted until the plane is clear of all obstacles.

DON'T!

Three additional items are considered worth comment in a discussion of take-off distance. They are "killer" items that become a hazard only because of pilot contempt or neglect:

Don't attempt take-off with ice or frost on the wings. Frost, particularly, has a spoiling effect on airflow far beyond its insignificant appearance. A layer of frost can add as much as 20 knots to the stalling speed of an aircraft.

Don't exceed center-of-gravity limitations in loading the airplane. Center of gravity (c.g.) will be discussed in the next chapter, where it has particular application to larger aircraft. For the light-plane pilot's safety, he need only know and heed the placarded weight limits of the aircraft baggage compartment and fly in the recommended seat when solo.

Don't attempt take-off with a malfunctioning engine. A pilot spends considerable effort and all of 15 min. of time inspecting a light plane, entering it, starting it, and taxiing to run-up position. Returning to the parking line with an engine that does not quite check out is not an inviting idea, but it has been proven a safer one than "trying it out on take-off."

Characteristic Maneuvers

The maneuvers about to be described are known collectively as characteristic maneuvers. They are designed to teach the limits of safe flight and to give the pilot the confidence and skill necessary to avoid abnormal flight attitudes or to recover from them; they are fundamental equipment for any airman.

CLEARING TURNS

Before considering several maneuvers that require a large block of unoccupied air space for their safe completion, the subject of clearing turns seems appropriate. Most pilot training programs have hard-and-fast rules about clearing turns—thus, two 90° turns in opposite directions before stalls and acrobatics, and two 180° turns in opposite directions before spins. The essential idea of the clearing turn is to be certain that the next maneuver is not going to proceed into another aircraft's flight path. While complying with the letter of clearing-turn rules, the pilot should take special care to examine the area likely to be covered in the maneuver. He must be quite sure that the area will be clear when he gets there, regardless of the number of turns required to get that assurance. Clearing turns are good common sense in or out of a training program before any practice maneuver described hereafter, except for lazy-8's, maximum-performance climbing turns, and chandelles, which are clearing maneuvers in their own right.

STALLS

The term *stall* is familiar to anyone learing to fly. It refers to the pattern of airflow over the wing. The coefficient of lift (Figure 2.36) increases with increasing angle of attack until the critical angle of attack is reached. At that point the airflow over the wing breaks down and lift falls off abruptly. This is the stall. To restore lift, the angle of attack is decreased, and power may be increased to add airspeed. The increased airspeed also adds lift.

When an experienced pilot flies an aircraft new to him, he invariably will try a few stalls or approaches to stalls before landing. He is going to be landing in a stall, or very close to one. The landing pattern and final approach will require positive control at speeds not far above stalling, and the pilot wants to know how the aircraft feels and reacts at those speeds. He wants to know what kind of warning precedes the stall, whether the aircraft becomes unstable in a stall, and what kind of recovery techniques are most effective. All these objectives also apply to the practice of stalls by student pilots, who have assigned to them a practice routine of stalls covering several representative conditions of stall entry.

The Stall Curriculum. The exact procedures for practice stalls vary, but each stall is designed to represent a condition that might arise from the pilot's failure to do some normal maneuver properly. *Power-on stalls,* straight ahead and turning, simulate stall out of climbing flight. *Power-off stalls,* straight ahead and turning, simulate stall out of gliding flight and especially out of the turn to final approach. The *approach to landing stall* is a final approach, roundout, and landing touchdown, all done at safe altitude to see how the airplane will react during landings. Landing gear and flaps are down, and the power is at idle, as they would be during actual landing. *High-speed stalls,* or *"accelerated"* stalls, will be discussed at more length later. When included in the student stall series, they simulate stalls out of tight turns or abrupt pull-ups such as might occur during acrobatic maneuvers. *Demonstration stalls,* as the name implies, usually are demonstrated by the instructor and offered for dual practice to show the result of grossly abusing aerodynamic principles. The student should leave them alone in solo practice until he is competent at acrobatics. Demonstration stalls are as varied as the possible insults to safe flight, but may include: *top and bottom rudder stalls,* with severe lack of rudder coordination during turns; *secondary stalls,* resulting from excessive back stick pressure immediately after stall recovery; *vertical stalls* (if the airplane is not restricted from them) such as might be encountered in aerobatics not properly performed; and *elevator trim stalls* to show the results of not controlling nose-up tendencies created by the elevator trim being set for glide during sudden application of power for go-arounds.

Together, the practice and demonstration stalls should establish that the airplane will stall at any airspeed, any attitude, any power setting, any configuration, and at any weight or *g* loading.

All that is necessary is that the critical angle of attack be exceeded.

High-speed stalls (or *"accelerated stalls,"* as the aerodynamicist prefers to call them) are stalls incurred during centripetal accelerations, such as pull-ups or turns. More lift is required to sustain the acceleration than would be needed for stabilized flight, so stalling speed is proportionately higher. When the demand for lift causes the angle of attack to reach the critical angle, the stall will occur.

Stall Warning. Recognition of an approaching stall must become automatic, because most accidental stalls occur while the pilot's attention is distracted from his primary task of aircraft control. If acceleration is not involved, airspeed is still the most reliable indication of what is about to happen to the aircraft. Stalling speeds for different weights and configurations, power-on and power-off, are published in the aircraft flight handbook. If the airplane will stall at 70 knots in a certain condition, and the airspeed indicator shows 75 knots in that condition, a stall can be anticipated from almost any new inducement. It is, of course, no more necessary to stare fixedly at the airspeed indicator at this time than at any other time. The sound of diminishing slipstream and laboring engine becomes especially noticeable just prior to stall, and controls become quite "sloppy" and lack effectiveness. These are the same customary signs of diminishing airspeed used in all visual flight.

Airspeed is a less reliable warning of high-speed stalls, which can happen at any airspeed if the turn or pull-up is tight enough. Here we must depend mostly upon the feel of control pressures, and upon the feel of *g* loads that are equal and opposite to the lift used in acceleration.

How the Aircraft Acts in the Stall. When the stall actually occurs, there should be no mistaking the event. Airspeed and control effectiveness will drop sharply with the rapid drag rise. The nose will start to drop, and more back pressure will serve only to aggravate that condition. Turbulent air from the stalled airfoil will buffet the airplane structure, causing a shaking vibration that can be felt through the airframe, seat, and pilot's controls. If the entire airfoil were to stall at the same airspeed, these sensations would be even more pronounced; but it then would be too late to take any action except a full stall recovery.

The designer of the aircraft arranges for the stall to occur progressively from the wing roots out to the tips, to reduce the abruptness of the stall and provide some aileron control as the airplane begins to stall. This may be done by "washout," twisting the wing to provide less angle of attack at the wing tips than at the wing roots, or by adding spoiler strips to the leading edges of the inboard sections of the wings, causing the disruption of airflow at a lower angle of attack in these areas. The aircraft will begin to buffet as the stall begins to develop, alerting the pilot to the onset of marginal lift conditions. The stall warning light or horn will be activated even before the buffet, also to alert the pilot.

When fully stalled, the aircraft becomes a falling object with some residual forward velocity in addition to that provided by propeller thrust. It is a curiously shaped object, though, and is not by any means "uncontrollable." Control surfaces lose their effectiveness in a definite order: *ailerons, elevators,* and *rudder.* They regain effectiveness in just the reverse order, *rudder, elevators,* and *ailerons,* just as on takeoff.

Some aircraft have slots in the leading edge of the wing and restricted elevator travel, so that it is impossible to get a prolonged full stall of the airplane over the wing ahead of the ailerons. These aircraft give good aileron control in the "full" stall; but they are not often used as primary trainers, in which capacity they might engender bad habits.

The rudder control is the least affected in the stall and will assist in bank as well as directional control. The elevator remains effective to lower the nose, where it is only aiding gravity; it cannot raise the nose without the airfoil's lift.

The nose drops in a stall because the aircraft designer has located the center of gravity slightly ahead of the center of lift, this difference being compensated by airflow across the horizontal stabilizer during flight. If left to its own devices after a stall, the aircraft will accelerate like any falling body and the acceleration will be aided by any propeller thrust in the nose-down attitude. As soon as flying speed returns, the resulting lift and correct airflow over the tail will tend to raise the nose. If the stall-producing situation has been removed, the aircraft will have made its own recovery. If the stick is held back, however, as soon as the elevator becomes effective enough to impose too much load on airfoil lift, the stall will recur. That is what happens when a high-speed stall is not relieved by easing off the causative back stick pressure. The pronounced shuddering in continued high-speed stalls is the result of a continuing series of stalls, partial recoveries, and stalls again.

Stall Recovery. Although the trainer will effect its own recovery, even from such aggravated stall conditions as a spin, too much altitude is lost in the process for the pilot to rely too heavily on automatic recovery. The *standard stall recovery* procedure is designed to restore controlled flight with the least possible sacrifice of altitude. It consists of a *positive forward movement of the control stick,* accompanied by smooth application of *maximum allowable engine power.* The wings, if banked, are leveled with rudder and ailerons. Some caution must be used as to rate of roll attempted at that point.

The phrase "positive forward movement of the control stick" needs some qualification. The object here is to get the nose in a shallow or moderate dive attitude as quickly as possible, *unless* flying speed returns before that attitude is reached. In a high-speed stall, of course, flying speed was there all the time; and as soon as some back pressure is released, the stall will cease. No throttle is needed for recovery from high-speed stalls, unless the airspeed at entry was only slightly above normal straight-ahead stalling speed. Some aircraft respond well to "popping" the stick briskly to just forward of neutral for stall recovery. In other aircraft, with quicker recovery characteristics, a few items might be lifted from the floor by negative *g* forces as the stick is "popped." The best all-around technique is a moderately rapid forward pressure, with its effectiveness judged by rate of nose travel and the degree with which positive elevator control returns. If excessive negative *g* forces develop from the pitch-down movement, the pilot should reduce his amount of forward pressure. He is starting to fly again. The rate at which control returns will depend, to a large extent, on the pitch attitude at time of entry, and also on whether the stall was complete. If the exercise involved only an *approach to the stall,* a modified stall recovery will be made with full control of all flight surfaces available during the entire recovery. If the recovery is from a partial stall, control will return earlier than for a full stall.

During practice stalls, recovery is easier, and the airplane may be in a better position for the next stall of the series, if airspeed is built up in a moderate dive angle recovery. The student (and the instructor) should not forget that one of the objectives of the stall series is to learn to recover with minimum loss of altitude, and to become familiar with that type of recovery so that panic will not delay the pilot's reactions when the altitude actually is not there to spare. To recover with minimum altitude loss, the recovery dive angle should be as shallow as possible, the wings should be level as soon as possible, and the dive should be abandoned for level flight as soon as possible—all without incurring too much drag penalty due to acceleration loads.

The customary word of caution about engine torque seems necessary. Flight controls should be fairly well trimmed for the condition preceding a practical stall entry. A rapid application of throttle for recovery will produce yaw tendencies to the left. If yaw is allowed to develop at stall speeds, a rolling tendency will accompany. Rudder will be found to be the most effective control for both yaw and roll until the ailerons come into service.

SLOW FLIGHT

There are occasions other than a minimum-altitude stall recovery when the aircraft must be maneuvered at speeds just above stalling—late go-arounds, short-field landings, and maximum-performance aerobatics, for example. Practice of *slow flight* will ensure a good "feel" of the aircraft in this condition: the *minimum controllable airspeed.* The airplane is flying at the maximum angle of attack, just below the critical angle of attack. To enter slow flight, ease the throttle back so that airspeed decreases to a point slightly above the stall. Maintain altitude by gradually increasing pitch as the airspeed decreases. When just above the stall, readvance the throttle to stabilize the airspeed. Use elevator trim as needed. Experiment with gentle turns and variations of pitch and power, trying to keep just enough airspeed to avoid an actual stall. Lower the nose only as far as necessary to recover at the first appearance of stall buffeting. Flying speed should return by the time the nose reaches the horizon. If a partial stall is not encountered occasionally during slow flight, the airspeed probably is being kept too high to get value out of the maneuver. The stall warning horn or light can be expected to be on continuously.

SPINS

A spin is autorotation of a stalled aircraft around a vertical axis, with substantial loss of altitude in each 360° of turn.

The building of "spin-proof" airplanes has progressed to the point where applicants for private and commercial pilot certificates no longer are required to demonstrate their proficiency at spin recovery. The light-plane trainer must be forced into a spin, will recover unaided within a couple of turns if the controls are released, and will recover within one turn or less after the controls are neutralized. Most high-speed aircraft also have excellent spin recovery characteristics. Those that do not have good characteristics usually require a modified recovery technique. Yet, the teaching of spin recovery should remain a firm part of pilot training. In addition to the fact that the best-mannered light plane can be brought out of its spin a little faster assisted than unassisted, the spin recovery is the best possible training for remaining oriented under stress and potential confusion.

Practice Spin.   Before entering an intentional practice spin, clear the area below carefully. Close the throttle, and raise the nose above normal climb attitude during the last of the clearing turns. As the first stall warning buffeting occurs, apply rudder briskly as far as the control will travel and bring the stick straight and fully back in a smooth manner. Hold both stick and rudder fully with the spin until ready to start recovery, because partial release of controls may permit a partial recovery and a secondary stall. Take care to avoid introducing aileron, which may increase rate of rotation in the spin or cause erratic rotation. Because of the large fore and aft movements required of the stick during entry and recovery, the student pilot is often guilty of inadvertent aileron movement.

Some aircraft will oscillate during spins. The nose will rise during part of a revolution, and the rate of rotation will be slower at that time; then the nose will drop at an increased rate of rotation. Recovery can be started at any point of such oscillation, and the most effective point varies among different aircraft types. In most cases the oscillations will diminish after the first turn or two.

The standard method of recovery from spins (Figure 8.9) is:

First. Apply full rudder briskly against the direction of spin rotation.
Second. Move the control stick briskly straight forward of neutral.
Third. As the rotation stops and control feel returns, neutralize the rudder, correct the wings to level bank, and recover smoothly from the dive.

As in the case of stall recoveries, there are some finer points to spin recoveries. The most effective spin recovery for any aircraft is the one recommended in the appropriate flight handbook. Some aircraft respond best to rudder and elevator control applied almost simultaneously; in others, the rate of rotation should be perceptively slowed (or even halted!) by rudder action before forward stick is brought in. Normal control returns to most light planes almost immediately after the forward stick part of the recovery. The pilot must be alert for returning pressures on his controls, or else some uncomfortable negative g and yaw will result. But the part of the spin recovery requiring the most "feel" is the pull-up after flying speed returns. To recover from the dive with minimum loss of altitude, it is important that the airplane be kept just comfortably outside the accelerated stall condition through the entire pull-up. This is a bit of a touchy job when airspeed is low, but that is when the most good can be done. If the airspeed is allowed to build up, excessive *g* loads must be imposed later to save altitude. An occasional accelerated stall probably will be encountered during practice of minimum altitude recoveries. Relax back stick pressure at the first stall indication, and no ill effects other than slightly increased altitude loss should occur. Practice spins in light aircraft should be planned to leave at least 3000 ft. of altitude available after recovery.

Inadvertent Spin and Recoveries.   If the spin was accidental, circumstances may be quite different from those in the typical practice spin. Because most modern aircraft do give ample stall warning and do not spin easily, an accidental spin most likely will start from an extremely nose-high attitude with engine power or misguided rudder pressure supplying uncorrected yaw. From unusual attitudes of entry, and especially if the entry stall was an accelerated one, the start of the spin may be a gyration that is difficult to follow or analyze. The first step in such an event should be movement of the throttle to idle. To avoid aggravating the stall, all flight controls can be held in neutral until some pattern of rotation is apparent. As a general rule, retract landing gear and flaps as soon as there is time. If altitude permits (and altitude *should* permit when one goes about courting accidental spins), the controls can be moved to the normal spin entry position, with rudder applied in the direction of existing rotation. The gyration should quickly become a normal spin, from which recovery will be routine. And if altitude does not permit? If skilled in recognizing stalls, you can prevent an accidental spin.

Perhaps training programs develop a little too highly the "one turn or two turns and standard recovery" outlook of student pilots toward spins. The recovery method must be learned; but, as you gain experience at spins and various types of slow flight, you should acquire an ability to ease the aircraft out of an *incipient* accidental spin without going the whole route. During the first quarter turn or so, before the autorotation and the excessive angle of attack which characterize the spin are well developed, the stall probably can be broken by correcting yaw with opposite rudder and *easing* the stick toward a nose-down attitude. Rolling tendencies must not be resisted with ailerons. The roll usually helps to get the nose down for recovery. Throttle should not be advanced unless you know *exactly* what you are doing. In fact, the best thing to do with the throttle is to retard it to idle. The uncorrected yaw caused by a blast of throttle at this point may be all the help that the spin tendencies require to take over. The amount of airspeed that engine power can contribute during the recovery is negligible. The chances are that airspeed already is adequate; you probably had to "snap" your aircraft into the spin situation through accelerated and uncoordinated flight above the straight-ahead stalling speed. Under these conditions, there should be enough airspeed to enable you to regain control, although the aircraft may exhibit an apparent inclination to spin.

Inverted Spins.   These usually are not pleasant to perform and have been barred from most training curricula. Any modern trainer will have to be forced

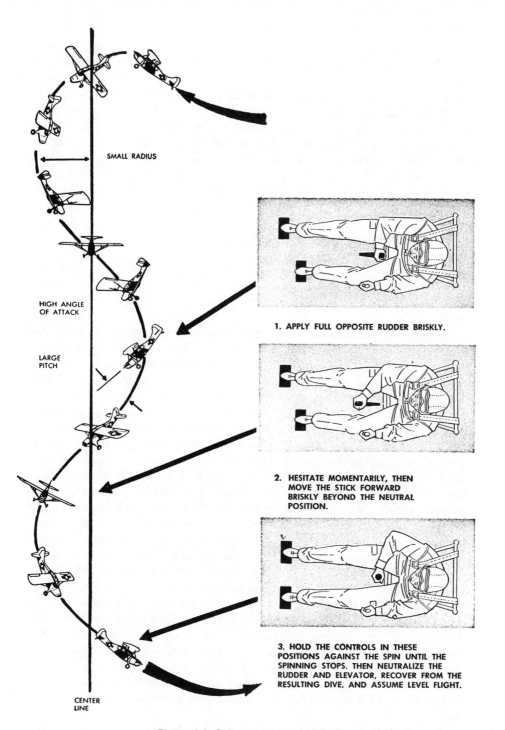

SMALL RADIUS

HIGH ANGLE
OF ATTACK

LARGE
PITCH

1. APPLY FULL OPPOSITE RUDDER BRISKLY.

2. HESITATE MOMENTARILY, THEN
MOVE THE STICK FORWARD
BRISKLY BEYOND THE NEUTRAL
POSITION.

3. HOLD THE CONTROLS IN THESE
POSITIONS AGAINST THE SPIN UNTIL THE
SPINNING STOPS. THEN NEUTRALIZE THE
RUDDER AND ELEVATOR, RECOVER FROM THE
RESULTING DIVE, AND ASSUME LEVEL FLIGHT.

CENTER
LINE

Figure 8.9. Spin recovery procedure.

into the inverted spin by holding full forward stick and either rudder as the aircraft stalls, preferably in an inverted or very nose-high position. The stick must be held forward to maintain the spin, or it will degenerate into a normal spiral or normal spin. If there is any doubt, the inverted nature of the spin can be recognized by the pressure exerted against the seat belt by negative g. Two or more negative g may be experienced in the spin, and this may lead to headaches and bloodshot eyeballs for the daring aviator. Recovery usually will result when the forward stick is released, but keeping the rudder in with the spin and pulling the stick smoothly straight back should immediately convert the spin to the normal erect type, from which the usual recovery can be made.

RECOVERY FROM HIGH-SPEED DIVES

Thus far, several undesirable results of too little airspeed have been considered. The airplane also has certain limitations at the high end of its speed range. These limitations are structural and are reached with excessive airspeed or excessive g loads.

The maximum permissible diving speed of any aircraft is listed in the flight handbook, possibly will be found on a placard on the instrument panel, and is represented on the airspeed indicator by a painted *red line* on the dial or a red needle on the instrument face. Beyond the listed airspeed, aerodynamic drag pulling at the flight surfaces creates dangerous loads on the aircraft structure. There is also a possibility that the propeller may overspeed because of the windmilling force of the airstream exceeding the engine load at the rpm setting.

Excessive g loads are caused by a too-rapid pull-up from a dive. The load of the airplane in acceleration must be borne by lift on the wing surface. Both wing and tail structure have certain carefully designed limits. The fully aerobatic military trainer is stressed to permit approximately 7.33 positive g at its maximum airspeed, with the ultimate failure point at about 11 g. This is more than adequate for any sensible maneuver, the average pilot being unable to withstand a sustained load of more than 4 to 4.5 g without impairment of faculties. The instrument that measures acceleration force (in centripetal acceleration only) is the *accelerometer,* or "g meter." The pilot also can learn to estimate g with fair accuracy by the pressure forcing him into the seat, and by an appreciation of his own tolerance for g loads. By any method of measurement, the maximum allowable g should be avoided by good margin during any flight in turbulent air. Sudden *gust loads* from turbulence will add sharp stresses of up to 3 or 4 additional g without warning. Another limit, not appreciated by many experienced pilots, is imposed if the pull-up also involves a rolling maneuver. For rolling pull-ups the stress limits are reduced to approximately two-thirds of their normal value, because the rising wing must bear the load imposed by the rotation as well as the pull-up.

The damage to the aircraft from excessive speed or excessive g load may not be immediately apparent if the wings and tail remain attached, but the damage may be found on postflight inspection, in the form of wrinkled skin, ''popped'' rivets, and twisted structure.

To avoid excessive *g,* the dive recovery should be smooth. Roll the wings level as soon as possible, and do not apply back pressure until the bank angle is 90° or less. Start the pull-out at least 10 to 20 knots before the limit airspeed is reached, depending upon dive angle, because the airplane will continue to accelerate in its nose-down attitude after pull-up is begun. The throttle should be idle if near limiting speed or if recovery with minimum altitude loss is the objective. The more airspeed the airplane has, the greater will be its centripetal acceleration during any degree of pitch change. Put even more simply, at higher airspeeds the pilot will have to ''pull'' more *g* to get out of the dive.

Ground Track Maneuvers

This group of maneuvers includes all those which require a specific pattern over the ground. They are important to give the airman a firm grasp of the effect of wind drift on his maneuvering judgment, particularly in maneuvering for landing.

WIND DRIFT

The airplane, moving through a mass of air, also shares the motion of the air relative to the earth below. Displacement of the aircraft's flight path due to its movement with the wind is known as *drift.*

Drift is not difficult to detect in level flight. In traffic pattern and ground track maneuvers, there is a runway or other line to be paralleled. If the aircraft heading is parallel, but the line moves closer or farther away, wind drift is the cause when the wings are level. To correct drift, make a coordinated turn of a few degrees into the wind. At the proper correction angle, the airplane will be pointing partially into the wind and away from the ground track. The point on the horizon toward which the aircraft is now traveling will appear off the normal straight-ahead position, as it would be in uncoordinated yaw. The airplane is not yawing but is crabbing into the wind to produce its straight ground track (Figure 8.10).

Drift correction grows more complicated during a turn, while the aircraft constantly changes its direction with respect to the wind. If you have had an opportunity to check the drift or other evidence of wind direction and velocity during level flight, you will know what to expect in a turn. That will help because there is no visible ground track to parallel while turning. You can visualize what that ground track ought to be and can observe the ground speed by watching the rate of travel past objects underneath. If you are not correcting for drift, the ground

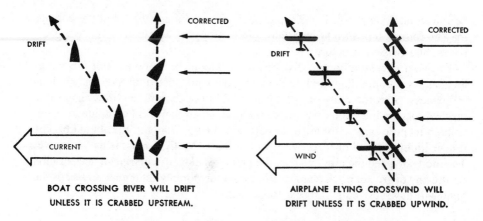

BOAT CROSSING RIVER WILL DRIFT
UNLESS IT IS CRABBED UPSTREAM.

AIRPLANE FLYING CROSSWIND WILL
DRIFT UNLESS IT IS CRABBED UPWIND.

Figure 8.10. Correcting for wind drift by crabbing.

track will be lengthened downwind and shortened upwind. The ground speed, if the wind is significant, will be appreciably faster downwind and slower upwind. When you start a turn in the traffic pattern or a ground track maneuver, you have an objective—a place that you plan to reach with a particular heading on arrival. Through experience you know that a steep, a shallow, or a moderate turn should enable you to reach place and heading at approximately the same time. If it seems that this is not going to occur, wind drift is the probable cause.

This subject is vital because many a crash described in the newspaper as "the plane, turning onto its landing approach, was seen suddenly to dive into the ground out of control . . ." is simply the result of the pilot not appreciating the effect of wind to the extent that preoccupied with making good a landing pattern with reference to the ground, he tightened the turn to the point where he forced the aircraft into a stall too close to the ground to recover.

SLIPS

The slip is included as a ground track maneuver because the principal use of the slip is to shorten the distance of a power-off glide on landing, final approach, or turn to final approach. The slip does this by the simple expedient of adding the drag of the aircraft side through a controlled yaw.

An intentional slip, just like those encountered accidentally in turns, results from an excess of bank (aileron) for coordinated flight. Opposite rudder is used to keep the flight path of the aircraft aligned with the runway or other ground reference.

A *side slip* and a *forward slip* actually are the same maneuver from an aerodynamic point of view. The difference is in the position of the aircraft relative to features on the ground. When you slip so that the aircraft moves toward a desired

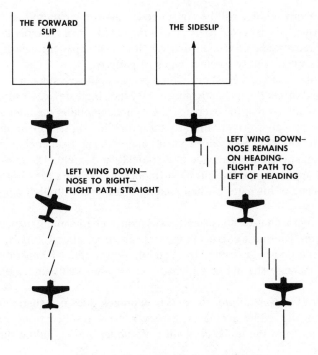

Figure 8.11. Forward slip and sideslip.

point, you are *forward-slipping*. Should you add additional bank, the aircraft will begin to move laterally from a track toward the point; you are then *side-slipping* (Figure 8.11).

The aircraft can be slipped in a turn as well as in straight flight. Excess aileron should be toward the inside of the turn, and compensating rudder toward the outside of the turn. If an inadvertent stall should then occur, any "snap" tendencies will tend to be toward the outside, reducing rather than increasing the bank angle.

The Rectangular Landing Traffic Pattern

A traffic pattern is a rectilinear course flown above the ground at a prescribed safe altitude, so that a landing approach may be accomplished in a systematic sequence. It is also a means of controlling aircraft returning to or leaving a field. There are many varieties of landing patterns. At this time only that most commonly used in light aircraft, the rectangular traffic pattern with a 45°

entry to the downwind leg, will be considered. This is the pattern that the light-plane pilot normally is expected to fly at a strange field, unless another type is announced to or requested by the control tower. It is a basic pattern, and the skill to fly it can be adapted readily to other types of patterns.

First to be discussed is the pattern with a power approach since it is the most widely used and offers the most advantages. By the single device of throttle manipulation, you can correct for overshooting or undershooting the intended landing point. For propeller-driven aircraft, the propeller slipstream improves the effectiveness of rudder and elevator control, and with any type engine, acceleration to go-around power can be made more quickly and safely. The use of power permits lower approach speeds, but most important, it allows the pilot to adjust the size and shape of his pattern when landing at airports with high-density traffic.

The power approach used consistently will reduce engine maintenance and extend engine life. The most excruciating punishment an aircraft engine takes in "normal" operation is a go-around from an idle-power glide. Temperatures over all the engine increase sharply and—what is worse—unevenly under such treatment.

In the event of power failure, the power approach does not guarantee a safe glide to the runway. More accidents, however, have resulted from misjudged approaches than from the unlikely chance of sudden power failure during the approach glide.

TRAFFIC INFORMATION

Before entering the traffic pattern to land at an airport, the pilot must know which runway is in use, what the wind conditions are, and which altitude to use. He should have reviewed airport data before beginning the flight, or en route, and already know the field elevation and the frequencies he may call on to obtain further information.

Traffic information is available through the control tower at tower equipped fields. They furnish wind direction and velocity, runway in use, altimeter setting, information on other traffic, and instructions for entering traffic. Those lacking towers are frequently served by Flight Service Stations (FSS), whose personnel can provide landing information, but may not direct traffic. At airports which have neither tower nor FSS, there are often Fixed Base Operators (FBO's) who operate a two-way radio called UNICOM and can furnish traffic information and other services, such as altimeter setting (and ordering a taxicab).

If the aircraft has no radio, or there is no radio equipped facility in service on the ground, the pilot must obtain his information from what he can see. Each airport has a traffic circle, which will have a wind sock, wind "tee," or tetrahedron, to tell the pilot the wind direction and the appropriate landing direction. If traffic is right hand on one or more runways, L-shaped markers within the circle will indicate it. If the field is lighted, the traffic circle is also illuminated at night.

A tetrahedron or wind "tee" may be locked to indicate desired landing direction, but a wind sock will always swing free, providing an indication of whether there is a crosswind, how strong the wind is, and whether it is gusty or steady. The wind at traffic pattern altitude may be quite different from that on the ground, particularly if there is a low level inversion. The pilot may find himself beginning final approach with a tailwind, which switches to a headwind for landing.

PATTERN ENTRY

Without radio communications with the ground, the pilot uses the last available altimeter setting, and flies across the field about 1000 ft. above the normal traffic pattern. He then maneuvers so as to enter traffic on a course to intercept the downwind leg at a 45° angle, at speed and altitude appropriate to his aircraft. If he has tower instructions or other radio information, he enters the traffic pattern accordingly, having completed his before-landing checklist.

The best assurance of a good landing is a good traffic pattern. It should be flown as nearly as possible the same way each time. Speeds, altitudes, power settings, and points for making changes should be standardized, so that the only variable left to contend with is the wind. Variables will therefore be drift corrections, angles of bank, and power settings. A typical traffic pattern, for the most typical of all light planes, is shown in Figure 8.12.

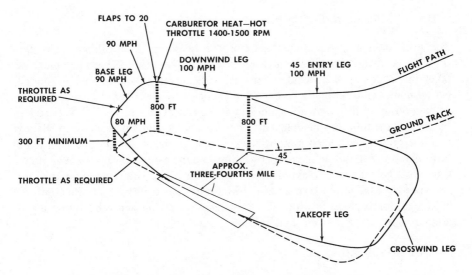

Figure 8.12. Rectangular traffic pattern, Cessna 172.

DOWNWIND LEG

The downwind leg sets up the landing. It should be far enough from the runway to allow comfortable turns to base leg and final approach. Prelanding checks are begun or continued. Airspeed is adjusted and power set to maintain the desired value. Course is maintained parallel to the runway. Landing gear is extended, power is reduced to a predetermined value opposite the intended landing point, altitude is held until speed is reduced to that desired for the descent, and then pitch is adjusted to maintain the glide speed. Initial flap setting is made, and elevator trim is adjusted. A sharp lookout for other traffic is maintained, and landing sequence is established.

BASE LEG

With calm or light wind, the turn to base leg is begun when the intended point of landing is about 45° behind the wing tip (7:30 o'clock). When a substantial headwind component is expected on the final approach, the turn is begun earlier, and rollout on base leg occurs with a drift correction established, so as to track over the ground perpendicular to the runway. Once rolled out on base leg, a judgment is made concerning the altitude, and the throttle may be adjusted to increase or decrease the rate of descent. Flaps may be further extended, in accordance with the manufacturer's recommendations and with need. Descent should be so controlled as to arrive at the turn to final approach with 300 to 400 ft. of altitude above the runway. Speed should be maintained constant at the recommended value.

TURN TO FINAL APPROACH

The turn to final approach has the purpose of aligning the aircraft with the runway, so that a straight-in final approach may be made. The wind must be taken into account. A tailwind on base leg indicates that the turn must be started earlier than would normally be expected, in order to avoid too steep a turn or an overshooting of the extended centerline of the runway. A headwind on base leg means that the turn may be delayed somewhat and need not be very steep. In any case the pilot should keep judging the outcome of the maneuver and "play the turn" so as to roll out on the proper heading, on the extended runway centerline. Care must be exercised to avoid steep turns, especially with crossed controls, at low speeds. If the turn does not place the aircraft in a position to allow a comfortable final approach to landing, the pilot displays his air sense by flying a go-around or missed approach, and coming around to try again.

FINAL APPROACH

The final approach turn positions the aircraft on a flight path aligned with the centerline of the runway, aimed at the selected touchdown point. Immediately after

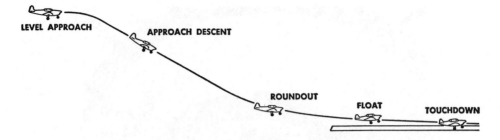

Figure 8.13. Final approach extends from line-up to touchdown.

rolling wings level, the pilot makes the final flap setting, and completes his before-landing checklist. The forward reference point in the windshield is placed on the desired landing spot (or even with it, if crabbed into a crosswind). From there to the flare point a nearly constant attitude is maintained, with the throttle used to control the airspeed.

With light or calm wind, maintaining the final approach is a matter of minor adjustments of pitch and power to maintain desired glide path. When the wind is blowing, the picture changes.

A strong headwind as the final approach begins may be cause for an immediate increase in power to carry the aircraft down the normal glide path. Somewhere in the vicinity of 100 to 200 ft. above the runway, the effect of friction at the surface causes a decrease (sometimes substantial and rather sudden) in the velocity of the air over the ground, often accompanied by turbulence. A sudden decrease in ambient headwind component causes an accompanying decrease in relative wind velocity, and a sudden decrease in indicated airspeed. The loss of speed causes a marked decrease in lift, and the aircraft may settle rapidly toward the ground or obstructions. The pilot responds quickly, increasing both pitch and power, to carry the aircraft to the selected landing point.

A crosswind component for the final approach and landing requires special skills also. The best approach is with the aircraft crabbed into the wind just enough to maintain the desired path over the ground. At 50 to 150 or 200 ft. above the runway, before the landing flare is begun, the pilot applies enough rudder correction to align the fuselage with the runway, then banks into the wind sufficiently to prevent drift across the runway. This is actually a very precise sideslip into the wind, with the objective of rolling the wheels on the runway in the direction of the runway, with no sideways skipping motion on the surface at touchdown.

A nearly constant attitude and constant airspeed on final approach help to assure a good landing. The point on the runway which does not move up or down in the windscreen is the point at which the aircraft will reach the ground if no changes are made in attitude and airspeed. Figures 8.14 and 8.15 show visual clues as observed from the rear seat of a tandem seat trainer. With side-by-side seating, or with a crosswind, the picture changes laterally. Constant attitude and

Figure 8.14. Runway appearance. *Upper:* For a constant-angle glide path, the aiming point and the horizon will appear a constant distance apart on the windshield. *Lower:* Distance between aiming point and horizon changes, as does the shape of the runway, for different approach angles.

airspeed permit trimming the airplane so that only light control pressures are needed.

Landing

Every landing is a challenge to a pilot, and no two are ever exactly alike. Even the most experienced and jaded pilots find landings enjoyable. To make the transition from flight to rolling on the ground smoothly and safely requires skill and frequent practice.

FLARE

A safe, smooth landing requires that the touchdown be made from a slow rate of descent, on the main landing gear, with the wheels aligned so as to roll straight

Figure 8.15. Obstacles in the glide path. If the trees appear to move up toward the aiming point, the airplane will not clear them.

down the runway. The transition from final approach to this condition, ready to touch down on the runway, is accomplished in the *flare*. This maneuver consists of applying exactly enough back pressure on the control stick or wheel to round out to a height just a few inches above the runway, applying whatever crosswind correction is needed, as described above.

Care must be exercised to apply enough back pressure to stop the descent at the desired height, but too much back pressure will cause the airplane to balloon, or start back up again (Figure 8.16). This tendency is most likely if the final approach is a very rapid descent, with pitch up for the flare delayed. Excessive *g* force is then required to avoid striking the runway, and when the descent is stopped, holding the back pressure will initiate a climb. On the other hand, allowing the aircraft to strike the runway prematurely will cause it to bounce back into the air, with the landing still to be made. A crosswind at this point will

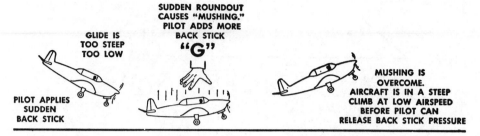

Figure 8.16. The story of a "balloon."

further complicate matters, so the best next move is to add full power and climb out for another try.

Once the flare is complete, a judgement about the actual height above the runway must be made, based on the visual clues. Height can best be judged by looking well down the runway ahead. If a bit too high, a little relaxation of back pressure will allow a slow descent to position closer to the runway. If much too high, a go-around is the recommended maneuver.

FLOAT

It is desirable to "float" some reasonable distance after the flare, before touchdown. This allows the aircraft speed to dissipate before the wheels touch and permits refining the attitude for landing. The desired landing attitude (something like that of a duck, to touch down in a nose-up attitude) is maintained. One viewpoint worth remembering is that the nose wheel is not for landing; it is a training wheel to be used for maneuvering on the ground. Power should be reduced slowly, remembering that increased back pressure on the elevator control will be needed to maintain desired height and attitude as thrust is reduced. Ground effect (see Chapter 2) plays a part in almost all landings, and prolongs the float. In no case should the pilot become impatient and force the airplane on to the runway with the elevator control. This will surely result in a bounce and a few more minutes flight time for another circuit of the pattern (even if it doesn't damage the nose gear or the propeller).

TOUCHDOWN AND LANDING ROLL

With the correct flare attitude established and power reduced almost to idle, the aircraft slows down so that in the desired tail-low attitude (could be three point in a tail dragger) the lift produced by the wings will no longer support the weight of the airplane. Consequently it settles to the runway. The final touchdown can be cushioned by continuing to increase pressure on the elevators.

Once the main wheels are on the ground, the throttle is closed and elevator

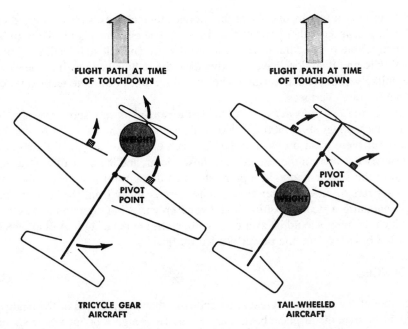

Figure 8.17. Why the tricycle gear is inherently stable at touchdown, touching down with a crab.

control is used to allow a gentle touch down of the tail wheel or nose wheel. Directional control is with the rudder, and crosswind correction (aileron up on the upwind side) may be required. The desired path on the runway is straight down the centerline. On some aircraft, the nose wheel steering is directly and inflexibly connected to the rudder pedals, so that if rudder were being held to correct for a crosswind, it must be returned to center at the moment the nose wheel touches down, to avoid an unscheduled lateral excursion of the airplane on the runway.

Directional control is critical in the initial portion of the ground roll, particularly on tail draggers. Figure 8.17 illustrates the tendency of a tail wheel aircraft to ground loop because of the location of its center of gravity, whereas the tendency of a nose wheel airplane is to straighten out when the main gear touches on landing. There are aircraft without rudders, which rely entirely on this straightening tendency in a crosswind landing. They are steered by a control wheel only, by ailerons in the air and the nosewheel on the ground.

STALL IT IN?

While it is desirable to land at a slow speed, in most cases it is not a good idea to land in a full stall. In a stall, the nose usually drops more or less abruptly, and even from only a few inches above the ground, unnecessary stress is placed on the landing gear and other parts of the airplane. The airplane also is likely to be less controllable in direction if landed in a full stall. The full stall landing from a normal flare and float does have the advantage that a bounce back into the air is quite unlikely. This characteristic makes it a sometimes desirable technique for landing on rough surfaces.

Planning the Touchdown Point.   Unless the runway is very short, you should plan to touch down about 500 to 1000 ft. from the approach end. This may be increased if there is an obstacle at the approach end. The aircraft is aimed at a point somewhat short of the desired touchdown point, and the float after flare will carry it to the selected area. The point at which the aircraft is aimed is the one which appears to move neither closer to nor farther away from the aircraft, but which maintains a steady position in the windscreen. Wind direction and velocity, which the pilot should be aware of, will determine the distance out at which to put the base leg, and the power needed on final.

BOUNCES

When rate of descent is too great at touchdown, the spring action of the landing gear will bounce the airplane back into the air. How high a bounce depends on descent vertical speed, margin above stall speed, attitude, and control positioning. High vertical rate, high approach or flare speed, and elevator-up inputs all contribute to the height of bounce (Figure 8.16).

The real problem with a hard bounce usually is that the aircraft ends up several feet above the ground at undesirably low air speed, with a surprised and somewhat slowly-reacting pilot at the controls. The safest action is to apply full throttle for a go-around, lower the nose just enough to start gaining airspeed, and fly another circuit for another try. The highly experienced pilot, with enough runway remaining, however, can recover from the bounce by judicious use of the controls (including throttle) to reestablish the correct flare, and continue for a safe landing.

Trying to land in too flat an attitude is the best guarantee of a bounce. The wings will probably be lifting yet, so that the bounce is accentuated. On a tail wheel airplane landing in a flat attitude, the bounce is almost straight back into the air. A tricycle gear airplane can make a more spectacular bounce. The nose gear hits first and bounces back into the air, just as the main gear hits. About that time (a bit too late) the pilot has raised the elevator, and aerodynamic forces (like lift) assist the upward leap. The pilot observes that he is too high for comfort and pushes over, then either flares too late again, or neglects to flare at all, and the second bounce is worse than the first. The nose wheel strikes harder each time, and will likely break off on about the fourth impact.

Bounces are minimized by proper flare and float, touching down above stall speed, but in a tail-low attitude. Going around is the best reaction to the first bounce.

POWER-OFF APPROACHES

The power-off approach is a good test of a pilot's feel for his airplane and his judgment of the effects of wind and attitude on his flight path.

When practicing power-off approaches, just as for every landing, standardizing the traffic pattern will help greatly. The power-off phase may begin at any selected point in the pattern, but for any given selected point, the airspeed, altitude, distance from the landing area, and course over the ground should be the same for each approach. The only adjustments necessary are then for wind.

At the power reduction point, the throttle is moved to idle, the altitude is held until glide speed is reached, and the elevator is trimmed for the new configuration. The nose is allowed to drop just enough to maintain the airspeed, which from then on to touchdown is controlled by pitch only. If flaps are to be used, they are extended according to when their effect is needed. Final setting is made as soon after rollout on final approach as possible, so that steady glide attitude and airspeed can be established.

If it appears that landing will be short, the flaps may not be fully extended, or perhaps not even be used. Glide distance can be extended by pitching down to increase glide speed, particularly with a headwind on final approach. Landing long can be counteracted by slightly reducing glide speed, earlier use of flaps, and forward slip. Less desirable are *S* turns on final approach.

GO AROUNDS

There are occasions when an approach may not terminate in a landing—there is an obstacle on the runway, there is another airplane too close ahead for safety, the crosswind is too strong, a bad bounce occurs, or for one of many other possible reasons. The approach must then be converted into a climb-out. The first step is to apply climb power, the next is to establish climb attitude, and then to configure the aircraft for climb (flaps to take-off setting, trim reset, wheels retracted when assured the airplane will not touch down). Important throughout the maneuver are maintaining attitude and airspeed, and avoiding obstacles. Once established in the climb, a normal departure is made.

ADVERSE LANDING CONDITIONS

Landing conditions are seldom ideal, but sometimes all a pilot's knowledge and skill are demanded.

Strong Gusty Winds.   Use a power approach. Add one-half the gust velocity to the minimum approach airspeed, for example, for a 20-knot wind gusting to 30, add 5 knots. Plan on a low roundout to keep the airspeed high until near the runway. Partial flaps, or no flaps at all, will reduce the ability of quartering gusts to raise the upwind wing. In extremely gusty winds, tail wheel aircraft should be literally flown onto the runway in approximately level flight attitude, using engine power until after touchdown. Just after touchdown in a wheel landing, slight forward pressure on the stick will hold the aircraft on the runway and prevent skipping. Raise flaps before lowering the tail and be ready for positive rudder action to retain control.

Strong Crosswinds. Crosswind landings are discussed on page 425. Each make and model of aircraft has a crosswind capability demonstrated by the manufacturer and shown on a placard in the pilot's position or published in the Approved Flight Manual. The major limiting factor for adequate control in a crosswind landing is rudder authority. Landing with partial power and minimum flaps is helpful in crosswinds. The pilot who knows his own and his airplane's capabilities can deal effectively with the crosswind problem, or know when not even to try.

Wet or Icy Runways. Expect poor braking and do not land long. Test the braking action by gradually increasing the strength of brake applications. On ice or snow, be alert for patches of dry runway. If one wheel is sliding when it reaches good braking surface, directional control suddenly will become difficult.

Soft Terrain. The problem of landing on a soft surface is to prevent the landing gear from digging into the surface. A tail wheel airplane could go over on its back; a nose wheel airplane could snap the nose gear off. The proper technique is to maintain power throughout the landing, touching down as gently as possible. The tail wheel airplane should make a three point landing. The tricycle gear airplane should be landed tail-low, holding the nose wheel off the ground as long as the elevator is effective. Power is taken off slowly after the main gear touches. With a soft surface, braking should be minimal or not used at all. This technique also applies to tall grass or rough or rocky surfaces.

Short-Field Landings. A runway may be marginal in length for a given airplane to begin with, or its effective length may be reduced by some sort of obstacle at the approach end. To cope with this kind of situation, a power-on approach is made, using full flaps and the manufacturer's recommended short-field approach speed (usually about 1.2 × stall in landing configuration). The approach path is established to provide a constant angle of descent which will clear any obstacles. The aiming point will be short of the touchdown point, to allow for flare and reduced float. Depending on the steepness of descent, it may be desirable to use a short burst of power to assist in the flare. When nose or tail wheel is lowered to the surface, flaps are retracted, and braking is used as required. Having the flaps up for braking reduces the effect of lift produced by them and places more weight on the main wheels, which can then brake more effectively. If distance is critical, any aerodynamic braking available, such as cowl flaps, open door or canopy, or even shutting off the engine, can be used to help slow the airplane.

No-Flap Landings. On runways of reasonable length, most light airplanes do not need landing flaps to land safely. There is one school of thought that they should be used for every landing, another that they should not. The larger general aviation aircraft, which have high approach speeds that can be significantly reduced by use of flaps, use them.

CROSSWIND LANDINGS

The wind pays little attention to a pilot's desire to land in a given direction (that of the runway), so crosswind landings are demanded of every pilot sooner or later. The direction an aircraft moves over the ground is made up of two vectors, that of the airplane through the air and that of the air over the ground. The pilot, for comfort, aircraft controllability, and airplane structural reasons, would like the landing wheels to roll straight along the runway immediately upon touchdown. In order to do that, he must steer the airplane so as to have the fuselage aligned with the landing direction, while preventing any drift across the runway during and after touchdown.

There are three schools of thought on how best to accomplish this feat. The first advocates holding a drift angle of correction, or crab, until just before the moment of runway contact, then applying sufficient rudder to align properly as the airplane lands. This has disadvantages in that a sudden gust can disrupt the plan, and most pilots do not know exactly when the aircraft will touch down anyway.

Another, and for some airplanes the only available method, is to land the airplane in a crab. With center of gravity forward of the main gear (tricycle gear airplane), the momentum will swing the craft into alignment with the runway immediately after touchdown. Tail wheel airplanes using this method should have crosswind landing gear. Otherwise a ground loop may be difficult to avoid (Figure 8.17).

The third method is essentially a sideslip into the wind. At some time in the latter stages of the approach, the pilot uses the rudder to swing the fuselage into alignment with the desired landing direction and then lowers the upwind wing enough to prevent drift across the runway. The aircraft will land first on one main wheel, then the other. Care must be taken to neutralize the rudder at the moment of contact when lowering the third wheel (nose or tail) to the runway, if it is steerable by the rudder pedals.

For any given airplane, there is a limit in the amount of crosswind that can be handled. It may be significantly greater than the value which is given as the demonstrated crosswind capability in the aircraft data furnished by the manufacturer.

It varies with the skill of the pilot and the rudder authority of the airplane. The value given by the manufacturer will be conservative to allow for a low skill level.

After all three wheels are on the ground, the airplane must be steered in the desired direction using the rudder and nose steering, assisted when necessary by braking on the downwind wheel. The upwind wing is kept from rising by the use of aileron.

Taxiing.   Nearly all light aircraft can be taxied safely in a wind up to 20 knots or so. With high wings it is prudent to keep the ailerons positioned so as to hold the upwind wing down. (With a quartering tail wind, put the upwind aileron *down*.) When the wind is stronger, it is often necessary to have wing walkers, and perhaps someone to hold the tail down as well, particularly when turning or taxiing crosswind or downwind. Holding the elevator down while taxiing downwind will help prevent nose over.

After the Flight

AFTER LANDING

A cockpit "cleanup" is due after the landing roll is complete and the aircraft is off the runway. If taxi distance to the parking area is short, the check might be delayed until then. The essentials of the "after-landing check" usually are:

1. Carburetor heat—*off*
2. Fuel boost pump switch—*off*
3. Propeller and fuel mixture controls—*full forward*
4. Wing flaps—*up*
5. Cowl flaps—*open*

PARKING AND STOPPING

Once off the runway of a tower controlled airport, the pilot calls ground control for instructions for taxiing to his parking spot. He may be assisted by a "follow me" vehicle, which will have a sign on the back (illuminated at night). Where there is no tower, assistance may be available on UNICOM radio frequency from a fixed base operator, or from the flight service station if one is located on the airport. Signals used by the ground crew, if any, are the same as those used for taxi out.

Engine temperature should be allowed to stabilize before shutdown. For safety, it is a good idea to perform a magneto grounding check, as follows:

1. Rpm—1000
2. Magneto switch—*Left only* (right grounded out, rpm drops)
3. Magneto switch—*Left off, right on* (rpm remains the same)
4. Magneto switch—*Off* momentarily, then *both on* (engine stops firing completely with switch in off position, returns to 1000 rpm with both on)

This procedure verifies correct operation of the magneto switch and assures that the engine will not fire, causing personal injury, if the prop is swung by hand with the switch(es) off.

Shutdown. Before the engine is stopped, all radios and electrical accessories should be turned off to avoid damaging surges and spikes from the electrical generating system as the engine stops. The engine should be set at 1000 to 1200 rpm for shut down. If it does not have a mixture control, it is stopped by turning the magneto switch off. If it does have mixture control, it is stopped by setting the mixture control to FULL LEAN or IDLE CUT-OFF position. In either case, the throttle is not moved as the engine stops. After the engine stops turning, the ignition switch and the master switch are turned off, and the control lock(s) installed. Chocks are placed and parking brakes released.

TIE-DOWN

For the airplane to be left outside a hangar, tie-downs should be used if winds are forecast or possible above certain levels. Table 8.3 provides recommendations for representative aircraft.

Tie-down attaching rings are installed on most aircraft. Ropes, chains, or cables are used to secure the aircraft, by means of permanent anchor points in the parking area, or, on grass, by anchors which can be driven or screwed into the ground. Hemp ropes shrink when they become wet, so some slack should be left in them when they are attached.

TABLE 8.3. Aircraft Tie-Down Guide.

Type of Aircraft	Maximum Wind Velocity (knots) for Parking Without Tie-Down
Average light plane (up to 2000 lb. gross weight)	20
Four-passenger, single-engine civilian types and military primary trainer (up to 4000 lb. gross weight)	35
C-46, C-47, DC-3, etc.	45
Modern tricycle-geared 2-engine	60

Advanced Maneuvers

A concert artist spends countless hours practicing scales and exercises that audiences never hear. For a similar reason, the conscientious pilot devotes effort to master certain practice maneuvers. Ability to perform them properly is a matter of craftsman's pride, but the important benefits are in the sharpening of fundamental skill to the degree that the pilot can cope with unusual or unforeseen circumstances occasionally encountered in normal flight.

These maneuvers are termed "advanced" because they require a degree of skill for proper execution that a pilot normally does not acquire until he has soloed and has obtained a sense of orientation and control feel in "normal" maneuvers.

The student pilot and the private pilot learn basic flying and maneuvering of the aircraft, including stalls, steep turns, slow flight, and other maneuvers related to simple, safe air transportation. The commercial student and pilot go on to more complex operations and uses, which require practice and demonstration of higher skills. These include such items as the maximum performance climbing turn (*chandelle*), the *lazy-8,* and *spirals around a point.* Some of these maneuvers are acrobatic by FAA definition (intentional maneuvers involving an abrupt change in attitude, an abnormal attitude, or an abnormal acceleration), but the attitudes, airspeeds, and accelerations are all within the capabilities of any certificated aircraft.

The pilot who wishes, or is required by military training, to develop the highest levels of skill learns to perform acrobatics or aerobatic maneuvers. The catalogue of acrobatic maneuvers is bounded principally by ingenuity and aircraft design limits. Stalls and spins, previously presented as "confidence maneuvers," fall within the definition of acrobatic flight. Acrobatic flight is defined by Federal Air Regulations as "maneuvers intentionally performed by an aircraft involving an abrupt change in its attitude, an abnormal attitude, or an abnormal acceleration." Other maneuvers currently taught in the Air Force and Navy courses of instruction are the *loop, barrell roll, aileron roll, half roll* and *split S, Immelman, Cuban eight,* and *cloverleaf.* Also to be considered in less detail are the *slow roll, inverted flight, snap roll, vertical reversement, vertical stalls, English bunt, outside loop,* and *inverted loop.*

Acrobatics are an essential part of military flight training. Through acrobatics, the pilot learns to control his aircraft precisely through its entire range of airspeeds and attitudes. The sense of confidence thus acquired is fully as important as the stick and rudder skills. And, certainly not the least of benefits is that acrobatics add immeasurably to the joy of flying. The thrill of mastering movement and acceleration in flying is the same as that which sends people down ski slopes and sailing a skiff close to the wind. Mount Everest was climbed and pilots do acrobatics because of the challenge to man's skill and daring.

Acrobatic maneuvers should be attempted only in aircraft properly stressed for them. An aircraft with limit load factors of 6 positive and 3 negative *g* is considered stressed for acrobatics, but the real limiting factor is pilot technique. A loop can be done with less than 3 positive and with no negative *g*. Even helicopters and four-engined bombers have been looped without fatal results. It is also quite possible to impose sufficient *g* loads during loop recovery to deform or detach the wings of any "fully acrobatic" trainer. Certain maneuvers involving inverted flight can be accomplished only in specially stressed and modified aircraft; these maneuvers will be described only briefly.

The relationship between limiting speed and *g* loads will be explored at greater length in the next chapter. Here it shall be noted only that maximum obtainable or allowable *g* is dependent upon aircraft weight, center of gravity, speed, and configuration; and that published *g* limits are based on straight flight. For a rolling maneuver, the normal *g* limits should be lowered by approximately one-third.

The reader who has yet to enjoy acrobatic flight should not infer that normal acrobatic maneuvers must be meticulously planned and executed to remain within a narrow safe operating range. No maneuver in the military acrobatic curricula with less than 4 positive or 1 negative *g* requires special talent to complete. Loads outside that range produce evident discomfort in blurred vision and pressures against the seat or seat belt. *The beginner should have dual instruction* on any maneuver to be attempted until he acquires a sense of control orientation and an appreciation of speed and load relationship, and until he can do fundamental acrobatics with reasonable precision and with confidence.

Because of the wide variation in performance of different aircraft, precise recommendations for airspeeds and power settings cannot be given. The best source for specific information is the manufacturer's or the military service's flight handbook for the aircraft. Lacking these, the capable pilot can find his own speeds and power settings by experiment, discontinuing any maneuver that threatens to lead to trouble. The range of satisfactory entry speeds for any maneuver except snap rolls is quite wide. The power setting used for normal climb will serve for most acrobatics involving vertical ascent.

Clearing Turns. Make at least two steep 90° turns before each acrobatic maneuver to permit careful inspection of surrounding air space. Begin maneuvers as soon as possible after the turns. The lazy-8 and chandelle are in themselves clearing maneuvers. Federal Air Regulations prohibit acrobatics in control areas (federal airways), control zones (airport), over populated areas or open-air assemblies of persons, and at less than 1500 ft. altitude above the ground. *Look around as much as possible during any of these maneuvers, especially watching behind during the starting turns.*

THE CHANDELLE

The chandelle is a maximum-performance climbing turn, with precise entry speed and a precise 180° of turn (Figure 8.18). It is entered in a shallow dive

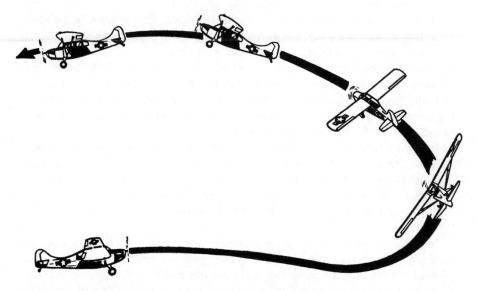

Figure 8.18. The chandelle.

about 20° to 30° below the horizon; select a road or other reference line on the ground, along which the maneuver may be oriented.

Engine rpm is monitored to insure that it does not exceed the red line; some power reduction may be necessary. The aircraft is banked 30° right or left after desired entry speed is reached; then back pressure is applied to make a coordinated pitch-up, which results in turning as well. The angle of bank is held with gradually increasing pitch until 90° of turn is completed. Power is increased as the red line limitation permits. From 90° to the 180° point, the angle of bank is smoothly and steadily reduced, and airspeed decreases constantly. The rollout is completed precisely at the 180° point, with airspeed just above stall. The nose is lowered while holding altitude, and normal flight is resumed.

This is only one of the ways of performing the chandelle. Other ways may be favored by the particular instructor or flight school.

LAZY-8

The lazy-8 (Figure 8.19) is a maneuver designed to develop the coordination of controls through a wide range of airspeeds and altitudes, so that certain accuracy points in the maneuver are reached with planned attitude and airspeed. It is a practice maneuver without peer and is one of the finest ways for a pilot to acquire the feel of an aircraft new to him. It consists of two 180° turns in opposite directions. During the first 90° arc of each turn, the nose of the aircraft describes a climbing and then descending path above the horizon. During the second 90° arc of each turn, the nose of the aircraft describes a descending and then a climbing path below the horizon. To an observer viewing the maneuver from the side and level, the total flight path traces a figure-8 lying on its side. The horizon bisects the 8 from one end to the other.

To execute the lazy-8, select a ground reference line with which to align the maneuver. If no ground reference is available below, it can be omitted, but you must align the entry so that a prominent feature on the horizon is at right angles to the aircraft heading. Normal cruise power is adequate for the lazy-8. Entry airspeed should be above normal cruising, about the same as used in the chandelle.

Obtain the entry speed in a shallow dive and then pull the nose *straight up* to the horizon. At the horizon, begin a gradual climbing turn toward the 90° reference point selected during entry. The initial degree of bank should be very shallow, because time must be allowed for climb and descent during the 90° of turn. The first part of the maneuver is mostly a climb, with rate of roll very slight. The maneuver should be "lazy," and *g* loads throughout should be negligible.

Make the climb and descent toward the first 90° point symmetrical in pattern, so that the steepest point of the climb will be after 45° of turn. As airspeed decreases in the climb, apply the usual correction for torque. If the turn is to the right, considerable practice probably will be required before adequate right-rudder pressure is maintained. Coordination may be checked by the rate of travel along the horizon as well as by sensing of slip or skid.

Bank angle at the 45° point should be about 15°. From that point continue to

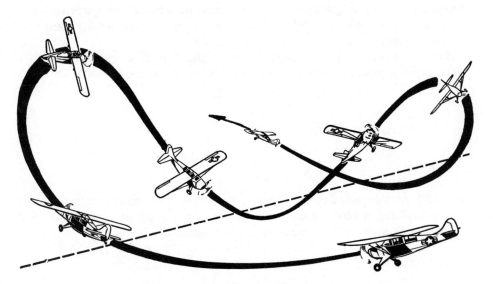

Figure 8.19. The lazy-8. *(Courtesy of US Air Force.)*

increase the bank angle and relax some of the back stick pressure to lower the nose toward the horizon exactly at the selected 90° point. Airspeed will continue to dissipate until just before the nose reaches the horizon.

Rate of roll and amount of back stick pressure together will determine where the nose of the aircraft crosses the horizon on its way down. Increasing the rate of roll or decreasing back stick pressure will bring the aircraft to the horizon earlier. Decreasing roll or increasing back pressure will carry the aircraft further along the horizon. Bank angle will be at its steepest on reaching the horizon. This bank should be 30° according to FAA standards. The bank angle should be predetermined as an accuracy condition for the maneuver. Airspeed as the nose passes through the horizon should be comfortably above stalling. It too may be an accuracy consideration, but too much attention should not be paid the airspeed until a good feel for the lazy-8 is developed and accuracy points can be made successfully.

As soon as the nose crosses the 90° reference point on the horizon in the descent, begin to roll slowly out of the turn. Again time the back stick pressure and rate of roll so that the arc below the horizon traces a pattern like that just completed above the horizon. When the aircraft reaches the horizon at the 180° point, the wings should just be coming to level, and the airspeed should be the same used for entry.

Precisely upon reaching the horizon, and without hesitation, begin a climbing turn in the direction of the original 90° reference point. Complete a second 180° pattern, identical to the first except that the turn will be in the opposite direction.

During the entire lazy-8 maneuver, the pitch and bank angle should constantly be changing. If a protracted pitch or bank attitude must be held in order to reach an accuracy point or to stay within airspeed limits, the planning of the maneuver has gone astray.

The lazy-8 may be flown with bank angles other than those specified above, again depending on the preference of pilot, school, or instructor.

Acrobatic Maneuvers

GENERAL

The maneuvers discussed in the following pages must not be attempted in aircraft not certified for acrobatic flight. They require the wearing of parachutes and should not be tried without instruction from a qualified instructor.

THE LOOP

The loop (Figure 8.20) is a 360° turn in the vertical plane. It is the most easily performed of all aerobatic maneuvers, for it involves a change only in the pitch attitude of the aircraft. The elevators are the basic control, and the ailerons and rudder are used only for maintaining a straight heading in coordinated flight. The power setting used for normal climb probably will fit the loop well. To insure precise control of heading, start the loop in alignment with a road or other straight line on the ground below. On reaching the correct entry speed, start the stick straight back with gradually increasing pressure. The wings should be level as the nose passes up through the horizon.

With the proper rate of pull-up, you will feel a definite but not uncomfortable seat pressure that results from *g* loads. A lack of seat pressure indicates the maneuver is proceeding too slowly, and too much airspeed may be lost in the climb portion. Any blurring of vision or evidence of high-speed stalling indicates an excessive stick pressure.

The decreasing airspeed in the climb will create a need for gradually increased right-rudder pressure to counteract engine torque. When the horizon ahead has disappeared beneath the nose, look out at the wing tips and keep them equidistant from the horizon. When a little past vertical flight, tilt your head straight back to use the approaching horizon for directional control and the planning of back pressure. Use rudder to keep aligned, and the ailerons to keep wings level.

Maintain back stick pressure through the entire loop maneuver, and keep the rate of travel of the nose around its 360° fairly constant. Some of the back pres-

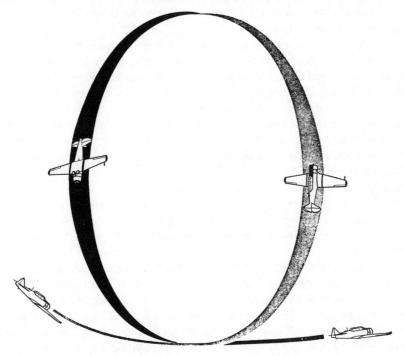

Figure 8.20. The loop. *(Courtesy of US Air Force.)*

sure exerted in the climb must be relaxed at the slow airspeed "over the top" of the maneuver. At any time that stall buffeting occurs, relax back stick pressure enough to stop the stall, but maintain positive seat pressure.

As the nose of the aircraft passes through the horizon from the inverted position, the airspeed will be increasing. Additional back stick pressure will have to be reapplied to keep the nose moving around its 360° at a fairly constant rate. A little extra back stick pressure—just comfortably outside the stall region—while below cruising speed will prevent acquiring too much speed in the recovery dive. As the nose nears level-flight attitude, ease off back pressure, and make a final check to insure that the wings are level. The loop should require 3 to 4 positive g and no negative g.

BARREL ROLL

The barrel roll (Figure 8.21) is a maneuver in which the aircraft rotates 360° around the longitudinal axis while the nose of the aircraft describes a circle about a real or imagined reference point on the horizon. It is a precision maneuver and should be smooth and coordinated. The circle described should be of constant radius around the reference point, and the aircraft should arrive with prescribed attitudes at accuracy points to the sides, top, and bottom of the center reference.

The barrel roll can be started from a wide range of speeds. The usual recommendation is slightly above normal cruising. Engine power should be at cruise or at the normal climb setting.

Reach the entry airspeed in a shallow-dive attitude about 20° below the selected center reference point. Bend stick and rudder to produce a coordinated climbing turn toward the accuracy point about 20° to either side of the center reference. On the way up to the horizon, roll out of the orignal turn so that the side accuracy point is reached with wings level.

With no hesitation, continue the climb and the new direction of the roll-up through the horizon and around the circle to the top accuracy point, about 20° above the center reference. At the top point, the rate of roll should have carried the aircraft to a vertical bank.

From the vertical bank position, relax some of the back stick pressure, and increase aileron pressure slightly. Time the rate of roll so that the aircraft will reach the horizon inverted, with wings level, at the same distance on the opposite side of the reference point as it was when passing through the horizon on the start of the maneuver.

As the roll continues down through the horizon, some increased top rudder pressure and back stick pressure will be required. The wings should reach the vertical position with the nose pointed at the same distance below the horizon from which the maneuver was started. Difficulties with "dishing out" of the maneuver in the bottom half usually are due to insufficient rate of roll and to insufficient elevator force shortly after crossing down through the horizon.

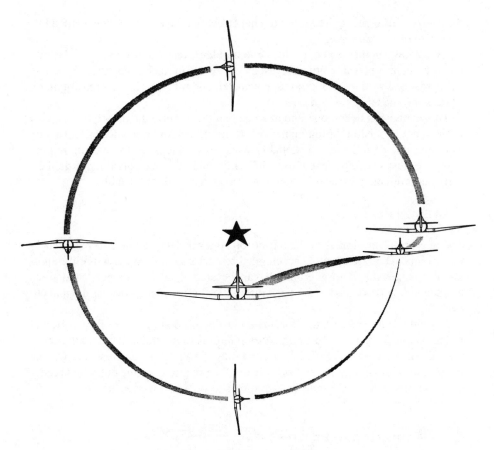

Figure 8.21. The barrel roll.

The barrel roll is completed when the aircraft, wings level, reaches the side where it first crossed up through the horizon.

Use the rudders only as necessary to keep the maneuver coordinated. Positive *g* forces should be slight throughout the barrel roll, but a definite seat pressure should exist at all times.

To coordinate the rate of roll with arrival at accuracy points, you must understand the effect of each control in any attitude. At the position in vertical bank above the center reference point, excessive back pressure between the vertical bank and the next 45° of roll can result only in crossing the horizon wide of the next side accuracy point. But as the aircraft continues its roll 45° past vertical bank, the effect of increased elevator pressure is directed predominantly down toward the horizon, thus decreasing the distance traveled along the horizon. If you accelerate the rate of roll proportional to the elevator presusre, you can meet the horizon in correct attitude, but the net effect will be a maneuver executed in a tighter than normal circle.

Examining another aspect of the same problem, one will see that a rate of roll too fast for the elevator pressure used will result in each quarter of the circle being reached with the bank too far advanced, and the "circle" decreasing in radius as the maneuver progresses.

This matter has been gone into at some length, because the principles are very basic to any precision rolling maneuver. If the student of acrobatics has difficulty in timing rate of roll, he can probably work his problem out by interrupting the maneuver just where he first senses difficulty, and there exaggerating or deliberately subordinating the use of the control that seems to be at fault.

AILERON ROLL

The aileron roll is similar to the barrel roll, but no effort is made to reach accuracy points. The rate of roll is accelerated by increased aileron pressure throughout the maneuver. The aileron roll can be started at any airspeed and pitch attitude that will not lead to airspeed difficulties before completion of the maneuver.

Normal rudder pressure to coordinate must be used at the start of the roll, and rudder also will be required as the aileron returns to neutral at the completion of the maneuver. If the aircraft has a good rate of roll, elevator pressures will be negligible. A small amount of back stick will be required during entry and recovery, and this is relaxed while the aircraft is at or near inverted flight.

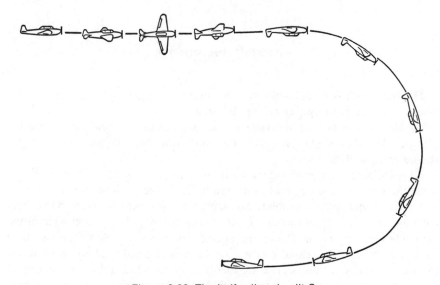

Figure 8.22. The half-roll and split-S.

HALF-ROLL AND SPLIT-S

The half-roll is just what the name implies. Neutralizing stick and rudder while inverted will halve any roll at the midpoint. A half-roll also can be executed from the inverted position to level flight.

The half aileron roll usually precedes the split-S (Figure 8.22). Start the aileron roll 10° to 20° nose-high and with 25% above stalling speed. As the rate of roll is reduced at midpoint, some forward stick may be necessary for a moment in order to prevent falling below the horizon before the wings are level inverted.

A split-S is exactly the second half of the loop. If you use throttle to enter the half-roll, retard it in the split-S to prevent accumulating too much airspeed. *Do not enter a split-S without plenty of altitude.*

VERTICAL ROLL

One or more aileron rolls (Figure 8.23) can be executed in a vertical climb, provided the aircraft has a good rate of roll, low power loading, and an adequate maximum limiting airspeed. Entry speed should be at least above that used for the loop.

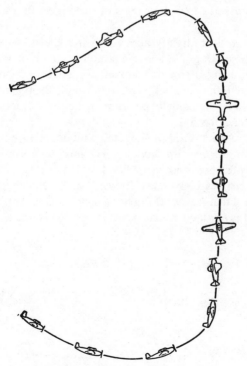

Figure 8.23. The vertical aileron roll.

As soon as entry speed is reached in a dive, come back smoothly but rapidly with the stick to establish a vertical climb. When vertical, start the roll with rather drastic aileron pressures and accompanying rudder. Release almost all the rudder pressure after entry.

The ground will be a little difficult to keep track of in this maneuver. Progress of the roll must be determined from a view of the horizon rather than the ground beneath.

To discontinue the maneuver, introduce back stick pressure and slow the rate of roll. The back stick will start the nose down toward the horizon. As the aircraft reaches level flight, decrease the rate of roll and the back stick pressure, or stop the roll as soon as the wings are level inverted; leave the maneuver with a split-S.

SLOW ROLL

The slow roll (Figure 8.24) is a maneuver in which the nose of the aircraft is made to rotate 360° *on* a point, rather than around the point as in the barrel roll. The maneuver has lost its popularity because it is uncoordinated and uncomfortable, and because modern high-speed aircraft do not slow-roll well. The slow roll does demonstrate very well the function of each flight control in rolling attitudes, because control pressures must be exaggerated to meet the requirements of the maneuver.

Begin the slow roll at slightly above cruising speed, with the nose of the aircraft pointed about 10° to 20° above a prominent point on the horizon. Start it to either direction, using aileron and a touch of rudder for coordination. As soon as the roll has started, the nose will want to move off the point in the direction of roll. It must be held on the point by pressure on the ''top'' rudder, opposite to the aileron that continues the roll.

As the roll progresses toward the inverted attitude, release back stick pressure and introduce a definite *forward* pressure. The nose will drop below the point if this forward stick pressure is not maintained while inverted. After passing the inverted position, progressively relax forward stick, and the top rudder again becomes the control that staves off gravity and holds the point. This bears repeating: The slow roll is *not* a coordinated maneuver. When the aircraft is at or near

Figure 8.24. The slow roll.

either vertical bank position, considerable top rudder will be needed to hold the point. You will definitely sense the skid that results. Complete the slow roll by neutralizing aileron and rudder as the aircraft again nears level flight.

As long as turns are not being attempted while inverted, do not worry about the controls "exchanging their functions" as the nose of the aircraft rotates in the roll. The elevator does not act like the rudder at any time. It moves the nose of the aircraft toward the top or bottom of the aircraft as the pilot sees top and bottom from within the cockpit. The left or right rudder will always tend to move or hold the nose of the aircraft to the pilot's left or right, respectively. How these movements affect the aircraft's relation to outside references is a matter for separate judgment.

IMMELMAN

The Immelman turn is a combination maneuver, consisting of the first half of a loop followed by a half-roll to level flight. The maneuver is named after a German World War I ace, who first perfected the maneuver to gain altitude and reverse direction by 180° simultaneously (Figure 8.25).

The Immelman requires a few knots more airspeed for entry than does the loop but otherwise proceeds in the same manner—to within about 20° of inverted flight. Before the nose of the aircraft reaches the horizon, inverted, relax back stick pressure. At the same time move the stick well to either side to start the rollout. The use of the word *move* in connection with aileron technique is significant. Because of the low speed and the fairly rapid rate of roll required, most aircraft require almost full aileron deflection for the Immelman rollout.

Figure 8.25. The Immelman.

The aileron executes the half-roll portion of the Immelman. Regard the rudder not as a coordinating control but as an independent control *to keep the aircraft headed straight* while the ailerons are rolling it to level. Rather extended rudder travel may be required near the completion of the maneuver.

The back stick pressure was relaxed when the rollout started, to forestall pulling the nose down below the horizon. As the aircraft nears level flight, add back stick again. Complete the maneuver at slow airspeed, and keep the pitch attitude well above the horizon to maintain level flight at the high angle of attack.

The Immelman requires approximately 4 positive *g* to complete. There should be no negative *g*. A common error in learning the maneuver is the use of forward stick while inverted to "hold the nose up." If the rollout is commenced above the horizon and is fairly rapid, the nose will not have to be "held up."

CUBAN-8

The Cuban-8 (Figure 8.26) is another combination maneuver, combining in this case the loop and a diving aileron roll. Each half of the "8" can be visualized as an Immelman turn, completed 45° below the horizon. Entry speed and power setting should be the same as used in the loop.

Start the maneuver like a loop, but relax all back stick pressure when the nose is about 45° below the horizon, in inverted flight after going over the top. Immediately perform a half-roll to place the aircraft in an upright 45° dive. Continue the dive until entry airspeed is obtained, and then repeat the maneuver.

Do the half-roll "on a point," a reverse heading from that used for entry. The half-roll can be a slow roll or, more frequently, a rapid aileron roll with some added rudder to stay on the point.

The Cuban-8, as described here, is known in Britain and on the Continent as a variant of "the spectacles," a combination of an erect and an inverted loop. Labels for various types of loop also are unstandardized.

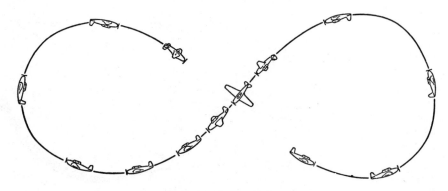

Figure 8.26. The Cuban-8.

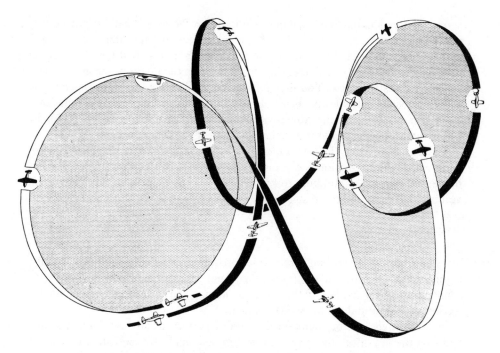

Figure 8.27. The cloverleaf. *(Courtest of US Air Force.)*

THE CLOVERLEAF

The cloverleaf (Figure 8.27) has in it part of the loop and part of the lazy-8. It is composed of four very steep climbing turns, with a change of direction of 270° in each turn.

Enter the cloverleaf from a speed a little less than that needed for a loop. From a dive, pull the nose of the aircraft straight up to about 70° nose-high pitch attitude. Then begin a roll in either direction, so timed that the horizon is reached in inverted flight after rolling 90° from the entry heading. During the 90° change of direction, continue the climb with back stick pressure, so that the climb attitude becomes almost vertical before the bank angle reaches 90° and the nose starts down toward the horizon.

The wings must be precisely level as the horizon is reached in inverted flight, and the rate of roll may have to be modified during the maneuver so that this will happen at exactly 90° from the entry heading. In this maneuver the same skills are required in timing of back stick and aileron pressures as are exercised in the barrel roll and the lazy-8. The cloverleaf is perhaps the most exacting test of the three.

At the horizon inverted, add back stick pressure sufficient to keep the aircraft just comfortably outside the high-speed stall region as you pull the nose through a vertical dive and into a normal climb attitude for the next ''leaf'' of the clover. During the dive, increased speed will demand increasing back stick pressure, and you will feel additional g loads as the pull-out continues. If the pull-out is not kept tight immediately after crossing down through the horizon, the aircraft will pick up excessive speed. You then must choose to accept the higher speed (and resulting loss of altitude) or incur additional g loads to complete the pull-out.

Enter the next leaf of the clover without pause after the dive recovery, with the roll to right or left as chosen for the first part of the maneuver.

The cloverleaf may be performed in any aircraft stressed for normal acrobatics. A dual introduction is especially advisable, because you can lose orientation easily during the climbing roll when learning the maneuver. An inadvertent vertical stall could be the result.

INVERTED FLIGHT

Inverted flight for more than about 10 sec. requires a specially equipped aircraft. It must have an extra oil sump or hopper and an extra fuel tank or hopper to supply lubrication and fuel to the inverted engine. The carburetor must be the injection type if altitude is to be maintained.

Inverted flight can be maintained for brief periods in any acrobatic aircraft. Half-roll the aircraft to the inverted position, and apply forward stick to keep in level flight or a shallow glide. If the carburetor is a float type, retard the throttle to idle position. If the carburetor is an injection type, retard the throttle toward idle to reduce the lubrication requirement. Watch the oil pressure gauge carefully, and recover when the indication drops below safe limits.

Maintaining altitude requires a steep negative angle of attack, for the wing does not produce lift efficiently off the camber of its undersurface. Hold the nose well above the horizon with forward stick. Turns while inverted require control pressures contrary to all habit, but a shallow turn produces no particular discomfort. To turn to the right (as you view the world) requires left aileron and right rudder with added forward stick.

Secure loose objects in the cockpit before inverted flying, and tighten the safety belt and shoulder harness.

SNAP ROLL

The snap roll (Figure 8.28) is a high-speed stall maneuver in which severe yaw from the rudder produces a roll. The roll can be held for 360° of rotation or stopped at half-roll by neutralizing controls before 180°.

The snap roll is peculiar to the light-plane trainer or special arobatic aircraft. Larger aircraft, such as fighters, are restricted from intentional snaps. Start the snap roll from a speed of about 30 knots above stalling. There is one best entry

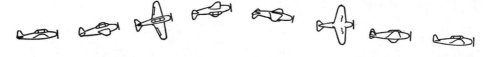

Figure 8.28. The snap roll.

speed for each aircraft at about that figure. A slower entry speed will not leave sufficient control to halt the maneuver precisely. A faster speed may result in a partial recovery while inverted.

To perform the full snap roll, bring the stick sharply back to its full rear limit of travel. As the stick comes back, introduce aileron in the desired direction of roll, and add full rudder travel briskly in the same direction. The result in some aircraft is quite breathtaking. The nose comes up sharply about 20°, and the roll occurs quickly. Hold entry controls. The rate of roll will slow as the roll nears completion. Near 360° of rotation, neutralize rudder and aileron, and retain only enough back stick pressure to keep level flight.

A half-snap can be done in many aircraft without using aileron. This makes the initial "snap" less violent and aids precise recovery inverted. Some aircraft will do an excellent half-snap to erect flight if back stick is accelerated at the top of a loop. Propeller torque effect supplies the necessary yaw.

VERTICAL REVERSEMENT

The vertical reversement (Figure 8.29) is a snap maneuver executed from a tight, low-airspeed turn. Added back stick and top rudder cause the aircraft to snap "over the top" into a turn in the opposite direction. Aileron only is needed to recover precisely in the new turn.

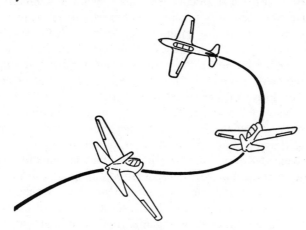

Figure 8.29. The vertical reversement.

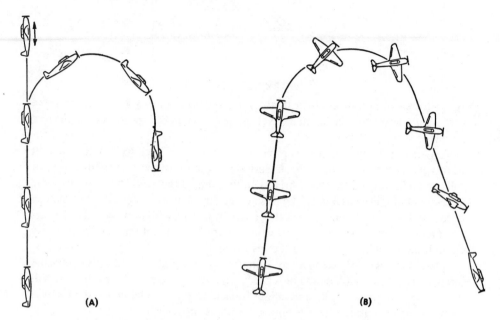

Figure 8.30. Vertical stalls. (A) A whip stall. (B) A hammerhead stall.

VERTICAL STALLS

Vertical stall maneuvers are of two types: the *whip stall* and the *hammerhead stall* (Figure 8.30).

The *whip stall* is not a recommended maneuver for any aircraft. It is sometimes encountered accidentally when the pilot exerts back stick pressure in a vertical or near-vertical climb near stalling speed for his aircraft. The aircraft slides tail-first momentarily, and then pitches violently forward and down. If you do not hold the stick firmly, reverse flow over the elevator may wrest the stick from your grasp and slam it forward. Should a whip stall be encountered, the best procedure is to do nothing except hold the stick firmly until the nose of the aircraft drops below the horizon. Then make a normal stall recovery.

The *hammerhead stall* also is entered from a vertical climb. It is not as violent as the whip stall and is a safe maneuver in an acrobatic aircraft if properly performed. If a whip stall appears imminent, it can be converted to a hammerhead. Near the stall, in vertical or near-vertical climb, apply almost full rudder travel in either direction. The rudder must be applied smoothly. As the nose drops toward the horizon, begin a quarter roll to level flight. Leave the stick about in neutral until sufficient airspeed is regained for the normal, nose-down, stall recovery.

Figure 8.31. The English bunt.

ENGLISH BUNT

The English bunt (Figure 8.31) is a maneuver for special acrobatic aircraft. It is a sort of inverted Immelman. The maneuver is performed by applying full forward stick in level flight or in a climb at near stalling speed. The forward stick is held until the aircraft is flying inverted on a heading 180° from entry. A half-roll is then made to level flight.

INVERTED AND OUTSIDE LOOPS

The normal loop starts with a climb from level and erect flight. As illustrated in Figure 8.32, the aircraft can be put through a "360° turn in the vertical plane" via three other routes.

The pilot can half-roll from level flight, split-S, and finish with an Immelman turn. This is an inverted loop. It is a practical maneuver for any acrobatic aircraft, the main precaution being not to develop too much speed in the split-S part of the maneuver.

The other two forms of the loop are strictly for the airshow artist and his specially constructed aircraft. That aircraft must be stressed for extra negative *g*

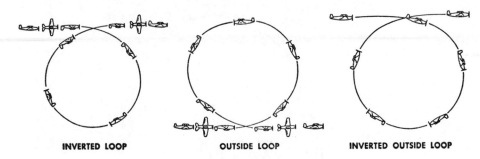

INVERTED LOOP   **OUTSIDE LOOP**   **INVERTED OUTSIDE LOOP**

Figure 8.32. Loops.

loads, must get fuel and oil to the engine while it is inverted, must have large control surfaces, a low power loading, and good "lift" from the wing when inverted.

The *outside loop* begins from level flight, inverted. The pilot applies forward stick pressure all during the maneuver, until he half-rolls to level flight again at completion. The *inverted outside loop* begins with a push-over from level flight, as for a steep dive. Both these maneuvers put stress on the pilot as well as the aircraft. He can tolerate about 3.5 negative g for protracted periods, although he is likely to get headaches and bloodshot eyes. If he suffers more than that figure for longer than about 10 sec., he will begin to lose vision.

IF THE ENGINE QUITS

Many thousands of flights are made every year in gliders and sailplanes, which of course have no engines. They make equal numbers of unpowered landings. Should the engine of a single-engine airplane fail to produce power in flight, it also becomes a glider and must make an unpowered landing.

Most light-aircraft engines are very reliable when properly maintained and operated. When such an engine stops running, it is more likely than not that the pilot failed, by allowing the engine to run out of fuel or oil, the fuel to become contaminated by water or dirt, the carburetor to ice up, or some similar problem to develop.

Except when a total mechanical failure occurs or fuel is exhausted, the loss of power is likely to be gradual rather than abrupt, or at least the impending trouble will be forecast by rough running or by indications of the engine instruments. The pilot should keep in mind that a landing is likely to be better and safer, even on an unprepared surface, if some power is available to allow choice of a landing area, and to maneuver to a final approach. A serious engine problem is his cue to find a safe landing area as soon as possible.

During training, the student has an opportunity to practice forced landings, which are quite similar to the real thing, until they are broken off by the instructor at a safe altitude, for a go-around. A student must never practice forced landings when he is flying solo. Even with an instructor on board, real forced landings grow out of practice forced landings embarrassingly often.

Unless the engine is shut off completely for the glide portion of the forced landing, the glide ratio will not be the same as for total power loss. Complete engine shut down is not recommended in practice. The small amount of power produced in idle will decrease glide angle by a small amount, but the student will still get good practice in estimating glide angle and distance, and develop judgment. "Clearing" the engine at intervals will assure that the go-around will be successful.

GLIDING DISTANCE

A beautiful landing area is no good to the pilot with a failed engine unless the normal glide of his airplane will carry him to it. Each light-airplane owner's manual provides a chart that shows how far the plane can glide power off at its optimum glide speed, for altitudes it can attain in normal flight. A glide ratio of 8 to 1, for instance, with a beginning altitude of 10,000 ft. above the terrain, would permit a still air glide in any direction of 15 miles, or a total area from which to pick a landing area of over 700 square miles. Even from 5000 ft., 180 square miles is available. It must be remembered that the area into the wind is much more limited; a downwind glide would carry one farther.

To determine for future reference the approximate distance which he can glide, power off, the pilot can proceed as follows:

1. Fly power on, level flight, constant altitude, at normal glide speed.
2. Note the point or line in the windscreen through which the horizon is observed.
3. Maintain normal glide speed with power off.
4. Again note the point or line in the windscreen established by (2) above. It rests on the ground point to which the airplane will glide, power off, with no wind.

The wind does not blow at constant speed and direction at all altitudes above a given area. Shifts and changes in velocity occur with changing altitudes. Vertical air currents also play an important part in glide distance. The pilot should never count on reaching a landing area at extreme range.

CHOICE OF FIELD

The first consideration in selecting an emergency landing area is minimizing the risk of injury to occupants of the airplane. Airframe damage is acceptable if personal injury can be avoided. This dictates selection of an area that can be readily reached in a normal glide, with adequate clearance over obstacles on the approach, sufficient length to at least get greatly slowed down before the end, as free as possible from gullies, stumps, and boulders. The approaches and orientation should allow landing into or across the wind, not downwind. Wires and fences are not especially visible from 2000 or 3000 ft. up, so initial planning should consider that they are there.

A prepared landing strip is best, of course, but knowing the terrain can often help in selecting good landing areas. Hayfields generally have a smooth surface, as do many pastures. A golf course fairway could be a good choice if there aren't too many players on it. A growing or freshly harvested grainfield is a reasonable choice, but land parallel to existing furrows. The same precaution applies to freshly cultivated land: Try to avoid plowed ground. In the desert, dry lakes are often the best areas available. In the cold season, especially in the northern climes, a frozen lake may present the best possible surface. Where there is nothing but forest, a thick growth of uniformly small trees may be the best available spot. Ditching in a shallow water near the shore of a lake is preferable to landing in an area of large trees or huge boulders, but pilot and passengers must be prepared to evacuate the plane immediately, before it sinks.

The single engine pilot, properly prepared for a forced landing emergency, constantly knows which direction the surface wind is blowing, and so which way he'd want to land in the event of engine failure. Surface wind of any consequence is often quite evident—drifting smoke, blowing dust, leaf color on poplar trees, "amber waves of grain," wind streaks on pond surfaces, flags, hanging clothes, cattle face downwind in a cold wind, etc. It is wise also to know the approximate elevation of the surface. In the absence of roads, fences, buildings, etc., it may be quite difficult to judge height above ground.

PLANNING THE APPROACH

Flying a precise pattern to a forced landing is not necessary or even desirable. The objective is to place the aircraft in a position at an altitude from which a safe power-off landing can be made. To be conservative, it is better to be slightly close or slightly high when ready to turn to final approach. You will then have some latitude for adjustment (Figure 8.33). Landing gear should be down unless the landing surface is either water or a very short field. On a smooth surface, aircraft damage will be minimized; on a rough surface the landing gear will absorb much of the energy of deceleration, protecting the aircraft occupants.

If you approach "key position" with too much altitude, you can lose it in a spiral, establish a downwind, or make S-turns approaching the key position. Try to avoid placing the base leg wide to compensate for extra altitude; the familiar pattern is easiest to fly.

At low altitude, it is too late to do much maneuvering. As a general rule, do

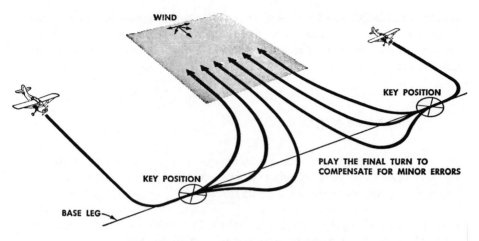

Figure 8.33. Play the turn to final approach.

not plan more than 90° of turn below 500 ft., and try to complete all turns with at least 200 ft.

"Play" the turn to final approach with a wide or tight turn. You can also "play" the landing area in an open field by planning originally on the center and changing to either edge of the field during final turn or on approach.

Lower landing flaps, if you need them, either during the turn or just after the turn to final approach, unless it then appears that the glide to the field may be short with the flaps down. The advantages of lowering them early are that they give a constant sink rate on final approach and indicate whether or not the flaps will operate in plenty of time to lose altitude by other means if necessary.

Pilots of fighter-type aircraft usually use a 360° overhead approach for forced landings. Their high sink rate makes it difficult to plan arrival at the "key point" by other means. The same pattern *can* be used in light aircraft, approaching the field in the direction of intended landing at an altitude 1200 to 1500 ft. above the terrain. Alignment slightly to one side of the runway centerline will permit you to keep the end of the runway in sight, so that you can make the first turn at the proper point. Delay the start of the 360° pattern until the aircraft has passed over the first third of the landing area if there is a significant headwind. Many pilots prefer to lower wing flaps completely at the start of the 360° overhead forced landing approach. This makes the pattern a bit tighter, but it provides a constant sink rate during the entire pattern and makes judgment of glide distance easier.

SETTING IT DOWN

Turn off master and ignition switches before touchdown. Jettison or open the canopy. Make the roundout and touchdown just as you would for a normal landing. If a crash appears to be coming, occupants should lean against the shoulder harness to take up any slack, protecting the head with some type of padding if possible. If the surface is at all level, the deceleration will probably be surprisingly slight.

Flight at Night

Full utilization of aviation requires that it operate at night as well as in the daytime. Private pilot, commercial pilot, and airline transport pilot certification all require night experience, as do the military pilot ratings. In some countries, single engine aircraft may not fly at night (or on instruments). That restriction does not apply in the United States. The military services require that aircraft used for night flying be equipped with the following:

1. All flight instruments and communication/navigation equipment required for flight under instrument flight rules.
2. Position, landing, instrument, and cockpit lights.
3. A dependable flashlight for each crew member.

Federal Aviation Regulations, for civil aircraft not to be used for hire, require the following for night visual flight rule operation:

1. Instruments and equipment only as required for daytime operation.
2. Approved position lights.
3. Approved rotating or flashing beacon.
4. Adequate source of electrical energy for all installed electrical and radio equipment.
5. One set of spare fuses.
6. Landing light if passengers are carried for hire at night.

The military services plan that aircraft may be operated at night over unfavorable terrain and in marginal weather for visual flight. The civilian pilot is allowed more discretion in the equipment of his aircraft, but should use that discretion and confine his night flying to local operations near well-lighted airfields and environs if the aircraft is equipped with the minimum list.

TECHNIQUE AT NIGHT

While many of your most delightful experiences in flying will come from flight at night, and while the ability to fly safely at night doubles the utility of your aircraft, there are some differences in the demands made upon the pilot. None of these demands is unusual, and all are well within the capability of the all-round pilot who is flying a properly equipped aircraft in good condition, and who has prepared carefully in advance.

Night flying does pose these problems:

1. The horizon is not visible, or is very indistinct.
2. Obstructions, clouds, and other aircraft are harder to see.
3. It is easier to get lost.
4. Forced landings on dark, unprepared surfaces require more luck than you can properly count on.

These problems all stem from man's limited ability to see in darkness, discussed in Chapter 7.

The Night Horizon. Remember, *you cannot fly by feel alone.* At night, the horizon must be provided by some small amount of light, or by instruments within the aircraft. Usually, pilots can rely on both. Because of the lack of depth perception at night, a good altimeter, properly set, is a primary requirement. An accurate airspeed indicator and a gyrocompass (or at least a well-dampened magnetic compass) are also necessities. There is no greater asset in the cockpit at night than an attitude indicator. (See Chapter 9.)

During takeoff, the problem is to remain on a straight line during the ground run. Look well ahead of the aircraft to the end of the runway or to a light at the end of the flying field. Landing lights, though not essential for takeoff, can be helpful and serve to identify your aircraft to others in the area. As the aircraft lifts off, airspeed, orientation, and direction are the vital concerns. Maintain a steady airspeed and a consistently steady rate of climb to insure clearing all obstructions and to prevent inadvertent flying back into the ground once the lighted airfield has passed behind. The horizon provided by a city's lights, by remnants of daylight, or by the night sky will not always be adequate. Inevitably at some time or other you will suffer disorientation or vertigo when there is no clear horizon in view. It need not be a problem with proper cockpit instruments and the ability to use them with confidence. Keeping cockpit lights low, using red lights for cockpit illumination and map reading, and careful adaptation of the eyes to darkness will all help.

Approaches and landings are much the same as in daytime, though you should avoid steep turns and make traffic patterns accurately and precisely. Some pilots prefer to land on lighted runways without using landing lights; when haze, smoke, rain, snow, or fog obstruct visibility, reflection renders landing lights more of a hindrance than a help. By maintaining accurate approach speed, and by looking well ahead of the aircraft to improve depth perception, roundout and landing will present no serious hazard. A little power carried through roundout and touchdown will insure more consistently smooth landings.

Obstructions, Weather, and Other Aircraft. When flying cross-country, observe IFR minimum en-route altitudes. Then the altimeter and good navigation will remove the hazard of obstructions. Careful planning, alertness, and the ability to do a 180° turn on instruments are the solution to encountering weather when VFR at night. As for other aircraft, remember that green lights are on the right wing-tip, red on the left, and white on the tail—so that you can tell the direction in which the other aircraft is flying. As long as there is relative motion between you and the lights of the other aircraft, you are not on a collision course. If there is no apparent relative motion, take evasive action immediately.

Getting Lost.   Lighted roads, cities, and villages look more alike at night, though aeronautical charts depict the shape of larger cities quite accurately. Careful preflight planning, using the basic principles of dead reckoning described in Chapter 11, is by far the most reliable means of knowing where you are at all times. Radio navigation can also be very helpful. Without careful observation of your position and headings after take-off, it is quite possible to get lost right over your own airport, particularly if it is near a large city with extensive suburban areas.

Forced Landings at Night. Engine failure is an extremely unlikely occurrence these days, but when it comes at night the only safe place for a pilot to put down his aircraft is on a prepared and lighted runway. Over any except the best terrain, a parachute is considered by some an absolute necessity during night flying.

Experienced pilots usually study the terrain past the ends of the runway of any field from which they operate. Such reconnaissance could save lives should an engine quit shortly after take-off, with no recourse but to make an immediate forced landing.

Roads and highways can be readily identified from the air at night, even when it is fairly dark. Once lined up for a landing, the landing light will be a big help. Remember, however, that autos and trucks not only have the right of way, but also are tougher than airplanes, and the drivers won't be expecting you. Moonlight is often sufficient for selecting fields in which a forced landing could be made at night, especially those containing ripe or freshly harvested small grain. In the high plains area there are many thousands of these strips of stubble visible from the air at night. Again, landing lights could be a big help for final approach and landing.

NIGHT LIGHTING OF AIRFIELDS AND TERRAIN

The following colors of lights are standard for airfield installations and surrounding obstructions.

Runway lateral limits	White (intensity may be adjustable on request)
Taxiways and ramp area	Blue
Threshold lights defining ends of usable landing area	Green

Obstructions	Red (High structures such as radio towers may have blinking lights that appear to climb and descend. Construction on ramp may be bordered with flare pots.)
High-intensity approach lights and lateral limits of runway overrun	Amber
Tetrahedron	Green outline and red center stripe
Wind sock and landing tee	White floodlight
Beacon lights	
Land	Rotating white and green
Attended water airport	Rotating white and yellow
Hazards (such as mountain tops)	Rotating red
White strobe lights	Flash rapidly toward approach end of runway
Approach lights	Variable intensity white or amber

Airfield beacon lights revolve at 6 rpm. They are located atop high structures on their airports. The white or yellow beam at a civilian airfield is a single beam. The white beam at a military airfield is split, giving the observer two successive flashes.

Flying the Light Twin Aircraft

By nature of its design, the thrust of a multi-engine airplane is divided into two or more packages. When one of those packages is not producing power, the performance of the aircraft is reduced accordingly. In the case of a light twin, cutting the available power by half reduces a snappy performer to a very sick machine demanding all the knowledge and skill the pilot can muster in some situations.

There is no requirement in certification of aircraft in the light twin category for any given margin of performance after engine failure. It therefore behooves the pilot to know his craft's capabilities in critical conditions and fly it accordingly. For instance, loss of only half the power on takeoff may just delay the inevitable forced landing a little longer than total engine failure, if one compares a twin powered by two 180 horsepower engines with a single having the same total horsepower. Given the reliability of modern light aircraft engines, such a situation occurs very rarely, but the pilot had better not have grown complacent.

Much has been made of the problems of "moving up to twins." With 100 hours or so—or less with good basic flying training, the "problems" are not unduly complex.

There are only three basic differences for the relative novice: (1) an increase in weight, with greater weight variation; (2) an increase in complexity with need to devote more effort to knowing the airplane and its systems; and (3) the management of unbalanced power, assuming engines are in wing nacelles.

WEIGHT VARIATION

With a gross weight of 5300 lb., the typical light twin is a light aircraft. Its maximum weight variation during flight due to fuel consumed with auxiliary tanks, allowing 10% for reserve, is about 14% of gross weight. There is also the problem of passenger and cargo weight location. With a roomier fuselage, it is critically important to be aware of the load distribution specified in the operator's manual so as not to exceed the limits, which can seriously affect controllability.

A greater weight problem, of course, is the difference between a light load in emergency and a heavy one—as is true in any airplane. There is no unusual flying problem associated with a larger weight variation. The principal problem is to be aware of it, and to know how it affects performance.

COMPLEXITY

There are two engines, controls, and engine instruments, instead of one. However, the throttle and propeller controls, and the engine instruments are designed to simplify the pilot's handling the engines individually or together. For substantial power changes, as in take-off, climb, descent, and landing, they are as one. The hydraulic, electrical, and fuel systems reflect dual or individual engine functions (e.g., on which engine is the hydraulic pump), but the problem is merely one of knowing the systems before flying the airplane.

UNBALANCED POWER

Airplanes have two engines to provide greater power rather than to provide greater reliability. But occasionally, engines do get sick. Being able to shut down one, or ease its load, is of course a distinct advantage.

The requirements for safe flight under one-engine-out conditions are similar to those for single-engine airplane flying: Know what the airplane will do, and practice enough to be able to get the performance built into it.

EMERGENCIES IN THE LIGHT TWIN

An abnormal situation is not always an emergency, depending on when it occurs. Failure of one engine as a twin breaks ground on take-off is an emergency, but if one fails in level flight it may mean only a decrease in cruise altitude and a little less fuel reserve on arrival at destination.

Engine Failure. In flying twins, this likelihood probably causes more pilot concern than any other, though each of the engines is just another engine. Sometimes a pilot will, and should, shut down an engine that is still producing power simply to save it from possible damage from low oil pressure, fuel or oil leak, high temperatures, etc.

Loss of Thrust. Loss of thrust on one engine is critical to a twin during and immediately after takeoff. Such a failure drastically reduces obstacle-clearing ability and increases time needed to accelerate to climb speed. Directional control is the principal problem, and the pilot must never allow the airplane to drop below *Vmc* (the *minimum control speed*) while establishing single engine climb speed.

Drag. A propeller being windmilled by the slipstream and thus turning a dead engine causes a very considerable drag. It is almost completely eliminated by feathering the propeller. A propeller which goes out of control and "runs away" produces the greatest drag situation, because its engine is driven at very high speed. There is also a possibility of destruction of the engine or dangerous loss of the propeller.

Effects of Unbalanced Thrust and Drag. The principal result is a rotational moment about the vertical axis (Figure 8.34). The loss of power of the no. 1

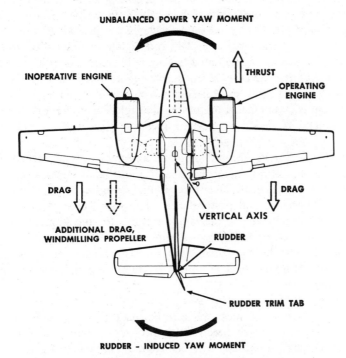

Figure 8.34. Forces and moments with engine failure.

engine causes an unbalanced moment acting counterclockwise around the vertical axis, assuming normal power from the no. 2 engine. The greater the power difference between the wings, the greater is the unbalanced moment.

The rudder counteracts the yaw moment. Right rudder in Figure 8.34 deflects it into the slipstream, which exerts a force that is translated into a clockwise moment to overcome the effect of unbalanced power. Trim reduces or eliminates the pilot's rudder effort needed. Placing the airplane in a slight bank toward the "good" engine also helps in maintaining directional control. An aft center of gravity (*cg*) shortens the moment arm of the rudder, reducing its effectiveness.

Minimum Control Airspeed (V_{mc}).   The maximum moment the rudder can exert depends upon the airspeed, since the vertical stabilizer and rudder are merely a flap-equipped symmetrically cambered wing that stands on end. The MINIMUM CONTROL AIRSPEED for any unbalanced power condition is the minimum airspeed at which the rudder can exert enough yaw moment, without stalling, to overcome the unbalance caused by the "good" engine putting out full power.

Critical Engine.   Each engine turning to the right is applying forces to turn the aircraft left. One of these forces is P factor, which effectively acts on the right side of the propeller disk. It is farther from the vertical axis for the right engine than it is for the left. The failure of the left engine produces greater unbalanced moment and therefore is more critical than failure of the right. So we say that the critical engine is the left one, and safety dictates that procedures used must account for its failure. The airspeed at which adequate directional control can be maintained with the left engine failed but not feathered, and with full power on the right, is therefore called the *critical single engine speed*, or V_{mc}. For engines turning left instead of right, critical engine would be the right one. The critical engine problem is counteracted by turning the right engine left and the left one right. Controllability versus airspeed is then the same for the loss of either engine.

Procedure with Engine Failure.   Should an engine failure occur before attaining V_{mc} on take-off roll, abort the takeoff immediately. If the decision point has been passed and it is possible to stop within the runway limits, or on the overrun, *STOP*. Experience and performance calculations before take-off dictate whether a stop can be made. If the failure occurs after breaking ground, above V_{mc}, maintain full power on the good engine, and fly the airplane level or with only enough climb to clear obstacles. Shut down and feather the faulty engine *as soon as it is positively identified,* in order to reduce drag and gain speed as rapidly as possible.

V_{mc} *must be reached and maintained* or the airplane cannot even climb safely over obstacles. The choice might have to be made to fly through the obstacle under control rather than lose directional control, or cut power and land short of the obstacle. During a go-around with one engine, the exact procedure may vary for different aircraft depending upon maximum power available, the drag of extended landing gear, and flap lift/drag characteristics. Generally, when going around with an engine failed, a safe climb is attained most quickly by this sequence:

1. Maximum power, maintaining Vx_{se}: or V_{mc}, whichever is greater, until past obstacles.
2. Adjust flaps for Vx_{se}.
3. Feather failed engine as soon as it is identified.
4. Retract landing gear as soon as it is evident that the wheels will not touch down on the runway.

Best Angle of Climb Speed. As discussed in Chapter 4, the best angle of climb speed is that speed at which the greatest ratio of vertical speed to horizontal speed is attained (V_x). This speed is desirable for clearing obstacles, but if it is less than V_{mc} in a twin, V_{mc} is used in the climb for safety in the event of engine failure.

Best Rate of Climb Speed. This is the speed at which the aircraft gains altitude at the greatest rate (V_y). It is usually somewhat slower for only one engine operating than for both. After obstacles are cleared with a failed engine, the V_y for engine out is used for continued climb. It is designated V_{yse}.

On some twins there is a single hydraulic pump, driven by the accessory drive of one engine. If the drag of the extended landing gear is greater than the windmilling prop, the prop should be allowed to windmill at high pitch (low rpm) long enough to retract the landing gear, before it is feathered.

A partial loss of power is not justifiable as a cause for shutting down an engine during the critical stages of initial climb and for avoiding obstacles after take-off. Any amount of power from an engine is better than none at this point, even at the expense of serious damage to the engine. Only complete loss of power or fire in the engine nacelle dictates immediate shut down.

Until you have V_{yse}, limit the climb to that necessary to fly safely over obstacles. When you have safe altitude and airspeed, retract any remaining flaps slowly to avoid stall or excessive settling. Reduce power to normal-rated or below if possible at the pattern altitude, establish a precise pattern if possible, and land—weather permitting.

Occasionally it happens that power loss occurs so late and is so extensive that the pilot can neither stop on the runway nor continue the climb-out. He then has no choice but to land, or continue to roll, straight ahead, using crash-landing procedures.

With engine failure in normal climb, cruise, or descent, first determine whether normal or nearly normal operation can be resumed by some simple measure such as changing fuel tanks, adjusting mixture, or reducing power. If power cannot be restored, or further operation will seriously damage the engine, or if there is danger of fire, shut it down. Make any necessary power adjustments on the other engine, and trim the aircraft for the new power and airspeed configuration. Change climb or descent rate and altitude as necessary in conformance with the operator's manual. Maintain at least V_{mc} even at the expense of altitude.

A twin-engine airplane with an engine out should land at the nearest suitable airport.

Landing procedure after engine failure depends upon the amount of power available. If you can maintain level flight, fly a normal pattern. When altitude cannot be maintained, a partial pattern is still desirable. Adjust the pattern so as to place the airplane at the proper position and altitude for as near normal a final approach as can be made. Use the flaps and lower the landing gear so that you will not have to use high power at slow speed to avoid undershooting, but with sufficient time remaining for the gear to extend, lock, and be checked. As you reduce power during final approach, take out rudder trim accordingly. Make the landing and rollout normal, remembering that the failed engine is not available for steering or reversing, and its opposite should be reversed only in dire emergency. If you must perform a go-around; do not hesitate. The earlier you decide, the safer and easier it will be.

Emergency Procedures in Light Aircraft

A flight emergency exists whenever the flight cannot be completed in a normal, safe manner. All the pilot's knowledge and skill may be demanded to achieve a safe outcome.

SYSTEMS MALFUNCTION OR FAILURE

Airplane auxiliary equipment may not cause as immediately severe an emergency by failing as does an engine, but it can nevertheless lead to serious trouble or disaster. There are of course many other failure possibilities, including those to electrical systems, pressurization, heat, oxygen, and fuel systems. However, the most common are those pertaining to flaps, landing gear, and fuel supply.

In any of these emergencies, it is generally better to declare an emergency and forget about possible later embarrassment. In the long run this principle will pay off; if the emergency requires no aid, the ground services will at least have had a drill that may save someone else.

Low fuel pressure on one of two or more engines, with the engine still running, may be a particularly critical problem. It may indicate a leak located so that any change in airflow in the nacelle may bathe the exhaust with fuel spray. With low fuel pressure, *never retard the throttle*. The best solution is to cut fuel off from the engine, then feather it. In any event, cut off fuel flow before landing. Retarding the throttle to land might change nacelle conditions just enough to result in explosion or fire.

FLAP FAILURE

While most airplanes have emergency systems for extending flaps when they will not extend normally, an occasional landing without flaps is necessary. Fly a normal pattern but make the turn to final approach farther out to allow for a flatter glide angle. Approach speed should be higher, by whatever percentage stalling speed is higher. It is important, too, not to overshoot. The nose will be higher than usual. On a big airplane, remember that as you pass over the end of the runway, the main landing gear may be much closer to the ground than is apparent from your position in the nose. Expect a longer landing roll because of higher speed and the absence of flap drag.

LANDING GEAR FAILURE

Landing with part of the landing gear extended is generally preferred to landing with all gear retracted. Less damage to the aircraft will result and there is less likelihood of injury to crew and passengers. Pattern, approach, and flare-out are normal. Plan to touch down on the extended gear, and hold the nose or the affected side off as long as possible. Lower the nose to the runway before you lose elevator control. There is much less danger of fire if the airport fire department is able to lay down a strip of fire-extinguishing foam on the runway before you land.

Brake failure is serious when it occurs on a short runway, or in conjunction with failure of drag devices during the landing roll. You may prepare to crash at the end of the runway, try to turn off on a taxiway or even try to ground loop to avoid going off the end, if brake failure becomes apparent during landing roll, and stopping appears critical. Some bigger aircraft have emergency brake systems which are independent of the normal systems. However, they provide only a limited number of brake applications.

Landing with a known flat tire or blowing a tire during landing roll can complicate directional control. Ordinarily you can maintain control with steering and braking on the side with good tires.

FIRE

Aside from mid-air collision, fire is the most disastrous accident that can happen in the air. Every possible precaution is taken while building and maintaining airplanes to minimize danger of fire.

Engine Fires. To control an engine fire in flight, first shut off the fuel and all electrical and oil supply that will not be needed for feathering. In a twin, as soon as the engine dies for lack of fuel, feather the propeller and shut off everything else connected with the engine's operation. If the fire persists at all, discharge the engine fire extinguisher for the affected engine, when so equipped. Continue to the nearest suitable airport for a single engine landing or establish a glide for a forced landing. *Never restart an engine that has been shut down because of fire.*

Engine fire is sometimes encountered during engine start. It is most likely to occur from backfire after overpriming or "pumping" the throttle of a carbureted engine. Keep the engine turning over for a start; fire in the intake should be drawn harmlessly into the engine. As with any engine fire, shut off the fuel supply. Use installed or available fire extinguishers, or the assistance of the fire department if necessary. Avoid discharging extinguishing agents directly into the intake, because severe engine damage will most likely result.

Fuselage Fire. Hand-operated fire extinguishers are maintained in occupied portions of most aircraft. If oxygen is available, everyone in a compartment where a fire occurs should wear masks and breathe 100% oxygen, to avoid injury from smoke and fumes.

In fighting a cabin fire, first make sure everyone is breathing 100% oxygen if possible; then determine the source of the fire. Cut off the source if possible; then try to extinguish the fire. Keep windows and hatches closed to reduce draft until the fire is out; then ventilate the cabin to remove smoke and fumes. Remember that the pilots' side windows will draw air from the cabin outward. They are seldom usable for fume evacuation. Use some other hatch if possible.

Electrical fire will be evidenced by smoke and the smell of burning insulation. Immediately shut off the master electrical power switch(es), and the fire should go out. Turn off all electrical equipment, and do not turn anything electrical back on unless it is absolutely necessary. If use of an "essential" item starts the smoke again, it is not that essential; leave it turned off.

CRASH LANDING AND DITCHING

A crash landing may be necessary either because the airplane cannot reach a suitable landing area owing to lack of fuel or engine failure, or because the landing gear cannot be extended for landing. Ditching is a crash landing on a water surface. If there is any choice in the matter, crash landing should always be made while there is still some power available, because this gives the pilot a better choice of landing area and direction, and permits a slower descent. It also provides control of the aircraft up to the moment of landing. All engines that can be used should be operating.

Landing with Gear Up on an Airfield. When it has been determined that part or all of the landing gear cannot be extended for landing, alert crash equipment, and prepare the airplane, crew, and passengers for the crash landing. If air-to-ground communications are available, give a report on fuel load and the location of crew and passengers, by compartment, to the ground agency for crash-crew use. If time permits and fire-extinguishing foam is available, the fire department may foam the runway to reduce fire danger during the landing.

Tie down or jettison all loose equipment in the airplane so that these items will not become projectiles to injure people on impact. Fasten all safety belts, and place any available padding material to reduce possibilities of injury. Facing aft, braced against a solid bulkhead is by far the safest position in which to survive the impact of a crash landing, if safety belts are available. Next best is an aft-facing seat with safety belt secured. Those in forward-facing seats should lean as far forward as their snugly fastened safety belts will permit. Those facing aft are instructed to lean back as far as possible, and place some kind of padding behind their heads. Crew members with shoulder harness should use it. Remember that a properly supported human body has withstood as much as 48 *g*'s of sudden deceleration. An improperly supported person cannot withstand even slight deceleration without injury.

Fly a normal landing pattern, except, of course, for the landing gear. Turn off all unessential electrical equipment.

Land on the runway, which has been covered with foam if possible, because experience has shown that less damage to the airplane and fewer injuries to personnel occur as compared with landing on an unprepared surface.

Landing on an Unprepared Area. When you cannot reach a prepared landing field, select the best area available for the landing. Send distress radio messages, giving position, difficulties, and intentions, if possible. In addition to the criteria for picking a forced-landing field, give consideration also to landing near a town where fire-fighting equipment and some rescue help may be available. Choose the landing direction and prepare the airplane, passengers, and crew for the crash landing, using flares and landing lights, if possible, at night. Highways are notoriously attractive yet likely to be unsafe for forced landings because of poles, culverts, signs, ditches, and vehicles.

With some airplanes, landing on an unprepared surface with landing gear up will lessen the danger of rupturing a fuel tank and increasing fire hazard. With others it is better to land with gear down, since even if it is torn off it will absorb some of the impact and lessen personnel injuries. Whether to use the landing gear or not is largely a matter of judgment in a particular situation and a knowledge of the aircraft manufacturer's recommendations.

Final approach and landing are the same as for crash landing on an airport runway. As soon as the airplane stops, able-bodied crew members and passengers aid the injured, if any, in getting out. Lose no time in getting clear of the aircraft, because it may suddenly burst into flame, even after an apparently safe landing.

Ditching. A water landing with a land-based airplane is a controlled crash landing with special problems. The preparation of the airplane, crew, and passengers includes making sure that everyone is wearing a life vest; placing first-aid kits, food, water, blankets, and other usable equipment where it will be readily accessible after landing but will not tear loose on impact; and preparing the life rafts for launching. The life vests and rafts must not be inflated inside the cabin.

Successful ditching without engine power to help control descent is very difficult. While the pilot certainly wants to avoid committing his airplane to a watery grave, he must make a decision early enough to retain some control over the rate of descent. If, for instance, a landing field is still an hour away, and he has only 20 minutes fuel, he must use the fuel for the safest possible water landing, rather than spend it on a futile effort to reach a land destination.

The landing direction for ditching must take into account not only the direction of the wind, but also the direction and size of any swells. The desired direction is upwind and downswell. Since this is usually not possible, a compromise choice is made, using the heading halfway between, but favoring the wind if it is strong. The best point to touch down is on the back slope of a swell just as it passes. This allows the aircraft to settle and decelerate before the next swell hits it. When two or more swell systems are present, the best touchdown point is in a fairly flat area where the systems cancel each other.

If at all possible, ditch near a surface vessel or a lee shore. Not only will the vessel be a valuable aid in rescue, but it also can give accurate wind and swell information. At night, a ship can lay down a row of flares to light up the surface and aid the pilot in lining up on the proper approach heading. The airplane may also carry flares which can be used to illuminate the surface.

An accurate position report transmitted as part of an emergency message is a vital element of success. Fly a descending pattern. During the first time over the prospective landing area, observe surface sea and wind conditions, dropping a flare if at night. On the second pass, get a better look, and at night launch a second flare to light the area for final approach and landing.

Make the approach and landing with power on and at a slow rate of descent, because it is almost impossible to judge accurately the height of an aircraft above water. Expect two impacts on landing; the first light, the second severe. The nose will probably go under water, then re-emerge. After the airplane has stopped, the ditching exits are opened, and the life rafts are launched. All useful equipment is taken along, and those unhurt aid the injured. Never jump into a life raft, because the rubber floor may be easily punctured.

Summary

The lore of flying is vast. To be truly professional, the pilot, like the doctor, can never stop learning. True competence demands an insatiable curiosity about the aircraft and the art of flying. Above all is being prepared for the unexpected—the contemplation, when things are going well, of what to do when they are not.

9

Instrument Flying

Transportation by air, like any other method, must be able to operate in foul weather as well as fair. Since its medium is the atmosphere, it must cope with natural phenomena occurring there, such as clouds, rain, snow, and other natural obstructions to vision.

Vision is man's primary means of keeping himself oriented with his surroundings. His body, however, senses the effects of gravity, and, even blindfolded, the normal person can get up from a prone position, sit, stand, walk, and even run.

Such body orientation without vision becomes impossible when the platform on which the body is supported is subject to varying accelerations, as is an aircraft in flight. Vision is, therefore, essential to aircraft control. When the natural horizon is visible, it is most often used as the primary reference for controlling an aircraft. When it is not, instruments must provide a substitute horizon.

With knowledge and practice, aircraft control by instruments becomes much more precise than is possible by reference only to the natural horizon. As a consequence, the more sophisticated aircraft are controlled more and more primarily by reliance upon the instruments.

Discipline

The pilot who flies an aircraft without reference to the natural horizon must develop the mental and physical discipline to ignore his body senses, which can be thoroughly confused by the varying accelerations of flight, and to rely implicitly on the sum total of the information provided by the flight instruments. He must discipline himself to do the hard, concentrated work of learning to fly by instruments in the first place, and to do the persistent practice to maintain a safe level of proficiency. He must learn to rely on *all* the instruments, to the degree that any bit of information he needs can come from the maximum number of instruments, and failure of a given instrument does not leave him unable to continue safe flight.

A high degree of discipline is also required to learn and use efficiently and easily the factors of time, communications, radio aids to navigation, and air traffic procedures needed to fit one's aircraft into today's dense air traffic flow. Fortunately this discipline can be acquired and improved under any flight conditions because the air traffic rules for instrument flying do not care whether the weather is clear or cloudy; an instrument flight plan can be filed and flown in any weather; and in clear weather the pilot can both initiate and cancel it at will.

Attitude Instrument Flying

The attitude of an aircraft is the relationship of its longitudinal axis (fuselage) and its lateral axis (wings) to the earth's surface or any plane parallel to the earth's surface. Attitude instrument flying means controlling the attitude of an aircraft by reference to flight instruments.

Both attitude instrument flying and visual flying use reference points to determine the attitude of the aircraft. While flying by visual reference to the earth's surface, the pilot determines the attitude of the aircraft by observing the relation between the aircraft's nose and wings and the natural horizon. While flying by reference to flight instruments, the pilot determines the attitude of the aircraft by observing indications on the instruments, which give him essentially the same information obtained by visual reference to the earth's surface. Because he uses exactly the same control techniques while flying by reference to instruments as he does in visual flying, the student of attitude instrument flying need not learn a different method of controlling the aircraft. Skill in basic instrument flying enables him to reduce the instrument flying problem to simple procedures in any aircraft. He is thus safely freed from concentration on the manipulation of controls to the extent that he can navigate, communicate with traffic control agencies, and still remain relaxed and deliberate.

TRIM

Proper trim technique is essential to smooth and accurate instrument flying. An inexperienced instrument pilot may think that he is too busy flying the aircraft by control pressure to devote time to manipulation of the trim controls; however, he cannot execute a maneuver precisely without trimming. *The degree of instrument flying skill which he ultimately will develop depends largely upon how well he learns to keep the aircraft trimmed.*

CONTROL AND PERFORMANCE INSTRUMENTS

Flight instruments by usage fall into three groups: *control instruments, performance instruments,* and *navigation instruments.*

Most instruments can provide more than one kind of information. For instance, both the attitude indicator and the turn indicator provide information on whether the aircraft is banked; the attitude indicator, the altimeter, the vertical speed indicator, and the airspeed indicator all provide information on pitch attitude. Failure of one instrument usually does not mean total loss of the information it provides.

Attitude and *power* indicators are the control instruments. By proper use of the attitude indicator and power setting you can obtain the performance you need to maintain level flight, turn, climb or descend, accelerate or decelerate, and even take-off and land.

Figure 9.1. Instrument or contact—the basic reference is the same.

CONTROL   INSTRUMENTS

PERFORMANCE   INSTRUMENTS

NAVIGATION   INSTRUMENTS

Figure 9.2. Functional grouping of the instruments. *(Courtesy of US Air Force.)*

ATTITUDE CONTROL

Proper aircraft attitude control is the result of maintaining a constant attitude, knowing when and how much to change it, and then changing the attitude smoothly a definite amount to obtain specific aircraft performance.

The attitude indicator provides an immediate and direct indication of any corresponding change in the aircraft's pitch or roll (bank) attitudes. Because small pitch or bank changes are easily seen and immediately evident, changes of any magnitude can be made smoothly and quickly.

Pitch Control for Level Flight. The pitch attitude of an aircraft is the angular relation of the longitudinal axis of the aircraft to the true horizon (Figure 8.4). In level flight the pitch attitude varies with airspeed. A change of several degrees in the angle of attack is necessary to maintain constant lift with changing airspeeds. The aircraft flies relatively nose-high at low speeds and relatively nose-low at high speeds. The pitch attitude of the aircraft for level flight also changes with difference in load, the angle of attack of the wings having to be increased with increased load and decreased with a decrease in load. It changes, too, with vertical velocity changes in the air: normal up and down drafts.

To attain the proper pitch attitude in visual flight, you raise or lower the nose in relation to the horizon. In instrument flight you adjust to proper pitch attitude by changing the "pitch picture" of the miniature aircraft or fuselage dot by definite amounts in relation to the horizon bar of the attitude indicator. These changes are called "bar widths," or "half bar widths," or degrees, depending upon the type of attitude indicator. Most modern attitude indicators have horizontal lines or dots (the pitch reference scale) in 5° increments.

Pitch changes of 5° or greater are called "degrees of pitch change," while pitch changes of less than 5° are normally called "bar widths" or "fractions of bar widths." Pitch adjustments can be refined in bar widths after establishing an

MINIATURE A/C ON BAR   ½ BAR NOSE HIGH   ½ BAR NOSE LOW

Figure 9.3. Pitch attitude indications. Though these indications seem small, they are characteristic of the degree of pitch change required for effective attitude control.

initial pitch change in degrees. In attitude indicators such as the J-8, pitch changes are made with respect to "bar widths." In instruments with clearly marked pitch reference scales (MM series), changes of any magnitude are made in terms of degrees.

PERFORMANCE INSTRUMENTS IN LEVEL FLIGHT

Before takeoff, if the "little airplane" is adjusted so as to be in line with the 90° indices at the sides of the instrument, it will usually be in the proper place for level flight. If not level with the horizon at normal cruising speeds in level flight, it is adjusted to show a level flight attitude at the preferred speed. Leaving it unchanged thereafter during varying airspeeds provides a constant true picture of the attitude relative to a familiar flight condition.

The attitude indicator has virtually no lag; it gives an immediate picture of a change in pitch attitude. By remembering that during rapid acceleration or deceleration the indication may be nose-high or nose-low, you can avoid the effects of the small misrepresentation due to momentary gyro precession.

When using the attitude indicator to make pitch corrections, apply light but positive control pressures. Up to true airspeeds of 260 knots, the normal movement of the "little airplane" should not exceed one width of the horizon bar. In level flight, as the true airspeed increases, pitch corrections should be smaller to prevent overcontrolling; use the vertical velocity indicator to determine if you are overcontrolling. Any movement of more than 200 fpm from the desired rate indicates overcontrolling. Usually, to restore the aircraft to a desired altitude after a small (100-ft.) deviation, a good pilot will use no more than 200 to 300 fpm. When it is obvious that this amount of correction is not enough, the best technique is to make a small correction on the attitude indicator and then observe performance instruments to determine if the correction is adequate. If not, make another correction of the same size immediately, and watch the result. Use prompt, light corrective pressures on the controls; this avoids later need for heavier corrective pressures. Although there is a small lag in the movement of the altimeter, for all practical purposes at lower altitude, it may be considered as giving an immediate indication of a change or a need for a change in pitch attitude.

Altimeter. The altimeter gives an indirect reading of the pitch attitude of an aircraft in level flight. The altitude should remain constant; any deviation shows a need for a change in pitch (and, possibly, power), and the rate of deviation shows the size of the pitch change needed. If gaining altitude, lower the "little airplane"; if losing altitude, raise it.

Vertical Speed Indicator. Correct use of the VSI is helpful for precise pitch attitude control. Although it gives an indirect indication of pitch attitude, it will, with smooth control technique, indicate by its initial movement the *trend* of vertical movement of the aircraft. Use it in conjunction with the attitude indicator and the altimeter for any flight attitude. In level flight, if you detect the movement of the needle immediately and apply corrective pressure to return the needle to zero, the altimeter will usually indicate that there has been no change in altitude.

The amount that the altimeter is off governs the rate to use to return the aircraft to the desired altitude. After applying corrective pressure to change the attitude indicator, cross check the altimeter and VSI to see if further correction is needed. Whenever you apply pressure on the controls and the VSI shows a rate which exceeds the rate desired by more than 200 fpm, you are overcontrolling. For example, if attempting to gain lost altitude at the rate of 300 fpm, a reading of over 500 fpm would indicate overcontrolling.

While the initial movement of the needle is instantaneous and indicates the trend of the vertical movement of the aircraft, there is a lag. The lag is proportional to the speed and magnitude of the pitch change. Train yourself to apply smooth control technique by using light pressures for adjustments in pitch attitude. Relaxation and crosschecking all the performance instruments to remain aware of the general pitch attitude will eliminate overcontrolling.

Airspeed Indicator. The airspeed indicator gives an indirect reading of the pitch attitude of the aircraft. With a given power setting and the corresponding attitude, the aircraft flies level and the airspeed remains constant. If the airspeed increases, the nose is too low and should be raised. A rapid change in airspeed indicates a large change in pitch, and a slow change in airspeed indicates a small change in pitch. There is very little lag in the indications of the airspeed indicator. When pitch changes are made, they are reflected rapidly by a change of airspeed.

WHEN AND HOW MUCH TO CHANGE ATTITUDE AND POWER INDICATIONS

The basic concept of aircraft control requires you to adjust the aircraft attitude and power to achieve the desired aircraft performance. Therefore, you must be able to recognize *when* a change in attitude and/or power is required. It is equally important to know *what* and *how much* pitch, bank, or power change is required.

Power setting, pitch attitude, altitude change rate, and airspeed are interdependent; attitude plus power equals altitude change rate and airspeed (performance). Through the use of power and attitude controls, altitude and airspeed are adjusted to meet the performance needs of the moment.

When to adjust attitude is determined by watching performance instruments. An indication, other than that desired, by the performance instruments, is the signal to change attitude or power.

Knowing *what* to change comes from knowing which control instrument adjustment will bring about the desired indications on the performance instruments.

The air is seldom so smooth that flight can continue for any extended period of time without making at least minor adjustments in attitude or power. Experience and judgment are the key to *when, what,* and *how much.*

INSTRUMENT CROSSCHECKING

It is quite possible to control an aircraft in instrument flight with only an altitude reference, a heading reference, and an airspeed reference. The altitude reference for any other than short periods of time must be the altimeter. Basic heading reference is the compass; to control heading, one needs a turn needle or turn coordinator, or an instrument that can substitute, such as the ADF needle. Airspeed reference can be a power setting that experience has shown will produce the desired airspeed in the chosen mode of operation.

With the full panel of instruments operative, the pilot has many interrelated clues concerning the performance of the aircraft. No change occurs without affecting several instruments. *Crosschecking* of the indications of all the instruments provides the best information about the airplane's performance.

The key to smooth, relaxed instrument flying is a continuous scan of all the instruments related to control and performance. Concentration on one instrument for any appreciable period of time leaves one unaware of the changes being reported by others. It takes only a glance at an instrument to get its story. If a change of power or attitude is required, it is made while the scan continues. The next glance at the particular instrument, and others telling the same story, will determine whether the applied change is effective, too large, too small, or just right. Attitude response should be immediate; performance instruments will lag to some degree because inertia is modified by force applied over some period of time.

The primary instrument for attitude determination is the attitude indicator. It is included in every crosscheck or scan across the instrument panel, because it is instantaneous in indication and accurately forecasts changes in other instruments—a bank forecasts a change of heading, a pitch-up forecasts a decrease in airspeed and an increase in altitude, etc., while a steady level flight indication predicts steady indications by other instruments.

Lag in the performance instruments should not interfere with maintaining or smoothly changing the attitude and/or power indications. When you control the attitude and power properly, the lag factor is negligible, and the indications on the performance instruments will stabilize or change smoothly. Making a flight control movement in direct response to the lag in indications on the performance instruments without first referring to the control instruments leads to erratic aircraft control and will cause additional fluctuations and lag in performance instruments. The sooner a deviation is noted and a correction applied, the smaller the required corrective adjustment of the control instrument will be.

The attitude indicator is the only instrument to be observed continuously for an appreciable length of time. Several seconds may be needed to accomplish an attitude change required for a normal turn.

The attitude indicator is the most frequently observed. For example, you may glance from the attitude indicator to the airpseed; back to attitude; then a glance at the altimeter; back to attitude, and so forth. This example of a normal crosscheck does not mean that it is the only method of crosschecking. It is often necessary to compare the indications of one performance instrument against another before knowing when or how much to change the attitude and/or power. An effective crosscheck technique may require that you note the attitude indicator, between glances at the performance instruments being compared.

Preponderance of attention to the attitude indicator is normal and desirable to keep the fluctuations and lag indications of the performance instruments to a minimum. This technique permits you to read any one performance instrument during a split-second glance and will result in smooth and precise aircraft control.

You must give a proper and relative amount of attention to each performance instrument. Pilots seldom fail to observe the one performance instrument whose indication is most informative in a particular maneuver, but they often devote so much attention to one performance instrument that they fail to crosscheck the attitude indicator for proper aircraft control. Another error is to become so engrossed with one performance instrument that the others are omitted from the crosscheck.

How can you recognize deficiencies in crosschecking?

Insufficient Reference to Control Instruments.   If you do not have in mind some definite attitude and power indication you want to establish or maintain, and the other instruments fluctuate erratically through the desired indications, you are not referring enough to the control instruments. Aircraft control will be imprecise, and you will feel ineffective and insecure.

Too Much Reference to Control Instruments.   This is rare indeed. It may be caused by your desire to maintain control indications within close tolerances. If you have smooth, positive, and continuous control over attitude and power indications, and are making small adjustments, but large deviations are occurring slowly in certain of the performance instruments, you need a closer crosscheck on the latter.

By crosschecking incorrectly, you will omit some performance instruments from the crosscheck, although you are observing other performance and control instruments properly. For example, during a climb or descent, you may become so engrossed with pitch attitude control that you fail to observe an error in the aircraft heading.

A 4° heading change is not as ''eye-catching'' as a 300- to 400-fpm change on the vertical speed indicator. Through deliberate effort and practice, you can insure that you include *all* the instruments in your crosscheck. Do this, and you will observe performance instrument deviations in their early stages, when corrections are easiest to make.

The habit of analyzing your crosscheck technique is very helpful in maintaining your skill. *Deterioration in effective crosschecking is the first evidence of*

Figure 9.4. Throttle movement and power change indications are simultaneous.

lack of enough practice; it is always a danger signal and requires positive effort to correct.

POWER CONTROL

Proper power control results from knowing *when and how much* to change the power indications. These power changes are established by throttle adjustment and reference to the power indicators. Power indications are not affected by such factors as turbulence, improper trim, or inadvertent control pressures as are other flight instruments. Therefore, once established, little attention is required to insure that power remains constant.

Adjusting the power to a desired amount is an easy step of power control. For all practical purposes, power indicators change simultaneously with throttle movement, and the manner in which they are calibrated permits adjustments to be made with a minimum of attention to the power instruments.

BANK CONTROL

The banking attitude of an aircraft is the angular relation of the lateral axis of the aircraft to the true horizon. To maintain a straight course in visual flight, you must keep the wings of the aircraft level with the true horizon. If the aircraft is trimmed properly and in coordinated flight, any deviation from a wings-level attitude produces a turn. For instrument flight the "little airplane" and horizon of the attitude indicator are substituted for the real aircraft and the true horizon, as in Figure 9.1, and the banking attitude is indicated accurately.

Changes are made by changing the ''bank attitude,'' or bank pointers, by definite amounts in relation to the bank scale. The bank scale is graduated at 0°, 10°, 20°, 30°, 60°, and 90°. Bank attitude control is used to maintain a heading or a desired attitude of bank during turns.

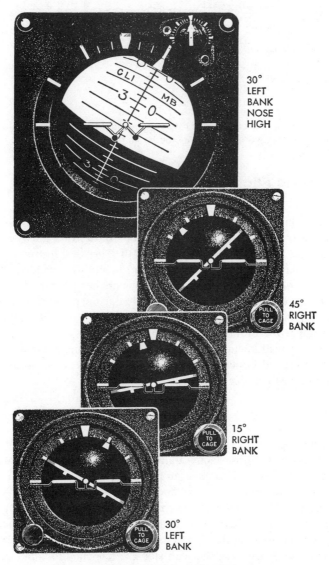

Figure 9.5. Bank attitude indications.

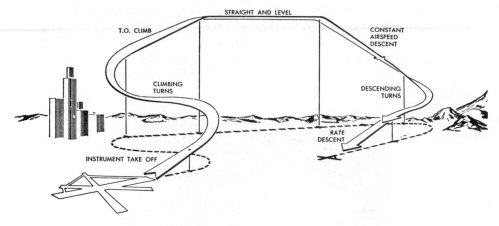

Figure 9.6. Typical instrument flight. Any instrument flight, regardless of how long or complex, is simply a series of connected basic flight maneuvers.

Basic Flight Maneuvers

For this portion of Chapter 9, the author is particularly indebted to the Air Force Instrument Pilot Instructor School which writes Air Force instrument flying manuals and develops a synthesis of the very best instrument flying techniques from Air Force and civil experience, involving a wide variety of pilots, flight conditions, and aircraft. The author has not hesitated to use verbatim parts of the Air Force text, except for occasional adjustments to suit the needs of the civil aircraft pilot. The techniques, however, are generally applicable to all aircraft equipped with full instrument panels.

Because any instrument flight, regardless of how long or complex, is simply a series of connected basic maneuvers, these maneuvers can form the basis for both training and practice.

TRIM TECHNIQUE

Proper trimming is very important in any attitude. In fact, improper lateral trim usually results in an unconscious strain on the pilot which can be distracting and dangerous. And if the airplane is properly trimmed, momentary diversions for navigation or other needs are easy and safe.

An incorrect setting of the aileron trim tab will lower one wing and start a turn. Incorrect rudder trim causes a skid. Skids result in banks, and therefore turns, because they increase the speed, and thus the lift, of one wing. Improper opposite aileron and rudder trim together can stop the tendency to bank, but result in a constant-heading skid. This costs airspeed. When trimming, use all available

Figure 9.7. Proper trimming. To trim properly, apply control pressure to maintain desired attitude. Then adjust trim until the control pressure is relieved.

bank instruments, but act on the principle that you are centering the ball with rudder trim and centering the needle with aileron trim. As trim is improved, speed will increase; small trim-tab deflections into the airstream will have greater effect. Continue the retrimming cycle, including elevators, until the needle and ball are centered hands-off, and the speed is stabilized. Use control pressure to get what you want, then hold it there with trim.

STRAIGHT AND LEVEL FLIGHT

Straight and level unaccelerated flight is a matter of holding a desired altitude, heading, and airspeed. Pitch attitude control maintains or corrects altitude; bank altitude control maintains or corrects the heading. Use power control to establish and maintain the desired airspeed.

Maintaining Desired Altitude.   The pitch attitude must first be held constant by reference to the attitude indicator and by keeping the aircraft properly trimmed. On leveling off, establish a pitch picture on the attitude indicator which will result in level flight. Then adjust the miniature aircraft reference of the attitude indicator to align it with the horizon bar. Readjust it as load and cruising airspeed change. This will simplify keeping a single cruise attitude picture and will make pitch control much simpler.

When correcting back to the desired altitude, make pitch attitude changes of a definite amount, such as ½- or 1-bar width. With practice and observation, you will come to know approximately how much these pitch changes will affect the vertical speed.

To avoid overcorrecting, change the pitch attitude just enough to give a vertical speed in fpm about double the altitude error in feet. For example, if 100 ft. off altitude, use a pitch correction that will give you a rate of correction of about 200 fpm. Remember that your initial pitch change is only an estimated amount; you must note the resulting speed and refine the adjustment as necessary.

When correcting back to an altitude, begin the level-off before reaching the desired altitude, and with a lead about equal to 10% of the rate of climb; that is, if climbing at the rate of 200 fpm, start leveling off 20 ft. below the desired altitude. With light, slow aircraft, such a lead is less needed.

This technique will also help to compensate for the lag inherent in performance instruments. With experience you can judge from the initial indication of the VSI whether or not your initial pitch attitude change was too large or too small and can correct before the altimeter has had time to react. Do not be led into overcontrolling pitch attitude by the oscillation of the VSI in rough air. This is a common mistake which can be corrected by more reference to the attitude indicator.

Maintaining a Desired Heading.   If the wings are level, the aircraft theoretically will not turn from the desired heading. Slight turbulence and slight banking will shift the heading, however, and the way to correct smoothly when the aircraft is properly trimmed is to refer to the attitude indicator and establish the amount of bank which will return the aircraft to the desired heading at a desirable rate. Most modern aircraft are designed with so little adverse aileron yaw that little rudder is needed, and correcting heading deviations with the rudder is a mistake, much less smooth than correcting by banking. Establishing an angle of bank on the attitude indicator equal to the number of degrees of heading correction needed will make smooth, precise corrections with minimum control effort.

Slight movements of the heading indicator are not particularly eye-catching; only the habit of regular and frequent crosschecking of the bank attitude and the heading indicator will develop skill in maintaining a constant heading.

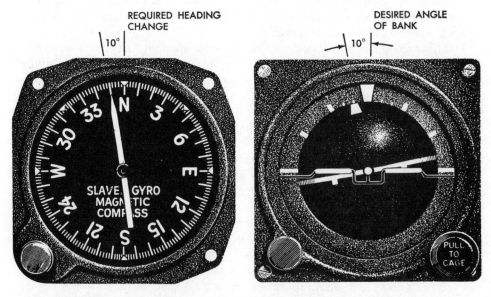

Figure 9.8. Limit the angle of bank to the number of degrees to be turned.

Establishing and Maintaining Airspeed.   To level off when the airspeed approaches that desired, it is helpful to know about what power will maintain that airspeed. Set the approximate value, continue the crosscheck of other instruments, and if the airspeed indicator and attitude indicator show that at the desired altitude the airspeed is high or low, make a minor power adjustment, establishing the pitch attitude which will maintain constant altitude. Make it a point to know what approximate power setting will give you a desired airspeed under the different conditions encountered in a typical flight.

When you observe a deviation in airspeed, it is generally wise to delay a power adjustment and check the altimeter and attitude indicator to see whether a pitch attitude change is needed. There are many times, *e.g.,* below the desired airspeed but above the desired altitude, when the required pitch change will correct both airspeed and altitude. Conversely, remember that changing pitch when the airspeed is correct may induce the need for a power adjustment. This is most likely at low airspeeds, particularly in jet or in large multi-engine piston aircraft.

Increasing Airspeed.   When controlling power to increase airspeed, advance the power beyond the setting required for the higher speed. As the airspeed increases, the aircraft will tend to climb. Crosschecking the altimeter and VSI with the attitude indicator will indicate the pitch change needed for the new airspeed. When the speed nears that desired, reduce the power setting to that which you estimate will support the new speed. Crosscheck, retrim, and readjust pitch and power until the airspeed is stabilized.

Decreasing Airspeed.   When establishing a slower airspeed, reduce the power beyond that which you believe will support the slower speed. Crosscheck pitch, altitude, and vertical speed with the airspeed. As it reaches the desired rate, bring up the power to support the slower speed. Crosscheck, retrim, and readjust pitch and power to stabilize the speed. If drag devices are used, it is good technique to reduce the power initially to that estimated for the new speed but not below it; as the speed slows to that desired, retract the drag devices and crosscheck pitch attitude, altitude, and vertical speed with the airspeed. Adjust pitch and power as necessary and retrim. Readjust pitch and power until the speed stabilizes. Retrim.

Extending or retracting drag devices on some aircraft will cause the aircraft to change pitch attitude. Be alert to this, and note the pitch attitude just prior to extending or retracting. Maintain this indication constant, and trim out the control pressures needed. Then proceed as described above.

LEVEL TURNS

Two facets of turns give pilots trouble. The first is the effect of loss of vertical lift during a turn. Compensation for this factor requires changes in pitch attitude, power, and elevator trim during the turn.

The second factor is the tendency of most attitude indicators, and particularly air-driven attitude indicators, to precess slightly during and immediately after turns, or during acceleration or deceleration. For example: Following a turn, the aircraft may actually be in a slight bank and turn even though the attitude indicator shows wings level. Or the aircraft may be actually climbing or descending with a level-flight pitch attitude indication.

These errors will disappear as the attitude gyro corrects itself; in the meantime, careful crosschecking of performance instruments and the awareness of these effects will permit using an attitude indication which will give the desired performance.

Establishing and Maintaining the Desired Bank.   Having determined the desired rate of turn, and what angle of bank is required at a particular aircraft speed and configuration to provide the rate of turn, enter the turn by reference to the attitude indicator. In normal instrument flying, you will seldom exceed 30° of bank.

Before entering a turn, make a definite decision on the angle of bank you

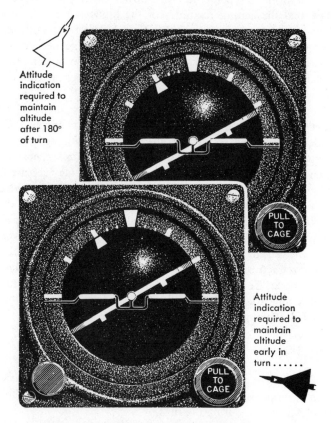

Figure 9.9. Effects of precession during turns.

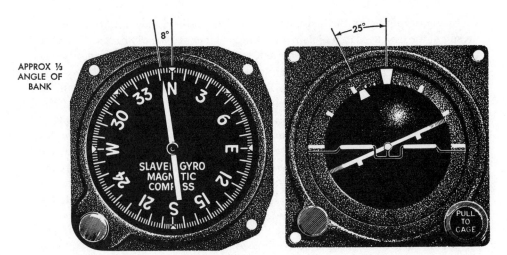

Figure 9.10. Leading the rollout of a turn. A lead of a number of degrees about equal to ⅓ the angle of bank is a good average, though it is somewhat different for all pilots.

require. You can then select a lead point in which to relax the control pressures to stop the roll-in smoothly. As the angle of bank increases, check the pitch attitude, changing it if necessary to prevent any change in performance instruments. Throughout the turn refer to the attitude indicator sufficiently to hold the bank indication constant, and use light pressures to readjust. It is important to use a precise, constant angle of bank.

To roll out of the turn on the desired heading, you must begin the rollout before reaching the new heading. The amount of lead you use will depend on the amount of bank, the rate at which the aircraft is turning, and the rate of roll you use in rolling out. An average amount of lead in degrees is ⅓ the angle of bank. Once you have developed a consistent rate of roll-out, a little experimenting will tell you what your individual lead should be.

Maintaining Altitude in Turns.   Maintaining the altitude constant during a turn requires the same pitch attitude control used in straight and level flight. Notice the effect of the loss of vertical lift, and anticipate the tendency of the aircraft to lose altitude by crosschecking carefully the altimeter and VSI. As the need for a pitch change becomes evident, apply back pressure to adjust the pitch attitude of the "little airplane" or nose dot on the attitude indicator. As you roll into the turn, crosschecking the pitch attitude of level flight will provide a reference for a slight increase in pitch attitude in the turn. Your principal error will be insufficient reference to the attitude indicator for pitch control.

During the turn, the horizon bar may precess up or down, requiring a change in the pitch attitude necessary to maintain altitude. Proper crosschecking and trimming in the turn will, with practice, make this process almost unconscious.

While rolling out of the bank, do not allow the nose of the aircraft to rise as a result of the trim used to maintain the altitude in the turn. Apply forward pressure during rollout to return the pitch attitude to that existing before the turn, remembering possible precession, and then trim out the forward pressure.

Maintaining Airspeed in Turns.   Power control is the same as in level flight, but the power also will have to be increased to maintain airspeed during turns. Compensating for the loss of vertical lift with a pitch attitude adjustment increases drag and reduces airspeed unless the power is increased. Anticipate this loss and be prepared to add the necessary power. At the higher airspeeds, the power needed may be negligible. At low airspeeds, particularly in jet aircraft and in larger piston aircraft in landing configuration, a considerable change in power may be required. If you are slow in applying the extra power as you roll into the turn, the airspeed may decrease rapidly until changes in power require a substantial change in pitch attitude, and the turn can become difficult and unstable. To avoid this, at low airspeeds add an estimated amount of power as you roll into the bank rather than having to catch up later; don't wait for the first indication of a loss in airspeed.

STEEP TURNS

Steep turns are those requiring more bank than is normally used in instrument flying in a particular aircraft. The techniques for turning steeply in level flight are the same for standard turns. You use pitch attitude control to maintain or correct back to the altitude, power control to maintain airspeed, and bank attitude control to maintain the bank attitude constant. Steep turns are an excellent practice exercise for improving your crosscheck in standard turns. In steep turns, hold the bank indication constant, and make all corrections back to altitude by changing pitch attitude. If control of the pitch attitude and altitude becomes erratic, or if excessive elevator control forces are required, roll to wings level and begin another turn. Perhaps you are using too much bank for your aircraft for your present crosscheck ability, or you are delaying too long using back stick pressure on rolling in. Power control in steep turns is also magnified, and power increases must be started earlier. Steep turns are excellent practice if you make them precise. The three primary factors affecting aircraft control during steep turns are: (1) rate of roll-in and roll-out, (2) use of trim and power, and (3) angle-of-bank control.

TIMED TURNS AND USE OF THE MAGNETIC COMPASS

With modern directional gyros, failure is unlikely, but it could happen, and turns using the magnetic compass are the only alternative.

The magnetic compass gives reliable information only during straight, level, and unaccelerated flight. It fluctuates widely, rushing ahead on turns to southerly headings, holding back on turns to northerly headings, and swinging during acceleration, as a result of the "dip effect." For this reason it is best to make turns at a standard rate, timing them to reach approximate headings.

A timed turn is one in which the angle of bank is established to give the desired rate, shown by the needle of the turn and slip indicator, and this bank is held for a definite period before leveling out. By holding the desired rate of turn for the proper number of seconds, you can make a turn of the desired number of degrees. For example, with the 2-min. needle, a full needle width turn at 3° per sec. will turn the aircraft 90° if held for 30 sec.

Start timing as you apply control pressures to begin the turn. Start rolling out when the time has elapsed. If your rates of roll-in and roll-out are consistent, the timing will be accurate; if the angle of bank is held constant and correct, the number of degrees of turn will be accurate.

As you practice holding definite amounts of bank by reference to the attitude indicator, note the turn needle to see what deflection is resulting. The rate of turn which results from any given amount of bank will vary with the true airspeed. By practicing turns at the various airspeeds normally used for your aircraft, you will soon become familiar with the angle of bank required for a desired rate of turn. Establish the initial bank, then crosscheck, trim, and adjust the bank indication on the attitude indicator to produce the turn needle deflection desired.

As an alternate to timed turns, you may refer to the magnetic compass to determine the lead point at which to roll out. This method can be effective in smooth air if you use angles of bank less than 15° in order to minimize dip error.

Because of the dip error, you must allow, in addition to your lead, a number of

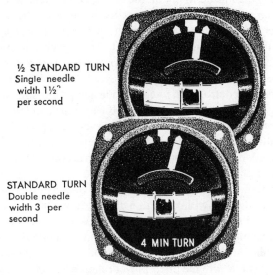

½ STANDARD TURN
Single needle
width 1½°
per second

STANDARD TURN
Double needle
width 3 per
second

4 MIN TURN

Figure 9.11. Turn indications.

degrees approximately equal to the latitude in which you are flying. When turning to south, turn past your normal lead point by this amount. Turning to north, turn short this number of degrees in addition to your normal lead.

Dip error is negligible when turning to east or west, and you can use your normal amount of lead if the turn is not too steep, so long as airspeed is held constant.

CLIMBS AND DESCENTS

In instrument flying, there are two general ways to climb or descend: constant *airspeed* and constant *rate*. In the former you establish a desirable power indication to be held constant throughout the maneuver, and then control the pitch attitude to give the desired airspeed, accepting whatever vertical speed results.

In the *rate* climb or descent, you control the pitch attitude as required to provide some desired constant vertical speed, and control the power as required to maintain the desired airspeed.

In either type of climb or descent, the bank attitude control used in straight and level flight or in turning applies, and during the climb or descent, you can hold a constant heading, or can turn.

Constant Airspeed Climbs and Descents.   Before entering the climb or descent, select the power setting you intend to establish, and estimate the amount the pitch attitude will have to change to maintain the airspeed you desire. Normally you make the pitch and power changes simultaneously.

If cruise airspeed is appreciably greater than that desired for climb, the pitch attitude should be adjusted to obtain the desired climb rate, and power should be increased to maintain the desired airspeed when it is reached. When cruise airspeed is higher than selected descent speed, the power is reduced, and level flight is held until the descent speed is attained. Then pitch and power are adjusted simultaneously to obtain the desired performance. Experience in the particular airplane makes possible very accurate selection of attitude and power settings.

It is important to control both pitch attitude and power smoothly and precisely, crosschecking to readjust pitch and power if necessary, and to see that the performance instruments are showing the airspeed, vertical speed, heading or heading change, and altitude change desired. Power changes should be smooth and uninterrupted, for good power control technique will simplify the coordination of pitch and power changes.

In some aircraft, particularly unsupercharged piston aircraft at lower altitudes, the change in air density will result in a change in manifold pressure, unless the throttle is adjusted to hold the power constant. With a moderate amount of practice in power and pitch change coordination, and with orderly crosschecking, the airspeed will remain within close limits as you enter the climb or descent. The more familiar you are with the pitch-power relationship of your aircraft, the more precisely you will be able to make the initial pitch and power adjustment.

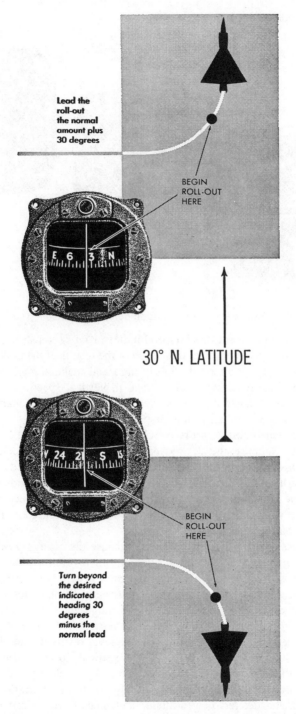

Figure 9.12. Use of the magnetic compass during turns.

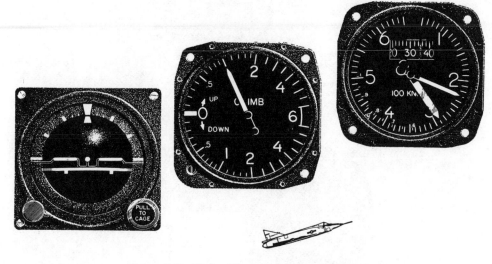

Figure 9.13. Pitch attitude determines airspeed.

After a pitch adjustment, the airspeed indicator will soon show a change, but the VSI will often give a quicker indication of the results of the pitch change. For example, you note the airspeed slightly high and make a small adjustment of pitch. If there is a resulting small change in vertical speed, you know, even though the airspeed may not have changed, that the small pitch change was effective. It will soon be evident in the airspeed.

The VSI can also point out that you have made an inadvertent change in the pitch attitude. Suppose the desired airspeed and vertical velocity have been remaining constant, but the pitch attitude has changed unnoticed. The VSI will usually show this pitch change more quickly than the airspeed indicator, and you can correct the pitch attitude to return to the desired vertical speed. The effect of the correction will be that the pitch attitude will have been corrected before the airspeed has had time to deviate.

This process is not ''chasing'' the VSI because the changes *are made on the attitude indicator,* and these result in a precise, positive, immediate change in the aircraft attitude. As long as the attitude indicator is kept constantly in the circuit of crosscheck, and in your consciousness, the rate of climb or descent will be stable in smooth air, and more nearly stable in rough air.

Leveling Off from Constant Airspeed Climbs and Descents.   As you approach your desired altitude, plan to lead the level-off by the number of feet on the altimeter about equal to 10% of the vertical speed. Adjust the power smoothly to that approximate setting required for level flight. Simultaneously change the pitch attitude of the miniature aircraft to the indication you believe will maintain the desired new altitude, and resume your normal level flight crosscheck. Retrim. Reset the height of the ''little airplane,'' if necessary.

In piston aircraft where the climb speed is substantially lower than cruise, good technique requires that you set a level pitch attitude as required to hold the desired level-off altitude and delay the change in power until the airspeed has increased to or slightly above that required. Retrimming is an important part of this process and begins before the power change. In heavily loaded aircraft, it may even be helpful to climb above the desired altitude, level off initially with climb power, and drift down to the desired altitude to insure that the airspeed reaches the optimum for the cruise power desired or available.

Rate Climbs and Descents.   The requirement is to execute constant airspeed and constant vertical speed, both predetermined, as might be used during the final portion of an instrument approach to landing.

As a general aerodynamic principle, pitch attitude controls airspeed and power controls rate of descent or climb. Within the fine limits of adjustment made during a carefully controlled rate descent, we reverse the rule. Having established airspeed and rate of descent, we, *for all practical purposes, control the airspeed with power and the rate of descent with pitch attitude.*

The reason is that the changes in pitch attitude and power should be minute; in fact, they will have to be minute if the approach is to be smooth. Only if a large adjustment is required, as, for example, when the aircraft has gotten dangerously low too early in the descent, will power govern descent. In this example, a substantial increase in power might be required to slow the rate of descent, or even climb.

Figure 9.14. Rate climbs and descents.

Also, aircraft of the present day are so clean aerodynamically, and jet aircraft, particularly, carry so high a proportion of power during a descent with gear and flaps out, that a pitch change, directly affecting the angle of attack and lift, can correct a small vertical speed error immediately without changing airspeed perceptibly. Similarly, the cleanness and high percentage of power held make the airspeed more responsive to small changes in thrust.

Just before initiating the climb or descent, recall the pitch change that will produce the desired vertical speed, and the power setting that will maintain the desired constant airspeed. Smoothly, and as simultaneously as practicable, establish the new power and pitch indications. Retrim. Crosschecking, adjust the pitch attitude to correct the vertical speed, and adjust the power to correct any deviation in airspeed. To level off, use the same procedure as for leveling off from constant air speed climbs or descents.

Bank Attitude Control During Climbs or Descents. By using the same bank attitude control for straight flight or for turns described for straight and level flight, you can turn or hold constant headings precisely as in level flight (Figure 9.5).

Remember the loss of vertical lift which occurs with bank. If you enter a turn while in a constant airspeed climb or descent, be prepared to lower the pitch attitude slightly to maintain the airspeed. If you enter a turn during a rate climb or descent, be prepared to raise the pitch attitude slightly to maintain the vertical speed, and to add power to maintain the airspeed.

Climb Schedules. For any one given load condition and altitude, there is only one airspeed which will produce the most efficient rate of climb. Because modern aircraft regularly climb through more than 40,000 ft. during their normal operations, using constant power indications or slowly deteriorating maximum power indications, climb schedules are used.

As the climb progresses, change the pitch attitude so as to adhere to the appropriate airspeed for the ambient altitude. The vertical speed is a help, because it will be the quickest indicator of the effectiveness of a small pitch change.

With a Mach meter, the climb is usually controlled by airspeed until a certain Mach speed is reached; thereafter the Mach meter is held constant throughout the rest of the climb.

INSTRUMENT TAKEOFF AND INITIAL CLIMB

Many an accident otherwise unexplained, and many a scare, have occurred right after takeoff in reduced visibility and at night in aircraft of all sizes in good weather and foul. The ability to shift completely to instruments after the beginning of the takeoff roll, and until the climb and initial heading are well established, is an essential of safe flight at night and in reduced visibility. The rapid acceleration and rapid rates of roll of modern aircraft make this particularly true.

Therefore, practice the instrument takeoff so that when the time comes, you can make the changeover successfully.

Preparation for Takeoff.   Make a complete cockpit check, as for visual flying. In addition, pay special attention to the gyro instruments and to trim tabs, because any irregularity could have a disastrous effect. Set altimeter setting in the Kollsman window. Check to see if reading varies more than ± 75 ft. If it does, have the altimeter changed, or don't take the aircraft into weather conditions.

After being cleared for takeoff, line up the aircraft with the centerline of the runway and allow it to roll straight ahead a short distance to be sure that the nose or tail wheel is properly aligned. Hold the brakes firmly to prevent creeping.

If you are using a vacuum-driven heading indicator, set it to the 10° mark nearest the published runway heading, and recheck it to be sure that it is uncaged. If using an electric heading indicator, set the heading needle and runway heading under the index at the top of the instrument.

Takeoff and Initial Climb.   Accomplish the takeoff roll, and establish the take-off attitude by combined visual and instrument reference. In the early part of your takeoff roll, you can keep the aircraft properly aligned with the runway by observing the centerline and runway lights. In extremely low visibility or hard rain, this may become more difficult as the speed increases. Therefore, make it a point to include the directional indicator in your crosscheck so you will know what heading will keep the aircraft on the runway. Remember that crosswind drift on the takeoff roll is recognizable only through outside visual reference.

In some cases, you must establish the take-off attitude almost entirely by reference to the attitude indicator. Make it a definite point to know the take-off pitch attitude required for your aircraft as indicated on the attitude indicator. Know also how much the attitude indicator can be expected to precess because of acceleration so that you can allow for the false nose high attitude it will show. The important point to remember is that you include the instruments in your crosscheck sufficiently so that a rapid or unexpected loss of outside reference will not cause difficulty.

Refer to the attitude indicator to hold the pitch-and-bank attitude constant as the aircraft leaves the ground. A small variation in pitch at this time may cause the aircraft to settle to the runway or stall. Wait for the altimeter to show an increase in altitude, and the vertical velocity indicator to show a definite climb before retracting the landing gear. Altimeters and vertical speed indicators sometimes indicate a descent during the takeoff roll and immediately after becoming airborne. Raise the wing flaps when appropriate for your aircraft, but after climb is established.

While the gear and flaps are being retracted, carefully maintain a climbing attitude on the attitude indicator. Refer to the VSI and altimeter to prevent leveling off, entering a descent, or allowing the vertical speed to increase or decrease excessively.

After the gear and flaps are retracted, control the pitch attitude to provide both a reasonable rate of increase of airspeed and a desirable rate of climb until you reach climbing airspeed. Control the bank attitude to maintain or correct back to the desired heading. Use trim as desired after takeoff, but avoid using trim to pull the aircraft off the ground. Doing so may result in an excessive nose high attitude as gear and flaps are retracted and the airspeed increases.

EXERCISES IN BASIC INSTRUMENT FLYING

These maneuvers combine climbs and descents at a definite rate with precision turns and require maximum speed of crosschecking for precise execution. They are real tests of basic instrument flying skill when executed precisely and checked by timing; they also simulate many of the problems encountered during instrument approaches. They should be flown with errors not exceeding 50 ft. altitude, 5 knots airspeed, and 10 sec. per min. of time.

The Vertical "S".   This is a continuing series of definite rate climbs and descents, made at constant airspeed and heading. Select a beginning altitude and heading, then climb or descend to another altitude, 500, 1000, or 2000 ft. different, and reverse. As you become proficient at coordinating pitch attitude and power changes, the airspeed will remain within close limits during climb, descent, and reversal. You will soon be familiar with the lead points needed for your aircraft and control technique.

The Vertical "S" in Constant Turn.   Use a bank angle suitable for your aircraft. Enter the turn simultaneously with the initial climb or descent, and maintain the bank constant throughout the maneuver. Make several successive and precise climbs and descents with the same bank.

The Vertical "S", Reversing Turn at Descent.   Add to the foregoing maneuver a reverse in the direction of turn each time you begin a descent.

The Vertical "S," Double Reverse.   In this exercise, reverse the direction of turn with each change in vertical direction.

Recovery from Unusual Attitudes

Situations can occur when you may become confused regarding the attitude and performance of the aircraft. Regard any position, even in a mild deviation from the attitude desired, *in which you are confused,* as an unusual attitude. These situations may result from turbulence, vertigo, wingmen getting lost from leader, carelessness in crosschecking, *and failure to crosscheck the instruments adequately when you suspect partial instrument failure.*

If the aircraft is in a moderate attitude, you can reorient yourself by establishing level flight and immediately resuming a normal crosscheck.

RECOVERY PROCEDURES WITH NONTUMBLING ATTITUDE INDICATORS

The reliability of these instruments is well proved. Using them properly will permit rapid, positive recovery; even from extreme attitudes that can be entered easily in large and small high-performance aircraft.

Bank control will aggravate the attitude if misused and will assist the recovery materially if properly used. Observe these principles:

In a DIVE, eliminating bank will aid pitch control.
In a CLIMB, using bank will aid pitch control.
Proper use of power and drag devices can aid airspeed control.

Recovery: First, verify attitude. Before taking any corrective action, verify the suspected unusual attitude by reference to the performance instruments and to the stand-by or copilot's attitude indicator, if available. Failure to take this step may create an unusual attitude where none previously existed, except in the indication of a faulty instrument.

Bank control by reference to the attitude indicator is of prime importance in recovery and is simple and positive. *The bank index pointer is the key to bank control.* Always consider it as pointing to "up"—a "sky pointer." To return to right-side-up, always roll toward the bank index pointer. Notice, in Figure 9.15,

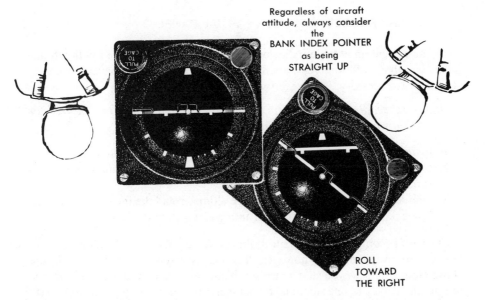

Regardless of aircraft attitude, always consider the BANK INDEX POINTER as being STRAIGHT UP

ROLL TOWARD THE RIGHT

Figure 9.15. Inverted attitude recovery. At left, the aircraft is fully inverted, in level flight. At right, it is inverted, diving steeply to the left.

that although the complete attitude may be difficult to detemine immediately, you can tell quickly from the bank index pointer which way to roll to right-side-up. Many attitude indicators, in addition to the bank index pointer, have the background divided by the horizon bar into light and dark halves, representing sky and ground. This presentation will also assist in rolling right-side-up.

When you have the bank index pointer in, or moving toward, the upper half of the instrument face, decide whether you are in a climb or a dive. The airspeed indicator is usually the best indicator, but a crosscheck of the altimeter, vertical speed indicator, and attitude indicator will help to make this decision. Having made it, return to level flight by reference to the attitude indicator.

If you are diving, reduce power and roll level and upright. Use drag devices if appropriate for your aircraft. When at a level flight indication on the attitude indicator, resume a normal crosscheck.

If you are climbing, add full power and roll, or continue your roll, until the bank index pointer *in the upper half of the case* approaches the 90° bank index mark. This will establish a banked recovery attitude which will bring the nose down to the horizon smoothly. Maintain the bank and just enough back pressure to keep seated comfortably. As the horizon and fuselage dot of the "little airplane" come together, roll the wings level, slightly nose-low, referring to the attitude indicator, and resume normal crosscheck and power. In some larger aircraft, to avoid undue stabilizer side loads, level the wings first, and then bring the nose down to the horizon smoothly.

RECOVERY PROCEDURES WITH PARTIAL PANEL

In the rare event an attitude indicator fails, or in aircraft equipped with attitude indicators that will tumble, a partial panel recovery with performance instruments is necessary. There are two cases:

High-Airspeed Recoveries: Airspeed indication is higher than that desired, or its trend is toward an indication which is higher than desired:

1. *Reduce power* to prevent excessive airspeed and loss of altitude.
2. *Center the turn needle and ball* with coordinated aileron and rudder pressures, in order to level the wings.
3. *Stop the descending indication* of the altimeter and the increasing airspeed by back pressure on the stick. If in a jet, use drag devices.

Change all components of control almost simultaneously with only a slight lead of one over the other; however, this lead is essential and *must* be in the above order. Suppose you start a turn and suddenly the airspeed is increasing, the turn needle is way to one side, and the altimeter is "unwinding." To apply back pressure to raise the nose would simply tighten the spiral. At this point, you would lose control, *unless you reduced power and centered the turn needle first.*

Use smooth control pressures. You are in level-flight pitch attitude when the movements on the airspeed indicator and altimeter stop before reversing their directions. Watch these instruments closely to avoid going into an uncontrolled climb. As their movements stop, observe the attitude indicator and VSI as a crosscheck, and as the airspeed indicator comes back to the desired airspeed, support that airspeed with the necessary power.

Low-Airspeed Recoveries: If the airspeed indication is lower than that desired, or is decreasing rapidly toward an indication lower than that desired, this is the procedure:

1. *Apply power.*
2. *Center the turn needle and ball* to level the wings by coordinated rudder and aileron pressure.
3. *Apply forward stick pressure* to increase the airspeed and prevent stalling.

Here, again, while the actions are almost simultaneous, they must be done in the above order. With a rapidly decreasing or low airspeed, a stall may be imminent. The first thing needed is airspeed; without it, a worse condition, from which lack of altitude may prevent recovery, may result. Use the same primary instruments as for high-speed recoveries. Make a smooth but large power adjustment. If in a jet, use 100% power because of the slow acceleration characteristic. In multi-engined aircraft, be sure that power is balanced between wings. Better no power than high power on one wing and none on the other if near or below the critical single-engine speed.

Use caution when applying the forward pressure. Avoid an excessive and prolonged amount of forward pressure which in certain jets could result in a flameout and, in any aircraft, could result in an unusually steep dive attitude which might require a subsequent high speed recovery. Allow the airspeed to increase to a safe value before you apply back pressure to level out. Make this back pressure smooth to avoid a secondary stalled condition. Pitch attitude is level when the altimeter stops. When the airspeed builds up to the desired value, reduce the power to that needed to support it.

Because of the very nature of the occurrence of an unusual attitude, make the recovery prompt. Excessive loss of altitude is undesirable and may be dangerous. Begin a climb or descent back to the original altitude and heading as soon as you have regained full control of the aircraft and a safe airspeed.

If your recovery from an unusual attitude occurs while making a low approach, consider going to the emergency altitude at once to collect your wits and make sure of your position before trying again.

RELAXATION

Instrument flying requires a state of mind and body best described as alert detachment. One is detached from natural references and stimuli and is virtually suspended solidly in another world in harmony with the stable, familiar references of known attitude and power. In this state of detachment, the mind is occupied with direction, speed, time, and useful radio voices. It can respond clearly and accurately without laboring unduly with the body over control manipulations, which most of the time require only fingertip pressures anyhow. Without relaxation, this state can be neither acquired nor maintained.

Instrument Flying Procedures

As shown in Figure 9.6, an instrument flight consists of a series of connected basic maneuvers. Instrument flying procedures connect these maneuvers, using the navigation instruments. Precise execution of these procedures is the means by which single aircraft use the system of airways and air traffic control facilities safely in marginal or adverse weather; precise execution also permits the orderly flow of the large volume of air traffic, particularly in congested areas.

Instrument navigation is essentially dead reckoning navigation using radio aids; the technique is discussed in Chapter 11 and brings us to the point of the approach and landing at destination.

There are various types of approaches. Air Traffic Control (ATC), in giving the approach clearance, will usually specify the type of approach. Pilots should indicate the type of approach desired on their initial radio contact with approach control.

PRECAUTIONS

Planning.   The best approaches are made when the pilot has studied the approach plate and possible traffic, radio, terrain, airborne equipment, or other pertinent factors before taking off. At that time he will fix in his mind the circumstances under which he will proceed to the alternate destination.

Radio Tuning and Checking.   Flying on the wrong radio signal has brought more than one pilot to grief, but certain habits will prevent it.

ALWAYS check the identification of any navigational aid station and monitor it during flight.

ALWAYS utilize *all* suitable navigation equipment aboard the aircraft and crosscheck heading and bearing information.

ALWAYS carefully crosscheck navigational aids and ground check-points if possible before overflying an ETA (Chapter 11.)

ALWAYS check NOTAMS and Flight Information Publications (FLIP) documents before flight for possible malfunctions or limitations on navigational aids to be used. Be sure the field and the approach runway you plan to use are available.

ALWAYS discontinue use of any suspected navigational aid and confirm aircraft position by radar or other means.

PLANNING

The best time for planning is before takeoff. Study the weather and airfield conditions which will govern the type of approach most likely to be made. Know what unusual conditions you may encounter by checking NOTAMS. A careful study of the terminal approach chart for the airfield of intended landing will show the airfield and runway layout, approach lighting, obstructions, and type of approach lighting. It will give a very important mental picture of the location of prominent landmarks at and near the airfield, the minimum altitudes, and the direction of traffic for circling approaches. With this information, and above all a clear orientation of direction, you will know what to expect during the transition from instrument to visual flight.

When you check the weather, or when you hear it in flight, remember that when the ceiling is reported as "obscured," "indefinite," or "precipitation," your visibility may be much less than reported. You are interested in the slant range from your eye to the nearest point of contact on the ground. This distance is a combination of many factors, such as approach and runway lighting, the shape, size, and slope of your windshield, its condition, your own adaptation to darkness, your fatigue, and the effects of rain or snow on the windshield or between you and the runway.

EN ROUTE DESCENT

When cruising at higher altitudes, you may request an en route descent for approach, either to the IAF or the FAF, or to some other point from which you can conveniently continue. You may request vectors en route. This manner of descent is customary and greatly expedites the flow of air traffic. The traffic at the time will govern. A transponder equipped aircraft has a great advantage during this maneuver because it is always clearly identified and does not need to make identifying turns. These and other time consuming maneuvers may be necessary to insure positive identification and separation from other aircraft.

The controller and the pilot must both understand the type of approach to be made on reaching the IAF. Range, weather, desired descent rate, and low altitude fuel consumption are factors to be considered before requesting an en route descent.

Normal ATC procedure where there is a Radar Approach Control (RAPCON) is to vector arriving aircraft to the final approach fix, assigning successively lower altitudes so that the aircraft arrive at final approach fix altitude before or at the fix.

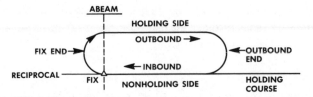

Figure 9.16. Holding pattern basic terminology. A standard pattern with right turns is illustrated. Nonstandard pattern is made with left turns. Check AIM for required airspeeds for specific aircraft. Generally, speeds are (IAS maximum) propeller, 175 knots; civil turbojet up to 6000 ft., 200 knots; 6000 to 14,000 ft., 210 knots; above 14,000 ft., 230 knots. Turboprop aircraft may climb in holding pattern at normal climb speed; turbojet aircraft may climb at 310 knots or less. (*Redrawn from Airman's Information Manual.*)

HOLDING

When traffic is heavy, it is occasionally necessary to wait for other traffic. This is *holding.* It is done over a radio fix specified by air traffic control. The fix may be the intersection of bearings from two stations, a DME distance and bearing from a station, or a station itself. While holding, the aircraft flies an elliptical pattern designed and oriented to facilitate the subsequent descent, the final approach, and the landing.

The pattern flown during holding receives airspace protection when entered and flown as prescribed in the Airman's Information Manual. The standard no wind length of the *inbound* legs of the holding pattern is 1 min. when holding at or below 14,000 ft. and 1½ min. when holding above that altitude. With DME, the maximum *outbound* leg is specified in miles.

Pattern Entry.   Enter the holding pattern at not more than the maximum airspeed, slowing down when 3 min. from the holding fix. When entering, use the pattern shown in Figure 9.17.

Timing.   The *initial* outbound leg should not exceed the appropriate time, depending on altitude, though aircraft holding at 100 KIAS or less may fly 2 min. on the initial outbound leg. The inbound timing of subsequent legs is maintained by adjusting outbound legs. Start the outbound timing when the aircraft is abeam the holding fix or wings-level, whichever occurs *last.* Start inbound timing whenever the aircraft crosses the inbound holding course or wings-level, whichever occurs *first.*

Wind Drift Correction.   Corrections are made on the outbound leg so that the aircraft will track the inbound holding course when inbound.

Constant rate turns are essential to accurate holding patterns. A standard rate turn (3° per sec.) is normally used at lower altitudes and airspeeds. Maximum bank angle is limited to 30°, which provides 3° per sec. at 210 knots. Strong crosswinds may be partly compensated for by shallowing turns upwind and

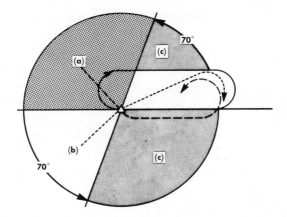

Figure 9.17. Entering the holding pattern. (a) Parallel procedure: Parallel holding course, turn left, and return to holding fix or intercept holding course. (b) Teardrop procedure: Proceed on outbound track of 30° or less to holding course, turn right to intercept holding course. (c) Direct entry procedure: Turn right and fly the pattern. *(Redrawn from Airman's Information Manual.)*

steepening turns downwind by 1° for each degree of inbound drift correction, but the minimum bank angle is 15° (3° per sec. at 90 knots, 1.5° per sec. at 185 knots).

An understanding of wind drift effect at various airspeeds is very helpful in maintaining courses in holding patterns. Table 9.1 is a helpful guide.

Meeting Expected Approach Time. The holding pattern may be shortened but never lengthened to meet approach times; plan to leave the holding fix to arrive over the Initial Approach Fix (IAF) at an "expect approach time." To make the adjustment, consider rates of turn and length of outbound and inbound legs.

TABLE 9.1. 90° Crosswind Effect.

TAS	30° Drift Correction Will Correct for a Wind of:	15° Drift Correction Will Correct for a Wind of:	10° Drift Correction Will Correct for a Wind of:
150	75 knots	37 knots	25 knots
200	100	50	33
300	150	75	50
400	200	100	67
500	250	125	83
600	300	150	100

EXAMPLES OF HOLDING

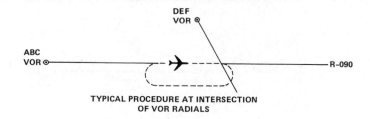

TYPICAL PROCEDURE AT INTERSECTION
OF VOR RADIALS

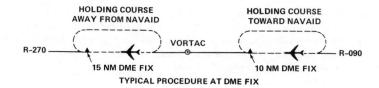

TYPICAL PROCEDURE AT DME FIX

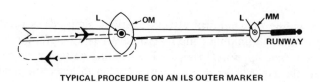

TYPICAL PROCEDURE ON AN ILS OUTER MARKER

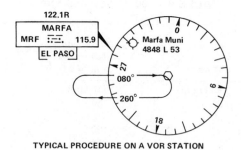

TYPICAL PROCEDURE ON A VOR STATION

Figure 9.18. Examples of holding patterns.

TACAN AND VOR/DME (VORTAC) HOLDING

Distance measuring equipment provides virtually an unlimited number of holding fixes and enables you to fly the holding pattern measured by actual distances rather than time. FAA Air Traffic Control maintains responsibility for defining holding patterns which provide vertical and horizontal separation and terrain clearance. With DME, patterns are different for various altitudes, speeds, and areas. The standard right hand pattern is usually used.

In ATC parlance, holding fixes are called *feeder points*. From these points aircraft are told to hold or approach. The fix is defined in holding instructions in these terms: "Hold southwest of the 30 nautical mile fix on the 045 degree radial, 8 mile legs, left turns."

Because it is not necessary to pass over an intersection or station before entering a holding pattern, entry is simply a matter of maintaining course, either inbound or outbound, on the specified radial. Then at the proper distance indicated on the DME range indicator, turn in the proper direction to enter a normal racetrack holding pattern, remembering that the nonstandard pattern has left turns. If a nonstandard pattern is required, the controller will specify "left turns" in holding instructions. The length of both outbound and inbound legs is the direct DME distance.

PROCEDURE TURNS

A procedure turn is a maneuver used to reverse course after passing a fix, to save time, restrict the maneuvering area, and insure return to the fix along a specified inbound course. It is primarily used at low altitude to place the aircraft on the final approach course. While the procedure turn is depicted on some approach charts by a specific pattern, the actual pattern to be used is at the pilot's discretion. It must, however, remain within a definite air space called the procedure turn maneuvering area, and must observe the minimum maneuvering altitudes published on the current approach plate.

The normal indication of a procedure turn is the barbed arrow (➤)located on the maneuvering side of the final approach course.

The procedure turn maneuvering area provides safe separation from other aircraft, as well as the necessary terrain and obstacle clearance. If it is in any way restricted, the approach chart will show it. For example, the chart may require completion of the turn in 7 nm. instead of the normal 10, or may prohibit procedure turns in certain sectors. The published altitudes provide a minimum of 1000 ft. clearance within the procedure turn area.

Figure 9.19 shows two recommended patterns for the turn. The one used depends upon the direction from which you approach the fix. Enter the teardrop or parallel turn in the same manner used for entering a holding pattern (Figure 9.17). Always set the inbound approach course in the course selector window

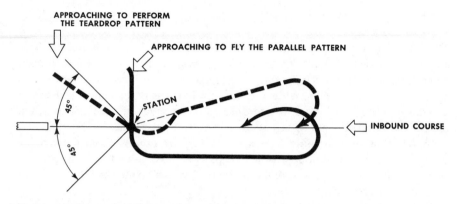

Figure 9.19. Standard flight patterns for performing the procedure turn. The 45° arcs are a general guide; entering the teardrop from outside this arc is likely to use too much maneuvering space. *(Courtesy of US Air Force.)*

while on the outbound leg. It is, of course, very important to keep the wind speed and direction in mind in order to avoid selecting a pattern in which you would fly outside the maneuvering area.

If you fly inbound with the outbound course set in the course selector window, FROM will, of course, appear in the TO-FROM window, and *you will have to fly away from, rather than toward, the CDI to center it.*

Flying the Teardrop Procedure Turn. Approaching the station, select an outbound course 20° from the inbound course on the side indicated for maneuvering on the approach chart: the holding side. Turn immediately after station passage to intercept the outboard course, level the wings, set the outbound course in the course window, and note the time. Complete the interception and maintain the outboard course, applying an appropriate drift correction and descending to the procedure turn altitude. The teardrop is flown as in the holding pattern. Normal time outbound is 2 min., not to exceed 10 nm.

Set the inbound course in the course window before turning inbound. When the time outbound has elapsed, begin the turn to the inbound course. During the latter part of this turn, adjust the rate of turn, if necessary, to roll out with the CDI centered. If the CDI is still fully deflected during the latter part of this turn, roll out 45° short and proceed with a normal course interception. If you fly through the course before completing the turn, continue turning and roll out to intercept from the other side. Remember that the procedure turn must be completed within the maneuvering area, above the minimum altitude. You may begin your descent to the FAF altitude when the aircraft is inbound and within 20° of the published inbound course, if you are cleared for the approach.

Flying the Parallel Procedure Turn. If you approach the station on a heading that is not within less than 45° of the reciprocal of the inbound course, plan to fly the parallel pattern entering as shown in Figure 9.19. Immediately after station

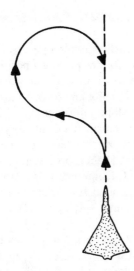

Figure 9.20. Ninety degree procedure turn.

passage, turn to the reciprocal of the inbound course. When the wings are level, note the time, and begin your descent to the procedure turn altitude. The length of the outbound leg depends on your ground speed. If TAS exceeds 180 knots while on the reciprocal of the inbound course heading on a parallel pattern, turn toward the reciprocal of the inbound course to avoid getting outside the maneuvering area. Turn inbound, and intercept and maintain the inbound course just as for the teardrop turn.

The 90° Procedure Turn.   This maneuver consists of a 90° turn from course followed immediately by a 270° turn back to the reciprocal course. If you have been maintaining course with a wind-drift correction applied, turn 90° from that heading, and roll out with drift on the opposite sign. (With a −10° correction outbound, use +10° inbound.) This procedure turn may be used for the turn phase of a low approach, though it is not as effective for checking drift, nor as smooth as the teardrop and parallel turns. It is important to set the inbound course in the course selector window before starting the turn because there may not be time to do it during the maneuver, particularly during turbulence.

Radar

*R*adio *d*etection *a*nd *r*anging (radar) equipment is based on the time differential between transmission of a radio signal and its reflection by an object, and on the ability of such a reflection to illuminate the face of a calibrated cathode ray tube, referred to as a "scope." The time lag provides distance information, and the illumination, or "blip" of light, on the scope face provides position information to the observer.

Ground Controlled Approach (GCA) equipment consists of two scopes: *search* or *surveillance,* and *final approach* or *precision.* The former provides position information, and the latter provides azimuth within narrow limits, and elevation. The GCA controller is in direct communication with the pilot and directs him to the landing runway by giving him heading and glide slope information.

GCA equipment has been made more precise and more capable of identifying individual aircraft by the use of airborne transponders which operate by transmitting a response when triggered by a ground station. The airborne station appears as a numbered bright spot on the radar scope. The equipment is adapted from a World War II method of identifying aircraft, called IFF (Identification Friend or Foe). It is known to the FAA Air Traffic Control System as ATCRBS (Air Traffic Control Radar Beacon System).

Ground radars equipped to interrogate the IFF can see responses from the beacon at much greater distances than is possible with the radar return alone. This is particularly true during periods of heavy precipitation when the IFF return will supplement the raw radar return and cause the aircraft to show more clearly on the scope.

VOICE PROCEDURES

Because the radar approach system is based entirely upon voice contact, you must use precise, crisp voice procedures. After the initial contact, identify your aircraft by the last three numbers of the serial number or the name and two numbers—"Smoky 26." During the approach repeat all headings, altitudes, and altimeter settings and acknowledge all other instructions.

Gyro-Out. If your directional indicator has failed, or if you are flying an aircraft without a stabilized direction indicator, you may fly a gyro-out pattern, using either precision or surveillance radars.

In this mode, the only difference is that the controller will advise you to make standard rate turns during the traffic pattern phase and half standard rate turns during the approach phase. Execute turns immediately on hearing the words "turn right" or "turn left." Stop the turn immediately on hearing "stop turn."

Published Instrument Approaches

TERMINOLOGY

A certain terminology is used in connection with instrument approaches. The terms are discussed and defined in the Instrument Approach Procedures publications discussed below. Some of the more important ones are given here.

Instrument Approach Category. Because aircraft differ in performance, the safe minimum altitude and visibility is not the same for all of them. The category to which an aircraft is assigned is determined by its approach speed. Safe approach speed is considered to be 1.3 times stall speed in approach configuration (Figure 9.21).

Obstructions and terrain features surrounding an airport and its instrument approach paths determine how low one may safely fly on an instrument approach without being able to see the surface environment. This minimum safe altitude also varies with the kind of surface guidance equipment and airborne guidance equipment being used.

Minimum Descent Altitude (MDA) is the lowest altitude MSL (above sea level) to which one may descend on final approach if the runway environment cannot be seen. The term applies to instrument approaches not using an electronic glide slope. It is higher for a circling approach (a turn of 30° or more must be made to align with the landing runway) than for a straight in approach. If the runway environment is not visible at the end of a precalculated time from the final approach fix or at an electronically established point, the pilot must execute a *missed approach* (climb back up again).

Decision Height (DH). This term applies when electronic (ILS or Precision Approach Radar) glide slope information is available. It is the aircraft altitude MSL at which the pilot must decide whether he can safely continue descent. If he has visual contact with the runway environment he continues; if not, he must stop the descent and prepare to execute a missed approach.

Height Above Aerodrome (HAA) is the height of the aircraft above the highest point on the airfield landing area, given for the MDA for each category of aircraft for a nonprecision or circling approach.

Height Above Touchdown (HAT) is the height of the aircraft above the point at which it will touch down on the runway if the glide slope is followed to touchdown, given for the DH for each category of aircraft.

Other definitions and data may be found in the published approach data, AIM, and other publications for pilots.

Thousands of instrument approaches to airports are currently published. Any airport with suitable nearby radio navigational aids can have an instrument approach. Minimum ceiling and visibility for some of them are above VFR minima. They range on down to ceilings as low as 100 ft. and visibilities less than a quarter of a mile (Category II). There may be only one instrument approach, or there may be half a dozen or more, for a given airport. (At last count, Chicago O'Hare International had 22.) Figures 9.26 through 9.36 include examples of various kinds of approaches. They are divided into two general classes, precision and nonprecision approaches. A precision approach includes a published DH, a nonprecision approach is made to a published MDA, shown on the approach chart.

Approach charts are published by National Ocean Survey of the Department of

Commerce, by the Department of Defense and by private firms, notably Jeppeson and Co. of Denver, Colorado. New approaches are being established, and old ones are being modified or discontinued, constantly. Consequently new revisions are being issued all the time. Government published charts are published in bound form every 8 weeks, with bound revisions at the midpoint between issues. New and revised charts are issued by Jeppeson immediately upon becoming effective, on single sheet inserts for loose leaf binders.

RADAR HANDOFFS AND COMBINED APPROACHES

IFR approaches are greatly speeded up where there is ATC approach control radar. Arriving traffic is identified, usually by transponder code, and vectored, with appropriate descent instructions, to the final approach course for the selected runway. The type of approach may be VOR, ADF, ILS, ILS back course, or precision radar.

Even when there is no radar approach control, ATC en route radar can frequently be used to vector an aircraft so as to minimize the time needed for an approach.

These procedures are helpful to both pilots and controllers, increase the flow of traffic, and help conserve fuel.

INSTRUMENT APPROACH CRITERIA AND SYMBOLS

Approach procedures are designed according to the FAA manual, *United States Standard for Terminal Instrument Approaches* (*TERPS*). These procedures are designed to provide a safe, orderly method for proceeding from the en-route structure to a point from which a visual approach to the runway can be made.

An instrument approach usually consists of four segments—initial, intermediate, final, and missed approach. The Initial Approach Fix (IAF), Procedure Turn (PT), the Final Approach Fix (FAF), and the missed approach itself, or transition to landing, are key points.

Figures 9.21, 9.22, and 9.23 are excerpted from the approach chart books and show the symbols used on the approach charts themselves.

Nonprecision Instrument Approaches

Nonprecision instrument approaches are those providing no electronic glide slope information. Instrument approaches using VOR, VORTAC, ADF, and RNAV are such approaches. Regardless of the type of approach, it is a blending of two skills: basic instruments flying and precise navigation. Both must be done regardless of the effect of wind or weather. A third element essen-

INSTRUMENT APPROACH PROCEDURES EXPLANATION OF TERMS

The United States Standard for Terminal Instrument Procedures (TERPS) is the approved criteria for formulating instrument approach procedures.

AIRCRAFT APPROACH CATEGORIES

Speeds are based on 1.3 times the stall speed in the landing configuration at maximum gross landing weight. An aircraft shall fit in only one category. If it is necessary to maneuver at speeds in excess of the upper limit of a speed range for a category, the minimums for the next higher category should be used. For example, an aircraft which falls in Category A, but is circling to land at a speed in excess of 91 knots, should use the approach Category B minimums when circling to land. See following category limits:

Approach Category   Speed

A. Speed less than 91 knots.
B. Speed 91 knots or more but less than 121 knots.
C. Speed 121 knots or more but less than 141 knots.
D. Speed 141 knots or more but less than 166 knots.
E. Speed 166 knots or more.

RVR/Meteorological Visibility Comparable Values

The following table shall be used for converting RVR to meteorological visibility when RVR is inoperative.

RVR (feet)	Visibility (statute miles)	RVR (feet)	Visibility (statute miles)
1600	¼	4000	¾
2000	⅜	4500	⅞
2400	½	5000	1
3200	⅝	6000	1 ¼

LANDING MINIMA FORMAT

In this example airport elevation is 1179, and runway touchdown zone elevation is 1152.

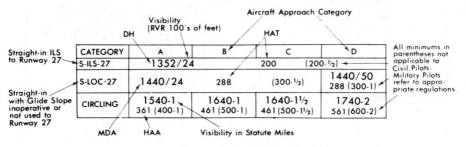

CORRECTIONS, COMMENTS AND/OR PROCUREMENT

Forward CORRECTIONS to and PROCURE from:

NATIONAL OCEAN SURVEY
DISTRIBUTION DIVISION, C44
RIVERDALE, MD. 20840

Figure 9.21. Approach-chart terms and procedures.

LEGEND
INSTRUMENT APPROACH PROCEDURES (CHARTS)

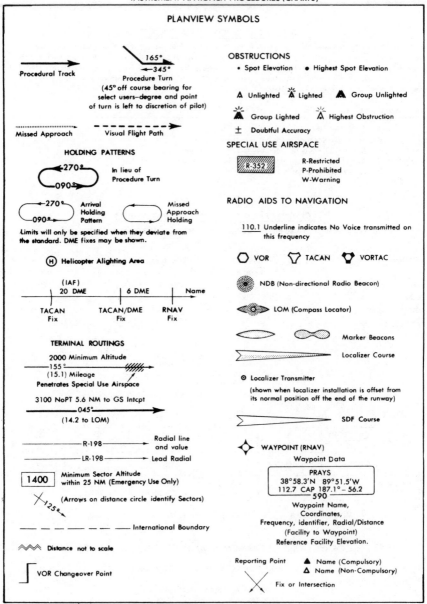

PLANVIEW SYMBOLS

Procedural Track

165°
345°
Procedure Turn
(45° off course bearing for
select users—degree and point
of turn is left to discretion of pilot)

Missed Approach   Visual Flight Path

HOLDING PATTERNS

270°
090°   In lieu of
Procedure Turn

270°   Arrival
090°   Holding
Pattern

Missed
Approach
Holding

Limits will only be specified when they deviate from
the standard. DME fixes may be shown.

(H) Helicopter Alighting Area

(IAF)
20 DME   6 DME   Name

TACAN   TACAN/DME   RNAV
Fix   Fix   Fix

TERMINAL ROUTINGS

2000 Minimum Altitude
155°
(15.1) Mileage
Penetrates Special Use Airspace

3100 NoPT 5.6 NM to GS Intcpt
045°
(14.2 to LOM)

R-198   Radial line
and value

LR-198   Lead Radial

1400   Minimum Sector Altitude
within 25 NM (Emergency Use Only)

125°   (Arrows on distance circle identify Sectors)

International Boundary

Distance not to scale

VOR Changeover Point

OBSTRUCTIONS
• Spot Elevation   ● Highest Spot Elevation

A Unlighted   A Lighted   A Group Unlighted

A Group Lighted   A Highest Obstruction

± Doubtful Accuracy

SPECIAL USE AIRSPACE

R-352   R-Restricted
P-Prohibited
W-Warning

RADIO AIDS TO NAVIGATION

110.1 Underline indicates No Voice transmitted on
this frequency

VOR   TACAN   VORTAC

NDB (Non-directional Radio Beacon)

LOM (Compass Locator)

Marker Beacons

Localizer Course

⊙ Localizer Transmitter
(shown when localizer installation is offset from
its normal position off the end of the runway)

SDF Course

WAYPOINT (RNAV)
Waypoint Data

PRAYS
38°58.3'N 89°51.5'W
112.7 CAP 187.1° – 56.2
590
Waypoint Name,
Coordinates,
Frequency, identifier, Radial/Distance
(Facility to Waypoint)
Reference Facility Elevation.

Reporting Point   ▲ Name (Compulsory)
△ Name (Non-Compulsory)

Fix or Intersection

Figure 9.22. Approach-chart symbols.

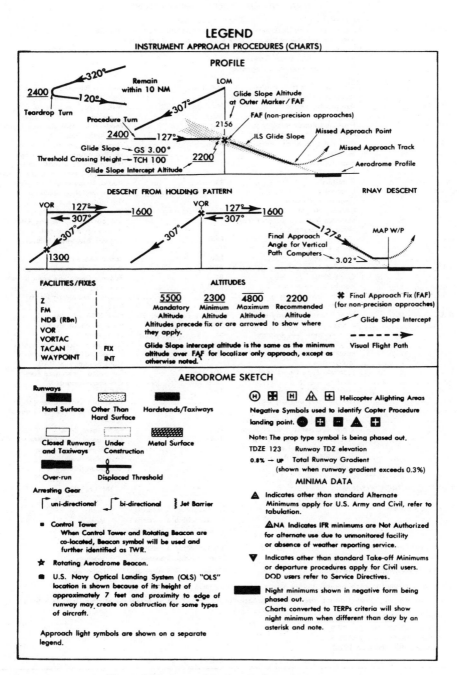

Figure 9.23. Approach-chart interpretation.

GENERAL INFORMATION & ABBREVIATIONS

★ Indicates control tower operates non-continuously
All distances in nautical miles (except Visibility Data which is in statute miles and Runway Visual Range which is in hundreds of feet).
Runway dimensions in feet.
Elevations in feet Mean Sea Level.
All radials/bearings are Magnetic.

ADF	Automatic Direction Finder
ALS	Approach Light System
APP CON	Approach Control
ARR	Arrival
ASR/PAR	Published Radar Minimums at this Aerodrome
ATIS	Automatic Terminal Information Service
BC	Back Course
C	Circling
CAT	Category
CHAN	Channel
clnc del	clearance delivery
DH	Decision Height
DME	Distance Measuring Equipment
DR	Dead Reckoning
elev	elevation
FAF	Final Approach Fix
FM	Fan Marker
GPI	Ground Point of Intercept (ion)
GS	Glide Slope
HAA	Height Above Aerodrome
HAL	Height Above Landing
HAT	Height Above Touchdown
HIRL	High Intensity Runway Lights
IAF	Initial Approach Fix
ICAO	International Civil Aviation Organization
Intcp	Intercept
INT, INTXN	Intersection
LDA	Localizer Type Directional Aid
Ldg	Landing
LDIN	Lead in Light System
LIRL	Low Intensity Runway Lights
LOC	Localizer
LR	Lead Radial. Provides at least 2 NM (Copter 1 NM) of lead to assist in turning onto the intermediate/final course
MALS	Medium Intensity Approach Light System

MALS/R	Medium Intensity Approach Light Systems/with RAIL
MAP	Missed Approach Point
MDA	Minimum Descent Altitude
MIRL	Medium Intensity Runway Lights
NA	Not Authorized
NDB	Non-directional Radio Beacon
NoPT	No Procedure Turn Required (Procedure Turn shall not be executed without ATC clearance)
RA	Radio Altimeter setting height
Radar Required	Radar vectoring required for this approach
Radar Vectoring	May be expected through any portion of the Nav Aid Approach, except final
RAIL	Runway Alignment Indicator Lights
RBn	Radio Beacon
REIL	Runway End Identifier Lights
RCLS	Runway Centerline Light System
RNAV	Area Navigation
RRL	Runway Remaining Lights
RTB	Return To Base
Runway Touchdown Zone	First 3000' of Runway
RVR	Runway Visual Range
S	Straight-in
SALS	Short Approach Light System
(S) SALS/R	(Simplified) Short Approach Light System/with RAIL
SDF	Simplified Directional Facility
TA	Transition Altitude
TAC	TACAN
TCH	Threshold Crossing Height (height in feet Above Ground Level)
TDZ	Touchdown Zone
TDZE	Touchdown Zone Elevation
TDZL	Touchdown Zone Lights
TLv	Transition Level
W/P	Waypoint (RNAV)

RADIO CONTROL SYSTEM

	KEY MIKE	INTENSITY
3 step light system	7 times within 5 seconds	High
	5 times within 5 seconds	Medium
	3 times within 5 seconds	Low
★2 step light system	5 times within 5 seconds	High
	3 times within 5 seconds	Medium
ACTIVATE (Rwy lights, Reil, or VASI)	5 times within 5 seconds	Lights on

★ Must be activated to High intensity before Medium may be selected.

This system will be indicated on the Instrument Approach Procedure (IAP) chart below the Minima Data as follows:

Examples of approach chart entry:

3 Step MALSR RWY 25-122.8, or
2 Step MALSR RWY 25-122.8, or
Activate MIRL RWY 25-122.8

Figure 9.24. Abbreviations and Radio Light Control Systems.

tial to instrument flying, and particularly to approaches, is the judgment and personal discipline required to decide if an approach and landing can be safely made or if the pilot should proceed to his alternate airport.

Be sure to study the published procedure and plan your approach *through the missed approach procedure* before you reach the IAF.

You may be required to hold at the IAF and will be given an expected approach time. The place to be at that time is determined by the type of aircraft, the time required to complete your prelanding check, and your proficiency, but the best place to be is inbound at the approach fix. You may make the approach from the holding pattern, but remain above the procedure turn altitude until you are inbound on the final approach course. If you are cleared to descend from an altitude above the initial approach altitude, descend in the holding pattern as quickly as your proficiency, aircraft, and load will permit. Be prompt, but be unhurried. Precision and safety are paramount.

VOR APPROACH

Obtain your clearance from Approach Control. At the IAF, turn immediately to the outbound course of the procedure turn you have selected. Note the time, set the desired outbound course in the course-selector window, and report your position, time, and altitude to Approach Control. The flight manual for your aircraft will specify when to accomplish the prelanding check; usually it is done on this leg. Complete the procedure turn, and establish your drift correction carefully on the final approach course.

On arrival at FAF, note the time carefully and report. If the station-to-field heading is different from the final approach course, turn to the former and reset the course selector. Maintain this course with the CDI centered. Station-to-field clearance permits you to descend immediately to the Minimum Descent Altitude. By doing this before the station-to-point-of-missed-approach time has elapsed, you place yourself in position to land more quickly. Watch time and altitude carefully, and keep the approach airspeed constant. The most common errors are to descend too late and to maintain excessive airspeed. As soon as you have the runway in sight, report to Approach Control in order that they may clear the next aircraft.

Missed Approach. If you have not established visual contact with the runway by the specified time from the FAF, or cannot for any reason make a safe landing, or are directed by Approach Control, you must execute the specified missed approach procedure.

The time-distance tables published on Approach and Landing Charts are based on zero wind. Therefore, you *must* consider the existing wind to determine the proper FAF-to-point-of-missed-approach time. Be prepared to execute a missed approach from every letdown; after you have started a missed approach and es-

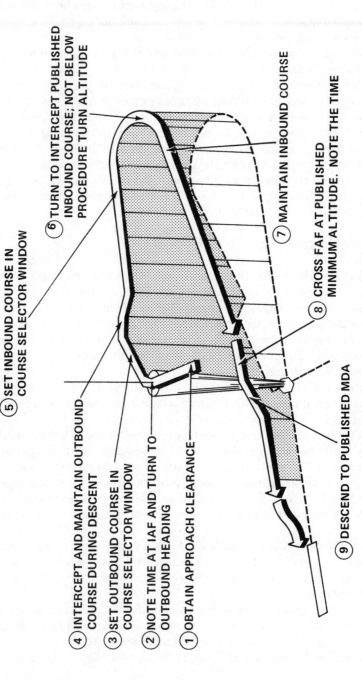

⑤ SET INBOUND COURSE IN COURSE SELECTOR WINDOW

⑥ TURN TO INTERCEPT PUBLISHED INBOUND COURSE; NOT BELOW PROCEDURE TURN ALTITUDE

⑦ MAINTAIN INBOUND COURSE

⑧ CROSS FAF AT PUBLISHED MINIMUM ALTITUDE. NOTE THE TIME

④ INTERCEPT AND MAINTAIN OUTBOUND COURSE DURING DESCENT

③ SET OUTBOUND COURSE IN COURSE SELECTOR WINDOW

② NOTE TIME AT IAF AND TURN TO OUTBOUND HEADING

① OBTAIN APPROACH CLEARANCE

⑨ DESCEND TO PUBLISHED MDA

Figure 9.25. Typical VOR low-altitude approach. (Courtesy of US Air Force.)

tablished your climb, call Approach Control and state your intentions, or request instructions.

TACAN AND VOR/DME APPROACHES

Because TACAN and VORTAC have range information as well as bearing information, procedures using this greater flexibility have been provided. There is one primary difference: The FAF or the IAF or both may be designated along a radial at certain distances. A procedure turn is usually unnecessary to execute a VORTAC approach. Because they are the same from this viewpoint, VORTAC is used to mean VORTAC, TACAN, or VOR/DME in further discussion.

During precision approaches, when shifting to the ILS frequency, distance information will not be available unless the DME tuner is separate or there is a dual installation.

The Arc Approach.   This approach requires that you fly the aircraft along an arc around the VORTAC station a specified distance out (Figure 9.27). The more smoothly one can turn from an inbound course to an arc, or leave an arc on a final approach course, the more precise arc approaches can be. To estimate the lead needed to begin a turn to or from an arc, remember this general relationship: At 10 nm., 1 nm. equals 6° of arc. At 200 knots and with a 3° per sec. turn, 1 mile (or 6°) lead will be required.

Most arc approaches guarantee obstruction clearance for only 2 nm. on either side of the arc. Do not become so engrossed with maintaining the arc that you miss crossing key radials related to altitude changes or missed approaches.

Under zero wind conditions, as the bearing indicator approaches the wing-tip position, you are as close to the station as you are going to get on that particular heading.

In theory it is a simple matter to remain on a particular arc, because maintaining the RMI bearing indicator at a relative bearing of 90° or 270° (wing-tip position) at all times would take you around the station in an exact circle. However, in practice, it is difficult to keep the station exactly off the wing tip, and wind effect further complicates the problem.

One system that works reasonably well is to turn very slowly, so as to maintain the relative bearing approximately 5° ahead of the wing tip, while noting the distance. Should the distance be decreasing, change relative bearing to the wing tip or a degree or two behind it until the distance increases to that desired. Should the distance be increasing, fly so as to put the relative bearing farther ahead of the wing tip. Another technique to use to remain near the desired arc is to fly a series of short legs. If you drift off the arc, make as small a correction as is practicable to return to it. *The size of your correction is continually reflected in the relationship of the bearing indicator to the wing tip.* In a crosswind you may find that it is necessary to keep the bearing indicator ahead of or behind the wing tip to remain on the arc. Because the graphic presentation of your position provided by

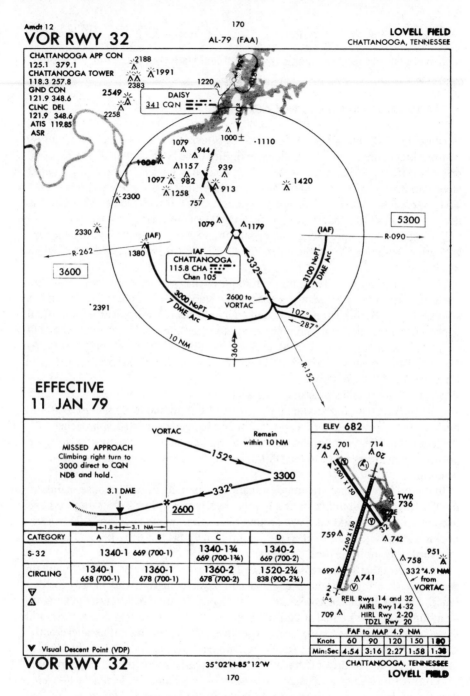

Figure 9.26. Typical VOR approach.

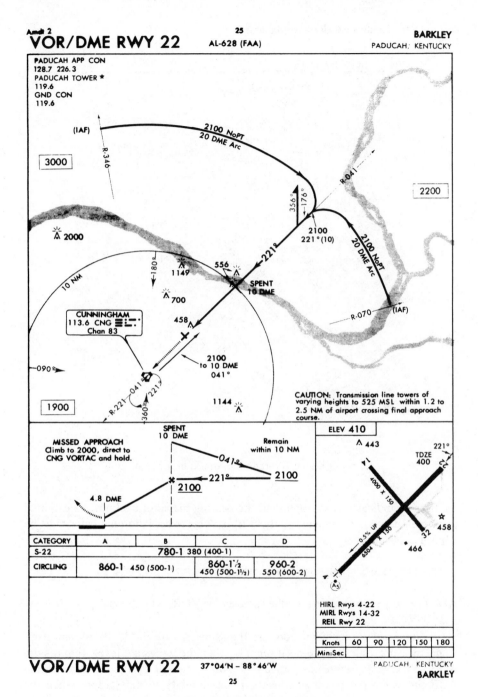

Figure 9.27. Typical VOR/DME approach.

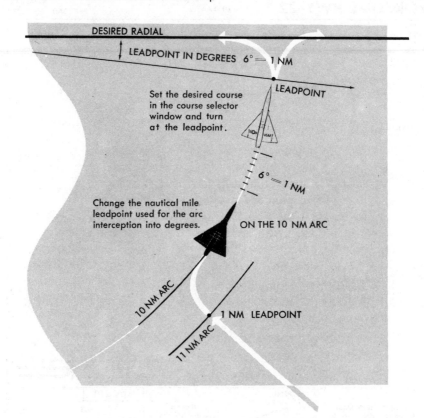

Figure 9.28. Intercepting an arc or a radial.

the bearing indicator is so important, the bearing indicator is almost an essential instrument for this maneuver. Without it, the course deviation indicator must be adjusted constantly.

ADF HOLDING, PENETRATION, AND LOW APPROACHES

ADF procedures for holding are the same as for VOR, with the exception that the CDI is not available.

Low approach procedures, too, are the same as with the VOR. When using ADF, measure your relative position only when the wings are level. In turns the ADF will give an erroneous indication. This is caused by the dip error in the ADF, in which the bearing indicator points directly to the station; when the aircraft is in a turn or descent close to the station, the direct line to the station may not be the actual bearing. Therefore, make precision turns to predetermined magnetic headings, and if in doubt, level the wings before reading the final approach course; then intercept the final approach course and proceed.

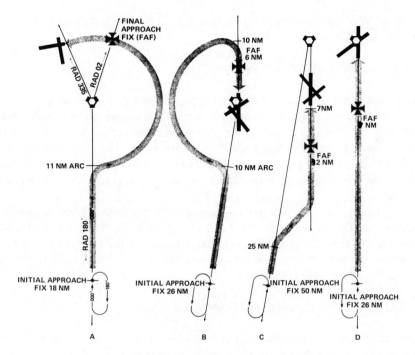

Figure 9.29. Typical VORTAC approaches.

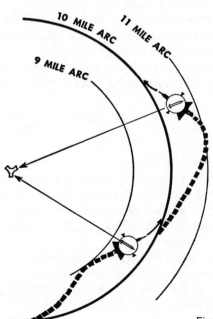

Figure 9.30. Correcting to maintain the arc.

As in VOR approaches, planning before the approach is a prime ingredient of a safe and efficient approach.

LOCALIZER BACK COURSE APPROACH

Many airports with ILS also have an approved and published back course approach. No glide slope information is normally available for the back course, unless it happens to coincide with a precision radar approach. In that case the radar controller can provide glide slope instructions. The normal operation of CDI will provide reverse sensing. The course needle becomes extremely sensitive as the end of the runway nears; the antenna will be reached before arriving at the runway.

The newer localizer receiver equipment often provides a selector switch, to reverse needle deflection in the localizer indicator for flying the back course inbound. It provides normal directional sensing for the pilot who is tracking opposite the normal approach course, and can also be "coupled" to the autopilot, for a back course approach.

RADAR SURVEILLANCE APPROACH

The surveillance (ASR) radar is not as precise as the PAR and cannot provide height information. It is used when precision approach radar is not available for the landing runway.

The traffic pattern and procedures are the same, except that instead of controlling your descent with reference to a precision glide slope, the controller will tell you when you are at the proper distance out on final to start your descent and then will direct you to descend to MDA or minimum altitude.

Surveillance approach minima are higher than those for precision approaches because of the decreased accuracy of the ASR scope and the lack of precise altitude information.

RNAV APPROACH

It often happens that, for a particular airport, a VORTAC station is not located so as to provide a good fix for a straight-in instrument low approach; it is too near the runway, too far away, or not aligned with the extended center line of the runway. The area navigation system of an aircraft so equipped can, in effect, place a VORTAC station anywhere desired, so long as line-of-sight "visibility" of the actual station can be maintained (Chapter 11).

It is a simple matter, then, to select a waypoint to serve as the station for the desired instrument approach. The waypoint is set into the aircraft's navigation equipment, and the approach is flown in the same manner as a VOR/DME approach, using the published RNAV approach plate.

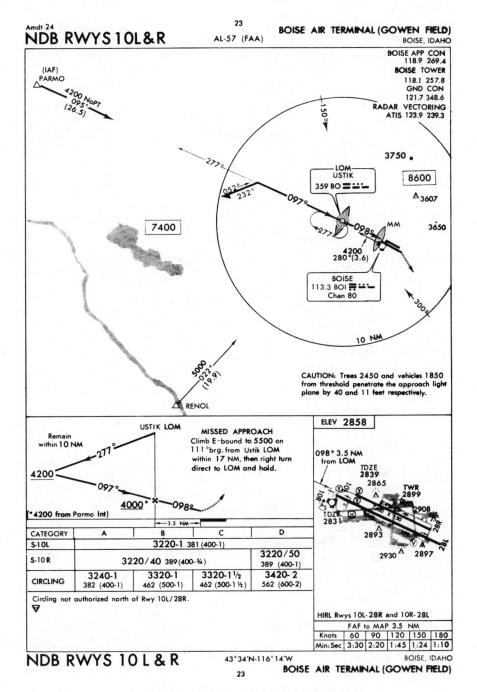

Figure 9.31. Typical NDB (ADF) approach.

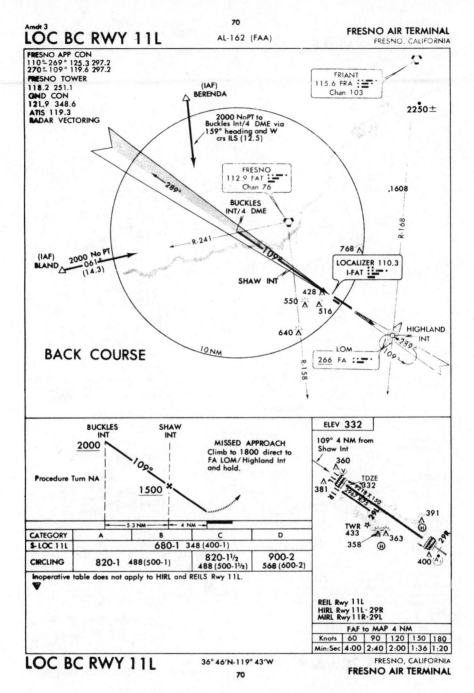

Figure 9.32. ILS localizer back course approach.

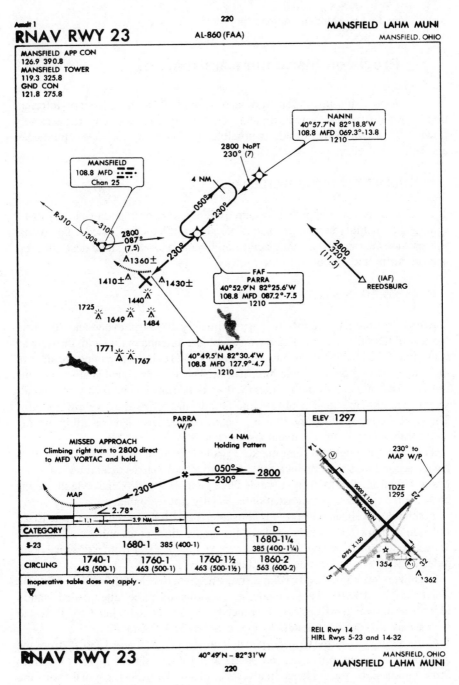

Figure 9.33. Typical RNAV approach.

Precision Instrument Approaches

A precision instrument approach is one for which glide slope information is furnished, all the way down to a DH. Altitude can thus be more accurately flown, and a lower safe altitude is possible. In this group are radar approaches and ILS approaches.

INSTRUMENT LANDING SYSTEM (ILS)

The ILS is a precision approach system designed to permit landings under lower ceiling and visibility conditions than is possible with other radio aids. To do this, ILS provides extremely accurate course alignment and glide slope descent information during the approach to the runway.

GROUND EQUIPMENT

Localizer transmitters are installed approximately 1000 ft. beyond and 300 ft. to the side of the far end of the ILS runway with the antenna in line with the runway centerline. The localizer transmits 90-Hz and 150-Hz signals on opposite sides of the centerline to provide course information. The 150-Hz signal is always on the right as you approach from the outer marker to the runway and forms the "blue" sector. The 90-Hz signal is always on the left and forms the "yellow" sector. The signals overlap along the runway centerline extended. The resulting overlap area forms a course line of equal signal strength.

The course through the outer marker is called the *front course*. Most localizers have a second transmitter which provides also a *back course* extending in the opposite direction. While many airfields have published approaches which use the back course, glide slope information is usually not available.

A reliable localizer will provide a usable signal for a distance of at least 25 miles in a sector 10° on either side of the course line, at an altitude of 1000 ft. above the terrain. You could expect to receive the localizer from a distance of 40 miles at 5000 ft., or 80 miles at 10,000 ft.

Localizers use the 20 odd decimal frequencies between 108.0 and 111.9 mHz (for example, 109.9). They broadcast a continuous three-letter coded station identification which will always be preceded by the coded letter *I* for ILS, and which may use the same letters as the local LF or VOR facility.

Glide slope transmitters, usually installed between 750 and 1250 ft. from the approach end of the runway, send 90-Hz and 150-Hz signals which overlap to form a glide slope for guidance in a vertical plane. The 150-Hz signal forms the lower portion of the glide slope, and the 90-Hz signal, the upper. The transmitter produces a signal which is normally usable at 10 to 15 miles in an 8° sector on either side of the localizer course line.

The glide slope beam width is approximately 1°, half above and half below the glide slope line. The glide slope line elevation varies, depending upon local terrain features. While 2.5° is desired, the angle may vary from a maximum of 4.0° to a minimum of 2.0°.

There are 20 glide slope frequencies, between 331.0 and 335.0 mHz. Most control boxes automatically crystal tune the glide slope when its companion localizer frequency is selected.

Marker beacons, used in conjunction with ILS systems, provide definite fixes along the approach. These markers transmit vertically on a frequency of 75 mHz. They appear on the ILS approach chart the same as any fan marker but are identified by the letters OM or MM for the *outer marker* or the *middle marker.* The outer marker transmits an audible series of dashes at 2 per sec. It is usually located so as to intercept the glide slope within ±50 ft. of the procedure turn altitude and is usually 4 to 7 miles from the end of the runway. The middle marker is identified by continuous alternating dots and dashes. It will normally be from ½ to ¾ mile from the end of the runway and intersects the glide slope 200 ft. above the terrain.

Radio compass locators, if installed, are located with the marker beacons but use separate transmitters. Nondirectional beacons, they operate between 200 and 410 kHz and transmit a continuous carrier wave and a keyed identifier. They are usually 25-watt facilities and appear on the ILS terminal chart in the same manner as any other radio beacons, except that the line forming a square around the radio information is broken by the letter *L*. Locations are identified by only two

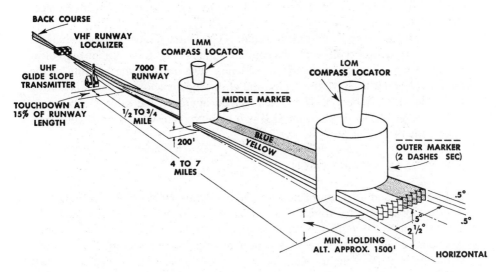

Figure 9.34. Instrument landing system. Many outer and middle markers do not have compass locators.

letters. At the outer marker, these are the first two letters of the three-letter ILS localizer identification. At the middle marker, they are the last two letters.

AIRBORNE EQUIPMENT

Although separate ILS receivers once were widely used, the principal one used now is the VOR receiver, tuned to the ILS frequency. When a glide slope receiver is installed, it is tuned automatically to the proper glide slope frequency when the VOR receiver is tuned to ILS. Localizer and glide slope indications are shown on the course indicator. The CDI indicates localizer course, and the horizontal GSI indicates glide slope. The warning flags appear whenever the localizer or glide slope is not being received. When tuned to an ILS frequency, the TO-FROM indicator and the course selector are inoperative. The heading shown in the course selector window is important. Set at the front course; the CDI is directional in relation to the heading pointer if installed. The RMI is also inoperative unless one needle indicates ADF, tuned to a compass locator.

The CDI, as shown in Figure 9.37, will indicate a full-scale deflection when the aircraft is displaced more than 2½° off the on-course line. Lesser displacement is shown by the number of dots over which the CDI is located away from center.

With a heading within 90° of the course, the CDI will be directional—that is, it will show the relative position of the course line with respect to the aircraft. To center it, fly toward the needle. However, when on a heading more than 90° off the course line—that is, outbound on the front course or inbound on the back course—the CDI is nondirectional, and to center it one flies *away* from the needle.

Glide Slope. The glide slope indicator (GSI) is much more sensitive than the CDI. It will indicate a full-scale deflection when the aircraft is .5° above or below the glide slope line. It is always directional, regardless of the heading of the aircraft. When it is above the center, the glide slope is above; when it is below, the glide slope is below. Therefore, as you approach the glide slope before beginning your final descent, the GSI will be high, and will center as your level course intersects the glide slope.

FLYING AN ILS APPROACH

Refer to your terminal approach chart and plan your approach. Observe particularly the procedure for transitioning from the local navigation facility (VOR, TACAN, or NDB) to the ILS, and whether or not the ADF or RMI, if available, can be used with a beacon or VOR station to help you intersect the ILS localizer at the outer marker. Setting the localizer course in the course selector window will have no effect on the instrument, but it is a handy reminder of the proper no-wind course heading.

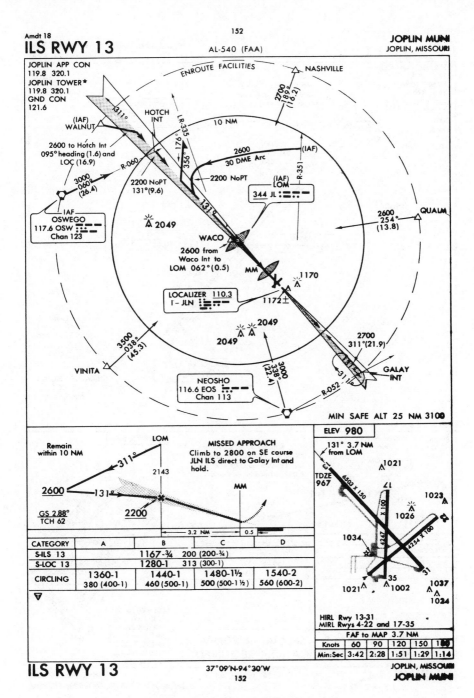

Figure 9.35. Typical ILS approach.

521

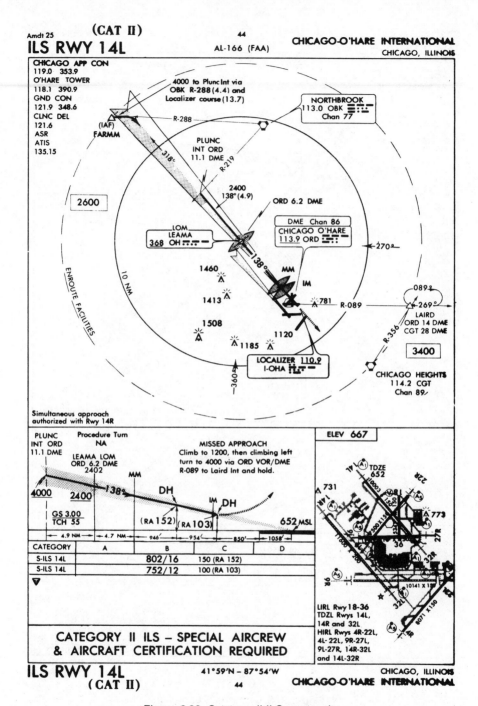

Figure 9.36. Category II ILS approach.

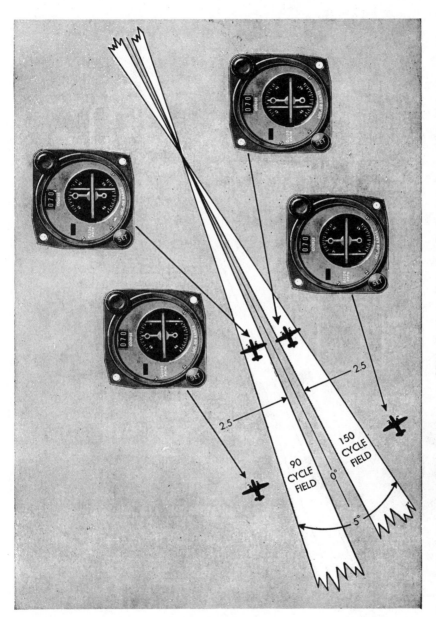

Figure 9.37. Relation of the localizer beam to CDI indications. Note that To-From indicator is blank, front course is set in course selector window, and CDI is directional.

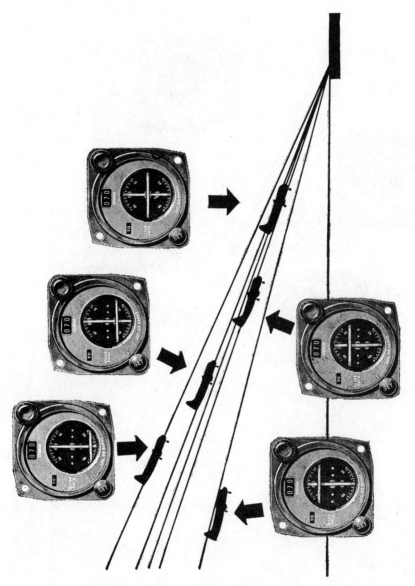

Figure 9.38. Relation of the glide path to the glide slope indicator.

As you intercept the course inbound, either from a procedure turn if one is required, or during your initial interception, the CDI will move from a full-scale to no-deflection rapidly, since it is fully deflected until the aircraft is within 2½° of the localizer course. As you approach the runway, the course will become narrower, and corrections will have to be smaller and more promptly executed. Make all corrections by reference to the control and performance instruments, crosschecking the CDI; a common error is to attempt to "fly" the CDI. Establish landing configuration according to the operating instructions for the aircraft, usually before reaching the outer marker.

Maintain the glide slope intercept altitude while reducing the airspeed to the final descent speed. At most installations, the GSI starts downward from full-scale upward deflection, approaching the outer marker or the compass locator. When the GSI starts to move, prepare to begin your rate of descent.

The angle of descent required is published on the ILS approach chart. Fly the descent with reference to the GSI. Hold your rate of descent constant until the GSI indicates a deviation; then correct by a change in pitch. Maintain airspeed with minimum power changes. During the final approach descent, you will have to scan and crosscheck your instruments with increasing speed. As you approach the runway, indications of deviation from the course and glide slope will occur more rapidly, demanding small prompt corrections to stay on course and on glide slope.

Be ready, as with other approaches, to go to the missed approach procedure immediately when you reach the DH unless you have established visual contact with the runway environment.

RADAR PRECISION APPROACH (PAR)

The only equipment required in the aircraft for a ground controlled precision radar approach is a radio transceiver for communicating with the controllers. In an emergency, once the controller realizes it is an emergency, a communications receiver in the aircraft is enough. Only the primary radar return from the aircraft appears on the precision radar scope.

The surveillance scope is used by the ground controller to direct the aircraft into the final approach, and the aircraft transponder is very useful in this phase, to both identify the radar return and make it more visible on the scope.

Your approach pattern may be either straight in from a fix, or rectangular as in a normal traffic pattern. The published approach chart will tell you whether radar approaches are available at your destination, and will give the frequencies needed. As soon as you have made radio contact with the approach controller, it is important that you follow his instructions precisely. He will first give you any significant changes in current weather, direction of landing, runway information, missed approach instructions, and lost communications instructions.

The controller will give you the headings and altitudes to fly. His reference is

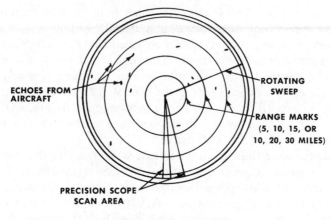

SURVEILLANCE SCOPE (ASR)

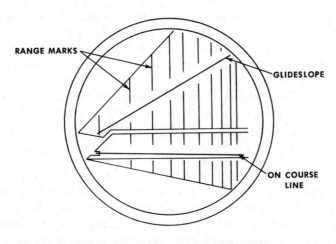

PRECISION APPROACH RADAR SCOPE (PAR)

Figure 9.39. Ground-controlled approach radar scopes.

the track you make after receiving initial heading instructions, so do not change the setting of your gyro-stabilized heading indicator, once set. Descend promptly to the altitudes he gives. You are assured safe terrain and traffic clearance while under radar control, and the controller will periodically report your position and other traffic as appropriate.

Maintain headings and altitudes accurately, and make your turns at standard rate. Though he will advise you when to perform the prelanding check, you may elect to hold final gear and flap settings until you reach the glide slope on final approach. Before reaching the glide slope, slow the aircraft to the final approach

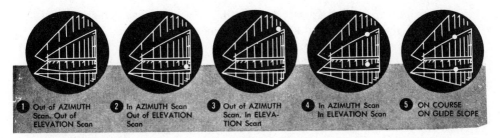

Figure 9.40. How aircraft look to the PAR controller. In normal use, scopes are more cluttered with ground reflections and other interference; however, if the controller cannot see you clearly, he will advise you of this fact and direct you to execute a missed approach.

speed specified in the aircraft operations manual; for a particular gross weight, final approach speed is roughly the stalling speed plus 30%.

When the controller directs you to begin your descent, reduce power or extend drag devices to keep the airspeed constant throughout your transition to the glide slope rate of descent. While this rate of descent is different for different ground speeds and airport glide slope angles, it will be near 500 fpm.

Establish pitch changes on the attitude indicator and crosscheck performance instruments to see if you are getting desired performance. Maintain airspeed with power, but avoid large changes by anticipating the need. This is where the power-speed relationship of your particular aircraft will come in handy if you know it well. Make corrections promptly after the controller says you are above or below the glide slope.

Maintain extremely accurate headings. When instructed to turn right or left to a new heading, turn promptly with coordinated turns. *Use an angle of bank no greater than the number of degrees you wish to turn.* Do not allow the aircraft to deviate from the heading, for the controller bases his corrections on the assumption that you are flying his last assigned heading. He cannot otherwise determine wind effect.

If you have not established contact with the runway environment at DH on a precision approach (PAR), execute the missed approach procedure.

During the entire approach, be particularly aware of the continuity of the controller's voice. *If you suspect that you have lost contact, query the controller, and act immediately to discontinue the approach.*

Landing from Instrument Approaches

The most critical part of an approach is the moment of transition from the instrument panel to the runway ahead. Your glimpse of it may be fleeting, and transition to ground contact under minimum conditions demands careful planning beforehand, as well as established cockpit procedures. *The principles you find here for low ceiling and visibility approaches are also applicable to approach and landing at night.*

TRANSITIONING FROM INSTRUMENTS TO CONTACT

You may have been on instruments for some time. You will be tired and inclined to rush the final part of the flight. Coupled with darkness and low ceilings and poor visibility, the brief time available for changing your complete frame of reference may make the required transition from instruments to contact very difficult, particularly if you are flying a high-performance aircraft.

Cloud bases may be ragged. As you near the end of the approach, you will get occasional glimpses of the ground or the runway ahead. The problem is solved by using your occasional ground reference as simply another factor to be cross-checked in an organized manner. Ground reference is no more important than your other control, performance, or navigation instruments. Approach lights, with which you are familiar as a result of prior planning, are a most important aid, but *continue to fly instruments*. If you have a copilot, have him watch for the runway and report when it is in sight.

Do not follow an urge to reduce power and dive to the runway, but continue a smooth, steady descent to normal flare altitude, transitioning to contact flight only when the visibility of the runway environment makes possible continuation to touchdown by visual references.

If ATC knows you will be able to see the airport before arriving on the final approach, you may be *cleared* for a *visual* approach, or you may *request* a *visual* approach, when you report the airport in sight. At any time during the approach that you are able to establish position and proceed by visual means to the airport, you may request a *contact* approach, even though the runway itself is not yet in sight.

CIRCLING APPROACHES

There are occasions when an approach does not culminate with the aircraft lined up for the landing runway, either because the wind does not favor landing on the runway aligned with the approach, or the approach path is more than 30° from the direction of the runway to be used. Some kind of circling approach must then be made. The approach plate will specify the Minimum Descent Altitude (MDA) and the required visibility.

A circling approach may be required under marginal conditions of ceiling and visibility. Altitude control can be very important, and the landing area must be kept in sight. Figure 9.41 shows some possible patterns for circling and low visibility approaches.

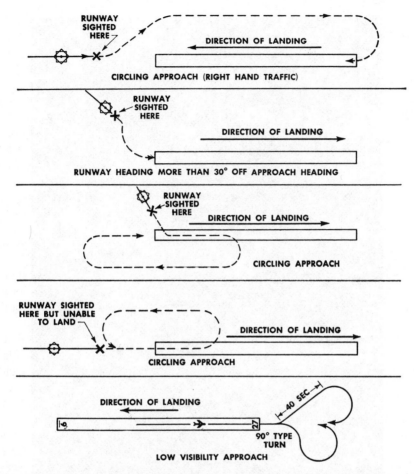

Figure 9.41. Circling and low-visibility approaches. Plan type to use well in advance, depending upon local obstructions and weather. Precision in altitude, airspeed, and heading is essential. These will usually be flown at the published circling approach minimum and below local VFR traffic pattern altitude.

APPROACH LIGHTING

Night landings are made much easier by a lighting system that aids the pilot in making a normal approach, without the usual cues available in daylight. A good approach lighting system also helps considerably in establishing visual contact with the runway environment during an instrument approach. Research and testing by the Air Force, FAA, and the airlines culminated in what is known as the Integrated Visual Approach and Landing Aids (IVALA) system (Figure 9.42).

The sequenced flashing lights are a key feature of this system and are used for

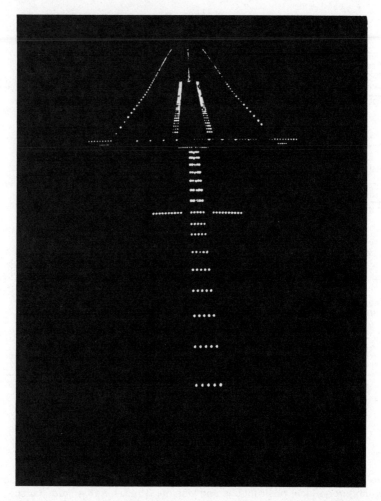

Figure 9.42. US Standard ALSF-1 lighting system with Touchdown Zone Lighting (TDZL). Sequenced flashing lights are not visible here because of short flash duration.

other purposes as well. Although they are of 30 million candlepower or higher, there is practically no blinding effect because of the short duration of light emission. In the IVALA system, they flash swiftly toward the runway, cycling twice per second. The effect is that of a tracer bullet fired along the approach to the runway threshold. Strobe lights also are used to identify the active runway, since their brightness and characteristic flash make them easy to pick out from among all the other confusing lights a pilot sees near the larger airfields at night.

Touchdown Zone Lighting (TDZL) is designed to provide depth information for touchdown and directional information during the runout. It is needed because on wide runways the side runway lights do not provide sufficient reference.

LEGEND
INSTRUMENT APPROACH PROCEDURES (CHARTS)
APPROACH LIGHTING SYSTEMS – UNITED STATES

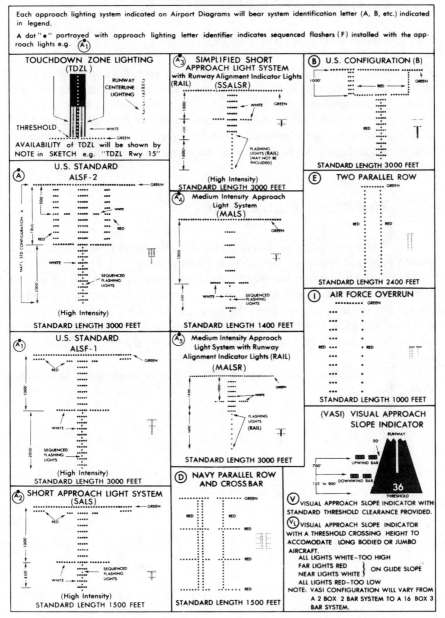

Figure 9.43. Approach lighting systems.

It is shown in Figure 9.42. Because each light fixture is covered by a steel grid capable of supporting aircraft weight, the light is not blinding. Heating elements in each light aid in ice and snow removal.

PRESENT APPROACH LIGHTING SYSTEMS

While the IVALA system has been adopted as standard, full conversion to it will not happen immediately. Its high cost means that it will be some time before existing systems are phased out in favor of it, or upgraded to its standards. Also, some airports, because of the nature of their traffic, do not need the full blown IVALA system. For these reasons, the other approach and runway lighting systems shown in Figure 9.43 will be seen for some time to come.

The Pilot's Burble Point

When a wing has too heavy a load, it stalls; the airflow departs from its smooth course at what may be called its "burble point."

When a pilot encounters weather beyond his normal capability, or when an unforeseen event occurs at a critical point in an approach, or when fatigue dulls his perceptions or slows his reactions, he must call on all his reserves of training, skill, and discipline to avoid reaching saturation, or his "burble point." Training, practice, and careful planning all prepare him for the time when "the chips are down."

The air, even more than the sea, is unforgiving of carelessness or incompetence.

10

Flying Higher-Performance Aircraft

The turbine engine in high powered aircraft eliminated the complications and vibration of the Otto cycle. Propellers have been very successfully combined with turbines to produce aircraft having both the short field capabilities of propellers and the smoothness and reliability of the jets. But for high altitude, speed, comfort, and the ability to carry heavy loads between big airports, the turbojet and the turbofan are unchallenged.

Jet Aircraft Characteristics

Jets differ from prop aircraft in take-off, climb, speed, and handling qualities; in the increased importance of the environmental factors discussed in Chapter 7; in the increased preflight planning needed; and in the fact that though most jet flying is above the weather, it is, even in aerobatics, largely attitude instrument flying.

It is important to realize that most light aircraft and heavy multi-engine prop aircraft, having straight wings and operating at lower speeds, are basically stable longitudinally and directionally. These aircraft move about their axes in a way that can be considered almost mutually independent, that is, they move in uncoupled modes of motion. Transonic and supersonic aircraft, on the other hand, can move in modes that result from coupling of longitudinal and lateral directional response. This coupling is caused by increased dominance of inertial and gyroscopic forces at high rolling rates over stabilizing aerodynamic forces, and while it becomes more and more pronounced as speed increases, it is more a function of configuration than speed. While relatively "stable" aircraft can be trimmed to fly "hands off," this is almost impossible to achieve in swept or delta wing jets. Stability augmenting systems are essential in most swept or delta wing jets to aid the pilot in his control responses. The Lear Jet is an exception. It is quite stable and can be flown "hands-off" at all altitudes including its maximum of 41,000 ft.

It is important to understand other characteristics in which prop and jet aircraft differ. In jet aircraft, level flight and dive speeds are higher, climb speeds are greater, but take-off performance is generally inferior; engine controls, however, are simplified and torque is eliminated. The jet's outstanding performance characteristics are not achieved automatically. In order to get the best results, the pilot must be willing to apply techniques different from those he would use under similar circumstances in prop aircraft. Further, he will be greatly helped by more training, particularly in theory, and he must "think ahead" of the aircraft he is flying.

Thrust characteristics of the turbojet engine are responsible for these airplane characteristics. Since power is the product of thrust and speed, the power available is directly proportional to the speed: low at low speed, high at high speed. This dependence of power on speed explains the inferior take-off characteristics of some jet aircraft as compared to their superior performance characteristics after accelerating.

The relationship of speed and power also explain the high climb speed of jets. Maximum rate of climb occurs at the speed of maximum excess power, as shown in Figure 10.1.

Note that maximum excess power for the propeller-driven aircraft is shown at A. The corresponding speed is the best climb speed for the propeller-driven aircraft. The situation in the jet aircraft is different. B indicates that the speed for maximum excess power is relatively high. Therefore, the best climb speed for jet aircraft is correspondingly high and produces much higher rates of climb. One can readily see how important it is to maintain proper climb speed in order to obtain optimum power. This higher jet climb speed and rate should not be confused with the very high angle of climb observed, for example, as a Boeing 727 takes off. This high angle of climb at the relatively low speed after take-off is largely the result of the high thrust-to-weight ratio of its three engines and take-off weight.

Another jet aircraft characteristic associated with the speed-power relationship is the jet's ability to obtain a variation of airspeed in level flight without any change in throttle setting. At certain cruise conditions (particularly at speed

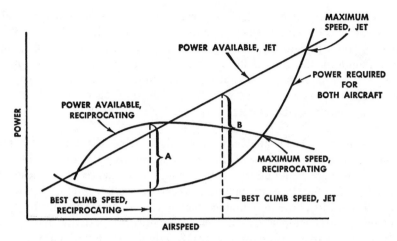

Figure 10.1. Approximate relationship between best climb speeds for reciprocating and jet engines.

ranges just below optimum cruise), as the speed increases—with no change in throttle—the power available also increases just enough to equal the power required. This characteristic makes it necessary for the pilot to accelerate *above* his cruise setting and then *reduce* power to recommended cruise settings. Theoretically, there is a proper power setting for every airspeed. At high altitude, only a slight increase in power is required to obtain a significant increase in speed.

Flying on the low-speed or "back" side of the speed-power curve (also called "area of reversed command") is most likely to occur when the airspeed is low, as in the traffic pattern or during a go-around, and under high-attitude, high gross-weight conditions, regardless of the type of aircraft (Figure 2.3). This is not an ideal situation, and may be very dangerous, because the angle of attack is high, approaching stall conditions. Full power may be inadequate to maintain level flight. The best, and often the only, solution for recovery is to put the nose down and sacrifice altitude, if available, to regain speed.

FLIGHT PLANNING

Most smaller jet aircraft have small cockpits. The extensive planning necessary if the flight is to be efficient and safe cannot be accomplished after the plane is airborne because of limited time and space. Therefore, preflight planning must be precise. The pilot, who usually does the high-speed navigation as well, must do it rapidly and accurately.

Weather Considerations.   In planning your jet flight, you should not be much concerned with en-route surface weather, except as it affects your choice of emergency en-route landing fields. The wind, however, is very important, because of the rate of fuel consumption, particularly if a jet stream should lie along the route. If it is blowing in your direction, plan to fly in it. If it is against you, plan to fly above it, below it, or to one side. The choice you make is determined by the altitude of the core of the jet stream, the velocities of the surrounding winds, and the aircraft performance at that altitude. Because of high cruise speed, wind has less overall net effect on the range of jets than on the range of conventional aircraft, since the jet is exposed to a particular wind a shorter period of time.

Fuel Planning. The available fuel for a flight includes, except in rare instances, the full capacity of the tanks. For planning purposes it is divided into three portions. These are the requirements for climb, cruise, and reserve at destination or alternate. The fuel for start, taxi, and take-off is included in the requirements for climb. For this phase of planning, use the cruise control data from charts in your Operations Handbook. Cruise fuel is that projected for use from the end of a climb to arrival over destination at cruise altitude. Reserve includes the fuel needed for descent, approach, and landing, as well as a safe margin to allow for adverse winds, landing delays, and flight to an alternate airfield if required.

TAKE-OFF PERFORMANCE CALCULATIONS

The take-off performance of jet aircraft is highly sensitive to temperature and pressure changes. The effect is illustrated by this tabulation from the Lear Jet Take-off Performance Charts, using a take-off weight of 12,500 lb.

TABLE 10.1 Take-off Performance Chart.

Pressure Altitude	Temperature	Take-off Roll (ft.) (no wind)
Sea level	59°F	3150
Sea level	95°F	4150
5000 ft.	41°F	4500
5000 ft.	95°F	6500

For every take off it is necessary that you not only compute take-off roll, but also the acceleration check speed and time, go-no-go speed, refusal speed, and decision speed. The commonly used terms are as follows.

Take-off Ground Roll: The distance required with existing conditions for the airplane to accelerate to take-off speed with all engines operating normally.

Rotation Speed (V_R): The indicated airspeed at which the nose wheel should be lifted from the runway to obtain the desired performance.

Take-off Speed (V_2): The indicated airspeed at which the airplane can be safely flown off the runway. This is also the minimum speed for directional control in the air in the event of engine failure—usually about 5 knots above rotation speed.

Critical Field Length: The length of runway needed under existing conditions for the airplane to (1) accelerate to critical engine-failure speed, (2) lose all power from one engine, then either (3) continue to accelerate with the reduced power available to safe takeoff speed, or (4) stop, using brakes, drag devices, and reverse thrust if available, within the remaining distance.

If critical field length is less than the runway length available, a safe take-off can be made allowing for the loss of one engine at or beyond critical engine-failure speed. If critical field length is greater than runway length available, the airplane cannot either stop or become safely airborne within the remaining runway, should an engine be lost at critical engine-failure speed. While marginal conditions may tempt one to take a chance, the solution, except in an emergency, is to off-load enough weight to make critical field length equal to or less than runway length.

Single-Engine Go-No-Go Speed: The minimum speed from which single-engine take off should be attempted.

Critical Engine Failure Speed (V_1):   The speed at which, under existing conditions, should one engine fail, the distance required to complete the take off exactly equals the distance required to stop.

Acceleration Check: The time required to accelerate between two preselected speeds—usually 80 to 120 knots (or V_1, whichever is lower). If at the expiration of the precomputed time, the 120 knots (or V_1) has not been reached, the take off should be discontinued.

Due to many variables, including wind, that enter into checking acceleration against a point along the runway, most acceleration checks in aircraft with more than one crew member are made using the time and speed method. This time may be quite short. For example in a Boeing 707 weighing 240,000 lb. on an 80° day at 1000 ft. pressure altitude and a flat runway, the time between 80 and 120 knots would be 10.4 sec.

STARTING YOUR AIRCRAFT

Starting usually requires a ground crewman in communication with the pilot by ground cord and interphone. Observe the normal safety precautions, particularly the distance from intakes to ground crewmen and tailpipe distances to other aircraft or persons. Safe distances are shown in the applicable Handbook listed under ground operation. Chocks and a suitable fire extinguisher are considered essential.

There is a certain sequence of events that occurs in the starting of jet engines. While the starting sequence may be manual or automatic and vary with the aircraft, the fundamentals are the same.

First: The engine must reach a high enough rpm to obtain a steady flow of air through it. Many jet aircraft will require auxiliary power units to obtain the proper start rpm. Most jets start using one of three types of systems: the explosive cartridge-operated starter, compressed air, or electric starter. Electric starters are most common for the smaller jets. Commercial-type business jets do not require the auxiliary ground equipment used and required for military and airline jets, because they require less energy to accelerate the compressor and turbine.

Second: An ignition spark must be introduced into the sequence in the burner cans, to ignite the fuel when it is injected.

Third: Fuel is sprayed into the burner cans through the fuel manifold and controlled through the fuel nozzles by use of the throttle. The proper sequence is absolutely necessary to prevent a fire on start. If the raw fuel were to enter the fire cans first, the introduction of the spark would result in a mild explosion and fire. Most jet starting systems are automatic and sequenced through the electronic fuel control. The primary instruments to watch during start are the exhaust gas temperature (EGT) gage and fuel flow meter.

Each jet engine has a maximum allowable temperature for starting. Temperatures increase very rapidly and must be monitored closely to avoid exceeding the limit. You should have your hand on the throttle ready to shut off the engine if the EGT should approach or exceed the start limits. The EGT gage is usually calibrated in 50°C increments from 0°C to 1000°C. Normal start range would be 500° to 700°C with an allowable instantaneous 800°C reading. If the temperature should exceed 800°C for 2 sec. or longer, it would be a *hot start.* With a hot start, shut down immediately. Allow a few minutes for fuel drainage, then attempt another start. Should the temperature exceed 900°C even momentarily, make no attempt to restart the engine because it will have to be removed for disassembly and detailed "teardown" inspection without further operation.

After completing the start, use your check list for all systems operation. The ground crewman will assist by visually checking that hydraulic-controlled rudder and elevator slab are in proper take-off position and that flap-to-stabilizer or similar control interconnect is correctly set for takeoff. Turn on your radio, call the tower for clearance, and proceed with your taxi.

TAXI AND TAKEOFF

The same taxi precautions applicable to prop aircraft apply to jet aircraft. In order to leave the parking area, it is necessary to increase power to approximately 60% to 70%, start rolling, check your brakes, and engage nose-wheel steering. Taxi out of the area to the taxi lane, complete pre-take-off checks as required, and obtain clearance to take the runway and line up. When cleared to take off, advance to 100% rpm or to limiting exhaust pressure (no warm up is required), release brakes, and you're on your way.

You will notice that there is no torque-produced yawing in a jet as there is in a propeller-driven aircraft. The takeoff distance and acceleration will probably be the first noticeable difference. You will feel you are not going to become airborne as the runway is rapidly used up. However, the aircraft accelerates slowly at first, and then more rapidly until you reach your precomputed rotation and lift-off speeds. This is where you must remember your takeoff performance calculations and use them. Upon reaching rotation speed, ease the nose up and the aircraft will fly off the ground. Do not attempt to pull the aircraft into the air because it may mush back into the ground. It cannot "hang on the props." The takeoff angles will be rather shallow initially. Crosswind effect on jet aircraft is not of much consequence, but during flight planning you should check the cross-wind component chart for your aircraft because in severe crosswind the nose wheel must be held on the ground longer than normal, and the aircraft crabbed into the wind immediately after liftoff to remain over the runway.

CLIMB

In the T-38, the initial climb is the same pitch attitude as takeoff (about 5° to 7° nose up) until reaching 300 knots. At 300 knots the throttle is retarded to shut off afterburners for a military power climb unless a continuous maximum power climb is desired. The aircraft is allowed to accelerate to the prescribed climb speed schedule, which is about 400 knots up to 10,000 ft., then to .78 M to 15,000 ft., .81 to 20,000 ft., .86 to 25,000 ft., .90 to 30,000 ft., .92 to above 40,000 ft. If a maximum power climb is to be maintained from takeoff, then the pitch angle is approximately 25°. The aircraft is allowed to accelerate to .92 M and holds this throughout the climb. This rate of climb will be in excess of 12,000 fpm.

Aircraft capable of high performance and high pitch angle climbs, such as the T-38 and Lear Jet, have visibility restrictions over the nose in this attitude. Therefore, it is highly desirable that in areas of air traffic concentrations, the climb be conducted under radar control or in assigned climb corridors.

Air Traffic Control centers are familiar with jet aircraft and the excessive fuel consumption they experience at low altitude. They will clear you to climb to altitude as rapidly as possible and try to avoid giving you involved low altitude departure procedures.

During the acceleration after takeoff and during climbout, the aircraft will feel stable and solid. The electric elevator trim control, usually under the thumb on the pilot's stick or yoke, permits continuous trim adjustments as the aircraft accelerates. Large changes in pitch and roll can occur fast and easily, however, unless one pays close attention to the pitch and roll attitude, to the airspeed, and to trim during the climb-out.

CRUISE

Fuel consumption is the primary problem in jet cruising. It is only the high speed obtainable that makes jet flying practicable and economical in cost, despite the fuel needed for acceptable ranges. The ideal flight profile in a jet aircraft is a climb to the most efficient initial altitude for the weight, followed by a slow climbing cruise as the weight is lowered by fuel consumption. In practical application, however, both from the standpoint of pilot technique and air traffic control, jets, like other aircraft, fly at assigned altitudes, which, if changed, are changed in steps.

In order to cruise efficiently, jets must cruise at a high percentage of maximum power. The thrust of a jet engine varies directly with power, roughly as follows:

$$100\% \text{ rpm} = 100\% \text{ thrust}$$
$$90\% \text{ rpm} = 75\% \text{ thrust}$$
$$80\% \text{ rpm} = 50\% \text{ thrust}$$

During cruise at altitude, a combination of fuel flow in pounds per hour, percent of maximum rpm, exhaust gas temperature (EGT), and engine pressure ratio (EPR), in some aircraft, is used to obtain optimum cruise. With proper preflight planning, it is seldom necessary to make corrections to power to maintain the flight plan and its schedule of fuel consumption. In large airline transports fuel is limited and carefully computed even over short routes because of the cost of carrying unneeded thousands of pounds of fuel. In the T-38 and Lear Jet, fuel is limited because the size of the aircraft limits the amount available. For this reason, some sort of meticulously prepared and meticulously kept fuel log is essential. While there are many types of logs designed to suit many different needs, one which has served well in the Air Force is shown in Figure 10.2.

While the difference likely to appear between forecast and actual winds may cause the computations to be off, the chances of their being either favorable or unfavorable are about equal; an appropriate adjustment can be made at the first check-point, and the cumulative effect can be detected early enough to alter the flight plan if necessary, insuring always the reserve needed to meet the unforeseen and land. The T-38 normal fuel flow at 40,000 ft. is 1400 lb./hr.; if, for example, you reached the cruise altitude with 2800 lb. of fuel remaining, you would have enough for only about an hour and 30 min. of flight with an 800 lb. reserve. Here again, Air Traffic Control is cognizant of this problem and will usually expedite your descent with minimum holding time at lower altitude, or provide you an en-route descent under radar control.

Charleston AFB - Grissom AFB

PILOT'S FLIGHT PLAN AND FLIGHT LOG

CLEARANCE M → 3000 expect	TAKE-OFF, CLIMB, CRUISE DATA
FL 390 10 min aft dep	FL 390
DC 319.8 Rwy Hdg 2 min	TAS 516 K
Sqawk 0300	I. M. N. .9

FREQUENCIES	Gnd 348.6	Cl. Del. 381.6
	Twr 239.0	

DEP FIELD DATA 2415				TOTAL DIST 577		TOTAL ETE 1 + 07		TOTAL FUEL 3790	
ROUTE	IDENT	LAT/IDENT	MAG CRS	DIST	GND	ETE	ETA/ATA	FUEL	ACTUAL FUEL
FIX	FREQ	LONG/FREQ			SPD			3790	
STTO								290	
								3500	
Vectors	CAE			77					
Columbia	94			77		↓			
J-47			331	10		11		880	
L/O				87		11		2620	
J-47	SPA		331	83		9		210	
Spartansburg	104		330	170		20		2410	
D→	TYS		300	110		13		310	
Knoxville	111			280		33		2100	
J-99	LOU		330	155		18		430	
Louisville	95		327	435		51		1670	
J-99	Romni		339	142		16		375	
GUS 228/30	112			577		1+07		1295	
Pen. & App, AFR 60-16 Res.						30		900	
						1+37		395	
Fuel Remaining						15		395	
						1+52		-0-	

AF FORM 70 OCT 78   PREVIOUS EDITION WILL BE USED

Figure 10.2. Pilot's flight plan and flight log.

MANEUVERING

While some jet aircraft, like the Lear Jet, have mechanical linkage controls, fighter types and large transports have hydraulically operated controls. Without this augmentation, it would be impossible to maneuver them easily and efficiently at high speeds. With augmented controls, fighters are highly maneuverable at both supersonic and subsonic speeds. Augmented controls are provided with an artificial feel system to indicate changes in pilot's control pressure.

The development of high-performance jets has been marked by problems in stability, and in the past most high-performance aircraft were very unstable. Either through design improvements in newer models, or modification of older ones, jet fighters and transports are now quite stable in the entire speed range.

These early stability problems were generated by lack of pilot understanding of dihedral effect and inadequate design compensation for it. Dihedral effect is manifest in two distinct flight characteristics. They are *Dutch roll,* or roll due to yaw, and *adverse yaw,* which is yaw due to roll. Both are primarily encountered while maneuvering at high angle of attack or in turbulence. Yaw dampers and stability augmentation systems of varying design are incorporated in most high-speed aircraft to minimize the effect.

If an aircraft is caused to yaw by rudder mismanagement, turbulence, or in multi-engine aircraft, unbalanced power, the wing away from the direction of yaw will swing forward, generating more lift and thus rise, causing a roll in the same direction as the yaw. When this occurs, it is relatively easy for the pilot to get out of phase with the oscillations and accentuate rather than dampen out the roll. In some aircraft it is possible to have the aircraft on its back in about five oscillations if yaw dampers and stability augmentation are not present or proper recovery technique is not applied.

With most aircraft, proper recovery technique is to neutralize rudder and level the wings with ailerons. In the Boing 707 the period of oscillation is relatively long (4 to 7 sec.) and as in all aircraft, can be controlled by stopping either roll or yaw. Dutch roll cannot exist without both roll and yaw.

Adverse yaw, yaw due to roll, is most pronounced at high angle of attack when an attempt is made to roll the aircraft with aileron. The yaw is produced by the drag of the down aileron; the dihedral effect, in turn, inhibits the roll. The yaw in this case is away from the intended direction of roll and the aircraft will in fact turn opposite to that intended. At very high angles of attack, that is, at or near stall, aileron inputs may cause rapid spin entry.

Delta wing aircraft which do not have conventional horizontal tail surfaces but depend on elevons for the elevator function may have *proverse yaw.* Proverse yaw is yaw in the direction of the turn. When the rudder is longitudinally located in the vicinity of the elevons (serving as both ailerons and elevators), displacing the rudder to the right and the right elevon up, as for a normal right turn entry, causes a build up of a pressure field on the right side of the vertical stabilizer. This pushes the tail left, which is a yaw to the right. Aircraft with this characteristic have built-in functions in the flight control system and stability augmentation system to compensate for this phenomenon. The rudder may even be deflected left for a right turn.

STALLS

The McDonnell Phantom II is typical in its stall characteristics of swept wing supersonic fighters. Stalls of 1g are preceded by a wide band of buffet warnings. Onset of buffet normally occurs at approximately 40 knots indicated airspeed above stall. A rudder shaker is activated by the angle of attack indicator. *Wing rock* is generally unpredictable but starts about 10 knots indicated air speed (KIAS) prior to stall and can progress to as much as 40° bank at stall. If there are no rudder or aileron inputs, wing rock may be delayed or absent. The stall is characterized by a yawing motion in either direction. The yawing is caused by a loss of directional control.

Stall recovery is effected by positioning the stick forward of neutral, while holding aileron and rudder neutral. Accelerated stalls are preceded by a moderate buffet, increasing progressively to heavy buffet just prior to stall. Prompt neutralization of controls will quickly effect recovery from accelerated stalls.

Clean configuration stalls are practiced only up to heavy wing rock or nose slicing (yawing). In the landing configuration the Phantom II has substantially the same characteristics as in the clean configuration. Wing rock is somewhat reduced, and a nose rise accompanied by lightening of aft stick forces is noted just prior to the stall. Even with maximum power, a sacrifice of up to 3000 ft. of altitude may be required to recover from the landing configuration stall.

The Boeing 707 is typical of large high performance jet transports in its stall characteristics. As the stall is approached in the clean configuration, a very gentle buffet may be felt in smooth air. At 15 to 20 knots slower, another buffet is felt but is still not strong and may be masked by turbulence or by the flaps and/or landing gear, if extended. Just prior to stall an airframe buffet is felt. This is a much stronger buffet and much less likely to be masked or confused with anything else. The buffet continues, and just at the stall a mild nose rise occurs. Following the stall the nose pitches down, and normal recovery procedures are easily executed. Due to the wide range of weights and configurations, altitude loss during recovery will vary considerably, but 1500 ft. loss of altitude is about average.

Stalls in the Lear Jet are not normally practiced in checking out the airplane because its stall warning devices make approaches to stalls an adequate preparation. From a moderately nose high attitude, the Lear Jet will stall and recover smoothly. In a straight ahead landing attitude stall, the nose will fall through with little or no roll. With power increased, the aircraft returns to straight and level flight.

The Lear Jet Model 23 has a *stick shaker* which provides a stall warning when the airspeed is 5% to 7% above stall speed. To assist in stall recoveries, and to inhibit further possibility of an inadvertent stall, a *stick pusher* is installed.

It is characteristic of T-tail airplanes to pitch up viciously when stalled in extreme nose-high attitudes, making recovery difficult or violent. The stick pusher inhibits this type of stall. About 1 knot above stall speed, an 80-lb. force automatically moves the stick forward, preventing the stall from developing. A *"g"* limiter is incorporated in the system. This prevents the pitch down generated by the stick pusher from imposing excessive loads on the aircraft. In addition, an angle of attack indicator is installed on the instrument panel, both to warn of near stall attitudes, and to serve as a checking reference for the stick shaker and stick pusher.

From a normal stall, no forward pressure is required for recovery; rather a release of the back pressure that was needed for the entry will suffice. The stall speed in the landing configuration at 12,500 lb. gross weight is 98 KIAS. The angle of bank increases the stall speed proportionately, as in any airplane. With the landing configuration and the bank indicated, the stall speeds are: 20°—102 KIAS; 40°—113 KIAS; and 60°—143 KIAS. A turning stall in the Lear Jet is accompanied by slight buffet; sufficient warning is given to the pilot during stall approach. At 12,500 lb. in a 1.5 g maneuver, a low-speed buffet will occur at 145 knots. High-speed buffet at 1.5 g does not occur until the speed is in excess of the maximum allowable Mach.

As in the straight ahead stall, a turning stall recovery is made by releasing the back pressure, increasing power, and leveling the wings. The recovery is very smooth. A crosscheck of the vertical speed indicator is necessary to determine sink rate associated with the complete stall.

Due to the extreme aerodynamically clean design and the great mass of jet aircraft, very rapid speed buildup can be anticipated in the nose down attitude. In view of this mass and rapidly building speed, a high sink rate should be anticipated during and immediately after rotation out of a stall recovery dive. A pilot's overreaction to this high sink rate can cause him to place the aircraft in a high angle of attack again and risk an accelerated secondary stall.

While all high performance aircraft will spin and will recover from spins, they are not practiced. Very great altitude losses are required in recovery. High-performance aircraft are also subject to another post-stall maneuver beside the conventional spin. This is simply called the *post stall gyration* and is described as a departure from controlled flight with random rotations about any or all axes. If the angle of attack is not reduced, these can be expected to develop into a spin. Spin recoveries are generally effected by neutralizing rudder and aileron and use of pitch control to reduce angle of attack. Drag chutes may also be used in this effort. In flat spins, if the drag chute is deployed early in the spin, recovery may be possible. Unfortunately, in fully developed flat spins, there is no known recovery technique for swept wing aircraft.

Aerobatics and High-Speed, High Angle of Attack Maneuvering.   Modern jets like the advanced trainers and jet fighters can perform the full range of aerobatic maneuvers easily and precisely. The lack of propeller torque makes only a little rudder control necessary, even in "over-the-top" maneuvers such as loops and Immelmans. There is, however, considerable altitude variation and lateral space required if these are performed at high speed. An F-4, with its wide range of speed and weight, may require from 3000 to 10,000 ft. or more altitude to execute a loop. Speed buildup in vertial descent is very rapid.

The use of speed brakes is one of several pilot techniques used in controlling acceleration in descending maneuvers. Speed brakes may be used at any airspeed, and though they produce some buffeting and vibration at higher speed, they affect trim and control characteristics very little.

The ability of the fighter pilot to maneuver his aircraft for long periods of time at high speed and at very high angles of attack is his stock in trade. It is immaterial whether he maneuvers his aircraft into an attacking position on an opponent or forces his opponent into uncontrolled flight. The end result is the same. In other fields of flying, except perhaps for crop dusting, high angle of attack maneuvering is generally confined to takeoff, approach, and landing, and during conditions of reduced thrust.

Because of the great amount of vertical and horizontal space required to maneuver, and the rapid, positive control response of high performance aircraft, a full understanding of high angle of attack maneuvering is essential.

The term *high angle of attack* is frequently confused with two other terms. They are "pitch angle" and "flight path angle." Pitch angle is the angle formed by the longitudinal axis and the true horizon. Flight path angle is the actual flight of the aircraft through space relative to the true horizon. Angle of attack is the difference between the two (see Figure 5.27). Angle of attack for all practical purposes is a function of gross weight, indicated airspeed, and *g loading*. Altitude is not a factor since we are concerned with indicated, not true, airspeed.

Of importance to the high performance aircraft pilot is the effect *g* loading has on his aircraft. There is no difference from the effect on light aircraft, but the greater potential for high values of speed, weight, and control response make *g* loading and unloading of the aircraft a prime control technique.

g loading is equal to the product of weight and the secant of bank angle, provided the aircraft is being maintained in a constant plane such as a constant altitude, *g* loading is always controlled by the longitudinal control (elevator or its equivalent). *g* loading and unloading can be used to accelerate or decelerate the aircraft as well as to increase or decrease its stall speed.

To illustrate, assume a 50,000 lb. aircraft in 1 *g* flight. Fifty thousand pounds of lift are required to sustain it. If the aircraft is placed in a 60° banked turn, 2 *g*'s will be generated to hold altitude in this turn, and now 100,000 lb. of lift are required and are acquired by increasing the angle of attack. The stall speed has now also been increased. Conversely, if the same aircraft is maneuvered into less than 1 *g* flight, such as .5 *g,* only 25,000 lb. of life are required, and the stall speed has decreased.

Theoretically an aircraft maneuvered into zero *g* flight requires zero lift and thus will not stall at zero IAS. The only problem here is that with aerodynamic controls there would be no way to maintain the zero *g.* As long as control can be maintained, *g* loading and unloading an aircraft is a highly effective control technique. For slowing down aircraft, a high *g* turn or roll will rapidly bleed off airspeed. For getting out of a high angle of attack condition, reducing *g* to as much as zero *g* will prevent a stall. A technique used by fighter pilots to accelerate rapidly is to use full thrust and then maneuver to near zero *g.* When this is done from a high *g* (4 *g* or 5 *g*) maneuver, it produces spectacular acceleration.

The Angle of Attack (AOA) Indicator.   This is a relatively recent development in aircraft instrumentation. The need for this instrument was recognized by Wilbur Wright as early as June 1907, but the production of an economical, reliable system was long in coming. Most modern fighter aircraft are now equipped with these instruments, and large jet transports are rapidly being equipped. Currently,

INDICATOR	INDEXER	ANGLE OF ATTACK UNIT	AIRSPEED	ATTITUDE
		20.3-30	VERY SLOW	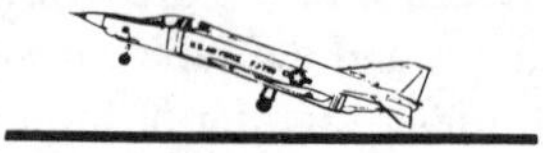
		19.7-20.2	SLIGHTLY SLOW	
		18.7-19.6	ON SPEED	
		18.1-18.6	SLIGHTLY FAST	
		0-18.0	VERY FAST	

Figure 10.3. Angle of attack conversion and displays. *(Courtesy of US Air Force.)*

standardization of presentation has not been achieved. Most indications are indexed in arbitrary "units" rather than specific angles, and some have only warning lights indicating *"slow," "on speed,"* or *"fast"* for final approach to landing. The required angle of attack will change with flap setting, and not all systems now in use compensate for this. Some systems are affected by airflow distortions caused by landing gear position and are inaccurate with gear either up or down, though in other installations the AOA indicator system compensates for change of gear and flap position. These installations usually consist of an indicator on the instrument panel, indexer lights on the wind screen, and a headset tone that increases in pitch as angle of attack is increased. A rudder or stick shaker or pusher may also be incorporated. As presently designed and used, the AOA indicator system is predominantly a stall warning system. With increased use, and since an aircraft wing has a desired angle of attack for any particular maneuver, that is, final approach, climb, maximum endurance, and maximum range, the AOA is rapidly relegating the airspeed indicator to a purely navigational instrument.

JET FIGHTER-TRAINER TRAFFIC PATTERN

Dear to the heart of the pilot of jet fighters and trainers is the "tactical approach" pattern shown in Figure 10.4. Originally designed to keep high-performance fighters within gliding distance of the field once the landing approach was started, it found use in early and present small jets because it does permit the rapid landing of many aircraft, as from a formation, usually all low on fuel. It is seldom used by other jets because of passenger discomfort and because jet transports and bombers are usually making some form of instrument approach.

The aircraft enters an initial approach about 3 miles out from the end of the runway, at 1500 ft. above the terrain. Airspeed on initial approach varies with

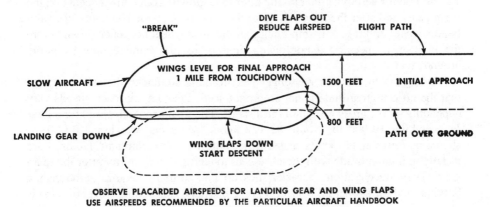

Figure 10.4. Typical jet fighter landing pattern.

the airplane, and is about 280 knots for the T-38. On the initial approach, the pilot lines up with the runway or slightly to the traffic side. The *pitch* or *break* point is approximately halfway down the runway. Retarding power slightly, the pilot makes a constant 60° bank to the downwind leg, reaching the gear lowering speed by exerting *g* forces, and the airspeed decreases proportionately to the *g* forces used in the turn.

APPROACH AND LANDING

From a tactical pattern the normal approach to the runway is from a point ¾ to 1 mile out from its end. At this point, the proper final approach speed, approach angle, and landing attitude are attained. The altitude is normally 400 ft. above the terrain. Now only minor throttle corrections are needed because with power well above idle, the newer jet aircraft have excellent acceleration characteristics, and minor throttle corrections can change airspeed rapidly. The approach in the T-38 is flat with almost a level flight attitude even though the rate of descent is 500 to 750 fpm. Final approach speed depends on gross weight of the aircraft computed from a standard minimum speed for normal landing weight. For instance, in the T-38, approach speed is 155 knots for aircraft with 1000 lb. of fuel or less. One knot is added for every hundred pounds over the thousand.

Extra airspeed on final approach requires more runway for the landing roll. A rule of thumb is: 1% additional airspeed on final approach requires 2% additional rollout on landing. Thus, the *proper* airspeed is important, especially when landing at an airfield with a relatively short runway. Too little airspeed results in getting behind the power curve and a dangerously high sink rate from which recovery is not possible with the remaining altitude.

Airspeed on touchdown in the T-38 will be approximately 140 knots. As airspeed decreases, the pilot brings the stick back, holding the nose at a 10° to 12° attitude for aerodynamic braking after touchdown. Touchdown occurs on the main gear, and after the nose lowers to the runway at about 100 knots, the pilot begins wheel braking. It is important that the nose wheel not touch down before the main gear, because a porpoising motion, usually curable only by going around, will result.

Landings in the smaller passenger jets are quite similar to the above, except that the nose high attitude is not so pronounced. Lear Jet approach speeds vary depending on the gross weight of the aircraft. The final approach speed for a Lear Jet weighing 10,900 lb. would be 122 knots, while one weighing 10,000 lb. should be flown at 117 knots. Passenger comfort being paramount, landings are usually in a flat attitude with the nose gear touching down shortly after the main gear. Their deceleration characteristics do not normally require aerodynamic braking, and the sooner all wheels are on the ground, the sooner wheel braking is effective.

Ground effect must be considered when operating jet aircraft. There is no more effect on jet aircraft than on any light plane, but size, speed, and mass make it more obvious to the pilot. Ground effect tends to reduce the angle of downwash and diminishes the effect of wing tip vortices. The result is a reduction of the aircraft's induced drag when near the ground. Generally ground effect is not a problem during takeoff in jet aircraft, since takeoff speed, distance, and acceleration are checked during the takeoff roll, and the rate of acceleration is so rapid as the jet aircraft approaches its takeoff speed and immediately after takeoff that it passes through ground effect quickly.

During landing the effect is more noticeable. Many jet aircraft do not change attitude (pitch angle) for landing. In the F-4 the same pitch angle as was used on approach is used during landing. Increased back stick is required to maintain the attitude constant since ground effect tends to force the nose down and angle of attack is rapidly increasing. In the Boeing 707 a reasonably good landing can be made out of a normal approach without changing attitude. In any aircraft if an excessive speed is maintained during approach, the aircraft will appear to, and may in fact, accelerate, if power is not reduced, as it comes into ground effect.

OTHER BUSINESS JET CHARACTERISTICS

The Lear Jet has certain interesting and informative flight characteristics somewhat different from the T-38.

It is usually taxied on one engine to conserve fuel and brakes. The location of the engines close to the fuselage centerline as opposed to being mounted in the wings facilitates single-engine taxiing and greatly reduces the problem of yaw during single engine flight. In fact, this yaw is very easily compensated for by rudder and trim.

Like the T-38, the Lear Jet accelerates spectacularly during take-off, despite the fact that it has no afterburners. During the takeoff roll, the nose wheel steering is disengaged at about 45 knots, when rudder control becomes effective. The aircraft rotation and liftoff speeds, precomputed, are precisely met, and gear and flaps must be retracted quickly to avoid acceleration beyond the limit speeds.

The Lear Jet accelerates to 300 knots for best climb, then climbs at 300 knots to .7 Mach; from that point it continues to climb at .7 Mach. It requires only 13 min. to reach 41,000 ft. However, the takeoff rpm and EGT limit is 5 min., so a power reduction is required during climb to remain within maximum continuous engine limits, which are 100% rpm and about 675° EGT.

Jet executive aircraft such as the Lear Jet are perhaps the most efficient form of high speed, highly reliable transportation for small groups whose time and flexibility is valuable enough to warrant the expense of operation. As the high performance aircraft they are, they require meticulous maintenance and professional operation within the performance limits which the designer intended. To a competent and proficient pilot, flying them is sheer delight.

FORCED LANDINGS

The original jet trainers and fighters required pilot techniques for forced landings. As performance advanced to that of modern fighters with performance comparable to and exceeding that of the T-38, forced landing became impracticable and ejection is now the rule. With the advent of business jets without ejection equipment, the requirement again exists, though forced landings are extremely unlikely because all are multi engined aircraft, and air start procedures have been made much more reliable.

It is generally recommended that a crash landing with a smaller aircraft be made with the landing gear extended. The gear will most likely tear off on rough terrain, but the process of tearing it off absorbs a great deal of the shock of deceleration, resulting in less serious injury to the occupants. This does not apply to ditching.

The crew of a jet fighter or trainer carries all emergency equipment attached to the parachute harness. Ejection is therefore preferred to ditching except at extremely low altitudes.

EJECTION AND BAILOUT

Military jet aircraft, particularly trainers, bombers, and fighters, are equipped with ejection seats. They are necessary because of the impossibility of climbing safely into the high-speed slipstream, and the probability of being unable to clear the empennage. Ejection seats are propelled by an explosive cartridge or by rockets. They are actuated by a trigger in the seat. The dangers of high altitude bailout or ejection—low temperatures, lack of oxygen, and parachute opening shock—are minimized by descending to lower altitude, if possible, and by automatic features of the parachute. They provide breathing oxygen and automatic free fall to below 14,000 ft., where temperatures are higher, exposure time is less, and opening shock is less. The greatest danger is from waiting too long and ejecting from too low an altitude.

With the advent of the rocket seats, the chance of survival from low-altitude and low-airspeed ejection has been greatly increased. Present seats can eject a man at zero altitude and zero airspeed to sufficient height for automatic parachute deployment if the pilot is using the seat properly. The absolute minimum *reliable* altitude for a controlled ejection is 2000 ft. Below that altitude, fatalities increase rapidly. The decision on when to eject must be firmly established in the pilot's mind before the emergency occurs.

The most advantageous aircraft attitude for ejection is a climb. When the aircraft is descending, the sink rate can be so great, and the direction of ejection so far from the actual vertical, that the capabilities of the ejection seat and parachute are greatly compromised. Successful ejections from modern high-speed aircraft require forethought and thorough training.

Figure 10.5. The Bell X-1. Flying high over the Mojave Desert, the Bell X-1 made history as the first aircraft to fly faster than the speed of sound. Now in the Smithsonian Museum. *(US Air Force photograph.)*

Supersonic Flight

FLYING THROUGH MACH 1

Since October 14, 1947, when Capt. Charles E. Yeager, USAF, flew the X-1 faster than the speed of sound, supersonic flight has become commonplace.

Pilots flying modern supersonic sircraft find the experience of supersonic flight unnoteworthy. Control and engine designs have so greatly improved that a pilot's first indication of having gone supersonic may be a glance at his Mach meter. Each aircraft has its own individuality in the transonic and supersonic regions, but the early experiences of control reversal, instability, and compressor stalls are largely history. Thermal limitations are and will be the limiting factors for atmospheric high-speed flight in the foreseeable future.

For many pilots, breaking the sound barrier in level flight also breaks a mythical bubble associated with this maneuver. The pilot can trim his aircraft for level flight, advance the throttle to the military power setting, and start the afterburner by moving the throttle outboard. Acceleration to Mach .95 is good. After this speed, acceleration is slower due to high drag rise. Then the pilot detects a hesitancy in his airspeed indicator for a second or two, followed by a more rapid movement of the indicator through Mach 1. Associated with this is a rapid change in altimeter reading, sometimes from 500 to 800 ft., as the shock wave passes over the static ports. Then all settles down again at Mach 1.05, and acceleration continues.

Meanwhile, controlling the airplane is quite normal. A slight increase in nose-down trim is necessary for level flight, indicating positive stability in this range. Lateral control remains as good as ever. The airplane can be turned and rolled, flying as if below the speed of sound. It must be remembered, supersonic flight is not without cost—in pounds of fuel consumed.

HIGH-SPEED STABILITY AND CONTROL

As noted earlier, most high-performance aircraft lack inherent stability. The amount or intensity varies with the aircraft and the particular flight condition. In any case, it makes the pilot's job more difficult.

Dynamic longitudinal damping is one instability characteristic. Longitudinal oscillations occurring from normal acceleration ideally damp to one-tenth amplitude in one cycle only, at 10,000 ft. in the power-approach configuration, gear and flaps down, and below Mach .89 at 25,000 ft. in the cruise configuration. Since good damping throughout the speed range of the aircraft is also necessary for ease of control, pitch dampers are designed and made part of the longitudinal control system, thereby artificially providing the desired stability.

The *dynamic lateral-directional characteristics* of most high-performance aircraft are a headache to the military pilot when maneuvering his airplane as a gun platform. When stability is good (positive), the aircraft resists change caused by rough air and maintains a steady desired path, without rolling and yawing objectionably. Transonic aircraft are equipped with a yaw damper to improve the characteristic tendency. With the yaw damper operating, lateral-directional oscillations dampen almost immediately.

The F-4 provides an example of adequate *static directional stability*, since rudder deflection produces a sideslip angle in the proper direction without lightening or reversal of rudder forces. Static lateral-directional stability is positive, with left aileron required to hold a straight path while in a steady left sideslip. From a pilot's viewpoint, the most important aspect is that there is no reversal of rudder or aileron controls throughout the speed range of the aircraft.

OTHER HIGH-PERFORMANCE CHARACTERISTICS

As a Mach 2.0 airplane, the F-4 is an excellent example of other high-performance phenomena.

Thermal Limitations.   Engine air-inlet temperatures are a recognized limitation of Mach 2.0 aircraft. In channeling the air into the compressor section, the inner chambers must dissipate large amounts of heat generated by the compressing function of the inlet and ducts. Many theories exist on how to cool this air, such as spraying water into the flow, improved duct design, etc.

Duct Problem.   Supersonic speeds require efficient engine operation and subsonic airflow to the engine. One of the problems is converting supersonic outside air to subsonic speeds and distributing it uniformly around the compressor section of the engine. In the F-4, automatically controlled variable inlet ramps and variable bypass bellmouths in the engine bay control the air volume and velocity delivered to the engine compressor face.

Negative-Dihedral Wings. When an aircraft is side-slipped, a rolling moment (dihedral effect) is induced. When left sideslip produces a right rolling moment, the aircraft possesses positive dihedral effect. Considering low-tail and high-tail configured airplanes coupled with any sideslip, the restoring force developed by the induced roll acts through the aerodynamic center of pressure (c.p.) of the aircraft. The high tail raises the c.p., increases the restoring capability of the tail, and in essence increases the positive dihedral effect. Too much positive dihedral introduces objectionable lateral-directional oscillations (Dutch roll). Therefore, wings are given negative dihedral to reduce the overall dihedral effect of the wing and tail combination.

Wing Fuselage Effects. Aerodynamic studies and flight-test results indicate that lift, in some cases, continues to increase above the wing stall, indicating that the fuselage itself provides lift. This characteristic can carry the aircraft into an extreme angle-of-attack condition where the horizontal stabilizer is acted upon by disturbed airflow from forward parts of the aircraft and wing downwash. The restoring moments of a high-mounted tail are negated under these conditions, and the aircraft will pitch up instead of down when the full stall condition is reached. Clearly this is not a desirable characteristic, since the aircraft usually enters a spin at that point. To prevent the condition from occurring, supersonic aircraft so affected have an automatic pitch-control system that senses pitch rate and angle of attack, warning the pilot of the condition and causing automatic control movement to keep the aircraft from entering the extreme attitude.

Ventral Fin. Supersonic aircraft may tend to lose directional stability above Mach 1.0 due to the characteristics of supersonic flow. To counter this effect, such aircraft require increased vertical fin area to maintain directional stability. Some have a ventral fin on the bottom of the fuselage, where it is always in the free airstream, for this purpose.

Thrust and Drag. Some of today's supersonic aircraft can exceed Mach 2.0 and altitudes above 100,000 ft. because their thrust-drag characteristics greatly vary with change in altitude. Figure 10.1 typifies the thrust-drag relationship at low altitude since the "power required" curve may also represent a "drag curve." The excess thrust available for developing rate of climb and accelerating to higher speeds is the important point. Where the thrust and drag curves cross, thrust and drag are equal; here speed and altitude stabilize. At low altitude and in the transonic speed range, drag rises steeply and requires a relatively high level of thrust to attain supersonic speeds. At altitudes around 35,000 ft., thrust exceeds drag over the complete speed range for a typical Mach 2.0 temperature-limited aircraft. The excess thrust over drag at this altitude results in higher rates of climb and aircraft ceiling. As altitude increases, the thrust-drag relationship changes and there is a new point where the thrust-drag curves meet. This is the supersonic ceiling of the aircraft, and its location depends on the particular aircraft configuration. Although today's supersonic aircraft have a supersonic ceiling, their speeds at this ceiling are well above their stall speeds. Consequently, this speed or energy can be exchanged at any time for altitude. This fact introduces the so-called *energy-transfer climb* or *zoom climb* technique. The amount of zoom capability available can provide a one-shot tactical advantage when additional altitude is needed.

COMPRESSOR STALLS

At some time during the jet pilot's career, he may encounter engine stalls. This will most likely occur when he initially advances the throttle, rather rapidly, while taxiing away from the line. It could also happen under low-airspeed maneuvering conditions at high altitude and high angle of attack, during rapid throttle movement, or under any conditions that will create an instability of the airflow over the blades of the engine compressor section. Stalls are characteristic of the twin-spool engine as well as other engine types. At times the stall is a very dramatic condition. It is variously referred to as a "pulsation," "chugging," "choo-choo," or "explosion"—and at times it certainly does sound like an explosion. Engine stall or breakdown of airflow over the compressor blades is sometimes compared to breakdown of airflow over a stalling wing. Stalls vary in severity and stem from marginally overrated engines and faulty duct design. Probably the worst type encountered is the "machine gun" variety. When this occurs, usually at high altitude, the engine cannot recover unless the throttle is retarded and the aircraft pitch attitude lowered to increase speed. If this is not done, rapid overheating of the engine occurs and engine failure follows.

SPECIAL DEVICES

The nature and complexity of high-performance aircraft demand special devices to provide satisfactory inherent stability for all flight conditions. For example, a vertical tail that gives satisfactory directional stability in the landing pattern and at high subsonic speeds may be entirely inadequate at supersonic speeds; damping in pitch may be satisfactory subsonically but intolerable supersonically. Considering these facts, it has been necessary to design stability devices that change or augment the inherent characteristics of the aircraft.

Yaw dampers on modern high-performance aircraft sense yaw and dampen reactions as a function of altitude and Mach number. They also keep the airplane in directional trim and give turn coordination automatically without the reactions being apparent at the pilot's controls. Yaw dampers are also designed to correct for either adverse or favorable yaw that develops when an aircraft is initially rolled. The yaw damper, based on a lateral accelerometer which senses the slip or skid of the aircraft, induces correction of infinitesimal yaw variations from the desired flight path. The end result is a better gun platform for military aircraft and a better ride for commercial aircraft.

Pitch Dampers. The pitch damper is similar in principle to a yaw damper. However, unlike the yaw damper—which operates on a control that is rarely used except for trim by the pilots of modern fighters—the pitch damper operates on a control which is continuously in use, frequently at maximum rates of displacement. As a consequence, the pitch damper is designed to eliminate residual pitch deviations without interfering with the pitch changes made by the pilot to maneuver his aircraft. The pitch damper is primarily needed for supersonic flight at high altitudes, where the air is thinner, and the characteristic restoring qualities of the aircraft to external or pilot induced motion are less, despite supersonic speeds.

Roll Dampers. Although, academically speaking, aircraft do not inherently produce roll oscillations directly related to the lateral axis of the aircraft, there is a need for roll stability devices in high-performance aircraft to assist the pilot in maintaining roll attitude during climbs and cruise, and to improve the spiral divergent qualities of some aircraft. (An aircraft with poor spiral tendencies tends to steepen rapidly in a descending turn.) These dampers sense roll rate from the roll-rate gyro of the autopilot, inducing small aileron deflections to compensate for this condition.

Mach Box. High performance aircraft require pitch trim changes throughout the wide range of their speeds, especially through the transonic speed range. The Mach box has been devised to reposition the horizontal stabilizer continuously as a function of Mach number, so that undesirable trim changes are minimized and necessary ones are in the proper direction. When speed is increased, the nose of the aircraft is trimmed down to maintain level flight, and vice versa.

Pitch "g" Limiter. The possible malfunction of automatic pilot systems can create control movements that will induce flight outside the structural limits of the aircraft. This is possible because these aircraft are equipped with irreversible hydraulic control systems. The pitch "g" limiter is a typical "black box" which senses a malfunction in an automatic control system and in microseconds disengages the system before structural damage can occur. To be effective, the device must be electronic (because of the time element) and reliable (to prevent false alarms). The g limiter utilizes pitch-rate gyro information, which when added to the output of a linear accelerometer, is amplified and used to trip a sensitive relay placed in the automatic flight system. Here then is an electronic device that prevents damage to an aircraft in a manner that could not normally be accomplished by the pilot.

Stick Shaker and Automatic Pitch Control Systems. The stall in some aircraft is a precarious maneuver. While in some high performance aircraft the nose drops in a stall, in others the aircraft pitches up with subsequent loss of control. Some aircraft that lack proper stall warning characteristics need stall-warning devices. The warning takes the form of a shaking control column, audible buzzer, or visual warning light. These devices are actuated prior to the stall, normally by sensing units that detect either angle of attack or a breakdown of the airflow pattern over the aircraft wing.

In aircraft that develop pitch up, a more complex system is involved. When the aircraft enters the pitch-up region, a push force of between 25 and 30 lb. is automatically applied to the control stick to return the aircraft to a normal flight attitude. The stick-pusher mechanism is actuated by angle-of-attack probes on the fuselage and a pitch-rate gyro which detects accelerated maneuvers that will result in pitch up. While the pilot can overpower the stick pusher, it is in some aircraft deenergized when the wing flaps are in the takeoff or land position to prevent a hard nose down movement at low altitude during takeoff or landing.

Unconventional Aircraft

Today's high performance tactical aircraft are the direct result of aerodynamic testing of the "X-Series" of experimental aircraft. In the short span of 10 years, these aircraft were used to penetrate the sound barrier at Mach 1.0 and to help solve the complex problems of stability, control, and aerodynamic heating associated with speeds of 2100 mph and altitudes up to 126,000 ft. The requirement for "X-Series" aircraft was visualized by Mr. John Stack of the Langley Aeronautical Laboratory, NASA, and made a reality by the military and industry. The combined efforts of many individuals—some of whom gave their lives—translated the idiosyncrasies of research aircraft into practical applications, resulting in the increased performance and improved stability of our present day service aircraft. The Bell X-2 and the North American X-15 are two outstanding examples. You may read about their exploits in earlier editions of *Modern Airmanship*.

Sustained Supersonic Flight

Two supersonic transports (SST), the British-French Concorde and the Russian TU-144, are flying regular routes. The B-58 flew for 10 years, frequently at a sustained speed of Mach 2.0. The North American B-1 bomber proved itself highly capable in the supersonic speed regime. But experience in sustained supersonic flight as the most economical speed of an aircraft is rare.

The outstanding example of very high cruise speeds, up to Mach 3.0, is the SR-71, a peerless reconnaissance aircraft, used by the US Air Force's Strategic Air Command.

Figure 10.6. Lockheed SR-71. Used by Strategic Air Command, USAF, in strategic reconnaissance. Mach 3.0 speed range, ceiling above 80,000 ft., range more than 2000 nm., and global with midair refueling. Two J-58TB engines rated at 30,000 lb. thrust each. "Double delta" wing, span 55 ft. Overall length, 107 ft., height of tail, 18½ ft. The YF-12, an experimental long-range interceptor, is one version of this aircraft, used largely for experimental purposes. *(US Air Force photograph.)*

DESCRIPTION

The airframe design, called a "double delta," offers an airfoil of remarkable lifting characteristics over its entire area. Though the "wing" shape is located rearward, the entire thin fuselage, throughout its length, generates lift. The lift developed by the forebody in particular is essential to the excellent stability characteristics shown in the entire speed range of the aircraft.

While many supersonic flight characteristics described earlier apply equally to the SR-71, its capability for economical cruise at supersonic speed is unique.

The SR-71 limit speed is above Mach 3.0, and it becomes more efficient as Mach number increases. In fact, it is above Mach 2.0 that range factors are superior to those of subsonic flight.

Efficiency for any supersonic aircraft, whether flown in a dash profile or a cruise condition, is dependent a great deal upon drag reduction. The dominant controllable source of drag in supersonic cruise is the trim condition of the elevons, a control surface peculiar to delta type aircraft.

As the aircraft moves into the supersonic ranges, the center of lift moves aft, and to avoid the use of excessive trim to counter this condition, the center of gravity must move aft also. This is accomplished by automatic control and distribution of aircraft fuel so as to complement the center of lift. Since the SR-71 is a military aircraft, designed primarily for a standard, one-profile mission, c.g. control can be automatically programmed with little or no attention required from the flight crew. The same system was used on the B-58.

Any time the standard profile is changed, such as for extensive subsonic cruise, the pilot must control his fuel distribution manually to provide optimum c.g. for the existing speed and center of lift location.

The pilot can use rule-of-thumb c.g. limits as a guide for varying speed ranges. A c.g. location indicator in the cockpit gives an easy reference for fuel transfer requirements. A pitch trim indicator permits monitoring the trimmed position of the elevon.

Center of gravity trimmed too far forward during any portion of flight is at best uneconomical. Trimmed too far aft, it can cause a dangerous condition of instability.

Speeds of Mach 3.0 and above create many departures from conventional flying procedures. Auto-pilot control is for the most part essential during supersonic flight. Although it would be perfectly safe for a pilot to hand-fly an entire mission above Mach 3.0, precision would suffer, and errors would be magnified.

A very sophisticated navigational system employing inertial guidance with star tracking updating is utilized to provide the accuracy and timeliness needed for inputs to the auto-pilot. Deviations from programmed track are absolutely minimal throughout global operations using this combination. Displays from the navigational system are also presented to the flight crew and, in the event of an autopilot malfunction, the pilot can fly the aircraft manually along the same course, though some degradation in track maintenance results.

Because of the automated navigation capability, the flight crew need not prepare a conventional type of flight plan. The entire mission is prepared and "canned" by a staff of technicians. The result is a computer tape which is fed into the astro-inertial navigation platform, and 35-mm. film strips installed in projectors in the crew compartments. The projectors automatically display a map presentation of the aircraft position at all times, regardless of speeds flown. They also show pertinent flight data such as fuel, times, heading, climbs, and descents. Mission planning for the supersonic flight crew then consists of studying the mission from back-up maps and cruise cards which are prepared for crew use in the event of display failure in flight.

PILOT AND RECONNAISSANCE SYSTEMS OFFICER (RSO)

The pilot and RSO constitute the basic aircrew. They fly in tandem, in separate, isolated cockpits, dependent on interphone for communications. They both wear the full pressure suit, (Figure 7.6) on all flights. Though these suits are encumbering and detract from flight comfort, they are considered economical and safe in comparison with the much heavier "capsule" which would provide "shirt-sleeve" flying. The military concessions to battle damage and crew ejection are of course not present in the pressurized civilian SST environment.

Crews have ejected safely, using the installed ejection seats, from ground level to the limit speed and altitude of the aircraft.

The bulk of pressure suit equipment makes cockpit layout very important, because it is very difficult to see or reach anything not in view with the head positioned forward. For this reason the forward part of cockpits are extremely "busy." There are conventional stick and rudders, twin throttles, and familiar high-performance cockpit instruments, modified as supersonic flight requires.

The RSO, a rated navigator, actually performs many duties that would fall to a copilot in a plane with a bigger crew. He has duplicate flight instruments (and must be proficient in interpreting them), fuel monitor systems, his reconnaissance system controls, and the astro-inertial navigation computer controls and displays. Crew coordination and standardization are extremely critical. There is no room for mistakes or confusion at Mach 3.0.

The RSO reads all checklists for both cockpits so as to free the pilot from distractions. The aircraft and crew are carefully monitored by a host of support technicians, including crew physical examinations prior to every flight. Groundcrew checks continue right up to takeoff time.

When aviation vaulted from propellers to jets, the compression of time was considered to be a near insurmountable obstacle for all but the most adept of crew members. It has proven not so. SR-71 crews feel they are now sampling another such quantum compression, and that only time will tell the ability of the average crew member to adapt successfully.

TAKE-OFF AND CLIMB

The aircraft uses afterburners during takeoff, allowing operation from conventional runways. Liftoff occurs at slightly over 200 knots. The best subsonic climb speed is 400 knots to Mach .90. Conventional FAA instrument departures are made after takeoff. Since the most damaging and annoying sonic booms, the nemesis of supersonic flight, occur at lower altitudes, the flight path is planned carefully over unpopulated areas for beginning initial acceleration to supersonic speeds.

When supersonic, the crew transitions to a "triple display indicator" (TDI) which presents a computerized, digital readout of equivalent airspeed (KEAS), Mach number, and altitude. Although conventional pitot static instruments are also in the cockpit, the TDI presentation is used for supersonic accuracy since it is corrected for temperatures, compressibility, and altitude.

Programmed turns during the acceleration phase are avoided because they seriously degrade climb and acceleration performance due to the increased angle of attack. Acceleration is also a time of very high fuel consumption because of power requirements and relatively low altitudes. Corridor type departures are very desirable after going supersonic. Because of the high fuel consumption, it is very critical to the success of the mission that, once begun, the acceleration to cruise speed and altitude not be interrupted. This proves unwieldy for traffic controllers but is an essential factor in SST planning.

SUPERSONIC EQUIPMENT AND TECHNIQUE AT CRUISE

The SR-71 has flight control damper systems in all three axes, a Mach trim system that automatically generates pitch trim during acceleration and deceleration, and a "stick shaker" system to warn of approach to limiting angles of attack.

High Mach cruise requires that external rotating beacons retract flush with the skin to avoid burning off. There is a tiny periscope mounted in the pilot's canopy to let him view the top of the fuselage and engines in flight. The blinding glare of high-altitude sunlight is muted by multi-positioned sunshades.

The extreme temperatures of high Mach cruise require special precautions. Titanium is used extensively, since it is both light and able to retain its shape despite the severely high temperatures experienced.

Yet there are limiting temperatures, most critical at the face of the engines, which the pilot must watch and control. On leading edges, temperatures of over 1000°F are sustained. To touch the cockpit windshield, even with an air-conditioned pressure suit glove, is like touching a hot stove barehanded.

On reaching cruise altitude and Mach, the pilot throttles back to about one-third of the fuel flow needed during the early stages of supersonic acceleration. The resulting range factors are most impressive and confirm the vital need for uninterrupted climb from low altitude.

Now the flight crew is primarily concerned with monitoring the aircraft condition, with navigation, and with the operation of required military equipment. The autopilot, placed in "Mach hold," or "KEAS hold" function, controls pitch to give smooth speed performance. In "Auto-Nav" (automatic navigation) position, the roll and yaw axes are controlled so that the aircraft precisely follows the computer fed flight plan. The pilot must be alert to increase power manually to hold speed in programmed turns. A Mach 3 turn of 360° requires a radius of about 90 miles depending on the angle of bank required and the angle of attack allowed.

The engine air inlet system works continuously to provide optimum control of inlet pressure, and makes constant, minute adjustments in air bypass doors and spike location. This compensates for altitude changes, speed deviations, turns, and out-of-trim conditions which might exist in yaw.

At high altitudes, though always in VFR cruising conditions, the pilot flies almost entirely on instruments.

External visual references are not only limited but can be misleading when maneuvers are being performed, such as turns or angle of attack changes with airspeed alterations. There is simply no time for sightseeing, although the view is breathtaking. The conventional pitot static instruments are of very limited value during supersonic cruise. Inherent lag, reversals, and inaccuracies make them useful only as a backup or emergency reference. Instrumentation for supersonic cruise is still in the infant stage of development, and numerous approaches are being tried.

The basic flight director system presents a reasonable navigation picture, but newer director models are in development. They display dual DME's and backup TACAN information for simultaneous selection and instantaneous presentations.

Flight reference platforms need major engineering improvements in their design to reduce inherent errors in heading, pitch, and roll axis during the vastly extended periods of acceleration, deceleration, and turns. The primary reference source for accurate attitude and heading displays falls on the astro-inertial navigation platform with the standard flight reference system serving as a poor backup.

Angle of attack information is one of the most critical displays available to the pilot. It is very conveniently displayed by a modified glide slope indicator or "bug" on the left of the attitude director indicator (Figure 5.42). This "bug" shows angle of attack information except when the pilot switches to ILS mode, when it reverts to its normal purpose, showing degrees above or below glide slope.

The angle of attack indicator is backed up by "shakers" and audio warnings as angle of attack limits and limit rates of approach to critical angles of attack are approached.

Pitch corrections require habitual precision and rapid crosscheck, because a ¼ bar width attitude indicator correction can result in pitch excursions that have devastating effect on speed and altitude. The "instantaneous" vertical speed indicator is a great help in resolving angle of attack, pitch, and attitude corrections.

Cruising. While cruising, SR-71 pilots usually maintain Mach constant on the TDI and allow airspeed to decrease slowly as altitude increases with fuel burnoff. Reversing this procedure, or establishing a different constant, requires considerable practice in both the aircraft and the flight simulator. A momentary excursion beyond the target or limit of one parameter is usually a "wipeout," or near loss of control, on the other two. From this it is evident that continuing research and development in high speed instrumentation is vital.

Monitoring the condition of the aircraft systems and engines fully occupies the remainder of the pilot's time. Many of the aircraft systems, such as the engine air inlets, must be operating at optimum at all times and demand immediate attention from the pilot if malfunctions occur. He must also show immediate response to a c.g. which moves 1° beyond desired, an inlet temperature which reaches the limit, a bank angle which may be a few degrees in error, or a simple FAA advisory of conflicting traffic.

The lack of timely response to the development of such conditions can result in exposure to conditions of instability, equipment damage, violent inlet shock wave expulsions, and gross course deviations. All of these conditions can develop rapidly from an otherwise serene and uneventful flight if crew attention wanders. Recovery from such conditions can be time-consuming and operationally degrading at best. Communications, which are frequently an untimely distraction, are handled by the RSO, whose duties are usually (though not always) less critical than the pilot's.

Clear air turbulence is occasionally encountered at high altitudes and can be very disconcerting due to the sharp, sudden impact associated with high speeds. This is essentially a matter of education, however, since the g forces exerted on the aircraft at such a high altitude are acceptable, despite the high Mach condition. They are usually minimal in duration because of the speed.

Present weather forecasting is becoming more accurate in determining the location and severity of CAT. Large air mass temperature changes are also quite noticeable in supersonic cruise. They are detected by rather subtle but sudden changes in the indications of Mach, altitude, and KEAS. These temperature deviations are also passed quite rapidly and usually constitute more of a nuisance than a discomfort or hazard.

A similar, subtle excursion is frequently encountered when cruising the aircraft at its ceiling altitude. A long period oscillation, known as a "phugoid," develops around the pitch axis. It is a common characteristic of longitudinal stability and constitutes a very gradual interchange of potential and kinetic energies about some target or equilibrium speed and altitude. It is usually handled adequately by the autopilot and creates only a minor physical sense of altitude and speed change that would hardly be discernible to a non-crew-member.

Collision Avoidance.   Generally it is considered a "see and be seen" environment for the few military aircraft that operate above Flight Level 600, but the traffic is a little more dense than one might suspect. B-57's and U-2's have been there for years. The FAA has become increasingly concerned over traffic separation above FL 600 and has put into effect a modified control system that places all traffic on a common, discrete UHF voice frequency and employs a common SIF code. Flight crews must give their altitudes in code to each new controller as they enter his area of responsibility.

Reporting altitudes in the clear will be a necessary change in the future. Controllers provide such traffic advisories as possible to minimize conflicting tracks and provide a 5000-ft. altitude separation. With a head on closure rate of over 4000 mph, it is easy to see why controllers are concerned about how to evaluate control problems and issue instructions to two SR-71's. Of course the capability for a Mach 3.0 aircraft to accomplish meaningful maneuvers is severely limited. For this reason slower high-altitude aircraft, such as the U-2, are asked to do most of the maneuvering involved. Pilots have also discovered that momentary use of the aircraft fuel jettison system will create an immediate "contrail" of many miles which substantially aids in mutual visual detection. In the future, hard altitude assignments for SST aircraft, in lieu of the cruise climb profiles now used by the military, will probably be necessary for safe traffic separation.

SONIC BOOM

One of the most controversial aspects of supersonic flying is, and will continue to be, the sonic boom. As mentioned before, the most damaging boom effects occur when the aircraft is at medium altitudes during the acceleration and deceleration phases. The overpressures at these times can be quite severe, and great care must be exercised in selecting routes of flight. Pressures from supersonic shock waves vary with the speed, altitude, and size of the aircraft. The SR-71 is relatively small (approximately 107 ft. long), and the overpressures that result from its sonic booms are reasonably light because of the altitudes involved. Even so, it can constitute a large nuisance factor.

DECELERATION

Deceleration from high Mach is generally a reverse of the crew procedures employed during the acceleration. The most critical phase of planning a deceleration is the accurate initiation of descent procedures a set distance from projected subsonic leveloff point. All understandings and clearances with traffic controllers have to be confirmed prior to initiating descent since, for all practical purposes, it is an irrevocable decision. Since altitude and equivalent airspeed (KEAS) are usually undergoing a constant change, descent timing must be continually updated. It will usually require in excess of 300 miles to descend and decelerate from above Mach 3.0 to an altitude from which a conventional instrument approach can be initiated, or to an altitude for air refueling. A delay of several seconds can result in missing the target by miles, just as in a space reentry and splashdown. Although some latitude for adjustment of the descent profile is available to the pilot, it is still rather limited because of the established inlet configuration, engine temperatures, and fuel consumption. Communications become extremely critical and crew actions must be paced well ahead of the aircraft throughout the entire flight.

LANDING

Upon reentering the subsonic speed regime, and the altitudes of conventional aircraft, the SR-71 fits beautifully into normal traffic molds. The maneuverability of the vehicle, wide selection of compatible speeds, and adaptability to standard instrument approach procedures all tend to minimize the dramatic differences that existed only moments before. The SR-71 pilot does have the difficulty of limited visibility in subsonic traffic. He is not encumbered by the restrictions of the full-pressure suit but has rather small windows out of sides of his canopy. Forward visibility is quite good.

Conventional TACAN/ILS and radar instrument approaches are suitable for the SR-71. An exception is the use of a "modified precision approach" at bases where the SR-71's are stationed. This provides for a glide slope which is flatter by ½° and has a glide path ground intercept point short of the runway. These approaches are restricted for use by the SR's and are a concession to their float characteristics when entering into ground effect. Conventional radar approaches are considered acceptable at alternate fields, if required.

Ground effect during landing is very noticeable, and speeds must be accurately controlled or excessive float can result before touchdown. There are no landing flaps or speed brake features on the aircraft. As soon as the aircraft touches down, a large landing parachute is deployed, resulting in a very dramatic deceleration. Differential braking is quite touchy, despite a good antiskid system, and care must be exercised to avoid excessive tire wear. The feel of the brake system is quite diluted because of the heavy pressure suit footgear, and a degree of trial and error exists in developing a good braking technique. Nose wheel steering visibility from the cockpit is adequate for conventional taxiing. The aircraft retains a very high residual heat from its exposure to the high Mach Cruise, and ground personnel must use care when handling it immediately after it completes a sustained high Mach cruise leg.

Several turns in a holding pattern or low-altitude air work before landing will result in adequate cooling.

Flying Large Multi Engine Aircraft

The complexity and cost of flying multi engine aircraft are such that their crews are professionals, especially in attitude and outlook. They must be experienced enough to be capable of transitioning to a new aircraft with a minimum of training time. In determining training costs, the commercial operator of large aircraft must consider not only the direct operating expenses, but also the loss of revenue caused by removal of an aircraft from revenue producing service and the increased insurance costs associated with training operations. As a result, most training in large aircraft is conducted in flight simulators. The most advanced of these simulators duplicate the aircraft in every way and include sophisticated motion and visual systems to provide sensory and visual cues. Presently up to 85% of the training time required to transition an experienced pilot into a new aircraft is performed in a simulator, and as advances are made in visual simulation of the real outside world, this will approach 100%.

What is different about multi engine flying? First, in all except the light twins, the plane demands a crew. One man cannot handle it alone. The crew members are specialists in engineering, navigation, radio, and in caring for passengers or cargo. The pilot is their captain. It is up to him to develop the closely integrated teamwork essential for efficient flying—efficient enough to provide either a profit (in commercial flying) or a measure of superiority against an enemy in wartime.

Second, there are from two to several engines. They eliminate the problem of the single-engine pilot—constant consciousness of a place to set it down—but they introduce complexities in systems and control.

Third, multi engine planes are heavy, and their load can change by 100 tons or more in a single flight. This great weight and inertia led to high control forces and sluggish response in early large aircraft whose control systems basically used the muscle of the pilot assisted by various forms of aerodynamic boost. Modern large aircraft, designed with redundant fully hydraulic control systems and artificial feel, may be comfortably flown with one hand and are as nimble as many small single engine aircraft.

FLIGHT PLANNING

In a commercial airline, a large business fleet, or a military organization, much of the burden of planning may be taken over from the flight crew by the operations section. Nevertheless, a flight plan must be prepared, and the crew must know the plan before takeoff. Even a flight over a route that the crew flies regularly requires detailed new flight planning to accommodate load, weather, wind, and runway limitations. While nowadays this process is usually a computer process, the steps are the same.

As the first step in flight planning, the crew or operations section assembles the known information concerning the flight to be made. This will include load to be carried, destination, route to be flown, en route stops, and time schedule to be followed—perhaps halfway around the world. They agree on a tentative general plan.

The pilots plan the takeoff and departure, fitting it into traffic procedures of the departure airfield. They study the weather forecast for the route and the destination, and the location of possible alternate landing fields. They plan for approach and landing at destination under any weather conditions. They may draw up a detailed communications plan for position reporting, and also do the specialized planning for any function or position not covered by other crewmembers.

If the crew includes a navigator, he assembles his materials and plans the flight in as great detail as necessary, generally following the procedures outlined in Chapter 11.

The flight engineer (or one of the pilots if the crew does not have one) obtains a loading list or manifest, to give him the approximate load so that he can estimate the weight and center of gravity. Using performance charts for the airplane, he determines the optimum altitudes and recommends them to the pilot. If these altitudes cannot be flown for reasons of traffic, wind, or weather, the best possible compromises are made. With altitudes selected, he can make final determination of the best long range airspeeds. He gives this information to the navigator, who develops a no wind flight plan, or a metro wind flight plan based on forecast winds. Fuel requirements are then computed, and the plan is complete.

Take-off Computations.   While the light-twin pilot need only glance at the wind sock or call the tower for his takeoff data while taxiing out, the multi engine plane pilot has a complex problem. He must consider airplane gross weight, engine thrust available, temperature at the runway, pressure altitude, wind direction and velocity, runway surface, and runway grade. The computations are made by the crew from charts showing takeoff performance in the aircraft operating manual. The primary results of these computations are the critical engine failure speed (V_1), takeoff ground roll, rotation speed (V_R), takeoff speed (V_2), critical field length, and acceleration check time, as discussed earlier in this chapter.

AIRCRAFT PREPARATION AND LOADING

Size makes this a tedious task. Necessary maintenance, inspections, loading, and servicing may be the responsibility of maintenance sections, but the pilot must make sure that they have been accomplished.

During flight planning, loading is itself a major problem. The load must be within the capacity of the airplane, must be so placed as to locate the c.g. within limits, and must be tied down so that it cannot shift or bounce around in flight, changing the c.g. or damaging the aircraft. Limitations from the airplane flight manual regarding the sequence of loading and using fuel from the various fuel tanks must be carefully followed to ensure that the aircraft c.g. will remain within limits at all times. Figure 10.7 shows a typical change of c.g. during flight as fuel is used.

Correct c.g. location also avoids drag due to poor trim. The most economical operation is obtained when the location of the c.g. is so established and maintained in flight that cruising speed trim requirements are as near zero as possible, that is, when the balancing force (normally downward in modern aircraft) produced by the horizontal tail is nearly zero.

Any aircraft can be loaded outside c.g. limits, and each is provided with a weight and balance handbook, or a section in the operator's manual devoted to

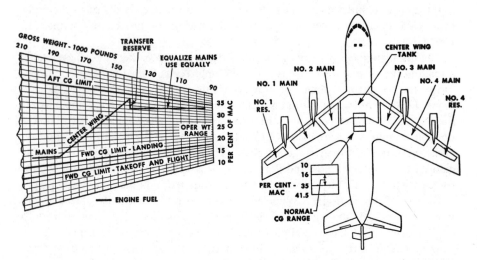

Figure 10.7. Center of gravity shift in flight. This schematic of the KC-135 shows the distribution of fuel, indicating its effect on lateral and longitudinal stability. Note that any load in addition to fuel would have to be placed so that fuel consumption did not exceed c.g. limits, and that uneven use of fuel in the wings would disrupt lateral trim.

loading, weight, and balance. Center-of-gravity limits are expressed in percentage of MAC and in moments about an arbitrary reference point, usually the nose, or near it. With many larger airplanes, the manufacturer supplies a specially designed slide rule for this c.g. computation, used to govern airplane loading.

Lateral location of the c.g. can be a serious problem in an airplane which carries fuel far out in the wing or in tip tanks; opposite wing tanks must be loaded equally. In flight, uneven use of fuel from outboard tanks, especially should one of them fail to feed, can develop into a real emergency. Uneven loading of the wings could produce an otherwise unbalanced moment so great that the ailerons could not counteract it, particularly at low airspeeds. If an outboard or tip tank cannot be emptied, the fuel manifold system will sometimes permit transfer of the remaining fuel sufficiently to get lateral c.g. within aileron control limits.

The wing bending moment is a special consideration in the loading of fuel on a multi engine airplane with a wide wingspan and tanks far out in the wings. The lift of a wing is distributed along its span. To avoid excessive spanwise bending moments, it is desirable to distribute as much weight as possible along the span also. If the fuselage is carrying a heavy load, fuel *must* be placed as far out in the wings as possible, and *must* be used first from the tanks nearest the center of the airplane. Doing so appreciably increases the positive-*g* acceleration safety margin.

GETTING AIRBORNE

The vast lore of flying is available to the pilot of any large airplane who will follow the prescribed procedures for that aircraft. Preflight inspection—a job in which all crew members participate with particular reference to their own specialities—follows careful checklists; a haphazard approach always causes trouble sooner or later.

Actually, the procedures of starting multi engine planes and getting to the end of the runway are little different from those for any other airplane; however, the increased size and complexity again require a passion for orderliness. Planning will avoid starting engines too soon and wasting fuel; it will also prevent a late start which results in rushing to make good a tight schedule. Whether the aircraft carries a load of revenue producing passengers who can choose another airline if you're not careful of their comfort and thrifty with their time, or whether it carries a huge and lethal bomb load, smoothness in engine starting, taxiing, stopping, braking, and turning is the true measure of piloting skill. No airplane is too small to be flown under the same principles.

TAKEOFF

Takeoff and landing are the two most hazardous parts of flying. Every takeoff must be made as carefully and skillfully as the experience of the crew and the quality of the airplane permit.

Runway acquisition rates are such at most large airports that the pilot will not be allowed the luxury of applying full power and checking his instruments before brake release. In fact some jet aircraft now have enough power available that full power with brakes locked will cause tires to skid, or worse, to rotate on the rims. For these reasons a running takeoff is usually made. This means that preparations for the takeoff must be completed in advance of the anticipated runway clearance. These preparations must include review of the planned departure route and rebriefing of emergency procedures and takeoff performance data.

When the pilot is cleared on the runway, if he is also cleared for takeoff, he will advance power far enough to roll smooothly onto the runway. He will advance throttles to takeoff thrust as he aligns with the runway centerline. He must be careful to prevent excessive side loading of the landing gear assembly during this turn onto the runway. There is no time for fumbling or misunderstanding on this type of takeoff roll. The crew briefing must have been explicit prior to the runway clearance, and each crew member must know his exact duties. The takeoff roll may last 40 sec. or less. Precise terms are used. A classic example of the failure to do this concerns a pilot who decided he could not land and would execute a go around. He commanded, "Takeoff power." The copilot complied by retarding power to idle! What the pilot wanted was "maximum power."

Takeoff Roll. When brakes are released and the takeoff roll begins, the first concern is directional control, which is maintained with nose wheel steering or differential use of throttles until the rudder is effective. Torque is less noticeable in multi engine aircraft, and of course not present in jets. The pilot avoids using brakes to maintain direction, except in emergency, because they lengthen takeoff roll.

The copilot calls out V_1, V_R, and V_2 speeds, and acceleration check time, if used. Acceleration check time is usually used only in military operations where loads are high or runways short, resulting in very slim performance margins. Usually the pilot reserves to himself the job of flying, and copilot and engineer observe engine performance. Close coordination is vital.

CLIMBOUT

As the airplane approaches V_2, the rotation of takeoff speed, the pilot exerts back pressure on the yoke, and the aircraft smoothly leaves the ground. Rotating too soon may lengthen the ground roll or cause the aircraft to leave the ground at too high an angle of attack; the increased drag delays reaching climb speed. Rotating too late wastes runway and may not permit safe obstacle clearance. Limit tire speed may also be exceeded.

The climbout attitude then depends on acceleration characteristics. With rapid acceleration, the pilot keeps the nose high to avoid passing the maximum gear-retraction speed; with slow acceleration, the nose stays low, but in a definite climb attitude, to accelerate to safe single engine climb speed as soon as possible. When the gear is up, flaps are retracted and power is set for the climb to cruising altitude. Particularly at night and in weather, it is very important to perform each of these operations in an orderly and smooth manner to avoid unusual deviations in rate of climb and possible disorientation.

If an obstacle has to be cleared immediately after takeoff, the pilot lets the speed build up after takeoff only to a value that will give the maximum angle of climb until the obstacle is passed.

Flying the heavier multi-engine aircraft requires use of the flight instruments for every takeoff—not just those under instrument conditions. A night takeoff is almost the same as an actual weather takeoff. The cockpit lights are turned up high because of the usually bright field lights outside at night. They are dimmed after the bright outside lights have fallen behind.

CLIMB AND LEVELOFF

For a climb of several thousand feet, pilots use the airspeed that will give the best rate of climb because it is most economical to establish cruise conditions in the shortest practical time.

Level-off from climb should be a smoothly coordinated maneuver. The technique is approximately the same in all aircraft. By delaying climb power reduction until a small excess over cruising speed is attained, the job of trimming at cruise speed is made easier.

Because of high climb rates obtainable in jet transports at low and intermediate altitudes, a leveloff required by Air Traffic Control in these lower altitudes must be anticipated and initiated with a 400 to 800-ft. lead on the altimeter. At higher cruise altitudes the rate of climb is considerably reduced, and no special techniques are required. Initial cruise speed is quite close to climb speed in most jets so establishing airspeed is easy.

CRUISE

Obtaining the most economical and dependable operation in a multi-engine airplane is both a science and an art. The fact that it requires a great amount of attention in flight is the chief reason for having flight engineers on the larger airplanes.

In large swept wing aircraft, careful lateral-directional trim is essential in order to obtain the maximum performance of the aircraft. After first determining that engine thrust is symmetrical, proper lateral-directional trim is most easily obtained by first carefully holding the wings level with aileron, then applying rudder trim to stop any heading drift, and finally applying aileron trim to maintain the control wheel position required for wings level. Since the needle-ball instrument is used only as a backup instrument in modern aircraft, it is not sufficiently accurate to be dependable for "fine tuning" trim.

An all moving horizontal stabilizer is the normal surface used for longitudinal trim. It may be moved either through an electrical switch on the control wheel or by engaging the autopilot, which can move both elevator and stabilizer.

ECONOMICAL ENGINE OPERATION

Maximum output, high altitude, and relatively high airspeed all improve turbojet economy. Airspeed increases the efficiency of the turbojet because the load of the compressors is reduced by the ram air effect.

With increasing altitude, fuel flow decreases while engine rpm and EGT are maintained, thus maintaining maximum efficiency while reducing thrust.

Reducing thrust without climbing reduces efficiency, because the throttle must be retarded to reduce thrust at a constant altitude, thus reducing EGT and rpm; the most economical jet engine operation is therefore obtained at high altitude.

MOST ECONOMICAL ALTITUDE

In Chapter 3 we saw that drag at a constant true airspeed decreases with reduced density at altitude. The power, and therefore the fuel required to overcome drag, is correspondingly less, and the airplane can fly the same true airspeed with less power, or a higher true airspeed with the same power.

Jets are equipped with Machmeters in addition to more or less conventional airspeed indicators. A Machmeter is a much more useful, quick reference for speed at high altitude, because the speed of sound varies only with absolute temperature. Experience has shown that best jet airplane economy is obtained at high altitude, flying a constant Mach number and a given power setting, and letting the airplane climb gradually as weight decreases with use of fuel. Fuel flow rate continues to decrease with the climb. This most economical operation occurs in current subsonic designs somewhere between Mach .75 and .84 depending on the individual design.

No attempt should be made to exceed the recommended altitude for a given weight for any jet aircraft. While the aircraft may be climbed above the recommended altitude for a particular weight, excessive fuel will be required in the climb and cruise, and the airspeed margin for maneuvering and penetrating turbulence will be greatly reduced.

CRUISE CONTROL

The performance section of the operator's manual for each airplane contains charts describing the performance under all conditions of range, altitude, airspeed, power setting, and gross weight. These charts are based on data obtained from test flying the airplane.

It is rarely, if ever, possible to make a flight under maximum economy conditions, because of weather, other traffic, or operational requirements. Since the great advantage of air travel is speed, it is generally desirable to cruise faster than the most economical airspeed, accepting a penalty of somewhat greater fuel cost but reducing indirect costs per mile. This tradeoff is greatly affected by availability and cost of fuel.

The problem of *cruise control,* then, is to obtain the maximum economy of operation under the existing conditions and requirements. There are three types of cruise:

Long range cruise consists of operating constantly at power settings and airspeeds, within engine limitations, which will approach most economical, no-wind, miles per pound performance. This results in a gradual reduction of power settings, gradually increasing altitude as the flight progresses.

Constant airspeed cruise requires adjusting of power as necessary to maintain a standard indicated airspeed once it is attained. Using this procedure, a specific power is used until airspeed gradually builds up to specified value, and then power is gradually reduced as necessary to maintain the desired airspeed for the remainder of the flight.

Standard cruise consists of establishing a specified power after leveloff and maintaining it while allowing airspeed to increase gradually as weight decreases.

DESCENT

Approaching destination, the crew can effect a substantial saving of both fuel and time by intelligent planning and skilled use of the airplane's capabilities, taking other traffic, weather, and Air Traffic Control clearance into account.

Because a jet airplane gets its best economy and minimum fuel consumption rate at high altitude, a long-range descent, with throttles in idle and the airplane "clean" is advantageous only when the descent can be timed so as to arrive at the final approach fix or the landing pattern entry point at the end of descent. Under VFR or under radar control, this is done.

Under IFR without an ATC radar controlled en route descent, it is usually more efficient to remain at high altitude until over the destination fix, and then to make a steep "penetration" pattern to the IAF (Chapter 9). This kind of operation is modified to shallower descents for airlines use.

LANDING PATTERN

Before entering the landing pattern, one of the pilots or the flight engineer determines the gross weight and the maneuvering speeds that will be used with the various flap settings. This step is vital since the gross weight of large airplanes may vary by 200,000 lb. or more between takeoff and landing. Modern sophisticated flap systems are so effective that in some cases the minimum speed "clean" may be nearly 100 knots greater than the final approach speed with landing flaps, and as many as five intermediate increments of flap may be selected between the cruise and landing configurations. Thus, initial flap extension is typically started several miles from the airport, and additional increments are taken as speed reduction is required to fit into the airport traffic flow. Both pilots must keep alert for other traffic throughout the let down, approach, and landing pattern.

Normal Pattern.   Every step of landing preparation should be planned by the crew for a certain point in the pattern. This eliminates variables that could adversely affect the pattern and presents the pilot with the simplest possible problem on the final approach.

A typical multi engine landing pattern is shown in Figure 10.8. With contact already established with the tower, the aircraft enters the downwind leg at a point even with the upwind end of the runway with maneuvering flaps (usually about ⅓) selected. Opposite the midpoint of the runway, lower the landing gear and adjust power to traffic pattern requirements. Complete the landing checklist except for final flap setting. Turn to the base leg when the end of the runway passes a certain reference point in relation to the airplane and select the next increment of flaps, coming to the speed and configuration specified in the airplane flight manual for circling approaches. Start the turn to final at a definite angle to the runway, select landing flaps and reduce to final approach speed as the desired glide path angle is intercepted. Throughout the maneuver, altitude and airspeed should be changing to conform to values predetermined by calculations and experience.

Admittedly, this is a mechanical approach to the landing problem. The key to

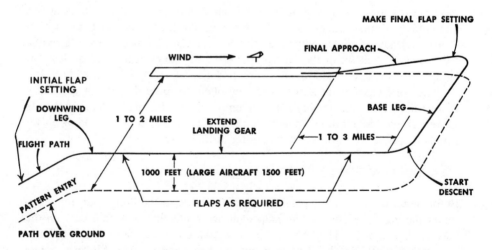

Figure 10.8. Multi-engine landing pattern. Size of the pattern depends on the size and speed of the aircraft. Large aircraft fly the pattern 500 ft. higher to allow more time to establish a good final approach. Descent may begin sooner than shown.

a good landing in any aircraft is a good pattern and approach. They can be made consistently smooth and safe only by an orderly, standard procedure.

Straight In Approach. This is a traffic pattern without any turns. It is frequently used with an instrument approach and quite often used in scheduled airline flying. All the normal traffic pattern checks must be made, and final approach must be established at the same point as in a rectangular pattern. A common mistake is failure to reduce airspeed, and to establish the correct attitude and airspeed for final approach before beginning the final descent. When VFR, some pilots prefer to establish the final approach configuration (except for power setting) somewhat early, then fly up to final approach at 500 ft. above the runway, or at whatever altitude is normally used for turning on final. When they reach the correct approach angle with the runway, they simply reduce power to obtain the necessary rate of descent, much as is done during an ILS approach.

Final Approach. Under visual conditions, the final approach descent begins as soon as the aircraft has rolled out on the runway heading. The final approach flap setting is made and attitude for the approach is established. Airspeed is held about 30% above stall. The correct speed varies with flap setting and gross weight and has been precomputed. From this point, except for the effects of sheer weight, the procedure is the same as described in Chapter 8.

Long flat approaches should be avoided. The pilot who enters final approach too far out or at too low an altitude will do best to maintain level flight and delay the final flap setting until he attains the correct final approach angle with the runway. If he finds himself too far out at too low an altitude, he must level off and maintain airspeed by adding power. Raising the nose to level flight without adding power will almost certainly result in getting on the back side of the power curve (Figure 2.37). This is dangerous when the airplane is heavy and has the high drag induced by gear and flaps. In this condition, the more the pilot raises the nose, the more power will be required to maintain level flight. Then the only solution is to lower the nose, add power, and sacrifice some altitude to regain operation on the front side of the power curve; or raise the gear, add power for go around, then raise the flaps cautiously as speed permits.

Should a go around be necessary, the throttles are immediately advanced to maximum power, the aircraft is rotated to climb attitude, and the flaps, if full down, are partially retracted. As soon as the pilot is certain that the aircraft will not touch down, he retracts the landing gear for the climb-out.

LANDING

Here again, inertia is an important factor. The maneuver may be divided into two phases: before and after touchdown.

Before Touchdown. Adjustment for crosswind is discussed in Chapter 8. However, inertia makes it essential that final corrections be made as early as possible, 2 to 5 miles out from the runway end.

A few aircraft have crosswind landing gear. With them, the best technique is to hold the crab right through the landing. At touchdown the wheels swivel enough to allow the airplane to roll straight down the runway while still in a crab. An interesting variation of crosswind landing gear is found on the Boeing B-52. Through hydraulic steering cylinders the pilot sets all four of its main gear at an angle, equal to and away from the angle of crosswind. He lands the airplane in a crab and reduces the setting to zero during the ground roll.

A properly executed approach has the airplane approaching the end of the runway, aimed at the selected landing point. Power, speed, and rate of descent are constant, and the pilot is ready to break the glide. At from 20 to 60 ft. above the runway, depending on the type of airplane and the angle of approach, he eases back on the control column, rounding out the glide with the main wheels a foot or two above the runway surface. Power is reduced to idle and the aircraft settles to the runway. A stall landing is not recommended, because experience has shown that control is better and landing shock is less if the airplane is flown on to the runway, generally using a little power when landing distance is not critical. This is particularly true with large aircraft which vary greatly in landing weight. The main gear must touch down ahead of the nose gear, and definite yoke pressure is necessary, once the main gear is on the ground, to lower the nose gear smoothly.

The Boeing 707 has excellent flight characteristics in the landing configuration and is typical of large jet transports. Approach speeds vary greatly due to the wide weight range possible. Full flaps (50°) are always used unless minimum in-flight control speed, crosswind, or other conditions dictate the use of the 40° flap setting. Full flaps are not set until landing is assured. Approach speed for a 150,000-lb. aircraft would be 141 knots. This would be bled off to 131 knots over the approach end of the runway for a touchdown of 121 knots. On an 80°F day, landing under these conditions at sea level would produce a 4550-ft. ground roll without reverse thrust, if maximum speed brakes were used just after touchdown. The approach itself is quite flat, and the rotation for landing is minimal. In fact, a fairly smooth landing is possible out of a normal approach without changing attitude due to the cushioning of ground effect. If, however, a high rate of descent is acquired during final approach, as the aircraft is rotated to break the descent a high sink rate may develop. Pilots must understand the inertia of large, fast aircraft.

After Touchdown.   The pilot maintains directional control immediately after touchdown by use of the rudder. He changes to nose steering and differential braking as the rudder becomes ineffective.

Modern large aircraft are universally equipped with antiskid systems, and if maximum braking is desired, the pilot applies full steady braking force to the pedals. The antiskid system will detect the beginning of a skid by any wheel and release pressure on that brake just sufficiently to allow the wheel to return to speed. Braking is more effective with flaps retracted, because there is more weight on the tires.

If the pilot uses reverse thrust, he applies it as soon as all the wheels are on the ground, removing it when the airplane has slowed to 50 or 60 mph. A four-engine airplane can reverse the inboard and outboard engines separately; the pilot may use only one pair if he desires.

The drag of the airplane is very effective in slowing it down. Landing gear, flaps, and spoilers all contribute to drag.

Jet bombers are fitted with drag parachutes stowed in the tail section. After the aircraft touches down, the pilot or copilot releases the chute by a cockpit lever. The pilot chute pops out and pulls the drag chute out. It blossoms and acts as a very effective brake. While very high landing speeds may cause it to fail, it must be used early in the landing roll to be effective. So, the chute is deployed as soon after touchdown as chute speed limitations permit.

EMERGENCY OPERATION

All emergencies discussed in Chapter 8 are applicable to multi engine jets, with the obvious exception of those peculiar to reciprocating engines and propellers. Certain others merit discussion here.

Modern technology, manufacturing specifications, and maintenance and inspection procedures have progressed to the point that abnormalities of operation are rather rare in multi engine jets. In reciprocating engine aircraft such as the older four engine transports and bombers, loss of one engine was not at all unusual and loss of two engines not uncommon. Today's jet engines make it highly probable that a pilot will fly many thousand multi engine hours without experiencing an engine failure. Strut mounted engines, either from the wing or fuselage, make engine fire a much less feared emergency than with the imbedded reciprocating engines. The lack of torque and prop drag makes the loss of a jet engine much less critical. Only loss of an engine at takeoff point, or after decision speed is passed, is likely to cause great concern. Even then, as long as minimum control speed can be maintained while clearing obstacles, loss of one engine is not likely to be the calamity it was at one time.

Flameout may occur in a jet engine anytime a fuel or air interruption occurs. Relight can usually be immediately effected by simply pushing the ignition button. If at high altitude an appreciable decrease in engine rpm has occurred before detection, descent to a lower altitude and an increase in indicated airspeed may be required to start the engine. If a relight does not occur, a check of the fuel system is made to insure that fuel is available to the engine. If the engine restarts and instrument indications are normal, it can be operated normally for the remainder of the flight. An engine is never restarted in flight if the cause of failure was fire or overheating.

Shutting down a jet engine is simply a matter of putting the throttle in idle cut-off, and if fire is indicated, cutting fuel and oil to the nacelle. This is usually accomplished with a single switch and need not distract the crew from flying the aircraft. Most large transports have a rapid fuel dump capability so that weight may be reduced to cope with a power loss emergency.

All jet aircraft have air-conditioning and pressurization systems. These may give false alarms about fuselage fires. This is caused by water separators being unable to cope with a high-humidity condition or a malfunction of the air conditioning system. Increasing temperature will usually clear up the vapor and relieve the crew and passengers of the anxiety of having seen "smoke." In the event of actual fire or heavy fumes, the air conditioning system must be shut down, and all persons on board must go to 100% oxygen. Once the source of the fire or fumes is located and eliminated, the air conditioner may be restarted.

Flying Turboprops

The turboprop airplane has a turbine power plant generating thrust for propulsion through a conventional propeller. The expanding gases of the turbine (gas generator) are converted to shaft power by turbine wheels. A reduction gear box reduces the high turbine speed (more than 40,000 rpm) to a workable propeller speed of 1500 to 2200 rpm.

Figure 10.9. Beech King Air Model 200 Turboprop airplane.

The turboprop airplane offers advantages of great reliability and high power-to-weight ratio. Poor low altitude efficiency and acceleration capabilities of the turbojet are effectively improved by the use of the propeller.

The turboprop airplane is a compromise, however, in its reduced cruising speed and altitude capability as compared to a turbojet. These are relatively less important when the aircraft is operated on short flights, at lower altitudes.

Although no more difficult to operate than a reciprocating or a turbojet powered aircraft, the turboprop aircraft has certain operational characteristics with which the pilot should be familiar.

COCKPIT INSTRUMENTATION

The instrument panel of a turboprop airplane has engine instrumentation similar to that used in both reciprocating and turbojet powered aircraft, yet uniquely different to meet the requirements of the turboprop power plant (Figure 10.10). The controls may vary from one type engine to another, and a given make engine may vary in its instrumentation from one airframe manufacturer to another. This discussion will deal with the Beech King Air Model 200 (Figure 10.9) powered by the Pratt-Whitney of Canada PT6A-41 (Figure 4.36) turbine engine.

The *torquemeter* indicates power output and is calibrated in pounds. It is used to set power for all modes of operation and corresponds to the manifold pressure gauge in a reciprocating engined aircraft.

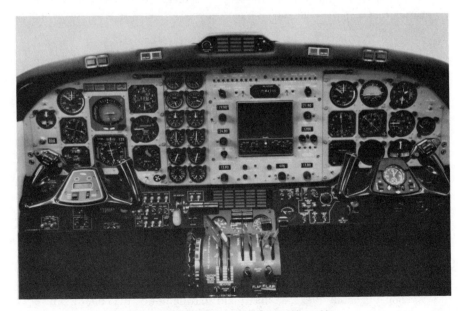

Figure 10.10. Beechcraft Super King Air

The *interstage turbine temperature* (*ITT*) gauge indicates the temperature between the first and second compressor stages in degrees centigrade. Knowing this temperature helps the pilot avoid severe damage to the engine that might occur if the manufacturer's limits are exceeded for only a few brief seconds.

The *turbine speed* (*N1*) gauge indicates turbine rotational speed and is expressed in percent, *i.e.*, 90% of maximum rated speed.

The *propeller tachometer* indicates propeller speed in revolutions per minute (RPM).

Fuel flow, fuel pressure, oil pressure, and *oil temperature* gauges indicate their respective data as in other types of aircraft.

TURBOPROP CONTROLS

Three different controls are located on the *power quadrant*. The *power levers* correspond to throttles on a reciprocating powered aircraft and are used to control turbine speed, hence power output. The power levers, when fully retarded and lifted over a "gate" or stop, provide *"BETA"* control and propeller reversing.

Propeller controls are used to set the desired propeller speed, as with any other propeller aircraft.

Condition levers control fuel shutoff for starting and stopping the engines and low and high idle speeds. The normal or low idle speed is 52% N1, and the high idle speed is 70% N1. Intermediate positions provide idle speeds between these two values. A higher than normal idle speed is selected for battery charging, high electrical or accessory load, and rapid response, when propeller reversing is used.

STARTING ENGINES

Starting a turbine engine is simpler than starting a reciprocating power plant. In some turbine engines the starting is totally automatic with only monitoring required by the pilot.

The PT6A in the King Air is started by selecting the "start position" on the starter switch. The engine will immediately start rotating and will stabilize at an N1 speed of 15 to 18%.

After N1 speed has stabilized for approximately 5 sec., the condition lever is advanced from the off position to low idle, introducing fuel to the combustion chamber. Igniter plugs, which are energized with the start switch, ignite the fuel. Lightoff is indicated by a rapid rise of N1 speed and ITT. The pilot must carefully monitor this portion of the starting sequence to preclude the ITT from exceeding the manufacturer's maximum allowable limit. The N1 speed should increase at a constant rate to low idle speed of 52% N1. If the ITT rises very rapidly and appears about to exceed maximum starting limits, the condition lever is returned to the off position, shutting off the fuel. The start switch is then placed to ''motor'' position to rotate the engine and purge the combustion chamber of fuel and gases and to maintain a flow of cool air through the engine for cooling purposes.

After N1 and ITT stabilize, the start switch is turned off. The condition lever is increased to high idle, and the generator is turned on to recharge the battery prior to starting the second engine. The second engine is started in the same manner as the first.

TAXIING

Taxiing a turboprop is little different from any other type aircraft except that perhaps it is easier. This is because of the feature called *"BETA."* In *BETA,* the pilot has complete control of propeller blade angle, providing very effective deceleration after landing and positive speed control while taxiing, all without brake application. *BETA* control is established by moving the power levers past the totally retarded position and lifting them over a stop or gate. A farther rearward movement of the power levers through the *BETA* range will activate reverse-blade angle for maximum stopping capability after landing.

RUNUP

The turboprop requires little or no runup except when making a check of engine and propeller governors, which is generally done only on the first flight of the day. It is important to make certain that fuel valves, trim, and power plant instrumentation are normal for takeoff. These items are included on the pre-takeoff checklist.

TAKEOFF

After he completes all pre-takeoff items on the checklist and receives takeoff clearance, the pilot taxies the aircraft on to the runway and aligns it with the runway centerline.

Prior to the takeoff roll, the pilot should mentally review critical speeds for the takeoff and climb. For example, if the aircraft has not reached its decision speed (101 knots) and an engine failure occurs, the takeoff must be abandoned.

Power levers are smoothly advanced to 2000 lb. torque. The ITT gauges are monitored to avoid exceeding maximum temperatures. At a speed of 90 knots, the nose wheel is lifted slightly from the runway. Passing a speed of 101 knots, the aircraft is further rotated and established in a climb attitude.

After liftoff, the landing gear is retracted, and best rate of climb speed is held to an altitude where an emergency is less critical. Propeller rpm is reduced to 1900 rpm for climb. During the climb, the power levers must be advanced peri-

RECOMMENDED CRUISE POWER

1700 RPM

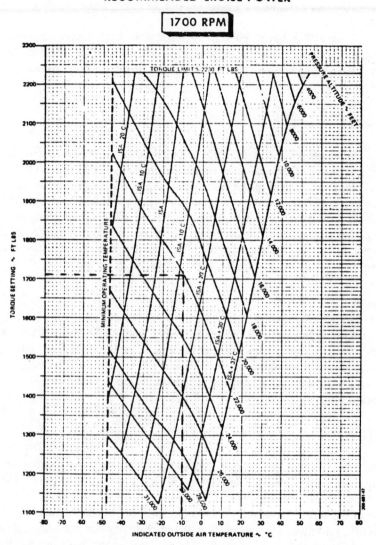

Figure 10.11. A cruise power chart must be used to select the proper power setting. For example, if the O.A.T. is −10°C and the aircraft is cruising at FL 220 (22,000 feet), the chart indicates a power setting of 1710 FT-LBS torque. *(Courtesy Beech Aircraft)*

odically to maintain torque setting. Engine instruments are monitored for readings within the manufacturer's specified limits.

CRUISE

As cruising altitude is approached, the aircraft's nose is lowered to a level flight attitude. When cruise speed is attained, propeller rpm is reduced to 1700 rpm, and torque is set for cruise. This setting is determined from the power chart (Figure 10.11).

DESCENT

Unlike descent in a turbojet aircraft, it is not good procedure to reduce power to idle during descent. Doing so will put the propeller blades in flat pitch, producing very high drag and a subsequent steep angle of descent, which may make passengers uncomfortable and uneasy.

LANDING

Preplanning establishes a descent profile to the runway about like an ILS approach, with a descent angle of approximately 3°. This profile requires about 400 to 500 lb. torque throughout the approach, ensuring near instant power response in the event of a go around or turbulence induced high rates of sink. If the approach is conducted with little or no power, the flat blade angle and the resultant propeller disc drag may produce a hazardous angle and rate of descent. Further, an immediate need for thrust is compromised because the inertial resistance of the propeller mass and the time delay for a significant blade angle change produce a lag from idle to high or full power.

Landing techniques are normal. After the aircraft slows and the nose gear is allowed to make runway contact, the power levers may be lifted over their stop gates and retarded into the *BETA* range for rapid deceleration. For maximum stopping effect, the power levers are moved farther aft through *BETA* into the reverse position, and aggressive braking action is applied. If he uses reverse thrust, the pilot should bring the power levers out of reverse at 40 knots, to minimize likelihood of engine foreign object damage (FOD).

SHUTDOWN

Taxiing to the parking ramp, the pilot should shut off nonessential electrical equipment to allow complete charging of the battery. After parking, he accomplishes the checklist and stops the engines by moving the condition levers to the off position. He shuts off the fuel pumps and battery switch and secures the aircraft.

11

Air Navigation

"..... the winds and the waves are on the side of the ablest navigators."

—EDWARD GIBBON

The art of navigation has challenged man from the beginning of time. Although little changed in its fundamentals, navigation has in recent years been greatly advanced through the use of electronic aids, to a degree of accuracy and simplicity undreamed of by navigators of the past.

Paradoxically, today's complex flight environment demands precision and flexibility at an ever increasing pace to give safety, utility, and economy.

Speed, time, distance, direction, and wind are fundamentals which today are adroitly coupled with electronic aids and the navigator's skill to make up the art of modern air navigation.

Like other topics in *Modern Airmanship*, this chapter is highly condensed; yet the principles, essential details, and standards of performance are provided.

The Earth's Surface and Mapping

A map of the earth's surface is our primary instrument. Although maps designed for navigation are known as charts, the terms *map* and *chart* may be used interchangeably. Any flat map is a compromise among the features to be portrayed, as no completely accurate map of a round object can be made on a flat surface.

THE SHAPE OF THE EARTH

A basic knowledge of the earth, upon whose surface the navigator works his problems, is important. Actually the earth is an oblate spheroid and only approximates a true sphere. The centrifugal force of rotation has expanded the earth at the equator, causing a flattening at the poles. If the earth were represented by a ball 25 ft. in diameter at the equator, the polar diameter would be approximately 24 ft., 11 in. Though small, this ellipticity concerns the map maker. For most navigation purposes, however, the earth may be considered as a perfect sphere.

CIRCLES ON A SPHERE

If a sphere is cut by a plane, the resulting intersection is a circle. If the plane passes through the center of the sphere, the circle formed is a *great circle* and is the largest circle that can be drawn on that sphere. Any circle formed by a plane which does not pass through the center of the sphere is a *small circle*. Segments of a circle, called *arcs,* are measured in degrees, minutes, and seconds. If the circumference of any circle is divided into 360 equal parts, each curved arc would be 1° in length. The number 360 is not a particularly convenient one with which to work in our decimal civilization, but it is historic, well-established, and probably beyond change. A degree further divided into 60 equal parts forms arcs of 1 minute each, and $1/60$ of a minute is a second. For very fine measurement beyond a second, decimal parts are used. Eight degrees, 40 minutes, and 7.18 seconds would be written: 8° 40′ 7.18″. Measurements finer than minutes are rarely used in navigation.

LATITUDE AND LONGITUDE

Determination of position requires a point of reference. It is sometimes satisfactory to pinpoint an aircraft as simply being 8 miles south of Flemington or over Darby at 12,000 ft. Generally this is not sufficient, and over water not even possible. There is, moreover, the problem of locating the towns of Flemington and Darby themselves in relation to some other fixed point. On the earth's surface a universal positional reference system has been established by arbitrarily drawn lines of latitude and longitude.

On any sphere, circles make the best lines of reference. The only problem is where to draw the circles. On an ordinary ball, any orderly system of lines would do. Since the earth is a spinning ball, the axis of spin itself forms the most logical and convenient starting place. The ends of the axis are called the North Pole and the South Pole. Midway between them lies the great circle known as the equator.

Latitude ranges from 0° at the equator to 90° north and 90° south at the poles. Any line of latitude other than the equator is a small circle parallel to it and is therefore known as a *parallel of latitude,* or more simply a *parallel.*

Half of a great circle passing through the poles is a *meridian of longitude,* or simply a *meridian.* All meridians intersect at right angles with parallels of latitude. Any number of meridians can be drawn, and each is exactly like the others. To be useful as reference lines, therefore, one specific meridian, a *prime meridian,* must be chosen as a starting line. Many such lines have been used in the past, and several are in use even now. The most commonly accepted prime meridian throughout the English-speaking world is the great circle passing through the observatory at Greenwich, near London, England. Longitude is measured around the earth both eastward and westward from this Greenwich meridian, through 180°.

Any point on earth can now be fixed by reference to the unique intersection of its parallel of latitude and its meridian of longitude. The numbers of these lines

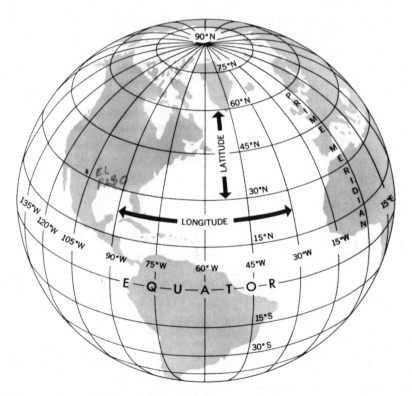

Figure 11.1. Meridians and parallels, the basis of measuring time, distance and direction.

are the coordinates of the point. By custom, when coordinates are given, latitude is named first, then longitude. Thus, the coordinates of El Paso are 31° 47'N,, 106° 27'W (Figure 11.1).

TIME

The fact that the earth is round and rotates on its axis is the basis on which our system of telling time is established. The earth makes one complete rotation of 360° with respect to the sun in 24 hours. The equator could, therefore, be divided into 24 hours as logically as into 360°, each hour being equal to 15° of longitude.

A day is defined as the time required for the earth to make one complete rotation of 360° on its axis with respect to the sun. Noon is the time when the sun lies directly above a meridian; to the west of that meridian it is forenoon, to the east, afternoon.

The earth is divided into 24 time belts of 15° longitude, which gives a difference of exactly 1 hour between belts. There are four such belts in the United States: Eastern Standard Time, Central Standard Time, Mountain Standard Time, and Pacific Standard Time. The dividing lines are irregular because communities near the boundaries often find it more convenient to use time designations of neighboring communities or trade centers.

When the sun is directly above the 75th meridian, it is noon Eastern Standard Time. At the same time, it would be 11:00 A.M. Central Standard Time, 10:00 A.M. Mountain Standard Time, and 9:00 A.M. Pacific Standard Time.

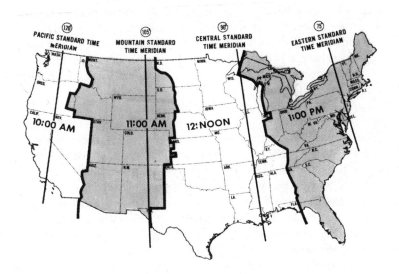

Figure 11.2. Time zones.

In reporting estimated time of arrival (ETA) and other time references to *Air Traffic Control* (ATC), pilots sometimes become confused and add an hour when it should have been subtracted, or vice versa. The speed with which an airplane can pass across time zones, and the confusion likely to occur, have resulted in the adoption of Z time for flying. Z time is local time corrected to *Greenwich Mean Time* (GMT). It is the time on the 0° or the prime meridian at Greenwich, England.

For example, 10:15 A.M. in New York (EST) would be 3:15 P.M. in Greenwich, England. Using four digits and the 24-hour clock (to eliminate confusion between A.M. and P.M.), the time would be 1515Z.

Conversion factors for time zones in the United States are:

Local Zone	Abbreviation	Conversion Factor to Z
Eastern Standard Time (EST)	E	Plus 5
Central Standard Time (CST)	C	Plus 6
Mountain Standard Time (MST)	M	Plus 7
Pacific Standard Time (PST)	P	Plus 8

MAP PROJECTIONS—THE ROUND EARTH ON A FLAT CHART

There are all sorts of maps, but most fall into one of several generic types called projections. Each has certain advantages and disadvantages, which may either permit or limit its use for a given purpose. To understand each, it would be helpful to consider the cartographer's basic problem. For convenience, the user wants a flat map. Yet one has only to try to flatten an orange peel to realize that a rounded surface cannot be flattened without some stretching, wrinkling, or tearing. The result on a map is called distortion. In making a flat map of any portion of the earth's surface, it is impossible for the cartographer to eliminate all distortion. He can only control or systematize it in such a way as to minimize those errors most detrimental to the purpose of the map.

The framework of meridians and parallels on any map is called the *graticule*. Since the configuration of the graticule determines the general characteristics and appearance of the map, proper selection of the graticule is the primary and most critical job the cartographer must face. In practice, the graticule is drawn mathematically by formula. For some maps, however, it can readily be shown visually by actually projecting a picture on a screen, hence the name *projection*. Consider a translucent globe, marked with opaque lines of latitude and longitude, with a small bright light inside. If a piece of paper is held anywhere near the globe, an image of the lines will be formed. The pattern of this projected graticule will be determined by the position of the light, the position of the paper, and the shape into which the paper is formed. For some maps, the paper is kept flat and held tangent to the surface of the globe, but for the most commonly used maps it is

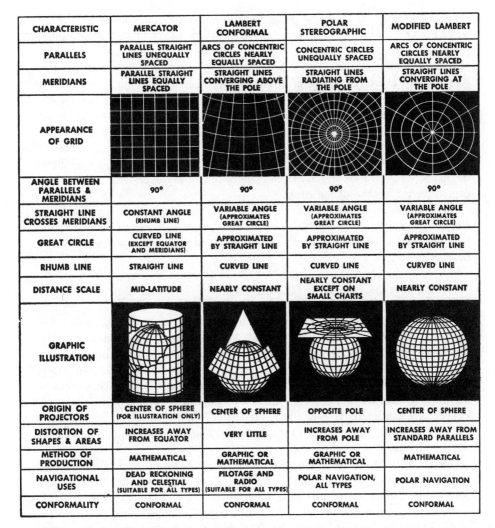

CHARACTERISTIC	MERCATOR	LAMBERT CONFORMAL	POLAR STEREOGRAPHIC	MODIFIED LAMBERT
PARALLELS	PARALLEL STRAIGHT LINES UNEQUALLY SPACED	ARCS OF CONCENTRIC CIRCLES NEARLY EQUALLY SPACED	CONCENTRIC CIRCLES UNEQUALLY SPACED	ARCS OF CONCENTRIC CIRCLES NEARLY EQUALLY SPACED
MERIDIANS	PARALLEL STRAIGHT LINES EQUALLY SPACED	STRAIGHT LINES CONVERGING ABOVE THE POLE	STRAIGHT LINES RADIATING FROM THE POLE	STRAIGHT LINES CONVERGING AT THE POLE
APPEARANCE OF GRID				
ANGLE BETWEEN PARALLELS & MERIDIANS	90°	90°	90°	90°
STRAIGHT LINE CROSSES MERIDIANS	CONSTANT ANGLE (RHUMB LINE)	VARIABLE ANGLE (APPROXIMATES GREAT CIRCLE)	VARIABLE ANGLE (APPROXIMATES GREAT CIRCLE)	VARIABLE ANGLE (APPROXIMATES GREAT CIRCLE)
GREAT CIRCLE	CURVED LINE (EXCEPT EQUATOR AND MERIDIANS)	APPROXIMATED BY STRAIGHT LINE	APPROXIMATED BY STRAIGHT LINE	APPROXIMATED BY STRAIGHT LINE
RHUMB LINE	STRAIGHT LINE	CURVED LINE	CURVED LINE	CURVED LINE
DISTANCE SCALE	MID-LATITUDE	NEARLY CONSTANT	NEARLY CONSTANT EXCEPT ON SMALL CHARTS	NEARLY CONSTANT
GRAPHIC ILLUSTRATION				
ORIGIN OF PROJECTORS	CENTER OF SPHERE (FOR ILLUSTRATION ONLY)	CENTER OF SPHERE	OPPOSITE POLE	CENTER OF SPHERE
DISTORTION OF SHAPES & AREAS	INCREASES AWAY FROM EQUATOR	VERY LITTLE	INCREASES AWAY FROM POLE	INCREASES AWAY FROM STANDARD PARALLELS
METHOD OF PRODUCTION	MATHEMATICAL	GRAPHIC OR MATHEMATICAL	GRAPHIC OR MATHEMATICAL	MATHEMATICAL
NAVIGATIONAL USES	DEAD RECKONING AND CELESTIAL (SUITABLE FOR ALL TYPES)	PILOTAGE AND RADIO (SUITABLE FOR ALL TYPES)	POLAR NAVIGATION, ALL TYPES	POLAR NAVIGATION
CONFORMALITY	CONFORMAL	CONFORMAL	CONFORMAL	CONFORMAL

Figure 11.3. Characteristics of common charts used in navigation. It is necessary to consider what purpose the chart is to fulfill, and to select the projection that most nearly affords the properties desired.

rolled into a cone or cylinder. Various positions and results achieved are illustrated in Figure 11.3.

DISTANCE AND DIRECTION ON A SPHERE

In plane geometry, a straight line is defined as "the shortest line that can be drawn between two points." On the surface of a sphere, the shortest distance between two points lies along the arc of a great circle. A great circle on the earth is a line marked out by the intersection of the earth's surface with any plane passing through the center of the earth.

Distance along a great circle is measured in the same way that latitude and longitude are measured along the great circles represented by the meridians and the equator: in degrees, minutes, and seconds of arc. For practical purposes the distance is converted into distance units of nautical or statute miles. A minute of latitude or a minute of any great circle is a *nautical mile*.

The *statute* mile established by law in English-speaking countries is 5280 ft. The nautical mile is approximately 6080 ft. or 1.15 statute miles. The length of the nautical mile varies slightly from country to country depending upon the accepted circumference of the earth.

On a sphere, *direction* is always measured and expressed in degrees clockwise from the North Pole (true north). For example, east is 90°, west is 270°. Complexities occur with various charts and compass equipment, as will soon be evident.

DEFINITION OF COMMON MAP TERMS

Map Scale.   Map scale is a ratio of length on a map to true distance represented. To express relationship of scale between maps, the terms *small scale* and *large scale* are used. These terms can be confusing. A map showing much detail, and therefore a small area, is a large-scale map; a map including a large area on the same size of paper in comparison must be small scale. Scale may be shown in several ways. On a road map it is most often given as "1in. equals 10 miles." On aeronautical charts it is shown as a ratio, called a representative fraction, as 1:500,000 or 1/500,000. In this system one unit on the map equals 500,000 of the same unit on the surface of the earth.

Great-Circle Distance.   The shortest distance between two points on the curved surface of the earth lies along the great circle passing through these points. The shorter arc of this great circle is the great-circle distance.

Rhumb Line.   A rhumb line is a line crossing all meridians at a constant angle. This is the line which an aircraft tends to follow when steered by a compass. It is easier to fly because the true course remains constant, though it is a greater distance than the great-circle route between the same two points. Under certain conditions it is advantageous to fly a rhumb line course instead of a great circle, because (1) in low latitude, a rhumb line closely approximates a great circle, (2) over short distances a rhumb line and great circle nearly coincide, and (3) a rhumb line between points on or near the same meridian of longidude approximates a great circle.

Conformality (correct representation of angles).   To be conformal, a chart must have uniform scale around any point, though not necessarily a uniform scale over the entire map. Meridians and parallels must intersect at right angles.

The two most useful navigational charts, the Mercator and the Lambert, are both conformal.

COMMON CHARTS USED IN NAVIGATION

In navigation, charts are used principally for two purposes, (1) map reading and (2) plotting and measuring course directions and distances. Some maps serve several purposes quite well. Other maps, such as loran charts or those used for flights in polar regions, are designed for one specific purpose. A summary of the various types follows, indicating appearance, characteristics, and principal uses.

Mercator Projection. The Mercator chart, one of the oldest precision charts, still serves an important navigational need. A Mercator chart is actually constructed mathematically, but an approximation of its graticule can readily be imagined by visualizing a cylinder tangent at the equator to a translucent globe with a light source at the center. All parallels and meridians on the globe will be projected on the cylinder as straight lines crossing at right angles. Meridians will be evenly spaced, whereas distance between parallels will increase rapidly with latitude. If the cylinder is then opened and laid flat, the graticule will appear as a series of rectangles, almost square at the equator and becoming increasingly elongated at higher latitudes.

Scale on a Mercator is true only along the equator. Elsewhere it expands as the secant of the latitude, so that at 60°N or 60°S, scale is twice that at the equator. When expressed as a representative fraction, scale can apply to only one latitude, and a variable scale is required to measure distances. This is a disadvantage for the navigator, but it is readily overcome by measuring along a mid-latitude scale.

The tremendous scale expansion in high latitudes restricts the use of the Mercator above approximately 70°. It is actually best suited for use within 25° to 30° of the equator. In this band of equatorial and temperate latitudes, the Mercator will remain a most useful and popular projection as long as magnetic reference is used for steering, for its outstanding advantage is the fact that on it a rhumb line course is a straight line. A line drawn between two points anywhere on the chart will thus have the same direction from true north at any point of measurement, an advantageous characteristic unique among all projections.

Except for the equator and any meridian, great circle routes on a mercator are curved lines, lying between the rhumb line course and the closer pole. In low latitudes, a rhumb line and a great circle will be close together; at the middle and upper latitudes, the amount of divergence becomes quite marked (Figure 11.4). The great circle route will always be shorter, but the ultimate choice of a route will depend on many other factors as well, as will be seen later in the discussion of flight planning.

Lambert Conformal Conic Projection. In any of the conic projections, the graticule is formed by placing a cone over the globe so that its apex lies along the extended polar axis. In a *simple conic projection,* the cone is held tangent to the

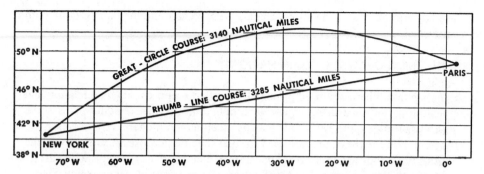

Figure 11.4. Rhumb-line and great-circle routes on a Mercator chart. Distance saved by flying a great circle instead of a rhumb line between New York and Paris would be 145 miles. At their maximum divergence, the great-circle route is 400 miles north of the rhumb line. If other considerations such as weather or favorable winds are of little consequence, transatlantic air routes follow a great circle.

globe along a line of latitude called the *standard parallel*. Scale is exact everywhere along this standard parallel but increases rapidly above and below it so the simple conic is not used. The earliest and most widely used modification of the simple conic is the Lambert Conformal Conic Projection, more generally known as a Lambert Conformal or simply a Lambert. Instead of holding the cone tangent to the globe, Lambert visualized the cone as making a secantal cut, thus giving two parallels. Scale along both is exact. Between them, scale is too small, and beyond them too large. The advantages over the simple conic are a reduction of total range of scale error and a more nearly homogeneous scale over the entire map.

On the Lambert, all meridians are straight lines that meet in a common point beyond the limits of the map, and parallels are concentric circles whose centers lie along a line containing the poles and the point of intersection of the meridians. Meridians and parallels intersect at right angles, and since scale is very nearly uniform around any point on a given chart, it is considered a conformal projection (Figure 11.3).

Any straight line on a Lambert chart very closely approximates a great circle, and the shortest distance between two points can therefore be drawn with a straight edge. The actual difference between a straight line on a Lambert and the true great circle course over small distances is negligible. Even between San Francisco and New York, a distance of 2570 miles, the two vary at mid-longitude by less than 10 miles. A rhumb line on a Lambert curves toward the equator and is difficult to plot. For map reading and radio navigation, the projection is unequaled, and most areas of the world through 80° latitude are covered by aeronautical charts with a scale of 1:500,000 and 1:1,000,000. Above 80°, scale on a standard Lambert is too inaccurate for navigational use.

Polar Charts.   Mercator and Lambert charts complement each other for all usual navigation needs except flights in polar areas, where distortion is too great. Therefore, true Mercator and Lambert charts are not used above 80° latitude.

For this specialized need, several types of aeronautical charts are now in common use, including the Polar Stereo, the Modified Lambert, and the grid overlay system.

In appearance, the Modified Lambert closely resembles the Polar Stereographic chart (Figure 11.3). On both, straight lines are approximately great circles. The Modified Lambert, however, is now used for polar navigation more than the stereographic because area of accurate portrayal can be extended farther from the pole.

In grid navigation a polar grid overlay is constructed on a polar projection chart as a series of straight lines parallel with the 180° meridian. The grid system enables the polar navigator to fly a constant heading from departure to destination (Figure 11.5).

Problems in polar navigation are not limited to charts alone. Direction and steering and position fixing are inherent problems in these latitudes. The pilot-navigator considering flying in the polar areas should consult a navigation manual for detailed polar procedures.

AERONAUTICAL CHARTS

The National Ocean Survey (NOS) publishes and distributes aeronautical charts of the United States. Charts of foreign areas are published by the Defense Map-

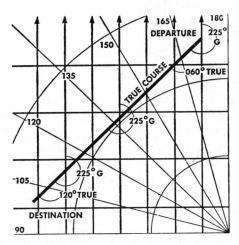

Figure 11.5. The grid system. The grid is drawn parallel to the 180° meridian. The grid system enables the polar navigator to fly a constant heading from departure to destination. As shown here, the true course measured at the local meridian changes from 60° at departure to 120° at destination, whereas grid course is a constant 225°.

ping Agency Aerospace Center (DMAAC) and are sold by the National Ocean Survey. The charts most commonly used in aerial navigation are:

(1) *Sectional*
Projection: Lambert Conformal Conic
Scale: 1:500,000 (1 in. = 6.86 nm.)
Purpose: Designed for visual navigation of slow to medium speed aircraft.
Description:Topographic information features the portrayal of relief and a judicious selection of visual check-points for VFR flight. The latter includes populated places, drainage, roads, railroads, and other distinctive landmarks. Aeronautical information includes visual and radio aids to navigation, airports, controlled air space, restricted areas, obstructions, and related data [Figure 11.6(A)].

(2) *Local and VFR Terminal Area Charts*
Projection: Lambert Conformal Conic
Scale: 1:250,000 (1 in. = 3.43 nm.)
Purpose: Same as sectional
Description:Same as sectional

This series provides an expanded scale of selected large metropolitan areas. They are highly detailed and are of great value for pilotage.

The VFR Terminal Area Charts depict the air space designated as *"terminal control area"* (TCA) which provides for the control or segregation of all aircraft within the terminal control area [Figure 11.6(B)].

(3) *World Aeronautical Charts (WAC)*
Projection: Lambert Conformal Conic
Scale: 1:1,000,000 (1 in. = 13.7 nm.)
Purpose: To provide a standard series of aeronautical charts, covering land areas of the world, at a size and scale convenient for navigation by moderate speed aircraft.
Description:Topographic information includes cities and towns, principal roads, railroads, distinctive landmarks, drainage, and relief. The latter is shown by spot elevations, contours, and gradient tints. Aeronautical information includes visual and radio aids to navigation, airports, airways, restricted areas, obstructions, and other pertinent data [Figure 11.7(A)].

(4) *En-Route Low Altitude Charts*
Projection: Lambert Conformal Conic
Scale: Varies
Purpose: En-Route low altitude charts, sometimes called *Radio Facility* or RF charts, provide aeronautical information for en-route navigation under *instrument flight rules* (IFR) in the low altitude stratum, up to but not including *flight level* (FL) 180 (18,000 ft.).

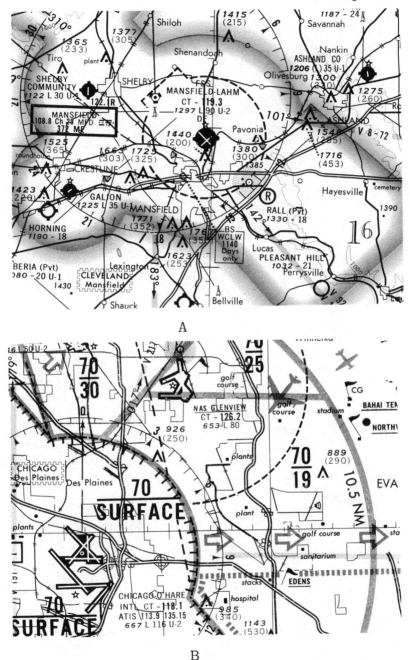

Figure 11.6. (A) The Sectional and (B) the Terminal Area Charts are commonly used charts for navigation by pilotage. The Terminal Area Chart offers large scale affording more detail in the vicinity of Terminal Control Areas. Boundaries of Terminal Control Areas (TCA) are depicted as appropriate. *(National Ocean Survey)*

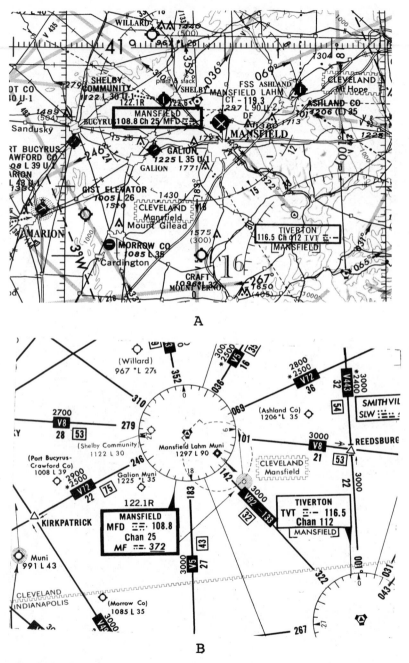

Figure 11.7. (A) World Aeronautical (WAC) chart is used for a combination of pilotage and radio navigation. Small scale affords long distance to be covered on one chart. (B) Enroute Low Altitude Charts are used for navigation under IFR conditions, or VFR using radio aids as the sole means of navigation. *(National Ocean Survey)*

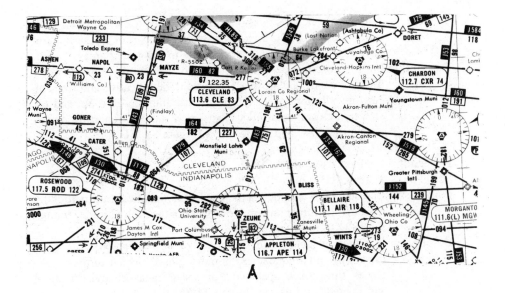

A

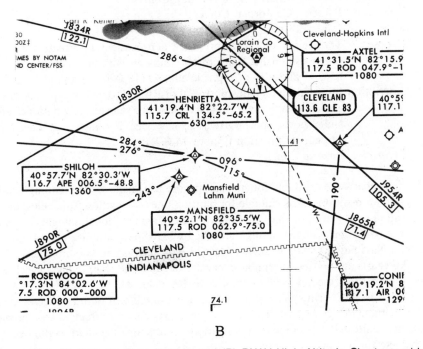

B

Figure 11.8. (A) Enroute High Altitude and (B) RNAV High Altitude Charts provide aeronautical information for instrument navigation (IFR) at FL 180 and above. RNAV routes are established and can be used only by those aircraft equipped with Area Navigation Systems. *(National Ocean Survey)*

This series of charts is also extensively used for VFR flying when using radio aids. The area chart, which is a part of this series, furnishes terminal data at a larger scale in congested areas.

Description: Information includes the portrayal of L/MF and VHF airways, limits of controlled air space, position, identification and frequencies of radio aids, selected airports, minimum en-route and obstruction clearance altitudes, airway distances, reporting points, restricted areas, and related data [Figure 11.7(B)].

(5) *En-route High Altitude Charts*

Projection: Lambert Conformal Conic

Scale: 1 in. = 38.5 nm.

Purpose: En-route high altitude charts provide aeronautical information for en-route instrument navigation (IFR) in the high altitude stratum, FL 180 (18,000) and above.

Description: Information includes the portrayal of jet routes, position, identification, and frequencies of radio aids, selected airports, distances, time zones, special use air space, radar jet advisory areas, and related information [Figure 11.8(A)].

(6) *Area Navigation (RNAV) High Altitude Charts*

Projection: Lambert Conformal Conic

Scale: 1 in. = 23 nm.

Purpose: This series of charts provides aeronautical information for air routes established for aircraft equipped with area navigation systems at altitudes of flight level 180 (18,000 ft.) and above.

Description: Information includes portrayal of RNAV routes, waypoints, track angles, changeover points, distances, selected navigational aids and airports, special use air space, oceanic routes, and other transitional information [Figure 11.8(B)].

(7) *Special Purpose Charts*

Various charts are available for special purpose use such as: (1) planning charts; (2) aircraft position charts suitable for plotting lines of position from celestial observations and electronic aids on long-range flights; (3) global navigation charts for use over long distances at high altitude by high speed aircraft; and (4) Global Loran Navigation Charts to be used with Loran navigation systems.

All charts are revised at regular intervals. Because the chart is one of the most important tools the pilot-navigator has at his disposal, he should consult the "Dates of Latest Prints," a list published and available from the same source as the charts. Most charts will list the date the chart becomes effective and the date the next edition is scheduled for publication. At this time the chart would become obsolete for use in navigation.

Commercial Sources. In addition to the previously discussed charts, commercial sources are available as well. Jeppesen and Company of Denver,

Colorado, is perhaps most noteworthy. The Jeppesen series of radio navigation charts is used exclusively by many airline, executive, and private pilots. They also offer specialized charts for special needs, tailored to any unique requirement.

Pilotage and Dead Reckoning

PILOTAGE

Flying an airplane from one place to another solely by reference to visible landmarks is known as pilotage. During conditions of good visibility, it is only necessary to plot the route desired to follow, noting from the chart prominent landmarks along the route and, in flight, to direct the airplane along the route by reference to the selected landmarks.

If it is seen from the chart that a prominent road, railroad, river, or such feature lies along or parallel to the route, the magnetic compass can be used to an advantage by heading the airplane so that its path or track is continuously over the landmarks selected. The compass heading thereby established is the correct heading to fly and should be maintained until it can be ver:fied again in a similar way. If the aircraft is accurately tracking the landmarks in question, we need not be concerned with magnetic variation, compass deviation, or wind drift, because all three would have been allowed for in establishing the compass heading.

Pilotage as a sole method of navigation is of very limited usefulness. In the industrial areas of the country, restricted visibility is commonplace. Therefore, the prominent check-points vital to this method of navigation often are ill-defined or obscured. Pilotage should be considered fundamental, however, as it is combined with other forms and techniques of navigation.

DEAD RECKONING (DR)

Determining position by means of direction and distance from a known position is known as dead reckoning. This method is used in determining where the aircraft should be, or will be, based on the wind applied to its true heading and true airspeed from the last known position. The accuracy of dead reckoning is dependent upon how closely the wind used in planning approximates the actual wind encountered and how closely the pilot holds the calculated compass heading. Headings and ETA's to reach check-points or destination must be determined by dead reckoning because there is no other practical method of fixing ahead of the aircraft. Dead reckoning encompasses many things such as: (1) plotting and measuring, (2) reading flight instruments to determine the aircraft's airspeed, heading, and altitudes, (3) flight planning with the forecasted winds, (4) calculating the winds encountered during the flight, and (5) altering heading and changing the ETA as new fixes and changing performance data reveal unpredicted changes in speed and path of flight. Whatever the fixing technique, dead reckoning is always the foundation, and an understanding of it is the most logical starting point in learning the art of navigation.

Fundamentally, there are two basic problems in navigation by dead reckoning. The first is the planning of the flight to determine, from the chart and weather data, the distance and the compass heading to be followed between two points. The second problem occurs during flight: to determine from the indicated compass heading, airspeed, and other observed data, if the aircraft is, in fact, making good the desired track.

These two problems are carried on continuously side by side. Before departure the pilot estimates his compass heading and ground speed from known data and predicted winds; then, once in flight, he begins to check his instruments and check-points in order to determine his actual track and ground speed. If he determines that actual track is differing from desired track, he must revise his original estimates accordingly. Again, he checks the results to determine if the revised heading is proper. This process will continue until he reaches the destination airport.

Flight Planning

COMPASS HEADING

In planning a flight a pilot will use charts, aircraft performance data, wind data, magnetic variation, and compass deviation to determine the "compass heading." The compass heading is the end product of navigation problems. To determine what the compass will read to make good the course desired, in view of the winds and all pertinent factors affecting the flight path of the aircraft is, in reality, what navigation is all about.

To determine a compass heading five steps are required:

1. Plot the true course (TC) on the charts.
2. Measure the true course on the chart.
3. Determine true heading (TH), by applying performance and wind data.
4. Determine magnetic heading (MH) by allowing for magnetic variation.
5. Determine the compass heading (CH) by allowing for compass deviation.

The true course and magnetic variation (Steps 1, 2, and 4) are obtained from the chart. Wind information (step 3) is obtained from the Weather Bureau or Flight Service station at the airport. Compass deviation (step 5) is obtained from the deviation card for the particular compass in the aircraft to be flown. Since the information for the five steps is obtained at different times and places, it is desirable that some form be used for recording the data as received in an orderly way.

TOOLS OF DR

The tools required for solving navigation problems are the chart, plotter, pencil, eraser, dividers, long straight-edge, and computer.

The uses of the chart, pencil, and eraser are self-explanatory. The dividers are used for the measurement of distances. Along a desired scale the dividers are spread to span a desired distance. This distance can then be transferred to another part of the chart. Conversely, an unknown distance on the chart can be spanned and the distance found by comparison with the distance scale of the chart. The straight-edge (yardstick) is also used in measuring distances and drawing course lines over great distances.

The principal use of a plotter, which is nothing more than a transparent straight-edge and protractor combined, is the drawing and measurement of "true course" (Figure 11.9).

It is also used for measuring distances from the convenient scales provided along its edges. The plotter is the most indispensable of the navigator's tools.

The computer is used to solve dead reckoning mathematics. The front face is essentially a circular slide rule; the reverse is a rotating compass rose used in conjunction with the sliding insert for solving wind and other vector problems (Figures 11.15 and 11.16).

TERMS AND DEFINITIONS

The following terms and definitions are commonly used in navigation. An understanding of each term and its application is essential for the airman to solve wind vector problems.

True course (TC)	Direction of intended flight.
True heading (TH)	The true course plus or minus any necessary wind correction.
Magnetic heading (MH)	True heading corrected for magnetic variation.
Compass heading (CH)	Magnetic heading corrected for compass deviation.
True airspeed (TAS)	The speed of an aircraft through a mass of air. TAS is determined by the correction of calibrated air speed for altitude and temperature.
Ground speed	The speed of the airplane with respect to the ground.

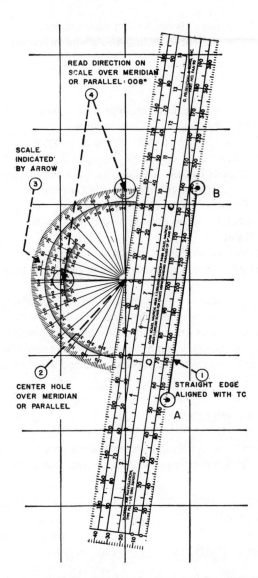

Figure 11.9. Using an aircraft plotter. The course from A to B is 008°. To measure direction from B to A, the plotter would be positioned in exactly the same way, but the reading would be made on the inner scale (188°). NOTE: The small scale is used in conjunction with parallels of latitude for measuring north-south directions, and the arrows on the large angular scale indicating the scale are used to coincide with the direction of flight. To avoid errors, direction should be estimated prior to exact measurement as a common-sense check.

Drift angle .... The angle between the true heading of the aircraft and its track.

Wind correction angle .... The angle, with respect to the true course, at which the airplane must be headed into the wind in order to counteract the effect of the wind, either to maintain the intended track or to return to it.

PLOTTING THE TRUE COURSE (Step 1)

Most plotting of courses will take place on sectional or world aeronautical charts (WAC's), which are Lambert Conformal Conic Projections. These charts may be considered as flight charts, in contrast with planning charts, which are intended primarily for ground study.

When the starting point and destination fall within the limits of a single flight chart, it is only necessary to draw a straight line between them. When the starting point and destination do not fall within a single chart, two or more charts can be joined together. A straight line can then be drawn between the two points, across all the charts involved, with a long straight-edge.

An alternate procedure is to plot the route first on a small scale planning chart and then transfer the route to a flight chart. Transferring a course from a planning chart can be easily done by using points of latitude and longitude along the route. These points are then located on the flight chart, and the route is plotted as a series of lines connecting these points.

MEASURING THE TRUE COURSE (Step 2)

The true course may be defined as the angle between the plotted intended track (course line) and the selected true geographic meridian printed in the chart. It is always measured in a clockwise direction from 0° at true north up to 360°. All courses are usually expressed in three-figure numbers: 000°, 090°, 180°, and 270°.

To measure true course, or the direction of any line on a chart, lay the plotter lengthwise with the outer edge or one of the black inscribed lines coincident with the line to be measured, and with the hole at the center of the plotter directly over a meridian (Figure 11.9). A course between 0° and 180° is measured along the outer scale; one between 180° and 360° is measured on the inner scale. In working rapidly, it is easy to make an erroneous reading by using the wrong scale. It is, therefore, wise to estimate the direction roughly prior to exact measurement as a common-sense check and to note the arrows at the 90° to 270° point indicating the proper scale for each direction.

The method of measuring course lines described above will give the direction from true north at any meridian chosen for the measurement. Because the graticule of a Lambert chart is characterized by converging meridians, a large deviation in measurement could exist over any extended course, depending on the location of the meridian selected for reference. This error will be at a maximum for east-west courses and at a minimum for north-south courses. You have two choices when the true course measurements at departure and destination are at variance. For distances of 200 to 300 miles, it is generally advantageous to make one course measurement for the entire route at mid-longitude. For longer distances (3 or 4 degrees of longitude), it is wiser to break the course into segments, changing to a new course with each new segment encountered (Figure 11.10).

It should be remembered from the previous discussion of the Lambert and Mercator projections that the methods of measuring distance and direction used on the Lambert chart will not apply to a Mercator. It will be recalled that the meridians on a Mercator are parellel to each other; therefore, any straight line cutting them will cut across each at the same angle, giving a rhumb line. Direction from true north can therefore be measured accurately at any meridian and not only at mid-longitude as on a Lambert.

However, on the Mercator, measurements of distance cannot be made with a fixed scale, as on the Lambert, because the scale of a Mercator is constantly changing. The true Mercator distance of the entire course can be measured only against a scale chosen along a portion of the course where an average scale exists. For ordinary navigational purposes, this average is assumed to be a section of the graduated meridian whose mid-latitude is the same as the mid-latitude of

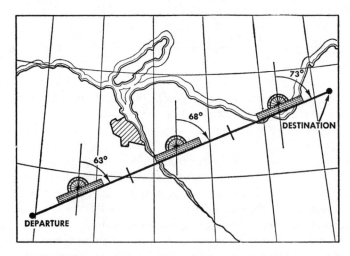

Figure 11.10. Measurement of courses on a Lambert chart. Flights over 300 miles are generally broken into legs, and course measurement of each leg is made at its mid-meridian.

the course. This mid-latitude scale is not precisely equal to the average scale, but for dead reckoning use the difference is undetectable.

DETERMINING TRUE HEADING (Step 3)

If air were absolutely motionless, aerial navigation would be simple. A pilot could easily set his aircraft on any desired course by merely correcting his compass for variation and deviation. He could determine his position at any time and compute arrival time at the destination by using true airspeed. His only errors would arise from unknown instrument errors, unsteady flying, and poor arithmetic. Such a condition rarely exists. The air mass through which man flies is also moving, a motion known simply as wind. It is wind that makes navigation, with its costly equipment and detailed procedures, necessary.

Wind has both direction and speed. As with other directions in navigation, wind direction is expressed in degrees from 0° to 360°. Thus, a wind blowing from the west is a west wind with a direction of 270° true. Wind direction differs from other directions used in navigation in that it indicates the direction ''from'' rather than the direction ''toward.'' This seemingly reverse labeling of wind direction is a traditional practice, and it actually causes little problem to the airman. It should be noted that wind direction is, with one exception, expressed in reference to true north rather than magnetic north. The exception is local wind reports by a control tower or other airport facility to departing and landing aircraft. These wind reports are referenced to magnetic north for convenient wind and runway relationship. Speed is expressed in knots (nautical miles per hour).

To understand dead reckoning, it is absolutely essential to understand that wind effect on any untethered object is completely independent of other motion. A toy balloon released into a steady 20-knot wind will in 1 hour be 20 nautical miles downwind from the release point. An aircraft flying in the same wind for 1 hour will also be displaced 20 miles regardless of its own airspeed. Flying directly into the wind at 200 knots, it will in 1 hour be only 180 miles from departure point; flying with the wind, it will be 220 miles; flying at right angles to the wind, it will be 20 miles right or left of its intended path, and ground speed will be the same on the northern course as on the southern course (Figure 11.11). The lateral displacement in this last instance is *drift,* and the angle between intended and actual track is *drift angle.*

Knowing that the aircraft is drifting to the right or left is not in itself enough to get to a desired destination. A correction must be made in the aircraft's heading in order to compensate for wind. The number of degrees that an aircraft is turned into the wind in order to fly a desired path over the ground is called the *wind correction angle,* or *crab angle.*

Wind correction angle is the same quantity as drift angle but is applied in the direction opposite to that of drift. When the drift angle is 10° left on a desired course, the drift correction is 10° right (the aircraft is steered 10° right of the orig-

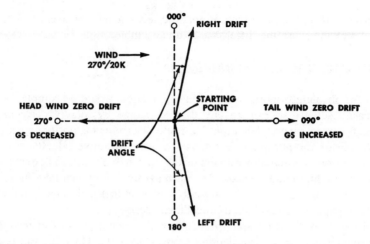

Figure 11.11. Wind effect on aircraft heading due north, south, east, and west. Circles represent what the position of the aircraft would be if there were no wind; arrowheads indicate the actual position if wind is blowing from the west at 20 knots (270/20).

inal course) in order to compensate for the wind. In making a drift correction, the pilot has not prevented drift but corrected for it. All he has done is to head off-course enough so that the wind will cause the aircraft to drift along the intended course to its destination.

WIND TRIANGLE

It has been shown that in order to proceed directly toward the destination, the airplane must be headed into the wind at such an angle that the effect of the wind is exactly counteracted. This angle can be determined graphically when the wind direction and velocity are known as follows:

Figure 11.12 illustrates the construction of a graphical wind triangle. This wind triangle was constructed from the following data:

Wind . . . . 240 at 30 knots   (From Weather Bureau reports)
True course . . . . 195°   (Measured on chart with plotter)
True airspeed . . . . 180 knots   (Calculated)

The true course from E to distant point D is determined to be 195°. From the departure point E, on any convenient scale lay off EW to represent the direction and velocity of the wind. From point W, with radius equal to the true airspeed of the aircraft, swing an arc intersecting the intended track at P. In this diagram, EW equals wind direction and velocity; WP equals true airspeed of airplane; and the wind correction angle equals angle EPW. It is *this angle,* when measured with a protractor (plotter) and applied to the true course, that gives the true heading. The fact that *WP* represents the true heading to be flown may be visualized more clearly if *EH* is drawn parallel and equal to *WP*. An E6B computer (Figure 11.16) solves the problem using *EE'P*.

The measurement of the wind correction angle with a plotter reveals the angle to be 7°. A study of the wind triangle shows wind drift is to the left. Therefore, to offset this drift it is necessary to correct (crab) to the right an amount equal to the drift. Applying this correction to the true course of 195° results in a true heading of 202° (195° plus 7° equals 202°).

The question of whether to add or subtract the wind correction angle can be easily remembered by this general rule: wind from the right, add; wind from the left, subtract. Perhaps the best technique of all is the common-sense approach in which one visualizes the direction of the airplane's path, the direction from which the wind is blowing, and the obvious drift that would result. The correction (crab) would necessarily be opposite the drift.

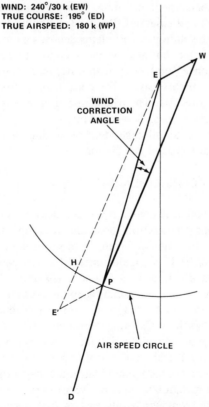

WIND: 240°/30 k (EW)
TRUE COURSE: 195° (ED)
TRUE AIRSPEED: 180 k (WP)

WIND
CORRECTION
ANGLE

AIR SPEED CIRCLE

Figure 11.12. A wind triangle is a simple graphic explanation of the effect of wind upon flight. It provides essential information about ground speed, heading, and time for any flight. It applies from the simplest cross-country flight to the most complicated instrument flight. In practice a navigation computer is used simplifying the task.

A pilot must be careful if he reverses his direction of flight and retraces his course. He must also reverse the correction for wind. If 15° were subtracted from the true course to compensate for drift on the original heading, 15° must be added to the new true course after turning back.

Ground Speed.   The wind triangle, in addition to revealing the wind correction angle, will also give the ground speed of the aircraft. With the same scale used in constructing the triangle, line *EP* is measured and will represent the ground speed. If the ground speed and the miles to be traveled are known, the exact elapsed time can be determined using the formula:

$$\text{Time} = \frac{\text{Distance}}{\text{Ground speed}}$$

Ground Preparation.   The construction of a wind triangle is obviously impossible in flight and cumbersome in planning. However, a clear understanding of the wind triangle is basic and essential. Practical methods are discussed later.

The accuracy of the solutions to problems depends a great deal not only on the accuracy of the navigator, but also on the accuracy of the information used in making the calculations. Winds vary with time, place, and altitude. The conditions experienced in flight may differ considerably from those indicated in weather reports and forecasts. The true airspeed of an airplane can vary greatly with altitude, temperature, and power settings.

Once the pilot is en route, pilotage, dead reckoning, and radio aids must all be used to verify or adjust ground calculations.

DETERMINING MAGNETIC HEADING (Step 4)

As previously stated, the true course is measured with reference to a true meridian printed on the chart (true north). The magnetic compass used in air navigation is referenced (points) to magnetic north. In most areas magnetic north does not coincide with true north. This angular difference between true north and magnetic north at any place is known as magnetic *variation* (Figure 11.13).

From this diagram it can be seen that magnetic north is located at a considerable distance from true north. Magnetic north, a gradually shifting area, at present is located in northern Canada, more than 1100 nautical miles from the true pole. Lines of equal magnetic variation, known as isogonic lines, are shown on aeronautical charts for each half or full degree of variation.

In the northeastern part of the United States, the magnetic compass points west of true north, and the variation is expressed as westerly. In the western part of the country, the magnetic compass points east of true north, and the variation is expressed as easterly. The dividing line between these two areas of opposite variations is known as the agonic line, with 0° variation. At all points along this line, the direction of magnetic north and true north are the same. Local irregularities

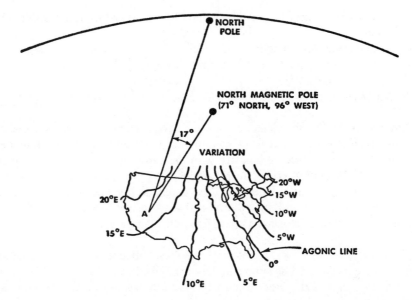

Figure 11.13. Variation chart of the United States. The variation at point *A* is 17°E, because the compass needle is deflected 17°E of the true North Pole at that point. Irregularities in the isogonic lines are caused by the attraction of local magnetic deposits. The agonic line is the line of no variation.

resulting from concentrated mineral deposits may cause minor bends and turns in the isogonic lines.

The compass is of limited value for directional information in the polar areas because of the tremendous variation potential of these areas. Steering in the polar regions is primarily done with gyros.

Incorrect allowance for magnetic variation has always been a prime source of error. Ships have piled against rocks and airplanes have crashed into the sides of mountains, or become completely lost, because of misapplication of variation.

Pilots need remember only one rule: ADD WESTERLY VARIATION. If westerly variation is added, it stands that easterly variation is subtracted. The rhyme, "East is least and West is best," may be helpful as well.

Magnetic heading equals true heading plus westerly, or minus easterly, magnetic variation. In the previous wind vector problem, we determined the true heading to be 202°. If this flight were taking place near Los Angeles, California (variation 15°E), the magnetic heading would be 187° ($202° - 15° = 187°$); near Boston, Massachusetts (variation 15°W), the magnetic heading would be 217° ($202° + 15° = 217°$); and near South Bend, Indiana (variation 0°), the magnetic heading would be 202° ($202° \pm 0° = 202°$).

Pilots often disregard variation when it is less than 5°. This may be satisfactory under some conditions, but is not good navigation and is not recommended. The fact that some errors may be present in a problem is no justification for introducing another.

A magnetic heading has no importance of its own. It is simply a necessary step in correcting a true heading into a compass heading.

DETERMINING COMPASS HEADING (Step 5)

In common with all instruments, compasses have both inherent and induced errors. Magnetic needles tend to point downward in middle and high latitudes, aligning themselves more completely with the earth's magnetic lines of force. This induces fluctuating errors in the compass during acceleration or deceleration, and during turning. Therefore, the magnetic compass should be relied upon only while flying straight and level and at a constant speed. During flight in turbulence it will be necessary to average out the reading of an unstabilized compass.

The directional gyro (DG) is not subject to these influences and is necessary to hold accurate headings. The simple directional gyro, it must be remembered, has no magnetic sense and therefore must be carefully set from magnetic compass readings.

The magnetic compass is often affected to a greater degree by magnetic attractions in the airplane itself. Metal parts, ignition system, electric lights, radios, etc., can disturb the magnetic compass, causing it to indicate incorrectly on many headings.

The angular difference between magnetic north and the north indication of the compass is known as compass deviation.

Swinging a Compass. Most light aircraft, including many light twin-engined aircraft, have no magnetic directional information other than the simple magnetic compass (Figure 5.24). It is this information that is used for setting the directional gyro. Thus, the need for accuracy is obvious.

Ideally, the magnetic compass is located as far away from the instrument panel as possible, such as the top center of the windshield. The electrical wiring located in and around the instrument panel and the resulting magnetic fields induced when electrical current flows can create large errors in a compass.

This error can be greatly reduced by proper adjustment. The magnetic compass is adjustable through two small adjusting slugs located on the face of the instrument (E-W and N-S). The process of adjusting the compass in a particular aircraft and recording it for future use is known as "swinging" the compass. When swinging the compass, all radios and other electrical equipment that would normally be in use on a cross-country flight should be in operation.

A compass can be either ground swung or air swung. Ground swinging is potentially more accurate, but it is often difficult to find a "compass rose" for accurate ground adjustments. Establishing the proper flight attitude and creating equivalent electrical loads can make ground swinging difficult. Therefore, air swinging is often the only means available.

For accurate air swinging, a good directional gyro is necessary. The aircraft is flown parallel to a surveyed straight line on the ground, of known *magnetic* direction, which of course takes *variation* into account. In areas surveyed under the Northwest Land Ordinance, section lines are the best reference. While the aircraft is flying parallel to such a line, the directional gyro is set to the line's magnetic heading. Then, using the gyro for heading reference, the airplane is turned to the nearest cardinal heading (N, E, S, or W.) The N-S or E-W adjusting screw, as appropriate, is adjusted to align the compass with the magnetic heading. The airplane is then turned 90° to the next cardinal heading, and the other adjusting screw is set so as to align the compass on the new heading. The aircraft is turned to the next cardinal heading. This time (and hereafter) the appropriate adjusting screw is used to remove one-half the error. The aircraft is then turned, with shallow banks, through at least one complete circle in each direction, stopping at each cardinal heading to adjust out one-half the compass error, and also stopping, each time aligned with the ground reference line(s), to check for and adjust out precession error in the directional gyro.

The next step is to turn at least one complete circle in each direction, recording the compass heading after the compass settles down, at stops of 15° or 30° intervals as desired. The directional gyro reading is again checked, and reset as necessary, each time the airplane is aligned with the ground reference line(s). The compass headings for each stop are then averaged. The final step is to record the averaged compass reading for each heading on a compass deviation card (Figure 11.14).

When air swinging without a directional gyro, the ground reference lines used must be aligned as nearly as possible with *magnetic* north and south, east and west. Accurate readings on intermediate headings will not be possible.

Many of the problems of deviation can be eliminated by a remote indicating compass (Chapter 5).

A remote system is essential in aircraft with electrically heated windshields. In operation, the strong magnetic field of the windshield may affect the compass, rendering it totally unreliable. These aircraft are usually equipped with one or two gyro stabilized remote compass systems, which must be ground swung if errors are to be adjusted out.

Accurate directional information is essential for navigation, yet the compass probably receives less attention than any other navigational device. Errors of 2° or 3° are usually insignificant when compared to the pilot's ability to hold a heading. However, errors of 15° or 20° are excessive for safe navigation.

The magnetic compass should be checked several times a year for accuracy. Changes in the aircraft's magnetism with the passage of time, severe landing shocks, and severe electrical storms can affect the compass. A recheck of devia-

__NAVIGATOR'S__ COMPASS SWUNG: 29 June 55 BY: H.C.N.			
TO FLY	STEER	TO FLY	STEER
N	000	180	177
15	016	195	192
30	031	210	208
45	047	225	223
60	062	240	238
75	078	255	254
90	092	270	270
105	106	285	285
120	120	300	301
135	134	315	316
150	149	330	331
165	163	345	345

Figure 11.14. Compass deviation card. Magnetic heading is listed on the left under TO FLY; compass heading is listed under STEER. Thus, to fly a magnetic heading of 215°, the aircraft compass should read 213°, or a correction of − 2° is applied to the magnetic heading.

tion should also be made where any considerable difference of latitude is involved. Checking the compass against the runway heading before starting take-off is an excellent habit.

Applying Deviation. The fifth and final step is to correct the magnetic heading for compass deviation and thereby realize our goal, a compass heading. Referring back to the wind vector problem with which we have been dealing, it was determined by step four that the magnetic heading was 217° (using Boston variation). Check the compass deviation card (Figure 11.14) and determine how much error exists for this magnetic heading. It should be noted that 217° lies between the card's 210° and 225°. It is evident that for 210° the compass would read 208° or 2° less, and likewise for 225°. Therefore, a compass deviation of −2° is applied to the magnetic heading of 217°, resulting in a compass heading of 215°.

If the data used are accurate, the wind conditions do not change while the airplane is en route, and no errors in calculating are made, flying this heading will track the airplane directly down the plotted true course to the destination at the predicted time.

A summary of the five steps and respective results in determining this compass heading follows:

195°	True course
+7°	Wind correction angle
202°	True heading
15°W	Variation (Boston)
217°	Magnetic heading
−2°	Compass deviation
215°	Compass heading

DEAD RECKONING COMPUTER

The preceding five steps involved in the solution to the navigation problem were presented to show the logical and orderly plan that must be followed for satisfactory results. Likewise, the solving of the wind vector problem by graphically drawing out the wind triangle was done to make clear the various vectors involved in solving a navigation problem.

It can be seen that graphic solutions are time consuming and would not be

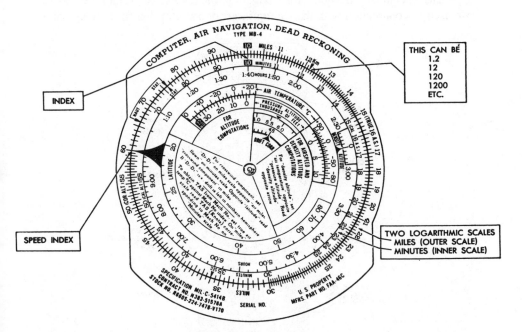

Figure 11.15. The MB-4 computer. The MB-4 is the high-speed version of the famous Dalton E-6B. The slide rule face above is used for basic mathematical computations of distance, rate, and time and also for correction of altitude and airspeed meter readings. *(Courtesy of US Air Force.)*

practical in flight. Problems that must be solved will appear in flight. For these in-flight calculations, a navigation computer is much more practical.

The computer is a quicker and less cumbersome method of solving the wind triangle or vector problems. The solution of all dead reckoning mathematics can be quickly and conveniently done on the computer as well.

There are many types of aircraft computers, but they all serve the same fundamental purpose: rapid solution of dead reckoning mathematics. The arithmetic of navigation is not difficult; it is just endless. How far? How fast? How many minutes? What direction? What is the wind? In a single flight a great number of individual computations might be made. A speedy mathematician might get the arithmetic done, but he might get hopelessly lost, too. Any airman given the responsibility for safe guidance of a high-speed aircraft finds some sort of mechanical computer (or pocket calculator) indispensable.

The most famous and popular of all computers is the Dalton E-6B. A newer version, the MB-4, is a modification for high altitude, high speed jet navigation (Figures 11.15 and 11.16). There are many variations of these, but all work essentially the same way. The front face is a circular slide rule with several auxiliary scales specifically designed for navigation; the reverse is a rotating compass rose used in conjunction with the sliding insert or a vector dial for solving wind

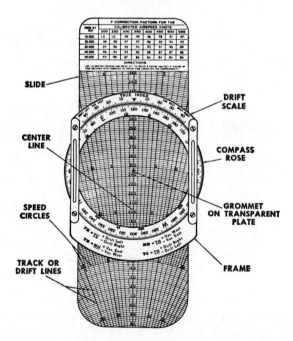

Figure 11.16. The wind face of the MB-4 computer is used for solving wind-triangle problems. The sliding insert can be turned over to provide a speed range from 70 to 800 knots. *(Courtesy of US Air Force.)*

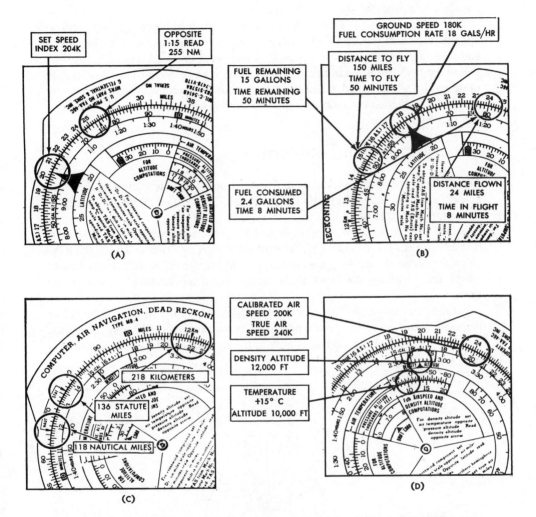

Figure 11.17. Representative problems solved on the slide rule face of the MB-4 computer. (A) Ground speed is known to be 204 knots. Find the distance traveled in 1:15. (B) The aircraft has traveled 24 miles in 8 min. Find the ground speed and time required to fly another 150 miles. Fuel consumed is 2.4 gal. in 8 min. Find the time remaining if you have 15 gal. left. (C) Distance is known to be 136 statute miles. Find the equivalent distance in nautical miles and kilometers. (D) The aircraft is flying at 10,000 ft. at a calibrated airspeed and density altitude. The temperature is + 15°C. Find the true airspeed and density altitude. *(Courtesy of US Air Force.)*

and other vector problems. Representative dead reckoning problems solved on the slide rule face are shown in Figure 11.17.

The slide, in the wind face of the computer, shown in Figure 11.16, is developed as shown in Figure 11.18. There are many problems which can be solved and there are several ways to do each. Some ways are preferred by navigators whose methods require more precision and whose workroom is larger; others are preferred by pilots. The detailed instruction booklet accompanying the computer should be consulted for more information on the wide variety of problems that can be solved.

WIND VECTOR ON THE COMPUTER

The problem previously solved by drawing a graphic wind triangle can be solved quickly and with little effort with the computer. You will recall the following information from this wind triangle.

Wind: 240° at 30 knots. True course, 195°. True airspeed, 180 knots. Referring to Figures 11.19(A) and (B), set wind direction (240°) under TRUE INDEX. Draw the vector down along the centerline to the center of the disc, to the proper length for 30 knots; the head of the vector is on the grommet (circle in the center).

Set the true course (195°) under the true index. Move the slide until the tail of the wind vector is on the speed circle for the true airspeed (180 knots). Read the drift, 7°. Since the heading line is now to the right of the track line, the drift is

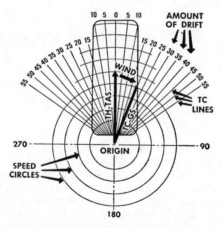

Figure 11.18. The wind triangle on the computer slide. Point of origin is omitted so that a higher range of speeds can be placed on the slide. The tip of the true heading arrow is positioned by setting the speed circle representing true airspeed under the center of the plastic window on the frame of the computer. The wind arrow is drawn from this same point.

left, and the drift correction is plus 7°. The true heading is therefore 195° plus 7°, equaling 202°. Read the ground speed, 158 knots, under the grommet.

The actual conditions in flight more than likely will not agree with educated predictions used in preflight calculations. The wind direction and velocity are perhaps most unpredictable. To find the wind direction and velocity while in flight, the computer can be used as follows (Figures 11.19 and 11.20).

Knowing the true course (track) to be 32° with a true heading of 20°, the wind is obviously from the left. With a true airspeed of 143 knots and a ground speed of 156 knots, it is also from the rear.

Place the true course, 32°, on the circular scale opposite the true index (Figure 11.20). Move the slide to set ground speed, 156 knots, under the grommet. Take the difference between true course and true heading to determine the drift correction angle. A minus drift correction angle is plotted to the left of the centerline, and a plus drift correction angle to the right. Mark the intersection of the wind correction angle and the true airspeed on the transparent disc.

Rotate the transparent disc until this cross is above the grommet, directly over the centerline of the computer. Read the wind direction on the rotating scale opposite the true index, 273°, and the wind velocity, 34 knots, on the slide from the marked cross to the grommet. Note, to avoid error, our original estimate of a wind from the left rear is confirmed.

CALCULATION OF AIRSPEED

The computer also serves as a rapid means of calculating true airspeed. One of the problems in navigation of any sort is a precise determination of speed. On land this presents little problem, but in the air it is more difficult because all mechanical connections with the earth are broken, except where distance measuring equipment (DME) is available. The airman must therefore rely on the air itself for his speed data. The basic reference value is indicated airspeed.

Only rarely is the speed indicated on the meter the true airspeed of the aircraft. The difference exists because the meter can give proper readings only at sea level with barometric pressure reading 29.92 Hg and the temperature 15°C. True airspeed at altitude is therefore generally higher than indicated by about 2% per thousand ft. of altitude. The correction is very easily found by use of the computer, as shown in Figure 11.17(D). The data needed are calibrated airspeed (indicated airspeed corrected for instrument and installation error), pressure altitude (indicated altitude with barometric scale set to 29.92), and outside air temperature.

The above examples of solving navigation problems on the computer indicate the convenience of this method of solution. The detailed instruction booklet accompanying the computer should be consulted for complete information on the wide variety of problems that can be solved.

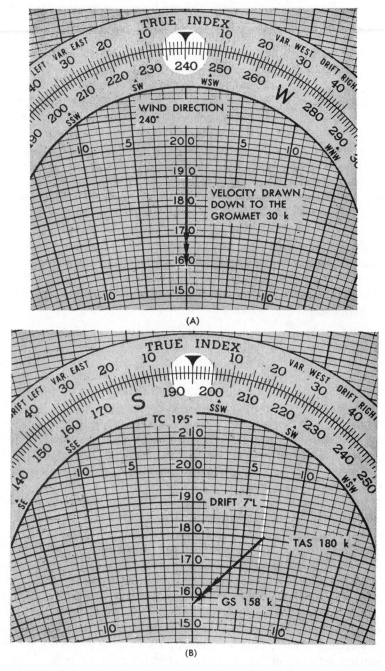

Figure 11.19. Finding true heading and ground speed. *(Courtesy of US Air Force.)*

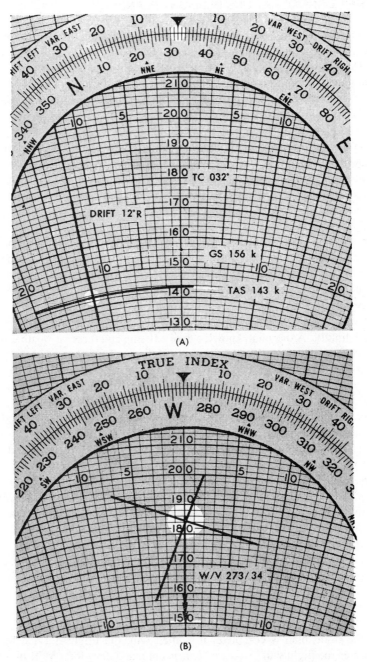

Figure 11.20. Finding wind direction and velocity. *(Courtesy of US Air Force.)*

OTHER CONSIDERATIONS IN FLIGHT PLANNING

There is no substitute for professional planning and preflight in insuring a safe, successful, and efficient flight. Ultimately the safety of a flight rests with the pilot. He must prepare not only for the usual events of an aerial journey, but also for the unusual and unexpected. The measure of a successful flight is not necessarily that it was accomplished safely, but that it would have been accomplished safely in the face of an emergency, or other contingencies, while en route.

The physical, mental, and emotional conditions of the crew members are of primary importance. Certainly an airman should not fly if he is ill, tired out, or emotionally distraught over personal problems. The Federal Aviation Regulations require that the crew member not only have an appropriate, current medical certificate, but also that he not fly as a crew member when he is subject to any potentially disabling physical condition.

Certainly the aircraft must be capable of the planned flight. It must be able to carry the planned load over the terrain to be crossed and have the necessary equipment for the planned operation, as well as for the unexpected. It must have adequate fuel reserves to meet unforecast winds and weather conditions. Load must be within limits, and c.g. located properly.

The weather remains an often present causal factor in accidents and must never be taken lightly. A thorough knowledge of existing and forecast weather is essential to proper planning. A good weather briefing is best obtained by a personal visit to a Flight Service Station or a National Weather Service Office. A telephone call to one of these activities is the next best thing.

Selecting a route is often influenced greatly by the weather and winds, existing and forecast, particularly over the greater distances. The route selected should take into account visual check-points, navigation facilities available, alternate airfields, obstructions, restricted areas, areas of high density traffic, and areas of bad or impossible weather.

The charts to be used must depict the information needed to navigate over the selected route. Sectional, World Aeronautical, and Terminal Area Charts meet the pilot's needs for VFR flight. For IFR he will need Radio En-route Charts, plus Instrument Approach Charts and Standard Instrument Departure Charts for any airports he may use.

Notices to airmen (NOTAMS) are checked in the Airman's Information Manual and on the Hourly Sequence Reports, to help minimize surprises in flight.

FILING THE FLIGHT PLAN

Once the flight plan is complete, it is filed with the Flight Service Station in whose area the flight will originate. This is not mandatory for a VFR flight, but is highly recommended so that someone will start looking for the aircraft if it does not arrive at destination. An IFR flight plan must be filed with a Flight Service Station or an agency of Air Traffic Control if it will enter controlled air space.

Aids to Dead Reckoning

Dead reckoning navigation can be used regardless of visibility outside the aircraft or radio aids available inside the aircraft; yet it cannot be used as an all-exclusive system of navigation. If the data used were always accurate, then its importance as a single navigation technique would be greater. However, the wind information is seldom accurate and is often in a constant state of change, particularly in and near weather frontal zones. Precise true airspeed information is often difficult to determine, and directional information from the magnetic compass is subject to errors. Human errors in computing, measuring, plotting, and instrument reading, and the fact that any errors are cumulative in effect can result in inaccuracies in the calculated position of the aircraft.

Consequently, dead reckoning is the basic foundation upon which navigation depends, but aircraft position must be verified and adjusted in flight by other aids. These aids can be visual or radio.

TERMS USED IN AIRCRAFT POSITIONING

Bearing.   The direction of an object from an airplane is called its *bearing*. Bearings are angular measurements, always measured clockwise, from 000° through 360°. They may be *true bearings, relative bearings,* or *magnetic bearings,* depending upon their point of reference.

A *true bearing* is measured in reference to true north (as a true course was measured); a *relative bearing* is measured in reference to the heading of the aircraft; a *magnetic bearing* is measured in reference to magnetic north.

Most bearings used in modern navigation are *radio bearings* received from a radio navigation facility, although *visual bearings* can be determined during flight and can be helpful to the pilot in position finding. The aircraft is the basic reference in estimating visual bearings; therefore, the bearing would be a relative bearing (relative to the aircraft's heading). To be useful in position finding, the relative bearing must be converted to a true bearing by adding the relative bearing to the true heading of the aircraft:

$$TH \text{ plus } RB = TB$$

An example of this procedure is as follows: true heading (TH) of the aircraft is 315°. An island is off the right wing tip (plus RB 90°). Its true bearing is 045°.

Note that when the sum of true heading and relative bearing exceeds 360°, the true bearing is determined by subtracting 360° from the total.

Sectional, World Aeronautical (WAC), and Radio En-route (RF) charts are designed with a compass rose, referenced to magnetic north, at each Omni station. A bearing plotted using this compass rose is a magnetic bearing, providing a convenient and accurate means of plotting bearings.

A compass rose is not provided around nondirectional radio beacons on these charts. In plotting accurate bearings to these facilities, the bearing from the *radio compass* must be converted to a true bearing before plotting on the chart.

The terms *bearing* and *azimuth* as used in air navigation are identical, but the former is generally used in radio navigation and the latter in celestial navigation.

Line of Position.   A line of position (LOP) is a line connecting all possible geographic positions of an aircraft at a given instant. LOP's determined by bearings, as in map reading or radio, cannot be plotted directly. A reverse bearing, or reciprocal, must be plotted from the known point to the aircraft. A reciprocal bearing is found by adding or subtracting 180° from the true bearing (the LOP from the island mentioned above is 225°) (Figure 11.21).

Fix.   A fix is a geographic position determined by visual reference to the surface or by reference to one or more radio navigation aids. It is a point, not a line. It is sometimes possible to establish a fix directly, as in flying over a landmark the exact location of which is known. Often, however, a fix can be found only by

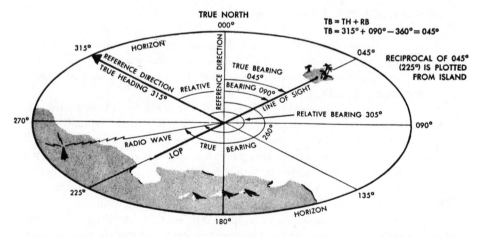

Figure 11.21. Lines of position. A relative bearing is obtained on the island. It is converted into a true bearing by adding it to the aircraft heading. Since the exact position of the aircraft is unknown, no origin for plotting the true bearing exists. A reciprocal bearing must therefore be computed by adding or subtracting 180°, and the LOP plotted from the known island. From any point along this line, the island bears 045° from true north. Bearings obtained from radios or radar must also be converted to true bearings and the reciprocals plotted from them. The exception is the OMNI bearings which are always magnetic bearings.

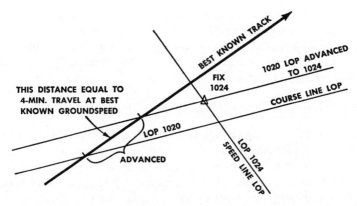

Figure 11.22. Advancing a line of position. At 1020 an LOP is obtained as shown. A second LOP is plotted at 1024. To rectify the time difference, the first LOP is adjusted to the time of the second by moving it up along the best known track a distance equal to 4 min. × best known ground speed. The adjusted LOP is always drawn parallel to its reference LOP.

crossing two or more LOP's independently determined (as being over a known highway intersection). Though any two LOP's intersecting at an angle greater than 30° will normally provide reliable fixing, the best results are obtained when they cross at right angles. When possible, a third LOP is used as a check on the point where the other two LOP's cross.

Using modern navigational equipment, and under favorable conditions, bearings or LOP's may be determined in less than a minute. In this case it may be considered for all practical purposes that the LOP's were taken at the same instant. If several minutes were required to obtain the LOP's, it would be necessary to advance or retard them to a common time to obtain a fix (Figure 11.22).

VISUAL AIDS TO DEAD RECKONING

Navigation by *pilotage* makes use of visual aids. Many landmarks are easily recognizable from the air, such as lakes, rivers, stadiums, race tracks, refineries, and cities and towns. Relief features such as hills and mountains can also be helpful visual aids. This means of position fixing is limited to conditions of good visibility.

MAP READING AS A VISUAL AID

In training and other smaller aircraft, the direct comparison of ground objects with corresponding portrayals on a map is very basic in navigation. The student pilot will use visual navigation as the primary aid in learning cross-country flying, supplemented by elementary dead reckoning. As his experience increases, more reliance is placed on dead reckoning position. Later, radio aids will be used for greater accuracy and increased confidence by the pilot.

Pilotage, or map reading, has little use in jet aircraft. Jet aircraft fly so high and fast that direct observation of the earth provides little navigational aid even during clear weather conditions.

ADVANTAGES AND LIMITATIONS OF MAP READING

Under ideal conditions, map reading is the best means of checking dead reckoning position. No fix is more precise than a positive identification of an accurately mapped landmark. If conditions were always ideal, there would be no need for any other aid to dead reckoning. But sometimes map reading is difficult or even impossible. It cannot be used when the ground is obscured by clouds or when flying over water. Even when the ground is visible, landmarks cannot always be readily identified on a chart. This is especially true at night and over uninhabited regions where the terrain is relatively uniform, or if the chart is incomplete or inaccurate. Excessive detail on a chart, as around large metropolitan areas, can also make identification difficult.

The most commonly used chart for visual navigation is the Sectional Aeronautical Chart [Figure 11.6(A)]. A series of 37 sectional charts covers the continguous United States. The World Aeronautical Chart (WAC) can also be used for visual navigation, but it is better suited for a combination of visual, dead reckoning, and radio navigation [Figure 11.7(A)].

MAP FEATURES USED IN MAP READING

Map reading is not always an easy job. No map can look precisely like the earth itself. The cartographer uses symbols to represent ground features, and the selection of symbols and features to show is not simple. Effort is made to represent those characteristics which have the most distinctive appearance when seen from the air, and to avoid those which would make the map a hopeless clutter. Successful map reading depends on correct interpretation of the symbols and, as in all navigation, success comes through practice.

Three general types of information appear on charts designed for pilotage purposes: (a) navigational reference data, (b) cultural features, and (c) terrain features. The symbols for all of these appear on the chart *legend*. The legend should be studied carefully to become familiar with the various symbols that are used.

Navigational Reference Data.   Every aeronautical chart contains a large amount of data which can never be seen from the air. Meridians, parallels, isogonic lines, VOR's, tower frequencies, airways, Air Defense Identification Zones, and terminal control area (TCA) boundaries fall into this category. All are necessary for safe, legal flight, but they may be confusing to anyone trying to follow on a map the path of an aircraft over the ground. Airways and radio aid data especially seem to clutter a chart in metropolitan areas where the clearest presentation of the ground is generally wanted. The cartographer shows these data in light overprint colors through which other features can be read, but only with practice can they be disregarded until needed.

Terminal area charts are provided for select large metropolitan areas. The large scale provided on these enables the cartographer to include many details not possible on the sectional chart [Figure 11.6(B)].

Cultural Features. These are man-made features such as cities and towns, roads, railroads, bridges, dams, race tracks, tower obstructions, and so on. They are not always the best navigational check-points, but they are generally the easiest for the beginner to use. Once properly identified, a cultural feature provides a pinpoint fix more readily than does any other map feature. In populated areas, only a few outstanding cultural features are shown, such as large towns, prominent highways, and individual buildings, only when they stand out clearly in contrast to surrounding terrain. In sparsely populated areas the reverse is true; every cultural feature possible is shown. In these areas small buildings, ranches, and even roads that are little more than wagon paths are charted.

One very important cultural symbol is the symbol for tower obstructions. Pilots utilizing map reading for navigation often fly at lower altitudes, and numerous high radio and television towers, around and near large cities, are a hazard to their aircraft.

It will be noted that the elevation of the top of tower obstructions above sea level is given in blue figures beside the obstruction symbol. Immediately below this set of figures is another set of lighter blue figures enclosed in parentheses, which represent the height of the top of the obstruction above ground level (Figure 11.23). Obstructions have some value as a navigational aid, but the primary reason for charting is to alert the pilot to the collision hazard.

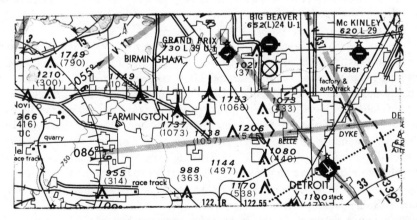

Figure 11.23. Tower obstructions are of great concern to the pilot. The elevation above sea level of the top of a tower is given beside the obstruction symbol; below, in parenthesis, is the elevation above ground level. *(National Ocean Survey)*

Terrain Features.   Rivers, mountain peaks, coastlines, lakes, and islands provide excellent check-points and are, therefore, prominently featured on aeronautical charts. Other topographic characteristics are also shown, but greater skill is required to use most of this information. Colored shading is used for terrain height, green reflecting low elevations and brown, higher elevations. Contour lines also reflect elevations and their relative shapes. These features are needed for safety, but only an experienced navigator can get useable navigational aid from them. Wooded areas are sometimes marked but are difficult to distinguish from the air. Dry river beds and dry lakes frequently blend into surrounding land areas, and marshy areas may look more or less wet with seasonal change.

MAP READING TECHNIQUES

Map reading is the best means of checking dead reckoning position. However, map reading as the only means of navigation is hardly satisfactory and is not recommended. Even in clear weather it is possible to become hopelessly confused without the supporting mathematics of dead reckoning. The small number of actual calculations that must be made is outweighed by the gains of safety, efficiency, and assurance.

More than one pilot has become disoriented while proceeding on course in anticipation of the next check-point, and, because of reduced visibility, passed the check-point without seeing it. Continuing, still in anticipation of this check-point, he becomes hopelessly lost with each passing minute. If he had, based on his last known position, calculated an estimated time of arrival (ETA) over the check-point, he would have been more likely to see it as it was approached and passed. *Therefore, if several minutes have elapsed since the ETA, without seeing the check-point, it should be concluded that it was passed, and preparation for observing the next check-point should be made.*

It is not necessary, and is inadvisable, to try to find every available check-point. So much time is consumed that long-range planning is impossible. Without long-range planning, a flight can quickly become hazardous if bad weather or aircraft emergencies are suddenly encountered. Good navigational practice consists of definitely locating the aircraft at a given instant and then preparing to establish another definite fix at some instant of time later. This may vary from 5 to 30 min., depending on many other factors. In good weather, where visibility is expected to remain clear and wind patterns are expected to remain constant, and at higher altitudes where an emergency is not as apt to require instant action, fixing every 20 to 30 min. is usually adequate. When approaching destination under poor visibility conditions, more frequent fixing is advisable.

A navigator prepares for his next fix by three steps: (1) he selects a good check-point some distance ahead of his present position; (2) he measures the exact distance and computes, on the basis of his best known ground speed, an estimated time of arrival at the check-point; and (3) at a reasonable period of time before this ETA is up, he watches for the check-point and notes the exact time of passage. Using this time, he can update his ground speed and again accurately estimate his ETA for the next fix.

MAP READING LINES OF POSITION

In map reading, fixes are generally obtained directly by either flying over a check-point or by estimating distance and direction from a check-point. Occasionally, however, it may be necessary to use lines of position (LOP) because the only available reference is too remote for an accurate estimate of distance.

In areas of the western United States, cities and towns and other cultural features are sparsely located. On a course of perhaps as much as 100 miles, no significant check-point may exist. Yet, lying left or right of course, at a considerable distance, a prominent check-point may appear. When this check-point is directly off the wing tip, a line of position can be easily determined, thereby fixing position on the course line.

Using the wing tip as reference is not always necessary but reduces the art of estimating the angle to its simplest degree, the wing tip being 90° to the longitudinal axis of the aircraft. As was shown earlier, the RB added to the TH of the aircraft will give the TB to the check-point. The reciprocal of this true bearing is plotted from the check-point to the aircraft's position (Figure 11.21).

NIGHT MAP READING

Map reading by night is more difficult than by day, but as an aid to dead reckoning it may be easier because there is less confusing detail. Even on moonless nights some ground features such as rivers can occasionally be seen.

In clear weather, lights from towns and cities can be relied upon as checkpoints. Highways can be spotted by automobile headlights, and even railroads can be found by the characteristic swinging headlights of locomotives. Terrain features should not be counted on unless they are very prominent or marked by some man-made feature, such as the lights of a dam or bridge across a river, or a beacon on top of a prominent peak (Figure 11.24).

The aeronautical beacon has value as a navigational aid, displaying flashes of white and/or colored light, which is used to indicate the locations of airports and landmarks, and to mark hazards. Airport beacons flash alternately white and green. Military airport beacons are differentiated from civil airports by dual-peaked (two quick) white flashes between the green flashes.

Often at night, the haze that limits visibility during the day appears to dissipate, increasing visibility. However, after sundown the resultant cooling of the earth often leads to the formation of ground fog. This can be difficult to detect at altitude and can be serious during approach and landing.

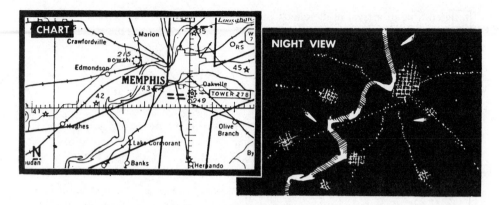

Figure 11.24. Night map reading. Not all features are distinguishable at night, such as the small streams, small towns, and country lanes. However, beacon lights (comet appearing in the figure) can be seen for considerable distances. Note the highways by the headlights of cars and the checkerboard pattern in the cities formed by street lights. On nights when the moon is bright, rivers and lakes will stand out clearly. *(Courtesy of US Air Force.)*

Although map reading at night can provide much navigational data, reliable radio aids should be available for night cross-country flights, and such flights should be attempted only after most careful planning. Flying at night is not inherently dangerous if you are prepared.

COMMON MAP READING ERRORS

There are two basic errors against which the inexperienced navigator should be cautioned. The first is a human failing that sometimes plagues even skilled navigators—the error of wishful thinking. It is easy under the pressure of rapid air work for a navigator to misidentify a check-point by fitting what he sees into what he wants to see. Since all features in an area are not always shown, he can readily convince himself that the map is in error. At any time of doubt, it is necessary to check back and forth from chart to ground, comparing every possible feature, before positive identification is assumed. It is usually best to select a feature on the map and then try to find it on the ground, rather than to work from ground to chart. A ground feature may be newer than the map or simply not have been selected for presentation.

The other error arises after the navigator becomes lost, or after a period of flying over an undercast which has obscured the ground. In either case, a checkpoint suddenly appears. The inexperienced navigator will search all over the map in a frantic effort to identify the check-point and reestablish his position. With a little experience, he soon learns to systematize his efforts and to search only in that portion of the map where he could possibly be. This area is roughly a circle

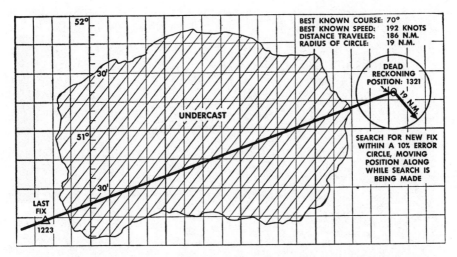

Figure 11.25. Establishing a position after being lost. While the aircraft is over the undercast, the best known course and distance traveled are continuously plotted. When the ground can again be seen, search is made within a 10%-error circle. Precise measurement is not needed. Approximate mathematics and a freehand circle are generally adequate.

with a radius in miles equal to about 10% of the distance traveled since the last fix. It is drawn around the dead reckoning position plotted from the best known ground speed and heading information (Figure 11.25). This is a good reason for flying steady, rather than wandering, headings.

EXAMPLE OF A VFR MAP READING FLIGHT

Figure 11.26 depicts a flight from Scottsbluff, Nebraska to Rapid City, South Dakota. It is planned direct, largely ignoring radio facilities plotted on the chart, in order to illustrate map reading principles. The return flight is planned by the way of Pine Ridge.

Flight Planning. Select the charts you wish to use, either Sectional or a World Aeronautical Chart (WAC) as used in the illustration. Check the date of the chart to determine if it is current or obsolete. Use only current charts for navigational purposes.

Visit the weather bureau of the Flight Service Station for a weather briefing or if a direct visit is not possible, call by phone. Give your intended route of flight, type of aircraft to be flown, estimated time of departure, estimated time en route, and the fact that the flight will be conducted under VFR conditions. With this information the meteorologist or flight service specialist will be better able to brief you in terms of your requirements. It is determined that the weather conditions will permit this flight to be conducted VFR.

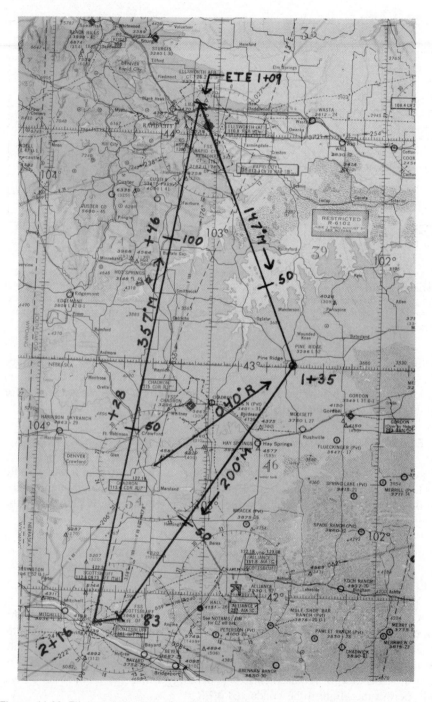

Figure 11.26. Flight pattern from Scottsbluff, Nebraska to Rapid City, South Dakota.

The altitude selected for this flight will be 8500 ft., which gives safe clearance over all terrain and complies with the FAA VFR altitude rules. The wind at 8500 ft. is forecast to remain constant from 300° at 25 knots.

The *Airman's Information Manual* (AIM) should be consulted to determine any information that may be pertinent to the flight. Part II, *Airport Directory*, should be checked to determine airport conditions, fuel availability, remarks, etc.; Part III, *Airport/Facility Data*, is checked for any recent changes in airport status. Airports along the route should be checked as well, to be prepared in the event of requiring an emergency alternate.

Draw the true course from Scottsbluff to Rapid City, to Pine Ridge, and to Scottsbluff VOR (which will be used to illustrate homing), and measure the mileage, dividing it conveniently into 50-nautical-mile segments.

Measure the true courses of each leg and subtract the 13° easterly variation taken from the isogonic lines on the chart. Write the magnetic courses thus determined along each leg, with an arrow indicating direction.

Good check-points lie along the route: Crawford, Nebraska, with the high ground to the west, Angostura Reservoir about 5 miles west of course, and the road network en route and near Rapid City. Coming back to Pine Ridge, note the Oglala Reservoir and the stream leading into Pine Ridge. Notice also Pine Ridge bears 040° from the Chadron VOR; if radio aids were being used, this would be a helpful additional check. The Box Butte Reservoir, the road and railroad pattern at Hemingford, and the water and road pattern at Scottsbluff are all examples of good check-points.

A Cessna Skylane will be used for this flight. It indicates 113 knots at our cruising altitude of 8500 ft.; *outside air temperature* (OAT) should be about 0°C. From the slide rule face of the computer you determine the aircraft's *true airspeed* (TAS) to be 129 knots. On the wind vector side of the computer, determine *ground speed* and *drift correction* required for each leg. Determine the time it will take to fly to each check-point, and note it on the chart. A log is generally better for this because it is more orderly and saves charts; but recording it on the chart helps visualize the problem and is one less piece of paper to worry about. The summarized data for this flight are as follows:

Check-point	Magnetic Course	Distance	Ground Speed	Elapsed Time	Total Time	Drift Corr.
Crawford	357	50	118	:28	:28	11°L
Reservoir	357	36	118	:18	:46	11°L
Rapid City	357	46	118	:23	1:09	11°L
Pine Ridge	147	63	147	:26	1:35	6°R
Hemingford	200	49	126	:24	1:59	11°R
Scottsbluff VOR	200	29	126	:14	2:13	11°R
Airport	240	5	112	:03	2:16	8°R

It is true that for this short flight this seems like a lot of trouble. There are many short cuts that can be taken, up to the point of gross carelessness. The process illustrates how it is done professionally, even though a large part of it may be mental. It would be very difficult to fly the magnetic courses, corrected for drift and compass deviation, closer than 1° in the Cessna Skylane. If speed and headings are held with care, the habit of holding careful headings will grow, resulting in a better check on the winds and the development of a good deal more confidence in basic DR.

Flying the Plan.   File a VFR flight plan with the nearest *Flight Service Station* (FSS). This will insure prompt search action if you should be forced down en route. It also provides one further opportunity to learn of any pertinent information vital to the safety of your flight that has developed subsequent to your original check.

After take off, and clear of the traffic pattern, turn and climb directly on course with a compass heading of 346°. This compass heading was determined as follows:

$$357°MC - 11° \text{ drift} = 346°MH$$
$$346°MC \pm \ 0° \text{ deviation} = 346°CH$$

(The compass deviation card [Figure 11.14] is checked, showing no deviation on this heading.)

The climb will slow TAS slightly on his leg, but not by more than 2 or 3 min. Two minutes will be added to the ETE to the first check-point to allow for this. By turning directly to a compass heading and holding it (the nature of the first check-point is such that it is not easy to miss), you will have a good check on both your compass and the wind.

Right after take-off, note the time of take-off on the chart or log. You may either hold the compass heading until the time has elapsed to Crawford to see just where you do come out, or you may "home" visually on Crawford, not trying to check the drift. When crossing Crawford, or a point near it, make a small triangle around the point (denoting the fix) and record the time; make whatever corrections appear appropriate and continue on course. When reaching the Reservoir, your next check-point, you will have a good check on your ground speed and drift and can make an accurate ETA to Rapid City. In making an ETA, whenever you pass a check-point, you will always know when you *SHOULD* be over the next check-point. This is a great help in finding one, particularly in congested areas. It is also vital in determining your fuel consumption. It is wise to note the fuel remaining each time you obtain a fix, and record it.

Though you can continue your map reading flying after leaving Rapid City, note also that by passing directly over the Rapid City VOR you can track outbound on the 147° radial, thus determining accurately your proper compass heading to get to Pine Ridge. If in doubt about Pine Ridge, it can be checked by intercepting the 040° radial from Chadron VOR, establishing a VOR LOP which would be over Pine Ridge.

As our cruising altitude is more than 3000 ft. above the surface after passing Rapid City, the cruising altitude should be corrected to an odd altitude plus 500 ft. (7500, 9500, etc.). This change is made to adhere to the altitude appropriate for the direction of flight as listed in the AIM (FAR 91.109). Likewise, after passing Pine Ridge an altitude correction will be required to an even plus 500 ft. altitude (8500, 6500, etc.).

In the vicinity of Hemingford, you may continue map reading direct to Scottsbluff, or home on Scottsbluff VOR if your equipment permits. If you have ADF, you can home also on Scottsbluff Radio Beacon.

Immediately before or after landing, cancel the VFR flight plan with the FSS. Forgetting to cancel a flight plan could lead to an extensive and costly search.

RADIO AIDS TO DEAD RECKONING

The *Federal Aviation Administration* (FAA) is responsible for insuring the safe and efficient use of the nation's air space, by military as well as civil aviation. The establishment, operation, and maintenance of a civil-military common system of air traffic control and navigation facilities is perhaps its most significant accomplishment.

A vast network of *airways* has been established covering the entire United States and beyond. These air highways connect all the principal cities and are continually being extended. Along the airways, radio navigation aids are appropriately spaced to provide navigation guidance and air-ground communication facilities. Additionally, *air route traffic control centers* (ARTCC), *flight service stations* (FSS), *control towers, radar facilities,* and *instrument approach and landing systems* are provided to supplement the en-route navigational aids.

A pilot on a *VFR* (visual flight rules) flight is not required to follow airways or use specific radio aids. However, he frequently will find them convenient as they have been specifically established to promote his confidence and safety. The increased use of radio equipment in aircraft makes it highly desirable, if not essential, for the pilot to have a thorough knowledge of the use of radio for navigation and communications.

TERMS USED IN RADIO NAVIGATION

Airways. An airway is a portion of air space of defined dimensions used for navigational purposes. Two airway route systems have been established for air navigation purposes: the *VOR system* below 18,000 ft. and the *Jet Route system* above 18,000 ft. The VOR system airways are called *Victor* airways and are number designated as V8 or V5, etc. [Figure 11.7(B)]. Jet airways are similarly numbered as J53 or J115, etc. [Figure 11.8(A)].

Radial. A radial is a line of magnetic bearing extending from a VOR. The 180° radial, for example, would be a line extending from the station directly south, magnetically.

It is customary, when reporting position in reference to a VOR station, to report the radial over which the aircraft is flying at that moment, regardless of the heading of the aircraft. If the aircraft's position were exactly south of the station, proceeding north toward the station, with the omni receiver reading 000° TO, the pilot would report his position as "inbound on the 180° radial." The term *bearing* is not used in VOR position reporting.

Radio Bearing. The direction of a radio navigation aid from an airplane is a radio bearing. As was discussed earlier, these bearings can be true bearings, magnetic bearings, or relative bearings depending upon what source of reference is used. Radio bearings can be used regardless of visibility outside the aircraft.

Arc. An arc is a *circular course* flown by course information from distance measuring equipment (DME). This procedure is called *orbiting* the station and may be used to intercept the final approach course during an instrument approach, to avoid protected air space (such as TCA's), and to separate aircraft under positive control.

Course Deviation Indicator. The CDI is located in the pilot's instrument panel displaying deviation left or right of a selected VOR or localizer course by deflection of a needle. During en-route navigation the pilot steers the aircraft to keep the needle centered, indicating an "ON COURSE" position. A TO-FROM indicator is also incorporated to show if the course selected will take the aircraft closer to or farther from the VOR station.

In its simplest configuration, the CDI has a vertical needle to indicate deviation from selected course, and a TO-FROM indicator. A horizontal needle can be built in. It displays deviation from glide slope on an ILS approach (Figure 11.27).

Pictorial-Symbolic Course Indicator. This instrument presents the same VOR/LOC course information as the CDI but is presented pictorially (Figure 11.27), showing the relative position of the course with respect to the aircraft. A compass card, slaved to magnetic north, is a part of this instrument, providing magnetic directional information integrated with desired course and course deviation. This unit is also called a horizontal situation indicator (HSI).

Constant Angular Course Width. Deflection of the left-right needle of the CDI is proportional to angle off course and is the display of standard VOR. The course width in miles varies with the distance from the VORTAC, equaling 0 miles at 0 distance, and ± 17 miles at 100 miles from the VORTAC. Thus, a standard VOR course is imprecisely defined at long range and extremely critical at short range. This latter factor causes the familiar large, rapid swings of the CDI as the aircraft approaches and passes over the station.

VOR Scalloping. Imperfection or deviation in the received VOR signal, which causes radials to deviate from a standard track, is called scalloping. VOR

Figure 11.27. CDI—Pictorial—ADF-RMI.

scalloping is generally the result of reflection from buildings, terrain, or other aircraft. This deviation or scalloping effect causes the Omni needle to slowly or rapidly shift from side to side.

Effective Range. VOR/DME facilities operate in the VHF/UHF bands respectively, and both are subject to line-of-sight restrictions. An aircraft 100 nm. from a VORTAC station at FL 180 may receive a reliable signal for both azimuth and distance indications. At a much lower altitude, signals received may be insufficient for reliable data; thus, the aircraft would be beyond the effective range of the station at the altitude. The effective range depends upon the class of VORTAC station, altitude above terrain, and the designed range of the receiving equipment.

VHF OMMIRANGE (VOR)

The VHF omnirange (VOR) operates within the 108-117.95 MHz band. This frequency band is relatively free from atmospheric and precipitation static, providing navigation and communication information when most needed. The VOR is not limited in the number of navigational courses available and provides accurate guidance "TO or FROM" the station on any course the pilot may select.

The word *omni* means "all" and is indicative of the infinite courses projected outward from the station like the spokes of a wheel. Each of these spokes, or *radials,* is denoted by the outbound magnetic direction of the spoke.

VOR's are subject to *line-of-sight* restrictions, and range varies proportionally to altitude of the aircraft (Table 5.2). Generally this does not create a serious problem in most parts of the country because of the great number of stations in use. In mountainous terrain some interruption of signal can be expected at lower flight altitudes.

Because of increased reception distance at high altitudes, an aircraft may receive erroneous indications due to interference between two different stations on the same frequency. Stations on the same frequency are spaced as far apart as possible. (There are more VOR's than there are frequencies available.) VOR's are, therefore, classed according to "useable cylindrical service volume." These classifications are: *terminal* (TVOR), *low altitude* (LVOR), and *high altitude* (HVOR), depending on their intended use.

Vagaries, such as brief course needle oscillation and flag alarm activity, similar to the indication of an "approaching station," may be experienced (usually in mountainous terrain). The pilot should be alert and use the "TO-FROM" indica-

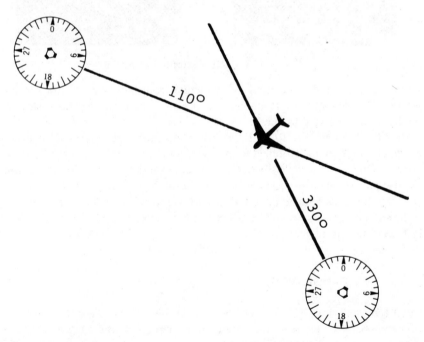

Figure 11.28. When in range of two or more VOR's a fix can be determined by taking and plotting a bearing (radial) from each station. The intersection of the LOP's is the position of the aircraft.

tor to determine positive station passage. This condition should create no major problem to the informed pilot.

When in range of two or more VOR's, a fix can be determined quickly and conveniently by taking bearings (radials) on the stations, thereby fixing position at the intersection of the radials (Figure 11.28). The compass rose printed on a chart around a VOR provides for quick, accurate, and convenient plotting of LOP's.

TACAN

TACAN or tactical air navigation was developed for reasons peculiar to military or naval operations, such as unusual siting conditions, like the pitching or rolling of a ship. As a result the FAA has integrated TACAN facilities into the civil VOR system. Though the two systems operate technically differently, and TACAN cannot be received on conventional VOR equipment, as far as the navigating pilot is concerned, the end result is the same. This integrated facility is called a VORTAC station.

VORTAC

A *VORTAC* station (VHF Omnidirectional Range/Tactical Air Navigation) provides distance information (DME) as well as azimuth. It consists of two components, VOR and TACAN, which provide three individual services—VOR azimuth, TACAN azimuth, and TACAN distance—at one site. Civil aircraft utilize the VOR azimuth and TACAN distance, when equipped with both VOR and DME receivers. With bearing and distance information, the pilot has an exact fix, eliminating the need for bearings from two or more stations. Most VOR stations are, or are being converted to, VORTAC stations. Throughout this chapter, VOR will be used to include both VOR and VORTAC stations. A discussion of VOR procedures will follow later in this chapter.

DISTANCE MEASURING EQUIPMENT (DME)

Distance measuring equipment provides the navigator with accurate knowledge of distance from the ground station to which his equipment is tuned. This equipment operates on an interrogation-response principle in the UHF (962–1213 MHz) band. Distance information is displayed in the cockpit by means of a dial or a digital indicator expressed in nautical miles (Figure 5.38).

DME is subject to the same line-of-sight restrictions as VOR. Reliable signals may be received at distances up to 199 nm. at line-of-sight altitudes with an accuracy of better than ¼ mile or 2% of the distance, whichever is greater. Distance information received is *slant range,* not true horizontal distance (Figure 11.29).

All values in nautical miles.   1 nautical mile equals **6076.1 feet.**

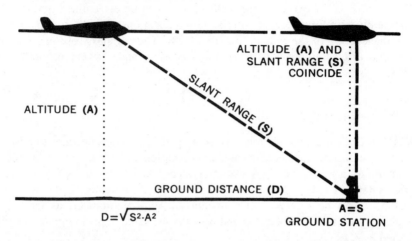

ALTITUDE **(A)** AND
SLANT RANGE **(S)**
COINCIDE

ALTITUDE **(A)**

SLANT RANGE (S)

GROUND DISTANCE **(D)**

$$D = \sqrt{S^2 - A^2}$$

A = S
GROUND STATION

Figure 11.29. DME distance displayed is slant range and may not be exactly the same as ground distance. Slant range equals altitude when the aircraft is directly over the station. *(Courtesy R. C. A. Corporation)*

The difference between slant range and ground distance is insignificant unless the airplane is flying at a high altitude close to the station.

Constant position information is provided the pilot when appropriate course and DME information are available. It eliminates the need of a second VOR or an ADF for crosschecking position. It can also be used to provide course information for operating over a circular course (arc) for purposes of traffic control or to avoid other areas as the pilot may deem necessary. Ground speed can be determined easily from the distance read-out information, by comparing distance traveled to elapsed time (Figure 11.44). Most newer DME's provide direct read out of ground speed also, valid when the aircraft is proceeding directly to or from the station. Some show time to or from, too.

Aircraft equipped with TACAN equipment (military) will receive distance information from a VORTAC automatically, while aircraft equipped with VOR must have a separate DME receiver. FAA navigation facilities designated VOR/DME, VORTAC, and ILS/DME assure reception of azimuth and distance information.

NONDIRECTIONAL RADIO BEACON (NDB)

A *nondirectional radio beacon* transmits normally in the 200–415 KHz band. To utilize these facilities the aircraft must be equipped with an *automatic direction finder* (ADF), also known as a radio compass. The ADF needle indicates a relative bearing or a magnetic bearing to the station, depending upon the type of indicator on board the aircraft (Figure 11.27).

It is possible to determine a radio fix with the ADF, although not as conveniently as with the VOR, by obtaining bearings from two or more stations, and then plotting radio LOP's on the chart (Figure 11.33). As in all types of radio bearings, the greatest accuracy will result if the LOP's cross at or near right angles.

Commercial broadcast stations can also be used by the ADF for homing and fixing. The principal disadvantage of the nonaeronautical station, such as a commercial broadcast station, is the difficulty of positive identification of the station.

ADF receivers are subject to disturbances that result in needle deviations, signal fades, and interference from distant stations during night operations. Unlike VOR, atmospheric disturbances (static) can render the ADF of guarded value for bearing and voice information. It does not account for wind drift, requiring more skill of the pilot. However, it is not limited to line-of-sight signals and perhaps it could be said that the nondirectional radio beacon is the *antithesis* of the VOR. Each has its advantages and disadvantages, and the adroit pilot will use both, one complementing the other.

In most parts of the United States, extensive VOR coverage has reduced the use of the radio beacon to terminal and approach aids. In other parts of the world, the radio beacon remains the primary en-route navigation aid. The cost of providing and maintaining a nondirectional radio beacon is only a fraction of that required for a VOR, and radio beacons can be located in terrain that would be unsatisfactory for VOR's. It can be expected that radio beacons will continue to provide valuable navigational information for years to come. A discussion of ADF procedures will follow later in this chapter.

RADAR

Radio detection and ranging (radar) is an electronic device, either ground or airborne, operating in the 3000 to 10,000 MHz band to provide a visual display of position information concerning objects within its coverage. Short bursts of radio energy are fired into space from a highly directional antenna system. These pulses are reflected from objects in their path and received at the radar site. The information is corrected and displayed on the radar scope as "blips" of light which represent the positions of the reflecting objects.

Ground-based radar can be of great value in fixing position, particularly during reduced visibility, or if navigational radios on board the aircraft have become inoperative. The only radio required is a *transceiver* to establish two-way radio communications with the radar facility. The radar operator will verify the aircraft's position by requesting identifying turns, or if the aircraft is equipped with a radar beacon (transponder), by asking for a reply on a *specific code*.

Airborne radar can be of some value as an aid to navigation. The weather avoidance radar in most large aircraft can be used for this purpose by tilting the antenna downward, revealing prominent terrain features such as lakes, rivers, and shore lines on the radar scope. Many military aircraft are equipped with specialized radar equipment which is better suited for this use.

VHF/UHF DIRECTION FINDER

The VHF/UHF *direction finder* (DF) helps the pilot without his being aware of its operation. The DF is a ground-based radio receiver used by the operator of the ground station where it is located. The DF display indicates the magnetic direction of the aircraft from the station each time the pilot transmits. It is used when a pilot requests assistance. Practice DF fixes and steers help keep the ground operator proficient. DF equipment is most commonly located at Flight Service Stations.

DF equipment is of particular value in locating lost aircraft. If the lost aircraft is within range of two or more DF stations, a fix can usually be obtained by the two lines of position obtained by the ground stations.

VOR Procedures

TUNING

To tune and identify a VOR station, turn the set on and rotate the tuning controls of the *frequency selector* until the published frequency appears in the appropriate window. Adjust the volume control to a level at which the coded identification or voice identification can be heard clearly and make positive identification of the station. It is possible to have read or to have set the frequency incorrectly, or the radio may not have channeled to the correct frequency.

ORIENTATION

The magnetic course to the station is read directly from the CDI by rotating tne OBS until the deviation needle is centered and TO appears in the TO-FROM window. This would be the no-wind heading to the station (disregarding compass deviation). The reciprocal of the TO indication will indicate the radial or position of the aircraft FROM the station. When plotting VOR LOP's on a chart, a FROM setting should be used. The radial thus determined can be plotted directly on the chart because all VOR compass roses are referenced to magnetic north.

TRACKING

After a VOR station is tuned and identified, and the position of the aircraft is determined by centering the needle of the CDI, the aircraft can be navigated on this course toward (to) or away (from) a VOR station by using the following procedure.

Tracking to the Station.   When the needle of the CDI is centered, check the position of the TO-FROM indicator. It must read TO; if reading FROM, rotate the OBS until the needle is again centered and TO is indicated. The aircraft is now turned to a heading corresponding to the magnetic course indicated in the CDI's course selector window. Maintain this heading. If the CDI needle begins to move from center, it indicates that you are drifting away from course, or that you are not holding an accurate magnetic heading. As soon as you determine that you are drifting, turn toward the needle of the CDI, usually an amount double the number of degrees off course. Make a firm and precise correction in order to return to course as soon as possible. Returning to course quickly will be helpful in establishing your correct drift correction promptly, which is particularly important in instrument approaches.

As soon as you have returned to course, establish an estimated drift correction. Any further corrections are made to this new heading as a base. Remember that a 30° correction is sufficient to counteract a direct crosswind of one-half the true airspeed of the aircraft.

If the aircraft again drifts off course in the same direction, after carefully holding the estimated drift correction, again make a correction toward the needle and return to course. Upon returning to course the second time, increase the amount of drift correction. When the correct amount of wind correction is determined by this trial and error method, the needle of the CDI will remain centered, indicating on-course tracking.

Because of *constant angular course width,* the needle will deflect rapidly when flying near the station and will move slowly when flying at a distance from the station. With experience the pilot will learn to recognize and react appropriately for this change in sensitivity.

Station Passage.   Upon reaching the VOR you will encounter the cone of confusion over the station. Since the width of the cone varies with altitude, the time spent in the cone will vary from a few seconds at low altitude to as much as 2 min. at 40,000 ft. As you enter the cone, the TO-FROM indicator may fluctuate between TO and FROM, and the CDI will move from side to side. The alarm flag also may appear.

After you have flown through the cone, the CDI will resume its normal indications, and the TO-FROM indicator will settle on FROM. The latter is the most precise indication of station passage. If you intend to maintain the same course from the station, no change is necessary. When you desire a different course from the station, turn immediately to that heading and set this new course in the course-selector window of the CDI. Intercept and maintain the new course.

Tracking From the Station.   Tracking from a VOR is done in exactly the same manner as tracking to a VOR, except that the CDI needle must be centered with a FROM reading in the TO-FROM window. Turn to the heading indicated in the CDI course window, and hold this heading until a positive needle deviation is evident. Then make a heading correction toward the needle and return to course. When back on course, return to the original heading corrected by an estimated wind correction angle. After flying this heading for a short period, it will be evident if the heading selected is correct or if additional corrections are required.

COURSE INTERCEPTION

In navigation it is generally desirable to track a predetermined magnetic course such as an established VOR airway. Course interception is done as follows.

Course Interception Inbound.   Turn the OBS knob until the CDI needle is centered and TO is in the TO-FROM window (Figure 11.30). The resulting heading in the course window is the direct course to the station. Turn the aircraft to this heading. Now set the course you wish to intercept in the course selector window. If the CDI is to the right, turn the aircraft right to the selected course by at least plus 30°; if the CDI is left, turn the aircraft by at least 30° left of the selected course; the result will be the intercept heading. More than 30° right or left of selected course may be necessary to intercept the course before reaching the station. Hold this intercept heading constant; the CDI will begin to center as you approach within 10° of course (when the CDI needle is exactly deflected to the outer dot, the aircraft is 10° away from the selected course). By watching the rate of movement of the CDI, you can regulate the steepness of bank needed to turn into the desired course without over or undershooting. This rate of movement is governed by the angle of interception, distance from the station, true airspeed, existing wind, and the rate at which you intend to turn on course. The angle of bank can be regulated during the turn so that you are directly on course when the wings are level. When on course follow the procedure for tracking as discussed above.

Course Interception Outbound.   The outbound procedure differs from the inbound procedure only in that FROM rather than TO should appear in the TO-FROM window (Figure 11.31).

Course Interceptions Near the Station.   Because of constant angular width deviation, the CDI needle will indicate an *exaggerated displacement* from the desired course when near the station. It is necessary to use smaller angles of interception when operating in close proximity to the station. Use an angle of interception which does not exceed the number of degrees off the desired course. The

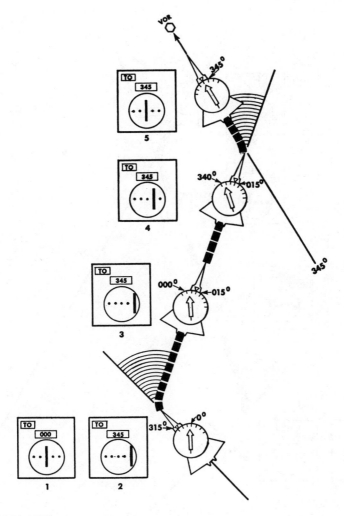

Figure 11.30. VOR course interception inbound, showing successive steps and action of CDI and RMI.

number of degrees off the desired course can be determined by centering the CDI if an RMI is not available.

RADIO MAGNETIC INDICATOR

A *radio magnetic indicator* (RMI) greatly simplifies orientation and course interception on the VOR. The RMI (Figure 11.27) combines compass, VOR, and

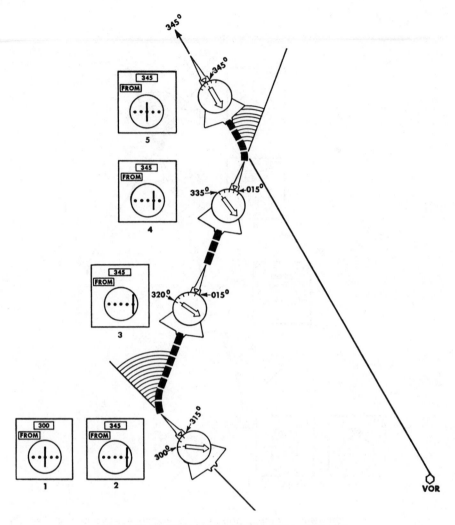

Figure 11.31. VOR course interception outbound.

ADF bearing information on one instrument. Aircraft heading, sensed by a remote compass system, is always displayed under the index at the top of the instrument. The radio pointers are synchronized with the revolving compass card so as to simultaneously indicate magnetic bearings to the stations and relative bearings.

When selecting a VOR station, the needle's head will always read the magnetic bearing to the station without pilot action as is required on a conventional

CDI. If the aircraft is turned until the head of the needle is directly under the top index, the aircraft is heading directly toward the station. The aircraft can *home* to the station by keeping the RMI bearing indicator under the top index (Figure 11.32). An aircraft can home on a VOR with a conventional CDI by centering the needle with a TO indication and turning the aircraft to this magnetic course.

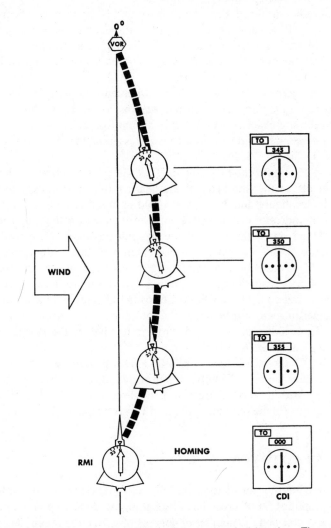

Figure 11.32. Curved flight path resulting from crosswind homing. The stronger the wind, and the greater the distance to the station, the greater the distance off course. The triangular index on the aircraft symbol indicates relative bearing to the station which, if read on the RMI compass card, is also a magnetic bearing.

If the needle moves off from the center position (due to crosswind), recenter with the OBS knob and turn the aircraft to this new magnetic course. Continue this procedure until over the VOR. This is not considered good procedure because any crosswind during homing will cause the aircraft to follow a curved path to the station. Neglect crosswind only when you are close to the station and the bearing on which you approach is unimportant.

The Omni radial on which the aircraft is located is the magnetic bearing from the station and will be shown under the tail of the needle. VOR magnetic bearings may be plotted directly on Radio Facility Charts and on maps as magnetic rather than true bearings because the Omni compass rose around each station symbol is oriented with magnetic north.

Course Interceptions with RMI. Course interceptions using RMI are greatly simplified because the pilot has a constant reading of bearing to the station as he flies his intercepting heading. This enables him to follow his progress and to anticipate the interception of the desired course, lessening the chance of overshooting (Figures 11.30 and 11.31).

Maintaining Course Inbound with RMI. After interception of the desired course, level off so that the head of the bearing indicator is over the desired course and under the top index. If, while holding the desired heading, you note that the bearing indicator drifts off this position, you are drifting opposite the direction of movement of the head of the bearing indicator. Turn toward it an amount equal to double the deviation, exercising the same judgment as used to determine the drift correction for the CDI. Fly this heading until the head of the bearing indicator is over the desired course. Now you are back on course; establish a drift correction. The head of the bearing indicator and the desired course will be together but will not be under the top index. The difference is the drift correction. If the indicator moves toward the top index, the correction is too small; if away, the correction is too large.

Maintaining Course Outbound with RMI. This procedure is the same as the inbound procedure but uses the tail of the bearing indicator in relation to the outbound course. After the drift correction is applied, if the tail of the bearing indicator moves toward the top index, the drift correction is too large; if away, it is too small.

DME AS AN AID TO VOR

Distance and azimuth information from the same station is of great value to the pilot. This simplifies VOR procedures by helping to decide upon the best angle to intercept a course and the number of degrees of correction required to return to course and to determine ground speed more accurately (Figure 11.44). It can also function as a navigation device when flying an arc (orbit) around a station.

ADF Procedures

The *radio compass,* or *Automatic Direction Finder* (ADF), is the most versatile low frequency radio navigation aid in use today. While it does not have the short-range flexibility or precision of the VOR, it may be used over much longer distances. It is so advanced that many recent long range flights by small aircraft have used ADF as their sole aid to dead reckoning; they have crossed the Atlantic and Pacific in both directions and continued on around the world with no other aid. The navigator uses the radio compass to obtain bearings which he can plot as LOP's.

RADIO COMPASS BEARINGS

When the radio compass needle points to a station, the indication is a radio bearing. If the instrument face is an RMI, the needle points to a magnetic bearing; if the card is fixed, or manually adjusted with 0° at the top index, it is a relative bearing. To be plotted as an LOP, either must be converted into a true bearing. With RMI, the true bearing can be read directly by applying variation and deviation. With the fixed or manually rotated compass face (when 0° is set under the top index), you obtain the true bearing by adding the aircraft's true heading to the relative bearing:

$$\text{TRUE HEADING} + \text{RELATIVE BEARING} = \text{TRUE BEARING}$$

With a rotatable compass card, you can do this mechanically simply by rotating the compass face until the true heading is under the top index, while allowing for variation and deviation. The true bearing to the station now lies under the head of the needle, and the true bearing from the station lies under the tail of the needle. Because bearings are always plotted from the station to the aircraft's position, read the bearing under the tail of the needle, or add 180° to the reading under the head of the needle (to determine reciprocal).

Plotting Radio Compass Bearings. To obtain an LOP, plot the true bearing like any other bearing by simply drawing a straight line from the station in the direction of the bearing. Unless the station lies directly ahead, behind, or is nearly off one wing tip, the LOP, to be useful, must be combined with some other bearing, either visual or radio. All LOP's must be advanced or retarded to a common time to obtain a fix (Figure 11.22).

ADF Position Fixing. The position of the aircraft can be fixed by plotting ADF LOP's from at least two L/MF stations. A third station is desirable to check the accuracy of the fix. The fix, as illustrated in Figure 11.33, is determined as follows.

Tune the ADF to each station, and record the magnetic bearing indicated by the

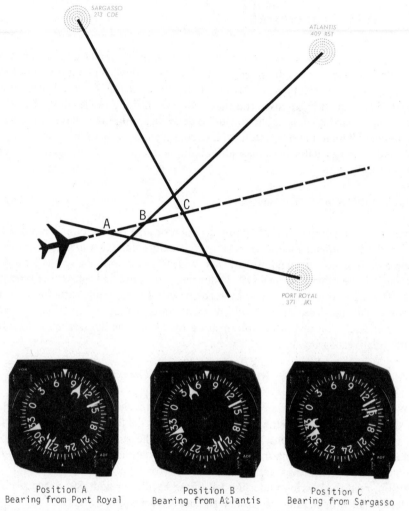

Position A
Bearing from Port Royal

Position B
Bearing from Atlantis

Position C
Bearing from Sargasso

Figure 11.33. Position can be "fixed" with ADF bearings when two or more stations are received. A third station is desirable to check the accuracy of the fix. Correct each bearing for local variation and plot from the station. The aircraft was within the triangle formed by the three bearings halfway between the times of the second and third bearings. *(Courtesy Collins Radio Company)*

tail of the bearing indicator (bearing *from* the station), noting the time as the last bearing is taken.

An RMI is used, in this example, indicating a magnetic bearing. Before plotting, magnetic bearings must be converted to true bearings by correcting for local variation. For the purpose of illustration, it will be assumed that the variation is 10°W and no deviation exists. The results are:

Station A   MB 281° − 10° = 271° TB
Station B   MB 221° − 10° = 211° TB
Station C   MB 150° − 10° = 140° TB

In actual practice the true bearing could be read directly from the RMI by subtracting (in this example) 10° from each station's magnetic bearing. The true bearings are plotted from the station (using a meridian as in plotting a true course). If the bearings were taken at equal time intervals, the aircraft probably was within the resultant triangle, at the time of the second bearing.

Note that the 10° westerly variation was subtracted, not added. This is because the magnetic bearing used was determined by an RMI which is oriented with magnetic north, and thus, variation was reflected in its indication.

If the fix in this example were determined with a fixed card indicator, the relative bearing read under the tail of the indicator is added to the true heading of the aircraft, giving a true bearing from the station for plotting.

It may appear that the fix determined in this example could hardly be done with a high degree of accuracy in the cockpit of an aircraft. This procedure was shown to demonstrate how a professional navigator on a large aircraft may determine a fix using the ADF. For increased accuracy all LOP's must be advanced or retarded to a common time (Figure 11.22). However, except in very high speed jet aircraft, with modern equipment the pilot can determine bearings in such a short time period that for all practical purposes, all LOP's were taken at the same time.

If an ADF bearing is plotted on an aeronautical chart (Radio Facility, Sectional, or World Aeronautical), the VOR compass rose can be useful for in-flight plotting even though the bearing is from a L/MF facility located some distance from the VOR. For example, if the bearing from a L/MF is 210° magnetic, plot an LOP from the nearest VOR on its 210° radial; from the ADF station plot an LOP parallel with this VOR radial (Figure 11.34). In this case the magnetic bearing from the station would be plotted directly without the need to convert to a true bearing because the VOR compass rose is oriented with magnetic north. Although this procedure is not as precise as more conventional methods, it is commensurate with cockpit capabilities.

Bearings over 150 miles long, plotted on Mercator charts, must be corrected because all radio bearings are great circles and all straight lines on a Mercator chart are rhumb lines. The rhumb line correction is obtained from the navigator's flight-planning documents.

Radio Bearing Inaccuracies.   While radio bearings over 150 miles long are not fully reliable, knowledge of the sources of error will help to use them intelligently. *Night effect,* indicated by a fluctuating needle, is caused by unpredictable sky-wave reflections; commercial broadcast stations over 1000 KHz are particularly vulnerable at sunrise and sunset.

Thunderstorms also will deflect the needle in the direction of the storm. *Shore-*

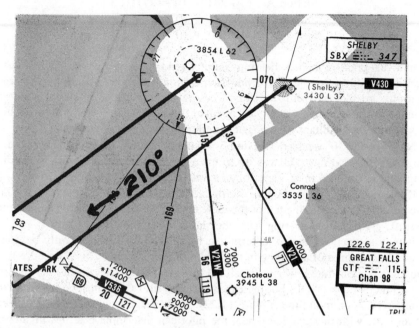

Figure 11.34. In-flight plotting of ADF bearings can be simplified by plotting the out-bound magnetic bearing on a nearby VOR, and then plotting a LOP from the ADF station parallel to this line. With this procedure a magnetic rather than a true bearing is used because the VOR compass rose is oriented with Magnetic North.

lines bend radio waves seaward when the bearing crosses the shore at angles less than 30°; mountains also reflect radio waves. Quadrantal error may occur, resulting from deflection of the waves when they strike portions of the aircraft, such as tail or engines, before striking the loop antenna.

TUNING THE ADF

The ADF must be tuned carefully; otherwise an incorrect station may be received or an inaccurate bearing may be indicated. Recent technology has been used to develop *crystal-controlled* ADF receivers, with digital frequency readout, greatly simplifying tuning. The tuning procedure varies with make and model of the ADF Radio.

BFO. In the United States all AM and FAA low frequency transmission is modulated. Some stations in other countries broadcast an unmodulated signal. To identify an *unmodulated* signal, switch to BFO mode, and the Morse code underlying the tone will be heard.

ADF Bearing Indicators. The ADF bearing indicator located on the pilot's instrument panel can be one of three types: an RMI which reads a magnetic bearing to the station; a fixed card indicator (0° is always at the top) which reads the relative bearing to the station; or a hand rotatable card which reads a magnetic bearing to the station if the card is rotated by its knob to correspond to the magnetic heading of the aircraft, or which reads a relative bearing if the card remains with 0° at the top index (Figure 11.27).

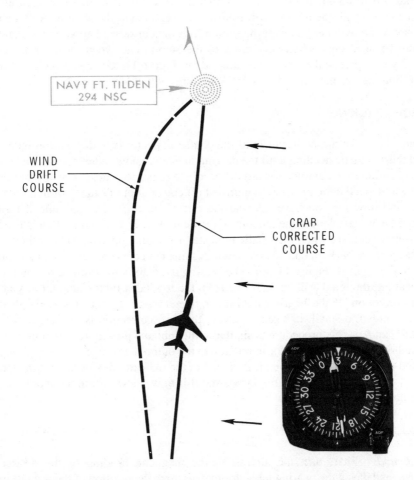

Figure 11.35. An ADF can be used to "home" on a station. The dotted line approximates the track, with a wind from the right of the aircraft, if the bearing pointer is maintained under the top index. To correct for crosswind, turn the aircraft into the wind to a heading that will maintain a constant bearing pointer indication as shown. *(Courtesy Collins Radio Company)*

HOMING

ADF homing is perhaps the easiest and most common use of the ADF. Tune in desired station frequency and identify the station; observe the position of the ADF bearing indicator, and turn in the nearest direction to place the head of the bearing indicator under the top index. The pilot flies to keep the indicator under the top index by changing the magnetic heading of the aircraft. If the magnetic heading must be changed to hold the bearing indicator under the top index, the aircraft is drifting due to crosswind. If no drift correction is made to compensate for the crosswind, the aircraft will follow a curved path to the station. Although this procedure will get you over the station, it is not considered good practice and should be used only when very close to the station and when the bearing on which you approach the station is unimportant (Figure 11.35). The principles for ADF homing are the same as for VOR homing with RMI.

ADF TRACKING

Inbound.   Turn the aircraft until the course indicator is under the top index; hold this magnetic heading until the bearing indicator shows a deflection. Make a turn of sufficient magnitude toward the needle to return to course. The size of this turn will depend upon how fast you drifted off course, your distance from the station, and how far you have allowed the aircraft to drift. The amount of turn should be at least double the amount of the deviation. If a 10° correction is used to return to course, when the needle indicates 10° on the opposite side of the top index, you are back on course and should resume the course heading with a wind correction applied (Figure 11.36). The bearing indicator will continue to point to the station, though it will be deflected from the top index by the amount of your drift correction. If the bearing indicator moves toward the top index, the drift correction is too small; if it moves away, the drift correction is too large.

Outbound.   This procedure is similar to the inbound procedure, but it utilizes the tail of the bearing indicator in relation to the outbound course. After applying a wind-drift correction outbound, if the tail of the indicator moves toward the top index, the drift correction is too large, and if it moves away from the index, the drift correction is too small.

COURSE INTERCEPTION

Inbound.   First, turn the aircraft to the magnetic heading of the desired course, and check the bearing indicator to determine the position of the aircraft in relation to the desired course. Unless you are exactly on the desired course, the bearing indicator will be deflected to the left or right. Turn toward the head of the indicator by the number of degrees of deflection plus 30°. The number of degrees turned is the *angle of interception*. Proceed on this heading; when the bearing in-

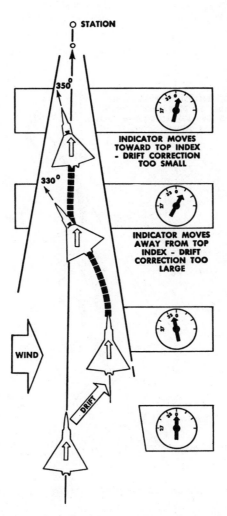

Figure 11.36. Maintaining course inbound with ADF. *(Courtesy of US Air Force.)*

dictor is deflected by the same number of degrees from the top index as the angle of interception, make the turn on course with an appropriate lead to avoid overshooting. If the angle of interception is left, the bearing indicator will indicate to the right of the top index upon interception (Figure 11.37).

The previous example used a fixed card for bearing information. If an RMI is used, the procedure is the same, but it is simplified because the bearing desired can be read directly under the bearing indicator as the desired course is reached. Likewise, if an indicator with a rotatable card is used and that card is rotated, setting the magnetic heading of the aircraft under the top index, the magnetic bear-

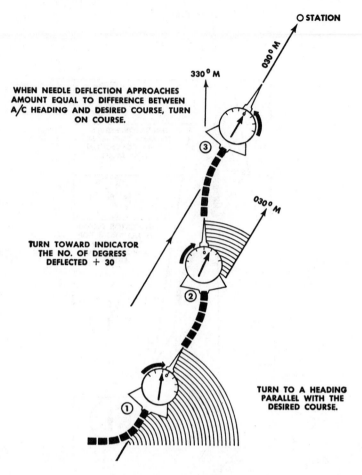

Figure 11.37. ADF inbound course interception. *(Courtesy of US Air Force.)*

ing desired can be read directly under the bearing indicator when intercepting the desired course.

Outbound. Turn the aircraft to the magnetic heading of the desired course and note the position of the tail of the bearing indicator in relation to the top index. If the tail of the bearing indicator is deflected to either side of the top index, turn away from the tail of the indicator by a sufficient number of degrees to insure course interception at a moderate rate. Maintain this intercept heading. When the bearing indicator (tail) is deflected by the same number of degrees from the top index as the angle of interception, the aircraft is on course. As the tail of the bearing indicator approaches the interception angle, lead the turn to roll out on the desired outbound course (Figure 11.38).

Immediately After Station Passage. Because you are quite close to the station, the angular amount you are off the new course is exaggerated, and you may use a simpler method. After definite station passage, immediately turn the aircraft to the magnetic heading of the desired course. Maintain this heading until the bearing indicator has stabilized. Then turn toward the desired course by the same number of degrees that the tail of the bearing indicator is deflected from the top index, not exceeding 45°. Turn to the desired heading with an appropriate wind correction as soon as you are on course.

TIME-DISTANCE CHECK

In normal cross country flying, the pilot will have information with regard to ground speed, distance to station, and ETA's. If the pilot is or has been lost, it

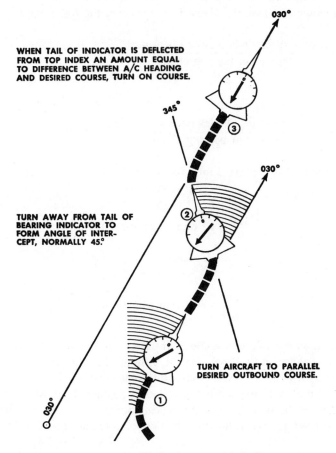

Figure 11.38. ADF outbound interception. *(Courtesy of US Air Force.)*

may be necessary to orient himself with respect to a radio facility. A bearing can be determined by use of an ADF or VOR receiver, but unless DME equipment is available, or a fix is determined by another LOP, the pilot has no information as to his position on the established LOP. This can be done by making a *time-distance* check, as follows.

VOR or ADF 90° Method.   Turn until the aircraft's heading is at right angles with a bearing to the station: (1) If using ADF, turn the aircraft until the bearing indicator is exactly at the nearest wing tip position, and (2) if using VOR, rotate the OBS, centering the CDI needle; turn quickly perpendicularly to this bearing (if necessary, recenter the needle and correct the heading); note the exact time. Hold this heading until you observe a 10° bearing change. This will be observed directly on the ADF or RMI indicator. If no RMI is available, or for better accuracy, set a 10° bearing change in the course-selector window (if the station is on your left, subract 10°; if on your right, add 10°). When the exact 10° bearing change is complete, note the elapsed time and use the following formula:

$$\frac{\text{time in seconds between bearings}}{\text{degrees of bearing change}} = \text{minutes to station}$$

or

$$\frac{\text{TAS}* \times \text{time in minutes between bearings}}{\text{degrees of bearing change}} = \text{miles to station}$$

For example, if it requires 2 min. to fly a 10° bearing change at a TAS of 200 knots, you are:

$$\frac{120}{10} = 12 \text{ min. from the station}$$

or

$$\frac{200 \times 2}{10} = 40 \text{ nm. from station.}$$

At the completion of the time check, if the aircraft is turned directly toward the station (90° change in heading), the ETA should be adjusted for the change in ground speed which will result from a different wind angle.

Another time-check method is illustrated by Figure 11.39. Maneuver the aircraft so that the station is at a relative bearing of from 15° to 45° on either side of the aircraft heading. When this has been done, note and record the time. Main-

---

* Use ground speed, if known.

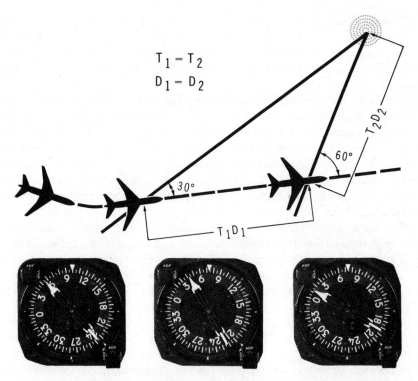

Figure 11.39. Time–distance to a VOR or LF/MF station can be determined by turning the aircraft so the station is at a relative bearing of from 15 to 45 degrees either side of the aircraft heading. Note the exact time for the relative bearing to double, while holding a constant heading. The time and distance to the station are equal to the time and distance between the two bearings. *(Courtesy Collins Radio Company)*

tain a constant heading until the relative bearing has doubled; again note and record the time. Time and distance to the station are equal to the time and distance between the two bearings. For accuracy, the ground speed and heading must be corrected for prevailing wind conditions.

Short and Medium Range Area Navigation

The electronic airways developed by the FAA have been laid out from one navigation facility to the next and often require other than a direct route between desired points. An economic need to shorten routes and increasing congestion over fixes (VOR's) have brought about a system of direct navigation. This system is called area navigation (RNAV).

RNAV permits aircraft operations on any desired course within the coverage of station-referenced navigation signals or within the limits of a self-contained system capability. This system makes it unnecessary to navigate from one navigation aid to the next, for the pilot can now fly the most direct route to his destination.

There are three generally accepted categories of RNAV equipment:

Mark I: A course-line computer, which converts bearings and distances from VORTAC stations into the heading required to maintain a course and a knowledge of current position, even though the aircraft does not fly over the fixes.

Mark II: A Doppler system which provides a heading reference, track, and current indication. Accuracy may be updated by VOR/DME, loran, or Omega.

Mark III: An inertial navigation system (INS). Accuracy may be updated by VOR/DME, loran, or Omega.

The course-line computer will be discussed first as it is the most common RNAV system in use. Its relatively low cost and weight make it the most practical system to acquire and maintain. The Doppler and inertial systems will also be discussed later in this chapter. Because their cost and weight are much greater, these systems are usually found only in larger aircraft.

TERMS USED IN RNAV

Waypoint.   A waypoint (WPT) is a predetermined *geographical position* used for route definition and progress reporting purposes that is defined relative to a VORTAC station or in coordinates of latitude and longitude. Two sequentially related waypoints define a *route segment*. Waypoints are often called *phantom* stations because they provide the same navigation information that a ''real'' VORTAC at that position would provide. A waypoint is defined by its *distance* and *radial* from a receivable VORTAC station; based on INS it is defined by latitude and longitude.

Waypoint Bearing.   The VORTAC radial on which the waypoint is set.

Waypoint Distance.   The distance in nautical miles by which the waypoint is offset from a VORTAC station.

Cross-track Deviation.   The number of degrees or miles off desired course as defined by the course set on the OBS. In the RNAV mode, course deviation is displayed in nautical miles shown by displacement of the left-right needle of the CDI. Each dot of displacement is 1 nautical; in approach mode left-right deviation needle sensitivity becomes ¼ nautical mile per dot. The pilot is, therefore, provided with accurate lateral position information.

Constant Course Width.   The deflection of the left-right needle of the CDI is proportional to distance off course. One dot deflection of the CDI would indicate 1 mile from the centerline course regardless of distance from the VORTAC. Constant course width is provided in RNAV mode.

RNAV COMPUTER

An RNAV computer allows the pilot to fly the most direct route to his destination, as easily as tracking a VOR when flying airways. Electronic waypoints, which the computer can use to direct the aircraft precisely on course, are established at convenient intervals along the course. Any departure from the intended path will immediately be displayed by the CDI needle. Distance to the selected waypoint is constantly displayed, providing a continuous fix (Figure 11.40). Additionally, this distance information provides an accurate and convenient method for checking ground speed by comparing distance traveled to elapsed time. The newer systems also provide a direct read out of ground speed.

In addition to the RNAV computer, the aircraft must be equipped with both a

Figure 11.40. RNAV units.

VHF navigation receiver (Omni) and a DME receiver. The information obtained by these two components, when tuned to a VORTAC station, enables the RNAV computer to solve simple geometric problems and display this course information on the CDI.

GEOMETRY OF THE RNAV COMPUTER

The computer uses relatively simple mathematical relationships to solve for the two fundamental elements of navigation information required by the pilot: distance and course to the waypoint. When you tune in a VORTAC, the computer knows side A of the triangle, illustrated in Figure 11.41, since this is the VORTAC radial and distance between you and the station, furnished to the computer by the VOR and DME equipment. When the waypoint bearing and distance are set into the computer, a second side, B, is provided in the triangle. Since radials A and B are related to magnetic north by angles 1 and 2, the computer can compare these two angles to determine angle 3. The computer has enough information to solve for side C which is the distance and the magnetic course to the waypoint from present position. The distance to a waypoint is displayed in nautical miles, and

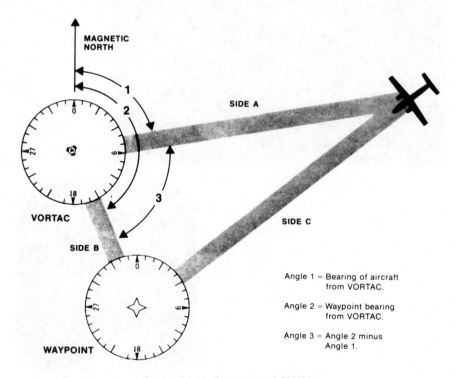

Angle 1 = Bearing of aircraft from VORTAC.

Angle 2 = Waypoint bearing from VORTAC.

Angle 3 = Angle 2 minus Angle 1.

Figure 11.41. Geometry of RNAV.

the bearing TO or FROM the waypoint is read on the CDI display after rotating the OBS knob until the needle centers.

PROGRAMMING THE COMPUTER

A VORTAC station wtihin the line-of-sight range of the navigational and DME receivers is selected and identified. If the DME receiver has a separate frequency selector, both receivers are tuned to the same station. A bearing from one VORTAC and distance information from another would provide sufficient data for the computer to solve a vector, but the resulting course information would be erroneous.

Set the waypoint location from the VORTAC facility in use by setting the bearing (radial) and distance in the appropriate positions on the computer (Figure 11.40); select RNAV on the *mode selector*. Turn the OBS knob until the left-right needle of the CDI is centered and the TO-FROM window reads TO. The magnetic course indicated by the CDI is the course, and the distance showing in the window marked "naut. miles" is the distance to the waypoint.

The aircraft is now navigated to the waypoint in the same manner as tracking to a VORTAC. When passing the waypoint, the distance will read 0 miles as the TO-FROM indicator changes to FROM. Tracking from a waypoint is done in the same way as tracking from any VOR; select the magnetic course desired and steer the aircraft to keep the left-right needle of the CDI centered. The distance display will indicate increasing distance as the aircraft proceeds outbound.

EXAMPLE OF PLOTTING DIRECT RNAV COURSE

An RNAV course can be plotted on any navigation chart in regular use, *i.e.,* Radio Enroute, Sectional, and World Aeronautical. The choice of which map to use depends, for the most part, on the type of flight to be made. For VFR operations the WAC series is convenient because map scale permits long distances to be covered within one chart, and visual landmarks are shown which can be helpful during VFR flight. For IFR operations the Radio Enroute chart should be used.

An example of plotting a direct RNAV course appears in Figure 11.42. In this example, the low altitude enroute chart is used; however, the procedure would be essentially the same using a Sectional Chart or World Aeronautical Chart (WAC).

A direct RNAV course from Oneida County Airport, Utica, New York to Olean Municipal Airport, Olean, New York, in this example, is plotted by first drawing a straight line from airport to airport. Next, select a series of VORTAC stations along either side of the course. It is not necessary to select a large number of stations. When flying high, fewer stations are required than when fly-

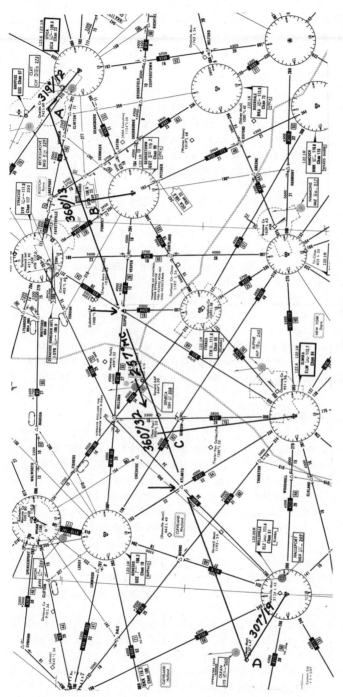

Figure 11.42. RNAV direct.

ing low for the usual line-of-sight reason. Above FL 180 choose VORTAC's intended for high altitude navigation.

The workload can be reduced by selecting all stations to either left or right of course if a single waypoint computer is used. If the course is generally west, and VORTAC's to the left (south of course) are used, plan to use the 360° radial of these stations. When the computer is programmed for the next waypoint, the only change required will be the mileage offset. If a multi-waypoint computer is used, this procedure is less important. When an examination of the chart reveals a suitable number of stations lying to the south of course, draw a line from each VORTAC to be used outward on the 360° radial until it intersects the course. Measure distance from VORTAC to point of intersection (using the appropriate chart scale) and jot it down beside the intersection (waypoint).

The direction of the course line to be flown is determined by measuring with a plotter, using a mid-course meridian (if the course is a distance of 2° or 3° longitude, break the course into several shorter legs). This measurement will give the true course. Because all Omni stations are referenced to magnetic north, it is necessary to convert this true course to a magnetic course. This is done by applying appropriate magnetic variation.

In the example, true course was found to be 247°. The magnetic variation for this route, as shown on the chart, is 12°W at the point of departure to 9°W at destination. An average variation of 10°W will be used. The magnetic course is determined: 247°(TC) + 10°W = 257°(MC).

In all navigation preflight planning, an orderly log of courses, distances, and estimates should be prepared. It is good technique to enter the magnetic course, waypoint distance, and radial on the chart for direct reference.

The Oneida County Airport, as measured, is 319°R/12 nm. from the Utica VORTAC (UCA 108.6 MHz). This will be waypoint *A*. The navigation and DME receivers should be tuned to this frequency, the RNAV computer programmed, and the OBS, rotated, setting a magnetic course of 257°. After takeoff the aircraft is flown to center the CDI needle with the TO-FROM indicator reading FROM, which places the aircraft on course toward waypoint *B*.

Approximately midway between waypoint *A* and waypoint *B*, the computer is reprogrammed, and the appropriate VORTAC frequency (GGT 115.2 MHz) is selected on the navigation and DME receivers. The aircraft is navigated toward this waypoint by keeping the CDI needle centered, with mileage readout to the waypoint declining as it is approached. This is an appropriate time to determine ground speed (Figure 11.44) which can be used to determine ETA's of waypoints. When waypoint *B* is passed, track outbound until approximately midway to waypoint C at which time this waypoint will be programmed.

When passing a waypoint, if the VORTAC used for the next waypoint is within service range, this waypoint can be programmed immediately and the aircraft flown "TO" this waypoint. If beyond the service range, it will be necessary to track "FROM" the previous waypoint until within range. When selecting the next waypoint, it may be necessary to center the CDI needle slightly. Because the average magnetic variation was used for the entire course, and it is difficult to measure distance closer than 1 mile, a slight error of 1° or 2° may occur in the selected magnetic course.

Olean Municipal Airport, as measured, is on the Wellsville VORTAC's (ELZ 111.4) 307°R/19 nm. This will be selected as the fourth and final waypoint, waypoint *D*. The mileage readout will now also indicate distance to point of landing and can be used as an aid in determining time for starting descent.

In this example the 360° radial was used for en-route waypoints for convenience. However, it can be seen that the direct course line passes across many victor airways, any one of which could be used as a waypoint radial. Distance for waypoint projection as well often can be read directly from airway mileage depicted on the chart. It will be noted that the plotted direct course passes over the "SCIPIO" intersection, made up by V423 and V14. Without measuring, a waypoint can be determined at this geographical point by any one of three VORTAC's: SYR 117.0 MHz, 221°R/24 nm.; GGT 115.2 MHz, 284°R/29nm.; ITH 111.8 MHz, 007 R/19nm. The "ATLANTA" intersection made up by V147 and V501 is another example.

Preflight planning, including drawing the point-to-point course line, determining its magnetic course, measuring and establishing waypoints, determining distance between waypoints, calculating estimates based on best known wind information, and calculations to possible alternate airports, as in all navigation, is good operating practice. Yet, it can be seen from the example that the flight could have been accurately flown with little more than drawing the course line from departure to destination, with all supporting data required readily available from airway radial and mileage information on the chart.

The step of measuring the true course with a plotter and correcting for magnetic variation can be eliminated for in-flight calculations by setting up a waypoint ahead and centering the CDI needle which will give the magnetic course to the waypoint. As each waypoint is passed, program the next waypoint ahead, center the CDI needle, and navigate directly to the waypoint. This latter technique often is the most practical procedure when it becomes desirable to proceed directly by RNAV to a distant destination when en route, and precise geographical position is unimportant.

PARALLEL TRACK NAVIGATION

In addition to providing the means to set up direct airways, the RNAV computer will also permit you to set up a course which parallels the published airways. This will permit you to keep well clear of most of the traffic, using high density airways. It will also permit you to bypass slower traffic flying at your altitude, or get you out of the way of a faster aircraft overtaking you from behind. While using the parallel track feature, you will still be able to take advantage of the communication, navigation, and emergency facilities that exist along the published airway system. The parallel track system can also be used during departure and approach to facilitate the flow of traffic in terminal areas.

The procedure for setting up a parallel tracking is as follows.

Select the airway along which you wish to fly, and note the published magnetic courses. In Figure 11.43 you can see that the magnetic courses for flying east on Victor 121 are successively 095° and 065°.

Determine the bearing of your waypoint by the following rule: To parallel an airway on its right side, always add 90° to the published magnetic course; to parallel an airway on its left side, always subtract 90° from its magnetic course. In this example, 90° was added to the published direction of leg 1, *e.g.,* 095° plus 90° equal 185°. This is the bearing of the first waypoint.

Select the waypoint distance for the parallel airway. This will be the second item in each waypoint input and will remain constant. In the example, the pilot elected to parallel the airway at a distance of 15 nm.

Set the proper VORTAC frequency on the navigation and DME receivers and the proper waypoint bearing and distances for the first waypoint. (These would be 115.4 MHz, 185°, and 15 nm., respectively.) Set the OBS knob to the magnetic course of the first leg (095°).

With the programming now complete, fly the aircraft into the parallel course by choosing a heading to intercept the course, and fly the left-right needle to center as during regular VOR navigation. Fly the heading required to keep the left-right needle centered to the waypoint.

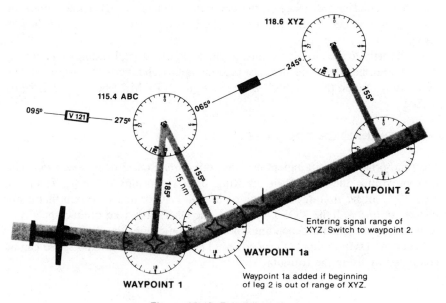

Figure 11.43. Parallel track.

After passing waypoint 1, hold the heading that was found to keep the aircraft on course to the waypoint, and set the frequency of the next VORTAC into the navigation and DME receivers. Set a bearing of 155° (065° plus 90°) into the computer; a 15-nm. setting would be retained in the distance window because the parallel track distance remains the same. Set 065° in the CDI, the new magnetic course to be flown. Continue to fly this heading until the needle of the CDI moves to the center of the indicator. The aircraft has now intercepted leg 2; turn the aircraft to a heading of 65° (assuming no wind), and proceed on course.

If, as in Figure 11.43, the beginning of leg 2 is out of signal range of VORTAC XYZ, the following step may be added to the above procedure.

Locate an additional waypoint (waypoint la) using the same VORTAC used on the previous leg. In this case, you would fly to waypoint 1, to waypoint la, from waypoint la until passing into signal range of XYZ, and then to waypoint 2.

LOCATION OF AIRPORTS

RNAV will aid in locating airfields with no navigation aids, particularly during low visibility. The procedure is as follows.

Locate a VORTAC facility near the airfield you wish to locate. The airfield must be within the VORTAC's effective range at the altitude you wish to approach the airport. Determine the radial and distance of the airfield from the VORTAC by measuring on the chart with a plotter.

Set the VORTAC frequently in the navigation and DME receivers and the waypoint radial and distance in the computer. Rotate the OBS knob until the "TO" flag appears on the display and the left-right needle of the CDI is centered.

Fly directly to the airfield on this magnetic course using headings which keep the needle centered. As the distance read out approaches zero, the airport should be in view. When the TO-FROM flag slips to "FROM," the airfield is being passed.

DETERMINING GROUND SPEED

Most airborne DME equipment provides direct read-out ground speed information providing the aircraft is tracking directly toward or away from the VORTAC. if the aircraft is tracking on a VORTAC or to a waypoint that is not directly in line with the VORTAC radial, the DME displayed ground speed will read low by an amount depending upon the geometry of the flight path.

When the DME computed ground speed is low, accurate ground speed can be computed by using the formula:

$$\frac{60 \times \text{miles}}{\text{time}}$$

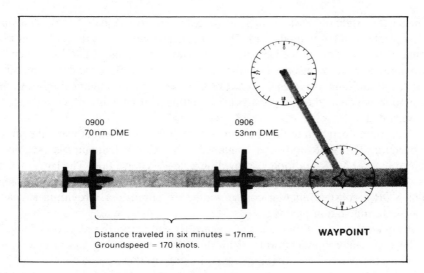

0900
70 nm DME

0906
53 nm DME

Distance traveled in six minutes = 17nm.
Groundspeed = 170 knots.

WAYPOINT

Figure 11.44. G.S. RNAV.

For example, if 17 miles were traveled (from waypoint distance display) in 6 min., the ground speed is 170 knots. If exactly 6 min. are used on each check, all that is required is to multiply the distance traveled by 10 (Figure 11.44). This is less cumbersome than using less convenient numbers.

If the airplane is one of high speed, 6 min. may be intolerably long. The pilot may prefer to clock the DME for 36 sec. and multiply the distance traveled by 100 to derive ground speed in knots. If 2.4 nm. were traveled in 36 sec., ground speed equals $100 \times 2.4 = 240$ knots.

Limitations in reading the mileage closer than about .1 nm. result in less accuracy when using shorter time periods. Other time periods and factors are: $20 \times$ distance traveled in 3 min.; $60 \times$ distance traveled in 1 min.

These methods of computing ground speed and position, while fairly accurate, should not be used if the aircraft is closer than 1 nm. to the ground station for each 1000 ft. of altitude if high accuracy is important, because the ground speed error equals the magnitude of slant-range error (Figure 11.29).

HELPFUL USES OF RNAV

Direct navigation, locating airfields, and parallel tracking are but several of the many uses the pilot will find for RNAV. Several other uses that can be of great value to the pilot are as follows.

Constant Course Width. RNAV can be used when navigating on regular VOR airways from VORTAC to VORTAC. The constant course width feature of RNAV will display on the CDI a positive indication of aircraft position in miles left or right of airway centerline (one dot deviation equals 1 nm). You simply operate in RNAV mode with 000.0 nautical miles set in the waypoint distance window. This procedure places the waypoint at the VORTAC, thereby providing navigation information directly to the station. Because the waypoint is located at the same place as the VORTAC by setting 000.0 nautical miles in the waypoint window, the waypoint bearing setting will have no effect and can be disregarded.

Auto-pilot operation in the navigation track mode is smoother using the above procedure. The VOR converter contained in an RNAV computer is capable of rejecting most VOR scalloping, providing a steady course line. The large swings of the CDI as the aircraft approaches and passes over the station, because of regular VOR's constant angular course width, are eliminated, providing a steady course during station passage.

Weather Breaks.   Often an ATC controller can report the location of breaks in lines of heavy thunderstorms. With the VORTAC bearing and distance supplied by the controller you can navigate directly to these breaks.

Time of Penetration.   Estimated time of arrival and actual time of arrival information for penetration of specially controlled air space areas can be determined by establishing a waypoint at the boundary. While navigating accurately to this waypoint, distance information will be available as well as the means to calculate ground speed and estimated time of penetration.

Mileage to Non-RNAV Fixes.   The RNAV computer can furnish computed distance to fixes used in conjunction with regular VOR and ADF navigation. Although the course guidance may not be important in this procedure, the distance read out can supply valuable information to the pilot for speed control and descent planning.

Bearing and distance information for this use is often readily available from the data on instrument approach charts. An ILS chart will show distance and bearing from one or more VORTAC's in the area to the outer marker (OM) or outer compass locator. The computer, thus programmed, will provide distance information to this fix.

RNAV Instrument Approaches.   The FAA is rapidly establishing RNAV instrument approaches at large and small airfields. These approaches are conducted like regular VOR-DME approaches.

If an airport is within the service area of a VORTAC, it is possible to establish an RNAV instrument approach to each runway. An important advantage of an instrument approach to each runway is that the pilot can select the approach which permits landing straight in. This eliminates the need to circle at low altitudes, during low visibility conditions, to reach the active runway, or to land with a strong crosswind.

The economic advantage is obvious. Numerous instrument approaches from *one* facility are possible, and that facility need not be located on the airport. Many small outlying airfields that could not afford the cost of establishing a local navigation facility, can have established instrument approaches using existing VORTAC facilities. Note: The pilot should not design his own instrument approach to an airfield while en route. Only FAA established approaches should be used. These insure proper signal reception and, most importantly, clearance from terrain and obstructions.

VERTICAL NAVIGATION (VNAV)

A third dimension can be incorporated into RNAV, providing a *vertical* reference. *Vertical navigation* (VNAV) provides computer pitch commands to the pilot, enabling him to maintain a vertical track angle (VTA), in ascent or descent, to an RNAV waypoint. The altitude desired over the waypoint is set into the VNAV computer; this information, along with data from a servo altimeter and the RNAV computer, is used by the VNAV computer to solve the geometric problem.

This is helpful if Air Traffic Control (ATC) directs the pilot to cross a given waypoint at a specific altitude. In making an RNAV instrument approach to an airport, the VNAV computer can provide vertical guidance like a glide path of an ILS system. It can be expected that VNAV will be integrated along with RNAV into the airway system.

Studies are being conducted to determine the merits of adding still a fourth dimension, *precision timing*. This dimension would provide for the metering and spacing of aircraft.

SLANT RANGE CORRECTION

DME distance is slant range distance, not horizontal distance (Figure 11.29). Although this generally is insignificant, it can result in an error in waypoint projection if the aircraft is flying at a very high altitude when near the VORTAC. Some newer RNAV computers have a provision for slant range correction. A servo altimeter provides altitude input with the pilot setting in the altitude of the VORTAC station. The computer, with this altitude information and DME distance, will adjust slant range distance to horizontal distance, providing increased accuracy.

Long-Range Navigation

The VOR/DME navigation system is accurate and reliable in all weather conditions (including thunderstorm activity) and is simple to operate. Cost of airborne receiving equipment is reasonable, and nearly all aircraft in the United States are equipped with one or more receivers. Other countries of the world have adopted the VOR system of navigation, if to a lesser degree.

Yet, the VOR/DME system of navigation has one major limitation which precludes its use as a "sole" method of global navigation—it is not useable over long distances and so is not practical in remote, inaccessible areas or over the vast ocean areas of the earth.

For long range navigation requirements, other types of electronic radio aids have been developed. These systems operate in the lower frequency bands, which have greater distance capabilities. (General characteristics of radio propagation were previously discussed in Chapter 5.) One such long-range system in use for many years is *loran*. A somewhat similar system is *Omega*.

The following discussion will provide a general understanding of the capabilities and limitations of these systems. For detailed operating instructions, consult the operator's manual for the particular system.

LORAN

Loran is a shortened name for *long range aid to navigation*. It was developed during World War II at the Massachusetts Institute of Technology Radiation Laboratory, to fill a need for an accurate method of determining position at long distances from land. Loran is the principal electronic navigation aid over the world's ocean routes, used by aircraft primarily to update self-contained navigation systems.

Two types of loran are in use throughout most of the ocean areas of the Northern Hemisphere: *loran-A* and *loran-C*. Both are based on the principle that radio frequency energy is *propagated* at a single velocity, and both make use of pulsed radio frequency signals transmitted alternately from fixed geographic positions.

Loran-A is the primary loran system used by the airlines and other civilian aircraft. Loran-C is used primarily by the Navy and Air Force. To prevent confusion, loran-C will be discussed separately.

Loran-A.   Loran-A has assigned frequencies of 1950 kHz, 1850 kHz, and

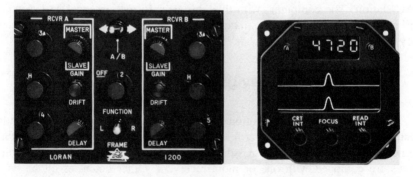

Figure 11.45. Loran receiver.

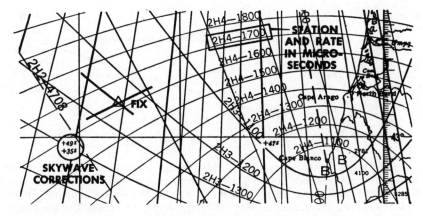

Figure 11.46. Fixing by loran. Section of a loran chart showing time-difference hyperbolas from four sets of loran stations. On the original chart, the lines from each station pair are in a distinct color to minimize confusion. The fix shown above represents the crossing of an LOP of 1620 from station 2H4 and an LOP of 1250 from 2H3.

1900 kHz for channels 1, 2, and 3 respectively. Range varies from about 700 miles in daytime (groundwaves) to 1500 miles at night (skywaves). In some areas, skywaves are used at a range of 3000 to 4000 miles. Loran requires two ground stations to furnish data for one LOP. The navigator uses a loran receiver (Figure 11.45) and a special loran chart (Figure 11.46). A list of standard loran charts is available in Publication No. 1-V, provided by the US Navy Oceanographic Office, Washington D.C.

The ground stations are called a *master* and a *slave* station. The navigator, using his DR position, selects and tunes his receiver to the most desirable pair of stations. The first signal is received from the master station and displayed as a *pulse line* on the scope face (Cathode Ray Tube). The same signal, also received at the slave station, triggers the slave station, which, after a set time delay, transmits a signal on the same frequency. The slave signal is received, and the time difference between the two determines an LOP (Figure 11.46).

The U.S Coast Guard now uses the term *secondary station* to denote the slave station. The ground stations have *atomic frequency standards,* permitting them to run independently with an occasional correction for drift. So the secondary station is no longer "slaved" to the master, causing it to follow the excursions of the master pulse under noise, as occurred formerly.

There are many points where the time difference could be the same. The locus of these points is a *hyperbolic* curve. Each loran chart is, therefore, printed with a series of hyperbolic curves, giving microseconds of difference for each pair of stations.

A station pair is identified by two numbers and one letter, *i.e.,* 1LØ. The first number indicates the channel and operating frequency (Channel 1, 1950 kHz).

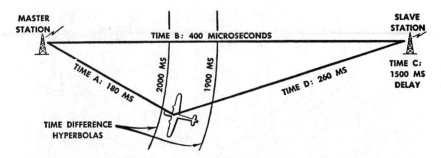

Figure 11.47. Measurement of loran rates. The loran set in the aircraft receives the master signal 180 microseconds (msec) after transmission and starts with this signal as a reference. This same signal takes 400 msec to travel to the slave station, which does not transmit until a controlled delay of 1500 msec has been added. The slave signal reaches the aircraft 260 msec after transmission. The loran set performs the following computation:

$$\text{Time B} + \text{Time C} + \text{Time D} - \text{Time A}$$
$$400 \quad + \quad 1500 \quad + \quad 260 \quad - \quad 180 \quad = 1980$$

The aircraft is therefore known to be on the 1980 msec hyperbola.

The letter L denotes the basic pulse recurrence rate (PRR). It could be Slow (20 pulses per second), Low (25 pulses per second), or High (33⅓ pulses per second). The last number indicates the individual station pair or specific PRR, which changes the number of pulses per second by $1/25$ for each Slow station, $1/16$ for each Low station, and $1/9$ for each High station.

Groundwave and Skywave. The range of loran stations, type of signal received, and accuracy of resulting time difference measurements are affected by the paths over which the radio waves travel. A portion of the radio energy travels out from the transmitter, parallel to the surface of the earth, and is called a groundwave. In addition, various signals are reflected back to earth by the ionosphere and are called skywaves (Figure 11.48.)

Skywaves cause two problems in loran navigation, recognition, and correction. The loran indicator will give a reading no matter what groundwave or skywave pulses are matched. To get the correct reading, proper pulses must be selected. This is the most critical part of loran operation.

Weak groundwaves from both stations of a pair are preferable to strong skywaves, because variations in the ionosphere cause skywaves to vary in timing and shape.

A special correction is required for matching the groundwave from one station to the one-hop-E skywave from the other. These corrections, where necessary, are tabulated on loran charts and marked "SG" if the master skywave is used, and "GS" if the slave skywave is used. When applying a correction (normally 70 microsec.) to a loran reading, remember "slave skywave subtract, master

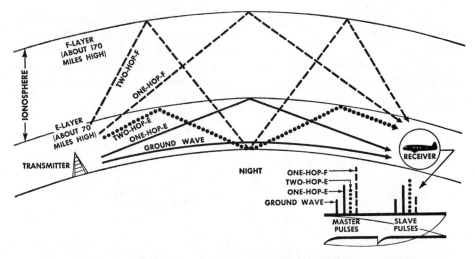

Figure 11.48. Sky wave reception. Sky waves are generally received only at night when the underside of the ionosphere is smooth enough to reflect energy striking it. Some energy may pass through the E-layer and still be reflected by the F-layer so that at times four or five signals are received simultaneously. The sketch of the cathode-ray tube at the right shows the order in which signals are received. It is absolutely essential to know whether signals being matched are sky waves or ground waves. At times it may be necessary to use a sky wave from a master station and a ground wave from its slave. Unless this is recognized and proper correction made for it, serious error may result. For this reason, identification of sky waves is the most important skill of a loran operator.

skywave add.'' To match skywaves, the correction appearing every 5° of latitude and longitude on the loran chart is used and may be interpolated if necessary.

Identification of Groundwaves and Skywaves.   Skywaves may build and fade, at times completing an up-and-down cycle in less than a minute, and splitting, which consists of breaking into two or more humps which fade more or less independently. In contrast, the groundwave is generally steady in amplitude and is always free from splitting, although if weak, it may flicker from noise. During the middle daylight hours, at ranges under 800 miles, groundwaves are normally the only ones present (Figure 11.49).

Operational Procedures for Loran.   Accuracy to a great degree depends upon care and judgment exercised by the operator. *Auto-tracking* loran sets now available are a vast improvement over the older manual sets, which require several minutes to obtain a fix. Modern loran sets require only 10–15 sec. for an initial set up, after which they will automatically and continuously provide the airman with one or two LOP's.

The following discussion is based on an *auto-track* loran system designed specifically for civil aircraft (Figure 11.45), capable of using either loran-A, or loran-C stations.

Loran-A acquisition.   Consult loran charts covering the area of the aircraft's assumed position, and select a station pair within range. Following the operating manual's instructions, set the station controls for the desired loran station pair. After adjusting as outlined in the operator's manual, an LOP in microseconds of the selected station pair will appear in the digital readout. To complete the fix, a

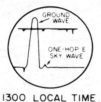

1300 LOCAL TIME
SKY WAVE VISIBLE BUT WEAK.

1500 LOCAL TIME

1600 LOCAL TIME
SKY WAVE BECOMING STRONGER.

1700 LOCAL TIME.
SKY WAVE HAS BECOME
STRONGER THAN GROUND WAVE

1800 LOCAL TIME.
SUNSET. GAIN REDUCED TO
SHOW TOP OF SKY WAVE.

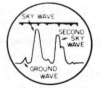

1900 LOCAL TIME.
LEFT EDGE
OF SKY WAVE IS FADING.

NOTE:
MASTER GAIN CONTROL ADJUSTED FOR NO SIGNALS ON TOP TRACE.

Figure 11.49. Typical groundwave and skywave presentations of loran-A are shown above. The identification of groundwaves and skywaves is the principal operational problem in loran, and must be solved correctly if satisfactory results are to be obtained. In making identification the aircraft's position, relative to stations; type of signals that might be expected; appearance of signals under observation; and spacing of pulse under observation, relative to other pulses in the train, are important considerations. The problem of identification requires that observations be made at frequent intervals to confirm the type of wave. *(Courtesy of Edo Commercial Corporation)*

second LOP, from a station pair having an intersecting LOP with the initial LOP, is determined.

These readings are plotted on a loran chart (Figure 11.46), interpolating as necessary, to draw a line parallel to the printed hyperbolic curves on the chart, near the DR position.

The auto-track loran receiver will track, and continually update, the LOP(s) of the selected station(s) to an accuracy of ± 1 microsecond and aircraft speed to 600 knots.

SUGGESTED OPERATING PRACTICES

Flying course line. Many times an LOP may be a nearly straight line from departure to destination. Examine the chart, determining if any loran lines run parallel to the intended course. If so, select one as a course line, and follow it. En route, the digital read out will indicate the time difference reading of the selected LOP. If the reading begins to increase or decrease, change course to the left or right as necessary to maintain the same reading.

Flying speed lines. Select labeled loran lines which cross the intended course at or near a 90° angle. Check the time between labeled loran lines. Measure distance between the two LOP's, and divide by the elapsed time, to obtain ground speed.

Position fixing. To reduce time required for position fixing, always keep the loran auto-tracking on one loran rate. When a position fix is desired, note the LOP reading, and switch to acquire the other desired rate. Thus, it is only necessary to acquire once to obtain a fix. Advanced loran receivers operate essentially similarly but provide a continuous fix, as two LOP's are displayed on the digital read out.

Loran-C. Loran-C, like loran-A, is a hyperbolic navigation system, operating on the same theory and principles. It extends operational range and accuracy, while using fewer transmitters. This is accomplished by using frequencies near 100 kHz, which are subject to less groundwave attenuation. A technique called "cycle matching" gives accuracy within several hundred feet. This accuracy is important for ships but not often required for aircraft during overwater flights.

Loran-C differs from loran-A in having more slave stations in a group. There is one master station, centrally located, and two, three, or four secondary stations (slaves). Each combination of the master station and any slave station of the same chain gives an LOP.

A second major difference is the signal format. Loran-C is transmitted as a series of pulses. The master transmits nine pulses in a single burst; eight at equal spacing and the ninth at a double space. The slave stations transmit a series of eight equally spaced pulses.

The ionosphere appears much more stable to signals at the lower frequencies of loran-C, with less fading and less splitting of skywaves. To determine whether the signal is a groundwave or skywave, the operator is guided by information appearing on loran-C charts. Only in areas of skywaves are correction factors printed on these charts. Here also, groundwaves are preferred to skywaves. Matching is done in the same manner as with loran-A.

If groundwaves cannot be received from the master and slave stations, first-hop skywaves are matched and skywave corrections applied. Only first-hop skywaves should be used.

The human factor is inescapably involved in the accuracy of a loran fix because of groundwave and skywave recognition and matching. Therefore, sufficient time must be taken to become thoroughly familiar with any loran receiver before attempting any long-range overwater flight.

OMEGA

Omega, like loran, is a hyperbolic radio navigation system. Initial work on Omega began years ago for ship navigation, at a time when there were no alternatives such as *inertial* or *satellite* systems. The concept is credited to Dr. J. A. Pierce of Harvard University.

Through the work of Dr. Pierce, the United States Navy, and the United Kingdom, a *very low frequency (VLF) long-range hyperbolic navigation system* has been developed. Some see it eventually replacing the VOR/DME system, but further development and refinement must come first.

The present status of Omega provides for long-range navigation with accuracy to within about 1 or 2 miles. A slightly degraded accuracy during darkness, characteristic of VLF radio navaids, results from *diurnal* changes in the ionosphere. This accuracy is acceptable for aircraft navigating over oceanic routes, and as a means of updating and improving accuracy of *self-contained navigation systems* (Doppler and inertial).

Omega operates in the VLF range of 10 to 14 kHz, which offers low *attenuation rates* and *stable phase velocities*. These signals are contained by the surface of the earth and the ionosphere (acting like a waveguide), providing for their reception over long distances (Figure 11.48).

Omega Concept. Because of their long range, a network of only eight 10-kw. ground stations is located around the world. Each transmitter station has a range of 6000 to 8000 miles, providing a world-wide navigation system. The ground stations transmit three frequencies sequentially, 10.2, 12.6, and 11.33 kHz; signals from all eight transmitters occur over a 10-sec. period. Each ground station has two transmitters plus other back-up systems to ensure a high degree of redundancy and operational availability.

The United States Navy is responsible, in partnership with six other nations, for installation and operation of these Omega ground stations. Two stations (North Dakota and Hawaii) are located in the US, and six are located in other countries: Norway, Trinidad, Reunion Island, Argentina, Japan, and Australia.

Airborne Receiver-Computer. The aircraft (or surface vessel) must be equipped with both an Omega navigation computer (ONC) and a VLF receiver. The computer and VLF receiver are usually contained in one unit (Figure 11.50). The computer measures the phase difference of the 10.2 kHz basic frequency, giving a phase difference "lane" 8 miles wide. The addition of 11.33 and 13.6 kHz frequencies increases lane width to 72 miles.

The phase difference of radio signals from any two pairs of ground stations is used to determine one LOP. The difference in phase of another pair of stations is used to establish a second LOP. The intersection of these two LOP's establishes position. Thus, a minimum of three stations is required to operate the system. With eight ground stations in service, four or five should always be received.

The Omega computer automatically selects the best combination of stations, based on station location with respect to aircraft and received signal strength. The additional stations provide useful redundancy. The computer controls the receiver as well as providing propagation corrections, phase tracking, signal to noise measurements, and other functions.

Unlike loran, where the navigator must carefully select appropriate stations and carefully interpret scope readings, Omega automatically selects the best

Figure 11.50. The Omega Navigation System provides a continuous position fix, in coordinates of latitude and longitude, with an accuracy of 1 to 2 nm. anywhere in the world. Since system is earth referenced, accuracy does not degrade with time. The control indicator (right), antenna (left), and computer-receiver (top) are shown. *(Courtesy of Canadian Marconi Company)*

combination of stations, and the computer provides useable information to the pilot directly.

In the event of signal loss, an automatic dead reckoning (DR) *reversionary* mode is incorporated in the system; return to Omega tracking from DR is also automatic. Thus, the computer-receiver provides the pilot with a continuous fix over any point of the earth, in coordinates of latitude and longitude. The Omega navigation system (ONS) may be interfaced with conventional HSI (CDI) instrumentation in an aircraft's instrument panel, and with the auto-pilot, for desired track guidance.

Control-Display Indicator.   The *control-indicator* (CI) is similar to *ARINC 561 INS* units (Figure 11.51). This unit provides for *data insertion* and related functions (lower portion), and a *data display section* (top portion), through left and right numerical display windows.

The *data keyboard* provides the means for inserting data into the computer,

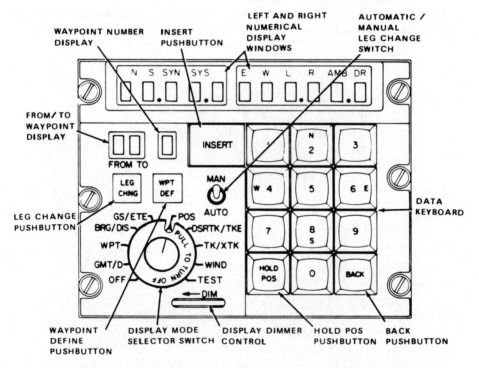

Figure 11.51. The control-indicator of an Omega navigation system is similar to ARINC 561 INS indicators. The Data Display Section is at the top where selected data are displayed in left and right numerical display windows. The Data Keyboard and operational controls are located in the lower portion. *(Courtesy of Canadian Marconi Company)*

such as *present position, Greenwich Mean Time and date,* and *latitude and lon-gitude* for up to nine waypoints. After the pilot enters data and initiates the *inital track leg,* the ONS will automatically navigate from waypoint to waypoint. Man-ual operation is also provided, allowing the pilot to bypass waypoints, or to change waypoints to allow for flight plan alterations. *Self-testing* is continually in operation to tell the pilot, through *annunciators,* of system performance.

Operational Procedures for Omega. The following discussion shows the general procedures for programming and navigating with the Omega navigation system. This discussion is general in nature and therefore should not be consid-ered as a pilot's manual.

After power is applied, Greenwich Mean Time and date (GMT/D) and present position (in coordinates of latitude and longitude) are inserted in the computer, using the data keyboard. The computer goes through a *synchronization phase,* usually within 2 min. under normal signal conditions. GMT/D and present posi-tion are used in the computer to select the appropriate part of the diurnal pro-gram. The position inserted must be within ±4 nm., and GMT within 5 min. GMT should be inserted as accurately as possible.

To enter GMT/D. Set the mode selector to "GMT/D" position, and enter GMT with data keyboard. Example: 09.30.5 (hours, minutes, and tenths of min-ute). Starting with the most significant digit (9 in this case), enter 9305 on the data keyboard. As each button is pressed, verify that the digit is displayed in the left display window as the least significant digit, and each preceding digit moves one place to the left. With the correct data appearing in the window (09.30.5), press the INSERT button.

Now enter the date. Example: 29.08.72 (day, month, and year). Starting with the most significant digit (2 in the case), enter the date in the same manner as GMT. When the correct date appears in the right display window (29.08.72), press the INSERT button [Figure 11.52(A)].

To enter present position. Example: With the mode selector on "POS" lati-tude and longitude are entered. Assuming latitude of 45.28.2 N, press "N" fol-lowed by numbers 45282 on keyboard. Verify the displayed data in the left window, and press the INSERT button. Longitude is entered in the same man-ner, in this example, W 73431 [Figure 11.52(B)].

During synchronization the pilot can perform preflight procedures for the sys-tem. By selecting "TEST" mode, the pilot can verify CI numerical displays, FROM-TO displays, waypoint displays, and that all annunciators are operating correctly.

Waypoint coordinates entry. After GMT/D and present position data are en-tered, the pilot enters coordinates for up to nine waypoints. The initial en-route waypoint would normally be entered as waypoint 1, and thereafter sequentially. In AUTO mode, leg waypoints are called up in numerical sequence, from 1–9.

Example: Set the mode selector to "WPT" position. Enter the latitude and longitude of the waypoint as follows: Press the appropriate waypoint number on

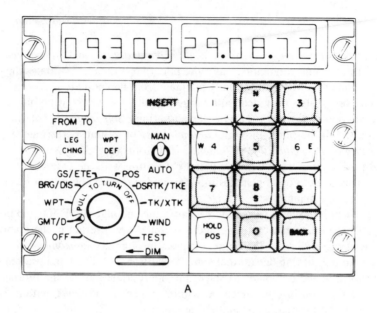

A

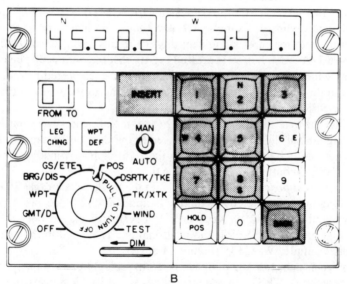

B

Figure 11.52. Greenwich Mean Time, date, present position, and one or more waypoints (in coordinates of latitude and longitude) must be set into the system before departure. (A) Greenwich Mean Time appears in left display window and the date in right window. GMT/D are entered as part of the diurnal program and can be used in flight as a time reference. Time shown is 30.5 minutes past 9:00 o'clock; date is 29 August, 1972. (B) Latitude and longitude to nearest tenth of an arc minute are shown in left and right display windows, respectively. Present position and selected waypoints are displayed in this manner. The display reads 45 degrees 28.2 minutes North latitude; 73 degrees 43.1 minutes West longitude. *(Courtesy of Canadian Marconi Company)*

the data keyboard, verifying that the figure *1* (example) appears in the waypoint display window. Press the WPT-DEF (waypoint define) button, illuminating the INSERT button. Enter the waypoint coordinates as previously done in entering the present position, verify the data displayed, and press the INSERT button.

Press the next waypoint number, *2* (example), to be inserted on the data keyboard; press the WPT-DEF button; enter the coordinates of the waypoint; verify; and press the INSERT button. The third and subsequent waypoints are entered in the same manner.

Initial track selection. After the initial en-route waypoint latitude and longitude are entered, initial track selection procedures can be performed. The initial track is the direct great circle route between the aircraft's present position (ground or in the air) and the initial en-route waypoint (WPT 1 in this case). The pilot selects and initiates the initial track as follows.

Progress the "LEG CHNG" (leg change) button and verify that the INSERT button is illuminated. Press the required leg waypoint numbers on the data keyboard. In this example, 0 and 1 should be pressed in this order; verify that the FROM-TO window displays 0–1. The bearing from the present position to waypoint "1" will now be indicated in the left display window.

The displayed track should be checked to see that it is reasonable. The initial track may be started on the ground or after take-off. In the latter case, data will automatically be from point of initiation, *i.e.*, present position. The mode selector does not affect this procedure (Figure 11.53).

En-route procedure. Prior to departure, or after departure, the pilot initiates the initial track as indicated above. When the bearing is displayed and verified for reasonableness, the pilot selects this course on the HSI (CDI) and navigates to waypoint 1 by steering to keep the needle centered as in VOR navigation. Passing waypoint 1, the bearing to waypoint 2 will be displayed and is set into the HSI. This procedure is followed as each succeeding waypoint is passed.

During automatic operation the ONS navigates through each entered waypoint in sequence and automatically changes track at each waypoint. During manual operation the ONS navigates from waypoint to waypoint, but the pilot must initiate a track change at each waypoint.

En-route Data Displayed. In addition to track guidance, the ONS provides the pilot with a multitude of information. Access to this computed information is gained by setting the mode selector to the appropriate position. Data selected will be displayed in the left and right display windows. Computer navigational information available to the pilot is as follows.

Greenwich mean time and date. Select "GMT/D" on mode selector. GMT is displayed in the left window to the nearest tenth of a minute, to the same accuracy as inserted in the initial programming. The date is displayed in the right window. In Figure 11.52(A), the time is 0930.5 and the date is 29 August 1972.

Waypoint display. Select "WPT" on mode selector. The waypoint coordinates, in latitude and longitude to the nearest tenth of an arc minute, are shown in

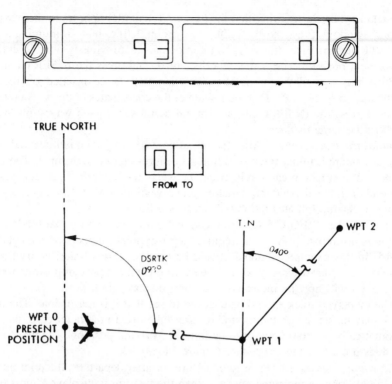

Figure 11.53. Initial track (first leg) is presented on left display window to nearest degree showing direct great circle route between aircraft's present position (ground or in the air) and the first waypoint. The initial track must be initiated by the pilot with subsequent tracks displayed automatically when operating in the AUTO mode. The display reads 093° (True Course); passing WPT 1, a track of 040° would be displayed. *(Courtesy of Canadian Marconi Company)*

the left and right display windows, respectively. Any previously entered waypoints can be displayed by pressing the desired waypoint number on the data keyboard [Figure 11.52(B)].

Bearing and distance.   Select "BRG/DIS" on the mode selector. Bearing from the present position to the next waypoint is shown in the left display window to the nearest degree; distance from present position to the next waypoint is shown in the right display window to the nearest nautical mile [Figure 11.54(A)].

Ground speed and estimated time en route.   Select "GS/ETE" on the mode selector. The aircraft's ground speed is displayed in the left display window to the nearest knot. Estimated time to go, at current ground speed, from present position to next waypoint, is displayed in the right display window to the nearest minute. In Figure 11.54(B), the aircraft's ground speed is 518 knots, and it is 6 min. from the next waypoint.

Present position. Select "POS" on the mode selector. The aircraft's present position in latitude and longitude, to the nearest tenth of an arc minute, is shown in the left and right display windows, respectively. The read out can be frozen in flight, to facilitate a position check/update, by pressing the "HOLD POS" button. Pressing the "BACK" button returns the actual present position to the display window. (The ONS continuously computes distance traveled while the display is frozen.)

Desired track and track angle error. Select "DSRTK/TKE" on the mode selector. The desired track is shown in the left display window and the track

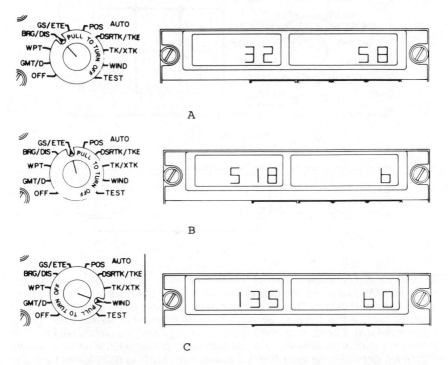

Figure 11.54. A multitude of navigational information is available to the pilot while in flight by selection of desired data on the mode selector. Examples are: (A) Bearing from present position to next waypoint, or any desired waypoint selected, is displayed in left window to nearest degree and corresponding distance in right display to nearest nautical mile. The display reads a bearing of 032° and a distance of 58 nm. (B) Ground speed and time are displayed by selecting GS/ETE. Ground speed is displayed in left display to nearest knot and estimated time to next WPT is displayed in right window. The display reads a ground speed of 518 knots and 6 minutes to WPT. (C) Actual wind encountered in flight can be displayed by placing mode selector on WIND position. The wind direction is displayed in left window to nearest degree, speed in right window to nearest knot. Wind information will be available only if an input from aircraft's true airspeed is included in system. Display reads a wind from 135° at a speed of 60 kts. *(Courtesy of Canadian Marconi Company)*

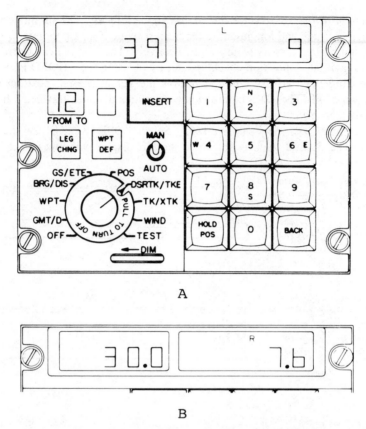

A

B

Figure 11.55. Aircraft position relative to desired track is displayed by Omega navigation system. (A) Desired track is displayed in left window and track angle error in right
window to nearest degree. The L or R annunciator indicates actual track is left or right
of desired track. In this example, aircraft is tracking 9° left (L) of desired track of 039°.
Further definition of position is displayed by selecting TK/XTK on mode selector. The
results are shown in (B) where track angle (030°) is displayed in left window to nearest
tenth of a degree (desired track 039°-9°L equals track angle of 030°). Cross track distance is displayed in right window to nearest tenth of a nautical mile. The aircraft is
currently 7.6 nm right (R) of desired track. *(Courtesy of Canadian Marconi Company)*

angle error in the right display window, to the nearest degree. The L or R annunciator above the TKE numerical display indicates whether the present track is left
or right of the desired track angle [Figure 11.55(A)].

Wind direction and speed.   Select "WIND" on the mode selector. Wind direction is shown in the left display window to the nearest degree, and wind speed
in the right display window to the nearest knot. (Note: Wind information will be
available only if the ONS has an input of the aircraft's true airspeed.) In Figure
11.54(C) the wind is from 135° at a speed of 60 knots.

True North Orientation.   The above examples reveal the great amount of information automatically available to the pilot at any time and position on earth. However, all displayed information in terms of bearings, track, wind, etc., is measured in reference to true north. Thus, if the pilot is using magnetic directional information, magnetic variation must be applied where appropriate.

Updating.   Omega can be used as a source of information for updating self-contained navigation systems such as INS, which accumulate error over a period of time.

Limitations of Omega.   Offering a *global navigation system* (for both ships and aircraft) with only eight stations and relatively little cockpit workload, Omega appears to be a practical solution to all navigation requirements. Unfortunately, certain limitations exist in Omega's present state of development, restricting its application as an exclusive navigation system.

Disturbances.   *Precipitation static* and heavy thunderstorms with resultant electrical activity create potential interference with VLF radio signals. Use of an *Orthogonal* loop antenna (H-field) eliminates precipitation static. Solar storms, although infrequent and predictable, likewise can interfere with VLF signals.

Antenna location.   Antenna location is of considerable concern because the frequencies of Omega are harmonics (multiples) of electric power frequencies used in ships and aircraft. It is necessary to skin-map each aircraft for optimum loop antenna location.

Ambiguities.   Omega has inherent *ambiguitis*. If power is lost temporarily, a position fix may be in error by 8 miles or by multiples of 8 miles. The Omega system experiences identical signal phase measurements every 72 miles, which requires the coordinates of the starting point to be inserted into the computer prior to departure, as discussed earlier.

The system must, therefore, keep track of its progress as the craft crosses from one 72-mile lane into the next, to avoid ambiguity uncertainties. For a jet aircraft traveling at high speed, a failure of a few minutes could result in lane loss, and an error of 72 miles in indicated aircraft position.

Cost of installing Omega.   The cost of Omega navigation systems, although reasonable in terms of capability for global jet aircraft, is excessive for small aircraft involved in domestic flying. This is the major obstacle to Omega's becoming a universal navigation system to replace the present VOR/DME system.

Accuracy.   The present state of development of Omega, with accuracies of 1 to 2 miles, is not adequate for terminal navigation aids for aircraft or ships. However, where more precise navigation accuracy is required in localized areas, such as harbors and airports, another technique, called *"differential Omega,"* can be employed. It employs a monitoring station detecting any discrepancies between predicted and actual propagation conditions, with corrections transmitted to users on a frequency other than those used for Omega.

Self-Contained Navigation Systems

A self-contained navigation system is independent of any externally derived information, such as visual or radio links, to determine position. It is, therefore, useable during any adverse weather conditions, not being affected to any degree by thunderstorms, propagation, precipitation static, or ionospheric disturbances.

The development of practical self-contained navigation systems is an outgrowth of space technology and an existing need. The computer, one of the key components of the system, is programmed with appropriate data prior to departure. Then, from extremely precise measuring devices, it senses and constantly calculates its present position, which is usually displayed through normal navigational instrumentation. The computer also provides access to a multitude of additional navigational information including wind direction and speed, distances and time to go, track and ground speed, heading and drift angle, cross track, and track angle error.

The present state of development of self-contained navigation systems dictates a high cost for acquiring, installing, and maintaining the equipment. So the system is normally found only on long-range jet aircraft operating over oceanic routes and in parts of the world where navaids are nonexistent.

The self-contained system has one major limitation: any system error will be "cumulative," resulting in a buildup of error with time, accumulating at the rate of approximately 1 nm. per hour (in inertial systems). To correct this error, self-contained systems have provisions for updating present position through radio fixes. The two most common self-contained navaids in use are Doppler and inertial (INS).

DOPPLER RADAR NAVIGATION

The phenomenon known as *Doppler effect* is familiar to anyone who has heard the change in pitch of a locomotive whistle as a train passes a stationary point, such as a railway station. This observation illustrates a principle described by the Austrian physicist, Christian Doppler (1803–1853). The Doppler principle states that whenever a sound source (or the observer) moves, the frequency of the sound heard differs from the frequency of the vibrating source from which it originates. This effect is also apparent in electromagnetic radiation, including light and microwave energy. A shift in observed frequency is noted when observer and source of radiation are moving toward or away from each other. In the case of light from a star, the frequency shift of certain recognizable spectrum lines is used to measure the speed with which the star is retreating from the earth. In the case of *Doppler radar* applications, frequency shift in the microwave echo is a measure of the relative motion of observer and reflecting object.

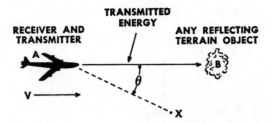

Figure 11.56. Measurement of velocity, Doppler principle.

In Figure 11.56 an airplane moving at V cm. per sec. transmits a radar signal and receives its reflection from B. If the transmitting frequency is F_t, the reflected returning frequency is F_r, and C is the speed of light, the relationship can be shown as follows: For aircraft approaching the reflection, the returned frequency for the forward beams is *upshifted*. Therefore,

$$F_r = F_t + F_t \frac{2V}{C}.$$

The relationship for the aft beams is

$$F_r = F_t - F_t \frac{2V}{C}.$$

If reflection is from any other point, such as X, which is not directly in the aircraft's path, the values become a function of cosine θ, or

$$F_r = F_t + F_t \frac{2V \cos \theta}{C},$$

and similarly

$$F_r = F_t - \frac{2V \cos \theta}{C}.$$

Therefore, the difference between the transmitted and received frequencies, called the Doppler shift, is $F_r - F_t = D$, and

$$D = \pm F_t \frac{2V \cos \theta}{C}$$

Thus, the Doppler frequency shift is directly proportional to the aircraft's velocity and cosine θ and is positive for forward beams and negative for rear beams. Obviously, when θ equals 90, $D = 0$.

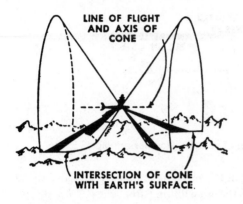

Figure 11.57. Diagram showing radar beams as a segment of an imaginary cone.

A Doppler radar navigation system, in order to measure D and at the same time average terrain effects from an irregular surface, transmits signals in forward and rear conical patterns, as shown in Figure 11.57.

If the axis of the cones is the longitudinal axis of the aircraft, and the aircraft is flying with no crosswind, *i.e.*, aligned with its ground track, radar beams will intersect the surface at four points as shown in Figure 11.58(A).

The Doppler shift in this illustration is the same for all four beams, except that the sign of the aft beams is negative, because the received frequency is less than transmitted frequency. Thus, the sum of the Doppler shift for each AB pair equals $2D$, and is proportional to the aircraft's velocity along the track. The hyperbolic curves, called *isodops*, are the locus of all points at which the Doppler shift is equal, for each AB pair.

When the antenna is not aligned with the track, as would be the case with a crosswind, the antenna structure is rotated so the A pairs [Figure 11.58(B)] move toward the zero shift line and the B pairs move away. Because these pairs are pulsed alternately, the receiver can sense the magnitude and direction of the Doppler shift difference between pairs, and this signal is used to drive a servo which rotates antennas until they are again aligned. The angle of deviation from the aircraft's axis is the drift angle and is displayed in degrees, right or left on the Control Indicator (Figure 11.60).

The antenna pattern described here has certain advantages over a simpler pattern. It is insensitive to changes in vertical velocity because an increase of decrease in vertical velocity changes the frequency shift of each pair by the same amount. For improved accuracy the Doppler antenna is *gyro stabilized* in the *pitch and roll axes* to maintain the antenna radiating elements in the horizontal plane. With a *fixed antenna*, which pitches and rolls with the aircraft, the beams do not exactly compensate, and errors in velocity and drift angle measurements are induced. These errors are small with small pitch and roll angles but can be considerable for large pitch and roll angles.

Doppler Radar Components. A Doppler radar navigation system is a basic

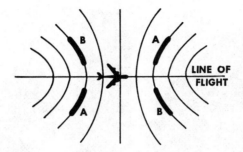

Figure 11.58(A). Illumination pattern, antenna aligned with ground track.

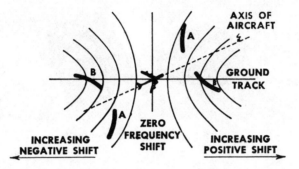

Figure 11.58(B). Illumination pattern, antenna not aligned with ground track.

dead reckoning (DR) device consisting of three major components: *Doppler radar, compass system, and computer*. The Doppler radar provides data for ground speed and drift information, and the compass system provides heading reference. The computer processes this basic information into position and guidance data.

Limitations. Doppler radar systems are self contained units, requiring no external references from radio signals except the transmitted microwave signal and reception of its reflection from the earth's surface. Error is cumulative with time. So Doppler position is correlated with a position fix by other means whenever possible. If Doppler error is indicated, results of the fix are entered into the computer manually, correcting the position indicator.

The primary source of error in a Doppler system is *heading reference* from the aircraft's magnetic compass system. Any error, even though small, will be integrated into the computer's navigational output.

Components Functions

Doppler radar. The radar transmitter/antenna generally transmits four symmetrical beams, one in each of the four antenna quadrants: forward left, forward right, rear left, and rear right. If the aircraft has a drift angle, the antenna is

Figure 11.59. Doppler components.

driven to azimuth until the Doppler shifts in the left beams are equal to the Doppler shifts in the right beams. When these shifts are equal, the antenna centerline is pointed along the aircraft's ground track, and the angular difference between the aircraft and Doppler antenna centerlines is equal to drift angle. The Doppler shift, D, is proportional to ground speed. The drift angle measured by the Doppler is added to the compass heading to get the actual track angle, which is sent to the computer. The Doppler converts Doppler shift to ground speed which is also sent to the computer.

Navigation computer. The computer first integrates Doppler ground speed to get distance travelled along the actual ground track. Second, it determines track angle error, the angular difference, if any, between the actual ground track (Doppler drift angle plus compass heading) and the desired ground track. Third, using the track angle error, the computer resolves distance travelled along actual ground track into distance travelled along and across the desired track, giving the distance made good along the pilot's preselected track and the distance right or left of the preselected track.

Control-indicator. The operating controls are located on this unit as follows: OFF, all power is removed; Standby (STBY), power is applied to warm up the system without the transmitter in operation; ON LAND, for use over land areas, and ON SEA, whenever flight path is over water, providing a special calibration circuit to minimize ground speed errors; TEST, a signal is generated, ensuring

Figure 11.60. Doppler control—indicator.

that the transmitter and receiver are functioning, and ground speed and drift indicators are driven to test values.

Drift is displayed on this unit, left or right, to a maximum of 40°. Ground
speed read out is displayed in the lower portion of the drift meter in nautical
miles per hour (Figure 11.60).

Computer controller. The desired magnetic course, miles to go, and miles
off track are set into this unit. Dual selectors are provided, allowing the next leg
to be programmed in advance. An infinite number of legs can be programmed
into the computer control by switching from stage 1 to stage 2 to stage 1, etc.
(Figure 11.61).

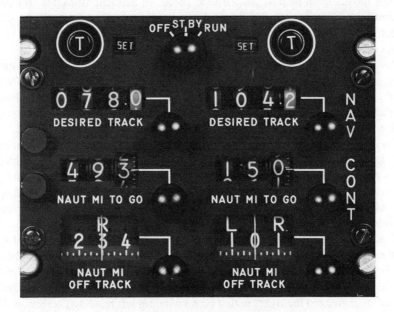

Figure 11.61. Doppler computer control.

Operational Procedures for Doppler.   Following is a description of a typical flight profile with three-stage legs over land and sea conditions. The procedures are based on the Singler Kearfott SKK-1000 Doppler Radar and SKQ-601 Computer. Doppler navigation systems by other manufacturers are operated in essentially the same manner.

Referring to Figure 11.62, the planned flight is from a land position, then over water, and again over land to the destination airport. The route is divided into three legs as follows:

Leg	Desired Track	nm. to go	nm. off track
1	058.3°	514.4	0
2	077.8°	425.9	0
3	073.4°	420.3	0

The aircraft is navigated along the desired track by steering it to maintain a *O NAUT MI OFF TRACK* indication on the Computer Control. If the aircraft is so equipped, *NAUT MI OFF TRACK* information can be fed to a flight director or HSI (CDI). After selecting a magnetic course of 058.3°, the needle is steered to center as with VOR tracking. A full-scale deflection occurs when off track deviation reaches 3 miles.

En route to waypoint 2, the Control Indicator is moved to SEA when the aircraft moves from land to sea. Ground speed, in nautical miles per hour, is indicated in the lower portion of the drift meter, allowing for accurate calculations of estimates. The *NAUT MI TO GO* displays remaining miles to the second waypoint.

Passing the second waypoint (0 miles to go), Computer Control automatically transfers from leg 1 to leg 2. Leg 2 is flown in the same manner as leg 1. The data for the third and final leg can be set in during this portion of the flight. Passing the third waypoint, Computer Control again automatically transfers from stage 2 back to stage 1.

Passing over land once again, the Control Indicator selector is placed on LAND. In the event of a solid undercast, this position would be determined by miles read out information corresponding to predetermined measurements.

The fourth and final waypoint is the destination airport. Descent and speed control can be planned, based on the *NAUT MI TO GO* readout, to bring the aircraft to the desired altitude and speed at the appropriate location.

Wind correction.   The Doppler system will automatically compensate for any change in wind direction by maintaining a *O NAUT MI OFF TRACK* indication and HSI needle centered (if equipped). For example, a wind from the left during this flight would require a correction to the left. Drift indication would be displayed in degrees to the right, with a corresponding left heading correction. If the magnetic course is 77.8° and the drift angle is 10° right, to maintain "on course" the magnetic compass would read 067.8°, reflecting the 10° drift correction.

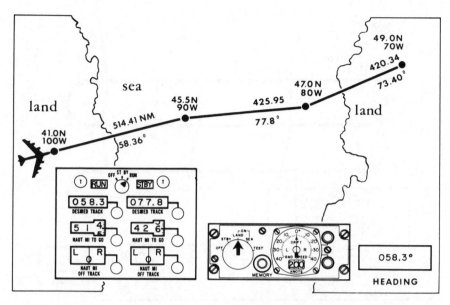

Figure 11.62. A typical flight profile is illustrated with 3 stage legs over land and sea. The computer control is programmed for stage 1 with DESIRED TRACK (0.58.3°), NAUT MI TO GO (514.4), and NAUT MI OFF TRACK (0). Stage 2 is programmed 077.8°, 425.9, and 0 respectively. At completion of stage 1 (0 miles to go) transfer to stage 2 is automatic. Data for 3rd leg would now be programmed in stage 1 position. Note, aircraft heading and desired track are identical. This is verified by 0° drift on drift meter. *(Courtesy Singer Kearfott)*

Track deviation. Deviation from the desired track for *weather avoidance* or other reasons can be made with position constantly indicated. Assume that thunderstorms lie ahead and the pilot deviates right of course. The miles right of course and miles to go to the next waypoint will constantly be indicated. After passing the weather area, the pilot steers to intercept the original track. When he is back on course, the *NAUT MI OFF TRACK* indication is 0 and HSI needle is centered.

Survey operation. Doppler navigation is of great value for unique operations such as *aerial survey,* providing: (1) accurate ground speed to assure proper overlay for photographs, (2) drift information to align the camera to the aircraft track, (3) parallel survey lines, and (4) a self-contained system requiring no ground based aids in remote parts of the world.

A typical survey program is displayed in Figure 11.63. Note that the Doppler system is magnetically referenced. Therefore, the desired true course must be corrected for magnetic variation, and it is this magnetic course that is set into the *DESIRED TRACK* window. In this example, a *magnetic course of 096°, 125 nautical miles, and 0 miles off track* are set in the appropriate windows of the Computer Control (leg A). The reciprocal course, 276°, 125 miles, and 5 miles

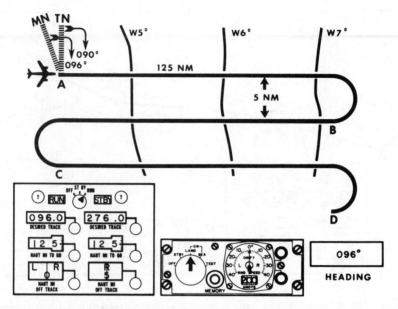

Figure 11.63. Aerial survey operations are enhanced by Doppler navigation. In this example, parallel tracks, 5 miles apart, are accurately flown by programming stage 1 of computer control with DESIRED TRACK (096.0°), NAUT MI TO GO (125), and NAUT MI OFF TRACK (0). Return leg, stage 2, is programmed with 276.0°, 125 nm, and 5R (right) respectively. At point A, computer control would be turned to RUN position and aircraft steered to maintain 0 NAUT MI OFF TRACK. Ending 1st leg (0 miles to go) computer control automatically transfers from stage 1 to stage 2. A 180° turn to the right is effected in a manner to ensure a reading of 125 NAUT MI TO GO and 0 NAUT MI OFF TRACK at position B. The 2nd and subsequent legs are flown in a like manner. *(Courtesy Singer Kearfott)*

(right) off track (in order to move the entire pattern by 5 nautical miles) are set for stage 2 (leg B).

At point A the Computer Control is turned to RUN position. The aircraft is steered to show *O NAUT MI OFF TRACK* to make good the desired track. If no crosswind is encountered, the drift angle would read 0°. In the event that drift is encountered, a correction is applied. Drift angle read out will provide accurate information for alignment of the camera.

Completing leg 1, the Computer Control will change to leg 2 automatically. A 180° turn is made to the right to intercept the second leg. The *NAUT MI TO GO* read out will show miles of overrun during the turn, and *NAUT MI OFF TRACK* will display closing distance on the next leg. At point B the aircraft is positioned on track with *O NAUT MI OFF TRACK* and 125 miles to go indicated. While en route, leg 3 (C) is programmed in stage 1 position, etc.

For a specialized aerial survey, a *long array* antenna is available as an option, providing a significant improvement in Doppler velocity accuracy and fluctuation, increase in overall antenna gain, and a reduction in overwater shift. Additionally, a cross track modification is available, increasing cross track resolution by a factor of 10 and allowing off track position accuracy to within 120 ft.

INERTIAL NAVIGATION SYSTEM

Improvements in recent years in *gyroscopes* and *accelerometers* have made it possible to design and fabricate a self-contained navigation system offering exceptional accuracy and flexibility. Miniaturization through solid state electronics has reduced the size of the inertial navigation unit (INU) to approximately 1 cubic foot in space with a weight of 57 lbs.

The following discussion is based on the Litton Aero Products LTN-72 Inertial Navigation System. Other units by other manufacturers are essentially similar in construction and operation (Figure 11.64).

Inertial Navigation Concepts. *A body at rest tends to remain at rest, and a body in motion tends to remain in uniform linear motion unless acted upon by an external force.* This, Newton's first law of motion, is the basic foundation upon which inertial navigation rests. While totally reasonable to Newton and his contemporaries, his unquestioning use of the term *at rest* eventually came under heavy criticism. The first seeds of doubt were planted by Einstein when he published his paper on the special theory of relativity in 1905, a theory in which the

Figure 11.64. INS components.

premise of *absolute motion* is shattered. The substance of this new view of things is, "nothing is at rest." The body that had been previously defined as being stationary was, in reality, only sharing a velocity with some other object and its coordinate system, all of which was being recorded by an observer who was also going along for the ride, whether he was aware of it or not.

The primary measuring device in an inertial navigation system, the accelerometer, bears out the theory. It makes no distinction between "at rest" and any other fixed velocity, because there is none. On the other hand, the accelerometer makes a disturbing distinction between truly fixed velocities and those that we like to think are fixed, but in reality are only fixed speeds along curved paths (all paths being curved, inertially speaking). Velocity is a vector quantity, made up of speed and direction. If direction is changing, velocity is changing.

The nature of the material of which a body is composed (matter) is to preserve its existing state of motion, *i.e.*, a body resists changes in its state of motion. This universal property of matter is known as inertia and is shared by all bodies.

In the development of an inertial system, primary interest is in Newton's second law which states, *"the acceleration (rate of change of velocity) of a body is directly proportional to the force acting on the body and is inversely proportional to the mass of the body."* For this law we can mechanize a device which is able to detect minute changes in velocity, an ability necessary to develop an inertial navigation system.

Based upon these laws, an inertial system, in simplest terms, includes a *stable platform* composed of accelerometers and gyroscopes suspended in a four *gimbal*

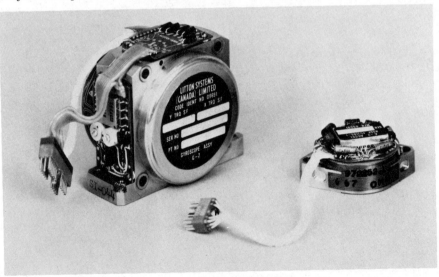

Figure 11.65. Gyro-accelerometer.

system. The gyros maintain the platform's accelerometers in a level position, in relation to the earth's surface, so that they may accurately measure aircraft accelerations, which are converted into velocities. These velocities, when modified by the computer with necessary correction terms, provide extremely accurate navigational information. A detailed discussion of these components and their functions follows.

Terms and Components of INS. An *accelerometer* is an instrument which senses and measures the change in motion, or vehicle acceleration. Acceleration is the rate of change of velocity per unit of time. The accelerometer is the basic measuring instrument of the inertial system (Figure 11.65).

It is basically a pendulous device tending to swing off the null position when the aircraft accelerates. The accelerometer has a signal pickoff device that tells how far the pendulum is off the null position. The signal from this pickoff device is sent to an amplifier and back into a torquing device which tends to restore the pendulum to the null position. The amount of current going into the torquer is a function of the acceleration which the device is experiencing. The output of the amplifier is also sent to an integrator which is a *time multiplication* device. Starting with acceleration, in feet per second squared, the output is multiplied, in the integrator, by time, and the result is a velocity in feet per second. It is then sent through a second integrator with an input of feet per second, multiplied by time, resulting in distance in feet or miles. Three accelerometers are mounted in the aircraft to measure acceleration along three perpendicular axes. One of these accelerometers measures acceleration in the north-south direction, and one measures acceleration in the east-west direction.

The computer associated with inertial systems knows the latitude and longitude of the take-off point (set in by the pilot), how far the aircraft has traveled in a north direction, and how far in an east direction (for example) from its accelerometers. With this data it is a fairly simple matter for the computer to compute the new present position of the aircraft.

The platform. The accelerometers are mounted on a gimbal set, commonly called the platform. The platform is a mechanical device which allows the aircraft to go through any attitude change while the inner element of the platform, on which the accelerometers are mounted, is maintained level by gyroscopes (Figure 11.66).

This is necessary to prevent the accelerometers from being affected by the changing attitudes of the aircraft. The effect of attitude changes of the aircraft, if allowed to influence the accelerometers, would put out an erroneous acceleration signal and would end up integrated into distance traveled. For example, at take-off the aircraft would pitch up, making the pendulum swing off the null position, due to gravity.

Gyros. A gyro (gyroscope) is a device which provides stable reference. It may be defined as a spinning mass, usually a wheel or disc, turning about an

Figure 11.66. Platform.

axis. The gyro is supported by a gimbal system which allows the wheel to rotate about one or two axes perpendicular to each other and to the spin axis, when disturbed by an external force or torque (Figure 11.65).

Gyros are required in an inertial system to control the level and heading of the platform stable element. These gyros, extremely precise in their design, manufacture, and operation, along with the accelerometers, are mounted on a common gimbal. When this gimbal tends to tilt from the level position, the spin axes of the gyros remain fixed. The case (housing) of the gyro tends to be moved from level, and the amount the case is tipped will be detected by the signal pickoff device in the gyro. The signal is then amplified and sent to a gimbal drive motor which restores the gimbal to the level position. Thus, the accelerometer is kept level, and the pendulum does not sense a component of gravity, but only the horizontal accelerations of the aircraft as it travels across the surface of the earth. Tilting tends to occur in the pitch, roll, and yaw axes of the platform; all are almost instantly detected by the gyros, preventing the gyro case from tipping off level, thereby keeping the accelerometer level.

Earth-orienting the platform. The gyro spin axes remain fixed in space, but the aircraft is not operating in space; it is operating on an earth which is "round and rotating." In order to keep the accelerometers level with respect to the earth, for sensing acceleration of the aircraft in a horizontal direction only, some compensation must be made for these two facts.

Figure 11.67 (left) shows what would occur if compensation were not made

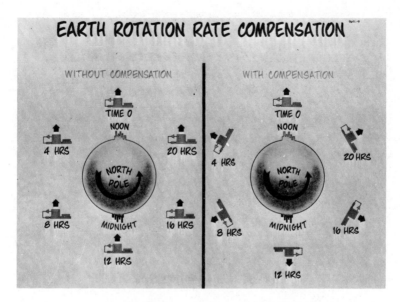

Figure 11.67. Earth rate compensation.

for the earth's rate of rotation. This example, looking down at the earth from a North Pole vantage point, shows that the platform would maintain the same orientation in space as the earth rotates. However, from an earth vantage point, it would appear to tip through 360° every 24 hours. To compensate for this apparent tipping, making the accelerometers sense only horizontal acceleration, the platform is forced to tilt in proportion to the earth's turning rate. Figure 11.67 (right) shows the platform with the earth rate compensation added. From a North Pole vantage point, it appears to tilt through 360° every 24 hours, while from an earth vantage point it remains fixed and level as required for proper system operation.

Transport rate compensation. Transport rate is compensated for in much the same manner as earth rate. In Figure 11.68 (left) what would happen without any compensation can be seen. Leaving the take-off point and flying to the landing point, the aircraft is actually traveling in an arc rather than on a flat surface. Since the platform maintains a space orientation, it would appear to tip with respect to the aircraft from take-off to landing. Shown in the same diagram (right), the platform is forced to tilt, maintaining the accelerometers level during the entire flight. Notice, at take-off and landing the platform has the same orientation with respect to the surface of the earth.

Torquing the gyros. Both earth rate and transport rate compensations are implemented by torquing of the gyros. Two integrators function in this regard as follows: integrate once to get velocity, and a second time to get distance; the distance traveled is summed with the initial position, in coordinates of latitude and

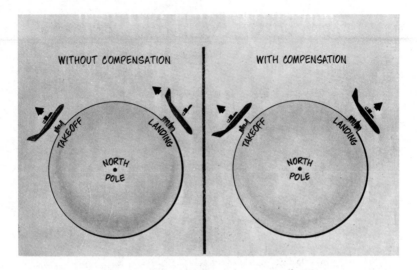

Figure 11.68. Transport rate compensation.

longitude, inserted into the system during the initial programming. This generates the present latitude, signaling some electronics to develop the earth rate torquing corrections.

Earth rate torquing is a function of latitude because what is being compensated for is the horizontal component of earth rate, which varies with latitude. At the equator, this value is about 15.04° (902.4 nm.) per hour; traveling farther north or south, it reduces until it becomes zero at the poles.

The velocity signal is also sent through some electronics to develop the transport rate torquing terms. The transport rate and earth rate terms are summed and sent into the gyro torquer, causing the rotor of the gyro to tilt with respect to the case. An output signal, thereby generated, is amplified to drive the gimbal drive motor, causing the gimbal to tilt in proportion to the two inputs, earth rate and transport rate torquing.

There are a number of other compensations generated within the system. The two examples were shown to demonstrate how a simplified inertial system works. Other compensations are necessary because the earth is not a perfect sphere; coriolis and centrifugal effects must also be compensated for within the system.

Aligning the system. When the system is first turned on, two things are required before becoming operational; the accelerometers must be leveled, and the platform must be oriented to true north.

During leveling the output of the accelerometers is put through some electronics and into the gyro torquer, causing the signal to come out of the gyro at the signal pickoff end and the gimbal drive motor to move the gimbals until the pendulum of the accelerometer is lined up with the gravity vector. At this point, there is zero output from the accelerometers.

During the orientation of the platform with true north, called *gyrocompassing*, it is assumed that the platform is aligned with true north and all the earth rate compensation is fed into the Y gyro. The Y gyro should be sensitive to all the earth rate if the platform is in fact aligned with true north, with no compensation being sent to the X gyro. However, if the platform is misaligned with respect to true north, the X gyro senses some earth rate, and the wrong compensation is sent to the Y gyro. Therefore, as the earth rotates, the platform will begin to tilt from level, which is detected by the accelerometers. A torquing signal is thereby developed and sent to the gyro that controls the platform in the azimuth or heading axis. This gyro control loop physically reorients the platform toward north. Eventually, when the platform is pointed to true north, the X gyro senses no earth rate, and the Y gyro is properly compensated for all of the earth rate it senses. Because the gyros are correctly compensated, the platform will not tilt as the earth turns, and the platform will remain aligned to north. As the system is flown in the navigation mode, north alignment of the platform is maintained by the computer torquing the azimuth gyro, using a combination of earth and transport rates, similar to the way platform level is maintained.

Wander azimuth inertial system.   The system just described is called a north pointing inertial system. A north pointing system has the disadvantage that it cannot be operated in polar regions. Because the platform must always be physically pointed north, when a system is flown directly over the pole, the platform would have to rotate 180° at the instant it crossed the pole. This is physically impossible, and in fact most north pointing systems cannot be operated within several hundred miles of a pole because of the high platform rates necessary to maintain north pointing. This problem is solved with a wander azimuth inertial system.

The basic fundamentals of a wander azimuth system are identical to a north pointing system except that during gyrocompassing the platform is allowed to take an arbitrary angle with respect to true north, *i.e.*, the initial wander angle. The platform is leveled as before and it is initially assumed that the wander angle is zero; therefore, all the earth rate compensation is sent to the Y gyro. Because this assumption is not correct, the platform will tilt from level in exactly the same manner as it did in a north pointing system. This off-level condition is detected by the accelerometer as before, except that now the accelerometer signal is split up between the two gyros to compensate for earth rate. Instead of continuing to send all the earth rate compensation to the Y gyro and orienting the platform to north to satisfy that condition, some earth rate compensation is sent to the X gyro. If the platform were pointed to north, the Y gyro would sense the entire horizontal component of earth rate. If the platform were pointed east or west, the X gyro would sense all the earth rate, and the Y gyro, none. Anywhere between, each gyro would sense a portion of earth rate. Eventually the right combination of earth rate compensation to the gyros is determined for the particular wander angle, and the platform will remain level as the earth rotates. The ratio of earth rate compensation to the gyros is then used to compute the initial wander angle.

As the system is flown in the navigate mode, the wander angle will change (*i.e.*, wander) as a function of longitude due to the convergence of the longitudinal meridians, which is taken into account by the computer.

Operation of a wander azimuth system is the same as a north pointing system, except that the wander angle is taken into account in all computations. For example, the accelerometers are not oriented along north-south and east-west directions but are offset by the wander angle. The computer knows this wander angle and can easily compute N-S and E-W accelerations using the sensed acceleration and the wander angle.

Terminology.   Terminology used in INS navigation is similar to other forms of navigation. However, inertial systems are referenced to true north rather than magnetic north as with VOR navigation.

Operational Procedures.   A typical INS consists of three units: Control Display Unit (CDU), Mode Selector Unit (MSU), and Inertial Navigation Unit (INU). The CDU and MSU are located in the cockpit for use by the pilot, while the INU is usually located in the electronics bay (Figure 11.64).

System alignment.   The INS must be aligned on the ground before precise navigation data are available. The process requires approximately 17 min. and is largely automatic. Alignment procedures are as follows.

Set the mode selector to standby (STBY) (Figure 11.69). This applies power to the system, starting the environmental control cycles for system warm up. It allows for the present position to be inserted, and for testing of the CDU, by setting "TEST" on the display selector (Figure 11.70). All segments of the numerical and FROM-TO waypoint displays, ALERT, BATT, WARN, degree, decimal, and minute displays, TK CHG, INSERT, and HOLD push-buttons, should be verified for illumination.

For the present position, set the display selector to position (POS), preparing the system for entry of the present position in coordinates of latitude and longitude. Example: present position, 33 degrees 56.4 arc minutes, North, and 118

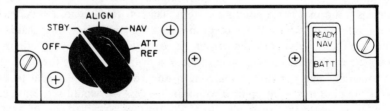

Figure 11.69. Mode selector-INS.

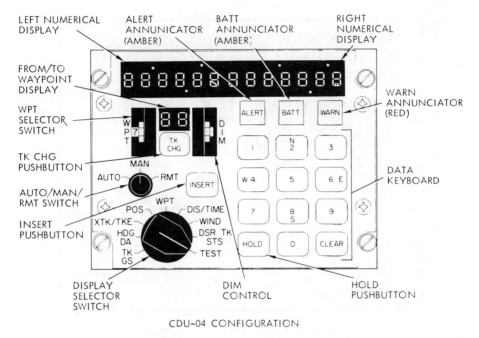

CDU-04 CONFIGURATION

Figure 11.70. Control-display Unit.

degrees, 24.4 arc minutes, West; press the *N* key for proper hemisphere, press 33564 buttons in that order, and verify that the left display reads the correct latitude. Press the INSERT push-button; the data is thereby transferred to the computer. Longitude is inserted similarly by pressing *W,* 118244, in that order, and verifying that correct data appear in the right display. Press the INSERT push-button (Figure 11.71).

After inserting the present position, set the mode selector to ALIGN. Receipt of the present position coordinates and the selection of ALIGN enables the system to continue the alignment sequence with no further action required by the pilot. The aircraft should not be relocated when the system is in the ALIGN mode. Gusty wind conditions and movement caused by fueling or cargo and passenger loading do not significantly affect alignment. When the alignment is completed, the READY NAV indicator, located in the MSU, is illuminated, at which time the mode selector may set to NAV (Figure 11.69). After selection of NAV the aircraft may be taxied or flown.

Integrity monitoring. During alignment, automatic monitoring of the system is conducted. If performance degradation occurs or a malfunction has been detected by ..he *integrity monitoring* circuits, the WARN annunciator will flash. In this event the pilot should set the display selector to Desired Track/Status (DSR TK/STS) and note the recommended action code displayed in the right numerical

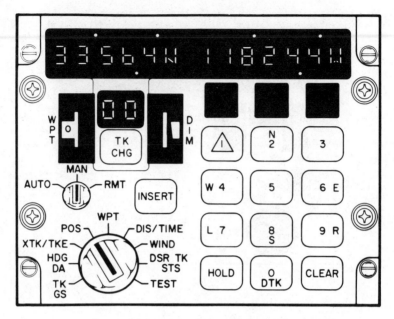

Figure 11.71. Latitude-Longitude display.

display. As shown in Figure 11.72, the numeral *2* is indicated. Referring to the listing, item 2, the pilot is instructed to ''Cycle system through off/on sequence; reenter present position.''

System detected malfunction codes are also available for display on the CDU by selecting DSR TK/STS and pressing the HOLD push-button. The maintenance code displayed should be recorded to inform maintenance personnel of the nature of the repair required.

There are three annunciator lamps—ALERT, BATT, and WARN—on the CDU front panel. The amber ALERT light signifies that the aircraft is within a preset time of the selected waypoint, usually 2 min. The ALERT annunciator will not come on when ground speed is below 250 knots.

The amber BATT light indicates that the primary AC power has failed and that the system has reverted to the back-up DC power source. In the event a back-up DC bus is not available in a given aircraft installation, a battery unit should be provided as a back-up power source. The battery source is capable of sustaining normal INS operation for a period of 30 min. A BATT warning lamp (red) is also located in the MSU to indicate that the back-up power source has been activated but is inadequate for system operation.

The current alignment process status may be monitored by selecting DSR TK/STS on the display selector; status, displayed in a numerical manner, will be displayed in the right display window. The number will start out high and gradu-

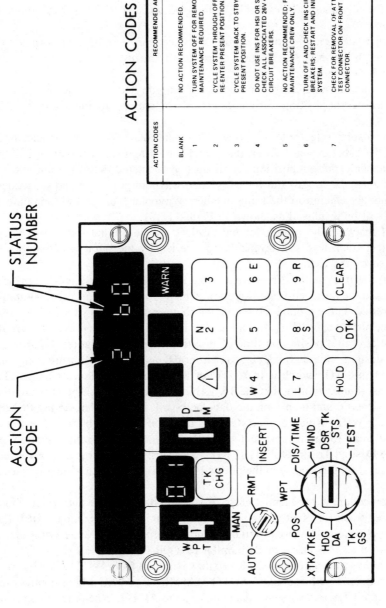

ACTION CODES

ACTION CODES	RECOMMENDED ACTION
BLANK	NO ACTION RECOMMENDED.
1	TURN SYSTEM OFF FOR REMOVAL. MAINTENANCE REQUIRED.
2	CYCLE SYSTEM THROUGH OFF-/ON SEQUENCE. RE-ENTER PRESENT POSITION.
3	CYCLE SYSTEM BACK TO STBY. RE-ENTER PRESENT POSITION.
4	DO NOT USE INS FOR HSI OR STEERING. CHECK ALL ASSOCIATED 26V 400 HZ CIRCUIT BREAKERS
5	NO ACTION RECOMMENDED. FOR MAINTENANCE CREW ONLY.
6	TURN OFF AND CHECK INS CIRCUIT BREAKERS. RESTART AND INITIALIZE SYSTEM
7	CHECK FOR REMOVAL OF ATTITUDE SLEW TEST CONNECTOR ON FRONT INU TEST CONNECTOR

STATUS NUMBER

ACTION CODE

Figure 11.72. Integrity monitoring.

ally decrease. A high number (90) indicates that alignment has just begun; a low number (02) indicates that the system is aligned to specifications (Figure 11.72).

Waypoints. Waypoints must be inserted into the system with the data keyboard. Up to nine waypoints can be inserted at a time; a good time is while waiting for the alignment process to be completed and the READY NAV light to illuminate.

Distance between waypoints should not exceed 7999 miles. In normal INS navigation, because of Air Traffic Control requirements, distance between waypoints will be considerably less.

Waypoints are entered in coordinates of latitude and longitude in the same manner as present position was entered for alignment. Waypoints are inserted as follows.

Set the display selector to waypoint (WPT) and rotate the waypoint selector to *1*. With the data keyboard, insert the latitude and longitude of waypoint 1 in the same manner as present position coordinates were entered. Set waypoint selector to 2 and enter the coordinates for the second waypoint; the third and subsequent waypoints are entered in the same manner. Waypoint latitude and longitude as entered will be displayed as shown in Figure 11.71.

Initial track. The pilot pilot must select and insert the *initial track* leg. Subsequent track legs can be manually inserted by the pilot or automatically sequenced by the INS.

The initial track is the direct great circle route between the aircraft's present position and the initial en-route waypoint, which is WPT 1 in this case. An initial track inserted on the ground represents the route from the aircraft's present position at the airport to WPT 1. An initial track inserted in flight represents the route from the aircraft's position at time of insertion to WPT 1 (Figure 11.73).

The desired initial track leg is inserted as follows: ensure that the AUTO/MAN/RMT switch is in AUTO or MAN; press the Track Change (TK CHG) push-button, noting that the TK CHG and INSERT push-buttons are illuminated. Press 0 and then *1* on the data keyboard, in that order, noting that the FROM/TO WPT display reads 0–1. Press the INSERT push-button. The initial track will read to the nearest tenth of a degree in the left display window by selecting Desired Track (DSR TK/STS). After verifying this true course for reasonableness, set this course in the pilot's HSI (CDI) and steer to center the needle.

This procedure is for manual entry of steering data into the auto-pilot. Since the desired track for great circle navigation is continuously changing, HSI data updating is required. Most installations provide for direct INS steering of the auto-pilot in an INS NAV mode (similar to VOR NAV mode).

Automatic waypoint change. En-route navigation from WPT to WPT is normally conducted with the CDU AUTO/MAN/RMT switch, which controls the method of WPT sequencing and data entry, set to AUTO. When the switch is set to automatic (AUTO) position, track leg changes are automatically sequenced in

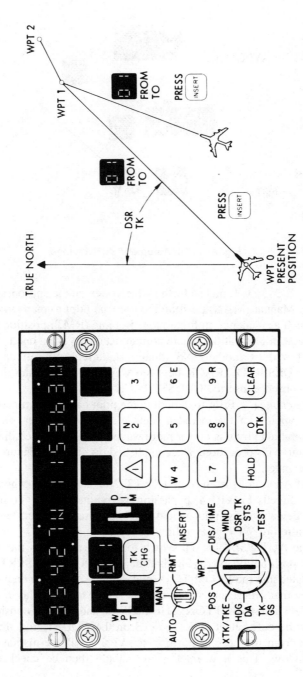

Figure 11.73. Initial track—INS.

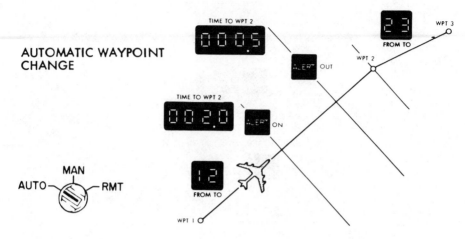

Figure 11.74. Automatic WPT change.

order, as 1–2, 2–3, 3–4, and so forth by the system at the appropriate times (Figure 11.74). Manual (MAN) position requires the pilot to manually initiate track changes, such as a change in flight plan. Remote (RMT) position permits semiautomatic auto-fill operation, where normal data entry from one INS is automatically transferred to another INS (dual installation), to call up DISTANCE, TIME, and DESIRED TRACK between waypoints other than the active track leg, and to observe inserted CROSSTRACK OFFSETS.

Track leg change. Prior to departure the pilot has programmed the computer with the waypoints of his intended route of flight. However, en route it may become desirable to bypass one or more waypoints and proceed directly to a distant waypoint. This procedure is accomplished, as shown in Figure 11.75, in the following manner.

Example: An aircraft is en route to WPT 2 on a 1–2 track when a decision is made to go directly to WPT 4, bypassing waypoints 2 and 3. To implement this flight plan deviation, the pilot would perform a manual track leg change from present position to WPT 4, as follows.

Select the AUTO/MAN/RMT switch to MAN, and press the Track Change (TK CHG) push-button, verifying that the TK CHG and INSERT push-buttons are illuminated and the FROM/TO display is blank. Press 0, then 4, on the data keyboard in that order, and verify that the FROM/TO display now reads 0–4. Press INSERT PUSH/BUTTON: THE INSERT and TK CHG push-button lights should now be out. Select DSR TK/STS on the CDU slector; the new desired track angle from the present position to WPT 4 will be displayed in the left display window. This new track angle should then be checked for reasonableness.

Data display selection.   Beyond basic track guidance from waypoint to waypoint, a multitude of navigation information is available for display in the Data Display portion of the CDU. The following discussion describes the data available and its selection [Figure 11.76(A)].

Track-ground speed: Select TK/GS on the Display Selector. The left display reads the aircraft track angle, with respect to true north, from zero to 360.0°. The right display reads aircraft ground speed, from zero to 3999 nm. per hour.

Heading-drift angle: Select HDG/DA on the Display Selector. The left display reads aircraft heading, from zero to 360.0°. The right display reads the aircraft's drift angle: L (left) or R (right), from zero to 180°.

Cross-track distance and track angle error: Select XTK/TKE on the Display Selector. The left display reads the aircraft's cross-track distance, L or R, from zero to 399.9 nm. The right display reads track angle error, L or R, from zero to 180.0°.

Present position: Select POS on the Display Selector. The left numerical display reads latitude, and the right display reads longitude of the aircraft's present position.

Waypoint position: Select WPT on the Display Selector. The left display reads latitude, and the right display reads longitude of up to 9 stored waypoints, corresponding to the digit on the waypoint (WPT) selector.

Distance and time to waypoint: Select DIS/TIME on the Display Selector. The left display reads distance to go, from zero to 7999 nm., to the waypoint currently selected for navigation. The right display reads time to go, from zero to 799.9 min., to the selected waypoint.

Distance between waypoints: During the flight it may be desirable to know the distance between two waypoints in the event a change in flight plan is desired.

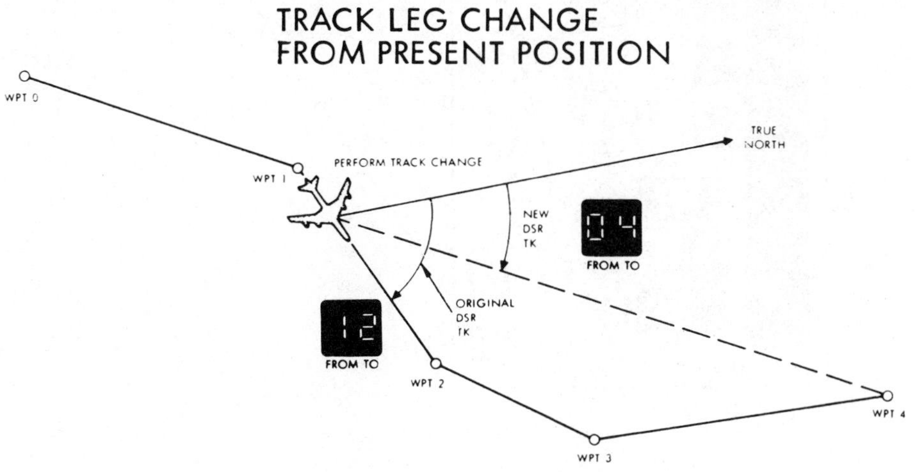

TRACK LEG CHANGE
FROM PRESENT POSITION

Figure 11.75. Track leg change.

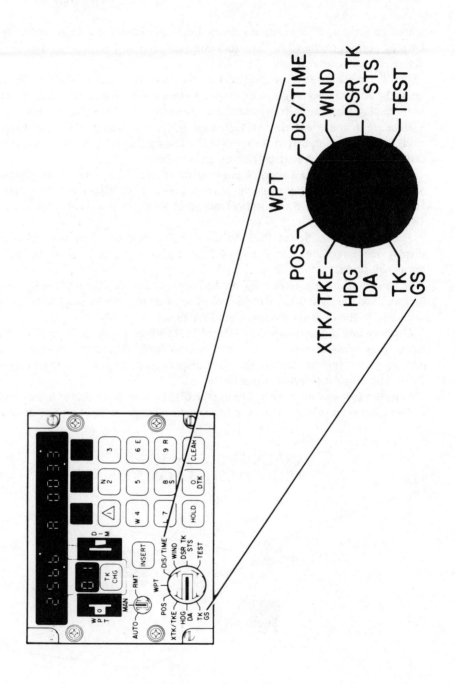

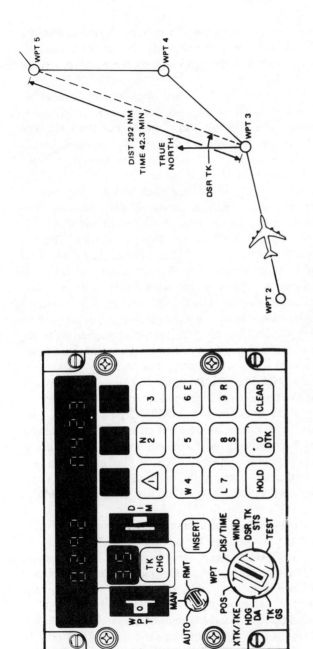

Figure 11.76. CDU display—selector.

For example, if the aircraft is between waypoints 2 and 3 and the direct distance between waypoints 3 and 5 is desired, the INS will read the great circle distance, time, and initial desired track between these two points when the procedure below is followed [Figure 11.76(B)].

Set the AUTO/MAN/RMT switch to RMT and the Display Selector to DIS/TIME. Press the TK CHG push-button and push-buttons 3–5, in that order. Press the INSERT push-button. The distance, time, and desired track between the selected waypoints (3–5) are now available for display. The left display reads distance in nautical miles. The right display reads the time between selected waypoints, based on present ground speed.

Desired track: In the above example, the desired track between waypoints 3 and 5 is displayed by selecting DSR TK/STS on the Display Selector. The desired track will be displayed in the left window, from zero to 360.0°.

Wind direction and speed: Select WIND on the Display Selector. The left display reads wind direction, from zero to 360.0°. The right display reads wind speed, from zero to 799 knots. A true airspeed (TAS) input from the aircraft data system is required for this function; otherwise, the wind display will be blank.

Rhumb line track. Track angles displayed by the INS are great circle courses from present position or from waypoint to waypoint. However, in addition to waypoint oriented navigation, the INS can produce a selected rhumb line mode of operation, which can be useful in flying legs of a holding pattern or a rhumb line course. A preselected rhumb line track is initiated as follows.

Set the Display Selector on DSR TK/STS. Press the *O* key on the data keyboard, and enter the track angle (rhumb line) to the nearest tenth of a degree with the data keyboard. To activate the track, press the INSERT push-button.

Position check and update. A significant concern with any self-contained navigation system is that any error entering into the navigation data is cumulative with time. Therefore fixes by other means, such as loran, Omega, or VOR/DME, are compared with corresponding INS position, and the INS position is corrected as necessary. A position check and update using VOR/DME is conducted as follows.

Select a VORTAC station to one side of the course, preferably within 30 to 60 nm. of the track. In this example, assume that the aircraft is eastbound and will pass north of the Cleveland VORTAC. Convert a true bearing of 360° to a magnetic radial by adding or subtracting the variation at the station. This is necessary because INS direction is referenced to true north, while VORTAC radials are referenced to magnetic north. A variation of 3°W at Cleveland VORTAC is added, giving a magnetic radial of 003°. Set the computed magnetic radial in the HSI (CDI). As the HSI cross-track deviation bar centers, depress the HOLD push-button and simultaneously read the DME distance.

To compute the aircraft's actual latitude, add the recorded DME distance as minutes to the latitude of the station. In this example, the DME distance (35.1 nm.) is added to the station's latitude, 41°21.5′N, because the aircraft is north of the VORTAC. The result, 41°56.6'N, was the actual latitude of the aircraft at the time of the position freeze (Figure 11.77). Use the VORTAC station longitude as the aircraft's because the aircraft was directly north of the VORTAC at the instant of comparison; therefore, longitude would be identical.

Compare the arithmetically computed latitude and longitude with the frozen coordinates, and determine if an update is required. If the comparison is acceptable, press the HOLD push-button again to restart the display. If an update is required, use the Data Keyboard to enter the computed coordinates data into the present position latitude and longitude displays. The present position can be updated within 30 arc-minutes of both latitude and longitude and is entered in the same manner that initial present position was entered for alignment.

When both coordinates are updated, the present position display will restart upon pressing the INSERT push-button the second time. If only one display is updated (latitude for example), press the HOLD push-button to restart display.

Position changes that occurred during the display freeze and updating procedures are automatically compensated for, and the position display reflects the new present position, referenced from updated coordinates.

Hybrid INS Navigation Systems. Hybrid INS is a combination of an inertial navigation system with some other type of navigation system, for the purpose of updating or improving the accuracy of the INS. Hybrid systems are typified by the following.

Radio inertial is a navigation system employing an INS which is updated or dampened periodically by a positioning system based on radio fixes. Radio and INS have complementary characteristics that can be integrated to provide performance characteristics and capabilities beyond those of either system. The inertial is used to carry the radio system through areas of poor or difficult reception, and the long-term errors of the inertial system are taken out by the radio system. Typically, loran, Omega, or VOR/DME is used to dampen the inertial velocities and to update the present position.

Doppler inertial is a navigation system employing an INS which is dampened by the accurate ground velocity signals developed by a Doppler Radar system. The ground speed is computed by utilizing the apparent frequency shift of reflected electromagnetic radiation due to vehicle velocity. Depending upon the velocity of the vehicle and type of terrain, accuracy may be in the order of a couple of feet per second. The advantage of this source of velocity error is that it is not cumulative and can be used to null those errors in the inertial system which are cumulative. Velocities obtained from Doppler Radar are referenced to the vehicle axes, usually designated heading and drift. In order to develop north-south and east-west velocity correction terms, the Doppler terms are resolved by the true heading angle of the vehicle and compared with the inertial velocities.

Stellar inertial is a navigation system employing an INS and optical star tracker system, used to eliminate the effects of gyro drift on the inertial platform. In this system, stellar observations are used to update any present position error

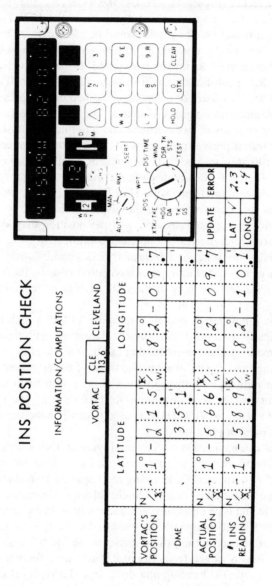

INS POSITION CHECK

INFORMATION/COMPUTATIONS

	VORTAC	CLE 113.6	CLEVELAND			
	LATITUDE			LONGITUDE		
VORTAC'S POSITION	N/$\cancel{S}$	1°	– 2 1 5 . 1'K/$\cancel{W}$	8 2	– 0 9 . 7'	–
DME			3 5 . 1			–
ACTUAL POSITION	N/$\cancel{S}$	1°	– 5 6 . 6'K/W	8 2	– 0 9 . 7'	
#1 INS READING	N/$\cancel{S}$	1°	– 5 8 . 9'K/W	8 2	– 1 0 . 1'	

UPDATE	ERROR
LAT ✓	2.3
LONG	.4

Fkgure 11.77. INS updating.

in the inertial system. Accuracy of stellar alignment and platform stability are accuracy limiting factors.

Grid Heading Display. The areas north of the Arctic Circle (66°30′N) and south of the Antarctic Circle (66°30′S) are generally considered the polar regions. Navigating in these areas has certain inherent problems, due to geographical location, of which a navigator must be aware. The most significant problems are direction, steering, maps, and fixing, due to the rapid convergence of meridians and proximity of the magnetic pole. To overcome these problems, a grid overlay system has been devised. Thus, polar navigation is often referred to as grid navigation. The polar grid overlay is constructed on a transverse Mercator or other polar-projection chart as a series of straight lines parallel with the 0° to 180° meridian. The parallel lines portray Grid North, with Grid North in the direction of the 180° meridian (Figure 11.5 and 11.78). With the 180° meridian as 000° grid and the Greenwich meridian as 180° grid, the other directions are easily recognized, *i.e.,* 000° is at the top of the chart, 090° is at the right of the chart, etc. All courses, headings, winds, and azimuths should be converted to grid directions.

Steering in the polar regions is done by gyros, since magnetic compasses become unreliable, and positioning the aircraft by pilotage, without adequate

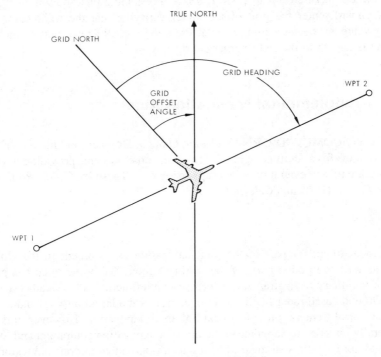

Figure 11.78. Polar grid heading.

landmarks and mapping, is very difficult. Other methods of fixing must be used, such as radio, inertial, or celestial.

Inertial navigation is invaluable for polar navigation. The precise gyros afford accurate steering and directional information, while the computer provides a continuous fix. A grid heading display can be selected and used as a compatible steering reference. A grid offset angle is inserted into the INS, and the aircraft's grid heading is calculated and displayed. The grid offset angle may be obtained from either a grid overlay map or a grid offset angle table.

INS displays and steering commands, except the grid heading display, are referenced to true north, and calculations are based on a great circle path. An INS course plotted on a map appears as a slightly curved line as opposed to a navigator's plot, which appears as a straight line. The rate the curved line deviates from the straight line is referred to as grid transport precession (GTP) and is not considered when the INS calculates the aircraft's grid heading; no correction for GTP is made. If a GTP correction is necessary, or if the map used to determine the original grid offset angle is changed, a grid heading update is made by determining the grid offset angle at the aircraft's position at the time of correction and by repeating the grid heading display selection procedures. Grid heading is selected as follows.

Set the Display Selector to DSR TK/STS. Press the 5 push-button; using the data keyboard, enter the grid offset angle, verifying that the right numerical display value has been entered ($\pm 0.5°$). Press the INSERT push-button; the grid heading is shown in the right numerical display.

Supplemental Navigation Aids

VOR/DME, ADF, RNAV, loran, Omega, Doppler, and inertial are the major systems for aircraft navigation. However, other systems providing navigation data are in use, even if only in a supporting role. These are *Radar, Radio Direction Finder* (DF), and *Celestial*.

RADAR

Radar is an important part of the total navigation environment in the United States as well as in other parts of the world. Although airborne radar has great value as a military navigation aid, its primary civil function is for weather avoidance. Ground based radar facilities, in operation at the larger airports and at Air Traffic Control Centers, provide a vital link in air navigation. However, this aid is generally intended to supplement the airborne navigation equipment and to facilitate the orderly flow of air traffic, not as an individual aircraft navigator.

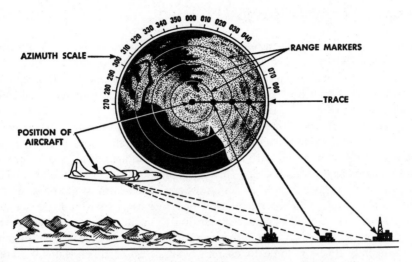

Figure 11.79. Navigation by radar. The trace moves in a clockwise direction around the scope face, painting blips every time the antenna receives a return from energy it has sent out. The navigator identifies the blip by reference to his maps. He then determines the distance the aircraft is from the object by reference to the range markers, and the bearing of the object by reference to the azimuth scale. This bearing and distance plotted from the object locate the aircraft in relation to it.

RADIO DIRECTION FINDER

MF-, VHF-, and UHF-DF stations are located primarily in airport control towers (civil and military) and in FAA Flight Service Stations (FSS). These facilities are used principally for emergency homing and as a simple and readily available aid to the pilot who needs navigation assistance. Two bearings on an aircraft can provide a fix.

CELESTIAL NAVIGATION

Celestial navigation is the art of determining position from observation of celestial bodies (sun, moon, stars, and planets). Celestial, once the only system of long-distance navigation, is now little used, except by long-range military aircraft. It would become of vital importance in wartime, over remote areas of the globe, and if ground-based aids were shut down or destroyed.

For applications requiring celestial fixes, a manual on celestial navigation should be consulted.

SATELLITE NAVIGATION AIDS

Development is currently in progress by the US Air Force and Naval Research Laboratory (NRL) to establish a *Spaceborne Global Positioning System*. This system would place in earth orbit a series of navigation satellites (NAVSATS) from which accurate position fixes and velocity data could be determined at any point on earth (Figure 11.80).

Three-Dimensional Fixes. It is projected that to assure four satellites would always be within view of every user on any spot on earth, a total of 24 satellites, eight in each of three orbital planes, would be required. This would provide three-dimensional position fixing (including altitude). With 18 satellites, six in each of three inclined planes, a user anywhere on the earth would be assured of finding at least three spacecraft in view at all times. This would provide two-dimensional (no altitude) or three-dimensional position fixing, depending upon the type of clock used. A user with a precision *atomic clock* could determine his position and velocity in three dimensions, while a less costly crystal clock would give position and velocity in two dimensions.

Anticipated position accuracies in the horizontal and vertical axes are 24 and 29 ft., respectively, 90% of the time, and 16 and 20 ft., respectively, 50% of the time. The accuracy of velocity measurements is expected to be within .2 ft./sec.

Position is determined by measuring range and range rate to the four broadcasting satellites of known position. A *computer-processor* in the user equipment solves four simultaneous equations to determine position and the time error in an on-board reference clock needed for synchronization.

A *master ground control station* and a number of *monitoring stations*, located in Guam, Hawaii, Alaska, and the northeast United States, will be employed. The monitor stations will receive navigation signals from satellites and format the data for transmission to the master station, where the data are processed. The master station determines satellite ephemerides (calculated, predicted position), ionospheric propagation, and clock bias errors with each pass. They are retransmitted to the satellites on their next passes for subsequent use. Consequently, a navigator or pilot could read his position, altitude, and speed very accurately at all times.

The first applications of navigation satellites will be by strategic and tactical military forces. The global system potentially could replace many existing positioning sensors employed by the military. Eventually, civil uses will be developed. One probable civil use would be updating inertial navigation systems. Ultimately all navigation guidance for all kinds of users may be derived from satellite navaids.

Many problems have yet to be satisfactorily solved. Extremely high costs of development and purchase of user equipment are the most serious. But the advantages of a global positioning system are too great to ignore.

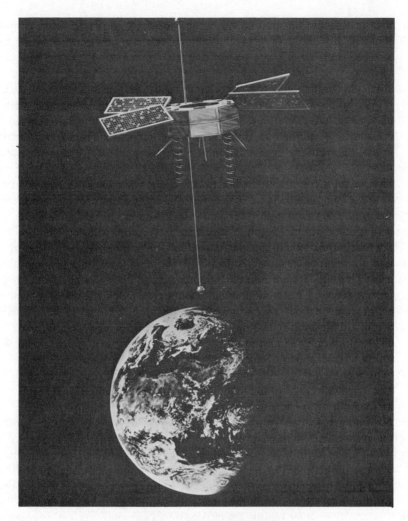

Figure 11.80. Satellite Navaid.

ERRORS IN USING RADIO AND ELECTRONIC AIDS

New and improved types of radio and electronic equipment are being developed to make flying safer, easier, and more precise. But as complexity has increased, the age old problems of malfunctioning equipment and human error are still present.

Examples of possible errors in the use of radio aids are: (1) Inaccuracies of receiving equipment: All electromechanical devices are subject to malfunctioning which can result in erroneous navigation signals. Therefore, navigation equipment should be checked periodically for accuracy. Many late model navigation receivers have self-testing circuits which can be used for verification of performance. (2) Improper selection of frequency or incorrect data programming: If an improper frequency is selected or inaccurate data entered, erroneous navigation information will be displayed. This can result from carelessness, or setting in data from an obsolete chart. Once a station is selected, it should be verified by checking the Morse Code or voice identification or by other appropriate methods, and the navigation data displayed should be checked for reasonableness. (3) Improper or incomplete understanding of how the equipment is operated: Accurate navigation equipment is of little value unless the airman knows how to use it. He must clearly understand the equipment's capabilities and limitations, correctly program it, and properly interpret its indications and readings.

The reliability of electronic equipment has increased at an accelerated rate with the advent of solid state technology. Continued improvement in reliability will undoubtedly continue. However, the *human factor*, although improved by "engineering out" many of the predictable human errors, will always be an important factor with which to reckon. Education of the pilot-navigator can make a significant contribution to keeping human error at a minimum.

Simplifying Navigation

Modern long distance flying requires the utmost in accuracy and efficiency if it is to meet the requirements of economy, reliability, safety, and dependable scheduling. It is these factors that have led to the development of sophisticated systems of aerial navigation. These systems, however, have multiplied the workload of the airline pilot and navigator. It has therefore become necessary to automate as much of the navigation and flight operation as possible.

The necessity to enter multitudes of data into a complex navigation computer prescribes a data entry keyboard (Figure 11.70). Programming by this method requires time and is subject to human error. The faster the airplane flies, the more time presses on the crew, further increasing the possibility of error. This situation leads to automatic programming of an entire flight.

The new navigation systems have a common characteristic: the need to convert visually recognizable *alphanumeric* information to *digital electrical signals* useable by the new computers. The primary means for this conversion has been the familiar manual entry data keyboard (Figure 11.70).

Data keyboards on the flight deck have increased the already high cockpit workload, during both the preflight and the en-route phases of operation. Increased workloads result in increasing human error and a reduction in safety. The coming of RNAV, to which the FAA is firmly committed, will place even greater loads on the pilot, particularly when using *standard instrument departures* (SID's) and *standard terminal arrival routes* (STAR's).

To compensate for additional workloads and to reduce errors, automation is desirable, to minimize the requirements for manual data insertion. It would be very helpful to preprogram the entire flight before take-off to avoid errors and simplify the pilot's task during flight, especially for that portion of the flight where several waypoints are needed in rapid succession (for departure path, climb-out path, final approach path, and alternates). The need for automatic data entry increases for jet aircraft.

CARD READER

One automatic data entry system is a card reader unit known as an *Automatic Data Entry Unit* (ADEU), shown in Figure 11.81. It is cockpit-mounted and eliminates the need for manual insertion of such navigation data as waypoints (distance, bearing, frequency, and station elevation) and en-route bearings.

Automatic start and shutoff are accomplished by the single act of insertion of

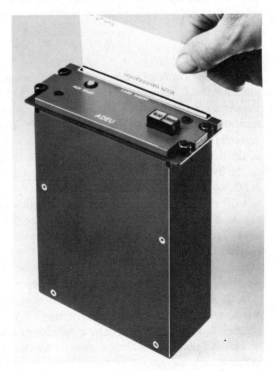

Figure 11.81. Automatic data entry unit.

the data card. The card is electro-optically scanned for pickoff of the printed digital data. A *READ* mode lamp remains illuminated throughout the *data scanning sequence* (approximately 10 sec. duration), automatically extinguishing after completion of the cycle. Error in the insertion of the card, or mistakes in actual data registration on the card, are automatically detected by the ADEU, triggering the illuminated *ERROR* indicator. The use of the card reader can reduce cockpit workloads by 90% and reduce the chance of human error to near zero. Existing navigation systems can take advantage of the ADEU's accuracy and convenience.

DATA CARD

Flight data cards (4⅝″ × 9¾″) used in the ADEU may be programmed well before take-off by a dispatcher, the pilot, or a navigator for "nonstandard" flights, or preprinted for regularly scheduled or frequently flown routes (Figure 11.82). Commercial sources distribute cards paralleling distribution of en route radio navigation charts.

Input data are in digital language, "ones" being black squares, "zeros" being blanks. Data for as many as 20 waypoints can be accommodated by one card. The reverse side of the card can be used for SID's, STAR's, flight charts, and other route identification information to help select the correct card.

DATA LINK

The ADEU is only one example of automation to reduce cockpit workloads and human error. Ultimately, Air Traffic Control (ATC) will use a data link for communications between the ground and the aircraft, reducing the endless voice transmissions between pilot and controller. Computer programmed commands and advisories could be transmitted automatically to the aircraft, with agreement of the Air Traffic Controller. In the cockpit, the pilot would have a digital display of the data linked information. He would signal "will comply" or "unable" to ATC by pushing appropriate buttons.

Conceivably, a data link may usher in almost totally automated control, joining Air Traffic Control directly with the aircraft. The total navigation scene from data "present" to en route and temporal changes for traffic sequencing all would be computer controlled from the ground. The pilot and controller would become systems monitors, whose functions would be to either allow the computer to do the flying, or to take over command as necessary. The data link would initially provide messages such as *notices to airmen, terminal conditions,* and *en route* and *terminal traffic control information.* With inertial and *aircraft parameter systems* linked to the on board data control unit, information such as position reports, winds, turbulence, avionics status, and fuel state would be transmitted automatically on ground command with no crew action required.

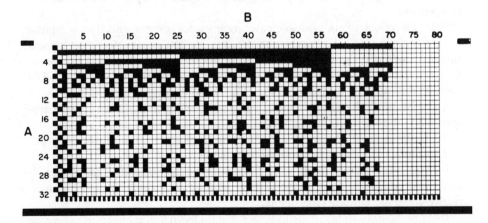

Figure 11.82. ADEU cards.

AUTOMATIC NAVIGATION SYSTEMS

Further relief for the cockpit crew is provided by an automatic navigation system (ANS). It recognizes the economic importance of reducing flight time; even 5 min. saved improves the profit picture. Horizontal planning as well as vertical (altitude) scheduling are essential.

A unit such as Collins Radio Company's ANS-70A provides such automatic navigation, increasing accuracy and flexibility, making use of vertical and time schedules, reducing workloads in planning and in flight with attendant reduction of human error, and reducing voice communications.

The heart of ANS is a general purpose digital computer which processes the outputs of multiple *navigation sensors*. The computer computes steering, controls aircraft equipment, automatically prepares flight plans, gives the pilot instant access to and display of flight plan and performance data, and automatically selects and tunes navaids.

Information Storage.   An on board magnetic tape cartridge, in addition to providing rapid and economical revisions of route structure, is the basic instrument of *system modification.* Change in operating functions, and interfacing with additional inputs such as satellite navaids, can be performed by new instructions on the tape. The system hardware remains unchanged, with no lost aircraft modification time.

Equipment.   An ANS is composed of a *Navigation Computer Unit* (NCU), *Flight Data Storage Unit* (FDSU) including a *Magnetic Tape Unit* (MTU), and a *Control Display Unit* (CDU). The NCU computes required navigation parameters for flight crew use by utilizing the best available inputs from the navigation sensors.

Selection of one of six available navigation modes is automatic. The computer checks to determine what navigation data are available and selects the *highest ranking* mode that can be used with the current data, providing increased accuracy and unparalleled redundancy of navigation data. Annunciators on the CDU indicate the mode of operation in use. The mode ranking is as follows.

Radio-inertial—Requires inertial, navigation radio, true airspeed, and barometrically corrected altitude data inputs.

Radio-air data—Requires navigation radio, magnetic heading, true airspeed, and barometrically corrected altitude data inputs.

Inertial—The inertial mode indicates that no navigation radio data are available. The system requires single or dual inertial, true airspeed, and barometrically corrected altitude data inputs.

Dead reckoning—This mode indicates that no inertial or navigation radio data are available. The system requires magnetic heading, true airspeed, and barometrically corrected altitude data inputs.

Radio—Navigation radio data are available, but true airspeed, magnetic heading, or barometrically corrected altitude data have been lost.

None—Insufficient navigation data are available to provide any of the above modes of operation.

Control display unit.   The CDU provides a display of navigation data stored and computed in the navigation computer and provides the means to alter, add, or delete information in the NCU memory (Figure 11.83).

At the bottom of the CDU screen is a *scratch pad line* where data typed into the system are displayed for inspection before insertion. Incorrect data entered on the scratch pad line are removed by pressing the clear (CLR) key. The three levers on the right side of the CDU screen are used to scroll the screen display, a line at a time vertically, or to change complete pages.

FAULT and *ALERT* annunciators are located on the right side of the CDU to indicate system malfunctions or other significant changes in status. If the FAULT

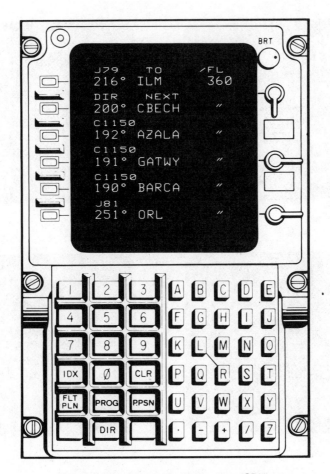

Figure 11.83. Automatic Navigation (CDU).

or ALERT lights are blinking, the PROG push-button is pressed, and the fault or alert message will be displayed on the screen.

Displays: Twelve pages of navigation data can be displayed. Several examples are:

Index Page—The index page, called up by pushing the IDX push-button on the CDU, provides a listing of six categories of information in the computer: RTES (routes), SID's, Airways, STAR's, Holding Patterns, and System Status. (Figure 11.84A). Pressing the line select key opposite a particular category calls up a page, listing all the stored items in that category. Selecting RTES, for example, displays all the stored route identifiers. Then pushing the button next to the desired route calls up its flight plan page. Figure 11.83 shows such a flight plan page.

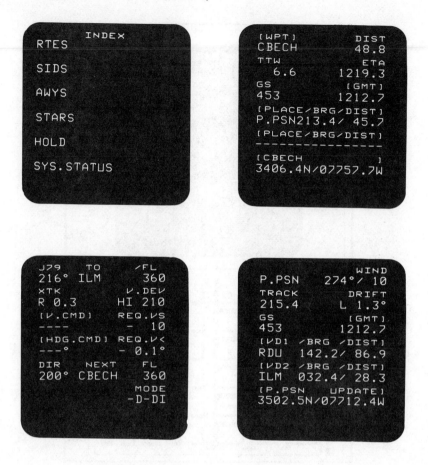

Figure 11.84. Pages of data display.

Flight Plan Page—The flight plan page shows waypoints, courses, and desired altitudes for each leg. Pressing the line select key next to a waypoint calls up a page showing detailed information on the chosen waypoint (Figure 11.83).

Waypoint Data Page—The waypoint data page shows time, distance, and bearing to the chosen waypoint, and its geographical position (Figure 11.84B).

Flight Progress Page—Pushing the PROG key on the CDU calls up the flight progress page. It shows present position relative to the present programmed leg and the next following one, and indicates the mode of navigation being used. "D-DI" indicates that data from DME-1, DME-2, and inertial sensors are in use (Figure 11.84C).

Present Position Page—Pressing the PPSN button calls up the present position page, which shows the airplane's present computed position and a listing of currently tuned radio stations. Also shown are wind direction and velocity, drift angle, ground speed, bearing and distance of selected VORTAC's, and present latitude and longitude (Figure 11.84D).

Pressure Pattern and Jet Stream Flying—Minimal Flight

Sometimes called Optimum Flight, this is the path between two points that can be flown in the shortest time. Under zero wind conditions, the great circle route is the shortest route in both distance and time. In many cases a deviation from a great circle route, while longer in distance, would result in a "lesser time" through minimal flight. Minimal flight planning requires multiple computations and is based on going around pressure systems on the side that gives best tailwind components or least headwind components. Likewise, locating the jet stream and using its high velocity winds (often in excess of 100 knots) to advantage when possible will result in less time, even though a distance greater than a great circle route is flown. Knowing the location of the jet stream and associated wind shears is also essential for avoiding clear air turbulence (CAT).

Minimal flight procedures are normally most productive on long distance flights, such as transcontinental or oceanic routes. Flights of relatively short distances are usually flown at lower altitudes, and a deviation from the great circle route would be of limited or no advantage. Because of the exhaustive calculations required to select and verify a minimal route in pressure pattern flying, ground based computers are frequently used, greatly simplifying the task. The potential economic gains resulting from minimal flight routes have influenced scheduled airlines to use all available resources, such as computers, dispatcher's expertise, and pilot reports, to ascertain what route of flight will result in minimal time. Atmospheric phenomena relative to pressure pattern and jet stream flying are discussed in Chapter 6 ("Circulation of the Atmosphere").

With the sky becoming increasingly crowded, ATC has established specific routes for more positive control of aircraft, to some degree restricting the free selection of the most advantageous route. The flight planner works out a minimal flight path and picks the established route closest to this minimal course. Although this is a compromise, the established route selected may be shorter in time than a great circle route.

Bibliography

Area Navigation System (KNC-610), Pilot's Guide. Olathe, Kansas: King Radio Corporation (1972).
Area Navigation System (LTN-104), Pilot's Guide. Woodland Hills, California: Litton Aero Products 1973).

Automatic Direction Finder (DF-203), Pilot's Guide. Cedar Rapids, Iowa: Collins Radio Company (1966).

Automatic Direction Finder (DF-206), Pilot's Guide. Cedar Rapids, Iowa: Collins Radio Company (1971).

Automatic Direction Finder (ADF-T-12D), Pilot's Handbook. Ford Lauderdale, Florida: Bendix Avionics Division (1972).

Automatic Flight Control System (FCS-810). Fort Lauderdale, Florida: Bendix Avionics Division (1972).

Automatic Flight Control System (M-4D), Pilot's Handbook. Fort Lauderdale, Florida: Bendix Avionics Division (1973).

Automatic Navigation System (ANS-70A). Cedar Rapids, Iowa: Collins Radio Company (February, 1974).

Computer Radio Navigation System (VLF). Costa Mesa, California: Communications Compo nents Corporation (1972).

Distance Measuring Equipment, Pilot's Handbook. Los Angeles, California: RCA Corporation (1966).

Doppler Radar for Aerial Survey & Enroute Navigation, Air Crew Manual. Singer Kearfott Division. Little Falls, New Jersey (1969).

Doppler Radar Navigation Computer (SKK-1000 and SK-601), Air Crew Manual. Singer Kearfott Division. Pleasantville, N.Y. (1968).

Doppler Radar Principles. Vol. 19: Singer Kearfott Division. Pleasantville, N.Y. (1970).

Edo Auto-Trak 800 Min-Loran, Operator's Manual. Melville, New York: Edo Commercial Corporation. (1968).

Edo 1200 Dual Lok-Trak Loran, Operator's Manual. Melville, New York: Edo Commercial Corporation. (1967).

Fundamentals of Inertial Navigation (TT 104). Woodland Hills, California: Litton Aero Products (1971).

How to Fly Loran. Melville, New York: Edo Commercial Corporation. (1967).

Inertial Navigation System (LTN-72), Pilot's Guide. Woodland Hills, California: Litton Aero Products (1973).

Inertial Navigation Systems (LTN-51), Pilot's Guide. Woodland Hills, California: Litton Aero Products (1973).

Inertial Navigation System (INS-61B), Pilot's Guide. Cedar Rapids, Iowa: Collins Radio Company (1972).

Introduction to Inertial Navigation (TT 100). Woodland Hills, California: Little Aero Products (1969).

Lyon, Thoburn C.: *Practical Air Navigation* (Civil Aeronautics Bulletin No. 23). Washington, D.C.: Superintendent of Documents, US Government Printing Office (1945).

Manual of Flight. Wichita, Kansas: Cessna Aircraft Company (1970).

Omega Navigation System (CMA-719), Pilot's Guide. Montreal, Canada: Canadian Marconi Company (1973).

Omega Navigation System—Summary of Evaluations in Progress. Montreal, Canada: Canadian Marconi Company (1973).

Omega Systems Status (No. 30). Washington, D.C.: US Naval Observatory (1973).

Ontrac II, Pilot's Manual. Costa Mesa: Communications Components Corporation (1973). California

Pictorial-Symbolic Navigation System, Pilot's Guide. Cedar Rapids, Iowa: Collins Radio Company (1968).

Pilot's Handbook of Aeronautical Knowledge (AC 61-23A). Washington, D.C.: Superintendent of Documents, US Government Printing Office (1971).

Weathervision Systems (RDR-1200), Pilot's Handbook. Fort Lauderdale, Florida: Bendix Avionics Division (1973).

Periodical Articles

"Airborne Navigation Gives 1 nm Accuracy," *Canadian Electronics Engineering:* (May, 1972).

Klass, Philip J., "Navigation Competition Grows Keener," *Aviation Week & Space Technology:* (December 10, 1973).

Klass, Philip J., "Omega Aviation Navaid Use Studied," *Aviation Week & Space Technology:* (November 29, 1971).

Schiff, Barry, "The Magic of Inertial Navigation," *The AOPA Pilot:* (September, 1973), pp. 35–37.

12

Air Traffic Control

When the first aircraft flew, there was no problem of encountering other aircraft and hence no concern about minimizing chances of a midair collision, about who had the right of way, and so on. A pilot simply took off when he could and had the sky to himself except for insects and birds of various sizes. This happy situation continued until aircraft proliferated during World War I. Then it happened—the first recorded midair collision of aircraft occurred between two Jennies in pilot training. Since then, of course, the dangers, and the problems, of keeping aircraft from striking each other and the ground, have multiplied many times over.

Control of Air Traffic

What are the basic requirements and objectives of controlling air traffic? Quite simply, rules, have to be devised, and tools designed, to prevent two or more aircraft from trying to occupy the same space at the same time, and to prevent aircraft from being a hazard (or sometimes a nuisance) to persons and property on the ground.

Coincidentally, the smooth, efficient flow of air traffic must be promoted, so that the aircraft's special advantages of speed and shortest possible routing are used to the fullest extent.

These objectives impose grave responsibilities on both pilots and controlling agencies. It is worth noting here that the pilot always is responsible, for during a major portion of flying no outside controlling agency is involved. For that reason this chapter approaches its subject primarily from the pilot's point of view.

Each nation has an agency in its government which establishes rules governing air traffic and which specifies the qualifications and certification of airmen and aircraft. In the United States it is called the Federal Aviation Administration (FAA) and is a part of the Department of Transportation. By international agreement there is an International Civil Aviation Organization (ICAO, usually pronounced I-K-O), which establishes rules for international air commerce.

Rules established by the FAA are called Federal Aviation Regulations (FAR). While any rule-making agency must be somewhat arbitrary, the FAA's rule-making process considers those who may be affected by any change to, addition to, or deletion from the rules. A proposed change is publicized, and all interested parties are invited to comment within a certain period of time. The comments received are considered very seriously and quite often drastically affect a proposed change. Occasionally a change is obviously required for reasons of safety or is directed by a higher authority (*e.g.,* the Congress). In these cases, aircraft operators do not get an opportunity to comment.

CERTIFICATION OF CREW MEMBERS

It was recognized quite early in the history of aviation that one could not walk up to an aircraft the first time he had ever seen one, get in it, and safely fly it away. The armed forces showed the way to safe flight by establishing formal training programs leading to special pilot ratings. From these beginnings the present rules concerning pilot qualifications have evolved. These are published in FAR Part 61.

A pilot-in-command of an aircraft, charged as he is by responsibilities to his passengers, if any, to people and property on the ground, to other people flying, and to the owner of the aircraft, must meet certain minimum training and experience requirements and observe particular rules.

STUDENT PILOT

The neophyte begins as a student pilot. His instructor is responsible for the mission whenever he accompanies the student in flight, and supervises the student's solo flights. The student must pass a flight medical examination and satisfy his instructor that he is competent to fly alone before his first solo flight. After this magic event, the student is more and more on his own. He has certain restrictions, however: He may not carry a passenger, he may not use flying to further a business, he may fly only under the supervision of an instructor, he may not cross an international border (except between Alaska and Canada), and he must have demonstrated solo capability, within the preceding 90 days and in the airplane he proposes to fly, to an authorized flight instructor. He must follow a training plan, develop his skills, gain specific knowledge, and pass special oral, written, and flight examinations as he progresses to the private pilot rating.

PRIVATE PILOT

Once he has completed his training requirements, and successfully passed his written and flight examinations, the new private pilot may spread his wings a bit. There is no longer an instructor looking over his shoulder, he may carry passengers, providing he meets recency of experience requirements, he may go when he pleases if he observes the rules, and he may even share expenses with those with whom he travels. He may not fly for pay and is restricted to visual flight operations until he qualifies for an Instrument Rating, discussed later in this chapter. He, like all other pilots, must have had a biennial flight review every 2 years by a certified flight instructor, or have passed some kind of FAA flight exam within the same period. If he wants to carry passengers, he must have flown and logged at least three landings within the preceding 90 days. If the passenger flight is to be at night, the landings must have been at night to a full stop. If it is to be in a tail wheel airplane, they must have been in a tail wheel airplane to a full stop. If it is to be in a multi engine airplane, they must have been in a multi engine airplane. The recent experience must also have been in a land or sea plane as appropriate.

To fly as a crew member, he must remain medically qualified. This is evidenced by having on his person, whenever he flies, an FAA medical certificate, issued by an FAA designated medical examiner within the past 2 years.

COMMERCIAL PILOT

The commercial pilot must have pursued a further course of training and have passed more stringent medical, oral, written, and flight examinations. He must have gained a good deal of flight experience, so that his total flight time (in powered aircraft for instance) is at least 200 to 250 flight hours. He enjoys the same privileges as a private pilot and may also fly for pay. But he cannot be paid for flying at night or more than 50 miles from home base until he has an Instrument Rating. He also must meet the same recency of experience requirements as the private pilot. His medical certificate must show that he has met the standards for a Second Class medical certificate within the past year. If he flies air taxi, he must have a flight check by an FAA designated examiner each year [every 6 months if he flies air taxi under Instrument Flight Rules (IFR)].

The pilot who plans to fly in weather conditions requiring control of the aircraft solely by instruments must qualify for and receive the Instrument Rating, and must also meet certain recency of experience requirements. These requirements are further spelled out later in this chapter.

OTHER CREW MEMBER RATINGS

Special training and experience requirements apply to Flight Instructors, Airline Transport Pilots, Flight Engineers, and other crew members. All these requirements are dictated by knowledge and skills needed to fulfill their duties. Also, there are specific requirements for training and specific ratings for each category and class of aircraft.

CERTIFICATION OF AIRCRAFT

Even the best qualified pilot could not be expected to fly safely without a safe vehicle. So, there are regulations concerning airplanes, too. No aircraft may fly in the United States, or under United States registry, unless it is certificated by the FAA or its counterpart in the country where the aircraft is built. Aircraft manufacturers must put each new model through a certification process. (Home built aircraft must be inspected periodically by FAA personnel as they are being built.)

Rules for certification and maintenance of aircraft are given in detail in FAR Parts 23, 25, 27, 29, 31, 39, and 43.

A detailed testing program of each new model is carried out by both the manufacturer and the FAA, before certification is issued. This test program both determines the capabilities of the new aircraft and uncovers any unsafe characteristics which must be corrected before it may be issued a type certificate.

EQUIPMENT REQUIRED

An airplane that is to be flown visually in daylight must have certain minimum equipment and instrumentation. These items include magnetic compass, altimeter, airspeed indicator, engine tachometer, manifold pressure indicator if it has a variable pitch propeller, oil pressure indicator, and oil (or liquid coolant) temperature indicator.

Aircraft to be flown at night, in addition to the equipment specified above, must have approved position lights on wing tips and tail, an approved anticollision light system, an adequate source of electrical energy for all installed electrical and radio equipment, and spare fuses. If it is to fly at night for hire, it must have at least one electric landing light.

To fly under Instrument Flight Rules, gyroscopic instruments and radios, suitable for communications with Air Route Traffic Control facilities and for navigation by use of the planned radio aids to navigation, must be added.

AIRCRAFT MAINTENANCE

The aircraft must also be maintained in safe flying condition. Its airworthiness must be attested to annually by an FAA authorized and licensed aircraft and engine inspector, and the pilot must perform a preflight inspection to assure himself that it is airworthy. An aircraft used for hire must also be inspected by an authorized aircraft and engine inspector every 100 hours of operation, or under a progressive maintenance and inspection plan approved by the FAA. Any discrepancies affecting safety, discovered by the pilot or an authorized inspector, must be corrected before the aircraft may fly. Separate aircraft and engine logs must be maintained to record 100-hour and annual inspections, repair and replacement of vital parts, overhaul, and compliance with special instructions affecting safe flight issued by the manufacturer or by the FAA. All of this is to assure that the aircraft will not be a hazard to its occupants, to other aircraft, or to persons or property on the ground.

Operating Under Visual Flight Rules

To be completely useful as a means of transportation, aircraft must be able to fly in all normal environments, including flying at night and in the clouds and other conditions of poor visibility. In order to consider the simpler situation first and then the more complex, however, we shall first take a look at flying when there are no obstructions to visibility. Rules for visual flying are detailed in Parts 91.61 through 91.109 of the FAR.

When an aircraft's position, course, and clearance from other aircraft and from ground obstructions are maintained by the pilot's looking outside the aircraft, it is being operated under Visual Flight Rules, or in common usage, VFR. When aircraft control relies solely on instruments, it flies under Instrument Flight Rules (IFR).

Visual flight rules are designed to minimize interference of aircraft flight paths with each other, to establish right of way when aircraft do approach each other, to minimize risk and nuisance to persons or property on the ground, and to provide for a safe, orderly flow of traffic around airports.

RIGHT OF WAY

Right of way rules for aircraft take into account the relative maneuverability of the airdraft concerned and apply principles used for surface traffic on land and sea. Aircraft having both directional and altitude control are considered most maneuverable. Those having only directional control are considered more maneuverable than those with only altitude control or altitude control and very slow directional control. So, right of way for aircraft, by category, is as follows:

1. Free balloon
2. Airship
3. Glider
4. Powered Airplanes and Rotorcraft

An aircraft experiencing an emergency has right of way over all other aircraft. An airplane towing or refueling another has right of way over any other aircraft not experiencing an emergency.

At an airport, a landing aircraft has right of way over one taking off, and if two aircraft are on final approach simultaneously, the lower one has the right of way.

When two aircraft approach each other head on, each is required to give way to the right so as to pass at a safe distance. An aircraft overtaking another must turn out so as to pass the overtaken aircraft at a safe distance to its right. If two aircraft converge on crossing paths, the one to the other's right has the right of way.

In spite of these rules, it must always be assumed that the pilot of the aircraft required to give way has not seen the other aircraft. A safety minded pilot never contests the right of way.

UNCONTROLLED AIR SPACE

In uncontrolled air space, the pilot is fully responsible for the conduct of safe flight. Uncontrolled air space is that over which no ground agency exercises control.

Many airports are in uncontrolled air space part or all of the time. Each pilot arriving, departing, or operating in the vicinity of an uncontrolled airport must observe certain rules.

Traffic Pattern.   Traffic around an airport normally is to the left, with straight legs on departure, crosswind, downwind, base leg, and final approach. A typical traffic pattern is depicted in Figure 8.12. Figure 12.4 illustrates traffic pattern terminology.

Sometimes traffic will be to the right; in that event markings in the airport traffic circle, visible from the air, will apprise the pilot of the correct direction to fly.

The standard entry into the VFR traffic pattern is to approach so that a 45° turn is made to the downwind leg. The normal traffic pattern altitude for light aircraft is 800 or 1000 ft. above the airport elevation; for larger multi engine aircraft and jets, it is 1500 ft. above the airport elevation.

Traffic Advisories.   Quite a few uncontrolled airports have a radio service known as Unicom. A fixed Base Operator (FBO) on such an airport operates a two way radio on a published VHF frequency for the convenience of its known and prospective customers. Over the Unicom the pilot can obtain traffic advisories concerning wind direction and velocity, other known aircraft traffic, favored runway, and any hazards or restrictions in effect. Such advisories are just that, not directive in nature. A pilot is free to land on another runway than the one recommended, for instance.

Many airports are served by FAA Flight Service Stations. During periods of VFR weather, these Flight Service Stations also provide traffic advisory service. This service is not directive, either.

In departure from an uncontrolled airport the pilot observes the normal direction of its traffic pattern. For example, if the normal pattern is to the left, the first turn after take-off is to the left.

Away from Airports. In uncontrolled air space away from airports, the general VFR rules and rules of right of way apply, but there are also specified minimum and cruising altitudes, and there are restrictions on certain maneuvers and airspeed. In remote, uninhabited areas and over water, there is no specified minimum altitude. In normal open country, flight must be maintained at 500 ft. or more above obstructions. Maximum speed below 10,000 above mean sea level (MSL) is 250 knots indicated air speed (IAS). Over cities, crowds and congested areas, minimum altitude is 1000 ft. above the nearest obstruction within 2000 ft. horizontally, or sufficiently high for the aircraft to glide outside the congested area in the event of engine failure. Acrobatic maneuvers must not be performed over cities or crowds, on an airway, or in other controlled air space, and must be completed at or higher than 1500 ft. above the surface.

VISIBILITY AND CLEARANCE FROM CLOUDS

Seeing and being seen are the basics of avoiding other aircraft in flight. Clouds and other atmospheric conditions affecting visibility play a major part in this. Therefore, the rules say that a VFR pilot must maintain specified distances from clouds and must have certain minimum forward visibility. The FAR require that when flying VFR at less than 1200 ft. above the ground, the aircraft must be kept clear of clouds and the pilot must have at least 1 statute mile forward visibility. In areas where air traffic is controlled from the ground, the visibility minimum is 3 statute miles.

When clouds are more than 1200 ft. above the ground, the VFR pilot must remain at least 500 ft. below, 2000 ft. horizontally from, or 1000 ft. above them. When flight is 10,000 ft. or more above sea level, the VFR minimums increase to 5 miles visibility, 1000 ft. below, 1 mile horizontally from, and 1000 ft. above clouds, except when at less than 1200 ft. above the ground.

Altitudes. In VFR cross-country operations, all flights at more than 3000 ft. above the surface are required to be made at an altitude governed by magnetic course. A flight whose magnetic course is between 0° magnetic and 179° magnetic must fly at an odd altitude (*i.e.,* 5000, 7000, 9000, etc.), plus 500 ft., above sea level. A flight whose magnetic course is between 180° and 359° magnetic must fly at an even altitude (4000, 6000, etc.), plus 500 ft., above sea level. The even thousand foot altitudes are reserved for aircraft on instrument flight plans.

Up to 18,000 ft. MSL, altimeter settings are used in establishing altitude. Above that, altimeters are set at 29.92 in. Hg, and pressure altitude is flown. The indicated altitude is then called a flight level, dropping off the last two zeroes. An indicated altitude of 20,000 ft. is flight level 200.

Above flight level 290, VFR and IFR flights are separated by 1000 ft. VFR uses even flight levels, IFR, odd ones. Thus a VFR flight making good a course between 0° and 179° would fly at flight level 300, 340, 380, or 420; one making good a course between 180° and 359° would use 320, 360, 400, etc.

Uncontrolled air space over the continental area of the United States extends upward to 14,500 ft. MSL. Above that altitude, all air space is controlled during IFR conditions.

RESERVED AIR SPACE

There are special areas of reserved air space, determined primarily by usage, in which flight may be restricted. One example is wild-life areas, over which a minimum altitude for flight is established. Another example is a temporary minimum altitude restriction over a disaster area, a sports event, or the scene of an aircraft accident.

Certain kinds of activities on the earth's surface project hazards to aircraft which may fly through the air space above them. An example is artillery fire. An area over which a hazard to aircraft may exist is designated a Restricted Area if over land, or a Warning Area if over a large body of water.

In cases in which aircraft may pose a special hazard to persons or property on the surface, the area in which the hazard is deemed to exist may be designated a Prohibited Area, and no one may overfly it below a designated altitude, or in some cases at any altitude. Perhaps the best known Prohibited Area is over the White House and Capitol areas in Washington, D.C.

Airborne Military Activities. Military aircraft fly at such speeds and often are so difficult to see that they may pose special hazards to other aircraft. Such activities are restricted to certain areas. An area in which a substantial amount of test flying is done or gunnery or bombing training carried out is designated a Restricted or a Warning Area.

Areas in which jet training is conducted are called Military Operational Areas (MOA). Routes along which high-speed low altitude navigational operations are practiced are called Olive Branch Routes. Locations of these areas are shown appropriately on aeronautical charts and in the Airman's Information Manual. The nearest Flight Service Station (FSS) has or can obtain information on when these are in use.

Flights Through Reserved Air Space.   Restricted and Warning Areas may be overflown with permission of the controlling agencies. The permission is automatic if the hazardous activity is not in progress. The Flight Service Station nearest the area will usually be able to tell a pilot whether the area is in use or not, or to obtain the information by contacting the Army, Navy, Air Force, Marine, or Coast Guard unit nearest the area.

CONTROLLED VFR TRAFFIC

Certain kinds of air space are established in which VFR traffic is controlled. The most common of these is the Airport Traffic Area around an airport, at which an air traffic control tower is operating. An Airport Traffic Area is 5 statute miles in radius from the center of the airport concerned and extends upward to but not including 3000 ft. above the runway surface. Within this area all air traffic is monitored and directed by personnel or "operators" (usually FAA employees) in the control tower, using radios. Certain limited control can also be exercised by use of light signals.

Traffic on the airport surface, excluding the runways in use for take-off and landing, is under the direction of Ground Control, a function of one or more operators in the control tower. A pilot desiring instructions and clearance to taxi for takeoff, or to taxi to a parking area after landing, calls "(Name of airport) Ground." Traffic on the runways in use and in the air is controlled by operators answering to the call sign "(Name of Airport) Tower." When ready for take-off, or approaching the field for landing, the pilot calls "(Name of Airport) Tower" for instructions and clearance.

One may fly in an Airport Traffic Area only for the purposes of proceeding to or from an airport in the area. If the airport concerned has the control tower, then radio communications must be maintained with it.

Within an Airport Traffic Area, a reciprocating engine aircraft may not exceed 156 knots (180 mph), while a turbine driven aircraft is restricted to 200 knots IAS (230 mph), except that an aircraft which cannot safely fly that slowly may exceed limiting speeds as is necessary for safety.

At many airports with control towers, there is not enough traffic during the late evening and nighttime hours to justify having tower operators on duty, and the tower is manned less than 24 hours a day. When there is not a tower operator on duty, the airport traffic area does not exist. In that situation the pilot of an aircraft operating to or from the airport must clear himself, assisted by whatever information he can obtain from Unicom, a Flight Service Station, or other agency on the airport.

TERMINAL CONTROL AREAS

Some big metropolitan airports have such a large flow of heavy jet traffic that the FAA considers special traffic control areas and procedures necessary. Terminal

Control Areas (TCA) are established around these airports. Those with the heaviest traffic are designated Group I, somewhat less busy ones, Group II. Operation in either area requires certain pilot qualifications and aircraft radio equipment, with Group I requirements being higher. Control of traffic in these areas is primarily by radar, even under VFR conditions. A typical TCA depicted in Figure 12.1.

Within a TCA, one may not exceed 250 knots IAS, and beneath its lateral overhanging limits, IAS may not exceed 200 knots.

SPECIAL VFR

A pilot approaching or departing a tower controlled airport or other controlled area with less than 3 miles visibility may (except in Terminal Control Areas) request special VFR clearance from the controlling agency, to fly with visibility as low as 1 mile. He must also be able to remain clear of clouds while maintaining a safe height above obstructions. A special VFR clearance at night will be issued only if both the aircraft and the pilot are qualified and current for instrument flying.

FLIGHT SERVICE STATIONS

One of the best aids to the pilot as he flies is the network of Flight Service Stations (FSS) which covers the 50 states in the US, in conjunction with the Federal Airways System. They are manned by highly trained specialists, who provide weather briefings and information on flight facilities and monitor the navigational radio net. They also assist in flight planning and accept formal VFR flight plans. A VFR flight plan is forwarded from the station with which it is filed to a Flight Service Station in the area of the flight's destination. This action assures a pilot that if he does not arrive at his destination within a short time after he has scheduled, someone will begin looking for him. Filing of a VFR flight plan is not mandatory, but it certainly is good practice. Periodic radio position reports should be made as one flies along.

Worth noting here is the necessity for the pilot who has filed a flight plan to notify Flight Service, through one of its stations, of his arrival at destination, and of any delay, change of flight plan route, or change of destination. This action will preclude wasting concern and search efforts for an aircraft and occupants who are not in any trouble at all.

For flight under Instrument Flight Rules (IFR), Flight Service serves as a channel of communication with Air Route Traffic Control. This function is discussed in more detail later in this chapter.

Certain communications radio frequencies are standardized and used by all stations. In addition each station may be assigned local use frequencies.

A pilot wishing to use any of the services available from Flight Service should visit the nearest FSS before flight if circumstances permit, to obtain a briefing

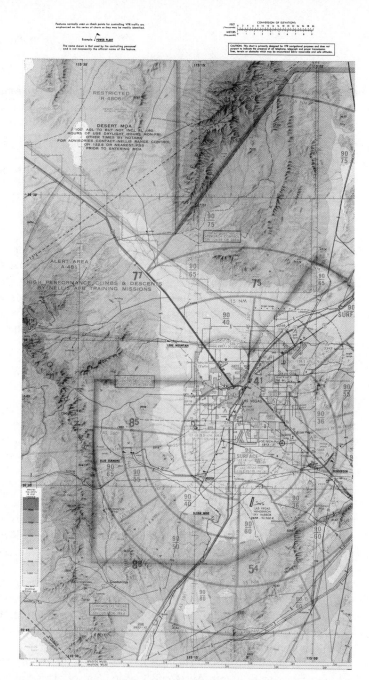

Figure 12.1. Las Vegas, Nevada Terminal Control Area. At a city/airport complex where there is a large amount of airliner traffic, a Terminal Control Area may be established. It is centered on the major airlines airport, extending in the area of the airport from the ground upward to a selected level about 7,000 feet above the surface. The base of the area steps to successively higher levels as one proceeds outward from the main air-

port. The area resembles an upside down stepped layer cake. Within the Terminal Control Area all airborne traffic is radar controlled, with the objectives of expediting traffic flow and providing safe separation of traffic. As produced and sold by National Ocean Survey, the scale of this map is 1:250,000, one inch on the map = 3.95 miles.

739

DEPARTMENT OF TRANSPORTATION— FEDERAL AVIATION ADMINISTRATION					Form Approved OMB No. 04-R0072		
FLIGHT PLAN							
1. TYPE ☑ VFR / ☐ IFR ☐ DVFR	2. AIRCRAFT IDENTIFICATION N9803U	3. AIRCRAFT TYPE/ SPECIAL EQUIPMENT Gr. Am. AA-1C /T	4. TRUE AIRSPEED 100 KTS	5. DEPARTURE POINT RAP	6. DEPARTURE TIME PROPOSED (Z) 1500 / ACTUAL (Z)		7. CRUISING ALTITUDE 8500

8. ROUTE OF FLIGHT

D→ GCC   D→ SHR

9. DESTINATION (Name of airport and city) Sheridan Sheridan, Wyo.	10. EST. TIME ENROUTE HOURS 2 / MINUTES 15		11. REMARKS Operative E.L.T. on board Passenger stop at Gillette

12. FUEL ON BOARD HOURS 3 / MINUTES 15	13. ALTERNATE AIRPORT (S)	14. PILOT'S NAME, ADDRESS & TELEPHONE NUMBER & AIRCRAFT HOME BASE Welch, J. Pilot and aircraft on file, RAP FSS	15. NUMBER ABOARD 1 to GCC 2 to SHR

16. COLOR OF AIRCRAFT White and Blue	CLOSE VFR FLIGHT PLAN WITH Sheridan FSS ON ARRIVAL

FAA Form 7233-1 (5-72)   *1976-G.P.O.-1703M-674-857-197

Figure 12.2(A). VFR flight plan filed with FAA Flight Service.

and to file a flight plan. If an FSS is not available nearby, the pilot may call the nearest one by telephone before his flight, for a briefing and for filing a flight plan. When neither a visit nor a phone call is feasible, he may call on his aircraft radio after he is airborne.

Future planning by the FAA is directed toward centralizing the location of Flight Service personnel and toward providing briefings, communications, and other services by remote electrical and electronic facilities, including televised information.

Instrument Flight Rules (IFR)

There are many times when an aircraft cannot be controlled by reference to the natural horizon, nor safely flown clear of other aircraft, because of clouds, rain, snow, fog, or other restrictions to visibility. To realize full utility of aircraft as a mode of transportation, provisions must be made, in the design of aircraft, in the rules of flight, and in ground facilities, to permit flight solely by reference to the aircraft's instruments, while maintaining safe separation of aircraft. Aircraft instruments and radios, electronic navigational aids, and flight rules for their use fulfill these requirements.

FLIGHT LOG

DEPARTURE POINT	VOR IDENT. FREQ.	RADIAL TO FROM	DISTANCE LEG REMAINING	TIME POINT—POINT CUMULATIVE	TAKEOFF	GROUND SPEED
RAP					1455	
CHECK POINT Gillette	GCC 112.5	265	102 67	1:12 1:12	ETA 1607 ATA 1610	93
Passenger Stop		↓ 273	↓ 67	:10 1:22	↓ 1625	∨
Sheridan		273	67 0	:45 2:07	1708 1718	89
Traffic Pattern, Taxi				:08 2:15	1718 1718	
DESTINATION Sheridan			TOTAL 175	2:15		

PREFLIGHT CHECK LIST	DATE 12 June 1979

EN ROUTE WEATHER/WEATHER ADVISORIES

Clear, good visibility

DESTINATION WEATHER Clear; 3010; 60 miles	WINDS ALOFT 9,000
ALTERNATE WEATHER	2710 +15°C

FORECASTS

VFR

NOTAMS/AIRSPACE RESTRICTIONS

Watch for antelope on GCC runway

Figure 12.2(B). VFR flight log on back of flight plan shown in Figure 12.2(A).

Again, the smooth, orderly flow of traffic must be promoted, in order to obtain the best efficiency and economy of aircraft operation. Radar makes minimum spacing between aircraft possible, permitting the movement of more aircraft through the same amount of air space in the same length of time. Its use will be discussed in more detail later in this chapter.

Rules for instrument flying are found in FAR 91.115 through 91.127.

PILOT QUALIFICATIONS

So long as a pilot can rely on visual references for controlling his aircraft and avoiding others, his visual training and skills are adequate. But before he may safely fly when he cannot see the horizon or the surface, he must develop his knowledge and skill in operation with reference to only the aircraft instruments and must learn the rules which apply to instrument flying.

To qualify for an instrument rating, a pilot must hold at least a private pilot certificate, have at least 200 hours total flight time, 40 hours of it in instrument training and practice (186 hours and 35 hours in an FAA Approved Flight School), and pass written and flight examinations. To remain qualified to file and fly under IFR, he must maintain a flight experience record of at least 6 hours instruments (weather or hooded) with at least six instrument approaches, in the last 6 months. Half the requirement must be met in an aircraft, the remainder may be accomplished in an appropriate aircraft simulator.

Category II.   Normal weather minimums for precision instrument approaches may be lowered if the ground equipment, the aircraft, and the pilot all meet certain standards established for Category II operations. The pilot must qualify once each 6 months in order to be authorized to exercise the privileges of this special rating.

AIRCRAFT QUALIFICATIONS

A pilot wants to be sure that the aircraft which he flies into clouds and conditions of poor visibility has all the instruments and equipment necessary to fly by reference to instruments only. He also wants to be assured that other aircraft sharing the murk with him are properly equipped and capable, too.

An airplane to be flown under IFR must have at least the following equipment relevant to flying instruments:

1. A gyro stabilized attitude indicator.
2. A gyro stabilized heading indicator.
3. A slip and skid indicator.
4. A sensitive altimeter.
5. A vertical speed indicator.
6. A clock with sweep second hand.

7. Navigation radios capable of providing navigation using the planned radio aids, including a marker beacon receiver if ILS is to be used.
8. A communications radio capable of maintaining communications with Air Traffic Control.

The equipment to be used for instrument flight must be inspected or checked at intervals depending on the specific system. The pitot and static systems must have been checked for leaks, and the altimeter must have been calibrated, by an authorized mechanic or shop, within the last 24 months. The VOR navigation indicator(s) must have been checked for accuracy within the past 30 days.

An aircraft to be flown IFR within a TCA or above 14,500 ft. MSL is required to have a radar transponder. For a Group I TCA, it has to be paired with an encoding altimeter, which transmits the aircraft altitude to Air Traffic Control Control Radar. A transponder must be checked and certified every 2 years by an FAA certified radio mechanic or shop.

Category II. Category II requirements for the aircraft, in addition to those above, are:

1. Appropriate approach coupler if automatic approach is to be used.
2. Radar altimeter if approach to less than 150 ft. altitude will be made.

CONTROLLED AIR SPACE

A vital part of IFR flying is someone to direct traffic. That responsibility is given to an entity of the FAA called Air Route Traffic Control (ARTC). The FAA has been given authority to control certain portions of the air space above the states, possessions, and territories of the United States, whenever the weather is below the minimums required for VFR flying. This mandate is exercised through Air Route Traffic Control Centers, which are located at major air route hubs around the country. Each center is responsible for a specified geographical area, which is in turn divided into sectors. Each sector is manned at all times by a controller, who has radio communications and radar to assist him. Radio transceivers and radar antennas are located at appropriate places in each sector and connected by ground lines to the controller's position. Nearly all aircraft flying under IFR now have radar transponders, which can be set to respond on a specified code. Many also have altitude encoding capability, so that the transponder, each time the radar beam sweeps past it, responds with both its assigned code and its altitude.

Thus the controller has geographical location and altitude immediately readable on his radar screen. He can maintain safe spacing of aircraft much closer together than with communications only, and the need for communications is much reduced. The newest equipment, by use of computers, can also sound a warning when two aircraft get too close together for safety.

Low Altitude Routes. The present primary means of navigation under IFR is the network of Visual Omni Range Stations which covers the country. The stations are described in Chapter 5.

Certain routes connecting these VOR stations are designated as VOR Low Altitude Airways. Each airway is assigned a number and establishes an aerial highway over a desired course. An airway is 8 nautical miles wide, and its center is defined by given radials of the VOR stations it passes over. Unless otherwise specified, it begins at 1200 ft. above the ground and extends upward to 18,000 ft. MSL. Should the stations defining a segment of a route be more than 102 nautical miles apart, the lateral limits of the airway are established beyond 60 nautical miles from each station by the radials 4.5° on either side of the one used to define the centerline. More details are found in FAR Part 71.

For use of crew members who are planning and flying IFR, Low Altitude En-Route charts are published every 60 days. Any changes between publication dates are published in the NOTAMS (Notices to Airmen).

Figure 12.3 shows a portion of a typical Low Altitude En Route Chart.

The chart depicts the VOR stations, showing locations, navigation and communications frequencies, the low altitude airways, minimum altitudes, ARTCC sector boundaries and frequencies, and much other useful information.

Altitude flown in the low altitude route structure is maintained using the nearest available altimeter setting.

High Altitude Routes. Many modern aircraft fly routinely above 20,000 ft. MSL. The limited number of VOR frequencies available, and the line of sight characteristics of VHF radio, make the low altitude route structure unusable at high altitudes. VOR frequencies must be repeated at different locations, two or more of which could be received simultaneously at high altitudes. Furthermore, since high altitude flights are generally longer, and between larger cities, they profit from more direct routes. For these reasons, high altitude routes have been established. The stations they use are chosen for their directness of routing and are separated much more distantly from other stations using the same frequencies.

Altitude in the high route structure is an assigned flight level, with the altimeter set at standard barometric pressure, 29.92 in. Hg. High Altitude or Jet Route Charts are published each 60 days, with interim changes published in the NOTAMS.

CONTINENTAL CONTROL AREA

The continental control area is the air space over the 48 contiguous states, the District of Columbia, and Alaska east of 160° west longitude, at and above 14,500 ft. MSL. It does not include air space less than 1500 ft. above the surface, or prohibited and restricted areas. Air Traffic Control Radar can "see" all this air space and thus control IFR traffic in it.

Aircraft flying above 12,500 ft. MSL, except gliders below 18,000 ft. and

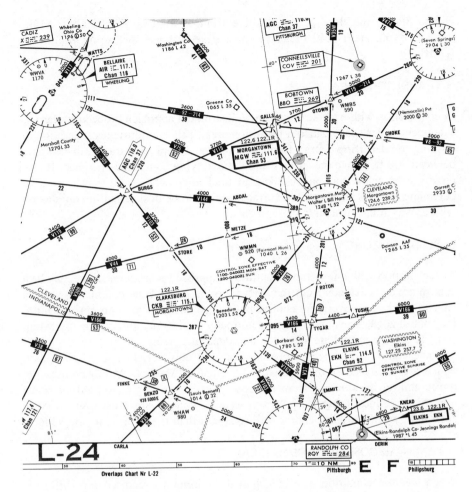

Figure 12.3. Portion of low altitude en-route chart.

others less than 2500 ft. above ground level (AGL), are required to have radar beacon transponders, with altitude encoding.

POSITIVE CONTROL AREA

Certain portions of the air space are under Air Traffic Control at all times. These are positive control areas and route segments. In them, the pilot must operate under IFR at an ATC assigned flight level, and pilot and airplane must be equipped for IFR flight, including altitude encoding transponder and two-way radio permitting direct communications with the controller. Over most of the

United States, the floor of the Positive Control Area is at 18,000 ft. MSL. Aircraft flying at FL 240 or above must have distance measuring equipment (DME).

TRANSITION AREA

In conjunction with each airport with an instrument approach procedure, there is a segment of air space called a transition area. This area provides separation of arriving and departing aircraft from each other and from other traffic, in the space between 700 ft. above the surface and the overlying controlled air space. It also provides air space for maneuvering to final approach(es). A transition area, beginning at 1200 ft. above the surface, may also be designated in conjunction with airway route structures or segments.

CONTROL ZONE

At each airport with one or more instrument approach procedures, there may be a control zone. It extends from the surface to the base of the continental control area. If there is no overlying continental control area, there is no upper limit to the control zone. The control zone is normally 5 miles in radius, with any extensions necessary to include approach and departure paths.

Some control zones are in operation during only certain hours of the day, which are published in the Airman's Information Manual. At other times, air traffic control is exercised from the base of the overlying controlled air space upward. Like most other controlled air spaces, control is exercised in a control zone only when visibility and/or ceiling is less than the minimums for VFR.

TERMINAL CONTROL AREAS (TCA)

When weather conditions are less than the minimums required for VFR, the operation of Terminal Control Areas changes very little. All traffic is still controlled, but of course a pilot operating within the TCA must be instrument qualified and current, and so must his airplane. These requirements do not apply for flight in a Group II TCA under VFR conditions.

UNCONTROLLED AIR SPACE

In some parts of the United States, over the oceans, and in the remote deserts and arctic regions, air space is largely uncontrolled. The airways serve to connect centers of population and aircraft activity. They are limited usually in width to 8 nautical miles. Instrument flight off the airways and outside controlled air space must abide by the usual Instrument Flight Rules, but there is no controlling authority to provide positive separation of traffic. "See and be seen" is not sufficient in clouds. The best protection in this situation is radar following, while maintaining altitude in accordance with magnetic course, safely above terrain and obstructions.

Before entering or crossing controlled air space with weather conditions below the minimums for VFR operation, the pilot must request and obtain an instrument clearance. The clearance given him will provide safe separation from other IFR traffic while he is in controlled air space. The clearance may be obtained by radio before takeoff or in flight, before entering the controlled airspace.

The air space below airways, TCA's, transition zones, and other controlled air spaces is, of course, also uncontrolled. It is therefore possible to do a great deal of flying in very poor weather conditions without a clearance. The pilot is ill advised to do so.

Electronic Aids to Air Traffic Control

Just as important as instruments used to maintain control and fly the airplane are the means of ascertaining and maintaining a desired course when the ground cannot be seen. Celestial navigation is one such means, but one must fly above the clouds to use it, and it is not precise enough to pinpoint position and course. Electronics has thus become the mainstay of aerial navigation, particularly when visibility is severely limited by clouds, fog, darkness, or other atmospheric conditions. This equipment is very important to control of air traffic, for both monitoring it and directing it. And without electronic communications the pesent system would be completely impossible.

RADAR

Radio direction and ranging (radar) was a development of World War II, used for detecting, identifying, directing, and intercepting aircraft and surface vessels. It has developed into one of the most useful tools man has.

Radar sends out, from a rotating antenna, pulses of electromagnetic energy which are reflected off certain kinds of materials. The reflected energy picked up on the rotating antenna is detected by appropriate electronic circuitry and displayed on a cathode ray tube or (oscillo)scope, as a light "blip." Aircraft, surface features, vehicles, and watercraft are good reflectors. So are rain, snow, hail, and other kinds of precipitation. From the scope can be read the bearing and distance of the reflector.

For defense identification and air traffic control, aircraft are equipped with radar transponders. These transmit a pulse of energy in response to each triggering by a radar antenna. This pulse is much stronger and makes a much brighter return on the scope than does the reflected energy. In addition, the response can be coded to provide identification and, very importantly to modern air traffic control, altitude of the responding aircraft. The radar operator thus has position and altitude of the aircraft displayed on his radar screen and can much more easily provide air traffic separation and vectoring.

Air Traffic Control Radar now covers the entire 48 contiguous states at high altitudes, and most of the low altitude route system. Many terminals are equipped with approach control radars, which assist immeasurably in monitoring and directing air traffic in the arrival and departure phases. Some airports also have precision approach radar, which is used to guide the pilot down a very precise final approach path. Military precision approach radar, for instance, allows approaches to minimums of 100 ft. ceiling and ¼ mile visibility.

NAVIGATION RADIO

Radio aids to navigation have developed apace with new concepts, materials, and electronic inventions. The old radio range system of 40 years ago has been replaced by much more precise and dependable equipment, and new ideas and systems continue to come along. These are discussed in detail in Chapter 11, and so will be reviewed only briefly here.

VOR and VORTAC. The old radio range system was subject to interference, beam bending and swinging, and other problems. When it was needed most, in severe weather, static was often so bad as to make it virtually unusable. The solution to most of its problems was the Visual Omni Range (VOR), which operates in the VHF frequency band, virtually unaffected by weather. The US Armed Forces were at the same time developing a system somewhat similar in concept and design, but in the UHF band, called Tactical Air Navigation (TACAN). TACAN provides not only bearing information from the transmitting station, but also distance, by means of interrogation from the aircraft to the ground station and measurement of time of response.

To make the two systems compatible, many VOR stations have been modified to add the same capabilities as TACAN, thus becoming VORTAC's. Each Vortac frequency is paired with a DME frequency, identified by a Channel number. Military aircraft equipped with TACAN obtain both navigation and distance information by tuning the Channel frequency. Civil aircraft obtain navigation information by turning the VORTAC frequency. Those civil aircraft equipped with DME tune the published VORTAC frequency, but actually operate on the paired TACAN frequency, for distance information. VOR stations, as distinguished from VORTAC stations, do not provide any distance information.

The VOR and VORTAC system is used for cross country navigation on or off the airways, and for nonprecision instrument approaches.

ILS. Another development of World War II was a precision Instrument Landing Approach System. It uses two transmitters located at the airport. One transmitter at the far end of the runway from the approach end sends a very narrow beam in the VHF band along the centerline of the runway. This beam is received in the aircraft and affects the positioning of the localizer or vertical needle of the VOR display instrument. When the frequency is tuned and the needle is centered, the aircraft is on the centerline of the runway. Lateral deviations from the centerline cause the needle to go off center.

A paired UHF frequency transmitter beside the runway near the approach end sends out a narrow horizontal beam at a precise angle above the surface in the approach area. It affects the positioning of a horizontal needle or pointer in the aircraft's navigation display instrument. It thus provides a glide slope for the pilot to follow on an instrument approach. When the ILS frequency is tuned and the horizontal and vertical needles or pointers are centered in the navigation display instrument, the aircraft is on a precise path, and by following it makes a precision approach.

ADF. Perhaps the most successful of earlier electronic navigation aids was the Radio Compass, operating on low and intermediate frequency bands, and incorporating Automatic Direction Finding (ADF). This system could use the carrier wave from the Adcock Ranges of the radio range system, as well as those of commercial broadcast radio stations. This equipment has been refined and improved and is now in wider use than ever. It is not subject to line-of-sight restrictions as is the VOR system, so can be used for long range navigation at low altitudes. It also can be used for nonprecision instrument approaches and is the only facility available at many smaller airports. It simply shows the relative bearing of the ground station from the aircraft's nose. Its major limitations are atmospheric interference and lack of height information.

RNAV. By incorporating a computer system in conjunction with VOR and DME receivers, a system of navigation off airways, not requiring the aircraft to pass over the ground stations, has been developed. Instead, waypoints, described by bearings and distances from VORTAC's, are programmed into the system. The course indicator shows the courses from waypoint to waypoint, and the DME measures distance, ground speed, and time to or from the selected waypoint. This system can significantly shorten a flight by providing precise navigation on a straight line. It also provides nonprecision instrument approaches to airports not equipped with electronic navigation aids.

Controlling Agencies

Much of the preceding discussion refers to the governmental agencies which play a part in the regulation and control of flight in the nation's air space. The primary agency is the Federal Aviation Administration, a part of the Department of Transportation. Others are the Civil Aeronautics Board and the National

Transportation Safety Board (NTSB). Still others, such as the National Weather Service, provide support but not regulation.

FEDERAL AVIATION ADMINISTRATION (FAA)

The FAA, headed by an Administrator, is charged by Congress with certification of airmen, aircraft, and airports; developing, publishing, and enforcing the rules of flight, establishing, maintaining, and periodically checking controlled air space, airways, and navigation facilities; controlling traffic in controlled air space when required; and promotion of air safety through all these activities. It also participates in the publishing of various kinds of information for airmen.

Control of air traffic is performed by Air Route Traffic Control Centers, Approach and Departure Control facilities, and Airport Control Towers. Electronic facilities are serviced and maintained by technicians located near or at each facility.

Accuracy and serviceability of electronic aids to navigation are verified at frequent intervals by FAA operated flight check aircraft and crews.

The country is divided into FAA Regions, which are subdivided into General Aviation District Offices. There is usually at least one General Aviation District Office (GADO) for each state.

Each GADO is charged with the certification and approval of flight schools and air taxi operators, examination and certification of airmen, certification of maintenance facilities, surveillance of maintenance of aircraft, promotion of aviation safety through various safety programs, and with assisting the NTSB in the investigation of aviation accidents. Surveillance of agricultural aviation is a little noticed part of all this activity.

Flight Service.   A very important part of the FAA is Flight Service. There are Flight Service Stations, discussed earlier in this chapter, at many airports around the country, generally in conjunction with VOR navigational facilities. These stations provide weather briefing and flight planning services, accept VFR and IFR flight plans, and provide traffic advisory services when their airports do not have control towers. By radio, they provide weather briefings, information on airports and other facilities, and communications links with Air Route Traffic Control Centers. They also provide valuable assistance for pilots who are lost or experincing emergencies, including VHF Direction Finding and radio fixes. They keep current information on the conditions of airports in their respective areas and can vector an aircraft with an emergency to the nearest suitable airport for landing. Flight Service provides assistance of all kinds to aircrews but does not control traffic.

Air Surgeon.   Considering the consequences of physical incapacitation while flying an aircraft, the major safety role played by the FAA's medical standards section is obvious. It has the responsibility of recruiting, training, and supervising FAA Medical Examiners, establishing medical standards for airmen, and deciding whether to issue medical certificates to applicants with certain physical problems. This section also establishes the physical standards that must be met by private, commercial, and airline transport pilots, those for the latter group being the most rigorous.

The exact organization of the FAA and the regulations it must develop to carry out its mission are constantly changing. So, no attempt will be made here to describe its organization and activities in exact detail.

CIVIL AERONAUTICS BOARD (CAB)

The scheduled major airlines had hectic and sometimes chaotic beginnings in the late 1920's and early 1930's. Congress, to bring order to the industry, established the Civil Aeronautics Board. It is charged with certifying and regulating the airlines so as to promote safe, orderly, and efficient air transportation linking the country's 500 or so major commercial centers. The Board also plays an important part in certifying and regulating the commuter airlines and air taxis which serve the smaller cities and towns.

The Board approves routes and rates for the scheduled airlines and monitors commuters and air taxis to assure that they operate safely and are properly insured.

NATIONAL TRANSPORTATION SAFETY BOARD (NTSB)

The national responsibility for maintaining and improving the safety of all forms of transportation rests with a section of the Department of Transportation called the National Transportation Safety Board. It is charged with determining the causes of accidents and searching for and recommending ways of improving safety. Most light aircraft accidents not involving loss of life are investigated by personnel of the GADO's within whose territories they occur. Their findings are forwarded to the NTSB, which publishes the necessary reports. The Board itself investigates all airlines accidents and most of those causing loss of life.

Operating in the System

The crop duster and the rancher out checking his cattle in his 30-year-old "Cub," while they and their aircraft must be certified by the FAA, seldom get involved in controlled air space. They also pose little hazard to other aircraft or to human activities on the ground. So, they are subject to comparatively little regulation: season, weather, and immediate need for the aircraft dictate most of their decisions.

The student training operation comes under substantially more regulatory control because of the interest of the public in adequate training, the safety aspects of flying in the vicinity of other air traffic, and fitting in with such traffic. VFR cross-country flying is a step up the ladder of the degree of regulation, and at the top is all-weather instrument flying.

VFR CROSS COUNTRY

The pilot who flies from point to point under Visual Flight Rules, whether for business, convenience, or pleasure, must obey the rules concerning airworthiness of the airplane and himself, the rules of right of way, altitudes to be flown, traffic in the vicinity of airports, air space restrictions, and the rules concerning ceiling and visibility restrictions. For his own safety he should obtain a briefing from FAA Flight Service, concerning the weather and any restrictions on the facilities he plans to use. He should also file a flight plan, report at intervals along his route, notify Flight Service of any changes, and not forget to close his flight plan when he arrives at destination. For his own VFR flight (unless he is a student pilot), he is his own clearing authority.

INSTRUMENT FLIGHT

The pilot who plans to operate within controlled air space during weather conditions less than the minimums for VFR is bound by the strictest regulations. He and his aircraft must be qualified for the flight to be undertaken, and a flight plan must be prepared and filed with Air Route Traffic Control. He must request clearance and obtain it to assure separation from other aircraft. He may not take off and enter the airways system until he is cleared, must maintain communications with Air Route Traffic Control, and must fly and report positions in accordance with clearances given him. If he wishes to change altitude, route, or destination, he must request and obtain clearance before he does so.

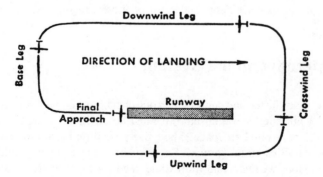

Figure 12.4. Standard traffic pattern terminology. *(From FAA Airman's Information Manual.)*

His departure, route of flight, minimum and cruise altitudes, maneuvering upon arrival, and instrument approach pattern are all spelled out by aeronautical charts and data he must consult before flight and should have in his possession, and/or by ground based traffic control facilities. The ground radar operators and Air Route Traffic Control personnel are in turn bound by regulations, standard operating procedures, and published material. Deviations must be approved before executed, except when dictated by safety considerations.

While closely regulated, full instrument flying with accuracy and precision is the most satisfying demonstration of a professional pilot's knowledge and skill. It is made possible by the combined efforts of countless people, whether they build the airplane, make the rules, man the radars, teach the crew members to fly, maintain the equipment, or sit in one of the crew seats.

13
Helicopters *

A powered, heavier-than-air vehicle which could ascend and descend vertically under precise control was among the earlier dreams of flying. After the successful application of internal combustion engines to airplanes, another 30 years was required to solve the problems of control for a zero speed takeoff and landing machine.

The more recent rapid development of the helicopter as a vehicle for air transportation has been brought about to a large degree by its usefulness to military forces. As has happened many times before, there has been a valuable fall out of civil benefits from military developments, and helicopters have become one of the most valuable tools of modern civilization. They have become indispensible in many fields, among them traffic control, emergency rescue, offshore oil drilling, setting up power lines, logging, transportation to remote areas, forest fire fighting, agricultural spraying and seeding, and many others. They come in many different sizes and models, with a large variety of special equipment, both determined by the purposes for which they are to be used.

Rotary-Wing Aerodynamics

The aerodynamics of rotary-wing aircraft and airplanes are basically the same. Both use airfoils to produce lift; both are subject to identical fundamental forces of lift, drag, thrust, and gravity. Because flight characteristics of the helicopter differ widely from those of the airplane, helicopter theory of flight requires separate discussion.

LIFT

The helicopter's airfoils are rotor blades which are turned at high speed. We have seen that for a given speed, an increased angle of attack increases lift until the stalling angle is reached; and for a given angle of attack, the greater the speed, the greater the lift. The helicopter obtains high speed airflow over the airfoils by high rotor rpm, rather than by high fuselage speed. In fact, it is normal for the tip speed of the rotor blades to be as much as 435 knots when the speed of the fuselage is zero. This explains why the helicopter does not require forward speed to produce lift, and why it can hover or fly backward, sideward, or forward.

AIRFLOW

During normal operating conditions, the direction of airflow is from the top down through the main rotor system. As the blades are rotated with a positive angle of attack, they accelerate air downward; thus, a downwash of air (Figure 13.1) is established through the rotor system. Notice that the leading edge of each blade bites into air throughout the complete cycle of rotation, producing a uniform downward flow of air.

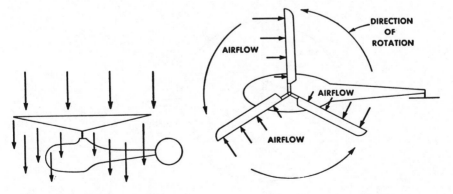

Figure 13.1 Downwash through the rotor system.

Figure 13.2. Airflow in the rotor system.

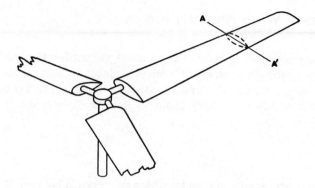

Figure 13.3. High-speed blade section.

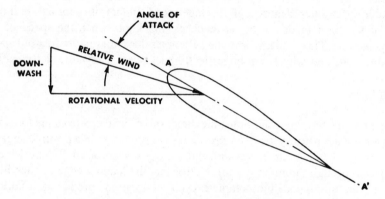

Figure 13.4. Relative wind components.

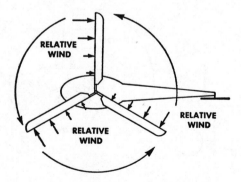

Figure 13.5. Direction of relative wind.

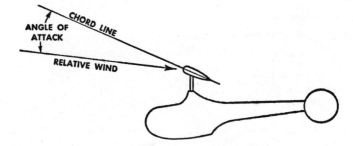

Figure 13.6. Rotor blade angle of attack.

At the root of the blade, airflow is slightly more than zero, but the velocity progressively increases throughout the length of the blades and at the tip may be 435 knots or higher. It is the blade velocity that determines the resultant strength and direction of the relative wind at a positive angle of attack. The helicopter changes the angle of attack by varying the pitch of the main rotor blades. The relative wind is developed throughout the complete cycle of 360° by rotation of the rotor system, and it usually varies considerably. This variation is dependent upon flight conditions.

ANGLE OF INCIDENCE

The angle of incidence is the angle formed by the chord of the airfoil and the longitudinal axis of the aircraft. The conventional airplane's angle of incidence is built into the aircraft by the designer and in most aircraft cannot be changed. The helicopter pilot, however, continually changes the angle of incidence during flight by increasing or decreasing the pitch of the main rotor blades.

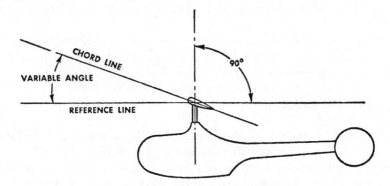

Figure 13.7. Rotor blade angle of incidence.

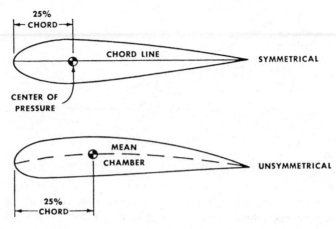

Figure 13.8. Airfoils.

AIRFOIL SECTION

The type of wing used on conventional airplanes varies considerably; the airfoils may be symmetrical or unsymmetrical, usually depending upon some specific requirement.

Historically, symmetrical airfoil sections have been used for rotor blades, primarily to restrict center-of-pressure travel and to permit easier blade construction. Increased testing of rotor airfoils by NASA and industry has resulted in use of unsymmetrical airfoils.

Many newer aircraft use "droop-snoot" blades with cambered leading edges and reflex curved trailing edges. These airfoils overcome center-of-pressure problems. They also give an increased stall angle of attack with generally improved stall characteristics, and improved hover and cruise efficiency. They postpone high-speed advancing blade drag and result in negligible pitching movements.

Both airfoils have a good lift-drag ratio throughout a wide range of velocities and spread lift forces over a wide area to equalize stresses. Usually a slight twist is built into the blade to help equalize these forces.

THRUST AND DRAG

As weight and lift are closely associated, so are thrust and drag. Thrust moves the helicopter in a designated direction and drag tends to hold it back.

The helicopter develops both lift and thrust in the main rotor system. In vertical ascent, thrust acts upward in a vertical direction; drag and weight, the opposing forces, act vertically downward. Lift sustains the weight of the helicopter, and excess thrust is available to give translation or vertical acceleration. During

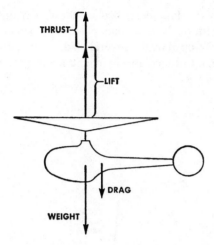

Figure 13.9. Forces in vertical ascent.

vertical ascent, drag is considerably increased by the downwash of the main rotor system striking the fuselage. Thrust must be sufficient to overcome both drag and downwash. The force representing the total reaction of the airfoils with the air is divided into two components: one is lift, the other thrust. However, drag is a separate force from weight, as indicated in Figure 13.9.

At all times the lift forces of the rotor system are perpendicular to the tip-path

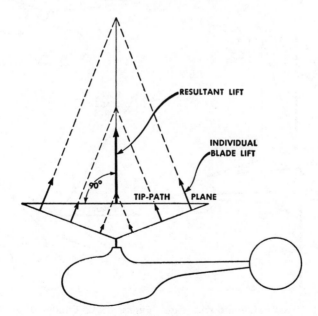

Figure 13.10. Direction of resultant lift.

plane. The tip-path plane is the imaginary plane described by the tips of the blades in making a cycle of rotation. The lift on the individual blade is perpendicular to the airfoil, but the resultant lift developed by the several blades is perpendicular to the tip-path plane (Figure 13.10). Lift increases in magnitude from root to tip of blade because of increase in velocity.

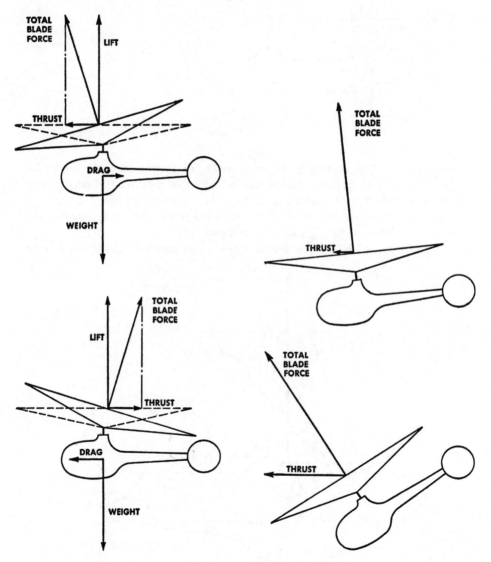

Figure 13.11 Forces in forward and rearward flight.

Figure 13.12 Effects of slow and high speed.

In vertical flight the tip-path plane is horizontal; and in forward, backward, or sideward flight, the plane of rotation is tilted off the horizontal, thus inducing a horizontal thrust vector in the direction of inclination. For example, to establish forward flight, resultant lift is inclined forward (Figure 13.11). Total force, being tilted off the vertical, acts both upward and forward; therefore it can be resolved into two components. One component is lift, and the other is thrust. Likewise, flight may be established sideward, or in any horizontal direction, by tilting the tip-path plane in the direction of desired flight. The rate of movement or speed depends upon the degree of tilt of the resultant lift force. Note the magnitude of thrust at the two speeds shown in Figure 13.12.

TORQUE

Torque effect is displayed in a helicopter by the turning of the fuselage in the opposite direction to the rotation of the main rotor system. This reaction is in accord with Newton's third law of motion, which states, "To every action there is an equal and opposite reaction." The engine is the initiating force that drives the rotor system in a counterclockwise direction, and the reaction to this driving force would cause the fuselage of the helicopter to rotate with an equal force in a clockwise direction (Figure 13.13). Torque is of real concern to both the designer and the pilot. Adequate means must be provided not only to counteract torque, but also to exert positive control over its effect during flight.

The designers of helicopters employ several methods of compensating for torque reaction. The tandem-rotor type helicopter turns the two main rotor systems in opposite directions, thus counteracting the torque effect of one rotor by the torque effect of the other. The coaxial configuration likewise turns its rotors in opposite directions to equalize the torque effect. In jet helicopters, if the engines are mounted on the tips of the rotor blades, no torque reaction is transmitted to the fuselage because the reaction is directly between the blade and the air. In the single main rotor helicopter, torque is usually counteracted by a ver-

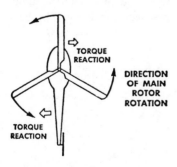

Figure 13.13. Torque reaction.

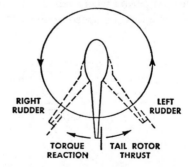

Figure 13.14. Torque correction.

tically mounted tail rotor which is located on the outboard end of the tail-boom extension (Figure 13.14). The tail rotor develops horizontal thrust that opposes the torque reaction. The pilot can vary the amount of horizontal thrust by activating foot pedals which are linked to a pitch-changing mechanism in the tail rotor system.

GROUND EFFECT

The high power cost in hovering is somewhat relieved when operating in ground effect. Ground effect is defined as the condition of improved performance encountered when hovering near the ground at a height no more than approximately one-half the rotor diameter. It becomes more pronounced as the ground is approached. Improved lift and airfoil efficiency while operating in ground effect are due to two separate and distinct phenomena: (1) First and most important, the rotor tip vortex is reduced due to the downward and outward airflow pattern. (2) Downwash angle is reduced. This reduces induced drag and permits lower angles of attack and decreased power requirements for the same lift (Figure 13.15).

TRANSLATION LIFT

Translation lift is the additional lift developed by a helicopter in horizontal flight. This additional lift becomes noticeably effective at an airspeed of 10 to 20 knots, and it continues to increase in magnitude as speed is increased. As horizontal flight is progressively induced, a higher inflow of air is established through the rotor disc, and greater lift is produced because of increased rotor efficiency. However, when a speed of 45 to 50 knots is reached, translational lift is canceled out by fuselage drag. Figure 13.16 illustrates a typical power curve indicating percentage of power required compared to translation velocity.

When hovering 3 to 8 ft. above the ground in a windless condition, the helicopter is aided by ground effect. But when the helicopter enters forward flight, it leaves ground effect, in effect sliding off a cushion and entering into a critical transition period where reduced lift initially causes the helicopter to settle.

When a forward speed of about 15 knots is reached, translational lift becomes

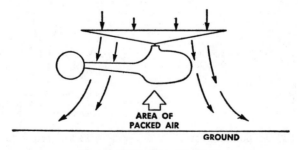

AREA OF
PACKED AIR

GROUND

Figure 13.15. Ground effect.

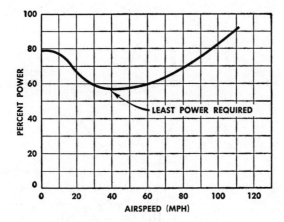

Figure 13.16. A typical power curve.

effective and the helicopter will gradually climb. As forward speed is increased, lift is increased, and less power will be required to maintain straight and level flight (Figure 13.16).

Under actual flight conditions, during the transition from 0 to 15 knots, the pilot obtains additional lift to avoid settling during the critical transition period, by increasing the pitch of the main rotor blades. When hovering in wind conditions of 10 knots or more, ground effect may not be present. The helicopter is in translational lift for all practical purposes, and forward flight is established without the settling effect.

DISSYMMETRY OF LIFT

Dissymmetry of lift is the unequal lift that develops between the advancing half of the rotor disc area and the retreating half of the disc area during horizontal flight.

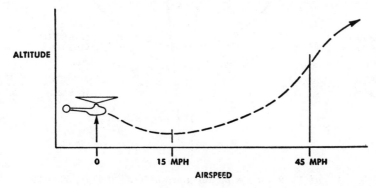

Figure 13.17. Translational lift.

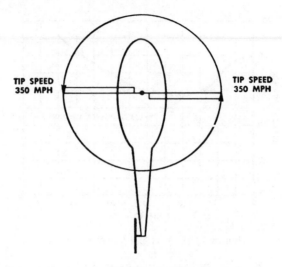

Figure 13.18. Hovering (zero airspeed).

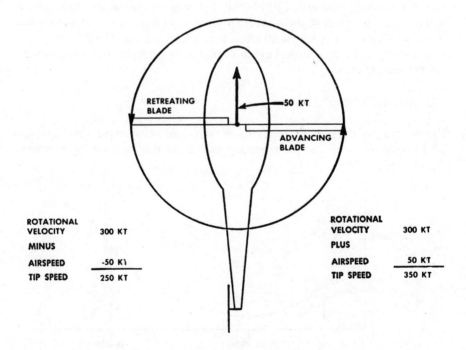

Figure 13.19. Differential tip speed in forward flight, 50 knots.

When the helicopter is hovering in a no-wind condition, the tip speed of about 435 knots, depending upon the type of helicopter, is constant throughout the complete 360° cycle, with a constant angle of attack on the blades. This creates equal lift and equal airflow velocity throughout the disc area.

When the helicopter enters horizontal flight, there will be a difference in tip airspeed between the advancing and retreating halves of the disc area (Figure 13.19). The forward speed is added to the rotational velocity on the advancing side of the disc area, and subtracted on the retreating side. Forward flight of 50 knots, therefore, would mean a differential of 100 knots between the advancing and retreating sides. This condition, if uncorrected, would develop unequal lift and the helicopter would turn over.

To counter this dissymmetry of lift, it is standard practice in rotor-head design to incorporate a flapping hinge, a device which permits the rotor blade to flap upward, or to design the blades to flex in the vertical direction. Under normal operational conditions, the high-speed rotation of the rotor system develops a centrifugal force of approximately 20,000 lb. on each blade. Centrifugal force holds the blades in a horizontal plane, and lift causes the blades to rise vertically. The rotor blades will take a resultant position between the effects of centrifugal force and lift and appear in a coned-up attitude (Figure 13.20).

Centrifugal force will be constant throughout the complete cycle of 360°. Lift will vary between the advancing and retreating portions of the disc area in forward flight. The advancing blade, with greater relative wind velocity, will lift more. This effect is reduced by the reduced angle of attack of the advancing blade due to forward tilt of the disc. Also, in developing greater lift, the advancing blade, hinged at the hub, is free to flap at a greater angle. In doing so, its effective span is reduced, reducing the effective lift area. Conversely, the retreating blade will have a greater angle of attack and more effective span and lift area. Thus the lift forces generated by advancing and retreating blades balance out.

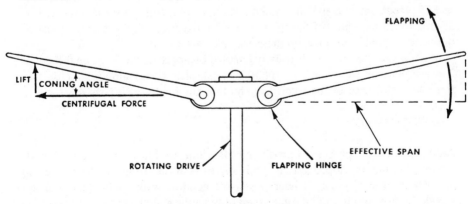

Figure 13.20. Flapping hinge device.

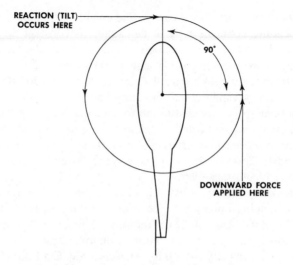

Figure 13.21. Gyroscopic precession.

GYROSCOPIC PRECESSION

Gyroscopic precession is the innate quality of all rotating bodies by which application of a force perpendicular to the plane of rotation will produce a maximum displacement of the plane approximately 90° later in the cycle of rotation (Figure 13.21).

Thus if a downward force were applied to the right side of a rotating disc, gyroscopic precession would cause the plane of the disc to tilt to the front, providing that the disc is turning from right to left. Maximum resulting displacement occurs about 90° further in the direction of turning, but speed of rotation, weight and diameter of the disc, and friction are factors which determine the actual displacement in a specific system. A main rotor follows these laws. If it is desired to tilt the plane of the rotor forward for forward flight, linkage must be provided to apply the force on the right side of the disc.

In helicopter design, this is usually handled by applying the force on the rim of a swash plate on which another rotating swash plate rides, applying the desired force to each sector of the rotor disc and thus the blade, as it passes.

AUTOROTATION

Autorotation is the process of producing lift with rotor blades that are kept rotating only by the developed aerodynamic forces resulting from the flow of air up through the rotor system. Under power-off conditions the helicopter will descend; the flow of air will be established upward through the blades. The rotor is

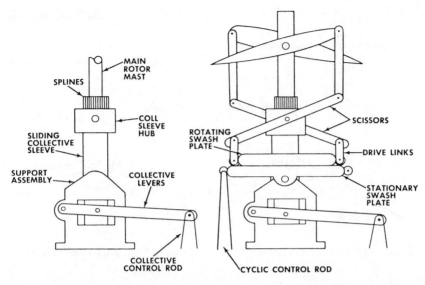

Figure 13.22. Cyclic and collective control head. Cyclic inputs tilt the swash plate, and as described in the text, the tip-path plane. Collective inputs act through the hub, with necessary flexibility provided by the drive links, to change the pitch of both blades equally, simultaneously, and in the same direction. This changes overall rotor lift.

automatically disengaged from the engine by a free-wheeling device, and the necessary power required to overcome parasite drag and induced drag of the rotor blades is obtained from the change of potential to kinetic energy produced by the helicopter's weight in descent.

During autorotation the pitch angle of the rotor blades must be reduced to minimum. The change in direction of airflow through the rotor causes a change in the direction of the relative wind which greatly increases the angle of attack at which the rotor blades are operating. If the pitch were not reduced, the blade would stall for much the same reason that an airplane's wing stalls when the nose of the airplane is pulled up too high. When the pitch angle of the blade is low and the angle of attack is small, the resultant lift force lies ahead of the axis of rotation of the blades, tending to keep the blades turning in their normal direction (Figures 13.23 and 13.24).

If, on the other hand, the pitch angle remains high, drag is increased and the resultant lift force lies behind the axis of rotation, tending to slow and stop the rotor. Autorotation is an emergency procedure that permits the pilot to land the helicopter safely in case of engine failure. It is necessary to maintain sufficient rotor rpm to provide both adequate airflow over the rotor blades and the required centrifugal force to hold the blades in an extended attitude; otherwise the blades would fold up and the helicopter would tumble out of control.

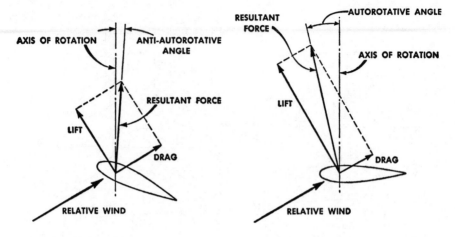

Figure 13.23. High pitch angle.   Figure 13.24. Low pitch angle.

PENDULAR ACTION

The fuselage of the helicopter is suspended from the drive shaft that mounts the main rotor head. Because the fuselage is bulky and suspended from a single point of attachment, it is free to oscillate laterally and longitudinally much like a freely swinging pendulum. As the rotor system introduces horizontal translation, the fuselage is dragged in the direction of flight. During established forward flight, the fuselage will assume a nose-low attitude. In effect, as the tip-path plane is inclined forward, resultant lift is inclined from the vertical, introducing thrust. The main drive shaft of the helicopter will have a tendency to align itself with the inclined resultant lift force (Figure 13.25).

The pendular action of the fuselage swaying is exaggerated by overcontrolling; therefore, control stick movements should be moderate.

RESONANCE

Resonance is the energetic vibration of a body produced by application of a periodic force of nearly the same frequency as that of the free vibration of the affected body; also, it is the condition of two bodies adjusted to have the same frequency of vibration.

The helicopter is subject to two types of resonance: sympathetic resonance and ground resonance. *Sympathetic resonance* is a harmonic beat which develops when the natural vibration frequency of one mechanism is in phase with another vibration. In the case of the helicopter, the rotor system has its own or some other

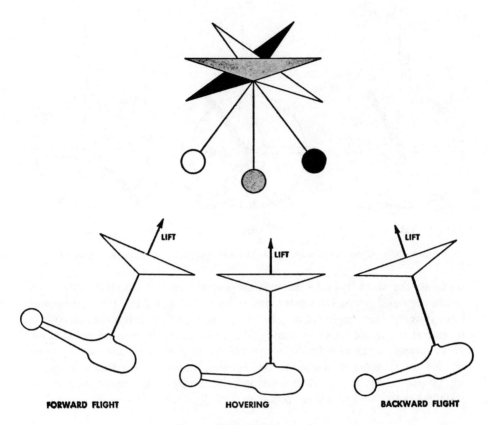

Figure 13.25. Pendular action.

mechanism's. Sympathetic resonance, however, has been successfully engineered out of most helicopters by controlling design features of gear boxes and other mechanisms.

Ground resonance, on the other hand, has persisted to plague helicopter designers, particularly during experimental stages. It is a self-excited vibration which develops when the landing gear repeatedly strikes the ground, thus unseating the center of mass of the main rotor system. The pounding effect of the loading gear is prone to occur during take-off and landing when the helicopter is 87% to 93% airborne. The aircraft, being light on the landing gear, bounces from one wheel to the other in rapid succession, setting up a pendular oscillation of the fuselage.

The succession of shocks is transmitted to the main rotor system, and the main rotor blades straddling the pounding wheel are forced to change their angular relationship. This condition unbalances the main rotor system, which in turn

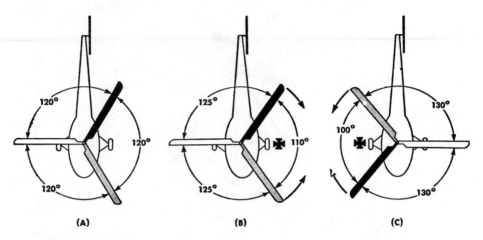

Figure 13.26. Ground resonance. (A) Normal; (B) and (C) successive shocks.

transmits the shock back to the landing gear (Figure 13.26). If this cycle of events continues, complete destruction is likely to occur. The fully articulated rotor system which employs a drag hinge for each rotor blade is more susceptible to ground resonance than the semi-rigid type of rotor. Present-day design employs various dampening devices to control the unbalancing of the rotor system, and helicopter pilots are trained to avoid critical maneuvers conducive to agitating ground resonance. Tandem rotor and skid landing gear helicopters do not usually have ground resonance problems.

WEIGHT AND BALANCE

The loading of all single-rotor helicopters is critical because of their lack of inherent stability. All lift is applied to the fuselage at a single point, the rotor mast; the fuselage is free to swing from this point like a pendulum. Loading forward of the center of gravity will introduce a nose-low attitude. When the fuselage tilts, it tilts the plane of rotation of the main rotor system. The degree of permissible tilt is dependent upon the amount of cyclic stick travel rigged into the controls of the particular helicopter.

If the nose-low attitude assumed by the fuselage, owing to faulty loading, is in excess of the rearward travel of the cyclic stick, the pilot is not able to stop the forward flight of the helicopter (Figure 13.27). It would be even more serious if faulty loading introduced uncontrollable rearward flight. Therefore the pilot must exercise care in loading the helicopter with respect to the weight of pilot, passengers, fuel, and cargo. Small helicopters are apt to be very critical as far as loading is concerned. However, in large helicopters the cargo compartment is usually located directly beneath the center of gravity, which facilitates loading. In the

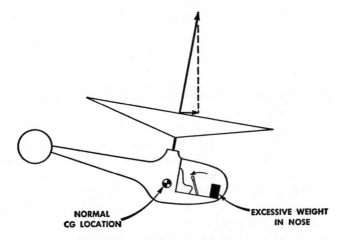

Figure 13.27. Excessive forward loading.

tandem rotor helicopters, loading normally is not as serious a problem because of the wide center-of-gravity range.

Helicopter Flying

The airplane pilot usually receives some surprises in learning to fly the helicopter. The controls are different in many ways from those of the airplane; and FAA check-out requires 25 hours of flight training. The airplane pilot is accustomed to changing from one type of airplane to another with a minimum of effort; but in a helicopter he must learn new and different control techniques peculiar to the rotary-wing aircraft. He must discard his valuable habits pertaining to slow and low flying.

He will probably overcontrol because of the control lag which is dominant in all helicopters. However, helicopter flying is not totally unrelated to airplane flying. The basic principles of flight are unchanged. Air experience and judgment are applicable in all types of flying, and understanding weather, navigation, instrument interpretation, radio, and related subjects is the same.

The marked difference is in high maneuverability, particularly at low speeds, beginning with the vertical take-off and departure in any direction, regardless of fuselage heading, from a standstill to 300 knots. But the helicopter is inherently unstable. Almost continuous control pressures are required to maintain position in flight, and to hold the aircraft in deft equilibrium about a given point, the center of gravity. Utmost dexterity is required, particularly while hovering. Many pilot trainees have compared hovering to balancing the aircraft on the point of a pencil. Coordinated control is required at all times about the roll, pitch, and yaw axes, if the fuselage attitude is to be held constant.

Power System

While reciprocating aircraft engines were adapted to early helicopters, many helicopters now use turbine engines.

These turbines are designed generally on the "free-turbine" principle. That is, the engine consists of two separate rotors, usually mounted coaxially. One is the *gas generator,* and the other is the *power turbine.* The latter is connected by a transmission and shafts to the main and tail rotors. The power turbine, whose speed is reflected in the pilot's power instruments, operates at constant speed. Power changes are automatically or manually made by changing the fuel-air ratio for the gas generator and thus its output.

In this way changes in load on the power turbine are met by an increase or decrease in the velocity of high-speed gasses from the gas generator through the blades of the power turbine. Because only the gas generator rotor mass must accelerate with throttle changes, the system is highly responsive to needed changes in power.

Flight Controls

The flight controls used in the single-rotor helicopter are the *cyclic stick,* the *collective-pitch stick* with its engine throttle handgrip, and the antitorque pedals.

THE CYCLIC STICK

The cyclic stick controls the horizontal direction of flight and the speed of the helicopter. It is connected by linkage to a control or "swash plate" (Figure 13.22), which in turn is connected to the main rotor blades. When the swash plate is level, pitch on the main rotor blades will be equal throughout the cycle of rotation; but if the swash plate is tilted, the pitch on the main rotor blades will vary throughout the cycle, proportional to the tilt of the swash plate.

Forward stick will cause the swash plate to tilt forward; the main rotor blade pitch will increase on the retreating blade and decrease the same number of degrees on the advancing blade. The equal but opposite change of pitch at points 180° apart will cause the tip path plane to tilt forward.

Resultant lift is always perpendicular to the tip path plane. When the tip path plane is tilted forward, resultant lift is inclined forward, and thrust is developed in the direction of tip path plane tilt. The rotor system will move rapidly in the

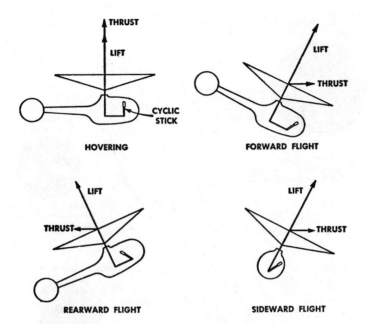

Figure 13.28. Directions of horizontal flight governed by cyclic stick.

direction of thrust and will drag the fuselage in that direction. The fuselage will pitch about the lateral axis, assuming a nose low attitude. The swash plate is free to tilt in the direction of movement of the cyclic stick—forward, backward, and sideward, through all 360°. As the cyclic stick is moved in any direction, the swash plate will tilt in that direction, thus establishing tip path plane inclination in the direction of cyclic stick movement. The directional speed of the helicopter is controlled by the degree of tilt of the swash plate. Figure 13.28 indicates the direction of flight in relation to cyclic stick movements.

THE COLLECTIVE PITCH STICK

This control regulates the pitch setting of the main rotor blades by a collective sleeve which passes through the swash plate on the rotor head. As seen in Figure 13.22, collective inputs are made through the scissors linkage to change blade pitch equally, simultaneously, and in the same direction.

The collective pitch stick is a lever with up and down travel located to the pilot's left and manipulated by his left hand. In conjunction with the collective-pitch stick movement, there is a synchronizing mechanism which automatically changes engine power output proportionate to the change in pitch. As collective pitch is increased, the engine is required to develop more power in order to maintain constant main rotor rpm. Almost all helicopters have a throttle to override

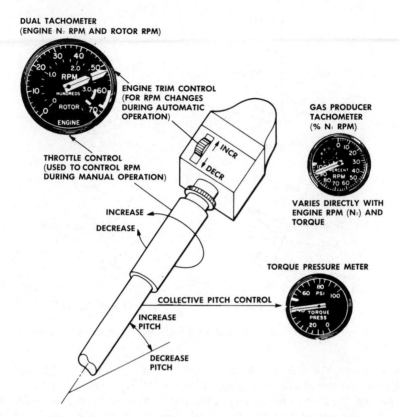

Figure 13.29. Rotor blade pitch and engine power coordination, collective stick.

the synchronization mechanism during emergencies. The throttle, or an additional electrical trim control, is also used to make fine power adjustments.

In order to climb, the main rotor blade pitch is increased; to descend, pitch is reduced. The rate of climb or descent is dependent upon the degree of movement of the collective pitch stick. Acceleration and level flight airspeed are also controlled by the collective-pitch setting.

THE ANTITORQUE PEDALS

The pedals control fuselage heading by changing tail rotor blade pitch. The primary purpose of the tail rotor is to compensate for torque, but fuselage heading is maintained by increasing or decreasing the horizontal thrust of the tail rotor. It is normal for the single main rotor to turn from right to left (viewed from the pilot's position), and torque would turn the nose of the fuselage to the right. The application of the left pedal increases the pitch on the tail rotor which

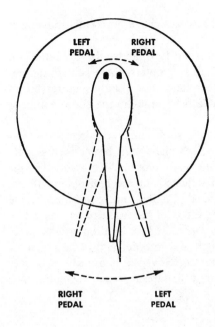

LEFT
PEDAL

RIGHT
PEDAL

RIGHT
PEDAL

LEFT
PEDAL

Figure 13.30. Fuselage heading control.

increases its sideward thrust to the right, thus establishing fuselage heading (Figure 13.30).

Flight Maneuvers

Helicopter flight maneuvers such as hovering, normal take-off, forward flight, climb, descent, and landings are basic. Advanced maneuvers include quick stops, backward and sideward flight, 360° hovering turns, maximum-performance takeoffs, running takeoffs, running landings, and steep approaches. The split second coordination of controls required in the performance of the advanced maneuvers is gained by experience; therefore no attempt will be made to discuss the advanced techniques. Understanding the basic maneuvers makes plain the basic principles of helicopter flight.

HOVERING

Hovering implies zero airspeed with a constant attitude and heading. It is sustained motionless flight requiring a high degree of coordination and concentration. The right hand controls the cyclic stick, the left hand manipulates the collective pitch stick and motorcycle type throttle, and the feet use the pedals to compensate for torque. At the same time the pilot must observe engine rpm, be cognizant of height above ground, and also anticipate horizontal movement of the aircraft.

Hovering is normally performed between 3 and 10 ft. above the ground. Since there is considerable lag in the controls of all helicopters, there is a strong tendency for the trainee to overcontrol. The airplane pilot usually requires 5 to 8 hours of flight training to achieve reasonable proficiency in the coordination of all controls essential for hovering.

The tip path plane, controlled by the cyclic stick, must be horizontal in no-wind conditions; if wind is present, the tip path plane must be inclined slightly into the wind to prevent drift. The pitch of the main rotor blade, controlled by the collective pitch stick, must be accurately adjusted, or the helicopter will not remain at a constant height above the ground. Engine rpm, controlled by the throttle, must be held constant. The turning effect of the fuselage introduced by torque reaction to engine power must be compensated for by proper pedal pressure. The coordination of cyclic stick, pitch and throttle, and pedals requires more than an average amount of concentration. Hovering is a basic maneuver because forward flight is started from the hover, and the approach to landing ends in the hover.

NORMAL TAKEOFF

The normal takeoff (Figure 13.31) employs a combination of two basic maneuvers: hovering, and climb with forward speed. Takeoffs should normally be made into the wind. To hold the desired heading, it helps to select a reference

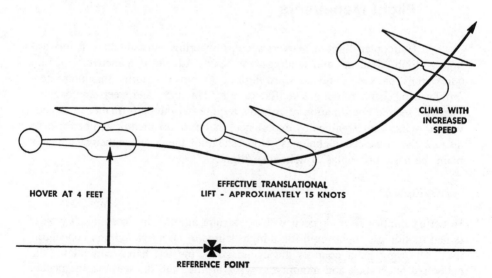

CLIMB WITH
INCREASED
SPEED

HOVER AT 4 FEET

EFFECTIVE TRANSLATIONAL
LIFT - APPROXIMATELY 15 KNOTS

REFERENCE POINT

Figure 13.31. Normal take-off.

point in front of the helicopter; if the nose deviates from that line of vision, appropriate pedal pressure will correct the change in heading.

Once the hovering attitude is established, apply forward pressure to the cyclic stick. As the helicopter moves forward, it has a tendency to settle. This loss of altitude occurs because the vertical lift vector is inclined forward, thus dividing total lift into two components: lift and thrust. To prevent settling, increase collective pitch, adding throttle to hold rpm, and apply additional left pedal to maintain heading. The coordination of cyclic stick, collective pitch stick, throttle, and foot pedals is essential to establish safe forward flight from the hover.

When the helicopter reaches a forward speed of approximately 15 knots, translational lift becomes noticeably effective. This is an additional lift that is developed because the velocity of airflow through the rotor system is more effective. Translational lift continues to increase with forward speed, but it is nullified at approximately 50 knots by parasite drag.

CLIMBS AND DESCENTS

To climb, raise the collective-pitch stick. This increases the pitch simultaneously on all the main rotor blades. As blade pitch increases, add throttle to provide the additional power needed to maintain constant rotor rpm. The synchronizing unit is activated by collective-pitch stick movement. If more power is required to hold rotor speed, coordinate necessary additional throttle with pitch. While climbing it is necessary to increase left pedal pressure in order to compensate for torque; otherwise the fuselage would turn to the right.

To descend, use the reverse of the climb procedure. Control airspeed throughout all maneuvers by the cyclic stick. Pitch and throttle coordination, particularly during climbs and descents, is the most difficult technique to master. Also, any change in power requires a coordinated change in collective pitch, throttle, and pedals.

STRAIGHT AND LEVEL FLIGHT

During straight and level flight, speed is determined by composite use of cyclic and collective stick. Altitude is controlled by the collective pitch stick and throttle. The foot pedals correct yaw trim and compensate for power adjustments. Streamlining of the helicopter in forward flight decreases the pedal input needed to correct for torque. If a constant airspeed is established, increasing pitch with a coordinated adjustment of throttle will result in a climb; decreasing pitch will cause the helicopter to descend.

The cyclic control is sensitive, but response to cyclic stick movement will lag, thus delaying control corrections for attitude and speed. If the nose of the helicopter should rise, airspeed would fall off; a slight forward pressure on the cyclic stick will bring the nose down. However, if this correction is held too long, or if it is a large one, the nose of the helicopter will continue to drop beyond the desired attitude. Neutralize cyclic stick corrections before reaching the desired attitude, or overcontrolling will result. The helicopter pilot must learn to anticipate the continued response to cyclic stick corrections even after neutralizing the control. If attitude is displaced by gust or turbulence, the helicopter may not return to straight and level flight as will an airplane. Therefore, maintain corrective control pressures at all times to maintain attitude.

NORMAL LANDING

The normal landing is performed by establishing an approach to the hovering attitude, followed by the vertical let down to ground contact. During the approach, maintain a constant ground speed and establish a constant glide angle. The cyclic control regulates speed, and the collective pitch control regulates rate of descent. On entering the approach, apply slight backward pressure to the cyclic stick to slow the speed, and decrease collective pitch to set up rate of descent and glide angle. Just before establishing the hover, relax the back pressure on the cyclic control so that the helicopter will assume a level attitude. Increase collective pitch to prevent loss of altitude, use left pedal to maintain heading, and use more throttle to hold rpm. From the hover, make the vertical let down by reducing collective pitch on the main rotor blades.

RECOURSE TO AUTOROTATION

Autorotation is an emergency procedure, used to make a landing in the event of loss of power, or in any other emergency when use of the engine might endanger the landing. During autorotation no torque is developed, for the only reaction is that directly between the main rotor blades and air. Should a tail rotor fail, for example, continued engine torque would cause the fuselage to turn in the opposite direction to the rotation of the main rotor blades. By cutting power and using autorotation, the helicopter pilot can prevent the turning of the fuselage and land in autorotation.

To establish autorotation, reduce the pitch angle of the rotor blades to minimum, for the reasons discussed on page 767. Place the collective pitch stick in the minimum pitch position. Set up a flight path into the wind, and establish a descending glide at best power off glide speed (approximately 60 knots). Use right pedal as needed to neutralize tail rotor horizontal thrust, since no torque is developed by the freely turning rotor system. Control attitude during the descent by manipulating the cyclic pitch control. Rotor rpm will be about the same as when the rotors are driven by the engine, or slightly higher. As the helicopter approaches the ground, slow the forward speed by the cyclic stick, and add sufficient collective pitch to slow the rate of descent. The simplest autorotative touchdown is the running landing type, wherein the craft is permitted to contact

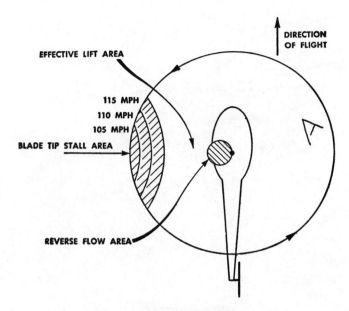

DIRECTION OF FLIGHT

EFFECTIVE LIFT AREA

115 MPH
110 MPH
105 MPH

BLADE TIP STALL AREA

REVERSE FLOW AREA

Figure 13.32. High speed blade stall.

the ground in a level attitude with a forward speed of about 15 knots. If the landing terrain is rough, use a flare type autorotation, with minimum ground run.

RETREATING BLADE STALL

Upon entry into blade stall, the first effect is generally a noticeable vibration of the helicopter, followed by a pitchup of the nose and a rolling tendency. The condition normally happens when operating at high speeds in conjunction with high blade loading, high density altitude, low rotor rpm, steep turns, or turbulent air and can best be corrected by reducing speed and bank and collective pitch, and by increasing rotor rpm.

SETTLING WITH POWER

Three conditions are necessary to produce settling with power. They are (1) airspeed below translational lift airspeed, (2) a high rate of descent, and (3) power applied. To recover from this flight condition, increase directional airspeed, reduce power, and obtain translational lift.

EXTERNAL LOADS

One great advantage of rotary wing aircraft is the capability to haul external loads. Two basic types of external load suspension systems are used: single-

Figure 13.33. Sikorsky CH-54 "Flying Crane" is designed for transport of large, bulky loads such as this pontoon bridge. A gross weight of 42,000 lb. permits a useful load of 20,170 lb. Its speed is 90 knots, endurance 1.5 hours plus 30 min. reserve, and its Pratt & Whitney JFTD 12A-5A engines are rated at 4800 shp. Detachable cabin carries 33 fully equipped infantrymen. *(US Army photograph.)*

point suspension, where the cargo hook is located near the center of gravity, free to swing fore and aft and laterally about a fixed point, and multipoint suspension through the use of load levelers which lift the load clear of the ground, dampen oscillations, and level the load.

Nylon or cable slings with clevises are used to attach the load to the cargo hook or attaching points. In some single point systems, the hook is free to swivel, has spring centering devices to limit cargo swing, or is hoist-mounted.

The same pendular actions that applied between the fuselage and the main rotor are present between the load and the fuselage. In addition, the slings tend to act like springs and can cause vertical oscillations that will make the helicopter bounce. Therefore, move the controls smoothly and avoid excessive speeds and speed changes. External cargo generally requires less loading time than internal cargo, is independent of fuselage dimensions, and allows the pilot to jettison the load in emergencies. Present day cargo helicopters are capable of hauling loads in excess of 20,000 lb.

OPERATIONAL LIMITATIONS

The performance limitations of a helicopter are determined primarily by the ratio of gross weight to available power. Many times the helicopter must operate only partially loaded in order to perform a particular mission. For the helicopter to be a useful aircraft, it must have both load carrying ability and range. Like any other aircraft, payload must sometimes be sacrificed for range, or vice versa, to reach a safe compromise.

Operational limitations are ultimately governed by atmospheric conditions, because helicopter performance is very sensitive to changes in air density. A decrease in atmospheric density brought about by an increase in altitude, decrease in barometric pressure, increase in temperature, or increase in humidity can prohibit hovering, as well as vertical takeoffs and landings.

Under such conditions the helicopter will have a certain minimum flying speed similar to the stalling speed of an airplane. A surface wind, sufficient to produce this minimum flying speed, will permit operation at zero ground speed. While increasing the minimum flying speed of the helicopter, this same decrease in atmospheric density will reduce the maximum safe flying speed, thus narrowing the range of safe operating speeds. A knowledge of these effects is essential to a rescue helicopter pilot who must frequently hover at altitude.

At the opposite extreme in atmospheric conditions, the performance of the helicopter is greatly increased by increased air density. This is not altogether a good effect, however, because it may give a false impression of the helicopter's load carrying potential. For example, a helicopter flying in very dense air might be able to hover, take off, and land vertically and have very good flight characteristics, while actually being dangerously overloaded from a structural standpoint. The helicopter pilot must consider both operational and structural limitations, and the effects of the load and existing atmospheric conditions on each, in planning a particular mission.

The Rigid Rotor Helicopter

Figure 13.20 illustrates the principle of the flapping hinge rotor which was a brilliant breakthrough in Cierva's autogyro design. Though then essential, this hinge caused instability, limited forward speed, lagging control response, and severely limited allowable center of gravity movement.

The rigid rotor concept uses the principle of gyroscopic precession 90° in the direction of rotation (Figure 13.21). A rigid rotor, with blades cantilevered from the rotor hub with freedom to rotate only about each blade's feathering axis, is itself a gyro. Mounted on the same mast with the rotor is a gimbal mounted control gyro consisting of spokelike flyweights, one for each rotor blade.

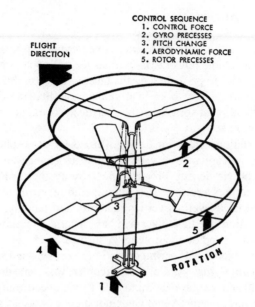

Figure 13.34. Lockheed rigid rotor concept and cyclic blade response. All parts above the "swash plate" at *1* rotate. A control force applied at *1* acts at *2*, 90° in the direction of rotation. *(Courtesy of Lockheed-California Co.)*

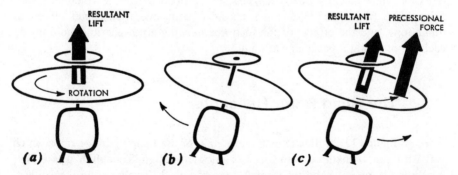

Figure 13.35. Gust reaction of rigid-rotor helicopter. (*a*) A strong gust may displace the fuselage and main rotor, but the control gyro remains in its plane of rotation. (*b*) The resultant angular variation of planes of the control gyro and main rotor changes blade pitch cyclically. (*c*) Resultant lift of the main rotor moves to a position forward of the rotor disk center, causing the rotor disk to precess or roll in a direction to restore the rotor disk and fuselage position. *(Courtesy of Lockheed-California Co.)*

When, during rotation, the pilot through his control stick applies a force to the control rotor through the "swash plate" (Figure 13.22), the force is felt 90° in the direction of rotation. This force displaces the rotors from parallel planes, individually changing blade pitch through shafts from the control rotor blades to respective pitch horns on the main rotor blades. The resultant differential lift on the main rotor disk causes it to precess 90°, tilting in the direction of pitch or roll indicated by the original control force.

This compensating and corrective effect may be induced by the force of the pilot's control stick, by gusts, or by the differing lift of the advancing and retreating rotor blades. The result of this system is a substantial improvement in helicopter design possibilities, enhancing positive control response, stability, maximum speed, allowable c.g. variation, and low vibration levels.

CURRENT DEVELOPMENTS

Power plant development is responsible for the major change taking place in rotary wing aircraft. This development is almost solely in gas turbine engines. The turbine engine apparently meets every requirement as the ideal power plant for the helicopter. Although fuel consumption of the gas turbine is greater than that of a comparable reciprocating engine, it is offset by consideration of engine weight, engine vibration, general maintenance, and performance at altitude.

In most cases, model conversion from reciprocating to turbine engines requires little design change, and provides greater speed, with greater load capacity and much lower noise level than the reciprocating engine versions.

Helicopter development at the present time is following these paths: turbine

Figure 13.36A. Aerospatiale Gazelle turbine driven helicopter. Note the inclosed tail rotor. *(Courtesy Aerospatiale Helicopter Corporation.)*

Figure 13.36B. Brantly 4 place helicopter. Note the full tricycle landing gear configuration. *(Courtesy Brantly–Hynes Helicopter, Inc.)*

powered helicopters for more efficient operation; multi engine helicopters for greater safety; crane type helicopters for the lifting of massive loads; larger helicopters with much greater load carrying capability; compound aircraft to provide greater range and speed; the capability of sustained automatic or manual instrument flight in adverse weather; and development of very capable light helicopters for general aviation use.

Very significant in the increased use of helicopters has been the growing number of light, economical, and relatively inexpensive helicopter designs. They provide versatility and utility not realized by any other form of transportation. Of all the many methods of transportation, they need the least real estate. And some of them even look like those conceived for movies 45 or 50 years ago! They are used for every imaginable purpose. Several examples are shown in Figure 13.36. About 900 civil use helicopters were delivered in the United States in 1977.

Vertical takeoff and landing aircraft, using rotors for vertical takeoff and then transition to forward flight where the rotors serve as propellers in a fixed wing configuration, have flown successfully. Bell Helicopter Company's folding proprotor concept is an approach to the possibility of even higher cruising speeds for vertical takeoff and landing aircraft (Figure 13.37). For cruise speeds above 260 knots, the rotors would fold, and fan jet engines would provide the thrust for high speed flight.

Figure 13.36C. Bell Helicopter's turbine powered 205A-1. This is the civil version of the U.S. Army's UH-1D "Huey." *(Courtesy Bell Helicopter Textron)*

Figure 13.36D. Enstrom Helicopter 280C. 500th helicopter produced by the Menominee, Michigan firm. Carries two. *(Courtesy Enstrom Helicopter Corporation)*

Figure 13.37. Bell Helicopter Textron's XV-15 tilt rotor research Aircraft No. 2 is shown making its first hover flight at the company's Arlington, Texas Flight Research Center. *(Courtesy Bell Helicopter Textron)*

14

Soaring [1]

The unavailability of lightweight engines led the early experimenters in heavier-than-air flying to concentrate their efforts on gliders. Among them were Sir George Cayley in the early 1800's in England, Lilienthal in Germany, and Montgomery and Chanute in the United States in the late 1880's. The Wright brothers first flew gliders at Kitty Hawk as a basis for their later powered flight. The success of the Wrights' first powered flights in 1903 caused interest in gliders to drop almost to the vanishing point until after World War I. The Wrights made some additional glider tests in 1911 at Kitty Hawk, where their best flight had a duration of 9¾ min. This record stood unchallenged for 10 years.

The Treaty of Versailles, ending World War I, prohibited Germany from building or importing powered aircraft. In other countries all efforts were concentrated on the development of powered aircraft, while the Germans threw all their very considerable technical talent into powerless flight. Their pilots and machines were in the fore of every gliding and soaring development in the 1920's and 1930's.

[1] By Harner Selvidge, S.D., Sedona, Arizona.

WORLD WAR II

The Germans again led in the application of gliders for military purposes, but late in the war the Allies' gliders made notable contributions, particularly in the Normandy invasion. Unfortunately, the technology developed for the large man and cargo carrying gliders turned out to have little application for peacetime use. In fact, the major wartime contribution to peacetime gliding and soaring was the surplus of training gliders and sailplanes released for public sale. These aircraft, and a few of the thousands of glider pilots who trained in them, formed the backbone of the soaring movement in the United States in the late 1940's and the 1950's.

RECENT DEVELOPMENTS

After about 10 years of doldrums, there was an almost explosive growth of soaring activities in the United States in the 1960's. The Soaring Society of America experienced an increase in membership of 20 times, from 1955 to 1970. American and foreign manufacturers now offer a wide variety of sailplanes for every need and purse. There are several hundred soaring clubs and dozens of commercial operators throughout the United States where instruction can be obtained, sailplanes rented, and where tows are available.[2] Today, soaring in America is the equal of that anywhere in the world, and a large proportion of the world's soaring records are held by American pilots.

Modern Sailplanes

The modern sailplane is as much of an improvement over the gliders of the 1920's as the modern business jet is over the Jennies of the same era. Despite the smaller market for powerless aircraft, there have been groups of very talented and dedicated designers, both in the United States and abroad, who have made great progress in the last two decades in increasing the performance and utility of sailplanes.

SPECIAL FEATURES

There are numerous special features of modern sailplanes which, while not entirely unknown on powered aircraft, are generally found only in sailplanes. The first is the provision for detachable wings and tail surfaces. The requirement for retrieving the sailplane in an auto drawn trailer from remote landings makes it

[2] Details on this and other soaring information can be obtained from the Soaring Society of America, Box 66071, Los Angeles, California 90066.

Figure 14.1. Schweizer Model 1-26. Over 400 of these Schweizer Model 1-26 sailplanes have been made. It has all-metal wings, monocoque nose, and fabric-covered chrome-moly tubing fuselage. *(Courtesy of Schweizer Aircraft Corp.)*

necessary to have wings and tail surfaces which can quickly and easily be removed by two or three persons and stowed for transit on the trailer. In the highest performance ships, the single wheel landing gear is retractable (by hand), but in most cases, whether retractable or not, there is no shock absorbing provision in the gear except for that furnished by the tire. The wheel is always provided with a hand operated brake, but some ships also have a drogue chute deployed from the tail cone when landing.

Spoilers are almost universally used for glide path control, but some sailplanes also have flaps. The spoilers also can be used as speed limiting dive brakes in many models so the nose can be pointed straight down without exceeding the red line speed. Despite their sometimes cramped cockpits, most sailplanes have superb visibility through their large canopies.

Figure 14.2. Schweizer Model 2-32. The Schweizer Model 2-32 is a high-performance all-metal two-seater with a 34 to 1 glide ratio. *(Courtesy of Schweizer Aircraft Corp.)*

One of the most outstanding features of modern sailplanes is their aerodynamic efficiency, which exceeds by far that of even the best performing military jet aircraft. They are also noted for their high structural strength, being designed for ultimate loads in the 8_g to 12_g range.

DESIGN TRENDS

In early sailplanes the main effort was to reduce weight to a minimum, since staying up was the main goal. Today, speed and distance are the principal objectives, and these call for the highest lift to drag ratio (L/D) at high speed ranges, combined with good circling performance at slow speeds. Weight is no longer as important. In fact, water ballast is sometimes added for contest and record flying. This added weight does not change either the lift or drag, so the ratio L/D is unchanged. This ratio is sometiems called the glide ratio or glide angle, since it gives the slope of the gliding flight path. For example, a sailplane with a maximum L/D of, say, 35 will be capable of gliding 35 miles from an altitude of 1 mile, in still air. The added ballast does not make the sailplane descend at a steeper angle, but merely makes it fly down the same slope at a faster speed. If lift conditions are strong, this more than counterbalances the slightly higher sink rate, which is a disadvantage when flying slowly. The water ballast can be dropped if lift conditions weaken.

Given a good airfoil and a smooth skin, the most important factor in obtaining a high L/D is the aspect ratio. Thus we see a trend toward very long narrow wings. Modern materials technology has permitted great strides to be made in this direction. The long wings reduce roll response greatly, and increase adverse yaw, but the advantages far outweigh the disadvantages. Drag reduction programs have also led to reductions in the cross sectional area of the fuselages. The

Figure 14.3. Libelle. The all-fiberglass Libelle, made by Glasflügel in Germany, is a high-performance sailplane widely used in the United States. *(Courtesy of Graham Thomson Ltd.)*

most important result of this is the necessity for the pilot to occupy a steeply reclining position in the cockpit. This reduces his visibility somewhat, and he can forget about elbow room, but the seats and controls can be arranged so that it is moderately comfortable.

Many of the best modern high performance sailplanes have either a "T" or "V" tail. The reason is partly aerodynamic, but more importantly, they are less likely to be damaged when landing in terrain covered with high vegetation.

As long as there have been sailplanes, pilots have dreamed of liberation from dependence on crews or outside mechanical help for launching into the air. The powered sailplane has been their goal, and many designs have been flown with power provided from every conceivable kind of small engine. Some featured fully retractable engines and propellers, while others merely feathered the propeller when the launch was finished. None of these ever went into production because of high costs and lack of large market interest. However, there was a rebirth of interest in powered sailplanes in the late 1960's, particularly in Europe, where at least two manufacturers had such models in limited production.

MATERIALS

Like powered aircraft, the early gliders and sailplanes were constructed of wood, wires, and cloth. Even as late as the mid 1950's, some of the best sailplanes were made entirely of plywood. Some fuselages were made of conventional metal tube and fabric construction, but the trend was to all metal structure and skin. Then in the 1960's plastic reinforced by fiberglass began to be widely used, particularly in European countries where labor costs were low. Fiberglass materials are strong, light, and can be formed easily into the necessary shapes with a very fine surface finish. Almost all of the high performance sailplanes coming out of European factories are all fiberglass, while American made ships are mostly all metal.

TYPICAL CHARACTERISTICS

Table 14.1 gives some of the significant characteristics of some modern sailplanes now flying in the United States. Some are of American and some of foreign manufacture. The examples are chosen to be representative of the various classes, such as training, medium performance, and high performance types.

Launching Methods

A typical modern high performance sailplane will weigh 660 lb. fully loaded and have a glide ratio (L/D) of 40. This means that it will take a pull (or thrust) of only 15 lb. to keep it airborne in level flight. Unfortunately, getting it

TABLE 14.1

Manufacturer	Model	Span (ft.)	Aspect Ratio	Gross Wt. (lb.)	Wing Load (lb./sq. ft.)	V_{stall} (mph)	V_{max} (mph)	$V_{L/D\,max}$ (mph)	L/D
Schweizer Aircraft Corp., U.S.A.	2-22E	43	9	900	4.8	35	90	47	17
Schweizer Aircraft Corp., U.S.A.	2-33	51	12	1040	4.7	35	98	47	22
Schweizer Aircraft Corp., U.S.A.	1-26D	40	10	700	4.4	30	114	49	23
Schleicher Aircraft, Germany	Ka-6CR	49	18	605	4.6	38	150	48	29
Schweizer Aircraft Corp., U.S.A.	2-32	57	18	1340	7.4	42	150	53	34
Glasflügel, Germany	Libelle	49	24	660	5.1	36	155	53	39
Bölkow, Germany	Phoebus C	56	21	825	5.4	36	124	56	42
Glasflügel, Germany	Kestrel	56	25	738	5.9	39	155	60	43
Schleicher Aircraft, Germany	ASW-15	60	26	860	6.2	40	150	62	48

off the ground and flying is not so easy. The power required to accelerate the mass from a standing start, to overcome the drag from the ground on the wheels, and to lift the aircraft to an altitude where sustained flight is possible is many times that required merely to sustain flight.

Man's earliest flights in heavier-than-air machines were in "hang gliders," a set of wings resting on his shoulders from which he hung in flight. Launching was by running downhill into the wind. Inclined tracks, catapults, and rockets have also been used. The only widely used launching means of real historic importance is the bungee, or shock cord method. It used the rubber rope made for landing gear shock absorbers and was developed in Germany and widely used in the early 1920's. The glider is hooked to the point of a "V" of shock-cord, and each end of the shock-cord is pulled by half the launching crew who stretch it tight by running while the glider is held in place. Then it is released and shot into the air as from a giant slingshot, usually from the brow of a hill. Although widely used for years, it is never seen today, having been succeeded by better means.

WINCH LAUNCH

The use of a winch to launch sailplanes and gliders has been very common in Europe, although it has been less popular in the United States, where the ready availability of automobiles and light aircraft has made the winch less advantageous.

In a winch launch, the aircraft is attached to the end of a long line which is rapidly wound up on a drum at the upwind end of the launching area. The aircraft climbs steeply at first, then levels off as it comes nearly overhead of the winch, where the pilot releases his end of the cable. The winch operator then continues to wind in the loose cable which is prevented from falling into a tangle by a parachute about 3 ft. in diameter attached to the end of the cable. A guillotine is provided at the winch to cut the cable if the sailplane cannot release its end. An automobile engine with an automatic transmission is the customary power for the winch.

All kinds of ropes, wires, and cables have been used for winch launches, with the best probably being the armored cable used by the military for towing aerial targets. It is strong, flexible, and will not kink. Since improper technique by the winch driver or carelessness on the part of the pilot can impart many g's load to the aircraft, the FAA requires that a weak link be used in the launching cable at the aircraft end. It should break with a pull of about twice the weight of the aircraft.

The winch launch has the advantage of being cheap ($.50 per launch being a common charge in club operations), and it can be used for takeoffs from a very short field, since the ground roll is only a couple of hundred feet. The winch cable can be several thousand feet in length with the winch located in rough ground beyond the takeoff area. The big disadvantage is that the pilot is always

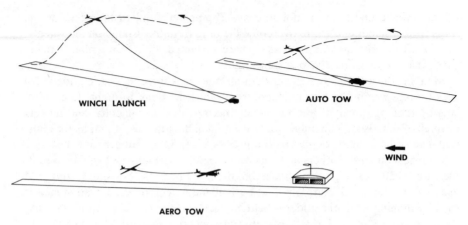

WINCH LAUNCH

AUTO TOW

WIND

AERO TOW

Figure 14.4. Launching methods.

released at the same spot, at a relatively low altitude, and has only a limited area and time in which to search for lift before he must land. The average altitude attained on a winch launch is about 40% of the length of the cable used.

AUTO TOW

If a long enough roadway or runway is available, a sailplane can be launched by towing it on a long line behind an automobile. While simple and cheap, this launching method suffers from the same disadvantages as the winch launch, including a limited height of release and a minimum chance to search out good areas of lift. If the field is small and the wind velocity low, a modification called an auto pulley launch can be used. One end of the line is attached to a stake, then over a pulley attached to the auto, thence back past the stake to the aircraft at the end of the field. This permits the car to move at half the speed necessary for the launch, a decided advantage in short or rough fields.

AIRPLANE TOW

When a sailplane is launched by aero tow, it can choose any altitude for release and can remain on tow until an area of lift is found before releasing. These advantages have made the aero tow the most widely used launching means by a wide margin, despite the somewhat higher cost. The requirements for the towplane are an FAA approved tow hook installation for attaching the towline, sufficient power to get a heavy sailplane out of a small field on a hot day, and the ability to climb at a slow speed (60 to 80 mph) indefinitely on a hot day without overheating the engine. A high horsepower to weight ratio is desirable; for example, a Piper Super Cub with a 150 hp engine makes a good towplane and is widely used for this purpose.

The tow hook on the airplane is made so that the tow pilot can release his end of the towline in an emergency, as can the sailplane pilot. Towlines are generally ¼-in. diameter braided plastic and range in length from 25 to 200 ft., 150 ft. being the length most commonly used in the United States. As in other types of launching, a weak link is required if the towline's breaking strength is more than twice the weight of the sailplane being towed. After the sailplane has released, the towplane can land with the towline still attached, or it may be dropped in some safe designated area before landing.

Basic Airmanship in Sailplanes

In Chapter 8 material was presented relating to the handling of light powered aircraft. Except where the use of power was required, all these techniques are equally applicable and valid for sailplanes. The operation of controls in flight is identical. In fact, after a couple of familiarization flights in a sailplane, the power pilot all too often feels that flying sailplanes is nothing new and there is little for him to learn. This is a dangerous fallacy and will surely lead to a serious accident.

It is true that the controls of a sailplane respond like those of a power plane, and the experienced pilot will soon handle them as effectively as those in any other aircraft. But in a powered aircraft he can buy time, altitude, and distance with his engine if he gets in a tight place because of carelessness or lack of experience. In the sailplane this crutch is not available. There are two things a pilot must learn before he can be considered a safe sailplane pilot: first, to plan ahead, particularly about his landing, to an extent far beyond that required of a power pilot; and second, he must learn the discipline which will always keep him within *easy and certain* gliding range of a landing spot suitable for (1) the conditions of the day, (2) his ability as a pilot, and (3) the capabilities of his aircraft. If he tries to cut his final glide to the airport landing pattern too fine, he will have no reserve if he encounters downdrafts en route, and he will be faced with an off airport landing for which he may not be prepared. If he tries to land on the first 20 ft. of the runway to show his great skill, a lurking area of sink can leave him with a cockpit full of fenceposts and barbed wire. The excellent safety record of soaring shows that the lack of an engine does not have to be a safety hazard, provided that the pilot plans ahead and observes proper soaring flight discipline.

The principal differences between soaring flight and powered flight are discussed in the following paragraphs.

TAKEOFF OR LAUNCHING

These are the regions where flight procedures differ the most from those of conventional powered aircraft.

Winch launch starts with the launching cable being hooked up to the sailplane. This should *never* be permitted unless the aircraft is occupied by a pilot, the canopy closed and locked, and everything ready for the takeoff. After the hookup, a helper at one wing tip signals the winch operator who slowly takes up slack until the cable is tight along the ground. The pilot then signals his readiness to take off by "fanning" the rudder, and this signal is relayed to the winch operator. The cable is only then rapidly reeled in. The acceleration of the aircraft can be quite surprising, and it will usually be airborne within 100 ft. or less.

The pilot holds the ship just off the ground until the airspeed builds up to about 1½ times stall speed and then firmly and deliberately pulls the stick completely back against the stop. This is where some passengers scream, and the pilot passengers who have never experienced a winch launch gasp audibly. As the climb progresses, the downward pull of the launch cable puts a positive *g* load on the wings as the nose is pulled down by the cable, and the pilot resists this force by full up elevator. Unlike normal pullups in free flight, the pilot does not feel this added *g* load in the seat of his pants. Nonetheless, it is there and increases the stalling speed proportionally. However, in this unusual flight configuration, the elevator stalls out first, resulting in a pitching down of the nose and an increase in airspeed. This in turn unstalls the elevator which then forces the nose up again. This cycle rapidly repeats itself and is called "porpoising." It is easily and instantly remedied by relaxing the stick pressure, thus shallowing the climb.

It is important to keep the airspeed well up during the climb so as not to be caught in this extremely nose high attitude at slow speed in the event of a cable break. If the cable does break, move the stick smartly forward to maintain normal flying speed and pull the release hook to drop the remains of the cable still attached to the ship. Land at whatever spot is appropriate for the altitude in hand.

As the launch progresses, the climb will flatten out until almost level flight is attained as the sailplane approaches a position almost above the winch. The winch operator will then cut the power. When the pilot feels this, he drops the nose slightly to take the tension off the cable and pulls the release. If he fails to release, or cannot, the winch driver will actuate the guillotine which cuts the cable at the winch.

Auto tows are quite similar to winch launches. The main difference is that the winch is capable of accelerating the sailplane much faster in the initial part of the launch. Auto speed will need to be 50 to 60 mph in light winds; less in stronger winds.

There is a standard set of signals used in sailplane launching, and every soaring pilot should be familiar with them. For example, it is up to him to let the winch or car driver know if the speed is too fast or too slow. The pilot rocks his wings to request more speed, and fans his rudder to request a slower speed.

Aero tows are the most common means of launching sailplanes in the United States. They are somewhat like a combination of power plane takeoffs combined with formation flying. Many soaring instructors find that learning to fly well on aero tow in turbulent air is the most difficult problem for their new students.

As in other types of launch, never hook up the cable until everything else is completely ready for takeoff. If it is the first flight of the day, hook up the towline and then release it while the helper is pulling on the towline. This confirms that the tow release mechanism will release under tension. Reattach the towline, and when the wing tip runner signals by leveling the wings, the towplane taxis slowly forward and takes up the slack.

The sailplane pilot then signals his readiness to take off by "fanning" the rudder. The tow pilot applies full power and the takeoff roll starts. In the absence of strong winds, the sailplane pilot will need to make vigorous motions of the control stick to keep the wings level and the nose skid from rubbing on the ground in the early part of the takeoff run when the two aircraft are accelerating slowly. However, normal control is quickly achieved, and the sailplane will shortly fly off the ground. Because the towplane will not yet be airborne, take great care to fly the sailplane only a few feet above the runway. Flying too high pulls the tail of the towplane up so far that its pilot never can get the tail down to a takeoff and climb attitude. Allowing the sailplane to balloon up too high suddenly could even dump the towplane up on its propeller. If the sailplane pilot holds his position properly, there will be very little drag on the towplane and it will quickly break off the ground.

There are two positions in aero tow: *high tow* and *low tow,* shown in Figure 14.5. The former is now used almost exclusively, but the FAA flight tests require that both be demonstrated. In high tow, the relative positions of towplane and

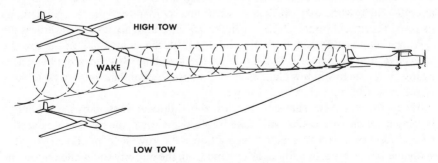

Figure 14.5. High and low tow positions.

sailplane on takeoff are maintained: The towed aircraft is directly behind and slightly above the towplane. This places it just above the wake turbulence caused by the wing-tip vortices of the towplane. In low tow position, it is just below the wake turbulence with the towplane well above it. The pilot can move directly from one position to the other by moving up or down through the wake. He will encounter moderate turbulence and a strong tendency to roll the sailplane, but this offers no problems to an alert pilot. The turbulence can be entirely avoided by going around it, by moving off to one side of the towplane (lots of rudder required), then moving down or up as the case may be, and then returning to the center position.

If the air is smooth, flying an aero tow is very simple. The problems arise when it is turbulent, which is always the case when instability and strong thermals are present.

The sailplane pilot usually tries to keep the towplane at some fixed spot on his windshield canopy. If he can do this precisely, he will never have any trouble with slack in the towline no matter how rough the air. A capable towplane pilot will carefully fly at a constant attitude, but when he encounters a thermal, the towplane will rise. The sailplane pilot should not wait to get to the updraft, but should immediately pull back on the stick to keep the towplane at the same position on the canopy. The reverse is true when the towplane sinks.

All this is easier said than done, and the common student fault of overcontrolling can create some difficult situations if he gets out of phase with the towplane movements. If the student gets too high, he can easily start to overtake the towplane when he makes his nose down correction, since he is much cleaner than the towplane. This creates a lot of slack in the towline.

Any large amount of slack is dangerous for three reasons: First, in extreme cases it can loop back over the sailplane wing and yank it off when the slack is taken out. This has happened, and any time a large loop of slack starts coming back, pull the tow release at once and dive away from it. Second, although the towlines are plastic and have some spring in them, when slack is suddenly taken out, there is a jerk which puts an unnecessary strain on the towline and the two aircraft. The towline may break at its weak link, perhaps leaving the sailplane on its own at a difficult time. Third, when the slack is suddenly taken out, unless the sailplane pilot is alert and holds his stick forward, the sudden pull of the towline will have a slingshot effect, ballooning the sailplane way up out of position again. When on tow in turbulent air is no time for the sailplane pilot to be looking at scenery. Constant alertness is required.

When the towplane banks for a turn, the sailplane does likewise, although lagging a couple of seconds will make it follow the towplane track more exactly. In the absence of other specific instructions, the towplane pilot will tow the sailplane into prospective lift areas upwind from the airport, and while on tow the sailplane pilot should make note of the position of any thermals he may be towed through so he can return to these areas if needed. Normal release is at about 2000 ft. above ground level, but it is usually poor economy to release too soon. The thermals are weaker and there is less time available to look for them at low altitudes. When the chosen release altitude is reached, or a thermal is encountered, pull the release knob and immediately start the sailplane into a *right turn*. The towplane pilot will feel the release and start a diving *left turn* as soon as he visually confirms that the sailplane is free. The sailplane pilot *must* remember one signal: If the towplane pilot rocks his wings, it is a *mandatory* release signal. The tow pilot will release his end of the towline if the sailplane pilot does not.

FREE FLIGHT

The sailplane in free flight is handled just like a powered airplane, except that no power is available on demand for climbing. The elevators control the airspeed. The rate of roll is slower because of the large wingspan, and the effectiveness of the rudder is usually less. In soaring, a much greater proportion of the time is spent in slow flight regimes than in powered flying. When circling in a thermal to gain altitude, it is common practice to fly only a few knots above stall speed. Thus the soaring pilot's experience with approaching stalls in turns is thousands of times that of the power pilot. Soaring pilots *must* be good at slow flight, using sound and feel as much as instruments.

Stall and spin recoveries are the same as for powered aircraft, but without power to help, the stick must be held forward longer. The spoiler or speed-limiting dive brakes help in holding down the excess speed which would otherwise be developed in the dive. While almost every aerobatic maneuver can be done after a fashion by sailplanes, they are very poorly designed for this purpose on account of the slow roll rate and small rudder. Thus the figures are sloppy, and except for the spin, inside loop, and chandelle, are generally hard to perform even poorly.

LANDING

Since the sailplane landing has to be right the first time, instructors give considerable attention to approach and landing practice. Actually, because of the excellent glide path control provided by the spoilers, it is very easy to land sailplanes on even a very small airport, provided the pilot has arrived in the traffic pattern with normal altitude. But the soaring pilot must be trained for the time when he may wish to land away from the airport in a small field, so landing drill is directed to precision spot landings. Power pilots and spectators are constantly amazed by the sight of good soaring pilots touching down time after time within a few feet of a given spot, and stopping with their nose a *few inches* from a target spot.

The *approach* is made in a normal rectangular pattern as with powered aircraft, and similarly the first key to a good landing is a good approach pattern. To be sure of reaching the field, the sailplane's pattern may be closer to the runway than the power plane's. This also keeps the two kinds of traffic separate in joint-use fields. *Pattern speed should never be less than 50% above stalling speed plus half the estimated wind velocity.* For example, if the sailplane stalls at 40 knots and the wind is estimated at 10 knots, the minimum pattern speed would be 65 knots. Hold this speed until flare-out, and you cannot have a stall-spin accident. If you are inexperienced, or the air is very turbulent and the wind gusty, it is well to add 5 or 10 knots to this minimum airspeed.

The *final approach* is also made as outlined in Chapter 8. Select the aiming point, or the flare out point, about 50 yards short of the touchdown spot. Soaring schools used to teach the final approach by having a pilot visualize a line to the aiming point and then adjust the position of the sailplane along this line by the use of spoilers. A new approach has been in use in recent years which has been found to give students much greater accuracy in a much shorter time. It sounds about the same, but is basically different.

In this method the same aiming point is still used, but the student is told to aim the ship at the flare out point, with the elevators as though he were strafing an enemy at that point. He is then told to use the spoilers for speed control to cancel out the effect of any change in pitch attitude caused by changes in aiming.

When the sailplane is almost at the chosen flare out point and at an altitude of 5 to 20 ft. depending on the angle of descent, the pilot smoothly rounds out the glide so that the flight path becomes parallel to the ground at an altitude of a foot or so. If the flare out point is reached with the spoilers closed, the glide path will be quite flat, and there will be little attitude change required for the flare out. On the other hand, full spoilers require a very steep glide angle to maintain airspeed, and the flare out is much pronounced and should be started higher up. Slowly closing the spoilers simultaneously with the flare will make this maneuver less critical.

With the flare out completed, you will be floating towards your chosen touchdown spot. Then imagine that the spoiler control is a throttle; if short, push it forward, closing the spoilers, extending the glide; if long, pull the control back, extending the spoilers to shorten the glide. When about 30 ft. from the chosen touchdown spot, open the spoilers smoothly all the way and allow the glider to touch down in a normal attitude without the use of the elevators. Never make a full stall landing intentionally in a sailplane because there is no shock capability in the landing gear except that in the sidewalls of the tire. Control the rollout by the wheel brake, and for a faster stop, push the stick all the way forward, digging in the nose skid in front of the wheel. Remember the landing is not complete until the aircraft has come to a complete stop safely. Plan to stop well short of obstacles such as other aircraft.

Sailplane Instrumentation and Equipment

While the instrumentation of training sailplanes is frequently a bare minimum, the instrumentation of high performance competition ships is quite complete. Conventional instruments are used for airspeed, altitude, acceleration, and turn and slip. The most important instrument for the soaring pilot is one which tells him whether he is going up or down. The regular rate-of-climb instrument used in power planes is much too slow and insensitive to be of much value in soaring, so a special instrument called a *variometer* is used for this purpose.

VARIOMETERS

When a soaring pilot encounters rising air, he needs to know it immediately, so he can take the proper action to stay in the area of lift. The variometer utilizes the same principle of operation as the conventional rate-of-climb (*i.e.,* measuring the airflow in or out of a small reservoir), but the soaring instrument has a very fast response of 1 sec. or less. Another feature widely used with this instrument is called "total energy compensation." If you push the stick forward, you will nose down and pick up speed. Your variometer will show a down reading. You have exchanged some of your potential energy (altitude) for kinetic energy (speed). The reverse would be true if you pulled the stick back. In the latter case you might think you had encountered rising air in a thermal, but it would only be what is called a "stick thermal."

Figure 14.6. Soaring instruments. Variometer instrument, *center.* Total energy compensation diaphragm, *left;* audio attachment, *right. (Courtesy of Rainco.)*

What you want to know is whether you have really encountered rising air, or have just inadvertently mishandled the controls. The total energy compensator tells you this. It utilizes pitot pressure to measure airspeed and to operate a diaphragm which changes the volume of the variometer reservoir. When this is properly done, you can push the stick forward or pull it back without changing the variometer reading over a surprisingly large range of airspeeds. The total energy compensation must be individually adjusted for each model of sailplane, but it is well worth the effort.

It is a very common experience to be spiraling in the same thermal with several other sailplanes. Under such circumstances you cannot spend much time looking at your instruments, yet if you do not pay close attention to the variometer, you may stray out of the best lift area. A solution to this problem is an audio attachment designed for use with any variometer. It generates an audio tone, whenever there is lift showing on the instrument. The pitch of the tone increases as the lift gets stronger. This permits you to spend more time looking out of the cockpit and still constantly monitor the strength of the lift where you are flying.

COMPASSES

Since while soaring you are likely to spend a large part of the time circling in thermals, a conventional magnetic compass is of little use, and special compasses have been designed for soaring use, such as the Cook compass, made in England. Its element does not float in a liquid, is light in weight for a fast response, yet is well damped. The Cook compass is almost as good as a gyro compass and can be used to roll out of a thermal, in or out of clouds, on a very precise heading.

Power sources for sailplanes are usually nickel cadmium batteries. Multi-channel VHF radios are commonly used for communication with ground crews and other sailplanes as well as FSS stations. Flights above 12,000 ft. are common in the western United States, so many sailplanes carry elaborate pressure oxygen systems for long duration flights at high altitudes.

Soaring Meteorology

To the power pilot, weather conditions are just one of the many considerations that are examined in planning and executing a flight. To the soaring pilot, favorable meteorological conditions are an absolute necessity for flight. If there is no lift, he cannot stay up. More than mechanical skill in flying his aircraft, the ability to understand the meteorological conditions which prevail during flight is the most important factor in making a top soaring pilot. In the broadest sense, the soaring pilot uses meteorological information in three ways:

first, to permit forecasting the expected conditions perhaps a day in advance, so that he can plan flying activity to take advantage of them; second, to understand the existing conditions which prevail at the start of the flight; and third, to permit recognizing and properly interpreting the dynamic changes in the meteorological situation which may occur during the flight.

KINDS OF LIFT

There are five kinds of lift which the soaring pilot may use. On local and many cross country flights, only one of these will probably be used, but on some long flights each kind may be encountered and used if the pilot is alert and knowledgeable enough to recognize them.

Slope or ridge lift is found when the wind encounters an obstacle such as a slope or ridge and is forced upward mechanically by the obstruction. It is the simplest kind of lift to recognize and understand and was the only kind used by the pioneers in gliding and soaring. A wind of about 10 knots at right angles to the ridge is required to keep a sailplane aloft.

The source of the wind may be the pressure gradient flow after the passage of a cold front, for example, or it may be a sea breeze caused by the heating of the land. In a 20 or 25 knot wind, it is possible to maintain an altitude one or two times the height of the ridge.

Convergence zones or *shear lines* are long lines of lift found when airflows moving in different directions collide. When the air masses come together (and they do not have to be in exactly opposite directions), the air is forced up along their line of intersection, and while this band of lift is often very narrow, it is soarable even when the winds causing it are very light, say 10 knots or so.

The convergence of the winds may be caused by local mechanical effects such as hills or valleys dividing or channeling the flows, or it may be caused by major air mass movements, such as cold fronts or their associated squall lines. One special case is the so called "sea breeze front" caused by stable air from the sea or a lake moving inland with the diurnal sea breeze and pushing against the less stable air over the land. This results in a miniature cold front roughly parallel to

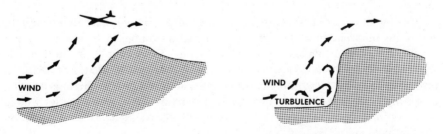

Figure 14.7. Wind flow at slopes or ridges.

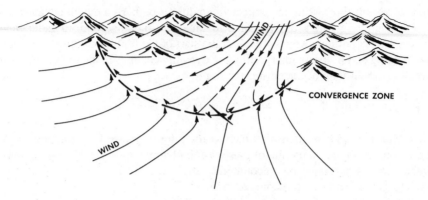

Figure 14.8. Lift at convergence zone.

the coast, sometimes moving a good many miles inland, depending upon the strength of the wind. It is frequently marked by a line of small cumulus clouds.

Thermals are the most commonly used kind of lift in modern soaring, although it was not until about 1930 that soaring pilots understood and started using thermals on a regular basis for staying aloft. The term *thermal* is used to describe lift resulting from unstable air rising in columns or bubbles. If the air is moist enough, the thermal will be marked by a cumulus cloud, providing the lift reaches high enough to condense the moisture. Lift will be found right up through the cloud to the top of the visible moisture. Stability concepts are discussed in detail in Chapter 6.

The soaring pilot is interested in the time thermals will start, their strength, and the height of their tops. The weather bureau forecasters will usually know the cloud base and the top of the unstable layer and can give a rough estimate of the thermal strengths expected, but they do not normally compute the starting time of the convection at ground level unless they are accustomed to preparing soaring forecasts. Since the soaring pilot will not usually have available the soundings of temperature and humidity which are necessary to make these computations, he will be forced to try to extract this information from some sympathetic weather man.

Once the pilot has his forecast, he is on his own. He will need continually to update his information in the light of the actual conditions as he observes them through his instruments and visual observations. For example, if the thermals start earlier than forecast, they may also be stronger than expected. The ability to recognize changing meteorological conditions in the course of a flight is an attribute which sets the top pilots apart from their fellows. The soaring pilot who is seriously interested in improving his flying will study all the written material on soaring meteorology he can find.

It is not necessary to have clouds to have thermals. Many long flights are made on days when there is not enough moisture to form clouds. The pilot then uses "dry thermals" which are just like those on moister days, but they are harder to locate since they have no clouds to mark their tops.

Another situation during which the soaring pilot can get information from the clouds occurs when they line up in rows or "cloud streets." These are frequently parallel to the wind, so the happy soaring pilot finds himself dashing downwind under a line of clouds, sometimes going miles without needing to stop and circle to gain altitude. A less pleasant sight is a high cirrus overcast or stratus deck. This will cut off the solar heating of the ground and spell a quick end to thermals and the flight.

Mountain waves or *lee waves* described in Chapter 6 give the soaring pilot the chance to climb to remarkable heights. In the updraft in *front* of the obstacle such as a ridge, a height of perhaps twice that of the ridge can be attained, but in the *lee* waves a height of ten times the obstacle height is not uncommon. Waves suitable for soaring can be expected when the wind at the top of the obstacle is about 25 knots or more, and the best conditions come with winds increasing with altitude. The occurrence of waves can usually be fairly well predicted by forecasters. Like slope or ridge soaring, the best conditions arise when the wind is blowing at right angles to the line of the ridge. Since wind direction and velocity are the primary criteria, the soaring pilot can do his own forecasting with winds aloft data prepared for other aviation use.

Slope or Ridge Soaring

The very first gliding and soaring flights were made by launching the aircraft from the top of a hill or ridge into a wind blowing up the slope. This gives a kind of lift which is easily visualized and measured and which provides the simplest kind of soaring. However, it is seldom possible to get very high by using slope lift, and unless the ridge is very long, flying an appreciable distance is also impossible. For this reason, slope soaring has lost most of its popularity except for training and duration flights.

For good slope soaring it is necessary to have a wind component at a right angle to a hill or ridge with a strength of about 10 to 15 knots. A suitable nearby launching site is required as well as a suitable emergency landing site at the foot of the ridge in case the wind velocity should drop so that the ship could not reland at the starting point.

LAUNCHING

Ridge lift is usually entered from a winch at the top of the ridge or from an aero tow from any nearby point. When launched by a winch from a field at the top of

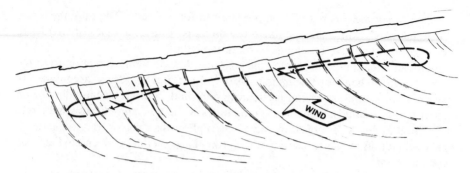

Figure 14.9. Flight paths in ridge soaring.

the ridge, the pilot will probably find himself 500 ft. above the ridge and can start one brief traverse along the edge of the ridge, looking for strong enough lift to remain airborne. If he finds none, he returns at once while he has sufficient altitude.

FLIGHT PATH

Fly a path parallel to the top of the ridge, moving carefully upwind or downwind to find the exact path which gives the best lift. When reaching the end of the traverse (for example, where the ridge ends or changes direction), turn *into* the wind away from the slope and complete a teardrop turning pattern back to the original traverse path, but now flying in the opposite direction (Figure 14.9). At the other end, repeat the process, always turning into the wind away from the slope. If other sailplanes are also flying in the same area, do not fly directly below one in the other pilot's blind sector. When overtaking a ship going in the same direction, always pass on the *inside* next to the slope. This may violate the usual passing rule in other air space, but it is necessary to avoid the possibility of the overtaken ship turning unexpectedly in front of the other. Since the name of the game is to stay up, in ridge soaring the pilot almost always flies at his minimum sinking speed, rather than the speed of best L/D. This means he is always on the edge of a stall and must be very alert at all times.

LANDING

If you wish to land back at the top of the ridge, never let yourself get too low to make that final turn in to land. Remember also that this landing will almost always be made downwind, and it is most important to maintain proper airspeed, no matter how fast the ground seems to be slipping by. If the wind weakens and you find yourself flying below the crest of the ridge, do not despair, but carefully fly in the best lift and wait in hopes that the wind will strengthen again, lifting you high enough to permit a landing on top.

Thermalling Techniques

Your ability to stay airborne and gain altitude above the launch point is the vital difference between soaring and gliding. You must be able to find the lift areas, and then stay in them until you have gained the maximum altitude they provide. Techniques for centering and staying in the lift areas can be taught, and most pilots become fairly proficient in this aspect of soaring. The thing which separates the good pilots from the ordinary ones is their ability to find the thermals in the first place. Yet this is one of the hardest things to teach. Some pilots seem to be born with a nose for thermals, and some apparently equally proficient ones seem never to acquire the knack.

CHARACTERISTICS OF THERMALS

Thermals are masses of air that are more buoyant than the surrounding air, usually because of an excess of temperature or moisture, or both. Some theoretical studies indicate that thermals are like a smoke ring blown upwards, with the rising air in the center returning around the outside of the doughnut. This would mean that no lift would be found either above or below the ring. The experience of soaring pilots indicates that most of the time they can find lift directly under and above other soaring aircraft almost all the way to the ground. This supports the theory that thermals are usually *columns* of rising air. In diameter they can range from a few feet (notice the small circles soaring birds sometimes use) to many hundreds of feet. In height they usually start at the ground and rise to the altitude where the air in the column is at equilibrium with the surrounding air. This can be as high as 50,000 ft., the top of some large storms. But most soaring pilots leave the lift at the cumulus cloud base, which will be 5000 to 10,000 ft. above the ground in most circumstances of good soaring weather.

Thermals drift with the wind, and since wind velocity almost always increases with altitude, the thermal will lean with the wind. When winds get to 25 to 30 knots near the ground, the thermals will be so badly blown apart that they are not useable for soaring. The air that goes up in a thermal must come down somewhere, and it does this in an area all around the rising air. However, since the descending air is spread over a much larger area than the rising air, the strength of the downdrafts is less than that of the updrafts. The soaring pilot soon learns that where there is strong lift, there is also strong sink nearby. This is important to remember in the final glide to a landing.

FINDING THERMALS

The variometer will tell you when you reach a thermal, but what direction should you take to have the best chance of finding one? Some of the common cues are listed below.

Cumulus clouds which are observed to be building, not decaying, are one of the most reliable indicators of lift. However, since the average life of a cumulus cloud is only about 20 min., they must constantly be observed to identify the growing ones. These are usually the ones with firm sharp edges. Top soaring pilots are always good cloud readers. On days when there is not enough moisture to make clouds, other indicators must be used.

Dust, scraps of paper, and other light particles can sometimes be seen carried up in strong thermals. The familiar dust devil, a common feature of the desert scene in summer, marks a very strong thermal lift area. But be wary of entering them at low altitudes as they are very turbulent and can easily upset a sailplane.

Birds and other sailplanes which may be observed circling are good thermal indicators. Go over and join them. If you are below, be sure to take into account that the wind will be causing the lift area to slant with its direction. Many a sailplane pilot has been saved from an undesired landing by spotting a hawk or buzzard circling in a nearby thermal and joining him, and many times the birds will come over and join the sailplane.

Surface features which are more prone to generate thermals should be searched for. These are large black areas such as parking lots and plowed fields and heat sources such as chimneys and fires. Avoid bodies of water and areas where rain has recently fallen, or which are irrigated. Dry ground is best for thermals. Usually forested areas are poor for thermal production because of their excess moisture content. Thermals will rise earlier from high ground than from the nearby valleys, so an important rule is "stick to the higher ground."

THERMAL CENTERING TECHNIQUES

Using the aids in the above paragraphs, or perhaps by luck, you encounter thermal lift. How can you best find the strongest part and stay in it? Normally the first indication of lift is shown by the reading of the variometer. This shows you are entering the area of lift which is usually roughly circular in shape, but you will have no idea if the center is directly ahead, or to one side. At this moment, notice that one wing or the other goes up. If this coincides with the increased reading of the variometer, it usually means that the lift is stronger on the side where the wing lifted. Count to perhaps 4 sec., and if the lift is still increasing, start a steep turn into the direction of the lifted wing. If there is no wing indication, turn either way.

A simple and effective way of locating the center of the thermal is to fly a complete 360° circle, noting the variometer reading at the four 90° points. This will give a good idea of the strength and distribution of the lift in this area. Then for the second turn, move the center of the circle over in the direction which shows the strongest lift. Another way of accomplishing the same thing is to shallow the angle of bank when the lift is decreasing. This process is shown diagrammatically in Figure 14.10.

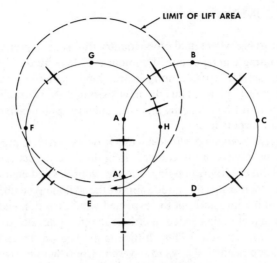

Figure 14.10. Thermal centering.

At point A' the variometer starts showing lift at the edge of the thermal outlined by the dashed circle. After 3 or 4 sec. with the vario reading still increasing, you elect to turn right at A. As you turn to an easterly heading at B, the reading has dropped to about zero rate of climb, and you strongly suspect that you turned the wrong way or that the thermal is a very small one. However, continue to turn, perhaps steepening the bank somewhat. At C and D, the two other 90° points, the vario shows "down." You know then that the strong core of the thermal lies to the west of the circle, so you roll out of the turn at D and fly straight for 5 or 6 sec. to point E where you roll back again into the right turn. The next circle will find small lift at F, strong lift at G, and small lift at H. Then shift the next circle northward so the entire circuit will be in the lift area.

When deciding which direction to turn in a thermal, remember the soaring rule-of-the-road: the first ship in a thermal sets the direction of circling. All others joining him later in the thermal (whether above or below) circle in the same direction. Further, anyone flying in a thermal should expect others to join him, so keep a sharp lookout.

Cloud flying is really thermal flying on instruments. It is difficult to fly in clouds legally in the United States, but it can sometimes be done. Since ice, hail, and turbulence may be encountered, there are two important cautions: First, choose a cumulus cloud that is in its early growing stage, so you will be in and out before it gets to thunderstorm size. Second, start to get out of the cloud when the icing or turbulence *begins* to get bad. It will always be more turbulent as you head for the side, and the ice will continue to build up during the exit time. Before entering the cloud get a heading firmly in mind which will take you into the clear air in the shortest time.

CROSS COUNTRY SOARING

Unlike power flying where dual cross country flights are a prerequisite for licensing, a glider rating can be obtained without ever leaving the familiar area of the home airport, and dual cross country instruction is seldom offered in commercial schools or in private clubs. Thus the new soaring pilot is usually on his own for his first away-from-home landing and is frequently apprehensive about the possibility of an off airport landing.

The first cross country flight loses much of its terror if preceded by careful planning. In most parts of the country, long distance flights can be made by the simple expedient of airport hopping. Lay out your desired course in a downwind direction, and then using the glide ratio of the sailplane (with a suitable safety factor), and taking into account the expected wind direction and velocity, calculate the minimum altitude needed in order either to glide back to the home airport or to go on to the next one. This altitude is marked on the chart, and you then know the altitude needed always to have an airport within easy gliding range if further lift fails to materialize. The same computation is made for the next leg, and so on. After a few such flights, you will have sufficient confidence and experience to face the possibility of an off airport landing without undue concern.

Maximizing distance flown or minimizing elapsed time will soon become your major concern as a cross country soaring pilot, since you will be endeavoring to surpass your own speed or distance records, will be trying for one of the FAI soaring awards, or may be entered in a contest with other pilots. Saving time is one of the most important factors, even though the objective of the flight is merely to attain the greatest distance without regard to speed. The reason is this: The hours of the day in which lift will be found are strictly limited. Wasted minutes mean wasted miles. The maximum glide ratio of the sailplane is a widely quoted figure of merit, and you will always know its magnitude and the speed at

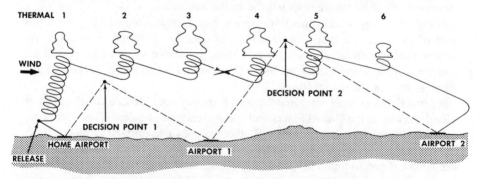

Figure 14.11. In planning flight to airport 2, the pilot has calculated the altitude of the two decision points shown. Staying above the dotted lines assures being able to land at any airport. Arriving above decision point 1, he continued. He was unable to get above decision point 2 in thermal 4, but saw the cloud marking thermal 5, so pressed on. Finding lift under cloud 5, no off-field landing between 1 and 2 was necessary.

which it is obtained. However, in cross country flying you will not fly at this speed except perhaps while making the final glide of the day to a landing.

The best speed-to-fly to maximize the distance flown on a given day can be computed from the performance curves of the sailplane if an assumption is made about the strength of lift in the next thermal. This is usually possible to do with considerable success, since on a given day the thermals are usually found to be of about the same strength. Details of how to make this computation can be found in most soaring books and will not be further described here.

Suffice to say, the speed-to-fly between thermals is considerably *greater* than the speed of the best glide ratio. You deliberately sacrifice altitude to obtain extra speed and save time, knowing you will be able to regain the altitude lost in the next thermal. At times when you get too low, or late in the day when the lift weakens, fly much slower. Here the problem is reduced to the simple one of just staying up. In any case, follow the general plan of flying fast when in sink and slow when in lift.

Navigation is pure VFR pilotage for soaring flights, using charts and landmarks. Your concern for a prospective landing place will usually keep you well aware of the terrain beneath. However, it is easy to get disoriented after circling for some time in a thermal, and it is wise to check the course heading with the compass upon leaving the lift.

Choice of Landing Field.   In most cases, the cross country soaring pilot faced with an off airport landing will have a choice of several fields. There are numerous factors which should be considered in making the choice of which field to use. The first is the question of adequate size. This, in turn, will be influenced by the wind direction and velocity, possible obstructions at the approach end, and the condition of the surface. Ideally the surface should be smooth, dry, and uphill into the wind. Except in the case of a very strong wind, make landings uphill no matter what the wind direction. The long wingspan of modern sailplanes makes road landings hazardous because of roadside marker stakes which are invisible from the air. Landings in even medium-high crops such as wheat can be quite hazardous, because if one wing gets a little low at touchdown, the vegetation will catch it and a violent ground loop will result. This seldom injures the pilot, but it frequently causes major damage to the sailplane. It is wise to avoid fields where livestock are seen. They may blunder into the landing path, or trample on the aircraft after landing when it may be left unattended. Lastly, if there is a choice, pick a field with easy access for the retrieve crew and trailer, but this should be the last in priority when choosing a site.

Wave Flying

When wind conditions are right, wave lift can be found behind very low ridges or obstacles, but for most pilots wave flying means going to 20,000,

30,000, or 40,000 ft. in the lee of substantial mountain ranges. Wave flying means flying in strong winds and at high altitudes. Do not undertake it lightly.

FORECASTS

Conditions which produce good waves can usually be forecast with good accuracy. In general a wind at right angles to the ridge and a velocity of 25 knots or more at the altitude of the ridge top is required. If the wind strongly increases with altitude, this is also advantageous, as is low level and high level instability with a stable layer in the middle altitudes. While lenticular clouds mark the wave when there is sufficient moisture present, dry and cloudless days can have just as good waves.

PREPARATION

Get ready for wave flying as carefully as if your life depended on it, which it does. Oxygen equipment suited to the task should be in first class shape, and it should be a pressure demand system if the flight is expected to go above about 30,000 ft. Carry an emergency oxygen supply as a backup. Frost from your breath will soon completely cover the inside of the canopy unless clear vision panels are provided.

Have batteries fully charged and insulated against the cold. Check the sealed recording barograph for operation in the cold. Dress warmly, particularly the legs and feet which will be in the shade with outside temperatures running $-40°$ or lower.

LAUNCHING

While it is sometimes possible to get into wave lift from a lucky winch launch, almost all wave flights start from an aero tow which takes the sailplane right into the best lift area. This frequently means towing through, or under, the rotor zone which lies in the lee of the ridge, resulting in an exciting time for both towplane and sailplane pilots. Beginners should not try it.

The severest turbulence known in the atmosphere is found in the rotor zone (Figure 6.26). As in flying powered planes in turbulence, careful attitude flying is the key to survival, together with a willingness to release the towline if dangerous attitudes result, or if there is excessive slack in the towline. Upon exiting from the rotor zone on the upwind side, you will suddenly encounter the wave lift, and within a few seconds, after hanging on for dear life, you will see the altimeter and rate of climb start winding up, in air so smooth as to seem uncanny. Pull the release immediately and head into the wind for the climb. Mark the spot on the ground below, so you can come back if you lose the lift. Adjust your speed to stay over the same spot, occasionally moving forward or back to be sure you are in the area of maximum lift, and bear in mind that the wind is probably increasing greatly with altitude. The world's altitude record for a sailplane is 46,267 ft., made in a wave behind the Sierras.

WAVE CLOUDS

When there is sufficient moisture present, lenticular clouds will mark the tops of the waves, and if there are several moisture layers, the clouds will also be stacked in layers. There will also be a ragged-looking cumulus cloud marking the rotor zone if there is moisture at that level. The rotor zone is always centered at the same altitude as the top of the obstacle causing the wave system. These clouds are most useful in showing the areas of lift and turbulence. Another cloud sometimes associated with waves is the "Foehn" or cap cloud which comes pouring over the top of the obstacle, propelled by the high winds (Figure 6.26). As the air flows down on the lee side of the obstacle, the cloud is warmed and evaporates. When the valley beyond is full of rotor clouds, or perhaps stratus, there will be a gap in the clouds where the air over the ridge is driven down and warmed. This is called the "Foehn gap" and sometimes provides a way to get on top, or back down, without going through the clouds. However, this can be hazardous, as the gap may suddenly fill with clouds, leaving you above a solid deck of clouds in mountainous terrain.

Soaring Safely

As a soaring pilot you start with many advantages which favor your safety. Unlike the power pilot you are not tempted into fog, rain, or snowstorms, because there is no lift there. The sailplane flies and lands slowly, its structure is much stronger than that of power planes, and with its single wheel and nose skid it can land safely in terrain which would seriously damage aircraft with conventional gear. Its spoilers give it an unsurpassed glide angle control on approaches. The obvious disadvantage of the sailplane is the lack of power to extricate it from a critical situation, particularly on landing. But trained in this problem from your very first flight, you can blame surprisingly few serious soaring accidents on a lack of power.

As with power planes there are too many stall-spin accidents close to the ground. The importance of maintaining adequate flying speed cannot be overemphasized. Many pilots who have just made the transition from power seem to feel that flying a sailplane is very simple and start doing things, such as flying low and slowly, that they would probably never dream of doing in a power plane. All the admonitions given to power pilots apply to sailplanes: Keep up your flying speed; don't fly into clouds unless instrument rated; don't try aerobatics unless qualified; keep a sharp lookout for other planes, etc.

All sailplanes are equipped with a shoulder harness in addition to seat belts, and most pilots wear parachutes when engaged in competitive flying or cross-country flights over rough terrain.

To the familiar takeoff and landing checklists, you must add another: *the assembly checklist*. Since the sailplane is frequently assembled on the field shortly before takeoff (having arrived disassembled on its trailer), use a checklist to insure that all assembly operations have been properly carried out. Sailplanes have been known to take off and fly without controls hooked up, or without wing pins in place, but never well.

There are special sets of signals used in winch launching and aero tow. Learn these and use them. In addition, there are three important conventions or rules of the road which are unique to soaring. They have been mentioned in previous sections but are repeated here: (1) The first ship in a thermal sets the direction of turns in this thermal. All newcomers must turn in the same direction. (2) In slope soaring all turns are made into the wind, and all overtaking and passing is done on the downwind side. (3) After release from an aero tow, the sailplane turns to the right, and the towplane makes a diving turn to the left.

15

Northern Wilderness Flying[1]

The extensive lore associated with the art of wilderness flying developed over the past 60 years has been built largely about northern flying experience. The effects of cold on men and materials warrant placing more emphasis on the problems of flight in cold climates rather than on the peculiar techniques which are, of course, associated with wilderness flying over jungle and desert areas, to say nothing of the vast water wilderness of the world.

The transportation problems of the subarctic are so similar to those of the arctic that the two are logically considered together. These problems also face the charter or sportsman pilot flying in the United States along the Canadian border, in mountainous areas, and in southern Canada.

Northern Flying Developed by the Bush Pilots

HISTORY

In 1920 the four open cockpit De Haviland biplanes of the Black Wolf Squadron flew

[1] By Terris Moore, Colby College; formerly President, University of Alaska.

815

from New York to Nome, Alaska, on the coast of the Bering Sea. This military flight, under the sponsorship of General "Billy" Mitchell, stands as history's first important northern flying operation. That achievement becomes more remarkable when one recalls that Alaska had then no airfields, little radio communication, and no weather forecasting.

Then as now, however, this military flight contrasted sharply with civilian flying in the extent of the US government resources which stood behind it. The "broken wings, smashed axles, busted tires and tail skids, broken oil lines and leaking tanks" suffered by the four planes on the 6 weeks' flight to Alaska represented expenses beyond the financial risks which civilian operations would undertake in 1920. Military interests in northern flying relapsed and became dormant thereafter with Mitchell's passing and remained so until the outbreak of World War II. During the years between, Canadian and Alaskan civilian bush pilots developed the basic techniques of northern flying.

By the time World War II brought military flying back into the arctic, the basic modern techniques of northern flying had been substantially developed. The airplane had become the automobile of the north, replacing, with its freighting capabilities, bus, truck, and even railroad. There are still thousands of northern native people who, though entirely familiar with the airplanes which bring them their mail, food, and medical help, have never seen an automobile or a railroad. In all the 5,000 miles of north country running from Newfoundland westerly to British Columbia and on to Bering Strait, Alaska, going north from this long line to the Pole, there are only four spindly north running railroad lines and two meager road stems, plus the Alaskan pipeline and its associated road system.

THE BUSH PILOT

Where the north begins, the bush pilot begins. His name incidentally derives from the Canadian word for the north woods, "the bush." Where the roads and railroads disappear as one goes north, the bush pilot takes over transportation. It is he whose self-reliant resourcefulness has developed the basic techniques of northern flying, which, with many minor variations and improvements, are the foundation of modern airmanship in the arctic, both military and civilian.

The term *bush flying* must be defined explicitly. It is the technique of airmanship involved in *wilderness* flying, where, the takeoffs and landings being away from airports, the pilot is confronted with the necessity of carrying them out in only modestly improved or even completely unprepared places, and where his ground handling and servicing must be frequently done without the usual airport, hangar, and shop facilities.

Originally all northern flying was bush flying. But as northern airports have come to be developed over the years, a growing part of northern flying has become simply conventional flying in the north, and not really bush flying. This is particularly true of military flying in the north. Military planes now do not lend themselves to the making of landings and takeoffs in unprepared or partially prepared places. They are airport-to-airport planes, designed for combat or transport at high altitudes and in ultracold atmospheres. And the techniques of using military planes for other than special rescue operations—even when the airports from which they must operate are in the arctic—have been standardized in Air Force Technical Orders. There is little need for the improvisation which to a considerable degree still characterizes northern bush flying. In fact, improvisation in northern military airport-to-airport flying, if it arises at all, is today correctly taken by the authorities to imply inadequate planning, improper operation, or inadequate maintenance.

The Role of Bush Flying Today.   Very few special arctic *flight* problems are now encountered by military aircraft which they are not already designed to meet. This is because most military aircraft, being designed for high altitude flying, where subzero temperatures are encountered in all parts of the world, come from the factory already equipped to cope with the problems of flight at arctic winter temperatures aloft. Thus, for military aircraft, special arctic techniques are limited as a practical matter to procedures for airport *ground* handling under extremely low temperatures, to search and rescue problems, and to liaison flying in support of isolated outposts or ground troops.

But in these aspects of military flying, and in the very considerable away-from-the-airport bush flying done by northern commercial and sportsman pilots, the resourcefulness, the ingenuity, and intelligent improvisation formerly demanded of all northern pilots should be familiar to every pilot flying in the north.

What may perhaps be called the classic description of bush flying, *The Flying North,* sets forth vividly the trials and tribulations of the early Alaskan bush pilots. Though describing incidents of the 1920's, '30's, and '40's, it provides modern and practical descriptions of situations with which the bush pilot may still have to cope. Many of the ways in which the early pilots solved their problems are still valid in principle. The methods of those years are now out of date only in that (1) modern civilian bush planes have much higher horsepower for the same weight, and therefore have greatly improved takeoff and rate of climb performance; (2) in today's engine the danger of forced landing due to power failure has been reduced from an average of one per several hundred hours of flight time to one per 5000 or more hours of flight time; (3) modern bush planes have highly effective wheel brakes, often combined with tricycle landing gear: (4) the problem of flying between dry ground and deeply snow covered landing strips has been solved by ski-wheel gear for even the smallest planes; and (5) excellent FAA facilities, and those of its equivalent, the Department of Transport, in Canada, provide reliable and continuous in flight radio communications and weather reporting not formerly available.

Winterizing

MODIFICATION OF AIRCRAFT FOR ARCTIC USE

Summer flying in the north does not require any special modification or adjustment of aircraft, but winter may; and "winterizing," in the sense of making adjustments, certainly will be required. Military aircraft produced after the middle 1950's have the necessary modifications for winter flying in the north in their design. Manufacturers of civil aircraft, particularly those like De Havilland of Canada and Cessna, whose aircraft are uniquely adaptable to northern flying, provide FAA approved winterization kits.

These kits generally consist of easily installed baffles designed to increase engine temperatures approximately 50°F by partially covering the engine's cooling air inlets. This is considered a good compromise between effectiveness and FAA removal requirements for operation in the United States and the lower provinces of Canada. Kits also include oil radiator shutters, intake system "lagging" (insulation), and carburetor intake air baffles designed to improve mixture distribution in cold weather.

The more extensive modifications and revised operational procedures for true arctic operation are left to individual preference.

Extensive modifications of some aircraft are still definitely necessary. Winterizing of the aircraft in the fall and winter as cold weather approaches, or when the aircraft is flown into the northern winter from temperate regions, not only includes making what modifications of the aircraft may be necessary, but may include minor structural changes as well as simple adjustments, in order to cope with extremely low temperatures. Shops with mechanics able to do this well are found at such air centers as Fairbanks and Anchorage, Alaska; Edmonton and Winnipeg, Canada; and other American and Canadian cities along the border.

MODIFICATIONS

1. Cabin Heating. A cabin heater which can maintain the temperature of instruments and cockpit accessories above 0°F and preferably above freezing, as well as warm pilot and passengers, is essential.

2. Defrosting. The cabin heater outlets must be designed to provide a blast of warm air onto the windshield in front of the pilot's face. At subzero temperatures the inside of the windshield will frost up despite adequate cabin heat, and planning to wipe the frost away will not work.

3. Baffle Plates. Baffles of proper size and shape are placed so as to restrict the flow of engine cooling air. This may be done by placing plates over the outlet at the rear of the cowling which stop the flow of air around the engine and minimize the cooling effect; however, in really cold weather, −30°F or lower, it is best to have plates on the forward side of the cowling and plates or "lagging" (asbestos sheathing) over the front oil sump of radial engines, over the front of the crankcase if exposed, and over the propeller governor. The external oil lines and oil-tank cooler may also have to be "lagged" with asbestos padding, and the oil-tank cooler vanes partly or entirely covered with shielding. All this is done in order to keep the engine oil temperature adequately high while the aircraft is in full flight through low subzero weather. Without this, it is quite possible for the engine oil to drop so low in temperature that partial congealing and inadequate lubrication develop, with the obvious danger of engine damage or even complete failure in flight.

4. Carburetor Air Intake.   In cold weather, fuel-air mixtures are inherently leaner due to increased air density. At −40°F, the equivalent sea level pressure altitude is −5000 ft. This effect, coupled with improper fuel vaporization and distribution to cylinders, can cause engine roughness. In extreme cold, without heating the incoming air or restricting its flow, takeoff power will never be developed. The effects of these conditions are especially noticeable while operating on one magneto during ground checks, though opening the throttle above idling alone may cause the engine to quit.

For this reason an adequate carburetor heat system is absolutely essential. During warm up and ground check, use of full carburetor heat will reduce the ram of air and thus enrich the mixture; it also improves fuel vaporization. Heat is also required for takeoff, climb, cruise, and descent into colder air near the surface, or for flight at reduced power.

Using carburetor heat requires close attention to engine temperatures; however, when operating in subzero cold, avoid using partial carburetor heat because this may raise the carburetor air temperature to the 32° to 80° range in which icing can be critical under certain atmospheric conditions. In addition to using heat, one can counter the effects of extreme cold by selecting relatively high manifold pressure and rpm settings, and by avoiding excessive throttle movements both in flight and on the ground.

Some older light aircraft air intakes simply cannot be redesigned for safe flight at extremely low temperatures, but most can. This should be discussed in detail with the A and E mechanic who does this part of the winterizing job. He must know the particular engine-cowling characteristics of the individual airplane in this regard, and if he does not, one should find a mechanic who does.

5. The Crankcase Drain.   The engine crankcase should be equipped with a quick-release drain of large enough diameter to allow cold or partially congealed oil to run through it, an extension tube to pass through the cowling, and also, ideally, a little hole or hook upon which to hang a clean oil drain canister into which to run the engine oil immediately upon landing, when staking out of doors for a northern winter night.

6. The Breather Tube.   If the engine breather tube extends more than ½ in. outside the engine wall, it should be cut off. Long breather tubes have a tendency to ice up and clog tight in flight in low subzero temperatures because of the condensation of combustion products going past piston rings and valve guides into the crankcase so that in the usual case where the oil tank has a tight fitting filler cap, pressure can build up and burst the tank or the oil cooler.

7. *Oil Dilution.*   In some airplanes an oil dilution system must be installed as a modification of the plane. This permits bleeding a standard amount of gasoline from the fuel line into the engine oil system as the very last step before shutting down the engine on nights when the temperature at starting time in the morning is expected to be in the 0 to +32°F range. This dilution system is not, however, much used on the smaller bush aircraft. In any case, the oil dilution system should not be relied upon at temperatures below 0°F, for at such temperatures full preheating becomes essential.

8. *Emergency Equipment.*   The winterizing procedure will include getting together and placing in semipermanent stowage in the plane the special winter forced-landing-in-the-bush emergency gear, which may save the lives of pilot and passengers. The list of equipment for this purpose is discussed in this chapter under the section "Search and Rescue."

9. *Auxiliary Equipment.*   If there is any possibility of intentional landings and takeoffs being made in the bush away from airports, the following equipment should also be stowed in the plane as a part of the winterizing procedure: tie-down ropes, the firepot, engine tent, and any other accessories necessary for preheating; and wing covers or a long-handled T-broom for brushing frost off the wings in the morning if wing covers are not used, or a rope sufficiently long to scrape the frost off the tops of the wings by sawing it back and forth.

Ground Handling

Some ground handling requirements have been implied in the presentation of the list of winterizing procedures outlined above. To it the following points should be added.

SUMMER

In summer the only peculiar ground handling problem which seems to distinguish operations in the north from those in temperate regions is the appalling number of mosquitoes, blackflies, and tabanids (popularly known as "bulldogs" or "mooseflies") which for some strange reason are strongly attracted by open aircraft cabin windows if the plane is standing in the sun. The numbers which may collect inside the cabin may be even more than a nuisance because of the serious distraction they create if the pilot takes off without killing them. If they are killed, usually with an aerosol insecticide, the numbers of the bodies which

Figure 15.1. A typical subarctic bush pilot's airfield. Hood Lake, just outside Anchorage Alaska, is a well equipped bush ski landing strip (1947). *(Courtesy of Bradford Washburn, Boston Museum of Science.)*

collect over a period of time in crannies and cracks around the instrument panel can become a menace to keeping the instruments clean. Keeping all cabin windows shut while on the ground seems to be the best solution.

WINTER

The special procedures of winter ground handling, however, are not so easily carried out. For military aircraft, procedures for northern air bases have been worked out in detail in Technical Orders. But the following observations on the winter ground handling of civilian bush planes can also serve as a checklist for military propeller planes, because general principles only are considered.

USE OF HEATED HANGARS

Heated hangars naturally solve most of the problems of winter ground handling of aircraft. But there is one danger with heated hangars which must be kept in mind. When a plane has been in a hangar all night at temperatures above freezing and then is brought outside in the morning into temperatures below freezing, any snow or freezing rain which is falling will at first melt on the wings, but subsequently will turn to ice. Crashes have occurred under such circumstances, where the pilot has started his takeoff run before the temperature of the wing surface has adjusted itself to the outside air; for by the time of beginning to be airborne, the wing surface had dropped below freezing and the resultant wing frost or frozen precipitation interfered with lift enough to prevent the plane from flying properly. Therefore, on bringing a plane out from a heated hangar, the pilot should, if there is any precipitation at all, delay his takeoff until the wing temperatures have become adequately adjusted.

Another important point in connection with heated hangars is the danger of condensation in fuel tanks when the plane is brought in. In the tank whose fuel selector valve is in the OFF position, water may collect which in summer would run off harmlessly into the carburetor water trap when the valve is turned to the ON position in flight. At extremely low temperatures in flight, this water freezes at the valve and obstructs fuel flow. Forced landings from this cause are quite frequent in winter. The best protection is to refill the tanks immediately after flight before leaving the plane for the night, so that there is less space in which condensation can occur. Also, after temperature adjustment in flight has occurred, it is well to test the fuel flow from all tanks before getting too far from the takeoff field to return to base on the tank used for takeoff.

PARKING IN THE OPEN

The real problems of ground handling in winter develop when the airplane is staked down overnight outside the hangar or out in the bush. To present the handling of this operation chronologically, it should be said that by far the most important part of getting the airplane under way on a subzero morning is the work done on it the night before.

Landing Gear. The first thing to do is to park the aircraft in such a way that the landing gear will not be frozen to the ground in the morning. Wheels are easily run onto a mat of insulating material of some sort (if indeed the wheels will be a problem at all). But skis require more thought, for when frozen down, not only are they harder to loosen, but the rough, freshly broken out under surfaces in the morning may set up a very heavy drag and create a takeoff problem. The usual bush pilot solution for this is to lay a short round log of firewood, or even spruce branches, in the snow at the place where the skis are to come to rest for the night, and simply taxi the ship onto this mat.

If ski friction on the snow is likely to create a heavily loaded takeoff problem in the morning, a quite effective and practical "ski wax" for airplane skis is arranged by getting a can of kerosene and a couple of burlap bags the night before. Before takeoff the bags may be soaked with kerosene and laid one in front of each of the skis so that the ski bottoms will get a good smearing of kerosene just at the start of the takeoff run.

The new plastic undersurface Teflon skis reduce but do not eliminate freeze-down of the skis. Ski-wheel gear (Figure 15.2), in which the skis can be retracted

Figure 15.2. A typical ski-wheel landing gear design. Dr. Terris Moore examines the ski-wheel landing gear on his Supercub. Taken on Kahiltna Glacier, Mount McKinley, 1951. *(Courtesy of Bradford Washburn, Boston Museum of Science.)*

a few inches above the tire tread, solves this problem: Merely retract the skis when the plane is parked.

Engine Oil. The second problem springs from the fact that the most engine oil congeals to the consistency of soft butter at temperatures slightly below 0°F, and to the consistency of laundry soap at temperatures far below zero. True, there are some engine oils now in existence which maintain the same free flowing viscosity at these temperatures as at full operating temperatures. But unless the engine is lubricated with these engine oils, time honored principles for handling congealed engine oils will have to be observed.

If the morning temperatures are going to be somewhere between 0° and 32°F, it may be possible to start the engine without preheating or draining the oil the night before—particularly if the engine is equipped for oil dilution. The smaller planes ordinarily do not have these oil dilution systems, but some aircraft are now factory equipped with them. Remember that shutdown time in the evening, not warmup time in the morning is the time to use oil dilution. Engine manufacturers' recommendations for dilution must be carefully followed. A common error is to start dilution while the engine oil is so hot that the diluting gasoline is immediately vaporized; the beneficial result of diluting engine oil is then of course nil.

Engine Heating.   If the morning starting temperatures are going to be below 0°F, the pilot should definitely plan to preheat the engine before trying to start. For at temperatures below 0°F, starting the engine without preheating puts too much of a strain upon the close tolerances in various parts of the engine, under the extreme temperature differentials which will prevail immediately after the start, to warrant such cold starts by anyone who values his engine. Moreover, in the cabin, gyro instruments and the tachometer may be damaged by making sub-zero starts without preheating.

In preparation for preheating the engine, the usual process is to put the engine tent over the cowl the night before, wrapping it around beneath to serve also as an engine cover during the night. This will greatly reduce the formation of frost on the engine parts within the cowl. Moreover it will protect the engine spaces and cracks from filling with any blowing snow which may fly during the night—a potential source of water in the engine during preheating.

Where electrical power is available from outlets, the engine(s) may be kept warm overnight by a special oil dipstick electrical heater, resistance heaters on each cylinder and on the oil sump, or an electric space heater. For heating cold engines and cabins quickly, the well-equipped fixed base operator has a motor driven hot air heater. Other less elaborate heaters using propane gas and battery power are equally successful, if not as quick, in getting the preheat job done.

If the only source of heat available is some kind of open flame device, such as a plumber's firepot or a blow torch, strong precautions must be taken to avoid fire, such as running the fuel system dry the night before at engine shutdown, keeping sheltering covers well clear of the fire, and keeping a fire extinguisher handy.

If temperatures below − 20°F are to be expected at starting time, it will probably be well to drain the engine oil immediately after shutdown, while it is still hot, into a canister kept for the purpose, and to take this oil into the pilot's night shelter. It is very easy to heat this oil to 150°F during the breakfast cooking operations the next morning. When poured back into the engine hot, this oil will aid a quicker preheating and more uniform distribution of heat throughout the engine. One caution, if it has been heating rapidly over a hot flame, the oil in contact with the sides and bottom of the canister can be boiling hot while there is still a large lump of congealed oil in the center.

Engine heating is all-important for cold weather starting. To take short cuts at this time will result only in extensive delay because of engine damage, or because of the need to repeat the process until done properly.

Wing Frost or Ice.   When the aircraft is to be left out overnight in below freezing temperature with little or no wind, wing covers should be used if available. Any coating of frost or snow must be removed from wings and control surfaces before flight is attempted, because the least roughness of lifting surfaces seriously degrades lift. Ice which may prevent full movement of control surfaces must also be removed.

The relative merits of using wing covers instead of brushing the wings in the morning with a long-handled T-broom is a question which only experience can answer satisfactorily for a particular aircraft and particular weather conditions. If the wing covers can be put on easily, then using them will always be the preferable course, because (1) they give more complete protection, (2) it takes less time to get the covers off in the morning than to do the necessary wing brushing, (3) actual ice may be almost impossible to brush off, whereas taking off the wing covers will usually break it up easily, and (4) if the type of wing covers which have built-in spoilers are used, the aircraft gets additional protection against being blown away by high winds during the night. Wing covers are bulky and heavy, and this is where nylon really comes into its own. Nylon covers are only a fraction of the weight of the canvas ones, not so bulky, and do not tear so easily. If carried in the fuselage on bush flights, the advantages of the smaller bulk and lighter weight of nylon wing covers will be especially appreciated.

Tying Down.   The fifth step in the process will be the tie-down arrangements. If the airplane is parked beside a hangar on an airport, the usual stakes and cables will serve; in fact they are usually much more dependably fast in the frozen ground of winter than in summer. On the other hand, if stakes must be driven, the frozen ground will present a real problem. The ordinary tie-down kit carried in the plane will be useless for a winter tie-down job in the bush. If the landing is on the ice of a lake or river, as is frequently the case in winter wilderness landings, the job is fairly simple; with an ax, or preferably a regular ice chisel (which will be easier), chop three "arches" in the ice, one under each wing and the tail. If these are cut so that the ice is at least 4 to 5 in. thick at its thinnest places, the strength of the arches will be easily able to hold a plane down even if the wind is lifting it up against the ropes.

If the landing is on a strip without tie-downs, or "in the rough" on a gravel bar or frozen tundra, the tie-down problem may still be solved if the air temperature is well below freezing. Tying down can be carried out by mixing water with snow into slush and cementing a branch or board or other object to the ground with the tie-down rope around it. A surprisingly strong tie-down can be constructed in this way. But if the air temperature happens to be above freezing, or only a little below, and the ground is still solid from the winter freeze, there will be a real problem.

The frequent attempts of pilots to make a secure tie-down by hanging weights under the wings of the plane are often based upon an illusion. Actually it will take a weight of many hundreds of pounds to give anything more than minor protection this way. It should be remembered that a weight hung under the wing does little more good than if it were simply placed in the cabin. How often has one seen airplanes weighted down in this way with 50 lb. or so hung on each wing, creating a false sense of security, when actually the weights had less effect than would that of a pilot and passenger in the cabin! The use of wing spoilers, the best of which is the type built right into wing covers, will do at least as much good as a great deal of weight hung around the airplane. Quite effective wing spoilers can also be rather simply arranged by tying two-by-four lumber, or even tree branches, along the tops of the wings just above the leading edge.

Snow Runway Preparation.   The sixth step will be to think about the snow runway in the morning. Usually there will be nothing to do. But if the airplane is being staked out in the bush, the following point should be noted: If the pilot is going to have to snowshoe-tramp a runway for the morning take-off on skis, it is much better to do it the night before, even if snow should fall during the night to the point where it has to be tramped again in the morning. The reason is that a snow runway tramped the night before (or taxied over by using the plane's power—an easier way of doing it than with the feet on snowshoes) will always be much firmer than one so made in the morning immediately before use. It is one of the curious phenomena of snow that even if the temperature is below freezing and the snow dry at the time of tramping the night before, the runway will be much "tighter" after standing overnight than if tramped just before use in the morning. And of course if the temperature is above freezing the evening before and below freezing in the morning, the evening is certainly the time to do the runway tramping.

Snow is most slippery at melting temperatures. Drag increases noticeably with subzero temperatures, and longer ski runways become necessary.

Batteries.   Before leaving the airplane, the pilot must give thought to the storage battery. If the engine is small enough to be started easily by hand, and if the battery is fully charged and will only have to stand out for a few nights, there will be nothing to do beyond turning off the master switch. But the following facts about lead-sulfuric acid batteries are worth remembering, for such batteries are greatly weakened by cold. A fully charged lead-sulfuric acid battery will produce 100% output at 80°F, 65% output at 32°F, 40% output at 0°F, and 10% output at −30°F. At 0°, the electrical load needed to crank an engine may be as much as 250% of the load at 80°F. The combined effect of these oppositely moving curves is something to ponder! It means that one must thoroughly preheat to get the engine load back to normal levels; and it also means that one must either take the battery indoors to keep it warm in subzero weather, or put it on charge for some hours prior to use, or "trickle-charge" (a slight charge of long duration) it overnight right in place in the plane's battery box. Charging warms it up chemically right at the point of need: in the plates and acid.

Another fact to remember about lead-sulfate lead-sulfuric acid batteries is that although a fully charged battery will not freeze until −95°F, a half-discharged battery will freeze at −40°F, and a battery fully discharged to a specific gravity of 1160 will freeze at 0°F. Conclusion: If the pilot is going to leave lead-sulfate batteries in an airplane parked outdoors, he should expect them to freeze and break unless they are not only left but kept in a fully charged condition. The pilot should also remember that lead-sulfate batteries slowly discharge themselves, so that even in a single month's time of standing they can become half or more discharged. And of course if the slightest electrical drain happens accidentally to be left on them, they could become fully discharged and freeze and break in a much shorter time.

Nickel-cadmium batteries are not subject to any of these failings, and moreover are lighter in weight. This is why they are increasingly coming into use in the North even though they are much more expensive than the lead-sulfate type.

Window Frost.   The last thing to do on leaving the plane is to open the windows very slightly, or open the cabin vents. Unless this is done, frost may form on the insides of the cabin windows because of the relatively warmer moist air left in the cabin. At very low temperatures this frost will not be easy to remove. Ordinary scrapers cannot be used because of the curved surfaces, and because some aircraft windshield materials scratch easily. If there may be blowing snow during the night, this ventilating of the cabin should of course be done in such a way as not to let blowing snow enter.

Refueling.   The pilot should definitely refuel the airplane before leaving it in summer or before placing it in a heated hangar in winter, in order to minimize condensation and collection of water in the tanks. But to insist on doing this at the end of staking an airplane out on a subzero day when all the preceding operations have been gone through would probably constitute a counsel of perfection! In the middle of the arctic or subarctic winter, when temperatures are virtually certain not to rise above freezing in the low sun of a short winter day, the danger of condensation in the partially empty gas tanks for a few nights and even days is small enough to ignore for a short time. But there may be a slight steady growth of frost crystals in the tanks if they are empty. Hence the refueling job should be done as soon as it can conveniently be arranged. Under these conditions refueling may be done the next morning when light has returned and the pilot can see what he is doing.

GETTING STARTED IN THE MORNING

Preheating.   Starting from a heated hangar presents no difficulties. On an active airport some type of aircraft engine heater is usually available, or it may even be possible to have plugged in an electrical engine compartment or oil sump heater. Out in the bush, the solution most often will be a catalytic heater, left under the engine overnight, a small propane heater running its ignition and fan from a 12 volt battery (perhaps the one in the airplane), or a firepot. An insulated engine cover or a tent (Figure 15.3) makes the operation much more efficient.

Since the preheating process will have to be in operation anywhere from 20 to 60 min., this is the first thing to get started, so that while this heating is going on, other preparations can be made. If the temperature is low enough to endanger the operation of the instruments in the cabin, it will be essential also to have a separate flow of heat in the cabin under the instrument panel, for in most airplanes the firewall prevents any appreciable amount of heat from under the engine getting to

Figure 15.3 "Firepotting" and firepotting equipment. The pilot, the Rt. Rev. William J. Gordon, Bishop of Alaska, is heating his Cessna 170 with a one burner gasoline heater. He has placed a can containing engine oil, which he drained on landing, on top of the heater. The patches in the engine hood attest to the need for vigilance to avoid disaster by fire.

the instrument panel. (BEWARE OF AN EXPLOSION in the cabin if the cabin firepot goes out with the lower firepot still burning.)

Refueling. Refueling should not be done at all while the engine is being heated with an open-flame type of firepot, and this is an additional reason for doing it at night. The center tank should certainly not be filled just above where the engine is being firepotted with an open flame.

The most important thing to remember about refueling from drums is *always* to strain the gas through a *chamois*. Canadian bush pilots seem to prefer felt to chamois; the writer always uses *both* a felt layer and a chamois in his gasoline funnel. Even the finest metal-mesh screens that will not pass water that may be in the fuel will, however, pass the very fine rust particles which (bush pilots soon learn) infest all gasoline that comes from drums. Only the filters and traps in use on the most active airports where gasoline is constantly being drawn through a pressure system can produce gasoline which may be safely run straight into the airplane's tanks. All other gasoline is dangerous unless run through a chamois. It is so because of the unavoidable condensation, with resultant rusting, of the drums or tanks in which gasoline is stored. A funnel with a good chamois is one of the most important pieces of equipment the bush pilot can carry with him.

While the preheating is going on, the daily preflight check can be made. For airplanes which have been staked outdoors in the snow country, always include in this preflight check a very careful examination of the interior of the fuselage of the plane near the tail. Wind-driven snow goes through the cracks and control openings of the tail assembly and easily fills the tail of the aircraft with snow, making it dangerously tail-heavy but leaving nothing outside to attract attention to this condition. Be sure to look for it. The same danger exists with some aileron designs, and these too should be checked for internal wind-packed snow.

How long should preheating go on? At least until the propeller can be moved freely by hand. And if the temperature is below zero, it should last long enough to be sure that some heat is distributed throughout the engine and that oil throughout the system has become warm enough not to congeal and stop the oil flow anywhere. With a standard plumber's firepot and a 150 hp engine at −20°F, this might take 30 min. If the temperature is down near −50°F, it will take at least a full hour with such equipment, and possibly more. But with the new viscosity-free-flowing oils, moving the propeller with the hand is no longer a valid test, and firepotting with such oils is done by time and by feeling of the engine, especially the magnetos, tachometer, and gyros, which might be damaged with too cold a start.

Immediately upon starting the engine, the oil pressure should be watched closely, and if it does not begin to rise within 30 sec. or so and remain at normal, the engine should be stopped at once to preheat some more. Sometimes congealed oil in some parts of the engine will prevent the oil pressure from coming to normal on the oil pressure gauge; on the other hand, oil congealing in other parts of the system may cause the pressure to go far above normal—in fact right off the pressure-gauge scale; and unless this is immediately stopped, the oil lines may burst at the fittings.

Ice-bridging of the spark-plug points is another phenomenon with which the northern pilot should be thoroughly familiar. It usually occurs when the engine has not been preheated, or when heating has been inadequate, and the pilot is trying to start it cold at temperatures near zero. Under these conditions, if the engine fires a few times and possibly makes 6 to 10 revolutions, stops, and then cannot be started at all, the chances are good that the spark plugs have become ice-bridged. This happens when a few explosions warm the interior of the cylinders and cause condensation just above freezing. Then if there is no firing for some moments, the spark-plug points cool below freezing and ice forms. If this takes place across the spark-plug points, the ice will probably short them out, and no amount of cranking thereafter will cause the engine to fire unless most of the other cylinders come on sufficiently to run the engine, in which case the repeated compression will melt the ice. But if most of the plugs in the engine have become ice-bridged, the engine cannot be started no matter how much it is cranked. The only thing to do in this case is to remove one plug from each cylinder and put them in a stove or under a blowtorch. The practice of some pilots of laying a blowtorch right on the heads of the plugs while they are in place is simpler and quicker and will get rid of the ice, but it is rather rough treatment and may scorch the spark-plug harness.

Taking off with flat struts is of course very hard on the struts and on the aircraft, both at takeoff and at the following landing. Proper attention to the oleo seals and constant watching of strut pressure will cure this trouble. A pilot flying from a warm airport to a cold one, or moving his aircraft from a warm hangar out into the cold, should expect trouble with oleo struts unless the struts have been properly winterized and unless air pressures have been increased to compensate for contraction of the air in extreme cold.

After taking off in slush or wet snow, landing gear should be left extended for a short time and then retracted and lowered two or three times to prevent thrown-up slush from freezing them immobile in the retracted position.

Condensation of water in oil tanks at very low temperatures means that oil sumps must be frequently drained. Therefore tank vents and crankcase breathers should be inspected frequently to make sure they are free from ice and snow. In some in-line engines, the oil is carried in the crankcase. Thus it is left there during the night and heated by the firepot while heating the engine. But if the oil is not drained after each flight, condensation can cause water to accumulate. To prevent this it is well to give the quick-drain valve a brief turn after each flight to allow water in the drain pipe to escape while the oil is hot and the water in a fluid condition.

LOADING

No discussion of ground handling in northern flying would be complete without some mention of the overloading and the strange cargoes sometimes lashed to the outside of aircraft by bush pilots. The only valid comment which can be made is: "Don't do it." The practice is dangerous, for the following reasons: (1) Overloading raises the stalling speed of the airplane. (2) Unless the overloading is done with the proper balance, the flight characteristics of the aircraft may be considerably altered. An airplane that is "forgiving" in its behavior with normal load can become vicious around the stalling point, snapping into a spin easily and coming out only with difficulty. This becomes particularly true if a rearward center of gravity has been created by the overload. (3) If an overloaded plane is flown through rough air, there may be some danger from structural failure which would otherwise be absent. The g load safety factor built into the aircraft to handle turbulent air may be exceeded if the plane is flown through strong turbulence in an overloaded condition. In mountainous terrain this can of course be encountered in clear air.

It is a fact that canoes are quite frequently flown successfully, lashed to the struts of light seaplanes. The practice cannot be validly endorsed, other than by the comment that if this must be done, it should not be undertaken without careful briefing from a pilot with experience, because of the possible aerodynamic effects on the particular airplane.

Federal Aviation Regulation 91.38 authorizes up to a 15% increase above certificated gross weight for certain air taxi and commercial operations and certain Department of Interior activities, in the state of Alaska. These provisions do not apply to other operators. Special authorization must be obtained from the FAA.

In-Flight Problems

LIMITING LOW TEMPERATURES

One question that may arise is: How far down in temperature can one safely fly a plane? Metallurgists report that many metals lose their strength quite rapidly in temperatures of 40°, 50°, and 60° below zero. What is the danger of structural failure in flying a plane at these temperatures?

Bush pilots formerly made it a policy (and some still do) not to fly below 35° or 40° below zero. But there now seems to be an increasing amount of flying done at these temperatures, and few if any reports of structural failure due to extremely low temperatures seem to be resulting. There are at least two good reasons why this may be true. One is that at these very low temperatures the air is almost always dead calm; hence there is no strain from turbulence. Another is that pilots flying in these extreme temperatures usually fly in a thoroughly restrained manner. If an aircraft would normally stand 4.5 to 5 *g*'s in flight before failure, and a 60-below temperature reduced this strength by, say 50% to 2.5 *g*, probably nothing would happen in flight with a 1.5-*g* load in very smooth air and the complete absence of any acrobatic flying.

An important but quite different argument against flying at temperatures below − 35 to 40°F, however, is that a forced landing in the wilderness at such temperatures is a very serious matter.

Temperatures below − 40°F, except in the Polar Basin itself, usually occur only under conditions of dead calm and in a very thin layer of air only a few hundred feet thick, close to the earth's surface. Thus the aircraft under these conditions makes most of its flight aloft where the air is warmer than at the surface. An example of this temperature inversion occurred when a thermometer at the base of the airport tower on old Weeks Field, Fairbanks, Alaska, read − 30°F at the same time that the thermometer at the top of the tower read 0°F. A more normal inversion was one observed during the coldest flight the writer happens to have made. A standard thermometer in an official shelter at the base of the hill at College, Alaska, read − 64°F on the coldest day of the winter of 1952–53. At a house part way up the hill, 150 ft. vertically above the base, the temperature was − 54°F. At the top of the hill, where the aircraft was tied down, approximately 225 ft. higher than the base, the wing thermometer read − 52°F. Staked out and cold-soaked, the aircraft was firepotted and operating in about an hour and a quarter. Immediately after take-off the wing thermometer showed a steady rise in temperature, and at 1000 ft. above the bottom of the hill the air temperature was − 35°F; at 2000 ft. it was − 24°F. This sort of temperature gradient may be considered normal for extremely low continental temperatures.

ICE FOG

This is undoubtedly one of the most serious problems of northern flying. There are quite different types of ice fog. The supercooled water-droplet variety can exist down to at least $-30°F$ in dead calm air. In fact laboratory experiments have shown that droplets can actually exist in liquid form down to $-50°F$. These droplets solidify instantly on contact with any object, and an aircraft moving in this kind of ice fog is immediately glazed with ice. The first problem for the pilot is that his windshield picks up an ice glaze from outside and rapidly becomes opaque. Flight may be possible for some time, however, because the coating is light and takes some time to build up enough thickness to appreciably alter the airfoil section of the wing. Since most ice fogs lie in comparatively thin layers only one to several hundred feet thick, close to the ground, the usual remedy is to climb out of it. Failing that, a ''one eighty'' or immediate landing is indicated.

The other type of ice fog does not attach itself to aircraft, but seriously obscures horizontal visibility. Though airfield buildings or lights may be dimly or sometimes quite clearly visible from above, the final approach through the shallow fog layer will be virtually blind. This is particularly true at dusk. At night, strong runway lights are valuable, but aircraft landing lights are worse than useless, for they give only a blinding glare.

WHITEOUT

Whiteout is another phenomenon peculiar to northern flying which can be exceedingly serious. It occurs most frequently in treeless regions when the ground is completely snow-covered and in addition the sky is overcast, perhaps with oncoming twilight. If the light conditions are such that downward, horizontally, and upward there is the same milkiness and no horizon is visible, the phenomenon of whiteout may occur. In this case the pilot suddenly and completely loses his orientation. He suddenly has no sense of horizon or of up or down. He is quite as helpless as if he were flying in a dense fog, even though the visibility may actually be perfect. He is in instrument conditions. It is a terrifying experience.

Whiteout is much less detectable in advance and therefore more dangerous than ordinary fog, because the latter is easily seen at a distance and therefore avoided—not so with whiteout. Though an experienced pilot will begin to get a warning from the onset of the symptoms just mentioned, it is not possible, as with fog, to see ahead just where the whiteout begins and ends. It is quite possible to get into it without realizing what is happening, until suddenly the pilot finds himself in instrument conditions—and possibly without the necessary equipment and training. Fortunately whiteout is not common except in the far north beyond the limit of trees, and in the Antarctic.

What to do about whiteout is easy to state but may be difficult to carry out.

This is definitely and insistently to avoid whiteouts unless one is completely prepared for instrument flight.

If unexpectedly caught in a whiteout without blind flying instruments, the situation is exceedingly serious, and not unlike the question of what to do if caught in fog or cloud of unknown depth and extent without blind flying instruments: a neat little problem in the theory of escape. The best advice probably is not to try to make a 180° turn without instruments, but instead to cut the power, lower wing flaps, take hands off the stick, and hope somehow with the rudder to keep the plane out of a fatal spiral dive and let it settle to earth in a mushing glide at 10 knots above stall and, say, 100 fpm. The pilot should keep his eyes open for a glimpse of some land object which will restore his horizon and depth perception, and permit him to carry out a landing.

If the pilot has smoke bombs with him and is only a few hundred feet above the terrain, he may, by throwing out a bomb and reducing the speed to no more than a safe margin above stall, be able to glimpse the bomb behind him on the snow, regain his orientation, and make a landing. Old time bush pilots sometimes carried a sack of spruce boughs to throw out for this very purpose. The spruce boughs have an advantage over smoke bombs in remaining visible indefinitely rather than burning out in a few minutes. It seems a little excessive to counsel carrying a sack of spruce boughs in the plane. But this might be a good thing to take on an essential flight in which whiteout appears a possibility.

CARBURETOR ICING

It might be supposed that carburetor icing is especially serious in northern flying. Though serious enough, it is not more of a problem in the arctic than it is in winter, early spring, and later fall flying in the temperate zones. The standard discussions of carburetor icing as they apply to a particular aircraft are equally applicable to the same plane in northern flying.

Most carburetor icing occurs when the air has a high relative humidity and the temperature is between 45° and 60°F above zero, because under such conditions refrigeration effects inside the carburetor create condensation and also drop its internal temperature below freezing. Application of carburetor heat raises the temperature at the points where freezing occurs, and melts out the ice; better yet, if used in time, it prevents ice formation.

But with some carburetors there is a zone where in very low temperatures if carburetor heat is applied, this will *cause* icing. Under these conditions (usually 0° to − 10°F) the carburetor will be operating at temperatures far below freezing. Now if the pilot applies the carburetor heat, he may raise the temperatures at critical points within the carburetor to just above freezing and cause melting water to run to colder areas and freeze. Thus at low temperatures the pilot should be cautious about applying carburetor heat until he knows the exact pecularities of his particular engine in this respect.

A very neat little problem can arise with some carburetor and engine cowling combinations. Some engines must, in extreme subzero air, have some carburetor heat on during throttle back when descending, in order to prevent their dying in flight. How likely is this carburetor heat to cause icing within the carburetor, as just described? In most cases it will not, but there may be some of the older planes in which it will. One should discuss the particular aircraft with an experienced northern A and E mechanic who knows that type of aircraft well with regard to these details.

ICING OF THE AIRCRAFT STRUCTURE IN FLIGHT

The discussion of this subject in Chapter 6 will be equally applicable to northern flying. Two special conditions do, however, occur in the far north. One of these has already been mentioned: the supercooled water-droplet type of ice fog. The other is a rather rarer type of fine frost which occasionally occurs in very low subzero temperatures. In this case the frost collects not merely on the leading edges of the aircraft wings and structure, but rather evenly over the entire structure of the plane. Typically it builds up quite slowly so that the pilot can keep flying for some time; but eventually he has to land, perhaps on a frozen lake or river, and brush the frost off wings and tail assembly before taking to the air again.

NAVIGATION AND RADIO

These two factors become especially important in northern flying, and they are closely related. Because distances between towns are so vast in the tremendous wilderness reaches of the north, navigation becomes a very serious matter. Getting lost is never a joke in the far north, where habitations are exceedingly few and far between. No pilot should undertake a northern flight unless he is absolutely sure that his navigation is accurate and dependable.

The special navigational problems, arising out of the fact that the magnetic compass becomes increasingly unreliable and finally useless as one draws within a few hundred miles of the North Magnetic Pole, are discussed in Chapter 11.

The radio aids which make navigation so easy in the States are scarce in the far north. This subject becomes highly technical and complicated and is worth a chapter of its own.

In general the VHF and UHF radio systems so widely used "Down South" are used to only a limited degree in the Far North, because these systems cannot give the long range which is obtainable from the older aircraft MF/HF sets developed during the 1930's and '40's. This long range is especially necessary for an aircraft down on the ground sending out signals. Moreover, the electrical static which is such a handicap to LF/MF radio in the States, particularly in summer, does not exist to nearly the same degree in the Far North, because there are fewer thunderstorms within radio hearing range.

Discussion of in flight problems of the North should not be closed without adding that most experienced northern pilots in general prefer northern flying to "south 48" operations, largely because the air in northern flying is so often "smooth as cream." With the sun rolling low around the horizon and never getting high in the sky, there is on the average very much less thermal turbulence in northern skies than father south. The average precipitation also is considerably lower in the Far North. Both of these factors tend to make for less cloudiness, smoother air, and therefore pleasanter flying.

Landings and Takeoffs "In the Rough"

The phrase *landing in the rough* is used by northern bush pilots to mean making a landing in an unprepared place, relying solely upon observations made from above in flight to carry it off. In the early days of northern flying, landings "in the rough" were quite frequent. And though the need for such landings is becoming less as airports and small airstrips have been placed throughout the North, nevertheless the northern pilot will not be fully acquainted with his subject if he does not know what is involved, nor fully accomplished in his profession if he has not had experience of this sort.

In general, the big problem with a landing "in the rough" is the great decision: whether to commit the airplane to the landing or not. By and large the successful bush pilots are the ones who in the past decided "No" at the right times and places. But it is not enough always to say "No" as a matter of principle. Any good airport pilot can make it an invariable rule to say "No" to the question of landing in the rough, but he will not then be a northern bush pilot. The fact of the matter is that landings in the rough can be made in remarkable locations and still keep the risk within reasonable limits. The competent bush pilot will be able to identify the occasions when the answer can be "Yes," and be willing and able to carry the operation out successfully.

Estimating Landing Possibilities

The method can be stated quite simply. A strip of ground, snow, or ice must be located, long enough and firm enough to get the plane down and to a stop and free of obstructions on the sides to clear the wings. And if it is an aircraft requiring more length of runway to take off than to land, the terrain must also be such as to allow the pilot to sufficiently improve the strip after he is on the ground, so that he can get into the air again. To appraise these possibilities from the air with any accuracy, the pilot must first of all know exactly what his airplane can and cannot do in the way of short landings and take-offs, with different loads, at various altitudes and temperatures. Ordinary airport flying with civilian planes does not call for such precise knowledge on the part of the pilot. But the competent bush pilot, at the time of his first acquaintance with a new airplane, will make a point of carefully trying the plane out with just these questions in mind. This can be done easily at the local airfield by pacing off distances carefully to measure landing rolls and minimum take-off runs necessary to get airborne with different load and air conditions.

A flap-equipped airplane lends itself well to placement landings. They are safely practiced by getting down about 10 ft. off the ground at least 100 yards away from the landing line. Then with flaps down, just enough power to keep the airspeed reasonably above the stalling point, and "hanging on the prop" in this condition, the pilot comes in, very gradually easing down his altitude to about 3 to 5 ft. and slightly reducing power as he gets deeper into ground effect. As his wheels are approaching the line—experience will teach him just where for the particular airplane—he cuts the power; then as rapidly as he can move one hand (for the other will be busy holding the stick against air pressure) he raises the flaps. This move, spilling the lift out of his flaps, will, if he is flying slowly enough at this point, drop him onto the ground right at the spot where he wishes to land. He then immediately applies brakes and keeps them on as hard as he can without nosing up the airplane, throughout the landing run to full stop.

On skis he practices this with different kinds of snow and ice landing surfaces. The takeoff, and particularly the landing runs on skis, with no brakes possible, differ widely with varying snow conditions.

Having learned from this practice how many airplane lengths he needs for landing and for takeoff, he next practices these from the air in another part of the airport, with different markers and unknown distances, attempting to judge the distances from the air by measuring airplane lengths on the ground with his eye. In this way he will learn to estimate quite accurately from the air how much distance on the ground he will need for landing in an unfamiliar place.

In actual experience in the bush, the three principal uncertainties will be (1) just how rough the ground is (it will almost never be really smooth), (2) whether the ground is sloping, and (3) whether the surface is firm enough to keep the wheels from sinking in too deep to roll. "Dragging" the proposed landing place several times from about 10 to 20 ft. high, with sufficient airspeed to keep flight safe, and studying the terrain carefully constitute the best way to develop some answers to these questions.

During this appraisal operation, the pilot must remember that rocks and rough spots on the proposed landing area must be lower than the radius of the wheels; otherwise the wheels will catch and tear off the landing gear, instead of rolling over the obstacles. Also the pilot must remember that even a slight downhill grade will make it virtually essential to use the "spill the flaps" technique of placements landing; otherwise he probably will float an unpredictable distance down the hill beyond the point of intended touchdown. Also, even a slight downhill grade will very greatly increase the length of the landing roll once his wheels are on the ground. Because of these fundamental disadvantages in downhill landings, it is often considered preferable to land uphill, even at the cost of landing downwind. Whether the benefits of uphill will more than offset the disadvantages of downwind will be something for the pilot to judge in each particular case.

Also the matter of appraising the firmness of the ground by its appearance from 10 to 20 ft. in the air is something for the pilot's judgment, for which descriptions in a book cannot begin to provide an adequate substitute for practical experience.

Replacing the stock wheels on the aircraft with large oversize wheels will help enable it to ride over rougher ground than is possible with stock model wheels. This is because the greater radius of the oversize wheels enables them to roll over obstructions that would stop the shorter radius wheels. Tandem wheel gear, if the wheels are of standard size only and not oversize, will do little, if indeed anything, toward overriding rough obstacles because the radius of the leading wheel—the critical consideration—remains unimproved. Tandem gear will help greatly in running on soft ground without sinking in because the ground pressure of the wheel tread has been halved. But tandem wheels render taxiing turns more difficult.

WILDERNESS LANDING PLACES

River Bars.   To pilots who have never flown in Alaska, accounts of the early fliers who used river bars (Figure 15.4) throughout the interior almost as if they were ready-made airports seem at times inexplicable, for river bars elsewhere are not usually firm enough, large enough, or smooth enough for landing airplanes. For pilots who have not had practical experience with them, it should be said that river bars in the interior of Alaska, Yukon Territory, and parts of the Northwest Territories of Canada do possess certain qualities of firmness and smoothness not found in other rivers. The reason why becomes an interesting speculation in geology. This aspect of the subject does not seem to have been thoroughly investigated, but the writer notes that it seems to be a peculiar windblown loesslike material in these river gravels which gives them their firmness and smoothness and also that these qualities seem to exist in those valleys which have glaciers at their headwaters. The rock flour getting into the river water from the glacial action above seems to add a binder to the river bar materials. Speculative theory aside, it is a fact that river bars in this region do not in general have the soft-as-sand consistency found elsewhere in river bars, but are usually firm enough to support the tires of light aircraft. The seriousness of snags or sticks and logs which may be lying about, and the depth of gullies cut out by the shifting water, will be the principal questions to answer when contemplating landings "in the rough" on these river bars.

Figure 15.4. Natural landing fields of the arctic. *Upper:* Braided streams, like the Yukon River in Alaska, with their broad expanses of gravel bar or snow covered ice, or water in summer, provide a principal type of northern landing field. *Lower:* A gravel bar at Allakaket, Alaska, on the Koyukuk, a tributary of the Yukon. Such bars are frequently large enough and firm enough for safe operation of aircraft the size of the DC-3. *(Courtesy of The Rt. Rev. William J. Gordon.)*

Sea Beaches.   In the north, as elsewhere, sea beaches often offer opportunities for the pilot of a light aircraft to make successful landings "in the rough." Although in other places beach landings would probably be in violation of local law and of very little use since there would be other transportation facilities nearby, in the north sea beach landings may occasionally be useful.

There are two main considerations in making sea beach landings. The first is that on most beaches only the area between low tide and high tide lines will be firm enough to support the landing gear adequately. With no more knowledge than can be gained by looking down on it from above, the pilot must assume that the sand above high-tide mark will be so soft that ordinary wheel landing gear will sink in and the aircraft will nose over. Therefore if the pilot coming in to land finds the tide is at full high, the sound decision, if there is any appreciable amount of fuel left in the tanks, will almost certainly be "Don't land." He should remember that by throttling the engine back to the point where the aircraft is just kept airborne and then leaning out the engine mixture to the point of minimum gasoline ratio consistent with full power for this minimum throttle setting, the gasoline consumption per hour can be dropped almost to one third of what it is at normal cruise. Also the pilot should bear in mind that the interval between high and low tide is about 6 hours anywhere in the world. Thus if the pilot has only a 1-hour reserve left in his tanks as he comes in over the beach at high tide, he can stretch this hour to 3, thus giving time to uncover half the area of the beach between high and low tide marks. Even a half-hour's supply can be stretched to 1½ hours, sufficient in many cases to uncover a strip of beach firm enough to set the wheels down.

If the pilot should, however, be confronted with the necessity of making an immediate landing at full high tide, the chances are he would be better off to avoid the soft sand above high-tide mark, and to land instead between the waves below high-tide mark. If he comes in low with flaps, "hanging on the propeller," 3 ft. or so above where the waves are running on the beach, and "drops her in" by spilling the flaps just as a wave is running out, he will have a few seconds on firm sand between waves to bring the plane to a stop, or at least to swing it up into the loose sand above high-tide mark at considerably less than landing speed.

The other important point to remember in beach landings is that most sea beaches are sloping, and yet the landing must be made across and not up the slope. This means that the pilot will have to cope with the fact that aircraft rolling on sloping ground have an inherent tendency to turn into the uphill direction. This tendency is increased by the steepness of the hill and in practical terms is caused by the uphill wheel getting more drag as the aircraft rolls along. On both landing and takeoff, the pilot will have to apply opposite rudder away from the hill. At low speeds there may not be enough rudder control to hold the airplane straight. Only experience based on trials will give the pilot the necessary judgment to carry off landings on sea beaches with dependable success.

Tides in the Polar Basin are slight, amounting to but a foot on most beaches of the Arctic Ocean. In the subarctic, however, they may be considerable: Cook Inlet in Alaska has one of the world's higher tide ranges. Geophysicists tell us that in the Northern Hemisphere a narrow inlet on the west side of a north-south peninsula will produce very high tides. Examples are Cook Inlet, the Bay of Fundy in Nova Scotia, Inchon in west Korea, and presumably also Kamchatka.

Tidal Flats. In the early years of bush flying, slippery tidal flats were sometimes used for landings and take-offs with skis and with floats. The fully qualified bush pilot should be able to handle such situations if need arises.

Historically, the tidal flats have been used in two quite different ways. In Alaska, particularly near Valdez, they were used by bush pilot Bob Reeve, before the days of combination ski-wheel gear, to solve the problem of landing on mountain glaciers when take-off had to be made from a snowless field or airport. The ingenious bush pilot's answer was, "You land your wheel plane on a road beside a tidal flat, wheel it over to the tidal flat, put it on skis there, and then, because the mud is slippery, you just take off from the tidal flat on skis— mud flying in all directions behind you—and go on up to the glacier." And coming back the tidal flat is just as easy to land on as to take off from with skis.

In Portland, Maine, and no doubt in other places, the tidal flats between the airport and the city were in the past successfully used by floatplane pilots who did not have time to wait until the tide came in. Some floatplane float bottoms will slide almost as well as skis.

Roads.   In most states landing on automobile roads is against the law. But in remote regions it is not against the law; and under emergency conditions such landings may of course become necessary. In general the problem is quite simple. To pilots who have not done it and may some day need to, it can be stated from experience that the narrowness of the road will be less of a problem than might be expected. But if there are any telephone or power lines along the road, at the usual standard distance from it, this usually makes the landing so dangerous that it should not be attempted at all.

Frozen Lakes.   These make excellent landing fields under certain conditions, especially for ski-planes (Figure 15.5). The ice must of course be thick enough to support the aircraft. A minimum of 4 to 5 in. will suffice for most light civilian planes, particularly if they are on skis, provided the thickness is uniform everywhere that the plane will be moving on the ice. A simple rule for clear uniform

Figure 15.5. Sea ice landing strip at Kotzebue, Alaska. Though there is a paved landing strip at Kotzebue, the sea ice is more convenient in winter because it is adjacent to the town and provides a variety of takeoff and landing directions. The picture was taken in November, 1949.

lake ice is to allow 6 in. for the first ton of weight and ½ in. extra for each subsequent ton. For sea ice these figures should be doubled.

The mere fact that a lake is snow covered does not of itself indicate that the ice is sufficiently strong. Only a measurement of the ice by someone on the ground or a reasonably sure indication of the ice thickness given by a knowledge of the number of freezing days preceding is a safe basis for landing. Appearance of lake ice from the air can be quite misleading. Black-looking ice in the fall, following some days of severe cold with no wind or snow, can look very thin yet be sufficiently strong.

Glare ice can be more treacherous in a quite different and unexpected manner. A surprising number of aircraft are damaged each year in the early winter by inexperienced pilots landing with ski planes on the glare ice of newly frozen lakes. To their surprise they are unable to stop the airplane, and it slides right on, crashing into the shore and woods. The landing slide of a ski plane on glare ice (particularly if the pilot mistakenly believes that there is no wind and actually lands with a slight wind on his tail) can be an unexpectedly long one, especially if the skis are not equipped with "skegs."

A wheel landing on glare ice is subject to the same problem. Wheel brakes can be almost completely useless if the lake surface is glare ice; the plane will be almost as badly off in this respect as if it were on skis. A very smooth ice surface with a recent film of light rain on it will produce the ultimate in this type of difficulty. Far below zero the difficulty disappears, for then ice ceases to be slippery.

Another phenomenon of frozen lakes can be that of unseen "overflow water," under the snow but on top of the ice. This sometimes occurs at extremely low temperatures, when the inlet stream of the lake happens to freeze in such a way that the flowing water gets forced out and runs along on top of the ice and so out onto the top of the lake (often unseen beneath the snow cover and insulated from intense cold by it). Frequently this will flow along the shore line. Because the water is underneath the snow, it may not be seen until one is into it; and immediately after the plane's skis get into the water, rough ice forms on their undersurfaces. About the only helpful comment is that overflow is most likely to occur somewhere near the mouth of an inlet stream and along the lake's shoreline. Hence if the pilot does not know whether this overflow exists or not, it is best to land away from the inlet stream and well out from shore, and not to taxi near the shore until the pilot has gotten out and walked over to the shore to test for overflow water. Another way is to touch down but keep up a fast taxi and then take off without stopping. Any water coming up beneath the skis will get swept off by the snow before it freezes. Aloft again, from the air any water on top of the ice but beneath the snow is very likely now to show up in the taxi tracks. If there is still any doubt, a second or even third "press down" of this sort will make the true condition clear.

It should also be mentioned that many airplanes have been completely demolished because the pilot assumed that he could see the surface of the snow when trying to land under poor light conditions. This is difficult and many times impossible when landing in the middle of a lake during a northern winter, under an overcast sky in twilight. Under these conditions it may be utterly impossible to see the snow surface, and pilots have been known to fly right into the ice, thinking they had 200 to 300 ft. of altitude. The remedy for this is to land parallel and close to a straight shoreline where the brush or trees provide a contrasting reference line. There is another very good reason for not landing in the middle of a lake when the light is poor. Wind often piles up snowdrifts like ocean waves. These may freeze as hard as concrete over a period of time. They are usually found in the center of the lake which is exposed to the wind. Thus the snow is smoother along the shoreline and may be perfectly smooth on the side lee to the prevailing wind.

Frozen Rivers.   These are sometimes less suitable than frozen lakes for landings "in the rough," because of the effects produced by current. Currents are likely to create rough ice by breaking up the ice layers and running them up over each other. Current also may produce variable ice, which in one place may be quite adequate to support a ski plane, yet immediately nearby is treacherous ice which looks thick but is not. Also rivers, even more than lakes, produce overflow water which even in subzero weather may be running on top of the ice, but is invisible under the snow. Once a plane is safely down, so that the pilot has a chance to walk around, frozen rivers usually offer good opportunities for airstrips as long as freezing temperatures prevail.

Winter Tundra.   The tundra in winter offers many excellent places to land light ski planes, but the surface must of course be fairly well covered with snow. The small vegetation mounds which make some parts of the tundra rough in summer are usually well filled in after a few snowfalls. In general, less snow is necessary on the tundra or barren grounds north of the limit of trees than on the patches of tundra farther south, because these mounds of vegetation are more prevalent farther south. On the other hand, the "polygonal ground" found in the Far North may at times produce roughness, though in general it can be said that the summer appearance of these polygonal cracks from the air gives a greater impression of ground irregularity than is found when actually walking upon them.

Inviting though it may appear from the air to the inexperienced pilot, the summer tundra is no place to land a wheel plane "in the rough," relying solely upon observations made from the air. It is generally too soft and spongy to take the risk. Airstrips can sometimes be devised on the summer tundra by prior study and minor improvement on the ground, especially on the true barren grounds beyond the limit of trees. An example of this is the old airstrip beside the civilian community at Point Barrow.

Undoubtedly ski planes can be successfully landed upon the unfrozen summer tundra. And possibly the modern very high powered small planes which are still light in weight might even be flown off the summer tundra with skis.

Mountains and Glaciers. Occasionally airstrips can be worked out, with only minor improvement, on the rocks and ground of mountain ridge tops. But in general true landings "in the rough" can be made in the mountains only by float-planes on mountain ponds or lakes, or by ski equipped planes on the surface of mountain glaciers.

The most serious problem in all mountain landings and take-offs, whether "in the rough" or on slightly or even well prepared airstrips, will be that of altitude, and if there is any appreciable wind, of downdrafts and severe turbulence.

The effects of altitude are fairly simple and straightforward. They can be determined in advance by study of the engineering tables for the particular engine and aircraft to be used. The engine loses power as altitude increases, in general proportionately as the barometric pressure falls off. The aircraft true landing speed increases with the decrease in the density of the air. But the pilot can obtain the true landing airspeed for a given altitude if he merely applies the correction which translates indicated airspeed for a given altitude to actual ground speed. The usual rule of thumb is that to get the actual speed from the indicated airspeed, the pilot must add to the IAS 2% for each 1000 ft. of altitude. This rule is good only for altitudes up to about 10,000 ft. The correction is actually slightly more than 2% per 1,000 ft. for the lower altitudes, slightly less than that in the higher altitude layers, and still less above 10,000 ft. The point is that the change goes not in a straight line but in a smooth curve, with a steadily decreasing rate of decrease in pressure with rise of altitude.

The operation can be simplified by stating that in landing an airplane even at high altitude, the pilot makes a landing normal in every way, including the fact that the stalling speed as it is *indicated* on his airspeed instrument will indicate the point of his actual stall, in exactly the same place on his instrument whether he is landing at sea level, at moderate altitude, or at very high altitudes.

Takeoffs are, as a practical matter, much more affected by high altitude than are landings. Not only will the airplane require greater actual speed to get airborne at its normal IAS, but for this very takeoff the power of the engine will have been considerably reduced by the altitude. The amount of this reduction in power can be accurately ascertained only from the engineering tables for the particular engine. In general it may be said that at 10,000 ft. the power available (unless the engine has a supercharger, as few light civilian planes have) will be about one third less than at sea level. But the snappiness of the takeoff and the rate of climb immediately available to clear obstacles will be reduced by even more than one-third. The reason is that these depend on the *excess* of power available above that required to overcome the ground friction in the takeoff and to maintain level flight after the plane is airborne. This *marginal* power will be reduced by much more, percentagewise, then the engine's total power is reduced.

The adverse effects of high temperature, relatively unimportant with modern planes at low altitudes, can become a serious matter in takeoffs at high altitudes.

The reason is that the margin of excess power necessary to get airborne and to develop a rate of climb is so greatly reduced that the introduction of another depressing factor—high temperature—may spell the difference between being able to take off and not. This point is well known to Rocky Mountain pilots who operate from high altitude western airstrips. But high altitude mountain takeoffs in the north, particularly when made from glacier surfaces, are not usually burdened with the additional handicap of a hot summer temperature, so frequent in Rocky Mountain summers.

The factors of downdrafts and severe air turbulence in mountain landings and takeoffs are so complicated, yet so important, that the best advice which can be given is the following: Do not undertake to make any landings and takeoffs in the mountains—either in the North or anywhere else, for that matter—at other than regularly approved airports until you have made a special study of and mastered this particular subject of downdrafts and turbulence. The only possible exception to this rule would be: If the day is clear and also the general air mass around the mountains is virtually dead calm and not moving, this particular danger factor is greatly reduced and could be temporarily nonexistent.

Downdrafts and turbulence are caused by wind, usually not by local winds, but by the fundamental air mass movement against the mountain range. If this air mass movement gets up to 20 to 30 mph aloft, there will in general be moderate, smooth updrafts generated on the windward side of the range. On the leeward side of the range, but above the level of the crest, there will be standing waves of powerful updrafts running to increasingly high altitudes as one goes farther downwind. Very turbulent air and strong downdrafts will exist on the leeward side of the range below the level of the mountain crest.

If the air movement aloft rises to 40 to 50 mph or more, the turbulence and downdrafts on the leeward side of the mountains below the crest can get so wild as to make the aircraft practically unmanageable. Downdrafts (and updrafts) with speeds of up to several thousand feet per minute can be the rule on days with such air mass movements. Do not believe the old saw that "a downdraft can't blow an airplane down into the ground because the air has no place to go down any more when it gets to the ground." The pilot who still harbors this idea should sometime go through the instructive experience of sitting on a cliff looking over the lee side of a mountain range down onto a lake on a windy day. Carefully noting what the gusts show on the water below, he will sometimes see vertical jets of air blowing straight down onto the water, as evidenced by wind caused waves running outward in all directions from an area where a downdraft is striking the water. A light airplane flying into such a downward jet of air could obviously be plastered right down against the valley floor in full flight.

The existence of strong air mass movements aloft in the mountains is not always easily recognized from on the ground in the valley below. Parts of some mountain valleys, even on such days, can be deceptively calm. The speed with which clouds around the mountain crest level are moving will more dependably tell the story. If there are no clouds to give this indication, snow blowing from ridges as seen with field glasses from below can sometimes give a reasonably reliable indication of the speed of the air mass movement aloft.

Glaciers present many good landing opportunities for ski-planes in northern mountains (Figures 15.6 and 15.7). In general this is because snow fills crevasses and rough places in the upper parts of glaciers even during the summer. In winter the snowfall may develop not only the upper but also the middle and lower parts of glaciers into possible landing places. Fortunately the glacier crevasse systems, often so treacherous to persons down on the surface, are usually clearly revealed to the pilot above, in somewhat the same way that a pilot from aloft can see things underwater not visible to persons on the surface itself.

The two biggest problems in carrying out a glacier landing "in the rough" will be (1) which way the wind is blowing on the surface of the glacier (sometimes quite different from that a few hundred feet aloft), and (2) how steep the slope of the glacier is, and which way it is actually sloping.

Throwing out one of the small smoke bombs down onto the surface of the glacier is an excellent way of getting the wind direction. It is well worth keeping a few of these bombs, only about the size of a large sausage, in the airplane's emergency equipment but readily available to the pilot's hand for glacier flights. A satisfactory substitute can be to drop one of the standard aircraft message con-

Figure 15.6. Noordyn Norseman on Muldrow Glacier. The Norseman has been an outstanding bush pilot's airplane since early in World War II. Here it is unloading a one ton load of miscellaneous freight on Muldrow Glacier, Mount McKinley. Mount Brooks (11,940 ft.) is in the background. Despite dramatic surroundings, there was an excellent 4000 ft. landing strip—hard-packed snow over solid glacier ice with no crevasses. *(Courtesy of Bradford Washburn, Boston Museum of Science.)*

Figure 15.7. The author's Supercub-125 on Kahiltna Glacier. The photograph is taken at 10,100 ft. The camp was entirely stocked by air drops, and more than a dozen Supercub-125 landings were made here.

tainers with a streamer attached. At the moment of impact the streamer will trail downwind and thus give its direction clearly.

As for slope, the experience of over 200 glacier landings and takeoffs at altitudes running up to 16,200 ft. indicates that high altitude landings should always be made going up the slope, and high altitude takeoffs should be made going down the slope. It is recommended that for really high altitude landings and takeoffs—say at 10,000 ft. and above—no downhill landings or uphill takeoffs should ever be attempted, even into the wind. The increased landing speed caused by altitude and the decreased engine power caused by altitude completely rule them out.

For those reasons a dome shaped, smoothly glaciated mountain summit represents the ideal altitude landing surface, for all landings can be made uphill and all takeoffs downhill, yet all of them into the wind regardless of its direction. Second best is a saddle, and poorest is a simple slope.

The serious tendency of the older bush aircraft to ground loop during downhill takeoffs has been considerably improved through (1) better designed landing gear, more widely spaced, and (2) greater engine power and hence stronger prop wash, which gives more effective control to the rudder surfaces in keeping the nose straight during the takeoff run.

Arctic Ice Pack and Ice Islands.   The Arctic ice pack is usually so rough from the breaking of the edges of floes and the building up of pressure ridges, where the sea current and wind have driven the floes against one another, that the principal problem will be to find a large enough smooth area to land a plane on. For small, light ski planes with low landing speeds, landing strips can be picked out quite generally all over the Arctic ice pack, and at times landings can be made even by wheel planes.

Another problem will be that of drifting. Except for those parts of the ice pack that are frozen to the shore, any given landing place is likely to be "here today and gone tomorrow." Thus the pilot should choose places (except for emergency or purely momentary use) in those parts of the pack that are attached to the shore and not drifting.

The surface of the ice pack becomes very uneven and dangerously rotten in summer. What few areas there may be that are close to land in summer are likely to be quite unsatisfactory as landing places.

The problems of actual landings and takeoffs on the sea ice are completely overshadowed by the problems of survival on the Arctic ice pack. These latter problems, which are not really concerned with airmanship as such, require whole books for their treatment and are well set forth by Vilhjalmur Stefansson in *The Arctic Manual*.

The operations of the United States Air Force on the ice island known as T 3 are remarkable in that heavy, fast landing military aircraft have been used. This has been possible because the ice which forms these floating islands is apparently of glacial origin and thus hundreds of feet thick instead of the 6 to 10 ft. maximum thickness of the ordinary Arctic ice pack.

Floatplaning. Floatplane weather brings the time of year when "landings in the rough are made smooth." Almost everywhere in the north there are innumerable bodies of water, most of which are perfect landing and takeoff areas for floatplanes during the open water season. This lasts anywhere from 3 to 7 months, depending upon where in the arctic or subarctic one is. The specific stick and rudder techniques for handling floatplanes generally are available in the standard flight instruction manuals, and therefore will not be repeated here. *Seaplane Flying and Operations* by Fogg, Strohmeier, and Brimm (Pitman, 1949) is a superior text on the basic techniques. The special northern aspects of floatplane flying, however, can be summarized in the following paragraphs.

Takeoffs and landings in northern wilderness lakes and rivers raise the inevitable question: How about hitting unknown submerged rocks and ripping out the bottoms of the floats? This is really not a serious problem for alert pilots, because (1) light floatplanes have a very shallow draught, as little as 8 to 10 in. in some Cub floatplanes; and (2) any rocks dangerously close to the surface are distinctly visible from the air in the clear water of northern lakes, or are plainly revealed, at least in rivers, by the current eddies and turbulence even where the water is muddy.

A special problem frequently encountered, however, is the matter of how to tie up a floatplane to a wilderness shore. If the pilot is fortunate enough to find a shelving sand beach, the plane is simply paddled in and then turned around and the heels of the floats drawn up on the beach. But in the much more frequent case of a rocky shore, what to do? Here it may be said from experience that a snug and secure tie up can be contrived by simply laying a trimmed log—and it need not be a large one—along the rocks right at the water's edge. The floatplane is then drawn up onto this, facing outward with the heels of the floats resting upon the log. It should of course be tied with three lines run to the trunks of trees, or to rocks or bushes, one line straight from the tail and one diagonally out from each wing.

Essential floatplane equipment includes paddle, bilge pump for the floats, canvas bucket, and plenty of nylon rope. Without these no floatplane should leave home base for any northern wilderness flight. The canvas buckets can be quite useful in quickly flooding the forward float compartments in case it becomes necessary to ride out a gale.

There are two ways for floatplanes to ride out a high wind in the wilderness. One is to moor the plane out and partially flood the forward float compartments, until the total weight of the airplane empty is increased by 50%, with all of the weight well forward. This weight tends to hold the nose down and to prevent the aircraft from flying away, particularly since a moored floatplane normally rides at a very much lower angle of attack than that at which a land plane stands on the ground. In fact it is quite possible to set up a slight negative angle of attack for the wings, even with the floatplane swinging at her mooring. This position is fundamentally stable against a big blow because the airplane, being moored from the nose, tends constantly to weathercock into the gusts as they batter in from different directions. A further downward thrust from the mooring cable at a most effective place is obtained if the mooring cable is laid either in a bridle right on the mooring cleats at the tips of the floats, or from the propeller hub. If the wind is as strong as 60 to 80 knots, so that still further precautions must be taken, a very simple device is to hang floating logs or gas tins or drums, almost full of water, from the tiedown rings of the wings, so that the weights are just awash under the surface of the water. Because of their water buoyancy in this condition, they lay no serious weight directly on the wings; yet as the wings begin to fly and lift the aircraft upward in strong gusts, these logs or drums are immediately lifted upward in the water, lose their water buoyancy, and exert their weight in holding down the wings.

The principal weakness of mooring out in the manner described is that the mooring must be secured to an anchor that will not drag in the highest winds—something usually difficult to contrive on the spur of the moment on the shore of a wilderness lake. However, because the plane is streamlined and always weathercocking, the drag against the anchor will not be as great as might be supposed. This drag can be computed, for it is approximately equal to the horsepower thrust necessary to cruise the plane in flight at the same airspeed at which the gale is blowing. This is of the order of 100 lb. for a light aircraft of the Piper Cub type.

If it is impossible to set out an anchor which will not drag, then bedding the floatplane down on a beach will probably be the best way to ride out a big blow.

If the plane can be tailed up onto a beach, it then becomes possible to flood all the float compartments completely with water. The pilot should also be careful first to set up a negative angle of attack on the wings by wedging up the heels of the floats if the natural slope of the beach is not steep enough. This negative angle of attack is necessary because even with all float compartments flooded, the plane will still weigh something on the order of only about 250% of its rated gross weight; and it should be remembered that lift varies as the square of the velocity. Therefore it does not require a wind of 250% of the normal takeoff speed to start blowing the plane away; a wind speed of somewhere near the square root of the increase only, that is, something less than 60% above the normal takeoff speed, will be all that is necessary to bring the danger point. Hence this matter of the negative angle of attack should not be overlooked.

In the fall or spring of the year, flying onto and off open water with air temperatures below freezing will of course splash the tail assembly of the plane with water, which may soon freeze into ice. There is much more danger of freezing up the control wires—especially the water rudder wires and thereby the air rudder control—than there is of freezing the hinges of the control surfaces themselves. This is because the manual control for the elevators and rudder exerts such powerful leverage from the control horns against the fulcrum in the control hinges that the ice is rather easily broken off the control joints themselves. But where the water rudder cables run through their tiny cable guides is the spot where the water, draining backward immediately after getting airborne, is quick to freeze, locking the water rudder cables tight and hence immobilizing the air rudder. Therefore, after getting airborne, the pilot should immediately seesaw the rudder controls back and forth until looking back he can actually see that all water has drained away from the cable guides or that ice has formed with the water rudder cables running through it in a tunnel. The hand pull cable which lowers and retracts the water rudder must also be pulled back and forth at the same time until all water has drained away from its cable guides or frozen into an open tunnel around the wire.

The northern bush pilot will probably find he needs to know how to operate floatplanes during freezing weather, onto and off a thinly ice covered lake, and even onto and off snow covered ground. Occasions for handling these unconventional situations are almost certain to arise sooner or later in northern flying.

For most pilots there is normally a period of some weeks during freezeup when "you can't use floats on account of the ice, yet the ice isn't thick enough to be safe for skis." But sometimes during this freezeup period the northern pilot will unexpectedly find that he simply must fly a floatplane onto breakable ice for some essential mission if it can possibly be done. For those who know the technique and are willing to give their floats some rough treatment, it is possible to operate some types of light planes onto all conditions of breakable ice, up to and including the point where the floats ride completely on the ice. Because it is not a commercial operation, the subject has not been sufficiently investigated to enable one to state that it can be successfully done with all types of floatplanes. But it has been successfully done with some light floatplanes in all conditions of breakable ice, including fully supporting ice.

For a plane which has been through the operation before, the landing on breakable ice is in every way normal, except for the absolutely appalling noise of the breaking ice, which can best be likened to a landing on a greenhouse roof with masses of broken glass flying in all directions! The strong keels of the floats break the ice in such a way that typically the skin of the floats does not tear.

Turns while taxiing will create the biggest problem in a breakable ice operation. If the ice is thin, turns can be made slowly without damage or danger. But when the ice gets close to ¾ in. in thickness, taxiing turns with power cannot be made without imposing such stress from the ice edges on the sides of the rear of the floats that ripping of the floats becomes a possibility. If the pilot looks back and watches what is happening, this danger will become evident to him, and he can take the appropriate action. When the ice reaches this critical thickness, it becomes necessary to make all turns by stopping the engine, getting out on the floats and breaking the ice around with a paddle or other tool, and then hand turning the seaplane with the paddle, to head straight in the direction of takeoff before applying power again.

The takeoff run will be very considerably lengthened, and as the ice becomes more than ½ in. thick, may become impossible, because of the greatly increased drag, unless a taxiing run is first made to break up the ice in a lane for the takeoff. When this has been done, the broken up ice does not create much drag, and the takeoff is easily and normally made. It should be added, however, that the use of a floatplane into and out of breakable ice cannot be really recommended. It is certainly not a commercial operation, because even when properly done there is a cumulative hammering effect on the float skins right on the edge of each bulkhead joint, which tends slowly to deform this area of the float skin.

As the thickness of the ice approaches 2 in., a lighter floatplane will land right on the ice without breaking through and simply slide along on the float keels like a ski plane. WARNING: On glare ice there is so little resistance from these narrow-keel runners that the plane will slide an unexpectedly long distance and may crash into the shore unless the pilot makes his landing by heading out into the broad reaches of a large lake. Riding on the surface without breaking through, the pilot is now beginning to use his floatplane as a substitute for a ski plane; and this, within certain limits, is reasonably feasible.

The big disadvantage of using floats as a substitute for skis is that unlike the ski landing gear, the float strut system has no shock absorber whatever. Any roughness in the ice or drifts in the snow will be transmitted with direct hammerlike blows right into the frame of the aircraft. Also, if any rough hummocky ice has developed—and pressure ridges do form quite soon on northern lakes in subzero temperatures—these rough spots will project up higher than the ridges of the keels, and the float skins will be struck and easily torn. A floatplane may seriously tear or crumple its floats on moderately rough ice which would cause a ski plane no difficulty whatever.

The floats will also operate reasonably well as skis from deep, soft snow. But a floatplane will not be able to carry the load of a regular ski plane without bogging down quite easily, for the V shape and the "step" in the float bottom dig into the snow and create a much greater drag. In any event even this would never be done during commercial operations because of the risk to the thin skin of the floats.

Search and Rescue and Emergency Gear

The dangers of running out of fuel or of getting lost or of having a forced landing and thus being down on the ground far out in the northern bush are very serious indeed, especially in winter. Absolutely no flights whatever should be made beyond gliding distance from a northern airport without having made a comprehensive plan to cover these contingencies. Most of what is needed can be built into or placed in the aircraft and can stay there for the particular season and thus require no more than a brief check at the beginning of each flight. But the original list of equipment must be thought through slowly, with plenty of time to insure adequate coverage.

In planning the list of emergency equipment and procedures, the use of mathematical probabilities can be quite helpful. For example, modern aircraft engines will fail on the order of about once in 5000 hours of flight. Add to this hazard that the pilot or navigator may get lost, or that because of some miscalculation the plane may run out of fuel, or that weather may close in both ahead and behind, demanding a forced landing, and the chances become something on the order of one in several hundred, for inexperienced pilots, that on any northern flight lasting several hours their airplane will have to make a forced landing in the northern wilderness far from any habitation. Such a forced landing is a seriously dangerous matter if it occurs in such a way that the airplane cannot get off again under its own power. If not found and rescued promptly in wintertime, the occupants, unless they are exceptionally experienced northern travelers, will not last long in the subzero temperatures which they may encounter. And even in the summertime, it is highly possible that pilot and passengers may starve to death unless they are uninjured in the forced landing and are experienced in the ways of the northern wilderness. Adequate emergency survival equipment should be carried in the plane.

RADIO

If the plane is equipped with a powerful long range two way radio capable of reaching at least one radio range station throughout a long flight, the use of this radio will virtually insure the safety of the plane's occupants in the event of a forced landing, because the rescuers will know what has happened and exactly where to look. The chances of such a radio failing while in flight are small if equipment is properly maintained. But the chances of *both* radio failure and engine failure occurring in the same hour will be far less likely.

Some of the fundamentals for radio are discussed in this chapter under "In-Flight Problems: Navigation and Radio." To those comments may be added the following as being especially pertinent to search and rescue operations. *The one great problem* in search and rescue work after an aircraft is lost and down in the bush, is to *find* the aircraft.

Today, the most important single rescue precaution one can take is to file a flight plan in advance with the FAA (in Canada, DOT) and then stick to it or advise of a change.

After that has been done, the rest is usually relatively simple. For this reason, the critical moments in a forced landing, and those in which "something can be done about it," are those minutes, even seconds, while the plane is still in the air after the need for a forced landing has first impinged upon the pilot. Most light planes with the engine off have a rate of sink of no more than 500 to 1000 fpm at most, and some as little as 350 to 400 fpm. On the very long range flights usual in the North, the pilot will probably be flying at somewhere between 5000 and 10,000 ft. for the greater visibility and ease in navigation this altitude affords, and also for the greater flight efficiency most planes develop at these altitudes. Hence the pilot is quite likely to have several minutes of time, even after engine failure, while the plane is still in the air, to inform the nearest monitoring radio range station that the plane is being forced down and exactly where on the map this is occurring. It cannot be overemphasized that these minutes or even seconds are vitally important, because the range of the aircraft's radio transmitter is potentially *many times* greater while the plane is still high in the air than after it is down on the ground.

The writer submits, from experience, that probably the most important single piece of emergency equipment that the pilot can put into his plane for northern flying is the best LF/MF radio transmitter that money can buy, plus an oversize reel out type antenna with enough wire to go to the *second* point of resonance for the emergency frequency (2182 kHz and 8364 kHz), plus an intimate knowledge of how to tune up the transmitter in flight. An LF/MF set rather than VHF or UHF is desirable, for in northern flying the ability to reach out dependably over long range with the transmitter—at least several hundred miles—is essential. VHF and UHF will not do this, since they are limited to line-of-sight ranges. Few of the LF/MF sets that can be put into a light aircraft will do the job of reaching out properly if they are equipped with a *fixed* antenna only. Of the HF "short-wave" sets which can be put into a light aircraft, only a set with a trailing antenna reeled out to at least the first point of resonance of a Hertzian antenna system, and preferably to its second point of resonance, will really do an adequate job. To operate such a trailing reel out antenna, one must know how to tune the set up to it in flight and to pass the test for the special license required. Prior experience as a radio ham in invaluable for the purpose; but failing that, an intensive special course of a few weeks can do what is necessary for this purpose.

When reeled out to the second point of resonace, the aircraft Hertzian antenna becomes highly directional, in a cone about 15° off the direction of the length of the wire. The antenna then has to be "aimed," and this is not difficult to do. When all this is done, dependable clear two way voice communication becomes possible over a range of 200 to 300 miles, even with the type of small 20 watt radio set appropriate for light aircraft. Often it works well two way over a range up to 500 miles.

Obviously it is not feasible to reel out the trailing antenna and tune the transmitter to resonance *after* the plane has begun to come down in a forced landing. Therefore for safety's sake it is important that the antenna be reeled out and the transmitter tuned to resonance promptly after takeoff and the antenna left that way throughout the flight. An additional precaution is to leave the transmitter filament turned on hot so that in an emergency there will not be even a 20 sec. delay for the transmitter to warm up. Flying along with the antenna trailing, the

(B)

(A)

Figure 15.8. Emergency locator transmitters are a priceless asset to survival. They transmit a coded beacon signal on 121.5 or 243 mHz, may be started by the force of deceleration on crashing, or manually, and permit rescue forces to locate downed aircraft quickly. (A) This is a fixed or hand held transmitter. Omnidirectional antennas. Range, 100 miles at 10,000 ft. search altitude. Actuated by 6 g shock. Weight, 2.3 lb. (B) This is similar but is a fixed, integral transmitter and antenna. Actuates on 10 g shock. *(Courtesy of Leigh Systems, Inc.)*

set in resonance, and the filament turned on and hot, all the pilot has to do in the event of an emergency is merely to press the microphone button and start telling the facts to the nearest 24 hour monitoring radio range station, even though it may be a couple of hundred miles away. FAA and Canadian range stations monitor the 2182 kHz and 8364 kHz frequencies on continuous 24 hour guard.

Five hundred kHz is an international short range emergency frequency. Silence is maintained from 15 to 18, and from 45 to 48 min. past each hour. During this period there is a much better chance of being heard.

Emergency locator transmitters, which are independently powered, electronic line-of-sight transmitters, are a most valuable emergency asset. Nearly all light aircraft are now required to have them.

They transmit on 121.5 mHz and 343 mHz and should have a power source capable of transmitting continuously for 24 to 48 hours. Most models are suitable for light aircraft.

The pilot will of course have filed a standard flight plan; but the pinpointing radio procedure described above will enormously improve the chances of being located promptly by the search and rescue organization, whose planes will not then be forced to comb over hundreds of miles conducting a needle-in-a-haystack search operation.

Provision must also be made against the contingency that for some reason a pilot will not have completed radio contact before making his forced landing and hence will have to try to make contact after the aircraft is down on the ground. Although the range is very greatly reduced, nevertheless if the radio transmitter can be got to working on the ground, the chances of making radio contact with a nearby search plane in the air are good. For this emergency on the ground, the following pieces of radio equipment will be important.

1. A permanently installed variable loading coil in the antenna circuit of the transmitter, in order to tune up the set most effectively on the ground.

2. A small vial of sulfuric acid, or else battery tops which really are spillproof. In a large proportion of forced landings in the bush, the airplane will come to rest upside down, and the acid will promptly run out of the storage batteries. If the pilot carries a small quantity of sulfuric acid and the necessary instructions, he can reconstitute this into the proper strength for replacement in the battery. Nickel-cadmium batteries are free of these difficulties.

3. A small portable and separate engine driven generator. The capacity of most aircraft radio batteries to run the radio transmitter is limited to only a few minutes of transmission time. Hence if the plane gets down on the ground without having been able to get a radio message out, the pilot may need a separate engine driven generator in order to send repeated signals until they are heard. The writer, for example, has a complete generator set weighing no more than 18 lb. which will charge the aircraft's 12 volt storage battery at 4 amps—quite sufficient to keep the transmitter on the air for at least 10 min. out of each hour, since the transmitter draws no more than 24 amps. Such a portable generator

can scarcely be listed as essential, but it does add to safety and may be worth taking when relative values of weight and space and the uncertainties of an unusually remote and difficult terrain indicate it.

SURVIVAL EQUIPMENT AND TECHNIQUES

There are almost as many different lists compiled for this purpose as there are outdoor experts. And this is in fact not so unreasonable as it might at first appear. The reason is that the ability to survive in the wilderness depends much more upon the personal skill of the individual than it does upon which particular items of equipment he has taken with him. In Alaska one can get no better advice than from the nearest Air Force Rescue detachment stationed there. In Canada, one should consult the RCMP or the RCAF.

Probably the most misleading aspect of this subject for the relatively inexperienced outdoorsman will be the selection of guns, ammunition, and fishing tackle—equipment for "living off the country." But this writer submits that these are the least important in the list, except possibly for a really experienced outdoorsman. An inexperienced person planning to live off the country in the far north, even with modern arms and fishing tackle, when he suddenly has to start doing it finds it very difficult. A white man can learn to make a living in the north as well as the Eskimo, but the arctic is "friendly" only to those with years of acquired knowledge. For others, to be cast away in the northern wilderness will almost certainly mean starvation unless they bring their own food.

The emergency equipment list will differ with the seasons and also for different parts of the north. For example, mosquito repellents will be utterly useless in winter, essential in summer. A snow knife and a primus stove, both of which will be important for winter use north of the limit of trees, will be unnecessary further south where the snow is seldom sufficiently hard-packed to permit cutting snow blocks, and where wood and birchbark can be used to build shelters and to make fires. Here a good ax, of little use north of the limit of trees, will be essential.

Instead of offering a recommended list of equipment, or series of lists, the writer offers the following summary of needs which will have to be met. Specific lists are readily available from which to choose items to meet these needs, such as the list required by the RCAF controllers of the Northwest Staging Route (the Alaska Highway Route). The authorities at Thule, Greenland, issue their list for flights over the Greenland Ice Cap. FAA authorities at Anchorage have an excellent list appropriate to Alaska.

Any pilot contemplating a flight into the subarctic or arctic areas of Canada can and should obtain this information, as well as a wealth of other essential information relative to rescue, survival, and Canadian flight regulations, by writing to the nearest District Superintendent, Air Regulations, Department of Transport. These offices are located in Vancouver, B. C., Edmonton, Alta., Winnipeg, Man., Toronto, Ont., Montreal, Que., and Moncton, N. B.

The needs to be met are: repair tools, clothing, water, food, shelter, fire, sleeping gear, protection from insects (and in some cases from animals), signaling, and traveling. Comment will here be offered only on the less obvious but essential aspects of their selection and use.

Kit of Repair Tools.   This is placed at the head of the emergency equipment section because of the large number of cases in which a plane can be set down successfully in a forced landing, repaired by the pilot if he has tools, and then flown on to destination. The pilot should select the items for this tool kit in consultation with the A and E mechanic who does the work on the particular aircraft. The makeup of the tool kit will depend not only upon which tools will be most useful for the particular plane, but also on the pilot's ability to do repair work.

Clothing.   Books could of course be written on the subject of emergency clothing for northern use. The most important warning is: *Don't go off on a flight during the cold months without parka, mittens, warm comfortable visored cap, felt shoes, mukluks, and the rest, adequate for −50° and −60°F cold.* Several layers are warmer than one heavy garment; also layers may be peeled off to adjust to the temperature. Lightweight, loosely woven material for the under layers is warmer than closely woven material. A light windproof parka should always be carried to wear over woolen clothing for protection against the wind. Even a very light oilskin raincoat is handy to throw on, when one has to land and refuel somewhere where there is a freezing wind blowing. Leather footwear will never do in subzero weather, and the ordinary civilian rubber "pacs" are not much better. Mukluks are good, but felt shoes are more convenient if there is much walking or snowshoeing to do at subzero temperatures. The shoes should be large enough so that for −50°F weather a pair of soft wool dress socks can be worn next the feet and also two pairs of woolen loggers' socks. A pair of felt insoles inside adds greatly to the warmth. Woolen underwear is essential. If the skin is irritated by wool, the fleece lined sort can be worn, or a suit of very lightweight cotton long underwear next to the skin. Keeping the feet and the rest of body dry is vital to survival (Figure 15.9).

Water.   This is plentiful anywhere in the North in summer. But in winter north of the limit of trees an ax or ice chisel should be carried to chop through ice. Otherwise the only source of water may be to melt snow. The old argument still persists as to whether or not to eat snow to quench thirst. For what it may be worth, the writer of this chapter, over years of mountaineering and arctic travel, has always eaten snow when he felt like it and never suffered any ill effects, except that in subzero temperatures there is a very unpleasant and possibly dangerous chilling of the mouth. On the other hand, the traveler can maintain a much better condition by melting snow in a pot from his cooking kit. The cooking kit therefore becomes a necessary item of emergency equipment, together with waterproofed matches, and a small gasoline stove when north of the trees or flying over glaciers.

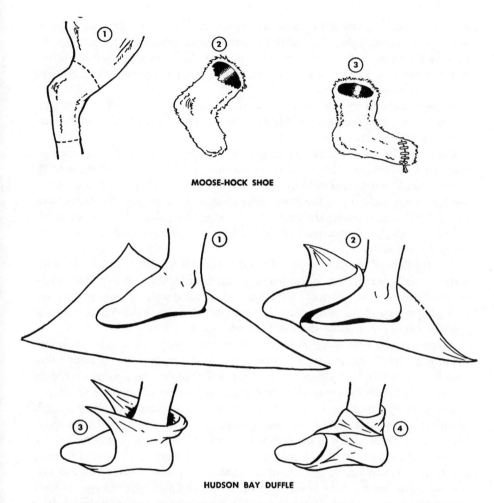

MOOSE-HOCK SHOE

HUDSON BAY DUFFLE

Figure 15.9. Emergency footwear. Care of the feet is vital to survival. The first principle is to keep them dry at all times. *(Courtesy of US Air Force.)*

Food. At least 7 to 10 days' supply of food per person should be carried in the emergency kit of airplanes on northern wilderness flights. Expressed another way, the plane should carry about 20,000 calories per person. The best form in which to carry these 20,000 calories is of course the subject for endless debate. The best calorie-weight ratios are obtained through extensive use of oils and fats. An investigation has shown that the body can utilize a higher proportion of oils and fats in the diet in cold climate than in warm. Of course protein is also important to rebuild muscle tissue. Hence the pemmican of explorers, which is a mixture of lean meat (protein) and fat, is a frequently mentioned emergency food. This writer has lived on pemmican on expeditions and from this experience believes that if weight and space permit, one should provide a food, for example, fat sausage, having nearly the same fat-protein and calorie-weight ratio as pemmican, with a taste more palatable to the inexperienced tongue. He would not unnecessarily impose the problem of adjusting to a strange and therefore unappetizing food like pemmican just at a time when other pressing and difficult adjustments have to be made.

In another recent investigation of survival foods in the arctic, it was brought out that a diet composed exclusively of a certain type of candy bar kept the survivors in surprisingly good condition.

Consequently the writer concludes that the time tested trinity of fats, proteins, and carbohydrates has been reconfirmed; and that if chosen as described, with an approximately equal proportion in bulk of the three, it is a simple diet, easy to assemble, store, and keep in the plane as emergency food supply. Small amounts of dried fruits, salt, sugar, and cereal can be added for palatability; but these additions do rapidly increase the bulk and reduce the calories-weight ratio and are not really necessary.

The problem of vitamin supply, if the individual has a full complement in his body normally, can as a practical matter be disregarded for a short period. Tests have indicated that in a normal individual well supplied with vitamins in the body at the beginning of the test, diets with even serious vitamin deficiency do not begin to cause real trouble short of a month's time.

A combination rifle-shotgun with ammunition, and fishing tackle, including nylon gill nets if practicable, will extend the food supply. But any notion that because one is supplied with these things, it will therefore be possible to obtain all one's food from the country is a very treacherous one. Living off the country *can* be done by experts and it *has* been done by novices who were lucky enough to have their forced landing occur near, for example, a migrating herd of caribou. But the pilot forced down in the northern wilderness should not count on obtaining more than perhaps 10% to 30% of the required food supply with firearms, fishing tackle, and snares. This continuing small amount of game and fish can become quite important, however, after the tinned and packaged emergency food supply has been used up. If the fish and game are eaten cooked only "rare," more vitamins will be retained. But bears and arctic sea mammals have been found to have trichinosis and must be thoroughly cooked; and polar bear liver is *poisonous* because of dense concentrations of vitamin A.

The smaller caliber rifle and small bore shotgun are generally preferable because so many more rounds of ammunition can be carried in a small weight; also it will be principally small birds and squirrels which the castaway will find to shoot, and for these the small ammunition will be more appropriate. Gill nets, except for emergency use, such as in a genuine forced landing, are illegal in most regions. But their great advantage is that when set, they are working all the time, require no bait, and are very effective, particularly if made of nylon, for this seems to be invisible to fish. The book *Edible Plants of the Arctic* by A. E. Porsild could prove of some help in the summer if carried along.

Shelter.   A tent with sewn-in ground sheet and tie-up tunnel entrance will provide protection against blizzards in winter and swarms of insects in the arctic summer. Such tents can be obtained with both cloth and mosquito-netting entrance, the unused one being tied back. The centerpole type is usually the easiest to pitch and the most comfortable to live in. An extra thin nylon poncho thrown over the top is the most dependable way of making a tent rainproof. On flight north of the limit of trees, one should add a collapsible or jointed aluminum pole and tent stakes. Remember that a fully waterproof shelter will frost heavily inside in very cold weather. This results in wet clothing and bedding.

Snow itself is the best survival shelter because of its excellent wind resistance, structural strength, and efficient insulating capacity. In deep, wind-packed snow, a snow knife, saw, or long bread knife can be used to cut structural blocks.

The best one man deep snow shelter is a single trench, made by cutting out blocks, then using them to make an A-frame roof. For several, an excellent solution is to dig a cylindrical hole of 6 or 8 ft. diameter, and cover it with tarps, parachutes, or other materials anchored with snow blocks. Then burrow spoke-like sleeping trenches into the sides.

Whenever possible in dug snow shelters, dig clear to the ground. Minimum ground temperature is about $+18°F$, and under shelter this heat will warm the sheltered air. With an outside temperature of $-40°F$, actual experiments have shown a temperature inside a snow shelter of $+18°$. This is warm enough to permit a downed airman to survive without a fire.

In light, fluffy snow, an adequate shelter strong enough to support the weight of several men can be made by piling up unpacked snow. With a 5-ft. stick, describe a circle on the snow. Erect the stick at the center, and place another from the center to the circle edge as a guide to the center. Shovel snow into a cone, one foot higher than the center pole, with its base filling the circle. *Work slowly enough to avoid sweating.* Allow the snow to set for 1 hour. Using the guide stick, burrow into the center. Remove the center stick carefully; evacuate the inside until the walls are 2 ft. thick. Then dig in the center to the ground surface to avail yourself to ground heat. A parachute pack, engine cowling, or piled snow makes a suitable door. The shelter is completed by providing some insulation between body and ground.

Fire. Matches in a waterproof container, or waterproofed by having the heads individually dipped in paraffin, a knife and an ax in timber country, or a gasoline pressure stove will provide fire. Starting fires in the woods is a special art, thoroughly treated elsewhere; we may mention the fact that birch bark or a handful of small dry spruce twigs easily solves most fire starting problems, even in the rain. A 2 or 3 in. length of miner's candle will save matches when starting a fire under difficult conditions.

Sleeping Gear. A sleeping bag, rather than blankets, is essential for adequate protection against cold; a double bag filled with down or feathers will be necessary for subzero temperatures. The sleeping bag also provides emergency protection against insects in summer if a tent is not available.

In flights over glaciers or the Arctic ice pack it is essential to add a light air

Figure 15.10. A lean-to shelter in the subarctic. A lean-to covered with a parchute, tar-paulin, bark strips, or brush provides excellent shelter and considerable warmth with a fire, but in extreme cold the fire, without a reflecting surface behind it, will warm the shelter only slightly. Keeping the fire can become a full time, energy draining chore.

mattress or some other insulating material, such as caribou hides or extra cloth-ing. Sleep will be impossible without this, for even in a tent with a sewn-in floor, when pitched on snow or ice, there must be an insulation layer to keep the sleeper in his bag off the icy floor. In the forested part of the subarctic, fir or spruce boughs can be cut and laid on the snow beneath the tent floor to do this job. A re-ally comfortable bed may be had by building a spruce bough mattress. Do this by cutting many small branches, not more than a foot long, avoiding thick, hard stems. Starting at the top, lay them like shingles with the stems toward the foot of the bed. The more layers, the better; if the mattress is thick enough, it will have a soft, springy action like a real mattress.

First Aid Kit.   Standard first aid kits are adequate for the north if dark glasses are added to protect against snow blindness. This can occur even on cloudy days on glaciers or snowfields and is most painful and disabling. If dark glasses are not at hand, one might follow the example of the Eskimos, who make snow-glasses by shaping a piece of driftwood to fit across the eyes and piercing a small hole in the slit center. This allows only a small amount of light to enter, but it is sufficient to see through. Almost any material can be used except metal. Never allow metal to come in contact with the skin in really cold weather; an incautious touch with lips or tongue can cause instantaneous "freeze-on" with painful results.

Frostbite is insidious because one may not be aware of its onset. Propwash at
$-40°F$ can cause frostbite on the cheeks or nose in a few seconds. Crew
members should watch each other's faces for white patches which indicate frostbite. Warming immediately with the hand will thaw out the flesh in light cases.
In more severe cases warm the flesh with lukewarm water. *Do not apply snow to
the place or rub the frostbitten parts with anything,* since this further damages
the affected tissue. Frostbite affects the flesh in approximately the same way as a
severe burn. Do not do anything to the affected part that you would not do to
such a burned area. It is permissible to massage the area next to the frozen part,
being careful not to touch the injured area. This will stimulate the blood circulation and expedite the thawing process. Really severe frostbite must be treated by
a doctor as soon as possible.

Protection Against Insects.   Mosquitoes and blackflies swarm in appalling
numbers in many parts of the arctic and subarctic in early summer. Though not
disease-bearing, their very numbers are a menace; they can kill horses. Headnets, gloves, trouser bottoms which tie around the tops of the boots, and clothing
that mosquitoes cannot bite through, such as a light nylon parka, are essential. A
fact seldom mentioned is that a very light nylon parka will give very convenient
protection against mosquitoes if its wearer draws the hood up close to expose
nothing but eyes, nose, and mouth. These and the backs of the hands are then
easily protected by repellents. Chemical repellents are convenient, but headnets
will of course outlast repellents and therefore provide a more basic protection.
There will be no sleep unless protection is arranged. If there is no tent, getting inside the sleeping bag and stuffing the headnet from the inside outward into the
breathing opening can do the trick.

Signaling.   The radio, already discussed, is of course the airplane's most important emergency signaling equipment. In addition a signaling mirror which can
be aimed accurately through a hole or cross and double mirror on its back should
also be carried, in order to be able to catch the eyes of scanners in searching
aircraft.

If forced down, a floatplane should be moored well out in the water away from
the shore, to catch the attention of searchers. Overhanging trees and branches
should be cut away from above any aircraft down among trees. Any fall of snow,
even an overnight frost, should be brushed off the wings and fuselage to make
the aircraft more visible to searchers from above. Choice of the airplane's color
is also important for wilderness flying. The new high visibility fluorescent paints
can be very helpful not only in the collision prevention role for which they were
primarily designed, but also in our problem of locating aircraft down in the wilderness.

One of the first things the pilot and occupants of the plane should do after a
forced landing is immediately to assemble the materials for touching off a large
smoky fire, so that this can be kindled without delay as soon as searching aircraft
appear in the sky. A rising column of smoke, even a small one, is very noticeable

from the air. If the plane is forced down in the snow, messages to search planes can be outlined with cut spruce branches, or tramped out in the snow (Figure 15.11). The aircraft should carry a flashlight with extra batteries to be able to flash an SOS (· · · - - - · · ·) should aircraft fly overhead during dusk or darkness.

Flares are usually listed as important emergency signaling equipment. Those

NO	MEANING	SYM-BOLS	NO	MEANING	SYM-BOLS
1	Require doctor Serious injuries	I	9	Probably safe to land here	△
2	Require medical supplies	II	10	Require fuel and oil	L
3	Unable to proceed	X	11	All well All is well	LL
4	Require food and water	F	12	No (negative)	N
5	Require firearms and ammunition	⩔	13	Yes (affirmative)	Y
6	Require map and compass	⊏	14	Not understood	⅃L
7	Indicate direction to proceed	K	15	Require engineer	W
8	Am proceeding in this direction	↑			

INSTRUCTIONS

1. Lay out these symbols using strips of fabric or parachutes, wood, stones, or other available materials.

2. Provide as much color contrast as possible.

3. If possible make symbols eight or more feet high.

4. Carefully lay out symbols exactly to avoid confusion.

5. Also make every effort to attract attention by means of radio, flags, smoke, etc.

AIRCRAFT ACKNOWLEDGMENTS—*Aircraft will indicate that ground signals*—
    ARE SEEN AND UNDERSTOOD—*By rocking from side-to-side or*
                               By making green flashes with signal lamp.

        ARE NOT UNDERSTOOD—*By making a complete right-hand circuit or*
                               By making red flashes with signal lamp.

Figure 15.11. International ground-to-air distress signals. *(Courtesy of US Air Force.)*

of the small, railroad type seem to combine maximum lasting time with minimum bulk.

Remember this jingle:

"Use smoke by day and flame by night. Your signal can never be too big or bright."

Protection Against Animals.   The polar bear will stalk humans on the Arctic ice pack not from anger, but strictly from hunger. The large brown and grizzly bears of Alaska are treacherous and definitely dangerous; although they usually retreat upon encountering a human, the fact that they have killed and wounded many humans over the years in unprovoked as well as provoked attacks makes it safer to assume that if encountered they will charge. But a rifle heavy enough to kill these bears outright is a cumbersome thing to carry in a light airplane and must be of a heavier caliber than is best for shooting game for the pot. Hammering on a dishpan is the time honored way of scaring off bears; they do not like that type of noise. One of the cooking pots struck repeatedly with a spoon is about as effective.

Traveling.   Snowshoes should always be carried in a skiplane so that the pilot can tramp out a runway in fresh snow for takeoff. The snow shoes can also of course serve as a means of travel if necessary. However, it cannot be emphasized too strongly that the almost invariable rule in a forced landing in the bush is "Don't leave the aircraft." An individual roaming about in the wilderness is infinitely more difficult for rescuers to locate than is the downed aircraft itself; in fact, it is virtually impossible to locate a lone man. And in a real wilderness forced landing, the chances of walking out without getting hopelessly lost are slim indeed for any but genuine experts in the particular type of country encountered.

For the same reason there is an invariable rule in northern wilderness flying: DON'T BAIL OUT (except over a settlement) unless absolutely forced to do so by imminent and certain destruction of the aircraft in the air.

Index

The letter "i" following page numbers indicates illustration.

865